Kimber-Gray-Stackpole's

Anatomy AND Physiology

Kimber-Gray-Stackpole's

Anatomy AND Physiology

17th edition

Marjorie A. Miller, M.S.

Associate Professor (Nursing), Cornell University–New York Hospital School of Nursing, New York / Formerly, Associate Professor (Science), Cornell University–New York Hospital School of Nursing, New York / Instructor of Nursing Education (Physiology), Teachers College, Columbia University, New York

Anna B. Drakontides, M.A., M.S., Ph.D.

Assistant Professor (Anatomy), New York Medical College / Formerly, Associate Professor (Pharmacology), Cornell University–New York Hospital School of Nursing, New York / Instructor (Science), Cornell University–New York Hospital School of Nursing, New York / Adjunct Assistant Professor (Biological Sciences), Hunter College of The City University of New York

Lutie C. Leavell, M.A., M.S., D.Sc. (Hon.)

Professor Emeritus of Nursing Education (Physiology), Teachers College, Columbia University, New York / Formerly, Consultant, College of Nursing, University of Iowa, Iowa City / Lecturer, School of Nursing, University of Pennsylvania, Philadelphia

Macmillan Publishing Co., Inc.
NEW YORK

Collier Macmillan Publishers
LONDON

Earlier Editions: Text-book of Anatomy and Physiology for Nurses *by Kimber, copyright,* 1893, *by Macmillan Publishing Co., Inc., and copyright,* 1902, *by Macmillan Publishing Co., Inc.:* Text-book of Anatomy and Physiology for Nurses *by Kimber and Gray, copyright* 1909, 1914, 1918, *by Macmillan Publishing Co., Inc.,* Text-book of Anatomy and Physiology *by Kimber and Gray, copyright,* 1923, 1926, *by Macmillan Publishing Co., Inc.;* Textbook of Anatomy and Physiology *by Kimber and Gray, copyright,* 1931, *by Macmillan Publishing Co., Inc.;* Textbook of Anatomy and Physiology *by Kimber, Gray, and Stackpole, copyright,* 1934, 1938, 1942, *by Macmillan Publishing Co., Inc.;* Textbook of Anatomy and Physiology *by Kimber, Gray, Stackpole, and Leavell, copyright,* 1948, 1955, *by Macmillan Publishing Co., Inc.;* Anatomy and Physiology *by Kimber, Gray, Stackpole, and Leavell,* © 1961, *by Macmillan Publishing Co., Inc.;* Anatomy and Physiology *by Kimber, Gray, Stackpole, Leavell, and Miller,* © *copyright* 1966, *by Macmillan Publishing Co., Inc.* Kimber-Gray-Stackpole's Anatomy and Physiology *by Miller and Leavell, copyright* © 1972, *by Macmillan Publishing Co., Inc.*

Copyright renewed: 1930, 1937, *by Mary F. Kimber;* 1946, 1951, 1954, 1959, *by Theresa Buell;* 1962, 1966, 1970, 1976, *by Lutie C. Leavell.*

Macmillan Publishing Co., Inc.
866 Third Avenue, New York, New York 10022
Collier Macmillan Canada, Ltd.

Library of Congress Cataloging in Publication Data
Kimber, Diana Clifford.
    Kimber-Gray-Stackpole's Anatomy and physiology.
    First–13th ed. published under title: Textbook of anatomy and physiology.
    Includes index.
    1. Human physiology.   2. Anatomy, Human.   I. Gray, Carolyn Elizabeth, 1873–1938, joint author.   II. Stackpole, Caroline Emorette, joint author.   III. Miller, Marjorie A.   IV. Drakontides, Anna B.   V. Leavell, Lutie Clemson.   VI. Title.   VII. Title: Anatomy and physiology.   [DNLM: 1. Anatomy.   2. Physiology. QS4 K49t]
QP34.5.K55   1977   612   76-28770
ISBN 0-02-381220-6

Printing: 1 2 3 4 5 6 7 8   Year: 7 8 9 0 1 2 3

Preface
to the Seventeenth Edition

The authors believe that real understanding of the functioning of the human body is based on knowledge of its structure—microscopic and macroscopic. Recognizing the difficulty of presenting many facts, we have integrated cellular and anatomic structure with functioning of the various systems (circulatory, reproductive, etc.) and have kept the unified aproach of the previous edition in presenting nerve and blood supply with the individual systems.

Although the book is geared to students of nursing and various paramedical disciplines and, therefore, includes some clinical applications, it is also appropriate for students taking courses in general anatomy-physiology.

As in the sixteenth edition, content is organized in seven units—The Body as a Whole; Locomotion and Support; Awareness and Response to Environment; Body Maintenance: Distribution of Energy Sources and Nutrients, Processing and Utilization of Nutrients, and Homeostasis of Body Fluids; and Perpetuation of the Species. Updating of the text has meant significant revision of portions of all of the 26 chapters and complete rewriting of the following chapters: "Cell Structure and Function," "Tissues of the Body," "Skeletal Muscle," "Fundamental Aspects of the Nervous System," "The Brain and Cranial Nerves," and "Dynamics of Circulation."

The authors have made every effort to help the student visualize cellular and an-atomic structure. To this end new halftones demonstrating surface anatomy, photographs of bone, line cuts, wash drawings, photomicrographs, electron micrographs, scanning electron micrographs (showing three-dimensional structure), and x-rays are correlated with textual material. Four full-page color-halftone plates depicting anatomic relationships in the thorax and abdomen, gastrointestinal system, endocrine system, and genitourinary tract have been added.

Tables simplify and highlight important points within the chapters. New pedagogic aids in this seventeenth edition are outlines at the beginning, and lists of additional readings at the end, of each chapter. Questions for thought and discussion by students, as well as summaries, are also included. A glossary of terms pertaining to structure and function apears in the back matter.

Revised editions (1977) of the *Teacher's Guide, Workbook and Laboratory Manual* (by Drakontides, Miller, and Leavell), and *Test Manual* accompany this seventeenth edition.

The authors are grateful to Mrs. Barbara Finneson for the quality of the wash drawings and line cuts (many of which appear in two, three, and four colors) and to Ms. Carol Donner for the color-halftone plates and the cover illustration. Although the majority of electron micrographs and photomicrographs are the work of one of the authors (A.B.D.), we also wish to thank the investigators who provided others, as indicated in the legends to certain figures: Dr. James L. German, III, Cornell University Medical College; Dr. George Palade, Yale University School of Medicine; Dr. Eduardo Nunez, College of Physicians and Surgeons, Columbia University; Dr. Martha Speigelman, Memorial Sloan-Kettering Cancer Center; Dr. Thomas F. DeCaro, PMC Colleges; Dr. Daniel Gomez, Cornell University Medical College; Dr. Donald Orlic, New York Medical College; and John Patrikes, College of Physicians and Surgeons, Columbia University. The Radiology Department of The New York Hospital supplied the x-rays. Appreciation is extended to Dr. George E. Mauriello, Pace University, and Vera Stolar, Cornell University–New York Hospital School of Nursing, who each assisted in the reading and correction of the galleys. Finally, the authors are particularly indebted to Miss Joan C. Zulch, medical editor of Macmillan Publishing Co., Inc., for her continued help and support.

M. A. M.
A. B. D.
L. C. L.

Contents

UNIT IV

Body Maintenance: Distribution of Energy Sources and Nutrients

UNIT V

Body Maintenance: Processing and Utilization of Nutrients

UNIT VI

Body Maintenance: Homeostasis of Body Fluids

UNIT VII

Perpetuation of the Species

UNIT I

The Body as a Whole

The human organism is able to perceive its environment and its changes, to respond purposefully on an unconscious as well as a conscious level, and to think and make judgments based on its perceptions. The body is made up of cells that are grouped together to form specific tissues. Tissues, in turn, make up the organs of the body. Each organ has a specialized function, and specific organs make up the systems of the body. The study of the general organization of the body, and of cells and tissues, is the topic of this unit.

CHAPTER 1

Structure of the Body

Chapter Outline

Both anatomy and physiology are divisions of a larger science, biology, which deals with the acquisition and organization of knowledge about living things, plant and animal. *Anatomy* is the study of the parts of the living organism and their relationship to each other; *physiology* is the study of the way these parts accomplish their functions—the multiple activities involved in the life of the organism. It is impossible to separate completely these two areas of study, and the fullest understanding of each comes from an understanding of the other.

A comprehension of these desciplines allows one to understand how and why the body performs its many functions and is also the foundation of the treatment of the individual who is ill.

The Disciplines of Anatomy and Physiology

Anatomy is concerned with all those divisions of knowledge that deal with the study of bodily structure and the relationship of one part to another. The term *anatomy* is derived from the Greek word meaning "to cut up"; hence with the use of the scalpel and forceps, one studies the gross or macroscopic relationships of the body. This means of study is termed *gross anatomy.*

With the invention of the microscope and other techniques, the details seen with the naked eye could be magnified, thereby elucidating and extending the knowledge gained by macroscopic observation. *Mi-croscopic anatomy* includes all those divisions of anatomy that study structural relationships with the aid of the microscope. Included under this heading of microscopic anatomy are *histology*, which deals with the study of tissues or organs and their organization or architecture, and *cytology*, which encompasses the study of the cells that make up tissues.

The living body is subject to structural changes from the time of conception until the time of death. The study of structural changes and relationships that occur during the time interval from the fertilized egg until birth is the science of *embryology.*

3

Developmental anatomy deals with both embryology and later development, including the postnatal period, infancy, childhood, adolescence, and early, middle, and later maturity. When structural changes occur that deviate from the normal, this study becomes *pathologic anatomy* or *pathology*.

Man is part of the Animal Kingdom, and his structure reflects similarities and changes that have occurred through evolution. *Comparative anatomy* and *comparative embryology* are the study of animal structure, the similarities and differences among various orders or species of animals.

Physiology is the study of function or how the parts of the body work. The discipline of physiology encompasses many areas of knowledge. For instance, in order to understand how something works, the structure or anatomy must be known. Since the body is composed of many chemical constituents, a knowledge of chemistry and biochemistry is necessary. The movement of the body and many of its functions are clarified by an understanding of physics.

The full understanding of both anatomy and physiology is therefore based on a broad knowledge of many sciences.

General Structure of the Body

Among the smallest units making up the body are *molecules*. These are compiled to form many billions of *cells*. Many cells having similar structure and function are put together to form *tissues*. Tissues, in turn, make up individual *organs* (e.g., stomach, liver, kidneys). Organs having similar functions make up the *systems* of the body and are found in given *regions*.

ANATOMIC TERMINOLOGY; PARTS OF THE BODY; TERMS OF DIRECTION; PLANES OF SECTION

In order to identify the various regions or locations of the body, the human is placed in the *anatomical position*—namely, standing erect, eyes looking forward and upper extremities at the sides of the body, with the palms turned out. The various regions or parts of the human body are indicated in Figure 1–1. To define the position of one part of the body in respect to another, or the location of specific structures or organs, a system of terms is universally employed.

Anterior or ventral means toward the front of the body.

Posterior or dorsal means toward the back of the body.

The head end of the body is referred to as *cranial*, and the opposite end is called *caudal*.

A part above another is described as *superior*, and a part below another is *inferior*.

Medial is defined as toward the midline or median plane of the body, whereas *lateral* means away from the midline.

Proximal means toward the point of attachment to the body, and *distal* is away from the point of attachment. For example, the hand is proximal to the wrist, whereas the elbow is distal to the shoulder. The terms proximal and distal are used primarily to describe the extremities.

Internal and *external* are defined as toward the inside or toward the outside of the body. These terms are used to describe the walls of cavities or hollow viscera.

Additional information as to the location of various parts of the body is acquired by cutting the body in planes or sections. Special terms are given to different cuts made in the body (Figure 1–2). *Sagittal section* is a longitudinal or vertical cut par-

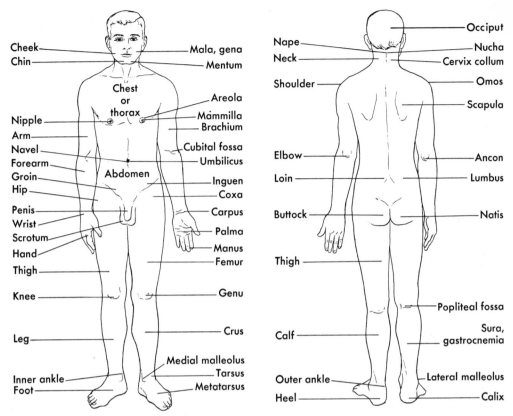

Figure 1–1. Anterior and posterior views of the body, the left side in the anatomical position. The regions are indicated; Latin terms or anatomic terminology given on right side of illustrations.

allel to the midline of the body and separates the body into right and left portions. The term *sagittal* comes from the sagittal suture of the skull and is in line with the midline of the body. A cut directly at the sagittal suture is *midsagittal*. A cut parallel to the sagittal suture but not at the midline is a *parasagittal section. Transverse* or *cross section* is a horizontal cut at right angles to the long axis of the body. This type of section separates the body into upper and lower portions. *Coronal* or *frontal section* is a vertical cut at right angles to the midline of the body and divides the body into anterior and posterior portions. The term *coronal* comes from the coronal suture of the skull, which runs at right angles to the sagittal suture.

The Cavities, Organs, and Systems of the Body

An anatomic characteristic of all vertebrate animals is the vertebral column, which supports the body. Another characteristic is that the body's organization is similar to a tube within a tube. In actuality the body wall is a tube that encloses another tube, the gastrointestinal tract. The cavity between these two tubes is the body cavity, or celom. This body cavity is divided into two large areas and several smaller areas. These cavities house specific organs and systems (Figure 1–3). The two major cavities are the dorsal (posterior) and the ventral (anterior).

The *dorsal cavity* is encased by the bones of the cranial portion of the skull

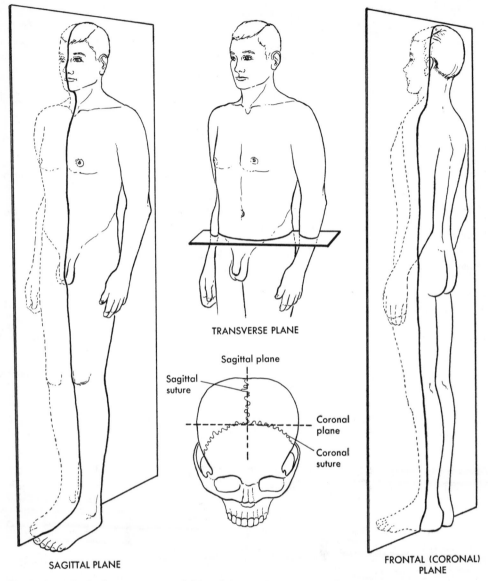

Figure 1–2. Sagittal, transverse, and frontal (coronal) planes of the body. The skull is shown, indicating location of sagittal and coronal sutures.

and the vertebral column. It consists of two parts: the *cranial cavity*, containing the brain, and the *spinal portion* of the dorsal cavity, containing the extension of the brain, the spinal cord.

The *ventral cavity* is located on the anterior surface of the torso and is divided by a muscular partition, the diaphragm, into an upper thoracic cavity and a lower abdominopelvic cavity (Plates I and II, inserted between pages 536–37).

The *thoracic cavity*, or chest, contains the trachea, bronchi, lungs, esophagus, the heart and great blood vessels connected to the heart, the thymus gland, lymph nodes, and nerves. The thoracic cavity is subdivided into two *pleural cavities*, one around each lung, and a central region, the *mediastinum*, which extends from the sternum to the vertebrae (Figure 1–4). Within the mediastinum is the *pericardial cavity*, which contains the heart. Each lung is covered by a thin serous membrane, the *pleura* (visceral pleura). The pleura ex-

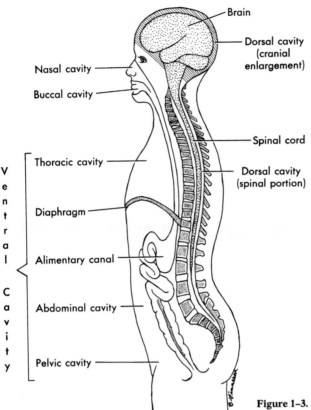

Figure 1-3. Longitudinal section of body to show dorsal and ventral body cavities.

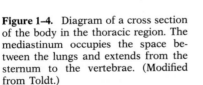

tends to and lines the inner surface of the chest wall (parietal pleura). The extension of the pleura from the lung to the chest wall creates the pleural cavity, which is a potential space filled with a serous fluid. A membrane that covers an internal body wall is called *parietal* (Latin *paries,* wall), and that covering the viscera (Latin *viscus,* an organ) is called *visceral.*

The *abdominopelvic cavity* is the second division of the ventral cavity. It is divided by an imaginary line across the prominent crests of the hip bones into the abdominal and pelvic cavities. The abdominal cavity contains the stomach, liver, gallbladder, pancreas, spleen, kidneys, and small and large intestines. The peritoneal cavity, a potential space like the pleural cavity, lies within the abdominal cavity and is formed by two layers of serous membrane: that

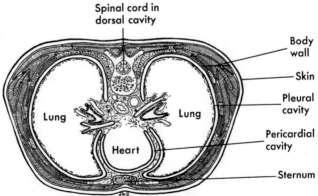

Figure 1-4. Diagram of a cross section of the body in the thoracic region. The mediastinum occupies the space between the lungs and extends from the sternum to the vertebrae. (Modified from Toldt.)

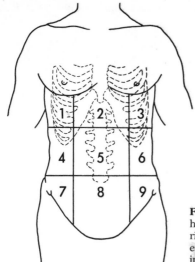

Figure 1–5. Upper horizontal line, transpyloric plane; lower horizontal line, transtubercular plane. Vertical lines, left and right lateral planes. Regions *1* and *3*, hypochondriac; *2*, epigastric; *4* and *6*, lumbar; *5*, umbilical; *7* and *9*, iliac or inguinal; *8*, hypogastric. (Modified from Drakontides *et al.*)

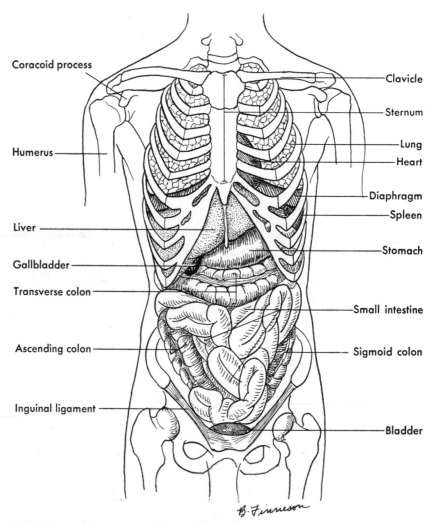

Figure 1–6. Diagram to illustrate the thoracic and abdominal viscera in their normal position and their relationship to the skeleton. Anterior view.

VISCERAL

covering the abdominal organs and that covering the inner surface of the abdominal wall.   *PARIETAL*.

The pelvic cavity is more completely bounded by bony walls than is the rest of the abdominal cavity. It contains the sigmoid colon, rectum, urinary bladder, and in the female the uterus.

In order to localize more specifically the organs of the abdominopelvic cavity, this area is often divided into nine regions, as indicated in Figure 1–5.

The skull contains small cavities in addition to the large cranial cavity.

The *orbital cavities* contain the eyes, the optic nerves, the muscles of the eyeballs, and the lacrimal apparatus.

The *nasal cavity* contains the structures forming the nose.

The *buccal* or *oral cavity* contains the tongue and teeth.

SYSTEMS

A *system* is an arrangement of organs closely allied to one another and involved with the same functions. Table 1–1 lists the systems of the human body, the component organs, and their functions; and Figures 1–6, 1–7, and 1–8 illustrate the position of the organs (refer also to Plates III and IV, inserted between pages 536–37).

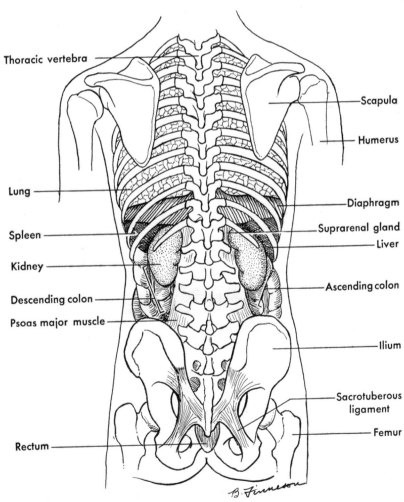

Thoracic vertebra

Scapula

Humerus

Lung

Diaphragm

Suprarenal gland

Spleen

Liver

Kidney

Ascending colon

Descending colon

Psoas major muscle

Ilium

Sacrotuberous ligament

Femur

Rectum

B. Finneson

Figure 1–7. Diagram to illustrate the thoracic and abdominal viscera in their normal position and their relationship to the skeleton. Posterior view.

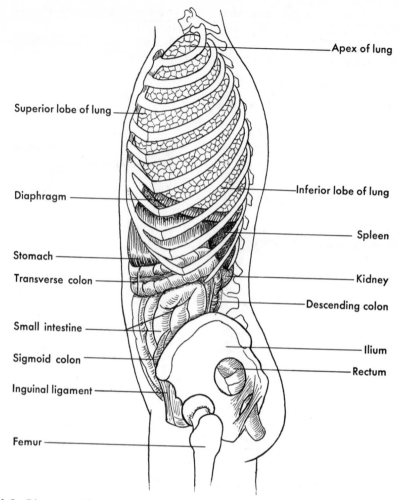

Apex of lung

Superior lobe of lung

Diaphragm

Inferior lobe of lung

Spleen

Stomach

Transverse colon

Kidney

Descending colon

Small intestine

Sigmoid colon

Ilium

Rectum

Inguinal ligament

Femur

Figure 1–8. Diagram to illustrate the thoracic and abdominal viscera in their normal position and their relationship to the skeleton. Lateral view.

TABLE 1–1
SYSTEMS OF THE HUMAN BODY

System	Main Organs and Tissues	Main Function
Skeletal	Bones and connective tissues that bind them together	Support, protection, and motion
Muscular	Striated skeletal muscle	To cause movement and to maintain static skeletal and postural support
Nervous	Brain, spinal cord, ganglia, nerves, organs of special sense	Reception of sensory stimuli or information from external and internal environment, integration of information, and an appropriate voluntary or involuntary response; integration of all bodily functions

TABLE 1–1 (*Continued*)

System	Main Organs and Tissues	Main Function
Circulatory	Heart, blood vessels, blood	Supplies the necessary nutrients and removes wastes from all tissues of body
Lymphatic	Lymph organs (tonsils, spleen, thymus, lymph nodes), lymph vessels, and lymph fluid	Lymph vessels return tissue fluid to blood vessels, thereby helping to maintain constancy of fluids around and inside all cells
Digestive	Mouth, esophagus, stomach, small and large intestine (all making up the gastrointestinal tract); and accessory organs—salivary glands, pancreas, liver, and gallbladder	Ingestion, digestion, and absorption of food and elimination of wastes
Respiratory	Nose, pharynx, larynx, trachea, bronchi, and lungs	Provides oxygen and eliminates carbon dioxide and maintains acid-base balance
Urinary	Kidneys, ureters, bladder, and urethra	Elimination of wastes in form of urine, maintenance of fluid and acid-base balance of body
Endocrine	Pituitary, thyroid, parathyroid, adrenals, endocrine portion of pancreas, ovaries, and testes; these glands secrete hormones into the blood	Hormones control many activities of body such as growth and development, metabolism, and fluid and electrolyte balance
Reproductive	Ovaries, uterine tubes, uterus, vagina, and vulva in the female Testes, seminal vesicles, penis, urethra, prostate, and bulbourethral glands in the male	Production of ova; development of embryo Production of sperm and secretion of semen; both male and female responsible for perpetuation of human species
Integumentary	Skin, nails, hair, glands of skin, sensory components of skin—i.e., nerve endings and specialized receptors	Protection, temperature control, perception of sensory stimuli—i.e., touch, temperature, pain

Questions for Discussion

1. While standing in the anatomical position, use each of the following terms in relation to the parts of the body.
   ventral–dorsal
   caudal–cranial
   superior–inferior
   proximal–distal
   medial–lateral
2. Define what is meant by a coronal and a sagittal section.
3. Name the body cavities and list the organs in each cavity.
4. What are the subdivisions of the thoracic cavity?
5. List the organs that make up the respiratory and digestive systems.
6. What is the function of the nervous system?

Summary

Biological Sciences deal with living things. Science is organized or classified knowledge.

Morphology, or Structure
{
Anatomy, macroscopic structure
　Gross Anatomy
Anatomy, microscopic structure
　Histology—study of tissues
　Cytology—study of cells
Embryology, early growth and development
Comparative anatomy { Study of similarities
Comparative embryology { and differences among
　　　　　　　　　　　{ various species of animals
Pathology, abnormal anatomy, both macroscopic and microscopic
}

Function—how parts of the body work {
Physiology
　Cellular—study of cells
　Comparative
}

Body cavities

Ventral cavity

Thoracic cavity
Two pleural cavities
　Mediastinum
　Contents: esophagus, trachea, lungs, heart, blood and lymph vessels, thymus gland, nerves
Diaphragm—separates thoracic and abdominopelvic cavities

Abdominopelvic cavity
Abdominal cavity
　Peritoneal cavity
　Contents: stomach, spleen, pancreas, liver, gallbladder, kidneys, small and large intestines
Pelvic cavity
　Contents: bladder, rectum, in female uterus

Dorsal cavity {
Cranial cavity—brain
Spinal canal—spinal cord
}

Facial aspect of skull

Orbital cavities {
Eyes, optic nerves,
　muscles of the eyeballs,
　lacrimal apparatus
}
Nasal cavity—structures forming the nose
Buccal cavity { Tongue, teeth, salivary glands

Body Regions

Anatomical position and terms

Dorsal	Superior
Ventral	Inferior
Anterior	Cranial
Posterior	Caudal

Anatomical planes
{
Medial
Lateral
Sagittal and midsagittal
Coronal (frontal)
Transverse
}

Tissue. A group of cells with similar structure and function
Organ. A physiologic unit composed of two or more tissues associated in performing some special function
System. An arrangement of organs, closely allied and concerned with the same function. Systems found in the human body:

Skeletal	Endocrine
Muscular	Respiratory
Nervous	Digestive
Vascular, or circulatory	Excretory
Lymphatic	Reproductive

Additional Readings

BASMAJIAN, J. V.: *Grant's Method of Anatomy*, 9th ed. Williams & Wilkins Co., Baltimore, 1975, Section 1.

HOLLINSHEAD, H. W.: *Textbook of Anatomy*, 3rd ed. Harper & Row, New York, 1974, Chapters 1 and 2.

PANSKY, B., and HOUSE, E. L.: *Review of Gross Anatomy*, 3rd ed. Macmillan Publishing Co., Inc., New York, 1975.

TOBIN, C. E.: *Basic Human Anatomy*. McGraw-Hill Book Co., New York, 1973, Chapter 1.

WOODBURNE, R. T.: *Essentials of Human Anatomy*, 5th ed. Oxford University Press, New York, 1973, Section 1.

CHAPTER 2

Cell Structure and Function

Chapter Outline

The human animal is made up of an enormous number of cells, thereby being referred to as a multicellular animal. This is in distinction to animals that consist of only one cell or are unicellular, such as the ameba or the paramecium. Each cell is a microcosmos, containing all the structures and machinery necessary for its survival; however, in higher organisms, cells usually function together as a group or tissue. All living matter, or protoplasm, is organized in a specific chemical, physical, and structural arrangement.

The term *cell* was first used by Robert Hooke[1] in 1665, when he described his microscopic examination of a piece of cork. The knowledge of structure and function of the cell has been expanded by the invention and adaptation of new methods and techniques. It is not within

[1] Robert Hooke, English experimental physicist (1635–1703).

the scope of this textbook to cover the numerous methods used in cell biology, and only a few are listed as examples.

In order to adequately visualize the parts and components of cells or tissues they must be stained. In the study of tissues and cells with *light microscopy*, many types of chemical stains are used. These stains react with specific chemical components of the cell; hence not only is structure clarified but the chemical nature of cells may be determined. For example, in a section of tissue stained with hematoxylin and eosin, the nucleus of a cell appears as a round to oval structure, which stains blue. The blue color is due to a concentration of the nucleic acid DNA. *Autoradiography* is a method in which radioactive substances are used and allows for the localization of metabolic events in cells and tissues; *immunohistochemistry* is a method in which specific proteins or polysaccharides may be localized in the cell. Although our information about the cell began with the light microscope and was expanded using other types of microscopy (i.e., phase contrast, interference, fluorescence), it was only with the invention of the *electron microscope*, which greatly increased the power of resolution, thus allowing for the magnification of structures many thousands of times, that we were able to observe the ultrastructure of cells and tissues. The *scanning electron microscope* is yet another instrument that has broadened our knowledge. Whereas the light and electron microscope give a one-dimensional view of structure, the scanning electron microscope reveals structure in three dimensions. The ultrastructure of the components of the cell is a foundation for understanding function; however, it is also necessary to define the chemistry, physical properties, and physiology to comprehend totally how cells work.

CHEMICAL COMPOSITION OF THE CELL

Chemical analysis of the human body has shown that it contains the following *chemical elements:*

Oxygen	(O)	Form 96% of total weight of body	65.0%
Carbon	(C)		18.0
Hydrogen	(H)		10.0
Nitrogen	(N)		3.0
Sulfur	(S)		0.25
Calcium	(Ca)		2.2
Phosphorus	(P)		0.8–1.2
Potassium	(K)		0.35
Chlorine	(Cl)		0.15
Sodium	(Na)		0.15
Magnesium	(Mg)		0.05
Iron	(Fe)		0.004
Iodine	(I)		0.00004
Silicon	(Si)	Very minute amounts	
Fluorine	(F)		

Traces of Cu, Mn, Co; perhaps traces of Ni, Ba, Li

In the human body free oxygen, hydrogen, and nitrogen have been found in the blood and intestines, but the bulk of these elements, as well as of all the others, exists in the form of complex compounds that are constituents of the cells and body fluids. The compounds are divided into two classes, organic and inorganic.

Organic Compounds. The *organic compounds* found in protoplasm are proteins, carbohydrates, and lipids. Wherever found in living organisms, these compounds are fundamentally similar.

In addition to this universal similarity, the proteins, carbohydrates, and lipids of each species of plant or animal possess distinctive characteristics. The essential chemical elements are present in varying absolute quantities, in varying relative quantities, and in varying combinations.

Proteins. All proteins contain the elements carbon, hydrogen, oxygen, and nitrogen, and some also contain sulfur and phosphorus. Proteins are molecules of large molecular weight, made up of chains of amino acids. Amino acids are chemical substances that contain the chemical group $-NH_2$; examples are glycine, alanine, serine, leucine, and valine. There are 20 kinds of amino acids in proteins. Not all of these amino acids are present in all proteins, and those present exist in proportions characteristic of each protein.

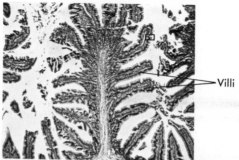

Light Micrograph (x70)

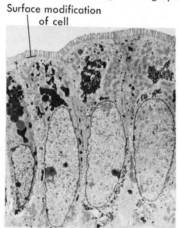

Surface modification of cell

Electron Micrograph (x4000)

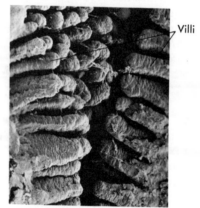

Villi

Scanning Electron Micrograph (x110)

Figure 2–1. Sections of the small intestine as visualized by three types of microscopes. The light microscope reveals overall pattern of the tissue; details are not readily apparent. With the electron microscope the detail of cells (boxed area of light micrograph), their internal structure, and the appearance of the surface can be seen. The scanning electron microscope reveals the three-dimensional nature of the projections (villi) of the small intestine. (Electron micrograph, courtesy of Dr. D. Orlic, New York Medical College, Valhalla, N.Y.; scanning electron micrograph, courtesy of Mr. J. Patrikes, College of Physicians and Surgeons, Columbia University, New York, N.Y.)

The amino acids of a protein are held together by peptide bonds (—CO—NH—). The product formed by the union of a large number of amino acids is called a *polypeptide.*

Proteins are among the most important constituents of cells. They act as catalysts (enzymes); they are essential for proper bodily functions (e.g., hemoglobin, insulin); they make up the contractile components of muscle tissue; they are responsible for the elasticity and tensile strength of tissues; and they have many other functions. About 3.5 per cent of the total protein present in the body may be destroyed and resynthesized. Protein in the diet is therefore essential to replace the continuous breakdown of body proteins.

Carbohydrates. Carbohydrates are substances containing carbon, hydrogen, and oxygen, the hydrogen and oxygen being present in the same ratio as in water. Thus, the general formula of carbohydrates is $C_n(H_2O)_n$. The simpler carbohydrates are known as sugars. A sugar structurally containing up to six carbon atoms is known as a *monosaccharide* (e.g., glucose, fructose). When two monosaccharides come together, we speak of a *disaccharide* (e.g., maltose, lactose, and sucrose); when a large number of monosaccharide units are present, it is a *polysaccharide* (e.g., starch, glycogen).

Carbohydrates are a chief source of energy for cells, and supply the energy needs of the body. Carbohydrates in excess of the body's needs are converted to glycogen and to fat (adipose tissue) and stored.

Lipids. The essential elements in lipids are carbon, hydrogen, and oxygen. Lipids are a heterogenous group of compounds such as fats, waxes, phospholipids, and sterols. This group of compounds is characterized by their relative insolubility in water. The most common lipids in the body are the neutral fats. These are triesters of fatty acids and glycerol.

Lipids are responsible for the structural integrity of the cell. They are part of the cell membrane and are involved in maintaining cell permeability. Fats are a source of energy and are stored in adipose tissue.

Inorganic Compounds. Inorganic compounds include mineral elements and water. Mineral elements exist in the form of salts or combined with proteins, carbohydrates, and lipids. A salt when dissolved in water dissociates into electrically charged parts called *ions*. If the ion carries a positive charge, it is a *cation;* a negatively charged ion is referred to as an *anion.* Salts are found throughout the body—in the cell (intracellular), in extracellular fluids, and in blood and lymph. The ions of these salts have essential roles in the proper maintenance and control of bodily functions, and are listed in Table 2–1.

Water. Water is the most abundant constituent of tissues and constitutes about two thirds of body weight and more than 70 to 75 per cent of nonbony body weight. It is difficult to obtain the normal water content of an isolated living cell. One estimate gives it at from 85 to 92 per cent of the weight of the cell. It is evident that water is by far the predominant constituent of protoplasm.

PHYSICAL CHARACTERISTICS OF PROTOPLASM

With the exception of dense connective tissue and bone, the intercellular, or interstitial, material is a viscous solution which is in general similar to cytoplasm—con-

TABLE 2–1
MINERAL ELEMENTS IN THE HUMAN BODY

Inorganic Element	Essential Function
Sulfur	Component of amino acids cystine and methionine, of certain vitamins, of coenzyme A
Calcium	For proper formation, maintenance of bone and teeth; necessary for proper activity of nerve and muscle; controls blood clotting and cell permeability
Phosphorus	In form of phosphate—for proper formation, maintenance of bone and teeth; component of many metabolic pathways and of many proteins, nucleic acids, and ATP
Potassium	Present in all cells; involved in conduction of nerve impulse and muscle contraction; participant in electrolyte and water balance
Chlorine	Important in controlling water movement in cells; makes up greatest anion of blood
Sodium	Important in controlling water movement in cells; in conjunction with K^+ involved in conduction of nerve impulse and muscle contraction; principal cation of blood
Magnesium	Component and activator of many enzymes
Iron	Component of hemoglobin (pigment in blood that carries oxygen) and of myogloblin (protein of muscle)
Iodine	Essential for the synthesis of hormone of thyroid gland

taining inorganic chemicals, proteins, carbohydrates, and lipids—the primary difference being one of kind of protein present and of amounts of the various chemicals present. Table 2–2 illustrates the difference in mineral makeup between the fluid within the cell and that without the cell. In some tissues fibers are embedded within

TABLE 2–2
CONCENTRATION OF MAJOR ELECTROLYTES IN CELLULAR AND INTERSTITIAL FLUID

	Cell Fluid	Tissue Fluid
Na^+	10 mEq/L	146 mEq/L
K^+	150	4
Ca^{++}	40	2.5
Mg^{++}	40	1.5
HCO_3^-		30
Cl^-		115
HPO_4^-	140	2
SO_4^-		1
Organic acids$^-$		5
Protein$^-$	40	1

[handwritten annotation: IDENTICAL TO SEA WATER]

the *matrix,* or background substance of intercellular material.

Physically cytoplasm and intercellular material are emulsions containing *ions* (electrically charged particles) and non-charged particles dissolved or suspended in water. Several advantages accrue from the fact that water is the universal solvent in animal tissue.

1. The solvent power of water is great. No other substance can compare with water in relation to the kinds of substances that can be dissolved in it and the great and varied concentrations that can be obtained.

2. The ionizing power of water is high; hence, the great number and many kinds of solute molecules yield large numbers of varied ions.

3. Water has a high specific heat. This means that it can hold more heat with less change of temperature than most substances; hence, the heat produced by cell metabolism makes comparatively little change in the temperature of the cell.

4. The heat-conducting power of water is high (for fluids). This means that the heat produced in the cells can pass to body fluids even if the temperature of the cell is barely above that of the fluid around the cell, which likewise can hold this heat with comparatively little rise in temperature and pass it on with little change in temperature to the blood and finally to the skin.

5. The latent heat of evaporation of water is high. Owing to the high latent heat of evaporation, a maximum of heat is taken from the skin for the evaporation of perspiration—about 0.5 large calorie per gram of water evaporated.

6. Water has a high surface tension. Because

of this, any immiscible liquid with which it comes in contact must expose to it the minimum of surface. The substances that dissolve in water lower its surface tension, since dilute concentrations especially have a tendency to lower surface tensions, and because of the great solvent capacity of water, the surface tension of water can be lowered greatly and by a great variety of substances. This permits the area of contact of the immiscible liquids to be enormously increased. Also, it is known that any dissolved substances that do lower surface tension will accumulate or be *adsorbed* at the surface of contact of the immiscible liquids.

The particles of material dispersed throughout protoplasm vary in size and may be dissolved, as in the case of glucose, or merely suspended in the case of fat droplets in the cytoplasm. Particles with diameters less than 0.1 mμ are dissolved in water and are said to form true solutions, whereas particles 0.1 to 1 mμ in diameter are termed colloidal particles and form colloidal solutions. Colloids may be individual molecules of large size, e.g., protein, or groups of molecules. Other characteristics of colloids that are important in cell physiology include:

1. They can take up large quantities of water and hold it within the cell.
2. Owing to their large size they do not diffuse readily.
3. They adsorb other substances at their interphase, or surface.
4. They possess electrical charges that contribute to chemical activity.

Sometimes in experimental work cells behave as *sols,* that is, as solutions in which the continuous phase is water, the colloidal particles of proteins and lipoids being dispersed in it. Sometimes they behave as *gels,* in which the protein molecules form networks enclosing areas of water between them. It has been said of protoplasm that its characteristics are like those of a reversible sol-gel colloidal system.

The cell is regarded as a highly organized unit engaged in ceaseless chemical activities. These activities are dependent on the continuous reception of substances from the so-called *internal environment* (tissue fluid) and the continuous elimination of substances to this tissue fluid. The circulatory liquids continually bring substances from the supply organs, which obtain them from the *external environment,* and continuously take eliminated substances to the eliminating organs for final removal to the external environment. Keeping the cell environment constant, within a narrow range, as regards oxygen, nutrients, acidity, and temperature is critical for optimum cell functioning. The term *homeostasis* was coined by Walter B. Cannon[2] and refers to the overall processes of maintaining optimum internal environmental conditions.

ENZYMES

Enzyme Action. Nutrients that enter the cell have potential value (1) as a source of energy, (2) as building blocks for the parts of the cell itself (especially if it is growing or dividing), and (3) as building blocks for products that the cell will secrete.

The utilization of nutrients is the process of *metabolism.* It is the step-by-step building of small molecules into large ones, referred to as anabolism, or the breaking down of large molecules to smaller ones, termed *catabolism.* These chemical reactions are many and varied and require the presence of enzymes. An *enzyme* is a catalyst and as such increases the rate of a chemical reaction without itself being used up. Enzymes are classified as *endoenzymes* (those that act inside of the cell) and as *exoenzymes* (those that act outside the cell—e.g., digestive enzymes).

Characteristics of Enzyme Action. In general, all enzymes are influenced by temperature and pH, and must have the appropriate substance or *substrate* to act upon. When the temperature or pH of the fluid in which the enzyme is found is not at optimum levels, the enzyme action is slowed down, or it may be completely ineffective. Endoenzymes act best at body temperature and at the pH of the cytoplasm and body fluids. A given enzyme can catalyze only one type of reaction, or

[2]Walter B. Cannon, American physiologist (1871–1945).

sometimes only one particular reaction. For the metabolic reaction to occur, the substrate molecule must come into contact with the specific enzyme. It is believed that there is a specific configuration for each enzyme and its substrate. Thus a particular enzyme and its substrate must possess shapes that complement each other. Inhibitors of enzymes, such as certain metals, act by interfering with this complementary shape, or by attaching to the site on the molecule that will react, e.g., to accept the hydrogen atom in the case of dehydrogenases.

Table 2–3 lists important classes of enzymes and their actions. Many of the biologic reactions are reversible, the enzyme influencing the speed of both the forward and the reverse reaction, operating to bring about equilibrium.

Endoenzymes are found in the active form within the cell. This is not true of *exoenzymes,* those that act outside the cell, for these are usually secreted in an inactive form and must be activated by another substance or the pH of the environment before they can catalyze the particular chemical reaction. The inactive form is known as *zymogen* or *proenzyme.* Activation of the proenzyme is believed to involve a chemical change, such as the removal of a small group from the molecule and exposure of the active site of the enzyme.

In some cases the action of an enzyme is helped by, or perhaps is dependent upon, the presence of some other substance. An example of this activity is the interaction of bile salts and lipase on fat digestion. The bile salts emulsify fat droplets, thereby increasing the surface area available for enzyme activity. These cases of *coactivity* are to be distinguished from activation by the fact that the combination may be made or unmade. For example, in a mixture of bile salts and lipase, the bile salts may be removed by dialysis. In activation, on the contrary, the active enzyme cannot be changed back to the inactive zymogen.

Nature of Enzymes. Most enzymes are proteins of high molecular weight, hence

TABLE 2–3
ENDOENZYME CLASSIFICATION

Name*	Action
1. Hydrolases:	Split molecules into smaller ones through utilization of H_2O
a. Esterases:	Split ester linkages of acids and alcohols
cholinesterase	Split acetylcholine to acetic acid and choline
lipases	Split fats to fatty acids and glycerol
phosphatases	Split phosphate group from phosphoric acid esters
pyrophosphatases	Split phosphate group from high-energy phosphate compounds
nucleases	Split nucleic acids to nucleotides
b. Carbohydrases:	Break down polysaccharides and other compounds with similar chemical bonding
amylases	{ Split glycogen to glucose { Split starch to maltose, then to glucose
hyaluronidase	Splits hyaluronic acid, intercellular material
c. Proteases:	Split peptide linkages of proteins and peptides
carboxypeptidases	Split terminal peptide bond to free amino acids from protein
proteinases	Split proteins by attacking interior peptide bonds
2. Phosphorylases:	Split molecule by addition of phosphate radical
muscle phosphorylase	Phosphate + glycogen $\rightleftarrows$ glucose phosphate
3. Oxidation-reduction enzymes	
a. Dehydrogenases	Oxidation of a compound by removal of $2H^+$
b. Oxidases	Addition of oxygen to a compound
c. Catalases	Remove atomic oxygen from H_2O_2; associated with cytochrome systems of mitochondria
4. Transferases	Transfer a radical from one compound to another
a. Transaminases	Transfer NH_2^-
b. Hexokinases	Transfer phosphate from adenosine triphosphate to glucose
5. Decarboxylases	Remove CO_2 from a compound without oxidation
a. Carbonic anhydrase	Carbonic acid $\rightleftarrows CO_2$ and H_2O
6. Hydrases	Remove H_2O from a molecule
7. Isomerases	Move radical from one part of molecule to another
8. Condensing enzymes	Transfer acetyl radical from acetyl coenzyme A into the citric acid cycle

*Note that the -ase ending indicates an enzyme and the prefix indicates its action (e.g., transaminase) or its substrate (e.g., amylase).

cannot diffuse across cell membranes. Some enzymes, such as pepsin and trypsin, appear to be simple proteins; others resemble the conjugated proteins in that they function with a nonprotein component known as a *coenzyme*. The importance of vitamins is becoming increasingly evident as more is learned about enzyme activity within the cells. The B-complex vitamins in particular are known to form parts of the molecules of various enzymes involved in energy release. These are discussed more fully in Chapter 22.

ENERGY SOURCES

In order to perform its varied functions, a cell must have a source of energy. Carbohydrates, lipids, and proteins are storage forms of energy. When these organic compounds are broken down to their simpler products in a test tube, heat is given off. Heat is a source of energy; however, the cell cannot transform heat energy into work. The energy released during the breakdown of organic compounds within the cell is captured by directly transferring chemical potential energy from one molecule to another. In all living cells, from bacteria to man, the major molecule that performs as an energy-carrier is *adenosine triphosphate* (ATP). This molecule consists of (1) a ring structure known as adenine, (2) a five-carbon sugar molecule, ribose, and (3) three phosphate groups linked in series to the sugar. When the chemical bond that joins the terminal phosphorus atom to ATP is broken in the presence of water, 7 kcal/mole of energy are released. This chemical bond is thus termed a high-energy phosphate bond. The resulting product is adenosine diphosphate (ADP) and inorganic phosphate (P_i). Conversely, when ATP is synthesized from ADP and phosphorus, 7 kcal/mole of energy must be added to the reaction.

There is a continual cycling of energy through ATP molecules in the cell. The energy released in the conversion of ATP to ADP is used to perform the many functions of the cell. In order to maintain a constant supply of ATP, ADP couples to energy-releasing reactions, namely, the decomposition of carbohydrates, lipids, and proteins. This transfer of energy to ADP takes place in an organelle of the cell, the mitochondrium.

PASSAGE OF MATERIAL ACROSS THE CELL MEMBRANE

The cell, as a living functional unit, must be able to both acquire materials from its surrounding medium and also either secrete substances it has manufactured or excrete waste materials. Therefore, it becomes obvious that materials must be able to enter and leave the cell. The physical processes that govern the movement of materials across the cell membrane include diffusion, osmosis, filtration, active transport, pinocytosis, and phagocytosis. The processes of diffusion, osmosis, and filtration are collectively referred to as *passive processes;* in other words, a source of energy is *not* required. *Active processes* are those that require a source of energy either as a specific molecule in the cell membrane, as in the process of active transport, or energy expended by the cell to carry out the process, as in pinocytosis and phagocytosis.

Diffusion. The term *diffusion* is applied to the spreading or scattering of molecules of gases or liquids. When two gases are brought into contact, the continual movement of the molecules of gas will soon

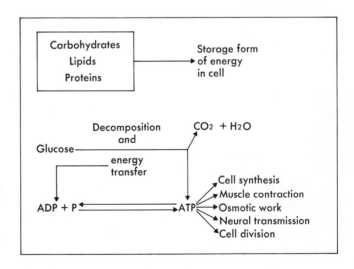

Figure 2–2. Energy sources and some cellular processes that require ATP.

produce a uniform mixture. If a solution of salt is placed in a receptacle and a layer of water poured over it, there will be a mingling of salt molecules and water molecules, producing a solution of uniform composition.

Molecular movement of particles is random; however, they will move in greater number toward the area where they are fewer in number, that is, from an area of greater concentration to one of lesser concentration (of that particular substance). Eventually an equilibrium will be reached in which all areas of the solution are identical. In the case of the salt solution, there will be the same number of salt molecules relative to water molecules in all parts of the solution. Oxygen moves from the blood into the fluid around the cell and into the cell as a result of diffusion. The amount of oxygen in the blood is much greater than that in the cell. Therefore, it moves to the area of lesser concentration. In this instance a static equilibrium is never reached, but a dynamic equilibrium is kept constant. Oxygen supply to the cell is continually replenished owing to movement of blood through the lungs and the circulatory system; the oxygen is continually used in metabolic processes once it enters the cell.

Osmosis. The usual definition of osmosis is the movement of solvent particles, such as water molecules, through a membrane. If a saline solution and water are separated by a membrane permeable to water, the water molecules will pass through the membrane to the salt solution, thereby raising the level of the latter. Theoretically, molecules of liquid are constantly in motion, a permeable membrane offering no resistance to their passage, and therefore the movement of water particles is in both directions; but the water molecules will travel in greater numbers per unit of time from the place where their number is highest to where it is lowest. If the number of the water particles on both sides of the membrane is the same, in a given time equal numbers of particles will travel in each direction, and equilibrium will be reached. Pressure is exerted by the flow of

these water molecules across the membrane and may be expressed as millimeters of mercury (mm Hg). Every solution then has a potential osmotic pressure.

Osmotic pressure is determined by the *number* of particles of solute dissolved in a particular solution. The more particles in solution, the greater the *osmolality* of that solution and the greater is its "pull" for water. In other words, water moves toward the area of greater osmolality. It is important to remember that the particles in solution are the critical factor in determining osmotic pressure. A solution made from an ionizing substance (e.g., sodium chloride) will have a greater osmolality than one containing an equal amount of nonionizing substance such as glucose.

In physiology the osmotic characteristics of different solutions are often determined by the way in which they affect the red cells of the blood. That is, their effect is compared with that of the blood serum. If red cells are subjected to contact with any fluid other than normal serum, they may remain unchanged or they may shrink or swell. If they remain unchanged, the solution is said to have the same osmotic characteristics as the blood serum and is called *isotonic.* If they shrink, the solution has higher osmotic characteristics than that of the blood serum and is called *hypertonic.* If they swell, the solution has lower osmotic characteristics than that of the blood serum and is called *hypotonic.*

Sometimes the word *dialysis* is used for the diffusion of molecules of the soluble constituents (solutes) through a permeable membrane. If two solutions of unequal concentration are separated by a membrane that is permeable to the solute, a greater number of solute particles will pass from the more concentrated solution to the less concentrated, per unit of time. The diffusing particles may be ions, molecules, or small molecular aggregates.

Filtration. Filtration is the passage of a substance in solution across a semipermeable membrane as a result of a mechanical force (e.g., gravity, blood pressure). The movement is from an area of higher pressure to an area of lower pressure, and the size of the membrane pores will determine which molecules will be filtered. Filtration, therefore, separates large molecules from small ones. The process of filtration that occurs in the kidney thus allows proteins, which are of a large molecular size, to be retained by the body, whereas waste materials of smaller molecular weight may be excreted.

Active Transport. The process of active transport involves the movement of materials across the cell membrane against a concentration or electrical gradient (if the

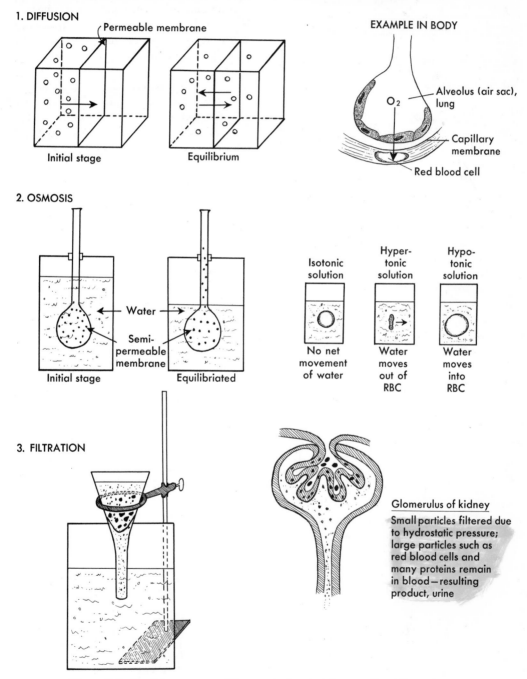

1. DIFFUSION

Permeable membrane

Initial stage　　Equilibrium

EXAMPLE IN BODY

O_2　Alveolus (air sac), lung

Capillary membrane

Red blood cell

2. OSMOSIS

Water

Semi-permeable membrane

Initial stage　　Equilibriated

Isotonic solution — No net movement of water

Hypertonic solution — Water moves out of RBC

Hypotonic solution — Water moves into RBC

3. FILTRATION

Glomerulus of kidney

Small particles filtered due to hydrostatic pressure; large particles such as red blood cells and many proteins remain in blood—resulting product, urine

Figure 2–3. Passive processes of transport of material across the cell membrane.

solute is a charged particle) and requires energy. The movement of material is in a direction opposite to what would be expected from the principles of diffusion and osmosis. In active transport, it is felt that the transported substance is attached to a component of the cell membrane. This may be a protein, a lipid, or most often an enzyme referred to as a "carrier" that picks up the substance for transport and carries it across the membrane. In the process of active transport the cell provides energy, usually in the form of ATP, to activate the carrier.

Pinocytosis. In the process of pinocytosis the cell membrane initially indents, then surrounds, and finally engulfs within a

vacuole, small molecules. The term itself means "cell drinking." Pinocytosis is not an alternative process of active transport, but rather a supporting one, since it provides a much larger area of cell membrane where both active and passive transport may be carried out. Molecules that may be used by the cell are transported in this manner and also noxious substances are ingested. If the substance is of a noxious nature, lysosomes become attached to the vacuole and cellular digestion occurs.

Phagocytosis. This process is quite similar to pinocytosis, the major difference being the ingestion of larger particles such as bacteria, cell fragments, or foreign material. The vacuole formed within the cell is

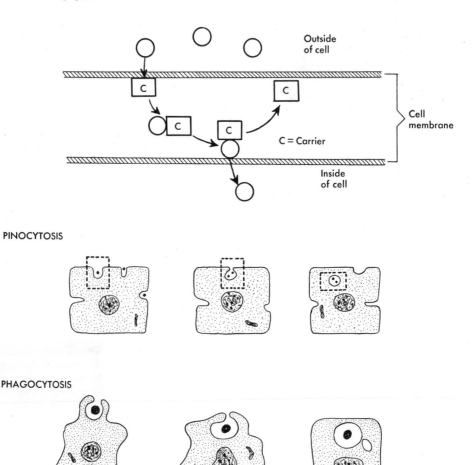

Figure 2–4. Active processes of transport of material across the cell membrane.

called a phagosome and attaches to a lysosome. The hydrolytic enzymes of the lysosome effectively digest the particulate matter.

CELL STRUCTURE, ORGANELLES, AND FUNCTION

Although the multicellular animal arises from a single cell, during the process of development many cells are formed and groups of cells differentiate. It is only by this process of differentiation, in which cells become different or specialized, that the organism can perform its various functions. In the field of biology, we attribute certain properties to protoplasm. All cells exhibit these properties to some degree; however, certain specialized cells are characterized by a particular property.

The physiologic properties of protoplasm include:

1. *Respiration*—taking in oxygen and using this for oxidation of food substances with the resulting liberation of energy.
2. *Excretion*—the ability to eliminate waste materials.
3. *Absorption and assimilation*—living cells can absorb food and other substances and utilize them.
4. *Secretion*—cells can synthesize useful substances from those that they absorb and give off these secretory products.
5. *Irritability*—the property that enables a cell to respond to a stimulus (change in its environment).
6. *Conductivity*—the ability to transmit a wave of excitation throughout the substance of the cell.
7. *Contractility*—the ability of a cell, on being stimulated, to shorten and return to its original length when the stimulus is removed.
8. *Cell division*—the ability of cells to grow to a limited extent and produce other cells.

Shape and Size of Cells. The cells that make up the tissues of the body differ in their shape and size. Figure 2–5 is an illustration of some cells of the body in reference to their shape and size. Usually the red blood cell, which is 7.5 μm in diameter, is used as a standard of reference to which other cells are compared. Some cells, such as the leukocyte, have a variable shape. Other cells have a typical shape, such as epithelial cells, erythrocytes, and sperm cells. Many of the cells of the body are either spherical or columnar in shape. Often the shape of the cell may be related to its function. The skeletal muscle cell, for example, is a long cell, sometimes inches in length. This shape is the best form for contractility, which is a characteristic feature of muscle cells. The neuron, the cell of the nervous system, has a number of processes; one of these processes, the axon, may be 1.2 to 1.5 m (4 to 5 ft) long. The axon is responsible for conveying information to various parts of the body. Table 2–4 lists relative dimensions and the various instruments used for the observation of cells and their component parts.

Although the size of cells may vary, they are all relatively small. One may ask, why are there so many small cells instead of larger and fewer cells? In the multicellular organism, most of the cells are not in direct contact with sources of nutrients, such as blood. This means that both nutrients and waste material must have an efficient means of entry and exit. A large size physically reduces the amount of surface area, whereas the smaller the cells, the greater the surface area (Figure 2–6) and hence the greater the area for transport of materials.

Certain cells (epithelial) exhibit a specialization on their apical surface, namely that surface facing a lumen. When viewed with the electron microscope, the surface of the cell is thrown into many folds called *microvilli*. These folds effectively increase surface area for absorption. Yet other cells have long filamentous protoplasmic processes called *cilia*. Cilia move, and the collective movement of cilia on the many cells lining the respiratory passageway helps to propel foreign material to the outside. The flagellum is a single long

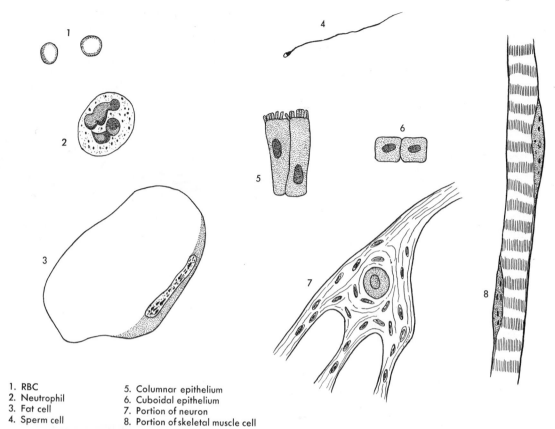

1. RBC
2. Neutrophil
3. Fat cell
4. Sperm cell
5. Columnar epithelium
6. Cuboidal epithelium
7. Portion of neuron
8. Portion of skeletal muscle cell

Figure 2–5. Diagram of a variety of cells from the human body showing variation in size and shape. Cells are drawn approximately to scale; the red blood cell (7.5 μm) provides a measure for other cells.

TABLE 2–4

Dimension	Discipline	Structure	Instrumentation
0.1 mm (100 μm) or larger	Anatomy	Organs	Eye or simple lenses
100 μm to 10 μm	Histology	Tissues	Various types of light microscopes
10 μm to 0.2 μm (2000 Å)	Cytology	Cells, bacteria	
2000 Å to 10 Å	Ultrastructure	Cell organelles, viruses	Polarization microscopy, electron microscope
Smaller than 10 Å	Molecules and atomic structure	Arrangement of atoms	Electron microscope, x-ray diffraction

1 meter (m) = 100 centimeters (cm); 1000 millimeters (mm); 39.37 inches
1 millimeter (mm) = 0.001 meter; 1000 micrometers or microns (μm or μ); $\frac{1}{25}$ inch
1 micrometer (μm) = 0.001 mm; 1000 millimicrons (mμ); 10,000 Angstrom units (Å)
1 millimicron (mμ) = 0.001 μm; 10 Å
1 Angstrom unit (Å) = 0.1 mμ

Figure 2–6. Diagrams to illustrate increased proportion of total surface area to total volume on fragmenting an object. *A* is a cube 5 cm (2 in.) on a side. *B* shows planes in which it may be cut to produce the eight cubes shown in *C*. Each of these eight cubes is 2.5 cm (1 in.) on a side. The total area of *A* is 60 sq cm (24 sq in.). The total area of the eight cubes in *C* is 120 sq cm (48 sq in.). The volume of material in *A* and *C* is the same, 20 cu cm (8 cu in.).

cilium, which in the case of the sperm cell, by its movement, allows this cell to swim.

Structural Components of the Average or Typical Cell.

All cells are essentially similar in their basic features; however, they differ in the details of organization. In the discussion of the cell, therefore, one describes the *typical* or *average* cell. The typical cell is a composite of many different cells and is illustrated in Figure 2–7.

The cell is a microscopic unit of *protoplasm* contained within a double-layered envelope, the *cell membrane* or *plasma membrane*. All cells of the body, with a few exceptions such as the mature erythrocyte (red blood cell), contain a more or less centrally located *nucleus*. Protoplasm is the living substance of the cell. Protoplasm

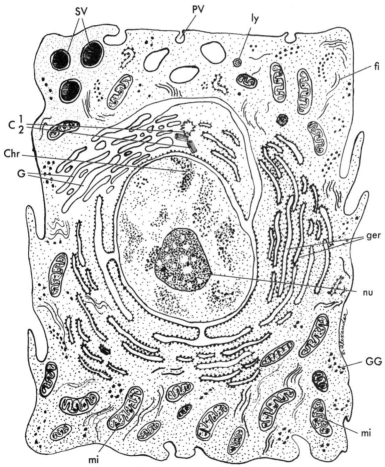

Figure 2–7. Diagram of typical cell illustrating fine structure. *C*, centriole (*1*, cross section; *2*, longitudinal section); *Chr*, chromatin; *fi*, filaments; *G*, Golgi complex; *ger*, granular endoplasmic reticulum; *GG*, glycogen granules; *ly*, lysosome; *mi*, mitochondria; *nu*, nucleolus; *PV*, pinocytic vesicles; *SV*, secretion vesicle. (From Drakontides, A. B.; Miller, M. A.; and Leavell, L. C.: *Anatomy and Physiology: Workbook and Laboratory Manual*, 2nd ed. Macmillan Publishing Co., Inc., New York, 1977.)

that surrounds the nucleus of the cell is called cytoplasm, and that which makes up the nucleus is termed nucleoplasm (karyoplasm).

Cytoplasm. Cytoplasm contains *organelles* and *inclusion bodies*. The cytoplasmic organelles are the *mitochondria*, the *Golgi complex* or *apparatus*, the *granular endoplasmic reticulum (ergastoplasm)*, *agranu-* *lar endoplasmic reticulum*, *lysosomes*, *peroxisomes (microbodies)*, the *centrosome* and *centrioles*, *filaments*, and *microtubules*. These organelles are organized living matter that are regarded as small internal organs of the cell, each having a specific function in the maintenance of the cell. Although many of these organelles can be visualized as minute particles with the

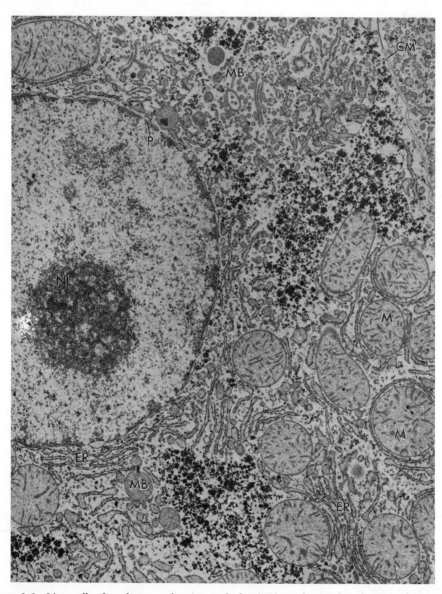

Figure 2–8. Liver cells of newborn rat showing nucleolus (*NL*); nucleus and nuclear membrane with several pores, one of which is labeled *P*; endoplasmic reticulum (*ER*); microbodies (*MB*); several mitochondria at lower right, two of which are labeled *M*; glycogen particles, which are visible as aggregated black dots. A portion of the cell membrane (*CM*) is visible at upper right. ×8400. (Courtesy of Dr. G. Palade, Yale University School of Medicine, New Haven, Conn.)

light microscope, the details or fine structure can only be studied with the electron microscope.

The inclusion bodies are lifeless and often temporary material of the cytoplasm, such as pigment granules, secretory granules, and nutrients such as lipid and glycogen.

The Cell or Plasma Membrane (Plasmalemma). The cell membrane has long been of interest since through it must pass all the materials the cell needs as well as any secretions of the cell. The cell membrane is selectively permeable (semipermeable); e.g., certain ions and molecules may enter or leave the cell, whereas others cannot. In addition, lipid-soluble substances can readily pass through the cell membrane. On the basis of the selective permeability of the cell membrane and other properties, Davson and Danielli in 1935 proposed that the plasma membrane probably consists of a bimolecular layer of lipids between two layers of proteins (Figure 2–10, *A*). Other proposals for the molecular configuration of the cell membrane are illustrated in Figure 2–10, *B* and *C*. When the cell membrane is examined with the electron microscope, it is visualized as two dense layers about 25 Å thick separated by a 30 Å light intermediate layer. This basic three-layered appearance is referred to as the *unit membrane*. Most of the organelles of the cell have unit membranes. The dense layers are thought to represent protein and the light area lipid. Since large-sized molecules can pass through the cell membrane, it is proposed that the membrane is not a continuous structure but that it contains fine openings or *pores*. Recently it has also been found that many animal cells have a thin external

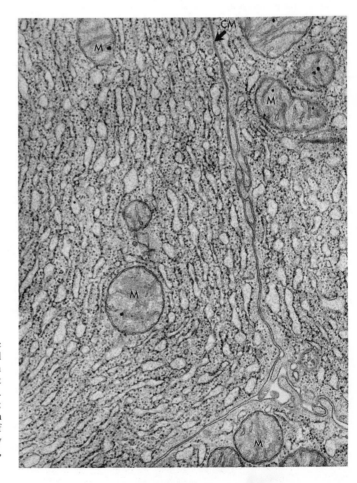

Figure 2–9. Portion of pancreatic exocrine cell (guinea pig) with cell membrane (*CM*) separating it from two adjacent cells visible to the right and in the lower right-hand corner. Endoplasmic reticulum is shown; several mitochondria (*M*) are seen in cross section. ×32,000. (Courtesy of Dr. G. Palade, Yale University School of Medicine, New Haven, Conn.)

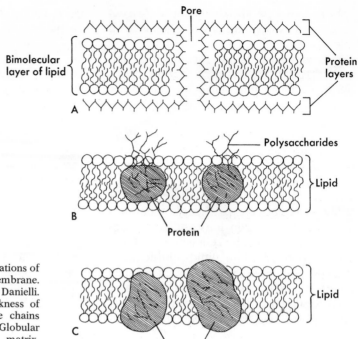

Figure 2–10. Schematic representations of the organization of the cell membrane. *A.* Classic model of Davson and Danielli. *B.* Globular proteins within thickness of lipid matrix with polysaccharide chains projecting above the surface. *C.* Globular proteins distributed in a lipid matrix. (Modified from Bloom and Fawcett.)

coating made up of polysaccharides (carbohydrates), called the *glycocalyx.* This coat may have an important function in the selective uptake of certain materials by the cell.

The Nucleus. The most notable structure within the cell is the nucleus. Most cells have one, some cells have two (liver cells), other cells have a multiple number (skeletal muscle cells), and some cells have none (the mature erythrocyte, platelets, cells in the lens of the eye). The nucleus is an essential organelle directing cell division, containing all the hereditary information in the form of genes, controlling protein synthesis and many of the metabolic activities of the cell. Those cells lacking a nucleus cannot undergo cell division, are not capable of protein synthesis, and are limited in their metabolic activities.

The nuclear membrane is a double-layered membrane that has openings or pores at intervals, through which materials can pass from either the nucleus to the cytoplasm or the cytoplasm to the nucleus. In the resting (nondividing) cell, clumps of dense granular material called *chromatin* are present. This chromatin is a combination of protein and deoxyribonucleic acid (DNA). The chromatin is transformed during cell division from its granular arrangement to one of long strands called *chromosomes.* Chromosomes contain *genes,* which control the transmission of characteristics from parent to child, or from one cell to daughter cells. Also present within the nucleus is a conspicuously rounded area of very dense material, the *nucleolus.* One or more nucleoli may be present. Nucleoli contain large amounts of *ribonucleic acid* (RNA). The chemical composition of DNA and RNA is discussed in a subsequent section in this chapter.

Mitochondria. In the light microscope mitochondria are seen as granules or filaments and are found in most types of animal cells. The number varies in different cell types from a few to several hundred. Mitochondria are mobile structures and capable of changing their shape. When examined with the electron microscope, mitochondria are seen to have a complex structure. They possess a double-layered membrane, each being a unit membrane. The inner membrane is extensively infolded, producing shelves called *cristae.*

The function of mitochondria is the production of energy, in the form of ATP, to support mechanical and chemical work performed by cells. The site of this energy production is felt to be within the inner cristae. Hence, the cristae not only increase surface area but also represent increased sites of energy production. It is not surprising that these organelles, often called the "powerhouse" of the cell, are most numerous in those cells whose energy requirements are high, such as cardiac muscle cells and liver cells.

Golgi Complex (Apparatus). The structure of the Golgi complex is best seen with the electron microscope. It consists of several flattened tubular membranes stacked upon each other, termed *cisternae*, and dilated terminal areas at either end of the cisternae, called *vacuoles*. The Golgi complex is most usually located between the nucleus and cell surface and is often connected to the endoplasmic reticulum. Protein secretory material formed in the endoplasmic reticulum passes to the Golgi complex. Here the secretory material is "packaged" (i.e., surrounded by a membrane). Often carbohydrates, which are synthesized within the Golgi complex, are added to the protein and the combination is referred to as *glycoprotein*. The packaged material buds-off the cisternae as presecretory vacuoles and eventually become secretory granules. These granules usually fuse with the cell membrane and their product is released from the cell. Thus, the Golgi complex has a major role in the synthesis of glycoproteins, and the transport of secretory materials in the cell and their eventual release at the cell surface. As might be expected from the preceding description, the Golgi complex is best developed in secretory cells found in glandular tissue like the pancreas. The Golgi apparatus is also well represented in many cell types that are not secretory in nature, and their function here still remains to be defined.

Endoplasmic Reticulum. The endoplasmic reticulum (ER) consists of a series of parallel arrays of membranes (unit membranes) creating canals (channels, or membranous tubules) and sacules or vesicles, which run throughout the cytoplasm of the cell. This system of canals, the endoplasmic reticulum, connects with the plasma membrane and the nuclear membrane. Two forms of ER exist: the *rough* or *granular* and *smooth* or *agranular*.

In rough ER the canals are studded with *ribosomes*, ranging in diameter from 120 Å to 150 Å. Ribosomes attached to the membranes are felt to be involved in the synthesis of protein that will be secreted from the cell by means of the canals of ER. Rough ER is seen to its best advantage, as is the Golgi apparatus, in cells that are active in the secretion of proteins, such as pancreatic exocrine cells and liver cells.

In smooth ER the same array of membranous canals or tubules is present but ribosomes are lacking (hence the term *smooth*). Smooth ER, which is the site of synthesis of steroid hormones such as that of the adrenal gland, is involved with lipid or fat synthesis and in striated muscle is concerned with the rapid transport of metabolites needed for muscular contraction. The ER may therefore be described as an intracellular transport system, conveying proteins through the canals either to the exterior of the cell or to the Golgi complex, where they are packaged and then released from the cell.

Lysosomes. Lysosomes are membrane-bound, dense-appearing structures that contain enzymes collectively referred to as *acid hydrolases*. These enzymes are capable of breaking down intracellular molecules and digesting foreign organisms such as bacteria, which may enter the cell. Indeed, the enzymes within lysosomes can digest and thereby destroy all the components of the cell; hence, these organelles have been given the name *suicide bags* (Figure 2–12).

Peroxisomes (Microbodies). These organelles are similar to lysosomes in that they are membrane-bound sacs containing enzymes. These enzymes are involved in either the production of hydrogen peroxide (product of reduction of oxygen) or the destruction of hydrogen peroxide to water. In addition to the oxidases involved in hydrogen peroxide, other oxidases are also present. Peroxisomes appear to be con-

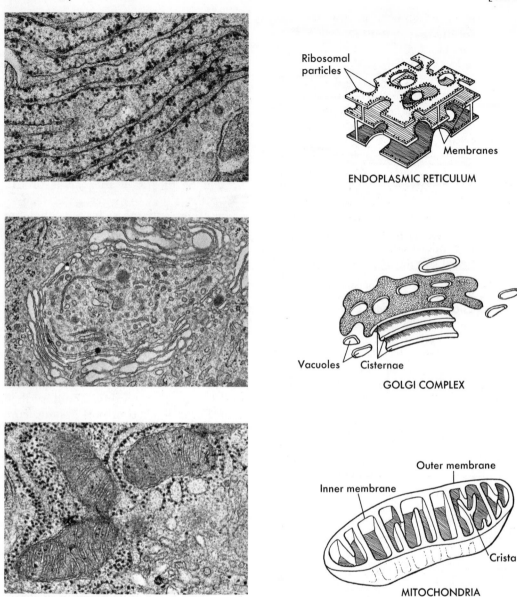

Figure 2–11. The endoplasmic reticulum, Golgi complex, and mitochondria as seen with the electron microscope on the left, and a schematic representation of these organelles on the right. (Electron micrographs, courtesy of Dr. E. Nunez, College of Physicians and Surgeons, Columbia University, New York, N.Y.)

cerned with purine catabolism, the breakdown of nucleic acids, as well as the conversion of fat to glucose. They are therefore active in the internal metabolism of the cell. Peroxisomes are most common in liver cells and the cells of the proximal convoluted tubules of the kidney.

Centrosome and Centrioles. The term *centrosome* means "cell center." Since the

nucleus has a central location within the cell, the region of the centrosome is quite close to the nucleus. Within this centrosome are a pair of small rodlike structures called centrioles. When centrioles are examined with the electron microscope, they have a very characteristic structure. They are minute cylinders whose walls, as seen in cross section, are made up of triplets of

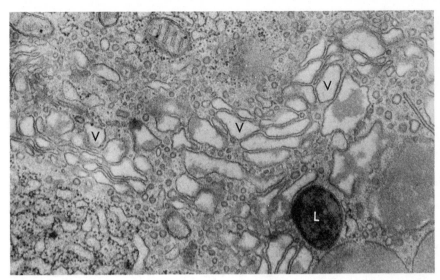

Figure 2–12. Golgi region of pancreatic exocrine cell (guinea pig). Clear areas are large vacuoles (*V*). A lysosome is visible as the dark granular body (*L*). ×30,500. (Courtesy of Dr. G. Palade, Yale University School of Medicine, New Haven, Conn.)

nine evenly spaced microtubules (Figure 2–7). Centrioles are active in the process of cell division (mitosis) in the formation of cilia and are self-duplicating organelles. If a cell lacks centrioles, it cannot undergo the process of cell division (e.g., the neuron).

Filaments and Microtubules. Filaments and microtubules are often considered to be part of the cytoplasmic matrix. All protoplasm has some degree of contractility, and the components responsible for this property are filaments. Filaments are most developed and organized in muscle cells.

In addition to the microtubules seen in the centrioles, microtubules are also evident throughout the cytoplasm. They are therefore important not only in cell division but also in the maintenance of cell shape and in the movements of inclusions and organelles within the cell.

CELL DIVISION

Both the growth and the maintenance of the multicellular organism are dependent on cell division. Some cells of the body have a limited life-span, yet others are subjected to continual wear and are destroyed; hence replacement of cells must occur. One form of cell division in which two identical cells are produced is called *mitosis*. The other form of cell division is meiosis, which will be considered in Chapter 26. Although many cells of the body undergo mitosis, not every cell possesses this ability (e.g., the neuron). Indeed, the rate (how often cells undergo mitosis) varies in different cell types.

During the process of mitosis (Figure 2–13) the chromosomes of the nucleus divide into two exact sets, and the cytoplasm is constricted to form two new cell bodies. The division of the nucleus is termed *karyokinesis*, and the partition of the cytoplasm is referred to as *cytokinesis*. Since the chromosomes of the nucleus consist of the highly coiled molecules of DNA housing the genes, the process of karyokinesis ensures that each new daughter cell will have the same number of chromosomes and genes. Mitosis is thus the means whereby all of the hereditary information can be duplicated in each new cell, thereby maintaining the characteristics of the species from generation to generation.

During mitosis the nucleus undergoes a number of structural changes. Although mitosis is a continuous process, for convenience of description it is arbitrarily di-

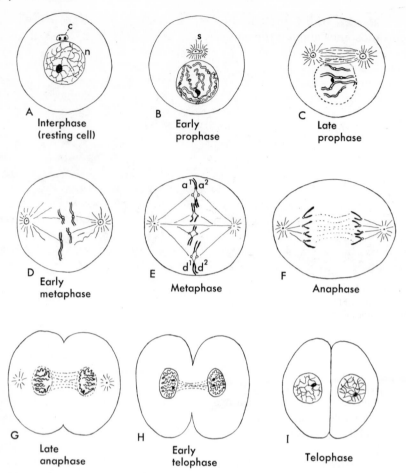

Figure 2–13. Diagrams to show mitosis. The cell illustrated has four chromosomes. *A.* Resting cell with nucleus (*n*) and centrosome (*c*); a nucleolus and network of chromatin are shown in the nucleus. *B.* Spindle fibers (*s*) forming; a chromatin thread, or *spireme*, is breaking into chromosomes. *C.* Nuclear membrane disappearing, *chromosomes* shown dividing *lengthwise* into halves. *D.* Chromosomes shorter and thicker, staining power increased. *E.* Chromosomes arranged on the *spindle*, the two halves of each chromosome opposite each other at a^1 and a^2, d^1 and d^2, etc. *F.* Chromosomes moving toward the poles of the spindle. *G.* Cell beginning to divide. *H.* Cell division continued. *I.* Cell division complete. (Modified from Wilson.)

vided into four stages—*prophase, metaphase, anaphase,* and *telophase.* Each of these stages can be further subdivided into either "early" or "late." The stage of *interphase* is also considered in the discussion of mitosis. No structural changes can be observed during interphase. However, during this interval replication of the chromosomes occurs, most of the DNA, RNA, and protein is manufactured, and it is a period when metabolic activity of the cell is at its greatest.

Prophase. In this phase, which is the first observable stage of cell division at a mi-

croscopic level, each *chromosome* becomes visible as two identical chromatids moving toward the center of the cell. The centrioles separate and begin to move apart, and *spindle fibers* extend from one centriole to the other. The nucleolus is no longer evident, and the nuclear envelope begins to fragment.

Early Metaphase (Prometaphase). This stage is the first part of metaphase. The spindle becomes larger, and the centrioles are clearly visible at its poles. The chromosomes move toward the center of the cell.

Metaphase. At this time the double-stranded chromosomes align at the center of the nucleus, and the spindle fibers can be seen attached to the centromeres. The array of chromosomes attached to the spindle is called the *equatorial plate.*

Anaphase. During this phase the chromosomes, which are now single stranded owing to the division of the centromere, move toward opposite poles of the cell.

Telophase. This stage is the reverse of prophase. The spindle dissolves, a new nuclear envelope forms, the nucleolus reappears, and the chromosomes return to relatively inconspicuous chromatin material. A cleavage furrow begins to form, which gradually divides cytoplasm and cell into two new daughter cells.

THE GENETIC CODE

In the framework of nature all living organisms differ from each other and are capable of reproducing only their own kind. Except for identical twins, each individual is unique in his features. There are no carbon copies. From bacteria to man it is the *gene* that acts as the biologic unit of heredity. Genes contain all the hereditary information and control protein synthesis of the cell. One gene is responsible for one protein. A protein can determine the structure of the cell or can be in the form of an enzyme. Enzymes control the chemical reactions of the body. In recent years it has become apparent that the information-containing portion of a gene consists of DNA. *DNA is capable of self-replication; it makes RNA. RNA, in turn, directs the synthesis of proteins.*

Chemical Composition of DNA and RNA. DNA and RNA are collectively referred to as *nucleic acids* because they were first identified in the nucleus. However, they are found in the cytoplasm as well. These molecules are immense in relation to others, e.g., a molecule of water or of sugar. Like a coiled ladder, the nucleic acid molecule is made up of repeating units that are basically almost identical. The unit is termed a *nucleotide,* each of which contains a sugar (with five carbon atoms), a phosphate group, and one of the following organic substances called nitrogenous "bases": adenine, guanine, cytosine, thymine, and uracil. The nucleotides are fastened to each other in a particular manner, the phosphate group of one attaching to the sugar of the next, forming the backbone of the molecule with the bases projecting outward from it.

A DNA molecule is a double helix (backbone of phosphate and sugar groups). The sugar is deoxyribose, and the bases are adenine, guanine, cytosine, and thymine. There is a selective binding of

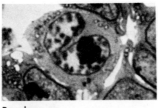
Prophase

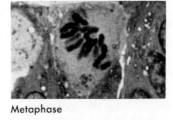
Metaphase

Figure 2–14. Cells of nine-day mouse embryo in stages of mitosis. (Courtesy of Dr. M. Spiegelman, Memorial Sloan-Kettering Cancer Center, New York, N.Y.)

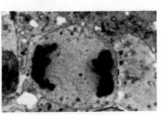
Anaphase

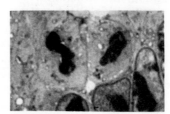
Telophase

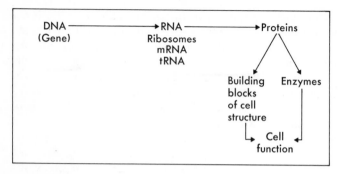

Figure 2–15. Summary of DNA function.

the bases, namely adenine-thymine (thymine-adenine), cytosine-guanine (guanine-cytosine). This is termed *specific base pairing.* The sequence of these base-pairs, however, can vary. A *gene* is a chain of approximately 1000 pairs of nucleotides, which appear in a very *specific sequence* on a portion of the DNA molecule.

RNA differs from DNA in that it is a smaller molecule and is essentially single stranded. In RNA the sugar is ribose, and the bases are adenine, guanine, cytosine, and uracil. Since thymine is lacking in RNA, adenine binds to uracil. RNA is found in the nucleolus and the cytoplasm. RNA transmits information coded by chromosomal DNA to the cytoplasm for use. The openings or pores of the nuclear membrane apparently allow for this transport. All RNA is formed under the direction of DNA. In other words, a part of the DNA molecule serves as a template for the formation of RNA. Generally the synthesis of RNA is similar to the duplication of DNA. The double helix of DNA separates, but only a given length of the DNA is involved—that part that carries the code for the particular kind of RNA to be formed. This synthesis of RNA on a DNA template is termed *transcription.*

There are three types of RNA, each having a different molecular weight, configuration, and function.

Ribosomal RNA (rRNA). This type of RNA represent 85 per cent of the cell's RNA. It is the main constituent of the ribosomes that line the endoplasmic reticulum and that are found free in the cytoplasm. The nucleolus appears to be a reservoir of rRNA and is probably the site of its synthesis. It is at the site of these ribo-

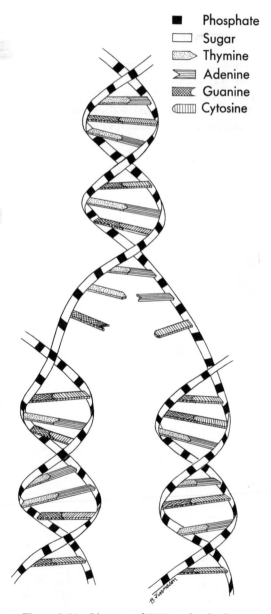

■ Phosphate
▢ Sugar
Thymine
Adenine
Guanine
Cytosine

Figure 2–16. Diagram of DNA molecule showing replication.

somes that protein synthesis under the direction of the other two types of RNA takes place.

Transfer RNA (tRNA). This RNA is the smallest of the three types. As its name implies, tRNA transfers or carries the appropriate amino acid to rRNA. There is a specific tRNA for each of the 20 amino acids that are used as the building blocks to make all the proteins of the cell and thereby the total organism.

Messenger RNA (mRNA). This form of RNA is smaller than rRNA, appearing as a long, thin strand, and it is formed in the nucleus by transcription of information encoded in DNA. mRNA carries the information that determines the *sequences* of amino acids in the specific proteins that are to be synthesized. Ribosomes become attached to the strand of mRNA, newly synthesized protein is released, and the ribosome is detached from the messenger. This process of assembly of protein governed by information encoded in mRNA is referred to as *translation*.

Each tRNA molecule contains within its nucleotide a sequence of three specific bases called a *codon*. This triplet of bases matches up with a specific codon (three base pairs) on mRNA. The sequence of three bases apparently forms a code word corresponding to a particular amino acid. In this manner the correct amino acids are assembled to form the appropriate polypeptide chain corresponding to a given protein. It is of interest that there are three bases in a code word and four different bases in either DNA or RNA. There are, therefore, 64 (4^3) different words available to encode the approximately 20 amino acids found in proteins. These 64 code words afford an infinite number of combinations for the many proteins of the body.

The events of protein synthesis are schematized in Figure 2–17.

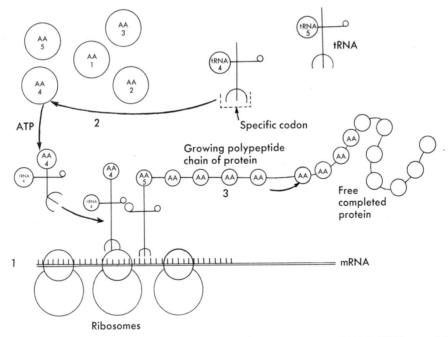

Figure 2–17. Schematic representation of protein synthesis. *1.* Messenger RNA (mRNA), containing coded information made from one strand of DNA molecule in nucleus, peels away from DNA and moves to ribosomes in cytoplasm. *2.* A specific transfer RNA (tRNA) molecule picks up a specific amino acid (AA) using ATP as an energy source and brings it to the appropriate site on mRNA, thus ensuring the proper sequence of amino acids for a given protein. *3.* As amino acids are added, they are linked by peptide bonds, a polypeptide chain grows, and a given protein is produced. This protein is now ready for appropriate packaging and eventual release from the cell.

THE CELLS AND TISSUE FLUID

All cells lie in a liquid environment called tissue fluid. This fluid serves as the only medium of exchange between blood plasma and the cells. Substances needed by cells for maintenance, growth, and repair diffuse from the plasma to the tissue fluid and on into the cell. The products of cell metabolism or other cell activity diffuse into the tissue fluid and enter either the blood or lymph capillaries.

The name *tissue fluid* covers all fluids *not* in the blood vascular system, the lymph vascular system, the great spaces of the body, or the cells themselves.

Points of view differ in regard to classifying the liquids concerned in the exchange of material between the blood and the tissue cells. In general, the lymph vessels form a closed system, and the name *lymph* should be applied to the fluid within the vessels only; the fluid outside the vessels, in the tissue spaces, should be called *tissue fluid.* In the different spaces of the body, e.g., the pericardial, pleural, and peritoneal cavities, it is serous; in the spaces of the cerebrum and spinal cord it is cerebrospinal fluid, and in joints it is synovial fluid. Lacteals are lymph vessels in the small intestine. During digestion, they are filled with *chyle,* a milk-white fluid composed mainly of emulsified fat.

Sources of Tissue Fluid. The walls of the capillaries are thin, and some of the fluid passes out into the spaces between the tissue cells. *Tissue fluid* is derived from the plasma of the blood mainly by diffusion and by capillary hydrostatic pressure. There is difference in pressure within the blood capillary and in the tissue spaces surrounding the capillary. For instance, at the arterial end of the capillary the hydrostatic pressure is about 30 mm Hg and in the tissue spaces surrounding the capillary the pressure is much lower. Since the pressure is highest within the capillary, fluid and other substances are driven from the capillary into the tissue spaces. Another force that must be considered is the protein osmotic pressure formed by the plasma proteins, which act as a "pulling" force to hold fluids within the vessels as well as to "attract fluids in." This opposing force prevents undue loss of fluid from the capillaries.

At the venous end of the capillary, hydrostatic pressure is about 15 mm Hg. This means that the difference in pressure within the capillary and in the tissue spaces is not as great as at the arterial end. As blood moves through the capillary network and fluid is lost to the tissue spaces, plasma protein concentration is slightly raised, and hence the "pulling force" is increased so that water and crystalloids reenter the capillaries readily.

Body Fluids	1. **Intracellular**	Protoplasm, a sol-gel system	
	2. **Extracellular**	Interstitial fluid	Around all cells / Tissue fluid
		Blood plasma	Within the circulatory system
		Lymph	In lacteals, lymph capillaries, nodes, ducts
		Serous fluid	Pericardial fluid / Pleural fluid / Peritoneal fluid
		Fluids in closed spaces	Cerebrospinal fluid / Endolymph and perilymph of inner ear / Fluid of eyes / Synovial fluid, and fluids of bursae, sheaths

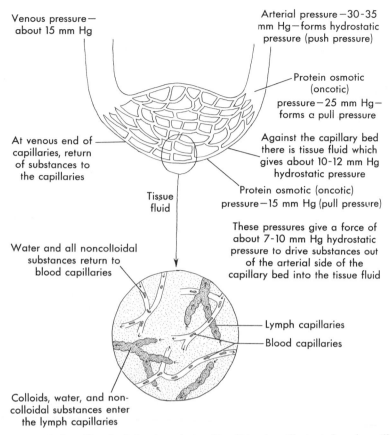

Venous pressure—about 15 mm Hg

Arterial pressure—30-35 mm Hg—forms hydrostatic pressure (push pressure)

Protein osmotic (oncotic) pressure—25 mm Hg—forms a pull pressure

At venous end of capillaries, return of substances to the capillaries

Against the capillary bed there is tissue fluid which gives about 10-12 mm Hg hydrostatic pressure

Tissue fluid

Protein osmotic (oncotic) pressure—15 mm Hg (pull pressure)

These pressures give a force of about 7-10 mm Hg hydrostatic pressure to drive substances out of the arterial side of the capillary bed into the tissue fluid

Water and all noncolloidal substances return to blood capillaries

Lymph capillaries

Blood capillaries

Colloids, water, and non-colloidal substances enter the lymph capillaries

Figure 2–18. Detail of capillary bed showing relationship of blood capillaries to lymph capillaries and return of substances to the bloodstream. By this process the amount of tissue fluid is kept constant.

Colloids, along with water and crystalloids, enter the lymph capillaries. Increases in hydrostatic pressure within the capillaries from any cause will interfere with return of materials to capillaries and will result in excess accumulation of tissue fluid, or edema.

There are two important exceptions to the capillary pressure figures used in the preceding discussion. These exceptions are the capillaries of the lungs and of the kidneys. Hydrostatic pressure in the lung capillaries is approximately 6 mm Hg; thus fluid does not move out of the capillary as it does in other tissues of the body. In the kidney, glomerular hydrostatic pressure is 60 to 70 mm Hg, which acts to force an increased amount of fluid from the capillary in the first step of urine formation.

Composition of Tissue Fluid. The composition is similar to that of blood plasma. It is a colorless or yellowish fluid possessing an alkaline reaction, a salty taste, and a faint odor. When examined under the microscope, it is seen to consist of cells floating in a clear liquid. Its resemblance to the plasma is indicated in the table following. In consequence of varying needs and wastes of different tissues at different times, both the tissue fluid and blood must vary in composition in different parts of the body. But the loss and gain are so fairly balanced that the average composition is constantly maintained.

Function of Tissue Fluid. The tissue fluid bathes all cells of the body. It delivers to the cells the material they need to main-

TABLE 2–5
COMPARISON OF BLOOD AND TISSUE FLUID

Blood	Tissue Fluid
Specific gravity about 1.055*	Specific gravity varies between 1.015 and 1.023
Contains erythrocytes	May contain a few erythrocytes
Contains white cells	Granulocyte count lower, lymphocyte count higher
Contains blood platelets	Does not contain blood platelets
A high content of blood proteins	A lower content of blood proteins
A low content of waste products	A higher content of waste products
A high content of nutrients	A lower content of nutrients
Normally—clots quickly and firmly	Clots slowly, and clot is not firm
Relatively high in colloidal protein	Relatively low in colloidal protein; globulin practically absent
Water, glucose, salts, same concentration in both	

*The specific gravity of a liquid is the ratio of the weight of the liquid (blood, urine, etc.) compared with the weight of an equal volume of distilled water at 15° C (60° F), the weight of water being considered 1.000.

tain functional activity and picks up and returns to the blood the products of this activity. These products may be simple waste or materials capable of being made use of by some other tissue. There is thus a continual interchange going on between the blood and the tissue fluid. This interchange is effected by means of *diffusion.* Some of the constituents of the blood pass into the tissue fluid; some of the constituents of the tissue fluid pass into the blood directly, and some into the lymphatics. Water and noncolloidal substances are returned to the blood capillaries; colloidal substances as well as water and noncolloidal substances enter the lymphatics. Diffusion of this kind is dependent on differences in concentration of diffusible particles of any substance at the two surfaces of the diffusion membranes.

As cells become active, varying needs must be met in relation to supplies and the products of metabolism. Experimental work has shown that as use of oxygen increases, the number of open capillaries in the muscle increases, the capillary diameter increases, the total area of the capillary bed increases, the volume of blood in the muscle increases, the distance of the farthest cell from a capillary decreases, and the difference in oxygen pressure inside and outside the capillary decreases. There are also other changes, such as increased temperature, and velocity of blood flow.

Since cellular metabolism is increased when muscular effort is increased, there is produced in the cells an increased quantity of metabolites, carbon dioxide, and other substances to be eliminated, as well as pH[3] change. Thus the chemical activity of the muscle cells may be the chief factor controlling the amount and distribution of blood through the muscle. This is an automatic control—the graded need to get and to give off brings about the graded means (variable blood flow) to do so accurately. If maximum blood supply does not bring sufficient oxygen and remove metabolites fast enough, oxygen hunger, followed by fatigue, results.

This automatic control for optimum distribution of blood in relation to muscular effort involves adjustment of pulse rate, pulse volume, general and local peripheral resistance, respiratory rate, and respiratory volume, and, in fact, an adjustment of all body functions. This control is brought about by the effects of variations in chemical equilibrium on the tissues themselves, or by the local changes in chemical equilibrium and the influence of the nervous system on the distant organs concerned.

This indicates the relative functions of blood plasma and tissue fluid. The blood

[3]pH is the measure of the relative acidity of a solution. Substances having a pH of 7 are *neutral;* above, from 7 to 14, are increasingly *alkaline;* and below, from 7 to 1, are increasingly *acid.* (See Chapter 24.)

brings (and takes away) substances. The volume of blood per minute in the muscle, the capillary area for diffusion, etc., are constantly varied. To the muscle cells the blood is the source of supplies, kept relatively high in concentration because circulation keeps blood in motion. Blood and tissue fluid are also the place of disappearance of products of metabolism, which are thus kept relatively low in concentration.

In the cells the supplies are used and wastes produced.

The tissue fluid stands between the two. It is a fluid that moves very slowly and is separated both from the blood and from the cell contents, which it closely resem-

bles chemically and physically, by diffusion membranes. In the cell the rate of change of chemical equilibriums, controlled by catalysts, constantly uses supplies and produces wastes; the blood constantly brings supplies and carries off wastes. The tissue fluid mediates this transfer and makes it possible for large amounts of substances to be transported and used with relatively small differences in concentration of soluble constituents in any of the body fluids.

This, together with the fact that blood returning from all tissues is *mixed in the heart* and hence all tissues receive the same blood, is probably the basis of all *physiologic integration.*

Questions for Discussion

1. Compare and contrast the physiologic activities of the human body and those of the individual cell.
2. List and give the function of the organic compounds that make up the cell.
3. What are organelles? Name them and give their function.
4. What are inclusion bodies? Give some examples and list their function.
5. What are the characteristics of water and colloids that make them so important in physiology?
6. Differentiate between protein osmotic and hydrostatic pressure.

7. Name and describe the processes that are included under passive transport across the cell membrane.
8. What are the processes of active transport across the cell membrane, and what is the source of energy used by the cell?
9. Discuss the nature and function of enzymes.
10. Briefly describe the events of mitosis.
11. Explain the structure and functions of DNA and RNA.

Summary

The cell is the unit of structure and function of the body. The protoplasm of cells has the same chemical composition and possesses similar physical characteristics. Although all cells possess similar physiologic characteristics, cells differentiate and become specialized in a particular function.

Constituents of Protoplasm: Chemical Composition

Organic compounds
{ Proteins
Carbohydrates
Lipids

Proteins contain
{ Carbon   Nitrogen
Hydrogen   Sulfur
Oxygen   Phosphorus

Proteins are the most important constituents of cells; act as enzymes; are essential for proper bodily functions; insulin and hemoglobin are examples of proteins; proteins make up contractile components of muscle

Carbohydrates contain
{ Carbon
Hydrogen
Oxygen

Constituents of Protoplasm: Chemical Composition (*cont.*)

Carbohydrates are chief source of energy for cells and supply energy needs of body

Lipids contain { Carbon, Hydrogen, Oxygen }

Lipids include fats, waxes, phospholipids, and sterols; are responsible for structural integrity of cell, making up part of the cell membrane, and are involved in maintaining cell permeability

Inorganic substances { Sulfur, Phosphorus, Chlorine, Sodium, Potassium, Calcium, Magnesium, Iodine, Iron, Traces of others }

Each inorganic substance has an essential role in the maintenance of the functions of the body

Water {
Most abundant constituent of { Cells, Intercellular material }
More than two thirds of weight of body
More than 75% of nonbony body weight
Estimate 85–92% of weight of cell
}

Physical Characteristics

Water {
Some physical characteristics having physiologic significance {
Solvent power is high { As to kinds of solutes, As to variable and high concentration of solutes }
Ionizing power is high
Surface tension is high
Specific heat is high
Thermal conductivity (for liquids) is high
Latent heat of evaporation is high
}
}

Colloids {
Importance in physiology {
Protoplasm is an emulsion of colloids
Ability to take up large quantities of water
Holds water within the cell
Adsorb other substances at interphase
Possess electrical charges that contribute to chemical activity
}
}

Cell organization {
Cells may act like liquids (sols)
Cells may act like semisolids (gels)
Body cells organized close to line of demarcation between sol and gel
}

Enzymes

Definition { Substances produced in living cells which act by catalysis }

Classification {
Endoenzymes—substances found within cells that promote chemical reactions in the cell.
Exoenzymes—secreted in an inactive form—activated by another substance—function outside of the cell
}

Nature of {
Are proteins of high molecular weight
Highly specific in action
Vitamins important in chemical action of enzymes
Inhibitors of enzymes are possible
}

Characteristics of {
Act best at body temperature
Require medium of definite pH
Action is specific and may be reversible
}

Passage of Material Across the Cell Membrane

Passive Processes

Diffusion—the movement of ions or molecules of a gas or liquid from an area of higher to one of lesser concentration, so that a uniform mixture is reached

Osmosis—movement of solvent particles across a membrane into a region in which there is a higher concentration of solute particles

Filtration—the process by which fluid passes through a membrane due to a difference in pressure on the two sides

Active Processes—require energy

Active Transport—involves the movement of materials across the cell membrane against a concentration or electrical gradient and requires a source of energy, ATP

Pinocytosis, or cell drinking—process in which cell membrane indents, surrounds, and finally engulfs molecules in a fluid-filled vacuole

Phagocytosis—larger particles such as bacteria ingested in similar fashion as in process of pinocytosis

Cell Physiology

Metabolism—life activities of cells
- Anabolism—building process—formation of larger molecules and their conversion into living substance
- Catabolism—forming smaller molecules from larger ones with release of energy

1. Motion
   - a. Ameboid movement
   - b. Ciliary movement
2. Irritability—ability to respond to stimuli and to *conduct* the stimuli throughout the cell
3. Respiration
   - Provides oxygen for oxidation
   - Liberates heat
   - Removes excess carbon dioxide
4. Circulation—streaming of the protoplasm in the cell
5. Use of nutrients—foods may also be oxidized to yield energy and regulate body processes
6. Excretion—discharge of waste substances
7. Cell division—mitosis, meiosis

Cell Structure

The cell is a microscopic unit of protoplasm contained within the cell membrane; protoplasm of the cell is called cytoplasm; cytoplasm contains organelles and inclusion bodies; most cells of the body have a nucleus; protoplasm of nucleus is called karyoplasm

Nucleus
- Nucleolus
- Chromatin
- Deoxyribonucleic acid (DNA)

Organelles of the Cell—are organized living matter regarded as small internal organs, each with a specific function in the maintenance of the cell

Mitochondria—energy production, "powerhouse" of the cell

Golgi apparatus—packaging of cellular products, proteins, and carbohydrates for export from cell as secretory granules

Endoplasmic reticulum—*rough*, studded with ribosomes, synthesis of proteins; smooth, synthesis of steroid hormones

Lysosomes—enzymes present, capable of breaking down intracellular molecules and digesting foreign substances, "suicide bags"

Peroxisomes—active in internal metabolism of the cell

Centrosome and centrioles—active in process of cell division

Filaments and microtubules—part of cytoplasmic matrix, maintain cell structure

Inclusion Bodies—lifeless and often temporary material in cytoplasm such as pigment granules, secretory granules, and nutrients such as lipid and glycogen

Cell Division: Mitosis

The formation of two new identical cells; during process of mitosis chromosomes of the nucleus divide into two exact sets, and the cytoplasm is constricted to form two new cells; each new daughter cell has the same number of chromosomes and genes

Stages
- Interphase
- Prophase
- Metaphase
- Anaphase
- Telophase

Genetic Code

Gene—pairs of nucleotides arranged in specific sequence on a portion of DNA molecule; the biologic unit of heredity

DNA directs formation of RNA

RNA—three types are ribosomal-RNA, transfer-RNA, and messenger-RNA; RNA responsible for protein synthesis

Body Fluids

1. Intracellular fluid
   - Protoplasm, a sol-gel system

2. Extracellular fluid
   - Interstitial fluid
     - Around all cells
     - Tissue fluid
   - Blood plasma — Within the circulatory system
   - Lymph — In lacteals, lymph capillaries, nodes, ducts
   - Serous fluid
     - Pericardial fluid
     - Pleural fluid
     - Peritoneal fluid
   - Fluids in closed spaces
     - Cerebrospinal fluid
     - Endolymph and perilymph of inner ear
     - Fluid of eyes
     - Synovial fluid, and fluids of bursae, sheaths

Location
- Surrounding all cells
- Occupies the spaces of loose (areolar) connective tissue and all other tissues

Body Fluids (*cont.*)	Source	Diffused from blood plasma Diffused from cellular materials Cell membrane is a two-way filtering and diffusion membrane between all cells and tissue fluid, including capillary cells
	Composition	Chemically and physically similar to blood plasma and protoplasm
	Function	Go-between for blood and lymph and cells

Additional Readings

BRILLEN, R. J., and KOHNE, D. E.: Repeated segments of DNA. *Sci. Amer.*, **222**:24–31, April, 1970.

BROWN, D. D.: The isolation of genes. *Sci. Amer.*, **229**:20–29, August, 1973.

COHEN, S. N.: The manipulation of genes. *Sci. Amer.*, **233**:24–33, July, 1975.

EVERHART, T. E., and HAYES, T. L.: The scanning microscope. *Sci. Amer.*, **226**:54–69, January, 1972.

FAN, H., and PENMAN, S.: Mitochondrial RNA synthesis during mitosis. *Science*, **168**:135–38, April 3, 1970.

FOX, F. C.: The structure of cell membranes. *Sci. Amer.*, **226**: February, 1972.

KURLAND, C. G.: Ribosome structure and function emergent. *Science*, **169**:1171–77, September 18, 1970.

LAMBERT, J. B.: The shapes of organic molecules. *Sci. Amer.*, **222**:58–70, January, 1970.

LOEWENSTEIN, W. R.: Intercellular communication. *Sci. Amer.*, **222**:78–86, May, 1970.

LURIA, S. E.: Colicins and the energetics of cell membranes. *Sci. Amer.*, **233**:30–37, December, 1975.

MANIATIS, T., and PTASHNE, M.: A DNA operator-repressor system. *Sci. Amer.*, **234**:64–76, January, 1976.

MILLER, O. L.: The visualization of genes in action. *Sci. Amer.*, **228**:34–42, March, 1973.

MIRSKY, A. E.: The discovery of DNA. *Sci. Amer.*, **218**:78–88, June, 1968.

NEUTRA, M., and LeBLOND, C. P.: The Golgi apparatus. *Sci. Amer.*, **220**:100–107, February, 1969.

NOVIKOFF, A. B., and HOLTZMAN, E.: *The Cells and Organelles*. Holt, Rinehart & Winston, Inc., New York, 1970.

RAFF, M. C.: Cell-surface immunology. *Sci. Amer.*, **234**:30–39, May, 1976.

STEIN, G. S.; STEIN, J. S.; and KLEINSMITH, L. J.: Chromosomal proteins and gene regulation. *Sci. Amer.*, **232**:46–57, February, 1975.

TEMIN, H. M.: RNA-directed synthesis. *Sci. Amer.*, **226**:24–33, January, 1972.

TONER, P. G., and CARR, K. E.: *Cell Structure: An Introduction to Biological Electron Microscopy*, 2nd ed. Williams & Wilkins Co., Baltimore, 1971.

WESSELLS, N. K., and RUTTER, W. J.: Phases in cell differentiation. *Sci. Amer.*, **220**:36–44, March, 1969.

CHAPTER 3

Tissues of the Body

Chapter Outline

Microscopic anatomy refers to the study of any structure under the microscope. *Histology* limits microscopic anatomy to the study of tissues. It is concerned with structural characteristics of cells and groups of cells as arranged to form tissues; hence a knowledge of the structure and activities of cells forms the basis of histology. The structure and function of tissues are closely related. An understanding of structure will clarify function and help to build the foundation for physiology.

In the previous chapter, we have considered the structure and function of the typical cell. Cells having similar characteristics and functions are grouped together

45

to form tissues. There are <u>four primary tissues</u>:

1. Epithelial
2. Connective
3. Muscle
4. Nervous

These tissues form the organs and systems of the body. The <u>cells forming tissues</u> have specific functions and are arranged in different ways. Often the <u>cells are packed very closely together,</u> and we characterize this type of arrangement as a tissue in which there is <u>very little extracellular (in-</u>tercellular) space, i.e., area surrounding outside of cells. The prime example of this arrangement is <u>epithelial tissue.</u> The reverse of the above arrangement is one in which the cells are not packed closely, and there exists an extensive amount of extracellular space. In this situation the extracellular space is made up of substances or materials that have been manufactured and released by the cells. The prime example of this arrangement is cartilage and bone. Indeed, in the case of cartilage and bone, it is the extracellular matrix that gives the tissue its characteristic texture.

Epithelial Tissue

CHARACTERISTICS

The cells of epithelial tissue are closely packed together; hence there is little extracellular space. The general functions of this tissue are *protection, secretion,* and *absorption.* Epithelial tissue or epithelium is arranged as follows:

1. Sheets or layers that *line* and *cover* the body.
2. Cords or clusters making up glandular tissue, which composes both the exocrine glands and endocrine glands.

Epithelium that lines and covers is arranged as a single layer of cells, called *simple epithelium,* or as many layers of cells, termed *stratified epithelium.* There is a third category of epithelium, described as *pseudostratified.* This is actually a single layer of cells; however, since the cells are arranged so that the nuclei occur at more than one level, it appears stratified. Epithelial cells have the following shapes: flat or *squamous, columnar,* and *cuboidal.* The arrangement and shape of the cells of epithelial tissue are the basis for the following classification:

Simple epithelium
  Simple squamous epithelium
  Simple cuboidal epithelium

Simple columnar epithelium
Pseudostratified columnar epithelium
Stratified epithelium
  Stratified squamous epithelium
  Stratified cuboidal epithelium
  Stratified columnar epithelium
  Transitional epithelium

Beneath all epithelial tissue is connective tissue. This connective tissue gives support and helps to keep the epithelium in place. Between the epithelium and connective tissue is a layer called the *basement membrane.* It is an extracellular condensation of mucopolysaccharides, proteins, and collagen fibrils. In addition to providing support, the basement membrane probably acts as a semipermeable filter under the epithelium.

Epithelial tissue is avascular (has no direct blood supply). Thus, the cells of epithelial tissue must derive their nourishment via diffusion from underlying connective tissue. Since epithelium lines and covers, it is subject to wear and injury. The cells of epithelial tissue undergo frequent cell division by mitosis and, therefore, new cells are continually available.

The function of epithelial tissue can be determined from its arrangement and location. Stratified squamous epithelium covers the surface of the body (skin) and has as its prime function protection against external mechanical damage and loss of moisture. As a covering membrane in contact with the external environment, epithelium also functions in *sensory reception*. Nerve endings in epithelium can receive and transmit external sensory information, such as pain and temperature. Taste buds in the epithelium of the tongue and cells of the olfactory epithelium in the nose become specialized epithelium, referred to as *neuroepithelium,* and receive sensations of taste and smell, respectively. Simple epithelium lining hollow organs, such as those of the digestive tract, and making up glandular tissue is best suited for the functions of *absorption* and *secretion*. The pathway of transport of substance either to the cell (absorption) or from the cell (secretion) is only one cell thick and thus the most efficient.

HOW EPITHELIAL CELLS ARE HELD TOGETHER

A fundamental property of epithelial tissue is the ability of the cells to maintain extensive contacts with each other. This close attachment was described by early workers as being due to the presence of an "intercellular cement." The electron microscope has elucidated structural modifications of the lateral cell membrane that apparently function to hold epithelial cells together (e.g., zonula adherens and macula adherens or desmosome). In addition, other structural modifications of the cell membrane have been implicated in functions such as transport and electrical coupling (e.g., zonula occludens and gap junction) (Figure 3-1).

Zonula Occludens, or Tight Junction. This is a region where the membranes of adjoining cells come together and the outer leaflets of their unit membrane apparently fuse. This region of membrane fusion is on the lateral surface but close to the outer surface of the cell. It extends in a belt around the perimeter of the cell. The zonula occludens seals the intercellular space and is thought to be important in the transport of water across epithelium.

Zonula Adherens. This region is found below the zonula occludens. The two opposing unit membranes are reinforced on their inner cytoplasmic surfaces by a

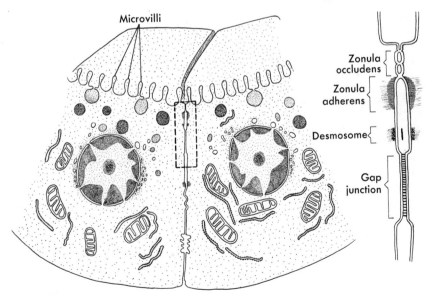

Figure 3-1. Diagram of two epithelial cells as seen with the electron microscope. The means of attachment of the two cells on the lateral surfaces is illustrated; the region within the dotted box is enlarged on the right.

dense mat of fine filamentous material that forms a continuous band around the cell parallel to the zonula occludens. The zonula adherens serves as a band of firm adhesion between adjoining cells.

Macula Adherens, or Desmosome. As with the zonula adherens, the cytoplasmic surface of the unit membrane appears thickened owing to the presence of a dense, filamentous material embedded in a dense matrix. Embedded within these dense plaques and apparently extending throughout the cytoplasm are individual cytoplasmic filaments called *tonofilaments.* Desmosomes, therefore, appear to be sites of attachment of the cytoplasm (cytoskeleton) to the cell surface as well as areas of cell-to-cell adhesion.

Gap Junction, or Nexus. This surface modification has the form of plaques of variable size where the intercellular space is greatly narrowed but not absent. The gap junction is felt to be the site of electrical coupling between cells. In other words, it is a region that is permeable to ions, thereby allowing ion flow and communication between cells.

Table 3–1 lists the principal types of epithelium, their characteristic location, and their function.

GLANDULAR EPITHELIUM

A gland is an organ that produces a specific product or secretion. All glands and their ducts (excretory ducts) are composed of epithelial tissue. Glands are divided by septa of connective tissue into lobes and lobules. Nerves and blood capillaries that supply the secretory units are located in the supporting connective tissue.

A gland that has a duct leading to the outside, to a cavity, or to a hollow organ is called an *exocrine gland.* For example:

1. Sweat glands: ducts leading to surface of skin, for collection of waste products and secretion of sweat.

2. Salivary glands: ducts that lead into mouth for secretion of saliva.

3. Pancreatic exocrine gland: duct that leads into small intestine for secretion of pancreatic digestive enzymes.

If the gland does not have a duct, it is called an *endocrine gland.* Endocrine glands have a rich blood supply and the secretory product of these glands (hormones) is carried in the blood. Endocrine glands and their specific hormones will be discussed in Chapter 14.

Histologic Organization. As with all structures, there exists a particular type of arrangement or organization of glandular tissue. Exocrine glands may be *unicellular* or *multicellular.*

Unicellular Glands. The mucous or goblet cell found among the columnar cells of the epithelium of mucous membranes is the example of a unicellular gland. This cell elaborates and secretes *mucin,* a protein-polysaccharide that along with water forms mucus. Mucus is a lubricating fluid and helps protect epithelial membranes.

Multicellular Glands. Multicellular glands are classified as *tubular, alveolar* (acinar), and *tubuloalveolar.* These terms refer to the arrangement of the secretory epithelium—e.g., in tubular glands in the form of a tube (Figure 3–10, *D, E, F, G*) and in alveolar glands in a cluster or saclike arrangement (Figure 3–10, H, I, J); tubuloalveolar refers to a combination both tubular and alveolar. There is yet a further distinction made on the basis of arrangement, and this is in particular referrence to the excretory ducts. If the duct to the surface does not branch, it is a *simple gland;* if the duct does branch, it is a *compound gland.* This histologic organization gives rise to the following types of multicellular glands.

Simple Glands. SIMPLE TUBULAR GLANDS. There is no excretory duct, and the terminal portion of secretory cells is a straight tubule that opens directly onto epithelial surface. Example: intestinal glands of Lieberkuhn.

SIMPLE COILED TUBULAR GLANDS. The

TABLE 3-1
TYPES OF EPITHELIAL TISSUE

Tissue	Characteristics; Location	Function
Simple epithelium Simple squamous epithelium	Consists of one layer of flattened cells. Edges are serrated and fitted together to form a mosaic. Nuclei centrally located, oval or spherical. Found lining air sacs (alveoli) of lungs, crystalline lens of eye, membranous labyrinth of inner ear, in glomerulus and thin segment of loop of Henle in kidney, in rete testis (male gonad), and in certain of smallest ducts of glands. Simple squamous lining heart, blood, and lymph vessels and forming capillary networks called *endothelium*. Called *mesothelium* when it lines serous cavities and covers visceral organs	Transport of material via diffusion, osmosis; ease of exchange because of one-cell-layer thickness
	Figure 3–2. Diagram of endothelium, surface and sectional view. Seen in the wall of a capillary.	
Simple cuboidal epithelium	Single layer of square, rectangular, or cube-shaped cells; nucleus centrally located. Found in many glands, such as thyroid, on free surface of ovary, choroid plexus, on inner surface of capsule of lens; lines kidney tubules, in ducts of glands	Secretion and absorption
	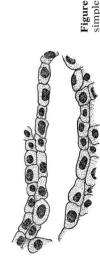 **Figure 3–3.** Schematic representation of simple cuboidal epithelium.	
Simple columnar epithelium	Cells are cylindrical and set upright. Oval nucleus located near base of cell. Modification of surface of columnar cells in form of *microvilli*, characteristic of epithelium of digestive tract, and *cilia*, characteristic of epithelium of respiratory system	Secretion and absorption

TABLE 3–1 (*Continued*)

Tissue	Characteristics; Location	Function
Simple columnar epithelium (*Continued*)	*Goblet* cells are specialized columnar epithelial cells in shape of goblet, manufacture and secrete mucin. Mucin component of mucus, which acts as protective film over epithelial surface. *Simple columnar epithelium* lines digestive tract—stomach, small and large intestine—and is found in excretory ducts of many glands. In small intestine epithelial cells with microvilli predominate, collectively termed *striated border*; in kidney called *brush border*. *Ciliated simple columnar epithelium* found in uterus and oviducts, in small bronchi of lung	

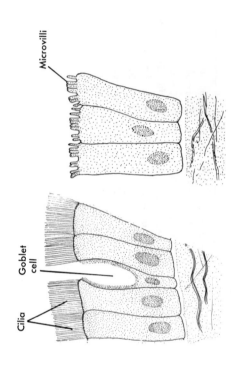

Figure 3–4. Diagrammatic representation of surface modifications of simple columnar epithelial cells. *A.* Sectional view showing cilia. *B.* View showing microvilli.

Pseudostratified columnar epithelium

All cells in contact with basement membrane, but not all of them appear to reach surface. Oval nuclei seen at more than one level. *Pseudostratified columnar epithelium* found in large excretory ducts and in male urethra. *Pseudostratified ciliated columnar epithelium* with goblet cells lines greater part of respiratory passages, including trachea, the eustachian tube, and part of the tympanic cavity

Secretion, movement by cilia of mucus with trapped dust particles and debris in respiratory passages

Striated border

Portion of two epithelial cells

Epithelium

A

B

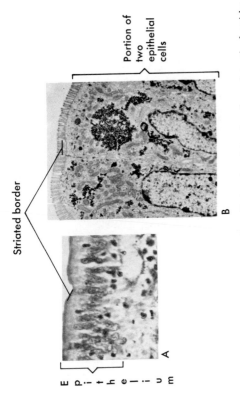

Figure 3–5. Microvilli, collectively called striated border, from intestinal epithelium. *A*, as seen with light microscope. *B*, as seen with electron microscope. (Courtesy of Dr. D. Orlic, New York Medical College, Valhalla, N.Y.).

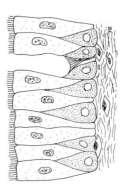

Figure 3–6. Pseudostratified ciliated columnar epithelium.

TABLE 3–1 (*Continued*)

Tissue	Characteristics; Location	Function
Stratified epithelium Stratified squamous epithelium	Several layers of cells, which vary in shape from base, where they are cuboidal or columnar, becoming flatter to scalelike near surface of epithelium. The most superficial layer of cells is composed of thin squamous cells. Basal cells constantly multiplying by mitosis. Thus, a continuing cycle of new cells from base to surface, surface cells sloughed off Two types: *nonkeratinized* and *keratinized* *Keratinized stratified squamous epithelium* occurs on exposed outer surfaces, such as skin. Superficial cells lose their nuclei and cytoplasm is replaced by keratin, a scleroprotein *Nonkeratinized squamous epithelium* is found lining surfaces such as mouth, esophagus, part of epiglottis, conjunctiva, cornea, vagina, and portion of female urethra	Protection, conservation of fluid loss from exposed surface

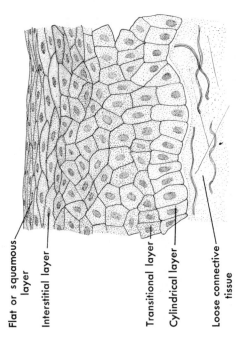

Flat or squamous layer

Interstitial layer

Transitional layer

Cylindrical layer

Loose connective tissue

Figure 3–7. Stratified squamous epithelium.

Stratified cuboidal epithelium	Usually two or more layers of cells, the superficial layer of cells being cuboidal in shape. Found in ducts of sweat glands. Epithelium of conjunctiva and female urethra often classified as stratified cuboidal. Relatively uncommon	Protection, secretion
Stratified columnar epithelium	Deep cells are small and do not reach free surface; superficial cells are cuboidal or columnar in form. Found in pharynx, on epiglottis, and in large excretory ducts of some glands. This type of epithelium is difficult to distinguish from the more common pseudostratified columnar epithelium. Relatively uncommon	Protection, secretion
Transitional epithelium	Varies greatly in appearance, depending on existing conditions. Lines hollow organs that are subject to great changes due to contraction and distention, such as the urinary bladder, ureters, and upper part of urethra. In contracted condition consists of many cell layers; superficial cells are large and have a characteristic rounded free surface. In stretched or distended condition usually only two cell layers can be distinguished, a layer of superficial squamous cells overlying a layer of cuboidal-shaped cells	Protection, allows for distention

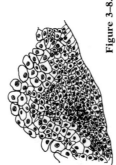

Figure 3–8. Transitional epithelium.

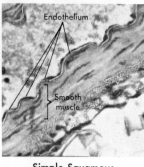

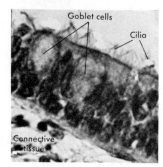

Figure 3–9. Principal types of epithelium as seen with the light microscope. The organ or site of the section is given.

Simple Squamous
(endothelium of blood
vessel)

Pseudostratified
Ciliated Columnar
(trachea)

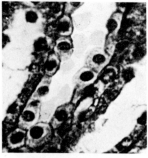

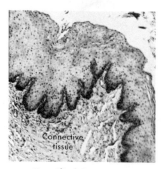

Simple Cuboidal
(kidney tubule)

Stratified Squamous
(esophagus)

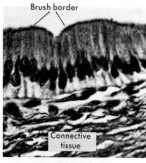

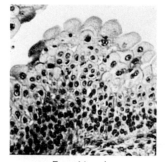

Simple Columnar
(gallbladder)

Transitional
(bladder)

terminal portion is a long, coiled tubule connected to the surface by a long, unbranched excretory duct. Example: sweat glands.

SIMPLE BRANCHED TUBULAR GLANDS. The terminal portion of tubules bifurcates into two or more branches. The excretory duct may be absent or may be short. Examples: glands of stomach, uterus, tongue, and esophagus; Brunner's glands in duodenum.

Compound Glands. COMPOUND TUBULAR GLANDS. These have a large number of distinct duct systems, which eventually open into a main or common excretory duct. The liver, kidneys, and testes are good examples of these glands.

COMPOUND TUBULOALVEOLAR GLANDS. These are numerous, and although the general principle of structure is about the same in all of them, there is considerable variation in their minute structure. These glands also have many distinct duct systems, which eventually open into a common duct. All of the salivary glands, the pancreas, some of the larger glands of the

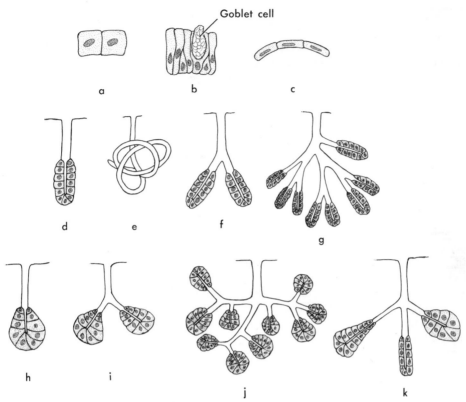

Figure 3–10. Diagram showing types of glands. (*a*) Plain cuboidal secreting cells. (*b*) Plain columnar secreting cells, one of which is a "goblet gland." (*c*) Plain flat secreting cells. (*d, e, f, g*) Tubular glands: simple, twisted, branched, and several times branched. (*h, i, j*) Saccular or alveolar glands: (*h*) simple, (*i*) branched, and (*j*) much branched. (*k*) Compound tubuloalveolar gland.

esophagus, the seromucous glands of the respiratory pathways, and many of the duodenal glands belong in this group.

COMPOUND ALVEOLAR GLANDS. These are very much like the other compound glands in general structure; however, the terminal ducts end in alveoli with a dilated saclike form. The mammary glands are good examples of this kind of gland.

Means of Secretion from Exocrine Glands. An exocrine gland whose secretory product consists of whole cells is called *holocrine* (e.g., sebaceous gland). If only a part of the apical cytoplasm is secreted, it is called *apocrine* (e.g., mammary gland); if no cytoplasm is lost, it is a *merocrine* type of gland.

Most glands are of the merocrine type. The secretory cells of merocrine glands are classified as serous and mucous.

Mucous Cells. These exist as unicellular

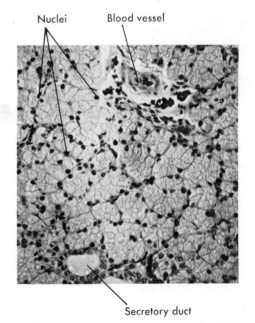

Figure 3–11. Histologic appearance of parotid gland, a salivary gland made up of serous cells.

glands found throughout the gastrointestinal and respiratory tracts and are prominent in many of the salivary glands. Such cells produce a secretion that is viscous, contains mucin, and is high in sugar.

Serous Cells. These are common in glands of the alimentary tract and are also found in salivary glands. Such cells produce a secretion that is watery but high in enzyme content.

Connective Tissue

CHARACTERISTICS

The fundamental characteristic of all connective tissues is the greater proportion of extracellular material in relation to cells. The extracellular material, described as an amorphous ground substance, contains cells, fibers, and tissue fluid. The chemical composition of the ground substance, the type of cells, as well as the type, number, and arrangement of fibers are the basis for the classification of connective tissue. Thus, under connective tissues we include:

1. The connective tissue proper
   a. Loose (areolar) connective tissue
   b. Dense (fibrous) connective tissue
      (1) Dense irregular
      (2) Dense regular
   c. Elastic connective tissue
   d. Reticular connective tissue
   e. Adipose tissue
2. Cartilage
3. Bone

The connective tissues provide the framework that supports epithelium and other tissues; they have an essential role in transport, protection, and repair. Adipose tissue functions as a storage site for fats. Cartilage and bone are the supporting tissues of the body. Many authors also include blood under the heading of connective tissue.

The derivative of all adult connective tissue is *embryonal connective tissue.* This exists in essentially two forms:

1. Mesenchymal connective tissue
2. Mucous connective tissue

Mesenchymal connective tissue or mesenchyme is a loose, spongy tissue that is found filling in the spaces between developing organs of the embryo. The cells of this tissue, mesenchymal cells, are stellate with long, slender extensions. The extracellular matrix is amorphous, with thin fibers. Mesenchymal cells are the precursors of many of the adult connective tissue cells, such as fibroblasts, fat cells, chondroblasts, and osteoblasts.

The most typical example of *mucous connective tissue* is *Wharton's jelly* in the umbilical cord. The intercellular substance is abundant and jelly-like. The cells include fibroblasts, a few macrophages, and lymphocytes.

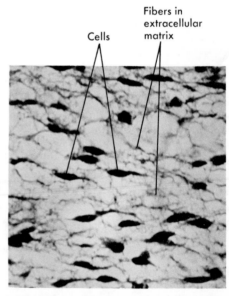

Figure 3–12. Mesenchyme embryonal connective tissue.

THE FIBERS OF
CONNECTIVE TISSUE

Three types of fibers are found in connective tissue: collagen, reticular, and elastic.

Collagen Fibers. These are the most abundant and characteristic of the fiber types. Collagenous fibers are flexible yet have great tensile strength. They are composed of an albuminoid protein *collagen* that yields gelatin when denatured by boiling. Collagen fibers appear as wavy interlacing fibrils that lie parallel to each other.

Reticular Fibers. When observed with the light microscope, these fibers are very thin and form delicate meshworks around cells. They make up the fibrous framework for the spleen, liver, pancreas, and lymphoid organs and are part of the basement membrane. It is currently felt that reticular fibers are very fine collagenous fibers that have the same properties as collagen.

Elastic Fibers. These fibers are homogeneous, branch freely, and are composed of the protein *elastin*. Elastic fibers possess the property of elasticity; namely, they have the ability to stretch easily when a force is applied, and return to their original shape when the force is removed, much like a rubber band.

TYPES OF CONNECTIVE TISSUE

Characteristics, Location, and Function
Connective Tissue Proper

Loose (areolar) connective tissue

This tissue has a semifluid ground substance, or matrix, in which are embedded many collagenous fibers, some elastic and reticular fibers. Many different types of cells are present in this tissue, as listed below. Loose connective tissue has many functions: (1) it is important in the repair of all tissue, (2) has the capacity to hold water in the tissues, and (3) is an important factor for changes in viscosity and permeability of the ground substance in the tissue. The matrix of this tissue is called tissue fluid. It is continuous throughout the body, under the skin, and in all mucous membranes around all blood vessels and nerves. This tissue forms the "internal environment of the body," delivers all substances to the cells, delivers products of metabolism to blood and lymph, and stores water, salts, and glucose temporarily.

Cells of loose connective tissue

Fibroblasts (fibrocytes)

These are large, flat branching cells with many processes. They play an important role in the formation of collagenous fibers and perhaps also form ground substance.

Macrophages (histiocytes)

These are irregularly shaped cells with short processes. Found in the sinusoids of the liver, lymph organs, and bone marrow. They are phagocytic and function in normal physiologic processes. They have great phagocytic capacity under such conditions as inflammatory processes.

Plasma cells

These are small round or irregular-shaped cells. There is evidence that they may be derived from special cells in the thymus and are distributed to other tissues shortly after birth. Found in greatest numbers in all connective tissue, especially that of the alimentary mucosa and great omentum. Plasma cells are the actual formers of circulating antibodies.

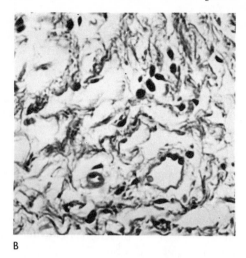

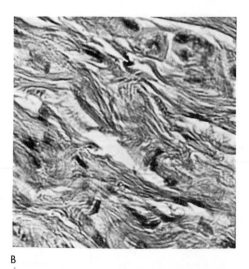

Figure 3–13. *A.* Diagrammatic representation of loose (areolar) connective tissue. *B.* Loose (areolar) connective tissue as seen with the light microscope.

Mast cells

The cytoplasm of these cells is filled with dense granules. Their function is poorly understood. Mast cells contain *heparin*, which is an anticoagulant, and *histamine*, which if released into tissues causes vasodilation and increases the permeability of capillaries and venules. Release of histamine occurs under various conditions of trauma and is a prime factor in tissue reaction to injury.

Adipose cells (fat cells)

These are found singly or in small groups, particularly near small blood vessels. They are specialized for the synthesis and storage of lipid.

Blood cells

Lymphocytes, neutrophils, monocytes, and eosinophils from the blood and lymph move in and out of loose connective tissue. Their numbers are variable, depending on physiologic conditions.

Dense irregular connective tissue

The elements found in this tissue are the same as loose connective tissue, but the collagenous bundles are thicker, more numerous, and *randomly* woven into a compact framework. The

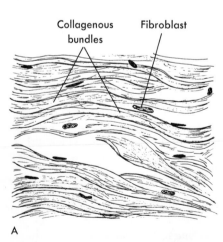

Figure 3–14. *A.* Diagrammatic representation of dense irregular connective tissue. *B.* Dense irregular connective tissue as seen with the light microscope.

amorphous ground substance is less, and the cells are more difficult to identify owing to the extensive fibrous components. This type of tissue is most typically seen in the dermis of the skin; it also makes up the capsules of many organs and the sheaths of tendons and nerves.

Dense regular connective tissue

In this connective tissue, collagen fibers are the predominant characteristic and are arranged in a *specific* pattern. The matrix between bundles of fibers contains fibroblasts arranged in rows, but the cells are not a prominent feature of this tissue. Dense regular connective tissue is part of the supporting framework of the body. It forms: (1) *Ligaments*, strong flexible bands or cords that help to hold the bones together at the joints. (2) *Tendons*, white glistening cords or bands that attach the muscles to the bones. (3) *Aponeuroses*, flat, wide bands that connect one muscle with another or with periosteum of bone. (4) *Membranes*, containing fibrous tissue found investing and protecting different organs of the body, e.g., the heart and the kidneys. (5) *Fasciae* (*fascia*, in Latin, means a band or bandage). It is found as sheets of fibrous connective tissue that are wrapped around muscles and serve to hold them in place. There are two groups: superficial fascia and deep fascia. Infection of superficial fascia is called *cellulitis*.

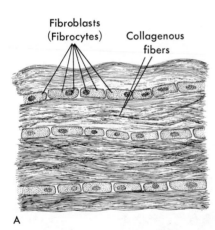
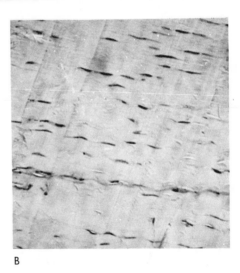

Figure 3–15. *A*. Diagrammatic representation of dense regular connective tissue, tendon. *B*. Dense regular connective tissue, tendon, as seen with the light microscope.

Elastic connective tissue

Elastic connective tissue is loose connective tissue in which the elastic fibers predominate. It consists of a ground substance containing cells with a few fibrous fibers and a predominance of elastic fibers that branch freely. They give it a yellowish color. Elastic tissue is extensile and elastic. It is found entering into the formation of the lungs and uniting the cartilages of the larynx. In the walls of arteries, trachea, bronchial tubes, and vocal folds; and in a few elastic ligaments and between the laminae of adjacent vertebrae (ligamenta flava, ligamentum nuchae).

Reticular connective tissue

The predominant fiber type is reticular. Small bundles of thin reticular fibers form complex three-dimensional networks. Within these networks are found a large number of cells. This

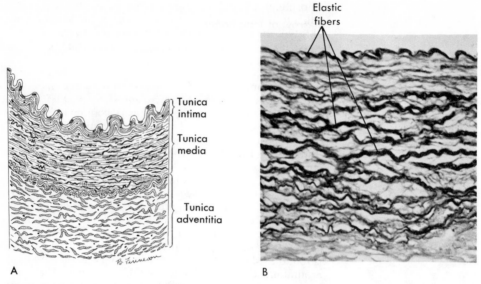

Figure 3–16. *A.* Diagrammatic representation of elastic connective tissue as seen in wall of a large artery. *B.* Section of large artery showing elastic fibers, as seen with the light microscope.

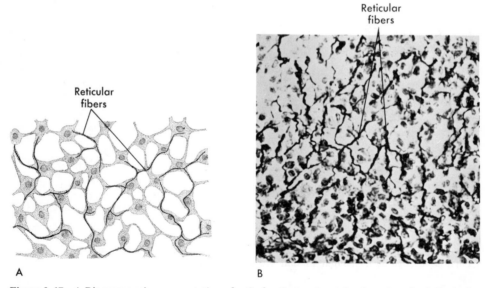

Figure 3–17. *A.* Diagrammatic representation of reticular tissue as seen in a lymph node. *B.* Reticular fibers as seen with the light microscope.

type of tissue makes up the stroma of the liver, spleen, and lymphoid organs.

Adipose connective tissue

This is loose connective tissue in which many of the cells are filled with fat. It exists throughout the body wherever loose connective tissue is found. It is an important reserve of food which may be oxidized, thus producing energy. It is a poor conductor of heat, thus reduces heat loss through the skin. Supports and protects various organs. Found generally throughout the body, under skin, in the subcutaneous layers. Large amount around the kidneys, covering the base and filling up furrows on the surface of the heart. Padding around joints. In marrow of long bones.

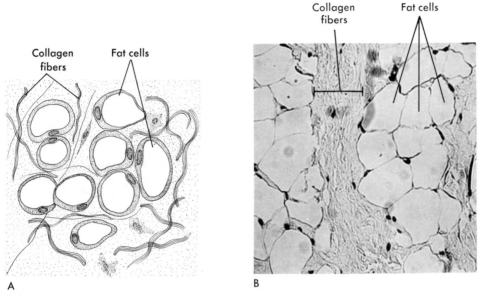

Figure 3–18. *A.* Diagrammatic representation of adipose tissue. *B.* Adipose tissue as seen with the light microscope.

Cartilage. Cartilage is firm, tough, and flexible. It consists of groups of cells called chondrocytes, in a mass of intercellular substance called matrix. Depending upon the texture of the intercellular substance three principal varieties can be distinguished: hyaline, fibrous, and elastic.

Hyaline cartilage

Comparatively few cells lying in fluid spaces or lacunae embedded in an abundant quantity of intercellular substance. This

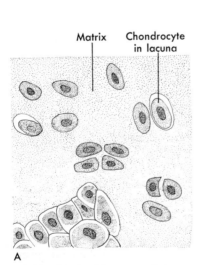
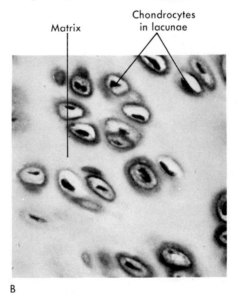

Figure 3–19. *A.* Diagrammatic representation of hyaline cartilage. *B.* Hyaline cartilage as seen with the light microscope.

substance appears as a bluish white, glossy, or homogeneous mass. It is made up of collagenous fibrils forming a feltlike matrix. Hyaline cartilage covers the ends of the bones at the joints forming articular cartilage; it forms the ventral ends of the ribs as the costal cartilage. It is frequently described as skeletal cartilage since it is in immediate connection with bone and may be said to form part of the skeleton. Hyaline cartilage enters into the formation of the nose, larynx, trachea, bronchi, and bronchial tubes. In covering the ends of the bones at the joints, it provides the joints with a thick, springy coating which gives ease to motion. In forming part of the bony framework of the thorax, the costal cartilages impart flexibility to its walls. In the embryo, a type of hyaline cartilage known as embryonal cartilage forms the matrix in which most of the bones are developed.

Fibrous cartilage

Intercellular substance is pervaded with bundles of collagen fibers, between which are scattered cartilage cells. The encapsulated cells frequently lie in rows with bundles of collagenous fibers between them. Fibrous cartilage closely resembles fibrous tissue. It is found joining bones together as in the round disks or symphyses of fibrocartilage connecting the bodies of the vertebrae and symphysis pubis between the pubic bones. In these cases the part in contact with the bone is always hyaline cartilage, which changes gradually into fibrocartilage. In the center of the intervertebral disks there is a soft mass called the *nucleus pulposus*. Herniation of this mass may occur into the spinal canal. Fibrocartilage serves as a strong, flexible connecting material between bones and is found whenever great strength combined with a certain amount of rigidity is required.

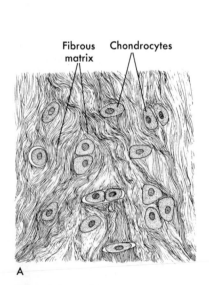

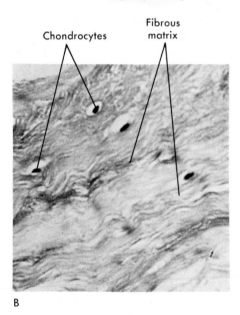

Figure 3–20. *A.* Diagrammatic representation of fibrous cartilage. *B.* Fibrous cartilage as seen with the light microscope.

Elastic cartilage

The intercellular substance is pervaded with a large number of elastic fibers that form a network. In the meshes of the network the cartilage cells are located. This form of cartilage is found in the epiglottis, the larynx, the auditory tube, and the external ear.

It strengthens and maintains the shape of these organs and yet allows a certain amount of change in shape.

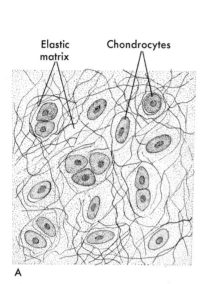

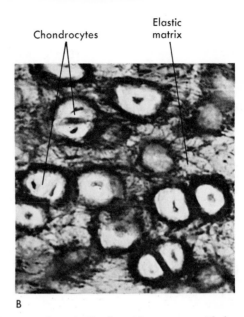

Figure 3–21. *A.* Diagrammatic representation of elastic cartilage. *B.* Elastic cartilage as seen with the light microscope.

Perichondrium: repair and aging

Cartilage has no direct blood supply. *Perichondrium,* a moderately vascular fibrous membrane, covers and nourishes cartilage except where it forms articular surfaces. Perichondrium also functions in the repair process of injured cartilage. When injured, the area is invaded by perichondrial tissues, which is gradually changed to cartilage. With the aging process, cartilage loses its translucency and bluish-white color and appears cloudy. Calcification may occur, with degenerating changes of the cartilage cells.

Bone, or Osseous Tissue. Bone is connective tissue in which the intercellular substance is rendered hard by being impregnated with mineral salts, chiefly calcium phosphate and calcium carbonate. This inorganic matter constitutes about two thirds of the weight of bone. Organic matter consisting of cells, blood vessels, and cartilaginous substance constitutes about one third. Bone freed from inorganic matter is called decalcified. It is a tough, flexible, elastic substance that can be tied in a knot. Bone free from organic matter is white and so brittle that it can be crushed in the fingers.

Forms of bony tissue
1. Cancellous or spongy

Has larger cavities and more slender intervening bony partitions and is found in the interior of a bone.

2. Compact

This bone has fewer spaces and is always found on the exterior of a bone. The relative quantity of spongy and compact bone varies in different bones and in different parts of the same bone, depending on the need for strength or lightness.

Structure of compact bone

The haversian[1] system consists of a canal containing blood

[1]Clopton Havers, English anatomist (1650–1702).

vessels, lymph vessels, and nerves, surrounded by concentric rings or *lamellae, canaliculi,* and *lacunae.*

Lamellae
: These are concentric rings formed by calcified intercellular substance, or bone matrix, that encircle the haversian canal.

Canaliculi
: A system of tiny canals that extend from one lacuna to another and to the surfaces of bone where there are many capillaries. They contain the processes of the osteocytes and form the "lifeline" of nutrition to bone cells.

Lacunae
: These are "little lakes" or spaces that contain the osteocyte.

Red marrow
: Consists of connective tissue that acts as a support for a large number of blood vessels and marrow cells: (1) *Myelocytes,* which resemble white blood cells, and a small number of fat cells. (2) *Erythroblasts,* from which red blood cells are derived. Giant cells called (3) *megakaryocytes* are found in both red and yellow marrow. Red marrow is found in the articular ends of long bones and in cancellous tissue. (See Chapter 15 for red and white blood cell formation.)

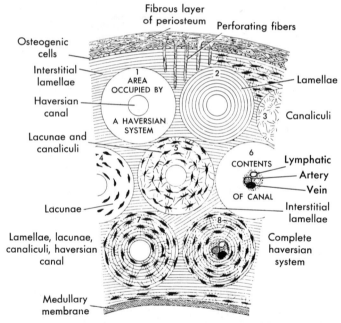

Figure 3–22. Diagram of a cross section of osseous tissue. Details are drawn to a very much larger scale than the complete drawing. A small part of a *transverse section* of a long bone is shown. At the uppermost part is the periosteum, covering the outside of the bone; at the lowermost part is the medullary membrane, lining the marrow cavity. Between these is compact tissue, consisting largely of a series of haversian systems, each being circular in outline and perforated by a central canal, left blank in the canals of five of the systems in this illustration. The *first* circle shows the area occupied by a system. The *second* shows the layers of bony tissue, or lamellae, arranged around the central canal. In the *third,* fine, dark radiating lines represent canaliculi, or lymph channels. In the *fourth,* dark spots arranged in circles between the lamellae represent lymph spaces, or lacunae, which contain the bone cells. In the *fifth,* the central canal, lacunae, and canaliculi, which connect the lacunae with each other and with the central canal, are shown. The *sixth* shows the contents of the canal: artery, veins, lymphatics, and areolar tissue. The *seventh* shows the lamellae, lacunae, canaliculi, and haversian canal. The *eighth* shows a complete haversian system.

Between the systems are interstitial lamellae, only a few of which show lacunae. The periosteum is made up of an outer fibrous layer and an inner osteogenic layer, so called because it contains bone-forming cells, or osteoblasts. (Modified from Gerrish.)

Yellow marrow

Consists of connective tissue containing many blood vessels and cells. Most of the cells are fat cells; a few are myelocytes. It is found in the medullary canals of long bones and extends into the spaces of the cancellous tissue and haversian canals.

Periosteum

Bones are covered, except at their cartilaginous extremities, by a membrane called *periosteum.* It consists of an outer layer of connective tissue and an inner layer of fine fibers, which form dense networks. In young bones, periosteum is thick, vascular, and closely connected with the epiphyseal cartilages. Later in life it is thinner and less vascular.

Endosteum

The marrow cavities and haversian canals are lined with a membrane called endosteum. Beneath it lies a layer of osteoblasts. After growth ceases, injury is the stimulus that activates the osteoblasts to produce new bone matrix.

Blood vessels of bone

Bones are plentifully supplied with blood vessels, which enter and leave the bones through canals (Volkmann).[2] These blood vessels proceed from the periosteum to join the haversian canals. The *lamellae* lie around the haversian canals, and arranged in circles are found the *lacunae,* which contain the bone cells (Figure 3–22).

Bone cells or osteocytes

Are found in the almond-shaped cell spaces, or lacunae. Their cytoplasmic processes project into the canaliculi. Before imprisonment in the lacunae, they were the osteoblasts, or bone-forming cells of the matrix. Osteoclasts are arranged in single rows on surfaces of growing bone. They are found on surfaces of bone where reabsorption takes place. They contain numerous mitochondria.

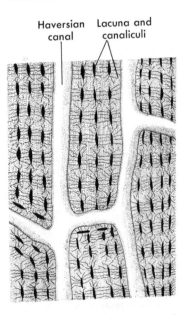

Haversian canal   Lacuna and canaliculi

Figure 3–23. Diagram of longitudinal section of bone.

Arteries to bone

The marrow of long bone is supplied by the medullary or nutrient arteries, which enter the bone at the nutrient foramen, located in most cases near the center of the bone, and perforate

[2] Alfred Wilhelm Volkmann, German physiologist (1800–1877).

the compact tissue obliquely. Its branches ramify in all directions in the marrow and bony tissue. The twigs of these vessels anastomose with arteries of compact and cancellous tissue. Veins emerge with or apart from the arteries.

Lymphatic vessels

Have been traced into the substance of bone and accompany the blood vessels in the haversian canals.

Nerve supply

The periosteum is well supplied with nerves, which accompany the arteries into bone. These nerve fibers form a plexus around the blood vessels. The nerve fibers include *afferent* myelinated and autonomic unmyelinated fibers.

Medullary canal

The shafts of long bone are formed almost entirely of compact

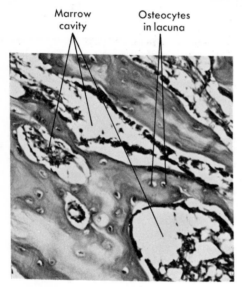

Spongy Bone

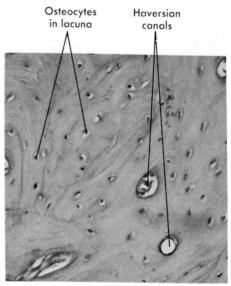

Compact Bone

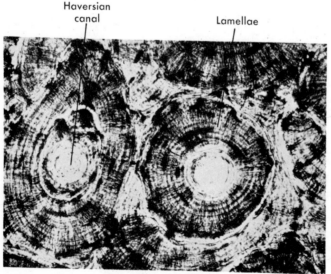

Ground Section of Bone
in Cross Section

Figure 3-24. Bone tissue as seen with the light microscope.

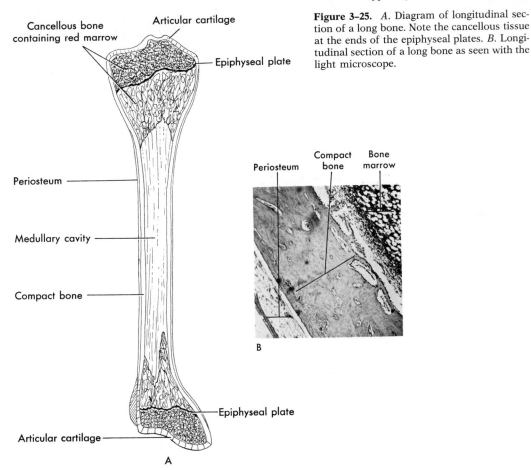

Cancellous bone containing red marrow

Articular cartilage

Epiphyseal plate

Periosteum

Medullary cavity

Compact bone

Articular cartilage

Epiphyseal plate

A

Periosteum

Compact bone

Bone marrow

B

Figure 3-25. *A.* Diagram of longitudinal section of a long bone. Note the cancellous tissue at the ends of the epiphyseal plates. *B.* Longitudinal section of a long bone as seen with the light microscope.

bone, except where they are hollowed out to form the *medullary canal,* which is lined by a vascular tissue called medullary membrane and filled with bone marrow.

Development of bone

In the early embryo the bones forming the roof and sides of the skull are preformed in membrane. Others are preformed in cartilage.

Intramembranous ossification

Before the cranial bones are formed the brain is covered by inner meningeal membranes, a middle fibrous membrane, and an outer layer of skin. Periosteum and bone are formed from the middle fibrous membrane. It is composed of fibers and bone-forming cells called osteoblasts in a matrix or ground substance. When bone begins to form, a network of *spicules* radiates from a point or center of ossification. The spicules develop into fibers. Calcium salts are deposited in the fibers and matrix, enclosing some of the osteoblasts in the lacunae. As the fibers grow out they continue to calcify and give rise to fresh bone spicules. Thus, a network of bone is formed. Successive layers of bony tissue are deposited under the periosteum, so that bone increases in thickness and presents the structure of compact bone on the outer and inner surfaces.

Diploe

With a layer of soft, spongy cancellous tissue between, the cancellous tissue between the layers of the skull is called the *diploe.* Ossification of bones of the skull is not complete at birth. At the site of the future union of two or more bones, membranous areas called *fontanelles* persist.

Intracartilaginous or endochondrial ossification	The skeleton of the embryo is preformed in cartilage by the end of the second month. Soon after this, ossification begins. Cartilage cells at the center of ossification enlarge and become arranged in rows. Then calcium salts are deposited in the matrix between the cells, first separating them and later surrounding them so that all nutriment is cut off, resulting in their atrophy and disappearance. *Perichondrium,* the membrane that covers the cartilage, assumes the character of periosteum. From this membrane cells grow, which are deposited in the spaces left by the atrophy of the cartilage cells. The two processes, destruction of cartilage cells and the formation of bone cells to replace them, continue until ossification is complete. The number of centers of ossification varies in differently shaped bones.
Ossification centers	In long bones, the center of ossification for the bone is called *diaphysis* and one or more centers for each extremity are called *epiphyses.* Ossification proceeds from the diaphysis toward the epiphyses. As each new portion is ossified, thin layers of cartilage continue to develop between the diaphysis and epiphyses; during the period of growth these outstrip ossification. When this ceases, growth of bone stops. Normal, healthy bone is under constant change as it is reabsorbed and repaired continuously.
Nutritional needs	Growth of bones is dependent upon (1) adequate amounts of calcium and phosphorus in the food, (2) chemical substances that enable the bone cells to utilize calcium and phosphorus, such as vitamins from foods, and (3) hormones. Low calcium content of blood may be caused by inadequate calcium in the diet, poor absorption of calcium, or too rapid excretion in the feces.

HORMONAL INFLUENCES ON BONE. *Somatotropin* (STH) influences growth of all tissues, especially bone growth. When the epiphyses of the long bones have united with the bone shafts, growth of bone ceases.

Thyroxin increases osteoclastic activity more than it increases osteoblastic activity.

Hyperactivity of *adrenocortical hormones* will cause protein mobilization from the organic matrix of bone, thereby decreasing it.

Parathyroid hormone acts directly on osteocytes and osteoclasts; it causes the release of calcium from bone mineral into the blood.

Calcitonin or *thyrocalcitonin* has the opposite effect of parathyroid hormone; it decreases bone resorption and thus lowers the blood calcium level.

Estrogens cause increased osteoblastic activity—for this reason, after puberty, bone rate of growth becomes rapid for several years. They also cause rapid uniting of the epiphyses with the shafts of long bone. Estrogens have a broadening effect on the pelvis.

Testosterone causes bones to thicken and deposit calcium salts, thereby increasing the total quantity of the matrix.

Ossification begins soon after the second month of intrauterine life and continues well into adult life. The sternum, the sacrum, and the hip bones do not unite to form single bones until the individual is well beyond 21 years of age.

The bones of the newborn infant are soft and largely composed of cartilage; since the process of ossification is going on continually, the proper shape of the cartilage should be preserved in order that the shape of the future bone may be normal. Therefore, it is obvious that a young baby's back should be supported, and a child should always rest in a horizontal position. The facility with which bones may be molded and become misshapen is seen in the bowlegs of children. However, the softness of the skeleton of a child accounts for the fact that the many jars and tumbles experienced in early life are not as injurious to the cartilaginous frame as they would be to a harder structure.

Rickets. Rickets is a condition in which the mineral metabolism is disturbed so that calcification of the bones does not take place normally. The bones remain soft and become misshapen, resulting in bowlegs and malformations of the head, chest, and pelvis. Liberal amounts of calcium and phosphorus in food, and vitamin D, found in fish-liver oils and in significant amounts of egg yolk, whole milk, butter, fresh vegetables, etc., are important factors in the cure and prevention of rickets. Exposure to sunshine, especially irradiation of the skin, helps also in the optimal use of calcium and phosphorus in the body.

Three types of rickets are recognized as due to chemical deficiency of the blood. The first, or so-called low-phosphorus rickets, is caused by a subnormal content of phosphate ions in the blood related to a lack of dietary vitamin D. This is the commonest type, sometimes called true rickets, and is characterized by histologic changes resulting in large joints, deformed bones of the cranium, chest, and spine, and a condition in which beadlike deposits occur at the ends of the ribs. The second type is characterized by deformed bones of the head, trunk, and limbs and is frequently accompanied by tonic muscular spasms (tetany) lasting for considerable periods of time. It is sometimes called low-calcium rickets or a "ricketslike condition." The third type is one in which both calcium and phosphorus are below

normal and is characterized by progressive porosity of the bones. A deficiency of vitamin C interferes with the function of the osteoblasts and their formation of organic intercellular substance. Individuals, especially women past middle life, frequently suffer from impaired skeletal maintenance, and the bones become relatively more fragile. This condition is called *osteoporosis*. There is some evidence that certain hormones may in part contribute to the condition.

Fracture. Fracture is a term applied to the breaking of a bone. It may be either partial or complete. As a result of the greater amount of organic matter in the bones of children, they are flexible, bend easily, and do not break readily. In some cases the bone bends like a bough of green wood. Some of the fibers may break, but not the whole bone, hence the name *green-stick fracture*. The greater amount of inorganic matter in the bones of the aged renders the bones more brittle, so that they break easily and heal with difficulty.

Regeneration of Bone. A fracture is usually accompanied by injury to the periosteum and tissues causing hemorrhage and destruction of tissue.

Fibroblasts and capillaries grow into the blood clot, forming granulation tissue. The plasma and white cells from the blood exude into the tissues and form a viscid substance, which sticks the ends of the bone together. This exudate into which the fibroblasts and capillaries grow is called *callus*. Usually bone cells from the periosteum and calcium salts are gradually deposited in the callus, which eventually becomes hardened and forms new bone. Occasionally the callus does not ossify, and a condition known as *fibrous union* results. The periosteum is largely concerned in the process of repair. If a portion of the periosteum is stripped off, the subjacent bone may die, whereas if a large part of the whole of a bone is removed and the periosteum at the same time is left intact, the bone will wholly or in great measure be regenerated.

Muscle Tissue

The cells of muscle tissue are long and often taper at their ends. Because of the length of these cells (1 to 40 mm, or as long as 40 cm), they are referred to as muscle fibers; hence when we talk of muscle tissue, the terms *cell* and *fiber* are synonymous. The length of muscle cells is one of the basic structural entities that defines one of the properties of this tissue, namely, *contractility*. A second property of muscle tissue is *irritability*.

There are three types of muscle tissue:

skeletal, smooth, and *cardiac.* The descriptions that follow concentrate on the histologic organization of muscle tissue. Additional details of structure that are concerned with the physiology of muscle (e.g., how does muscle contract?) will be discussed in Chapter 6.

CHARACTERISTICS, LOCATION, AND FUNCTION

Skeletal (striated) muscle

This tissue can be referred to by several terms: *striated,* because of transverse stripes or *striae* that run along the length of fibers; *skeletal,* because it forms the muscles of the body that are attached to the skeleton; *somatic,* because it helps to form the body wall; and *voluntary,* because the movement of this type of muscle is under conscious control. Skeletal muscle is capable of rapid powerful contraction and can maintain long states of partially sustained contraction. It has an extensive blood and nerve supply.

The cells have many nuclei, which are peripherally located. Individual cells and groups or bundles of cells are surrounded by loose connective tissue. The prefix *sarco* (Greek word for muscle) is used to refer to parts of the cell—hence the words *sarcolemma, sarcoplasm, sarcosome,* and *sarcomere.*

Sarcolemma

Cell membrane of muscle cell.

Myofibrils and myofilaments

Single muscle cells are made up of many fibers called *myofibrils* (1 to 3 μm in diameter). Myofibrils, in turn, are made up of smaller units, *myofilaments.* Myofilaments are of two types: (1) thick filament, 100 Å in diameter, which is the protein *myosin,* and (2) thin filament, 50 Å in diameter, which is the protein *actin.* The arrangement of myofilaments is responsible for striations and specific banding pattern, which under appropriate conditions can be seen with the light microscope as alternating light and dark bands.

Sarcoplasm

The cytoplasm of muscle cell. It contains the Golgi complex, numerous mitochondria (also called sarcosomes), endoplasmic reticulum (also called sarcoplasmic reticulum), glycogen granules, and lipid.

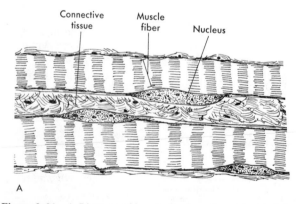

Connective tissue   Muscle fiber   Nucleus

A   B

Figure 3–26. *A.* Diagram of longitudinal section striated skeletal muscle. *B.* Longitudinal section of striated skeletal muscle as seen with the light microscope. *C.* Diagram of cross section of striated skeletal muscle. Three bundles (fasciculi) are seen. *D.* Cross section of striated skeletal muscle as seen with the light microscope. Arrows indicate one muscle cell.

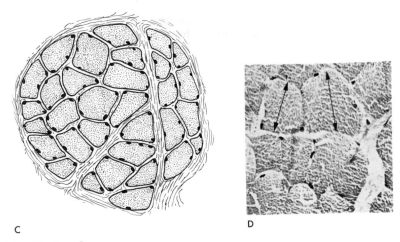

C   D

Figure 3–26. (*Continued*)

Smooth muscle (nonstriated)

Called smooth because striations are not visible with light microscope. Myofilaments are present and seen with the electron microscope, but not in the banding pattern of skeletal muscle. Cells are spindle shaped, 20 to 50 μm long, and have single, centrally located nuclei. Reticular fibers invest individual smooth muscle cells. Contraction of this muscle is involuntary. Smooth muscle makes up outer walls of gastrointestinal tract and walls of blood vessels. By contraction and relaxation, the size of gastrointestinal lumen and diameter of blood vessels can be changed.

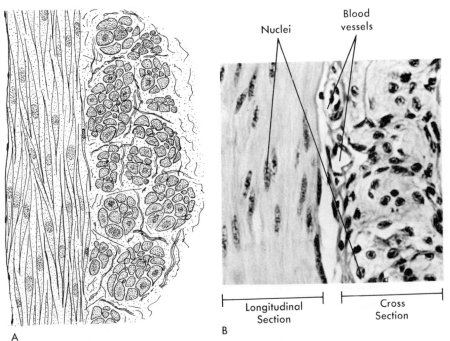

Figure 3–27. *A.* Diagram of smooth muscle as seen in gastrointestinal tract in longitudinal and cross section. *B.* Appearance of smooth muscle as seen with the light microscope.

Cardiac muscle

Tissue that forms heart. The cross-banding pattern is similar to that of skeletal muscle. Each cell has a large, oval, centrally located nucleus. The fibers branch and connect with adjacent fibers, forming a complex three-dimensional network. Where fibers connect, a special surface specialization is found, called the *intercalated disk*. These appear as dark-stained, steplike cross bands with the light microscope; with the electron microscope they appear typically as desmosomes. Cardiac muscle is involuntary and has little or no capacity for regeneration.

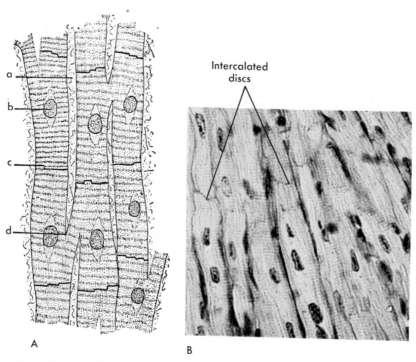

Intercalated
discs

A

B

Figure 3–28. *A*. Diagram of longitudinal section of cardiac muscle. (*a*) Connective tissue between cells, (*b*) centrally located nucleus, (*c*) intercalated disc, (*d*) branching of cell. *B*. Longitudinal section of cardiac muscle as seen with the light microscope.

Nervous Tissue

CHARACTERISTICS

Nervous tissue is the very specialized tissue making up the nervous system. The *neuron* is the cell of nervous tissue that possesses the properties of (1) *irritability* (excitability), the ability to respond to changes in the environment, and (2) *conductivity*, the ability to transmit impulses. The basic histologic organization of this tissue will be described here briefly; the function and detailed structure will be presented in Chapter 9, which deals with the nervous system.

Neuron

The neuron has a cell body or perikaryon and projecting from it numerous short processes called *dendrites* and one long process called the *axon*. The cytoplasm contains mitochondria, the Golgi complex, neurofibrils, and chromophilic granules called *Nissl bodies* (substance). The Nissl bodies contain ribonucleic acid and vary in quantity and distribution depending on the physiologic state of the neuron. Nissl bodies are not present where the axon begins to project from the cell body; hence this becomes a means of identifying the axon from the dendrites. Neurofibrils form a network within the cell body, projecting into the dendrites and axons. Both the shape and the size of the cell body are variable, as is the extent of dendritic branching. A given morphologic type of neuron is found in specific areas of the nervous system, and usually is associated with a specific function. Cell bodies are found in the central nervous system, namely the brain and spinal cord, and in areas outside of the central nervous system called *ganglia* (singular ganglion).

Neuroglia

The term *neuroglia* ("nerve glue") is used to describe cells found in the nervous system that do not have the ability to transmit impulses. These cells act as the specialized connective tissue of the nervous system. The neuroglia of the central nervous system include (1) *ependymal cells*, which line the ventricles of the brain and spinal cord, and (2) *neuroglial cells* proper, the *astrocyte*, *oligodendrocyte*, and *microglia*. The neuroglia of the peripheral nervous system include (1) the *satellite* or *capsular cells* of peripheral ganglia and (2) the *Schwann cells* of peripheral nerves.

Nerves

A nerve is a collection of many axons from many neurons.

Figure 3–29. *A.* Two cell bodies from anterior horn of spinal cord. *B.* Four cell bodies from dorsal root ganglion. *C.* Four nerve bundles seen, two of which (upper half) are sectioned longitudinally. Myelin is not stained; therefore, appearance of clear circular area with central dot. Dot is axon ⊙. *D.* A nerve between muscle fibers. Myelin is stained and appears as circular bands.

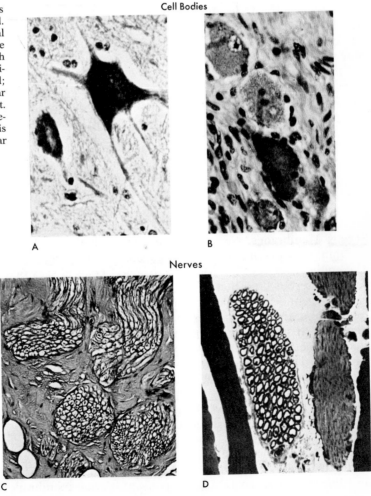

Cell Bodies

A   B

Nerves

C   D

Nerves extend over great lengths and make contact with skeletal, smooth, and cardiac muscle and glands. Since they are long, they are often called nerve fibers. Most of the nerve fibers found in the body are covered by a sheath of lipid material called *myelin* and are referred to as myelinated fibers; others are unmyelinated. Under suitable staining conditions the myelin sheath appears as a dark band encircling nerve fibers (Figure 3–29, *D*).

Membranes

In combination with the underlying connective tissue, epithelium forms *membranes*. The word *membrane* in its broadest sense is used to designate any thin expansion of tissues. In the commonest sense, the word *membrane* is used to denote an *envelope* or *lining* made up of tissues. The underlying connective tissue forms a stratum of closely woven fibers, which permits a variable amount of stretching but prevents an expansion that may separate the epithelial cells.

CLASSIFICATION OF MEMBRANES

The chief membranes of the body are:

1. Serous
2. Synovial
3. Mucous
4. Cutaneous

Serous Membranes. Serous membranes are thin, transparent, strong, and elastic. The epithelial cells of a serous membrane secrete a serous fluid that acts as a lubricating agent. Serous membranes are found lining the body cavities and covering the organs that lie in them. The part of a serous membrane that lines a body cavity is called the *parietal portion*, whereas the part covering the visceral organs in the cavity is the *visceral portion*. The serous fluid, as a lubricant, allows the organs to glide easily against the walls of the cavity or upon each other.

Under serous membranes are included (1) the two *pleurae*, which cover the lungs and line the chest cavity; (2) the *pericar-*

dium, which covers the heart and lines the inner surface of the fibrous pericardium; and (3) the *peritoneum*, which lines the abdominal cavity and covers its contained viscera and the upper abdominal surface of some of the pelvic viscera.

Synovial Membranes. Synovial membranes line joints everywhere except over the articular surfaces. They are richly supplied with blood capillaries, lymphatics, and nerves. These membranes consist of an outer layer of fibrous tissue and an inner layer of loose connective tissue with collagen and elastic fibers, fibroblasts, lymphocytes, macrophages, and leukocytes. There is not always a definite cellular layer on the inner surface; when present, it appears as a simple squamous or cuboidal epithelium.

There are three classes of synovial membranes: (1) articular, (2) sheaths, and (3) bursae. *Articular synovial membranes* line the articular capsules of freely movable joints. *Synovial sheaths* are elongated closed sacs that form sheaths for the tendons of some of the muscles, particularly the flexor and the extensor muscles of the fingers and toes. They facilitate the gliding of the tendons in the fibro-osseous canals. *Synovial bursae* are simple sacs interposed to prevent friction between two surfaces that move upon each other. An example is the subcutaneous bursae between the skin and underlying bony prominences such as the olecranon (elbow) and the patella (knee).

As with serous membranes, a fluid, acting as a lubricant, is associated with synovial membranes. This is synovial fluid, which is an ultrafiltrate of blood, some mucin, a variable number of white blood cells, and hyaluronic acid. Synovial fluid is the major source of nutrition for articular cartilage.

Mucous Membranes. A mucous membrane is composed of an inner epithelium and an underlying loose connective tissue. As the name implies, specialized epithelial cells of these membranes secrete mucus. Mucous membranes protect by forming a lining for all the passages that communicate with the exterior. These membranes are highly vascular and have an extensive lymphatic circulation.

The mucous membranes of the body are grouped in two divisions: (1) gastropulmonary and (2) genitourinary.

Gastropulmonary Mucous Membrane. This lines the alimentary canal, the air passages, and the cavities communicating with them. It is continuous from the edges of the lips and nostrils and extends through the mouth and nose to the throat, throughout the length of the alimentary canal to the anus. At its origin and termination it is continuous with the external skin. It also extends throughout the trachea, bronchial tubes, and air sacs. From the interior of the nose the membrane extends into the frontal, ethmoid, sphenoid, and maxillary sinuses, into the lacrimal passages, and also becomes the conjunctival membrane over the forepart of the eyeball and inside of the eyelids on the edges of which it meets the skin. A prolongation of this membrane extends on each side of the upper and back part of the pharynx, forming the lining of the auditory (eustachian[3]) tube. This membrane also lines the salivary, pancreatic, and biliary ducts and the gallbladder. The mucous membrane lining the gastrointestinal tract is thrown into many folds, thus increasing the amount of surface area for secretion and absorption by the epithelium.

Genitourinary Mucous Membrane. This lines the bladder and the urinary tract from the interior of the kidneys to the

[3]Bartolommeo Eustacchio, Italian anatomist (1520–1574).

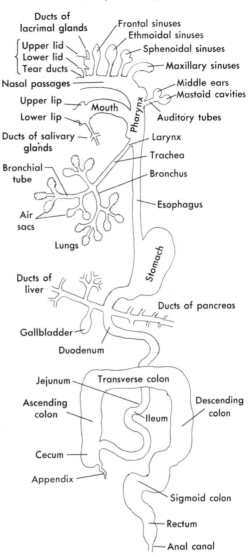

Figure 3–30. Diagram showing continuity of the gastropulmonary mucous membrane.

orifice of the urethra; it lines the ducts of the testes, epididymis, and seminal vesicles; it lines the vagina, uterus, and uterine (fallopian[4]) tubes.

Cutaneous Membrane. This is a very specialized membrane, more commonly referred to as the *skin;* indeed, it often is classified as an organ because of its many functions. The structure and function of skin are discussed in the section that follows.

[4]Gabriel Falloppius, Italian anatomist (1523–1562).

The Skin and Appendages

Most of our contacts with the environment are through the skin. Since living cells must be surrounded with fluid, the contact of the body with the air is made by means of dead cells. These dead cells form a protective covering for the living cells. The living cells of the inner layers of the skin are constantly pushed to the outside, shrinking and undergoing progressive chemical changes that cement them firmly together and render them waterproof. In this way a tissue-fluid environment is maintained for living cells although man lives in an air environment.

KAROTINIZED

SKIN

The skin has many functions. It covers the body and protects the deeper tissues from drying and injury. It protects from invasion by infectious organisms. It is important in many ways in temperature regulation. It contains end organs of many of the sensory nerve fibers by means of which one becomes aware of the environment. It also acts as an accessory mechanism for tactile and pressure corpuscles. In the skin fat, glucose, water, salts such as sodium chloride, and fluid accumulate in the tissues. The skin has excretory functions, eliminating water with the various salts that compose perspiration, and the dead cells themselves become an important way of eliminating many salts. It is an important light screen for the underlying living cells. It also has absorbing powers. It will absorb oily materials placed in contact with it.

The rule of nine gives the relative distribution of total body surface area.

Head and neck	9%
Anterior trunk	18%
Posterior trunk	18%
Upper extremity (9 × 2)	18%
Lower extremity (18 × 2)	36%
Perineum	1%
Total 100%	

Skin consists of two distinct layers: (1) epidermis, cuticle; (2) dermis, corium, or cutis vera.

Epidermis. The epidermis (cuticle) is stratified squamous epithelium, consisting of a variable number of layers of cells. It varies in thickness in different parts of the body, being thickest on the palms of the hands and on the soles of the feet, where the skin is most exposed to friction, and thinnest on the ventral surface of the trunk and the inner surfaces of the limbs. It forms a protective covering over every part of the true skin and is closely molded on the papillary layer of the corium. The external surface of the epidermis is marked by a network of ridges caused by the size and arrangement of the papillae beneath. Some of these ridges are large and correspond to the folds produced by movements, e.g., at the joints; others are fine and intersect at various angles, e.g., upon the back of the hand. Upon the palmar surface of the fingers and hands and the soles of the feet, the ridges serve to increase resistance between contact sur-

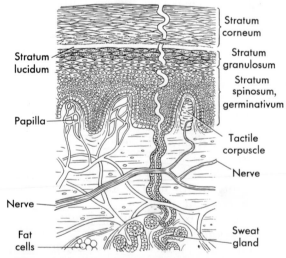

Figure 3–31. Diagram of a section of skin to show its structure. The epidermis consists of the stratum corneum, lucidum, granulosum, spinosum, and germinativum (mucosum). The dermis lies below the epidermis.

faces and therefore prevent slipping. On the tips of the fingers and thumbs these ridges form distinct patterns, which are peculiar to the individual and practically never change, hence the use of fingerprints for purposes of identification.

From without inward five regions of the epidermis are named: the *stratum corneum*, the *stratum lucidum*, the *stratum granulosum*, the *stratum spinosum*, and the *stratum germinativum*.

The three outer layers consist of cells that are constantly being shed and renewed from the cells of the *stratum germinativum*. In the *stratum corneum* the protoplasm of the cells has become changed into a protein substance called *keratin*, which acts as a waterproof covering. The reaction is acid; and many kinds of organisms, when placed upon the skin, are destroyed, presumably by the effect of the acidity. Underneath this is the *stratum lucidum*, a few layers of clear cells.

The *stratum granulosum* is formed by two or three layers of flattened cells that are cells in transition between the stratum germinativum and the horny cells of the superficial layers.

The *stratum spinosum* is of variable thickness and is composed of irregularly-shaped cells. This layer is also called the prickle-cell layer because the surface of the cells is covered with short cytoplasmic spines or projections.

The *stratum germinativum* consists of several layers of cells. The growth of the epidermis is by multiplication of the cells of the germinative layer. As they multiply, the cells previously formed are pushed upward toward the surface. In their upward progress these cells undergo a chemical transformation, and the soft protoplasmic cells become converted into the flat scales which are constantly being rubbed off the surface of the skin. The pigment in the skin is found in greatest amount in the cells of the stratum germinativum. No blood vessels pass into the epidermis, but fine nerve fibers lie between the cells of the inner layers.

Dermis. The dermis (corium) is a highly sensitive and vascular layer of connective tissue. It contains numerous blood vessels, lymph vessels, nerves, glands, hair follicles, and papillae and is described as con-

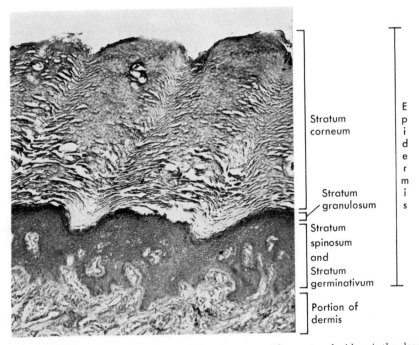

Figure 3–32. Section of skin as seen with the light microscope. The stratum lucidum is the clear area between stratum corneum and granulosum. The full extent of the dermis is not shown.

sisting of two layers: the *papillary*, or *superficial*, *layer*, and the *reticular*, or *deeper*, *layer*.

The surface of the *papillary*, or *superficial*, layer is increased by small conical elevations, called papillae, whence this layer derives its name. They project up into the epidermis, which is molded over them. The papillae consist of small bundles of fibrillated tissue, the fibrils being arranged parallel to the long axis of the papillae. Within this tissue is a loop of capillaries, and some papillae, especially those of the palmar surface of the hands and fingers, contain *tactile corpuscles*, which are numerous where the sense of touch is acute.

The *reticular*, or *deeper*, layer consists of strong bands of fibrous tissue and some fibers of elastic tissue. These bands interlace, and the tiny spaces formed by their interlacement are occupied by adipose tissue and sweat glands. The reticular layer is attached to the parts beneath by a subcutaneous layer of areolar connective tissue, which, except in a few places, contains fat. In some parts, as on the front of the neck, the connection is loose and movable; in other parts, as on the palmar surface of the hands and the soles of the feet, the connection is close and firm. In youth the skin is both extensile and elastic, so that it can be stretched and wrinkled and return to its normal condition of smoothness. As age advances, the elasticity is lessened, and the wrinkles tend to become permanent.

Blood Vessels and Lymphatics. The arteries that supply the skin form a network in the subcutaneous connective tissue and send branches to the papillae, the hair follicles, and the sudoriferous glands. The capillaries of the skin are so numerous that when distended, they are capable of holding a large proportion of the blood contained in the body. The amount of blood they contain is dependent on their caliber, which is regulated largely by the vasomotor nerve fibers.

There is a superficial and a deep network of lymphatics in the skin. These communicate with each other and with the lymphatics of the subcutaneous connective tissue.

Nerves. The skin contains the peripheral terminations of many nerve fibers and receptors. These fibers may be classified as follows:

1. Motor nerve fibers, including the vasoconstrictors and vasodilators distributed to the blood vessels, and motor nerve fibers distributed to the arrector muscles (arrectores pilorum) of the hair follicles.

2. Receptors concerned with the temperature sense, which terminate in *cold receptors* (end organs of Krause) and *receptors for warmth* (possibly the end organs of Ruffini).

3. Receptors concerned with touch and pressure, which terminate in *touch* (Meissner's corpuscles and free nerve endings around hairs and skin) and *pressure receptors* (pacinian corpuscles).

APPENDAGES OF THE SKIN

The appendages of the skin are the nails, the hairs, the sebaceous glands, the sudoriferous, or sweat, glands, and their ducts.

Nails. The nails (ungues) are composed of clear, horny cells of the epidermis, joined so as to form a solid, continuous plate upon the dorsal surface of the terminal phalanges. Each nail is closely adherent to the underlying corium, which is modified to form what is called the bed, or *matrix*. The body of the nail is the part that shows. The hidden part, in the nail groove, is called the root. The *lunule* is the crescent-shaped white area that can be seen on the part nearest the root. The nails appear pink except at the *lunule* because blood in the capillary bed shows through.

The nails grow in length by multiplication of the soft cells in the stratum germinativum. The cells are transformed into hard, dry scales, which unite to form a solid plate; and the nail, constantly receiving additions, slides forward over its bed and projects beyond the end of the finger. When a nail is thrown off by suppuration or torn off by violence, a new one will grow in its place provided any of the cells of the stratum germinativum are left.

Hairs. The hairs (pili) are growths of the epidermis, developed in the hair follicles, which extend downward into the subcutaneous tissue. The part that lies within the follicle is known as the root, and that portion which projects beyond the surface of the skin is called the shaft. The hair is composed of:

Cuticle, a single layer of scalelike cells which overlap.

Cortex, a middle portion, which constitutes the chief part of the shaft, formed of elongated cells united to form flattened fibers, which contain pigment granules in dark hair.

Medulla, an inner layer composed of rows of many-sided cells, which frequently contain air spaces. The fine hairs covering the surface of the body and the hairs of the head do not have this layer.

The root of the hair is enlarged at the bottom of the follicle into a bulb, which is composed of growing cells and fits over a vascular papilla that projects into the follicle. Hair has no blood vessels but receives nourishment from the blood vessels of the papilla.

Growth of Hair. Hair grows from the papilla by multiplication of its cells (matrix cells). These cells become elongated to form the fibers of the fibrous portion, and as they are pushed to the surface, they become flattened and form the cuticle. If the scalp is thick, pliable, and moves freely over the skull, it is favorable to the growth of hair. A thin scalp that is drawn tightly over the skull tends to constrict the blood vessels, lessen the supply of blood, and cause atrophy of the roots of the hair by pressure; in such cases massage of the head loosens the scalp, improves the circulation of the blood, and usually stimulates the growth of the hair. The hairs are constantly falling out and constantly being replaced. In youth and early adult life not only may hairs be replaced, but there may be an increase in the number of hairs by development of new follicles. When the matrix cells lose their vitality, new hairs will not develop.

With the exceptions of the palms of the hands, the soles of the feet, and the last phalanges of the fingers and toes, the whole skin is studded with hairs. The hair of the scalp is long and coarse, but most of the hair is fine and extends only a little beyond the hair follicle.

Arrector (Arrectores Pilorum) Muscles. — *imp.* The follicles containing the hairs are narrow pits which slant obliquely upward, so that the hairs they contain lie slanting on

involuntary muscle

Longitudinal section of hair follicle

Cross section of hair follicle

Figure 3–33. Section of skin showing hair follicles as seen with the light microscope.

the surface of the body. Connected with each follicle are small bundles of involuntary muscle fibers called the *arrector muscles*. They arise from the papillary layer of the corium and are inserted into the hair follicle below the entrance of the duct of a sebaceous gland. These muscles are situated on the side toward which the hairs slope, and when they contract, as they will under the influence of cold or fright, they straighten the follicles and elevate the hairs, producing the roughened condition of the skin known as "gooseflesh." Since the sebaceous gland is situated in the angle between the hair follicle and the muscle, contraction of the muscle squeezes the sebaceous secretion out from the duct of the gland. This secretion aids in preventing too great heat loss.

Glands in the Skin. *Sebaceous Glands.* These occur everywhere over the skin surface with the exception of the palms of the hands and the soles of the feet. They are abundant in the scalp and face and are numerous around the apertures of the nose, mouth, external ears, and anus. Each gland is composed of a number of epithelial cells and is filled with larger cells containing fat. These cells are cast off bodily, their detritus forms the secretion, and new cells are continuously formed. Occasionally the ducts open upon the surface of the skin, but more frequently they open into the hair follicles. In the latter case, the secretion from the gland passes out to the skin along the hair. Their size is not regulated by the length of the hair.

The largest sebaceous glands are found on the nose and other parts of the face, where they may become enlarged with accumulated secretion. This retained secretion often becomes discolored, giving rise to the condition commonly known as blackheads. It also provides a medium for the growth of pus-producing organisms and consequently is a common source of pimples and boils.

Sebum. Sebum is the secretion of the sebaceous glands. It contains fats, soaps, cholesterol, albuminous material, remnants of epithelial cells, and inorganic salts. It serves to protect the hairs from becoming too dry and brittle, as well as from becoming too easily saturated with moisture. Upon the surface of the skin it forms a thin protective layer, which serves to prevent undue absorption or evaporation of water from the skin. This secretion keeps the skin soft and pliable. An accumulation of this sebaceous matter upon the skin of the fetus furnishes the thick, cheesy, oily substance called the *vernix caseosa.*

Sudoriferous, or Sweat, Glands. These are abundant over the whole skin but are largest and most numerous in the axillae, the palms of the hands, the soles of the feet, and the forehead. Each gland consists of a single tube, with a blind, coiled end that is lodged in the subcutaneous tissue. From the coiled end, the tube is continued as the excretory duct of the gland up through the corium and epidermis and finally opens on the surface by a pore. Each tube is lined with secreting epithelium. The coiled end is closely invested by capillaries, and the blood in the capillaries is separated from the cavity of the glandular tube by the thin membranes that form their respective walls.

Perspiration, or Sweat. Pure sweat is very dilute and practically neutral. When gathered from the skin, it contains fragments of cells and sebum and has a pH

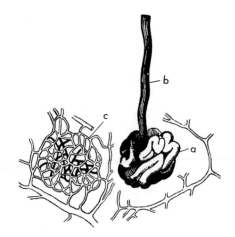

Figure 3–34. Coiled end of a sweat gland. *a*, The coiled end; *b*, the duct; *c*, network of capillaries, inside of which the sweat gland lies.

range of 5.2 to 6.75. Perspiration contains the same inorganic constituents as the blood, but in lower concentration. The chief salt is sodium chloride. The organic constituents in sweat include urea, uric acid, amino acids, ammonia, sugar, lactic acid, and ascorbic acid. Any factor that affects the composition of blood may also alter the composition of sweat. Sulfonamides are present after administration. Immune substances may also be present.

Under ordinary circumstances, the perspiration that the body is continually throwing off evaporates from the surface of the body without one's becoming aware of it and is called *insensible perspiration.* When more sweat is poured upon the surface of the body than can be removed at once by evaporation, it appears on the skin in the form of drops and is then spoken of as *sensible perspiration.*

The amount secreted during 24 hours varies greatly. It is estimated to average about 480 to 600 ml, but may be increased to such an extent that even more than this may be secreted in an hour.

ACTIVITY OF THE SWEAT GLANDS. Special secretory nerve fibers are supplied to the glandular epithelium of the sweat glands. The activity of these glands is supposed to be the result either of direct stimulation of the nerve endings in the glands or of indirect stimulation through the sensory fibers of the skin. The usual cause of profuse sweating is a high external temperature or muscular exercise. It is known that the high temperature acts upon the sensory cutaneous nerves, possibly the heat fibers, and stimulates the sweat fibers indirectly.

PHYSIOLOGY OF THE SWEAT GLANDS. Although perspiration is an excretion, its value lies not so much in the elimination of waste matter as in the loss of body heat by the evaporation of water. Each gram of water requires about 0.5 kcal for evaporation, and this heat comes largely from the body. This loss of heat helps to balance the production of heat that is constantly taking place. When the kidneys are not functioning properly, and the blood contains an excessive amount of waste material, the sweat glands will excrete some of the lat-

ter, particularly if their activity is stimulated. In the condition known as uremia, when the kidneys secrete little or no urine, the percentage of urea in perspiration rises.

Ceruminous Glands. The skin lining the external auditory canal contains modified sweat glands called *ceruminous* glands. They secrete a yellow, pasty substance resembling wax, which is called *cerumen.* An accumulation of cerumen deep in the auditory canal may interfere with hearing.

TISSUE REPAIR

Repair of tissues takes place continually under the normal process of living. Some tissues are subject to more wear and tear than others, such as, for example, those in the skin. The stratified squamous epithelium of the skin is constantly being subjected to varying degrees of friction and will respond to meet immediate needs. This might be shown by thickening, callus formation, or a blister, depending on the rapidity of the friction. When subjected to greater friction, sudden contact with sharp, dull, or blunt objects, the tissues are damaged or there may be tissue loss.

Some tissues are repaired easily and quickly, such as surface epithelium, connective tissue, and liver cells. Others repair more slowly; others, as typified by bone, must have parts kept in alignment and immobilized until repair is completed. Periosteum must be present for bone regeneration, muscle tissue has least ability for repair, and nerve cells destroyed by injury or infection do not regenerate.

Repair of Epithelial Tissues. Both the epithelial cells and underlying connective tissue cells have capacity for cell division and repair. The response to injury depends on the extent of the injured surface. There is usually rapid cell multiplication of the surface layer. Mitosis is decreased for the first few days but soon exceeds the normal rate. When injury is more extensive, the underlying tissue cells divide and migrate by a sort of ameboid movement into the injured area. There is formation of tonguelike processes, which grow into the area, and finally the surface layer of cells multiplies and they migrate over the injured surface, re-forming the surface layer of cells.

PRIMARY REPAIR. Primary repair takes place in "clean" wounds, such as, for instance, incisions, cuts, and the like, when infection is not present. If the injury simply involves the skin, the deep layer of stratified squamous epithelium divides longitudinally, the cells "push up," and the wound is rapidly and completely restored to normal. If the area is larger, the underlying connective tissue cells, the fibroblasts, take part in the repair process.

If the area of skin loss is great, fluid exudes from the capillaries; it dries and seals the open tissue, and a "scab" forms. Epithelial cells proliferate at the edges and continue to grow over the area until it is covered. If skin loss involves a large area, skin grafting is done to hasten the process.

When the deeper tissues are involved, if the edges of the wound are brought together with sutures, as, for example, in operative incisions, there is outpouring of serous fluid into the wound, and a coagulum is formed which seals the wound. The coagulum contains leukocytes and tissue fragments. In 24 to 36 hours, fibroblasts of connective tissue and the *endothelial* cells of the capillaries are multiplying rapidly. The newly formed cells remain along the edges of the wound, and by the third day new vascular buds are present. These grow across the wound along with connective tissue formation.

By the fourth or fifth day, fibroblast activity is markedly evident. Collagenous fibers are rapidly formed, and capillaries sprout and extend across the wound, holding the edges firmly together. Later the fibers shorten and scar tissue is reduced to a minimum.

SECONDARY REPAIR. In large open wounds with more or less tissue loss, the area is filled in by a process of building up "granulation tissue." Each granulation represents a minute vascular area consisting of newly formed, vertically upstanding blood vessels, which are surrounded by young connective tissue and wandering cells of different kinds.

The surface has a characteristic pebbly appearance. The connective tissue cells, the fibroblasts, increase in number; collagenous fibers are proliferated by them, and eventually the wound closes.

Granulation tissue secretes a fluid which has definite bactericidal properties. Its defense characteristics include hyperemia, which is more or less marked, active exudation, and marked local leukocytosis.

The amount of scar tissue formed is in relation to tissue damage. It is important that parts of the body undergoing extensive tissue repair (such as, for example, burned areas involving the chest and neck, the chest after breast removal, and the like) be kept in alignment, immobile at first, and in some instances stretched. Active movement should be encouraged early so that, as new tissues are formed, contracture from scar formation will not result. Every effort should be directed toward preventing or minimizing disfigurement.

At times, as when large areas of skin are destroyed by burns, considerable blood plasma is lost, with its contained proteins, electrolytes, and other substances. This may result in disturbance of fluid balance. It is, therefore, important that fluids be replaced by artificial means until the tissues themselves can prevent such loss by coagulation and initial wound healing response.

It is known that glucocorticoids have an anti-inflammatory action. However, the exact mechanism of action is not understood. If given in large doses, they diminish the inflammatory process and inhibit the formation of granulation tissue.

Conditions Favorable to Wound Healing. Nutrition plays an important role in the healing process. The tissues need plenty of protein for repair, hence the need for protein-rich diets.

The vitamins play an important part in wound healing, as well as in resistance to and prevention of infection. It is believed that *vitamin A* is important for repair of epithelial tissues, especially the maintenance of epithelial integrity of the respiratory pathway. There is no conclusive evidence that vitamin A has anti-infective or wound-healing properties when applied in ointments. However, reports of controlled experiments on animals show that ointments containing vitamins A and D shortened the healing time.

VITAMIN B. Thiamine, nicotinic acid, and riboflavin are important from the viewpoint of the general well-being of the individual as a whole, and specifically in relation to metabolism, vigor, appetite, relief of pain in some instances, and integrity of selective epithelial areas.

VITAMIN C. The normal production and maintenance of intercellular substance as well as cement substances of the connective tissues, especially collagen formations and the integrity of capillary walls, are directly dependent upon vitamin C. In wound healing by granulation, new capillaries must sprout and fibroblasts must grow; to meet the increased demands, vitamin C is essential.

VITAMIN D. Vitamin D is essential for the normal absorption of calcium from the intestine; so possibly it aids in the healing of fractures. A low serum calcium level stimulates parathormone production, which increases excretion of phosphorus by the kidney and tends to raise serum calcium.

VITAMIN K. Vitamin K functions in wound healing from the viewpoint of helping to maintain the normal coagulability of the blood.

On the whole, tissues heal faster and leave less visible scars in the young than in the aged. This perhaps is due to the fact that in the young the tissues are soft, pliable, and in a constant state of growth; and in comparison to the aged or older age group, cells multiply more rapidly.

Normally tissue repair takes place so readily and is so commonplace that thought is not given to it. Tissues may remain dormant for years, then suddenly become actively growing tissues in a response to stimulus. It is believed the stimulus is almost certainly chemical in nature, a substance liberated by the degenerating cells. When part of the tissue is removed, this in itself initiates regeneration and forms the basis of the theory that disturbance of *spatial equilibrium* is an important factor. Another theory suggests the coaptation theory— that cells of a tissue have common affinity due to highly specialized specific stereochemical bonds that exist at cell surfaces. If the bonds are disrupted, it provides a stimulus for proliferation of tissue activity.

The *leukocytes* have a specific function in relation to tissue repair. They are attracted to areas of injury, perhaps in response to the chemical liberated. The monocytes are especially active in the repair of tissue, possibly providing nutritive and building materials of protein and lipoid character.

Tissue Transplants. In the human, transplants of tissue from one part of the body to another are accepted because they are part of the same body and are said to be *autologous.* The tissues of identical twins are also autologous because they are from the same fertilized egg so that the tissues have the same tissue proteins. The transplanted cells survive only when nutrients and oxygen can reach the cells from the surrounding tissue fluid. Fraternal twins do not come from the same fertilized ovum and are genetically different; hence each

has different tissue proteins and they are said to be *homologous* or *allogeneic.*

Each individual has a different genetic structure and therefore different protein structure so that when tissues are transplanted from one person to another, *antibodies* are formed in the *recipient* against the foreign tissue proteins, which are antigenic, and destruction of the cells of the transplant results. However, the many tissues used for transplants from "tissue banks" such as bone, tendons, blood vessels, nerves, and fascia survive because of the large amount of connective tissue and the relatively small numbers of cells. The transplanted tissue serves as a temporary bridge which functions as a substitute for the destroyed tissue and stimulates regeneration of tissue in the recipient. In some instances the recipient for fresh tissue transplants is subjected to radiation to destroy his ability to form antibodies against the donor's tissue proteins. Antibody-blocking drugs, antimetabolites, and glucocorticoids are also used to prevent antibody formation by the recipient.

Successful corneal transplants have been done for many years. It has been thought that since the cornea is normally avascular and does not contain lymphocytes, it is different from other tissues. Recent studies show that the surface epithelium is regenerated by the recipient so that the transplanted cornea is covered in a few days. Regeneration of the endothelium takes longer, and the connective tissue cells are slowly replaced. Research today is vigorous in relation to *organ* transplants, and a certain degree of success is reported in the transplantation of the kidney and heart.

SUMMARY—TISSUE REPAIR

Name of Tissue	Repair Process
Stratified squamous epithelium	Thickens in response to slow friction. Forms blister in response to rapid friction. Readily repaired. Basal layer of cells divides by mitosis, the cells migrate upward, forming a tonguelike process which grows over the denuded area. If large areas are denuded, skin grafts may be necessary
Modified forms	
Cornea	When injured, the cells of the cornea form scar tissue.
Conjunctiva	Conjunctiva is repaired readily
Simple squamous	Considerable ability for repair
In kidney and small ducts	Repair in some areas may result in abnormal shape of cells
Mesothelium	
In serous cavities	Destruction by inflammatory processes frequently results in the formation of fibroblasts which form scar tissue
Endothelium	
All blood vessels and lymphatic capillaries	In capillaries, cells have ability to multiply rapidly. In all tissue repair, capillaries are formed by an outgrowth of former capillaries

SUMMARY—TISSUE REPAIR (*Continued*)

Name of Tissue	Repair Process
Plain columnar epithelium	Secretory cells are replaced occasionally. They multiply or the underlying cells replace surface cells. If large areas are destroyed, scar tissue results
Glands of stomach and intestine	
Ducts of many glands	
Goblet cells	Readily replaced
Cuboidal cells of liver	Liver cells have great power of regeneration
Ciliated columnar epithelium	Small injuries are repaired by mitotic division. Large areas repaired by underlying cell layers or connective tissue cells
Loose connective tissue	Blood oozes or flows into wound; clot forms; edges of wound are glued together, capillaries sprout, fibroblasts multiply rapidly. Collagenous and elastic fibers are formed quickly and invade area
Adipose tissue	Regeneration rapid
Elastic connective tissue	Good blood supply to tissue. Tissue repaired by fibroblasts and invasion of elastic and fibrous fibers
Fibrous connective tissue	Scant blood supply to tissue except in fibrous membranes. Repair process slow but complete
Ligaments, tendons, and fascial sheaths	Periosteum and dura mater—blood supply good, repair depends on the connective tissue
	Repair limited to a few cells
Muscle	Regeneration limited, healing takes place by scar formation. In pregnant uterus cells increase in size. Limited cell multiplication
Striated	
Smooth	
Cardiac	Do not regenerate. Repair by scar tissue
Cartilage	Healing of cartilage is by repair of the surrounding connective tissue called perichondium. Degenerative changes frequently occur as age advances
Hyaline	
Elastic	
Fibrous	
Bone	All bone is surrounded by periosteum, which is richly supplied with blood, lymph, and nerves. Bone is readily repaired if the periosteum is present. Deformations will not occur if good approximation is maintained. Cells under the periosteum proliferate and form a splint of cartilage around the fractured ends. Cells at the edges become osteoblasts and form bone
Nerve tissue	Nerve cells when injured are *not* replaced. Nerve fibers when severed from the cell body die. Myelinated nerve fibers will grow into the sheath if the neurilemma is present

Questions for Discussion

1. What is a tissue? List the primary tissues of the body and their functions.
2. What modification of the cell membrane of epithelial cells increases the surface area for absorption? Where is this type of modification found?
3. What are cilia and what is their function? In what areas of the body is ciliated epithelium present?
4. List the types and location of simple epithelium and stratified epithelium.
5. What are the differences between exocrine and endocrine glands?
6. Name three types of cells found in loose connective tissue and give their function.
7. Compare the types of muscle and indicate how their structure correlates well with their function.
8. Compare and contrast cartilage and bone in relation to structure and function.
9. List the areas of the body where the cell bodies of neurons are found.

10. Explain the relationship between tissue repair and nutrition.

11. Discuss bone healing in a young child as contrasted with that in an old person. In which is healing slower? Why? In which is a fracture more dangerous? What might be some of the complications, in both, of healing, immobilization, or future capabilities of the affected area?

12. Describe the structure and function of serous and cutaneous membranes.

Summary

Cells having similar structural characteristics and functions are grouped together to form tissues.

Primary Tissues

{ **Epithelial**
{ **Connective**
{ **Muscle**
{ **Nervous**

Epithelial Tissue

General Characteristics
1. Lines and covers body surfaces
2. Forms glandular tissue
3. No direct blood supply
4. Little surrounding extracellular space; cells closely packed together
5. Epithelial cells held together by specializations on lateral cell surface: zonula adherens and macula adherens or desmosome
6. Zonula occludens and gap junction specializations of lateral cell membrane implicated in functions such as transport and electrical coupling
7. Apical surface of columnar epithelial cells modified to form microvilli, characteristic of digestive system, and cilia, characteristic of respiratory tract
8. Basal layer of epithelial cells rest on a basement membrane
9. Three shapes of cells—squamous, columnar, and cuboidal
10. Arrangement—one layer or simple, many layers or stratified

Functions
1. Protection
2. Secretion
3. Absorption
4. Special sensation; specialized neuroepithelium; examples—eye, ear, nose

Most Common Types of Epithelium
Simple squamous—lining heart, blood, and lymph vessels and forming capillary networks—endothelium; lining serous cavities and covering visceral organs—mesothelium; ease of exchange of materials due to one-cell-layer thickness
Simple cuboidal—found in many glands, lines kidney tubules, adapted for secretion and absorption
Simple columnar—found lining digestive tract, stomach, small and large intestine; in small intestine epithelial cells with microvilli predominate; function of microvilli is to increase surface area for absorption
Pseudostratified—one layer of cells but appear as many layers; pseudostratified with goblet cells lines greater part of respiratory passages
Stratified squamous—many cell layers, adapted for protection; forms outer layer of skin
Transitional—many cell layers, lines hollow organs that are subject to changes due to contraction and distention, such as the urinary bladder

Connective Tissue

General Characteristics
1. Provides framework that supports epithelium and other tissues; essential role in transport, protection, and repair
2. Greater proportion of extracellular material in relation to cells

A. Connective Tissue Proper
Loose (areolar) connective tissue—*Composition:* many collagenous fibers, some elastic and reticular fibers; cells—fibroblasts, macrophages, plasma cells, mast cells, adipose cells, and blood cells
Function: connects, insulates, forms protecting sheaths, and is continuous throughout the whole body; fluid matrix is called tissue fluid; fluid matrix, often called internal environment, serves as a medium for transfer of supplies from blood and lymph vessels to cells, and wastes from cells to blood and lymph; stores water, salts, and glucose

Connective Tissue *(cont.)*

Dense irregular connective tissue—collagenous bundles thicker, more numerous, and randomly woven into compact framework; typically seen in dermis of skin

Dense regular connective tissue—collagen fibers predominant, arranged in specific pattern; this tissue makes up tendons, ligaments

Elastic connective tissue—elastic fibers predominate; tissue is extensile and elastic; found in walls of arteries, trachea, bronchial tubes, and vocal folds

Reticular tissue—predominant fiber type is reticular; this tissue makes up stroma of liver, spleen, and lymphoid organs

Adipose connective tissue—cells filled with fat; found throughout body wherever loose connective tissue found; important reserve of food, which may be utilized for energy; supports and protects various organs

B. Cartilage. Cartilage consists of a group of cells in a matrix. It is firm, tough, and elastic, covered and nourished by perichondrium

Varieties
- 1. **Hyaline cartilage** { Articular / Costal
- 2. **Fibrocartilage**
- 3. **Elastic cartilage**

C. Bone, or Osseous Tissue. Bone is connective tissue in which the intercellular substance derived from the cells is rendered hard by being impregnated with mineral salts

Bone or Osseous Tissue

Composition
- *Inorganic matter about* 67%
  - Calcium phosphate
  - Calcium carbonate
  - Calcium fluoride
  - Magnesium phosphate
  - Sodium chloride
- *Organic matter about* 33%
  - Cells
  - Blood vessels
  - Gelatinous substance

Varieties
- Cancellous, or spongy
- Dense, or compact

Canals
- Medullary—red and yellow marrow
- Haversian { Blood vessels / Lymphatics

Haversian System
- Haversian canals are surrounded by lamellae, lacunae, and canaliculi
- Lamellae—bony fibers arranged in rings around haversian canals
- Lacunae—hollow spaces between lamellae occupied by bone cells
- Canaliculi—canals that radiate from one lacuna to another and toward the haversian canals

Medullary Membrane. A vascular tissue that lines the medullary canal

Marrow
- **Red** — Consists of connective tissue supporting blood vessels, myelocytes, fat cells, erythroblasts from which red blood cells are derived, and giant cells; found in the marrow cavity at the ends of long bones and in cancellous tissue
- **Yellow** — Contains more connective tissue and fat cells than red marrow, fewer myelocytes, few if any red cells, and fewer giant cells; white cells of blood and lymph are derived from its myelocytes; found in the medullary canals of the long bones

Periosteum. A vascular fibrous membrane that covers the bones except at their cartilaginous extremities and serves to nourish them; important in the reunion of broken bone and growth of new bone

Blood Vessels. Twigs of nutrient artery in medullary canal anastomose with twigs from haversian canals, and these in turn anastomose with others that enter from periosteum; nerves accompany arteries into bone

Development of Bone
- In the embryo bones are preformed in membrane and in cartilage
- **Ossification** { Intramembranous / Intracartilaginous, or endochondral
- **Dependent on** { Adequate amounts of calcium and phosphorus in food / Vitamins and hormones

Rickets
- A disturbance of mineral metabolism
- **Prophylaxis, or Prevention** { Adequate amounts of calcium and phosphorus in food. Vitamin D supplied by fish oils, egg yolk, milk, butter, fresh vegetables, direct sunlight
- **Types** { True rickets / "Ricketslike condition"

Muscle Tissue

Types { Striated skeletal / Cardiac / Smooth

Muscle Tissue (*cont.*)

General Characteristics
1. Muscle cells long and often taper at their ends, called muscle fibers
2. Properties of contractility and irritability

Striated Skeletal Muscle
1. Forms muscular system, voluntary muscle
2. Single muscle cell is multinucleated, made up of myofibrils; myofibrils made up of myofilaments, actin, and myosin
3. Arrangement of myofilaments responsible for striations and specific banding pattern

Cardiac Muscle
1. Tissue that forms the heart, involuntary
2. Myofilaments create banding pattern similar to skeletal muscle
3. Fibers branch and connect at region of intercalated disc. Branching of fibers forms complex three-dimensional network

Smooth Muscle
1. Called smooth because striations not visible with light microscope; myofilaments are, however, present. No banding pattern
2. Cells spindle shaped, centrally located nucleus
3. Involuntary muscle, makes up outer walls of gastrointestinal tract and walls of blood vessels

Nervous tissue

General Characteristics
1. Specialized tissue making up nervous system
2. Tissue possesses properties of irritability and conductivity
3. The neuron is the unit of structure and function of the nervous system

Neuron
1. Cell body or perikaryon, from which project one long process, the axon, and usually many short processes, the dendrites
2. Shape and size of cell body and extent of dendritic branching vary
3. Function of neuron to receive and transmit impulses

Neuroglia
1. Cells of nervous system that do not have the ability to transmit impulses; act as specialized connective tissue of nervous system
2. Cells include astrocyte, oligodendrocyte, microglia, and Schwann cells

Nerves
1. Collection of many axons from many neurons
2. Nerves extend over great lengths and make contact with skeletal, smooth, and cardiac muscle and glands
3. Nerve fibers are myelinated or unmyelinated

Membranes

Definition — Any thin expansion of tissues that serves as a lining or covering

Varieties
1. Serous membranes
2. Synovial membranes
3. Mucous membranes
4. Cutaneous membrane

Serous Membranes

Consist of
1. Simple squamous epithelium
2. A thin layer of connective tissue

Derived from the mesoderm and called mesothelium

Found lining closed cavities or passages that do not communicate with the exterior. They are moistened by serous fluid

Three Classes

Lining the body cavities and covering the organs that lie in them
- Pleurae—cover the lungs and line the chest
- Pericardium—covers the heart and lines the outer fibrous pericardium
- Peritoneum—covers the abdominal and the top of some of the pelvic organs, lines the abdominal cavity

Lining the vascular system
- Heart
- Blood vessels
- Lymphatics

Forming the fascia bulbi and part of the membranous labyrinth of the ear

Functions—Protection
1. Furnishes a cover or lining for viscera and vascular system
2. Secretes serous fluid, a lubricant

Synovial Membranes

Consist of thin serous tissue associated with bones and muscles

Three Classes
- Articular synovial membranes — Surround cavities of movable joints
- Mucous sheaths—form sheaths for tendons
- Bursae mucosae — Sacs interposed between two surfaces that move upon each other

Functions

Furnish a lining or cover
- Joints
- Tendons
- Sacs under skin, muscles, and tendons

Furnish a secretion—synovia—which acts as a lubricant

Mucous Membranes

Mucous membranes composed of inner epithelium and an underlying loose connective tissue; specialized epithelial cells secrete mucus

Found lining passages that communicate with the exterior and are protected by mucus

Two Divisions

- Gastropulmonary
  - Alimentary canal
  - Air passages
  - Cavities communicating with both alimentary canal and air passages
- Genitourinary
  - Urinary tract
  - Generative organs

Functions

- Protection
  - Secretion of mucus
  - Action of cilia
- Support for network of blood vessels
- Absorption and secretion — Various modifications increase the surface

Cutaneous Membrane

- Forms the skin
- Covers the body
- Serves to protect underlying tissues
- Prevents loss of body fluids
- Contains structures for the reception of stimuli

Skin

Functions

1. Covers the body
2. Protects the deeper tissues from
   - Drying, injury
   - Invasion by infectious organisms
3. Important factor in heat regulation
4. Contains the end organs of many sensory nerves
5. It has limited excretory and absorbing power

Consists of

- **Epidermis** is a stratified epithelium
  1. Stratum corneum
  2. Stratum lucidum
  3. Stratum granulosum
  4. Stratum spinosum
  5. Stratum germinativum

  } Layers of epidermis

- **Corium** is a layer of connective tissue
  1. Papillary layer—papillae are minute conical elevations of the corium. They contain looped capillaries, and some contain termination of nerve fibers called tactile corpuscle
  2. Reticular layer — Bands of fibrous and elastic tissue that interlace, leaving tiny spaces that are occupied by adipose tissue and sweat glands

Blood vessels. The arteries form a network in the subcutaneous tissue and send branches to papillae and glands of skin. Capable of holding a large proportion of total amount of blood in body

Lymphatics. There is a superficial and a deep network of lymphatics in the skin

Nerve fibers

1. Motor fibers to blood vessels and arrector muscles
2. Fibers concerned with temperature sense
3. Fibers concerned with sense of touch and pressure
4. Fibers stimulated by pain
5. Fibers that are distributed to the glands

Appendages

- Nails, hairs
- Sebaceous glands, sudoriferous glands

Nails

- Consist of clear, horny cells of epidermis
- Corium forms a bed, or matrix, for nail
- Root of nail is lodged in a deep fold of the skin
- Nails grow in length from soft cells in stratum germinativum at root

Hairs (Pili)

- The hairs grow from the roots
- The roots are bulbs of soft, growing cells contained in the hair follicles
- Hair follicles are little pits developed in the corium
- Stems of hair extend beyond the surface of the skin, consist of three layers of cells: (1) cuticle, (2) cortex, and (3) medulla
- Found all over body, except
  - Palms of the hands
  - Soles of the feet
  - Last phalanges of the fingers and toes
- Arrector muscles are attached to corium and to each hair follicle. Contraction pulls hair up straight, drags follicles upward, forces secretion of sebaceous glands to surface, and forces blood to interior

Sebaceous Glands	Compound alveolar glands, the ducts of which usually open into a hair follicle but may discharge separately on the surface of the skin	
	Lie between arrector muscles and hairs	
	Found over entire skin surface except { Palms of hands / Soles of feet	
	Secrete *sebum*, a fatty, oily substance, which keeps the hair from becoming too dry and brittle, the skin flexible, forms a protective layer on surface of skin, and prevents undue absorption or evaporation of water from the skin	

Sweat Glands Tubular glands, consist of single tubes with the blind ends coiled in balls, lodged in subcutaneous tissue, and surrounded by a capillary plexus; secrete sweat and discharge it by means of ducts that open exteriorly

Sweat

Amount increased by
1. Increased temperature or humidity of the atmosphere
2. Dilute condition of blood
3. Exercise
4. Pain
5. Nausea
6. Mental excitement or nervousness
7. Dyspnea
8. Use of diaphoretics, e.g., pilocarpine, physostigmine, nicotine
9. Various diseases, such as tuberculosis, acute rheumatism, and malaria

Amount decreased by
1. Cold
2. Voiding a large quantity of urine
3. Diarrhea
4. Certain drugs, e.g., atropine and morphine
5. Certain diseases

Activity of Sweat Glands Due to
1. Direct stimulation of nerve ending in sweat glands
2. Indirect stimulation through sensory nerves of the skin
3. Influenced by external heat, dyspnea, muscular exercise, strong emotions, and the action of various drugs

Function of Sweat Importance not in elimination of waste substances in perspiration, but elimination of *heat* needed to cause evaporation of perspiration
When kidneys are not functioning properly, sweat glands will excrete waste substances, particularly if stimulated

Ceruminous Glands Modified sweat glands
Found in skin of external auditory canal
Secrete cerumen, a yellow, pasty substance, like wax

Additional Readings

Bickers, D. R., and Kappas, A.: Metabolic and pharmacologic properties of the skin. *Hosp. Prac.,* **9**:97–106, May, 1974.

Bloom, W., and Fawcett, D. W.: *A Textbook of Histology,* 10th ed. W. B. Saunders Co., Philadelphia, 1975, Chapters 3, 4, 6, 7, 9, 10, 11, and 12.

Greep, R. O., and Weiss, L.: *Histology,* 3rd ed. McGraw-Hill Book Co., New York, 1973, Chapters 3, 4, 5, 6, 7, 8.

Ham, A. W.: *Histology,* 7th ed. J. B. Lippincott Co., Philadelphia, 1974, Chapters 7, 8, 9, 14, 15, 17, and 18.

Junqueira, L. C.: Carneiro, J.; and Contopoulos, A. N.: *Basic Histology.* Lange Medical Publications, Los Altos, Calif., 1975, Chapters 4, 5, 6, 7, 8, 9, and 11.

Porter, K. R., and Bonneville, M. A.: *Fine Structure of Cells and Tissues.* Lea & Febiger, Philadelphia, 1973.

Ross, R., and Bornstein, P.: Elastic fibers in the body. *Sci. Amer.,* **224**:44–52, June, 1971.

Satir, P.: How cilia move. *Sci. Amer.,* **231**:44–52, October, 1974.

UNIT II

Locomotion and Support

Body movement is in essence movement of bones at their articulating surfaces as a result of skeletal muscle contraction. This unit describes the bones of the body, how they are grouped together at joints, and the skeletal muscles that move the body.

CHAPTER 4

The Skeleton

Chapter Outline

The bones are the principal organs of support and the passive instruments of locomotion. They form a framework of hard material to which the skeletal muscles are attached. This framework affords attachment for the soft parts, maintains them in position, shelters them, helps to control and direct varying internal pressures, gives stability to the whole body, and preserves its shape. The bones form joints, which may be movable. Here the bones act as levers for movement. Certain blood cells are formed in red bone marrow.

Divisions of the Skeleton

The adult skeleton consists of 206 named bones.

Cranium	8	
Face	14	
Ear {Malleus 2, Incus 2, Stapes 2}	6	
Hyoid	1	206
The spine, or vertebral column (sacrum and coccyx included)	26	
Sternum and ribs	25	
Upper extremities	64	
Lower extremities	62	

This list does not include the sesamoid[1] and wormian[2] bones. Sesamoid bones are found embedded in the tendons covering the bones of the knee, hand, and foot. Wormian bones are small isolated bones that occur in cranial sutures, most frequently the lambdoid suture.

The bones of the body may be divided into two main groups:

[1] Ses'amoid (Greek *sesamon*, a "seed of the sesamum," and *eidos*, "form," "resemblance").
[2] Olaus Wormius, Danish anatomist (1588–1654).

The Axial Skeleton	1. Head or skull	{ Cranium 8 Face 14
	2. Hyoid	
	3. Trunk	{ Vertebrae—child 33, adult 26 Sternum 1 Ribs 24
The Appendicular Skeleton	4. Upper	64
	5. Lower	62

Classification. The bones may be classified according to their shape into four groups: (1) *long*, (2) *short*, (3) *flat*, and (4) *irregular*.

A *long bone* consists of a shaft and two extremities. The shaft is formed mainly of compact bone tissue, the compact tissue being thickest in the middle, where the bone is most slender and the strain greatest, and it is hollowed out in the interior to form the *medullary canal*. The extremities are made of cancellous tissue, with a thin coating of compact tissue, and are more or less expanded for greater convenience of mutual connection and to afford a broad surface for muscular attachment. All long bones are more or less curved, which gives them greater strength. They are found in the arms and legs, e.g., humerus.

The *short bones* are irregularly shaped. Their texture is spongy throughout, except at their surface, where there is a thin layer of compact tissue. The short bones are the 16 bones of the carpus, the 14 bones of the tarsus, and the two patellae.

Where *flat bones* are found, there is need for extensive protection or the provision of broad surfaces for muscular attachment. The bony tissue expands into broad or elongated flat plates that are composed of two thin layers of compact tissue, enclosing between them a variable quantity of cancellous tissue, e.g., occipital bone.

The *irregular bones*, because of their peculiar shape, cannot be grouped under any of the preceding heads. A vertebra is a good example. The bones of the ear are so small that they are described as *ossicles*.

Processes and Depressions. The surface of bones shows projections, or *processes*, and depressions, called *fossae* or *cavities*. Qualifying adjectives or special names may be used to describe them. Both processes and depressions are classified as (1) *articular*—those serving for connection of bones to form joints, and (2) *nonarticular*—those serving for the attachment of ligaments and muscles.

PROCESSES

Process. Any marked bony prominence
Condyle. A rounded or knucklelike process
Tubercle. A small rounded process
Tuberosity. A large rounded process
Trochanter. A very large process

Crest. A narrow ridge of bone
Spine, or *spinous process.* A sharp, slender process

Head. A portion supported on a constricted part, or *neck.*

CAVITIES

Fissure. A narrow slit
Foramen. A hole or orifice through which blood vessels, nerves, and ligaments pass
Meatus, or *canal.* A long, tubelike passageway
Sinus[3] and *antrum.* Applied to cavities within certain bones
Groove, or *sulcus.* A furrow
Fossa. A depression in or upon a bone

---

[3]The term *sinus* is also used in surgery to denote a narrow tract through tissues, leading from the surface down to a cavity, and sometimes it refers to a large vein.

The Axial Skeleton

THE SKULL

The head, or skull, rests upon the spinal column and is composed of the cranial and facial bones. It is divisible into *cranium*, or *brain case*, and *anterior region*, or *face*.

The Skull as a Whole. The cranium is a firm case, or covering, for the brain. Four of the eight bones (occipital, two parietal,

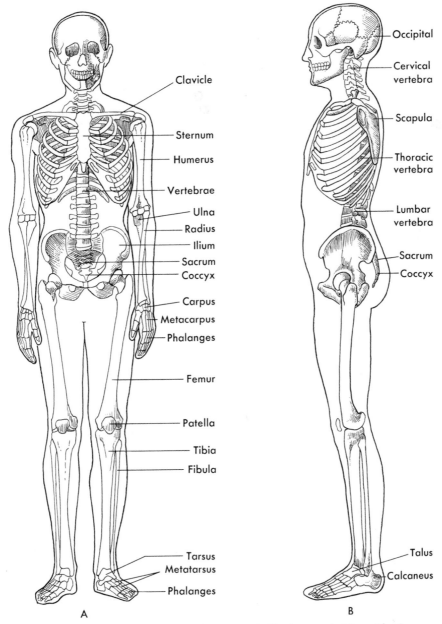

Figure 4–1. *A.* The human skeleton, front view. *B.* The human skeleton, side view.

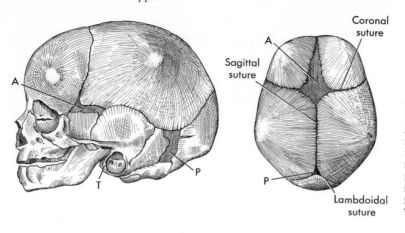

Figure 4–2. *Left:* Skull of newborn infant, side view. *A*, Anterolateral fontanelle; *P*, posterolateral fontanelle; *T*, tympanic ring. *Right:* Same, seen from above. *A*, Anterior fontanelle; *P*, posterior fontanelle. (Modified from Toldt.)

and frontal) which form this bony covering are flat bones and consist of two layers of compact tissue, the outer one thick and tough, the inner one thinner and more brittle. The base of the skull is much thicker and stronger than the walls and roof; it presents a number of openings, or foramina, for the passage of the cranial nerves, blood vessels, and other structures.

The bones of the cranium develop in early fetal life. Ossification of these bones is gradual and takes place from ossification centers, generally near the center of the completed bones. Ossification is not

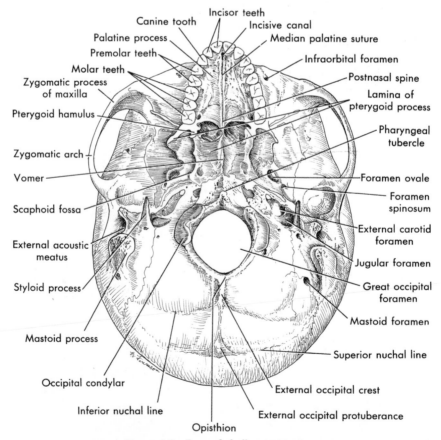

Figure 4–3. Base of skull, exterior view.

complete at birth; hence, membrane-filled spaces are found between the bones. These spaces are called fontanelles.

Fontanelles. At birth there may be many of these fontanelles. The shape and location of six of them are quite constant.

The *anterior,* or *bregmatic,* is the largest and is a lozenge-shaped space between the angles of the two parietal bones and the two segments of the frontal bone. Normally this fontanelle closes at about 18 months of age.

In abnormal conditions the fontanelle may close much earlier or much later. In cases of retarded brain growth, called microcephalus, it closes early. In hydrocephalus the increased internal pressure may cause it to remain open. In rickets and cretinism that are not yielding to treatment it may not close until much later.

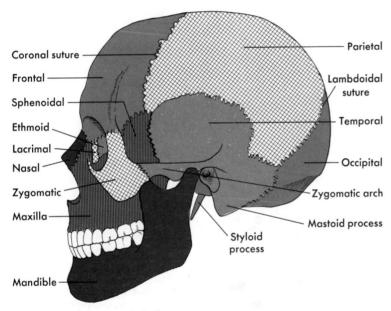

Figure 4–4. Side view of skull.

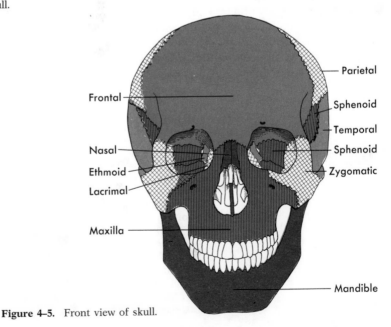

Figure 4–5. Front view of skull.

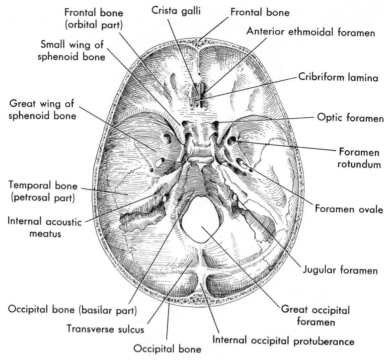

Frontal bone (orbital part)
Crista galli
Frontal bone
Anterior ethmoidal foramen
Small wing of sphenoid bone
Cribriform lamina
Great wing of sphenoid bone
Optic foramen
Foramen rotundum
Temporal bone (petrosal part)
Internal acoustic meatus
Foramen ovale
Jugular foramen
Occipital bone (basilar part)
Great occipital foramen
Transverse sulcus
Internal occipital protuberance
Occipital bone

Figure 4–6. Base of skull, interior view.

The *posterior* (or *occipital*) *fontanelle* is much smaller and is a triangular space between the occipital and two parietal bones. Usually this closes a few months after birth.

There are two *sphenoidal fontanelles* at the junction of the frontal, parietal, temporal, and sphenoid bones. They are quite small and usually close by the second month after birth.

There are two *mastoid fontanelles* at the junction of the parietal, occipital, and temporal bones. They decrease in size and usually close one or two months after birth.

The membranous tissue between the cranial bones at the sutures and fontanelles allows more or less overlapping during birth processes, thus reducing the diameters of the skull. This is called *molding* and accounts for the elongated shape of the head of a newborn infant, particularly if the labor has been long.

BONES OF THE SKULL

Name of Bone	Description
Occipital (os occipitale)	Is situated at the back and base of skull. *Internal surface* is concave and presents many eminences and depressions for parts of the brain. *External surface* is convex. The *external occipital protuberance* is a projection which can be felt through the scalp. From this median ridge the external occipital crest leads to the *foramen magnum,* the largest opening in the inferior part of the bone, for the transmission of the *medulla oblongata,* where it narrows to join the spinal cord. The protuberances and crest give attachment for the ligamentum nuchae. Other muscles are attached to the expanded plate of the bone. At the sides of the foramen magnum on the external surface, there are two processes called condyles, which articulate with the atlas.

Parietal bone, right and left (os parietale)	By their union they form the greater part of the sides and roof of the skull. The *external* surface is smooth and convex. The *internal* surface is concave and presents many eminences and depressions for reception of the convolutions of the brain and many furrows for the ramifications of arteries that supply blood to the dura mater, which covers the brain.
Frontal bone (os frontale)	Forms the forehead, part of the roof of the orbits, and the nasal cavity. The *supraorbital margin* is the arch formed by the part of the bone over the eyes. Just above the supraorbital margins are hollow spaces within the bone called the *frontal sinuses*, which are filled with air and open into the nose. The *lacrimal fossae* are located in the upper outer angle of each orbit in which lie the *lacrimal glands*, which secrete tears. At birth the bone consists of two parts, which become united soon after birth.
Temporal bones, right and left (os temporale)	Are situated at the sides and base of the skull. They are divided into five parts:

1. *The squama*, a thin expanded portion, forms the anterior and upper part of the bone. The supramastoid crest runs backward and upward across its posterior part. The *zygomatic process*, which projects from the lower part, articulates with the temporal process of the zygomatic bone.
2. *The petrous portion* is shaped like a pyramid and is wedged in at the base of the skull between the sphenoid and occipital bones. The internal ear, the essential part of the organ of hearing, is contained in a series of cavities in the petrous portion. Between the squamous and the petrous portion is a socket, the *mandibular fossa*, for the reception of the condyle of the mandible.
3. *The mastoid portion* projects downward behind the opening of the meatus. It is filled with numerous connected spaces called *mastoid cells*[4] or sinuses that contain air. These cells communicate with the middle ear.
4. *The tympanic portion* is a curved plate of bone below the squama and in front of the mastoid portion. It forms a part of the external acoustic meatus leading to the middle ear.
5. *The styloid* is a slender, pointed process that projects downward from the undersurface of the temporal bone. Some muscles and ligaments of the tongue are attached to its distal part.

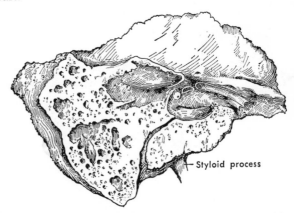

Styloid process

Figure 4–7. Right temporal bone, sectioned to show mastoid cells.

[4] *Cells.* Histologically, the word *cell* refers to one of the component units of the body, such as an epithelial cell. Occasionally it refers to such minute chambers as mastoid cells.

The ethmoid bone
(os ethmoidale)

Is a light cancellous bone consisting of a cribriform (horizontal) plate, a perpendicular plate, and two lateral masses or *labyrinths*. The cribriform plate forms the roof of the nasal cavity and closes the anterior part of the base of the cranium. There are many foramina through which *olfactory fibers* pass from the mucous membrane of the nose to the *olfactory bulb*. Projecting upward from the horizontal plate is a smooth, triangular process called the *crista galli*, which forms an attachment of the *falx cerebri* (page 251). The perpendicular plate, descending from the horizontal plate, helps to form the upper part of the nasal septum. On either side the lateral masses form part of the orbit and nasal cavity. The lateral masses contain many thin-walled cavities, the sinuses of the ethmoid, which communicate with the nasal cavity. On either side of the septum are two thin processes of thin, cancellous bony tissue, the superior and middle conchae.

Superior and middle conchae

The sphenoid bone
(os sphenoidale)

Is an important bone situated at the anterior part of the base of the skull that binds the other cranial bones together. It resembles an airplane with extended wings and consists of a body, two great and two small wings that extend transversely from the sides of the body, and two pterygoid processes (hamuli) that project downward. The body is joined to the ethmoid in front and the occipital behind.

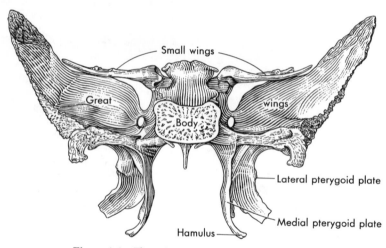

Figure 4–8. The sphenoid bone, seen from behind.

Sinuses of the sphenoid

The bone contains cavities called the sphenoidal sinuses that communicate with the nasopharynx.

Sella turcica

The upper portion of the body of the sphenoid presents a *fossa* with anterior and posterior eminences. This is called the *sella turcica*. The *hypophysis cerebri* is contained in the *sella turcica*.

BONES OF THE FACE

Name of Bone	Description
Nasal bones (os nasalia)	Are two small oblong bones placed side by side at the middle and upper part of the face, forming by their junction the upper part of the bridge of the nose. The lower part of the nose is formed by cartilage.

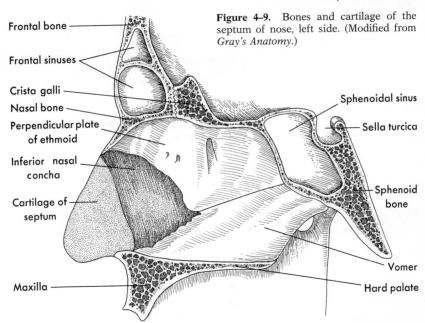

Frontal bone

Frontal sinuses

Crista galli

Nasal bone

Perpendicular plate of ethmoid

Inferior nasal concha

Cartilage of septum

Maxilla

Sphenoidal sinus

Sella turcica

Sphenoid bone

Vomer

Hard palate

Figure 4–9. Bones and cartilage of the septum of nose, left side. (Modified from *Gray's Anatomy*.)

Vomer	A single bone located at the lower and back part of the nasal cavity, forming part of the central septum of the nasal cavity. It is thin and may be deviated to one side, thus making the nasal chambers of unequal size.
Interior nasal conchae	Are located on the outer wall of the nasal cavity. They consist of a layer of thin cancellous bone curled upon itself like a scroll. They are below the superior and middle conchae of the ethmoid bone.
Lacrimal bones	Are located at the front part of the inner wall of the orbit and somewhat resemble a fingernail in form, thinness, and size. They contain part of the canal through which the tear ducts run.
Zygomatic malar bones (os zygoma)	These bones form the prominences of the cheeks and part of the outer wall and floor of the orbits. The temporal process (narrow and serrated) projects backward and articulates with the zygomatic process of the temporal bone, forming an arch on each side.
Palatine bones (2) (os palatinum)	These bones are shaped somewhat like an L and have a horizontal part, a vertical part, and three processes, the pyramidal,

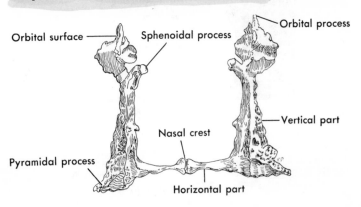

Orbital surface

Sphenoidal process

Orbital process

Vertical part

Nasal crest

Pyramidal process

Horizontal part

Figure 4–10. The two palatine bones in their natural position, viewed from behind.

orbital, and sphenoid. They are situated at the back part of the nasal cavity between the maxillae and pterygoid processes of the sphenoid. They help to form (1) the back part of the roof of the mouth, (2) part of the floor and outer walls of the nasal cavities, and (3) a small part of the floor of the orbit.

Maxillae, or upper jaw bones (2)

Form by their union the whole of the upper jaw. Each bone helps to form (1) part of the floor of the orbit, (2) the floor and lateral walls of the nasal cavities, and (3) the greater part of the roof of the mouth. Each bone contains a large cavity, the *maxillary sinus,* which opens into the nose. The *alveolar* processes are excavated cavities, which vary in depth and size according to the teeth they contain. Before birth these bones unite to form one. When they fail to do so, the condition is known as cleft palate.

The mandible, or lower jaw bone (os mandibula)

Is the largest and strongest bone of the face and consists of a curved horizontal portion, the *body,* and two perpendicular portions, the *rami.* The alveolar process border of the body contains cavities for the reception of the teeth. Each ramus has a *condyle,* which articulates with the mandibular fossa of the temporal bone, and a *coronoid process,* which gives attachment to the temporal muscle and some fibers of the buccinator. The deep depression between the two processes is called the *mandibular notch.* The *mental foramen,* which is just below the first molar tooth, serves as a passageway for the mental nerve, which is a terminal branch of the inferior dental nerve (of the trigeminal nerve). At birth the mandible consists of two parts, which unite at the symphysis in front to form one bone, during the first year.

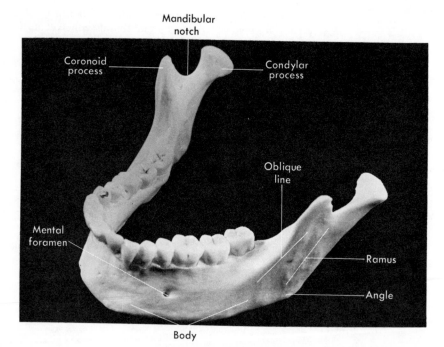

Figure 4–11. The mandible.

The hyoid bone (os hyoideum)

Is shaped like a horseshoe and consists of a body and two projections on each side called the greater and lesser cornua. It is suspended from the styloid processes of the temporal bones and

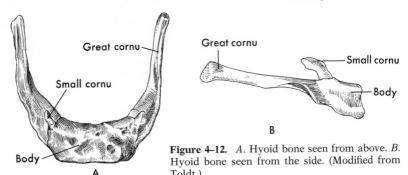

Figure 4–12. *A.* Hyoid bone seen from above. *B.* Hyoid bone seen from the side. (Modified from Toldt.)

may be felt in the neck just above the laryngeal prominence. It supports the tongue and gives attachment to some of its numerous muscles.

Sinuses of the head

There are four air sinuses that communicate with each nasal cavity: the *frontal,* the *ethmoidal,* the *sphenoidal,* and the *maxillary.* They are lined with mucous membrane. Inflammation of this membrane is called sinusitis.

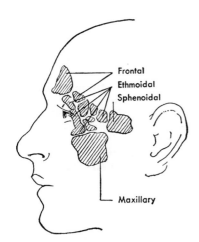

Figure 4–13. Sinuses projected to the surface of the face. All except the maxillary sinus are near the central line of the skull when viewed from the front. See also Figure 4–7 for mastoid cells.

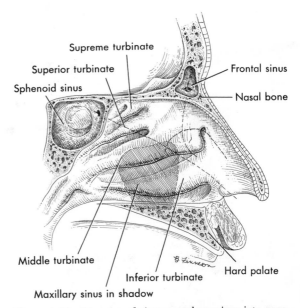

Figure 4–14. Diagram of sinuses and openings into nasopharynx.

The skull at birth

The skull is proportionately larger than other parts of the skeleton, and the facial portion is small. The small size of the maxillae and mandible, the noneruption of the teeth, and the small size of the sinuses and nasal cavities account for the smallness of the face. With the eruption of the first teeth there is an enlargement of the face and jaws. This enlargement is much more pronounced after the eruption of the second set of teeth. Usually the skull becomes thinner and lighter in *old age*, but occasionally the inner table hypertrophies, causing an increase in weight and thickness (pachycephalia). The most noticeable feature of the skull in the aged is the decrease in the size of the maxillae and mandible, resulting from the loss of the teeth and the absorption of the alveolar processes.

BONES OF THE TRUNK

Name of Bone	Description
The trunk	The bones forming the trunk are the *vertebrae, sternum,* and *ribs.*
The vertebral column	Is formed of a series of bones called vertebrae and in a man of average height is about 71 cm (28 in.) long. In youth the vertebrae are 33 in number:

Cervical, in the neck	7 ⎫	
Thoracic, in the thorax	12 ⎬	Movable, or true, vertebrae
Lumbar, in the loins	5 ⎭	
Sacral, in the pelvis	5 ⎫	Fixed, or false, vertebrae
Coccygeal, in the pelvis	4 ⎭	

In the cervical, thoracic, and lumbar regions the vertebrae are separate and movable throughout life. In the sacral and coccygeal regions they are firmly united in the adult, so that they form two sections, five bones entering into the sacrum and four into the terminal bone, or coccyx.

The vertebrae — Differ in size and shape, but their structure is similar. Seen from above, as in Figure 4–19, page 106, they consist of a body from which two short, thick processes, called the pedicles, project backward, one on each side to join with the laminae, which unite posteriorly, thus forming the vertebral, or neural, arch. This arch encloses the spinal foramen. Each vertebra has several processes: four articular, two to connect with the bone above, two to connect with the bone below; two transverse, one at each side where the pedicle and lamina join; and one spinous process, projecting backward from the junction of the laminae in the midline.

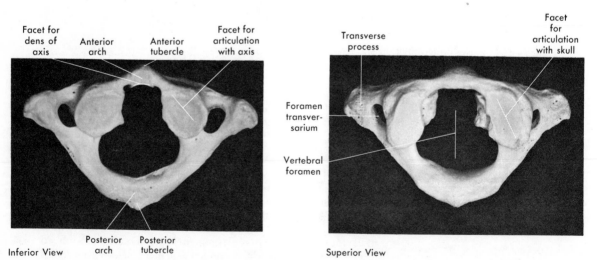

Facet for dens of axis   Anterior arch   Anterior tubercle   Facet for articulation with axis

Transverse process   Facet for articulation with skull

Foramen transversarium

Vertebral foramen

Posterior arch   Posterior tubercle

Inferior View   Superior View

Figure 4–15. The first cervical vertebra, the atlas. Inferior and superior views.

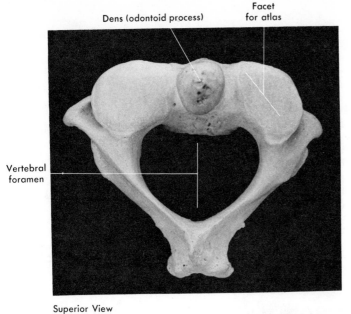

Dens (odontoid process)

Facet for atlas

Vertebral foramen

Superior View

Figure 4–16. The second cervical vertebra, the axis. Superior and inferior views.

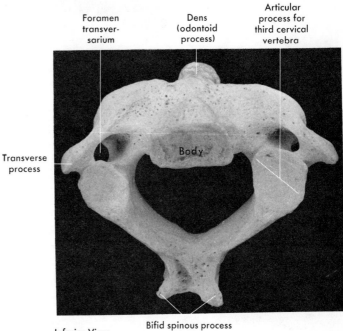

Foramen transver-sarium

Dens (odontoid process)

Articular process for third cervical vertebra

Transverse process

Body

Inferior View

Bifid spinous process

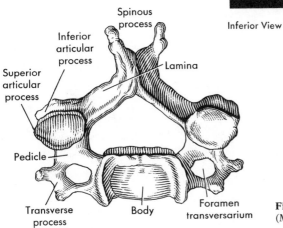

Spinous process

Inferior articular process

Lamina

Superior articular process

Pedicle

Transverse process

Body

Foramen transversarium

Figure 4–17. A cervical vertebra, viewed from above. (Modified from *Gray's Anatomy*.)

Cervical vertebrae

The bodies of the cervical vertebrae are smaller than the thoracic, but the arches are larger. The spinous processes are short and are often cleft in two, or bifid. Each transverse process is pierced by a foramen (foramen transversarium) through which nerves, a vertebral artery, and a vein pass. As Figures 4–15, 4–16, and 4–17 indicate, the first and second cervical vertebrae differ from the rest. The first, or *atlas*, so named from supporting the head, is a bony ring consisting of an anterior and posterior arch and two bulky lateral masses. Each has a superior and inferior articular surface. Each superior surface forms a cup for the corresponding condyle of the occipital bone and thus makes possible the backward and forward movements of the head. The bony ring is divided into an anterior and posterior section by a transverse ligament. The posterior section of this bony ring contains the spinal cord, and the anterior, or front, section of the ring contains the bony projection that arises from the upper surface of the body of the second cervical vertebra, the *epistropheus*, or *axis*. This bony projection, the dens (odontoid process), forms a pivot; around this pivot the atlas rotates when the head is turned from side to side.

Thoracic vertebrae

These vertebrae are larger and stronger than those of the cervical and have a facet or demifacet for articulation with the heads of

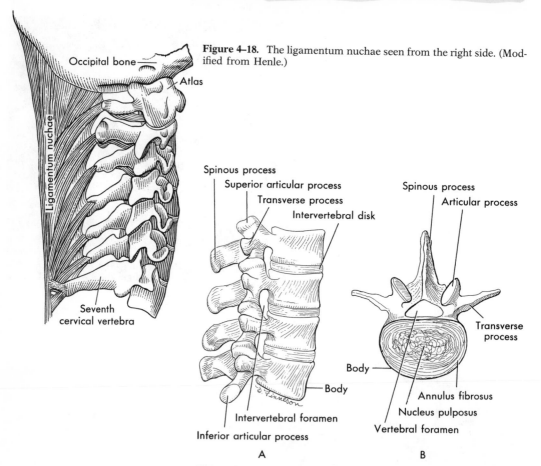

Figure 4–18. The ligamentum nuchae seen from the right side. (Modified from Henle.)

Figure 4–19. *A.* Diagram of several vertebrae with intervertebral disks between the vertebrae. *B.* Viewed from above; note the nucleus pulposus.

the ribs. The transverse processes are longer and heavier than those of the cervical, and all except those of the eleventh and twelfth vertebrae have facets for articulation with the tubercles of the ribs. The spinous processes are long and are directed downward.

The lumbar vertebrae

The bodies of these vertebrae are the largest and heaviest in the spine. The processes are short, heavy, and thick.

The sacrum

Is an important bone of the vertebral column. It is formed by the union of the five sacral vertebrae. It is a large wedge-shaped bone firmly connected with the hip bones. The pelvic side is concave and relatively smooth and the dorsal side is irregular. The sacrum is marked by four transverse ridges. At the ends of the ridges there are four pairs of pelvic sacral foramina, which communicate with the four pairs of dorsal foramina through which nerves and blood vessels pass. The sacral canal contains the lower part of the cauda equina of the spinal cord, spinal ligament, and fat.

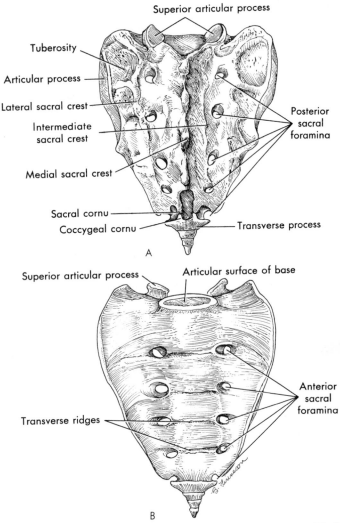

Figure 4–20. The sacral bone. *A.* Posterior view. *B.* Anterior view. (Modified from Pansky and House.)

The coccyx

Is usually formed of four small segments of bone and is the most rudimentary part of the vertebral column.

The intervertebral disks

Are disks of fibrocartilage interposed between the bodies of adjacent vertebrae from the axis to the sacrum. They vary in size and thickness in different regions, being thickest in the lumbar region. They are attached below and above by a thin layer of hyaline cartilage that covers the surfaces of the bodies of the vertebrae.

The nucleus pulposus

Is a soft, pulpy, elastic, and compressible substance centrally located within each disk. The disk as a whole permits flexibility of the vertebral column and the nucleus pulposus functions as an important shock absorber.

STRUCTURE OF VERTEBRAL COLUMN

Name of Bone	Description
As a whole	The bodies of the vertebrae are piled one upon another, form a strong, flexible column for the support of the cranium and trunk, and provide articular surfaces for the attachment of the ribs. The arches form a hollow cylinder for the protection of the spinal cord. Viewed from the side, the vertebral column presents four curves, which are alternately convex and concave. The two concave ones, named thoracic and pelvic, are called primary curves because they exist in fetal life and are designed for the accommodation of viscera. The two convex ones, named cervical and lumbar, are called secondary, or compensatory, curves because they are developed after birth. The cervical curve begins its development when the child is able to hold up his head (at about three or four months) and is well formed when he sits upright (at about 19 months). The lumbar develops when the child begins to walk (from 12 to 18 months). The joints between the bodies of the vertebrae are slightly movable, and those between the arches are freely movable. The *bodies* are connected (1) by disks of fibrocartilage placed between the vertebrae; (2) by the *anterior longitudinal ligament*, which extends along the anterior surfaces of the bodies of the vertebrae from the axis to the sacrum; and (3) by the *posterior longitudinal ligament*, which is inside the vertebral canal and extends along the posterior surfaces of the bodies from the axis to the sacrum.
The laminae	Are connected by broad, thin ligaments called the *ligamenta flava* (*ligamenta subflava*).
The spinous processes	Are connected at the apexes by the supraspinal ligament, which extends from the seventh cervical vertebra to the sacrum. It is continued upward as the *ligamentum nuchae*, which extends from the protuberance of the occiput to the spinous process of the seventh cervical vertebra.
Adjacent spinous processes	Are connected by interspinal ligaments, which extend from the root to the apex of each process and meet the ligamenta flava in front and the supraspinal ligament behind. The transverse processes are connected by the intertransverse ligaments, which are placed between them.
The spinal curves	Confer springiness and strength upon the spinal column, and the elasticity is further increased by the ligamenta flava and the

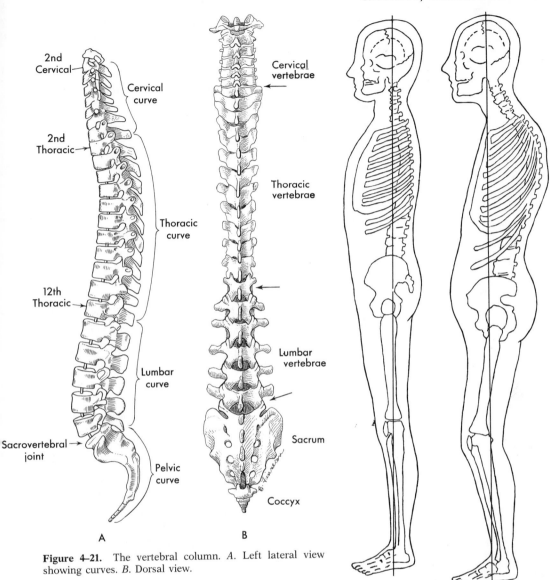

Figure 4–21. The vertebral column. *A*. Left lateral view showing curves. *B*. Dorsal view.

Figure 4–22. *A*. Skeletal form of a person with good body alignment. *B*. Skeletal form of a person with poor body mechanics. (Courtesy of the Children's Bureau, U.S. Department of Health, Education, and Welfare, Washington, D.C.)

disks of fibrocartilage. These pads also mitigate the effects of concussion arising from falls or blows. The vertebral column is freely movable, being capable of bending forward freely, backward, and from side to side less freely. Certain exercises increase the flexibility of the spine to a marked degree. In the cervical and thoracic regions a limited amount of rotation is possible.

Posture

The weight of the body should rest evenly on the two hip joints. A perpendicular dropped from the ear should fall through shoulder, hip, and ankle (Figure 4–22). In this position the chest is up, the head is erect, the lower abdominal muscles are retracted, and the body is well balanced and functioning efficiently.

As a result of postural habits, injury, or disease, the normal curves may become exaggerated and are then spoken of as *curvatures*. If the thoracic curve is exaggerated, it is called *kyphosis*, or humpback; if the exaggeration is in the lumbar region, it is called *lordosis*, or hollow back. If the curvature is lateral, i.e., toward one side, it is called *scoliosis*.

It occasionally happens that the laminae of a vertebra do not unite and a cleft is left in the arch (*spina bifida*). As a result the membranes and the spinal cord itself may protrude, forming a "tumor" on the child's back. This most often occurs in the lumbosacral region, though it may occur in the thoracic or cervical region.

THE THORAX

Name of Bone	Description
The thorax	Is a bony cage formed by the sternum and costal cartilages, the ribs, and the bodies of the thoracic vertebrae. It is cone shaped, narrow above and broad below, flattened from before backward, and shorter in front than in back. In infancy the chest is rounded, and the width from shoulder to shoulder and the

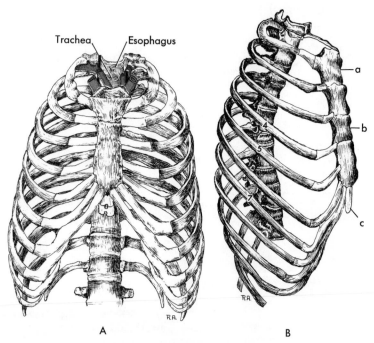

Figure 4–23. *A*. Bones of thorax, seen from the front. Lying in the superior aperture of the "thoracic basket," note esophagus (close to vertebral column), trachea, vagus nerves, arteries (*red*), veins (*blue*). (Modified from Toldt.) *B*. Bones of thorax, seen from right side. Between the fourth and fifth ribs note articulation of the head and tubercle of the seventh rib. *a*, Manubrium; *b*, body showing articular notches of ribs and lines of union of parts of body; *c*, xiphoid process.

depth from the sternum to the vertebrae are about equal. With growth the width increases more than the depth. The thorax supports the bones of the shoulder girdle and upper extremities and contains the principal organs of respiration and circulation.

The sternum

Is a flat, narrow bone about 15 cm (6 in.) long, situated in the median line in the front of the chest. It develops as three separate parts: the upper part, the *manubrium;* the middle and largest part, the *body,* or *gladiolus;* the lowest portion, the *xiphoid* or *ensiform process.* On both the manubrium and body are notches for the reception of the sternal ends of the upper seven costal cartilages. The xiphoid process has no ribs attached to it but affords an attachment for some of the abdominal muscles. At birth the sternum consists of several unossified portions, the body alone developing from four centers. Union of the centers in the body begins at about puberty and proceeds from below upward until at about 25 years of age they are all united. Sometimes by 30 years of age, more often after 40, the xiphoid process becomes joined to the body. In advanced life, the manubrium may become joined to the body by bony tissue. Posture, activity (play, work), and diet have much to do with shaping the sternum and the thoracic cavity.

The ribs (costae)

Twenty-four in number, are situated 12 on each side of the thoracic cavity. They are arches of bone consisting of a body, or shaft, and two extremities, the posterior (or vertebral) and the

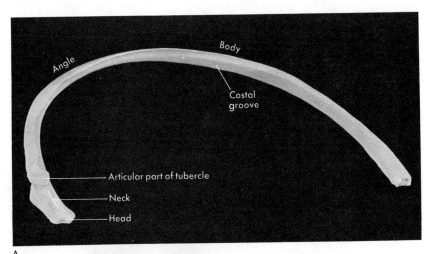

A

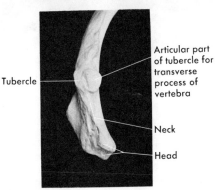

B

Figure 4–24. *A.* A typical rib. *B.* Detail of head and neck region of rib.

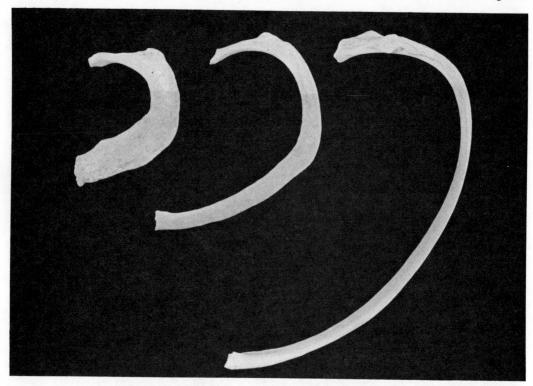

Figure 4–25. The first, second, and fourth ribs. Note difference in size, shape, and angle.

anterior (or sternal). Each rib is connected with a thoracic vertebra by the head and tubercle of the posterior extremity. The head fits into a facet formed on the body of one vertebra or formed by the adjacent bodies of two vertebrae; the tubercle articulates with the transverse processes. Strong ligaments surround and bind these articulations but permit slight gliding movements.

The heads of the first, tenth, eleventh, and twelfth ribs each articulate with a single vertebra. The heads of the remaining ribs articulate with facets formed by the bodies of two adjacent vertebrae. On the eleventh and twelfth ribs the articulation between the tubercle and the adjacent transverse process is missing.

The anterior extremities of each of the first seven pairs are connected with the sternum in front by bars of hyaline cartilage called costal cartilages. They are called *vertebrosternal* or *true ribs*. The remaining five pairs are termed *false ribs*. The cartilagenous attachment of ribs to sternum permits a degree of movement when pressure is applied. Pressure to the lower third of the sternum compresses the heart, which lies directly beneath and to the left (see Figure 17–1, page 389). Alternate pressure and release in a rhythmic fashion (cardiac "massage") simulates normal pumping action of the heart.

The convexity of the ribs is turned outward, giving roundness to the sides of the chest and increasing the size of its cavity; each rib slopes downward from its posterior attachment, so that its sternal end is considerably lower than its vertebral. The lower border of each rib is grooved for the accommodation of the intercostal nerves and blood vessels. The spaces left between the ribs are called the *intercostal spaces*. The red marrow of the sternum and ribs is one of the principal sites of red blood cell formation.

The Appendicular Skeleton

BONES OF THE UPPER EXTREMITY

Name of Bone	Description
The appendicular skeleton	Consists of the upper and lower extremities. The upper extremity consists of the *shoulder girdle* and the *upper limb*.
The shoulder girdle	Is formed by the two clavicles and the two scapulae. It is incomplete in front and behind. The clavicles articulate with the sternum in front but the scapulae are connected to the trunk by muscles. The shoulder girdle serves to attach the bones of the upper extremity to the axial skeleton.
The clavicle	Is a long bone with a double curvature, placed horizontally at the upper and anterior part of the thorax above the first rib. The inner end articulates with the sternum, called the *sternal extremity*. The outer, or *acromial*, end articulates with the scapula.

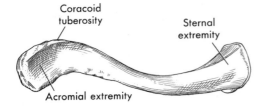

Coracoid tuberosity

Sternal extremity

Acromial extremity

Figure 4–26. The right clavicle, seen from above.

The scapula	Is a large, flat, triangular bone situated on the dorsal aspect of the thorax between the second and seventh ribs. On its dorsal surface there is a prominent ridge, the spine, which terminates in a triangular projection, the *acromion* process, which articulates with the clavicle. Below the acromion, at the end of the shoulder blade is a shallow socket, the *glenoid cavity*, which receives the head of the humerus.

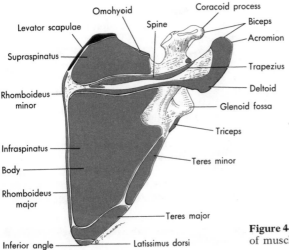

Omohyoid
Levator scapulae
Supraspinatus
Rhomboideus minor
Infraspinatus
Body
Rhomboideus major
Inferior angle
Spine
Coracoid process
Biceps
Acromion
Trapezius
Deltoid
Glenoid fossa
Triceps
Teres minor
Teres major
Latissimus dorsi

Figure 4–27. The right scapula, dorsal surface. *Red*, origin of muscle; *blue*, insertion of muscles.

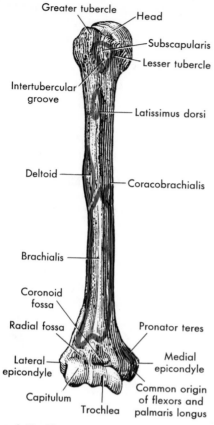

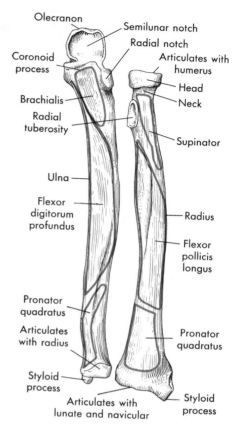

Figure 4–28. The right humerus, or arm bone, ventral view. *Red*, muscle origins; *blue*, insertions.

Figure 4–29. Anterior view of the bones of the left forearm. *Red*, muscle origins; *blue*, insertions.

The humerus, or arm bone[5]	Is the longest and largest bone of the upper extremity. The upper end consists of a rounded head, joined to the shaft by a constricted neck and two eminences, the *greater* and *lesser* tubercles, between which is the intertubercular groove.
The anatomical neck	Is the region below the base of the head.
The surgical neck	Is that part below the tubercles, so called because it is so often fractured. The head articulates with the glenoid cavity of the scapula, and the lower end is flattened from before backward and ends below in an articular surface which is divided by a ridge into a lateral eminence called the *capitulum* and a medial portion, the *trochlea*. The capitulum is rounded and articulates with the depression on the head of the radius. The trochlea articulates with the ulna. Above these surfaces on the lateral and medial aspects are projections called *epicondyles*.
The ulna	Is the largest bone of the forearm and is placed at the medial side of the radius. Its upper end shows two processes and two concavities; the larger process, called the *olecranon process*, forms the prominence of the elbow. The smaller process is called the *coronoid process*. The trochlea of the humerus fits into

[5] Anatomically, the word *arm* is reserved for that part of the upper limb that is above the elbow; between the elbow and wrist is the forearm; below the wrist are the hand and fingers.

the semilunar notch between these two processes. The radial notch is on the lateral side of the coronoid and articulates with the radius. The lower end of the ulna is small and ends in two eminences; the larger head articulates with the fibrocartilage disk, which separates it from the wrist; the smaller is the styloid process, to which a ligament from the wrist joint is attached.

The radius

Is placed on the lateral side of the ulna and is shorter and smaller than the ulna. The upper extremity presents a head, a neck, and a tuberosity. The head (caput radii) is small and rounded and has a shallow, cuplike depression on its upper surface for articulation with the capitulum of the humerus. A prominent ridge surrounds the head, and by means of this it rotates within the radial notch of the ulna. The head is supported on a constricted neck. Beneath the neck on the medial side is an eminence called the *radial tuberosity,* into which the tendon of the biceps brachii muscle is inserted. The lower end has two articular surfaces, one below, which articulates with the scaphoid and lunate bones of the wrist, and the other at the medial side, called the *ulnar notch,* which articulates with the ulna. Fracture of the lower third of the radius is called *Colles'* fracture.[6]

The carpus, or wrist (ossa carpi)

Is composed of eight small bones united by ligaments; they are arranged in *two rows* and are closely joined together, yet by the arrangement of their ligaments allow a certain amount of motion. They afford *origin* by their palmar surface to most of the *short* muscles of the thumb and little finger and are named:

Proximate, or Upper Row		*Distal, or Lower Row*	
1. Scaphoid	1	5. Trapezium	1
2. Lunate	1	6. Trapezoid	1
3. Triquetral	1	7. Capitate	1
4. Pisiform	1	8. Hamate	1

The metacarpus, or body of hand

Each metacarpus is formed by five bones (ossa metacarpalia), *numbered* from the lateral side. The bones are convex behind and

Figure 4–30. Bones of the right hand, volar surface.

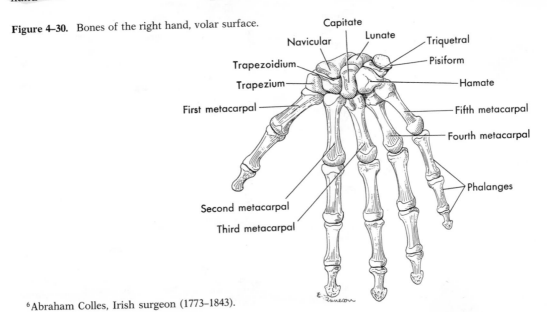

[6]Abraham Colles, Irish surgeon (1773–1843).

concave in front. They articulate at their bases with the second row of carpal bones and with each other. The *heads* of the bones *articulate* with the bases of the first row of the phalanges.

The phalanges (phalanges digitorum manus)	The bones of the fingers are 14 in number in each hand, three for each finger, and two for the thumb. The *first* row articulates with the metacarpal bones and the second row of phalanges; the *second row* articulates with the *first* and *third;* the *third* articulates with the *second* row.

BONES OF THE LOWER EXTREMITY

Hip bones, ossa coxae	2	
Femur, thigh bone	2	
Patella, knee cap	2	
Tibia, shin bone, 2		
Fibula, small bone of calf, 2 } leg	4	62
Tarsal, ossa tarsi	14	
Metatarsus, sole and lower instep	10	
Phalanges, 2 in great toe, 3 in others	28	

Name of Bone	**Description**
The two hip bones (os coxae)	Articulate with each other in front, forming an arch called the *pelvic girdle.* The arch is completed behind by the sacrum and the coccyx, forming a rigid and complete ring of bone called the *pelvis.* The pelvis attaches the lower extremities to the axial skeleton. The bones of the lower extremities correspond in general to those of the upper extremities, but their function is different. The lower extremities support the body in the erect position and are more solidly built. They are less movable than those of the upper extremities. The hip bones are large, irregularly shaped bones that form the sides and front wall of the pelvic cavity. In youth the hip bone consists of three separate parts. In the adult these have become united, but it is usual to describe the bone as divisible into three portions: (1) the *ilium* (pl., *ilia*), or upper, expanded portion forming the prominence of the hip; (2) the *ischium* (pl., *ischia*), or lower, strong portion; (3) the *pubis* (pl., *pubes*), or portion helping to form the front of the pelvis. These three portions of the bone meet and finally ankylose in a deep socket, called the *acetabulum,* into which the head of the femur fits. Processes formed by the projection of the crest of the ilium in front are called the *anterior superior iliac spine* and the *anterior inferior iliac spine.* The largest foramen in the skeleton, called the *obturator foramen,* is situated between the ischium and pubis. The articulation formed by the two pubic bones in front, called the *symphysis pubis,* serves as a convenient landmark in making body measurements.
The pelvis	Resembles a basin, is strong and massively constructed. It is composed of four bones, the two hip bones forming the sides and front, the sacrum and coccyx completing it behind, and is divided by a narrowed bony ring into the greater, or false, and the lesser, or true, pelvis. The narrowed bony ring that is the dividing line is spoken of as the *brim of the pelvis.* *The greater pelvis* is the expanded portion situated above the brim, bounded on either side by the ilium; the front is filled by the walls of the abdomen. *The lesser pelvis* is below and behind the pelvic brim, bounded on the front and sides by the pubes and ischia and behind by the sacrum and coccyx. It consists of

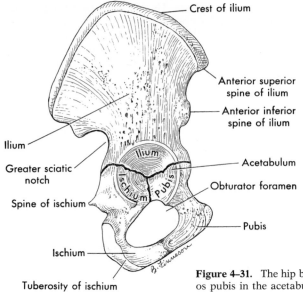

Crest of ilium

Anterior superior
spine of ilium

Anterior inferior
spine of ilium

Ilium

Greater sciatic
notch

Acetabulum

Obturator foramen

Spine of ischium

Pubis

Ischium

Tuberosity of ischium

Figure 4–31. The hip bone, showing the union of ilium, ischium, and os pubis in the acetabulum.

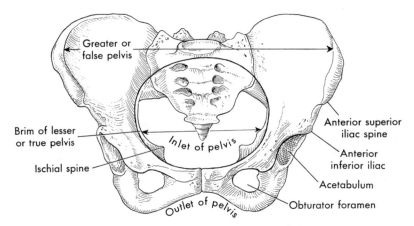

Greater or
false pelvis

Brim of lesser
or true pelvis

Inlet of pelvis

Anterior superior
iliac spine

Anterior
inferior iliac

Ischial spine

Acetabulum

Outlet of pelvis

Obturator foramen

Figure 4–32. The female pelvis, ventral view.

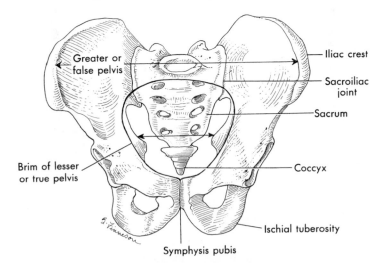

Greater or
false pelvis

Iliac crest

Sacroiliac
joint

Sacrum

Brim of lesser
or true pelvis

Coccyx

Ischial tuberosity

Symphysis pubis

Figure 4–33. The male pelvis, ventral view.

an *inlet*, an *outlet*, and a *cavity*. The space included within the brim of the pelvis is called the superior aperture, or inlet; and the space below, between the tip of the coccyx behind and the tuberosities of the ischia on either side, is called the inferior aperture, or outlet. The cavity of the lesser pelvis is a short, curved canal, deeper on the posterior than on its anterior wall. In the adult it contains part of the sigmoid colon, the rectum, bladder, and some of the reproductive organs. The bladder is behind the symphysis pubis; the rectum is in the curve of the sacrum and coccyx. In the female the uterus, tubes, ovaries, and vagina are between the bladder and the rectum.

The female pelvis differs from that of the male in those particulars that render it better adapted to pregnancy and parturition. It is more *shallow* than the male pelvis but wider in every direction. The inlet and outlet are larger and more nearly oval, the bones are lighter and smoother, the coccyx is more movable, and the subpubic arch is greater than a right angle. The subpubic angle in a male is less than a right angle.

The femur

Is the longest bone in the body. The upper end has a rounded head with a constricted neck, and two eminences, called the *greater* and *lesser trochanters*. The head articulates with the cavity in the hip bone, called the *acetabulum*. The lower extremity of the femur is larger than the upper, is flattened from before backward, and is divided into two large eminences, or *condyles*,

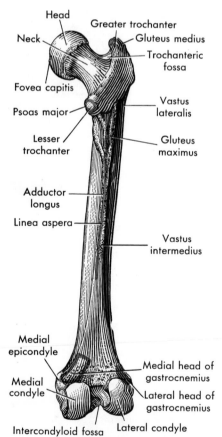

Figure 4–34. The right femur, or thigh bone, dorsal aspect. *Red,* muscle origins; *blue,* muscle insertions.

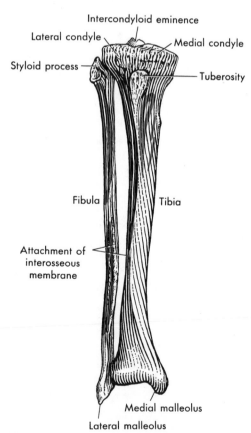

Figure 4–35. The bones of the right leg, ventral surface. *Red,* muscle origins; *blue,* muscle insertions.

by an intervening notch. The condyles are the lateral and medial, and the intervening notch is the *intercondyloid fossa*. The lower end of the femur articulates with the tibia and the patella. The bone inclines gradually downward and inward, so as to approach its fellow below to bring the knee joint near the line of gravity of the body. The degree of inclination is greater in the female than in the male.

The patella

Small, flat, triangular, sesamoid bone developed in the tendon of the quadriceps femoris muscle and placed in front of the knee joint. It articulates with the femur and is surrounded by large, fluid-filled bursae.

The tibia

Lies at the front and medial side of the leg. The upper extremity is expanded into two condyles, lateral and medial, with the sharp intercondyloid eminence between them. The superior surfaces are concave and receive the condyles of the femur. The lower extremity is smaller than the upper; it is prolonged downward on its medial side into a strong process, the *medial malleolus*, which forms the inner prominence of the ankle. There is also the surface for articulation with the talus, which forms the ankle joint. The tibia also articulates with the lower end of the fibula. In the male the tibia is vertical and parallel with the bone of the opposite side, but in the female it has a slightly oblique direction lateralward, to compensate for the oblique direction of the femur medialward.

The fibula (calf bone)

Is situated on the lateral side of the tibia, parallel with it. It is smaller than the tibia and, in proportion to its length, is the most slender of all the long bones. Its upper extremity consists of an irregular head by means of which it articulates with the tibia, but it does not reach the knee joint. The lower extremity is prolonged downward into a pointed process, the *lateral malleolus*, which lies just beneath the skin and forms the outer ankle bone. The lower extremity articulates with the tibia and the talus. The talus is held between the lateral malleolus of the fibula and the medial malleolus of the tibia. A fracture of the lower end of the fibula with injury of the lower tibial articulation is called a *Pott's[7] fracture*.

The tarsus (ossa tarsi)

Is formed by the calcaneus (os calcis), talus, cuboid (os cuboideum), navicular (os naviculare pedis), and the first, second, and third cuneiforms (os cuneiforme). The largest and

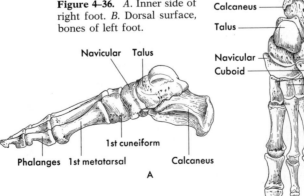

Figure 4–36. *A.* Inner side of right foot. *B.* Dorsal surface, bones of left foot.

A

B

[7]Percivall Pott, English surgeon (1714–1788).

strongest of the tarsal bones is called the *calcaneus,* or *heel bone;* it serves to transmit the weight of the body to the ground and forms a strong lever for the muscles of the calf of the leg.

The metatarsus, or sole and instep of the foot

Is formed by five bones that resemble the metacarpal bones of the hand. Each bone articulates with the tarsal bones by one extremity and by the other with the first row of phalanges. The tarsal and metatarsal bones are so arranged that they form two distinct arches; the one running from the calcaneus to the heads of the metatarsal bones on the inner (medial) side is called the *longitudinal arch,* and the other across the foot in the metatarsal region is called the *transverse arch.* The arches of the foot are completed by strong ligaments and tendons. The foot is strong, flexible, resilient, and able to provide the spring and lift for the activities of the body.

These arches may become weakened and progressively broken down, a condition known as flatfoot. This condition is thought to be due to prenatal conditions, dietary or hormone disturbances, improper posture, weight or fatigue conditions, or the wearing of shoes ill-fitting in last or size.

Phalanges

Both in number and general arrangement they resemble those in the hand, there being two in the great toe and three in each of the other toes.

Questions for Discussion

1. What are the functions of the bones?
2. Which bones articulate with the femur, humerus, scapula, and the atlas?
3. What is the anatomic relationship of the nerves, blood vessels, and lymphatics to the skeleton?
4. Discuss the structure and function of the intervertebral disk and nucleus pulposus.
5. Discuss the sternum, its position and flexibility in relation to heart massage.
6. What is the soft spot felt on a baby's head? What is its function? What becomes of this as the baby develops?

Summary

Bones

Functions
1. Organs of support
2. Instruments of locomotion
3. Framework of hard material
4. Afford attachment to soft parts
5. Help control internal pressures
6. Give shape to whole body

Classification
1. Long
2. Short
3. Flat
4. Irregular

Table of the Bones

BONES OF THE HEAD

Cranium		*Face*	
Occipital	1	Nasal	2
Parietal	2	Vomer	1
Frontal	1	Inferior nasal concha (inf. turb.)	2
Temporal	2	Lacrimal	2
Sphenoid	1	Zygomatic (malar)	2
Ethmoid	1	Palatine (palate)	2
	8	Maxilla	2
		Mandible	1
			14

Fontanelles	{ Anterior	1
	Posterior	1
	Anterolateral	2
	Posterolateral	2
Sinuses Opening into Nasal Cavity	{ Frontal	
	Ethmoidal	
	Sphenoidal	
	Maxillary	
Ear	{ Malleus	2
	Incus	2
	Stapes	2
	Described with ear (Chapter 12)	6
	Hyoid bone in the neck	1

BONES OF THE TRUNK

		Child	Adult
Vertebrae	{ Cervical	7	7
	Thoracic	12	12
	Lumbar	5	5
	Sacral	5	1
	Coccygeal	4 = 33	1 = 26
Ribs			24
Sternum			1
			51

Intervertebral Disks { Disks of fibrocartilage interposed between the bodies of vertebrae from axis to sacrum

BONES OF THE UPPER EXTREMITY

Clavicle	1	Pisiform	1
Scapula	1	Trapezium	1
Humerus	1	Trapezoid	1
Ulna	1	Capitate (os magnum)	1
Radius	1	Hamate (unciform)	1
Carpus		Metacarpus	5
Navicular (scaphoid)	1	Phalanges	14
Lunate (semilunar)	1		32
Triquetral (cuneiform)	1		32 × 2 = 64

BONES OF THE LOWER EXTREMITY

Hip bone (os coxae)	1	Third cuneiform	
Femur	1	(external cuneiform)	1
Patella	1	Second cuneiform	
Tibia	1	(middle cuneiform)	1
Fibula	1	First cuneiform	
Tarsus		(internal cuneiform)	1
Calcaneus (os calcis)	1	Metatarsus	5
Talus (astragalus)	1	Phalanges	14
Cuboid	1		31
Navicular (scaphoid)	1		31 × 2 = 62

Comparison of Female and Male Pelvis

	Female	Male
Bones	Slender	Heavier and rough
Sacrum	Broad, less curved	Narrow, more curved
Symphysis	Shallow	Deeper
Major pelvis	Narrow	Wide
Minor pelvis	Shallow and wide, capacity great	Deeper and narrower, capacity less
Great sciatic notches	Wide	Narrow
Superior aperture	Oval	Heart-shaped

Additional Readings

BASMAJIAN, J. V.: *Primary Anatomy*, 7th ed. Williams & Wilkins Co., Baltimore, 1976, pp. 21–75.

BOURNE, G. W.: *The Biochemistry and Physiology of Bone*, 2nd ed. Academic Press, Inc., New York, 1972.

CULLITON, B. J.: Repairing brittle bones. *Sci. News,* **97:**562–63, June 6, 1970.

GARDNER, E.; GRAY, D. J.; and O'RAHILLY, R.: *Anatomy—A Regional Study of Human Structure*, 4th ed. W. B. Saunders Co., Philadelphia, 1975, Chapter 2.

LOOMIS, W. F.: Rickets. *Sci. Amer.,* **223:**76–91, December, 1970.

ROMANES, G. J. (ed.): *Cunningham's Textbook of Antomy*, 11th ed. Oxford University Press, New York, 1972, pp. 75–206.

WARWICK, R., and WILLIAMS, P. (eds.): *Gray's Anatomy*, 35th Brit. ed. W. B. Saunders Co., Philadelphia, 1973, pp. 200–385.

CHAPTER 5

Joints

Chapter Outline

All bones have articulating surfaces so that various body movements may be accomplished. These surfaces form joints or articulations, some of which are freely movable, others slightly movable or immovable. All types are important for smooth, coordinated movements of the body.

Structure of Joints. The articulating surfaces of the bones are sometimes separated by a thin membrane, sometimes by strong strands of connective tissue, or fibrocartilage, and in the freely moving joints are completely separated. Strong ligaments extend over the joints or form capsules, which ensheath them. Tendons of muscles also extend over the joints.

Classification. Joints are classified according to the amount of movement of which they are capable and their structural composition:

1. Synarthroses (juncturae fibrosae), or immovable joints.

2. Amphiarthroses (juncturae cartilagenae), or slightly movable articulations.

3. Diarthroses (juncturae synoviales), or freely movable articulations.

Synarthroses, or Immovable Joints. The bones are connected by fibrous tissue or cartilage. The bones of the skull and face (with the exception of the mandible) have their adjacent surfaces in direct contact fastened together by a thin layer of fibrous tissue. The union is by a series of interlocking processes and indentations that

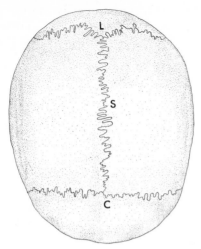

Figure 5–1. A toothed, or sagittal, suture, seen on the top of the skull. *L*, Lambdoidal suture; *S*, sagittal suture; *C*, coronal suture.

form sutures. (See summary for list of these sutures.)

Amphiarthroses, or Slightly Movable Joints. These include two varieties: (1) symphysis and (2) synchondrosis. *Sym-*

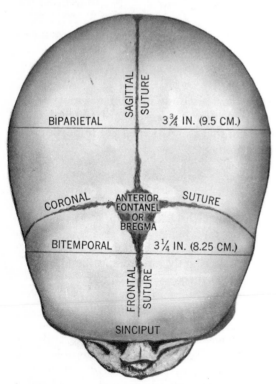

Figure 5–2. Diameters and landmarks of the fetal skull, upper surface. (Modified from Edgar.)

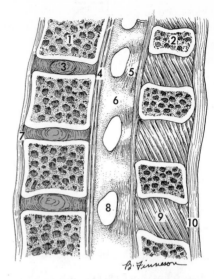

Figure 5–3. Diagram of vertebral symphyses, seen in longitudinal section through four segments of the vertebral column. *1*, Body of vertebra; *2*, spinous processes; *3*, intervertebral cartilaginous disk; *4*, posterior longitudinal ligament; *5*, ligamentum flavum; *6*, spinal canal; *7*, anterior longitudinal ligament; *8*, intervertebral foramen; *9*, interspinous ligament; *10*, supraspinous ligament.

physis is a joint where two long bony surfaces are connected by a broad, flat disk of fibrocartilage. *Synchondrosis* is a temporary form of joint. The cartilage is changed to bone before adult life, as found between the epiphysis and bodies of long bone.

Diarthroses, or Freely Movable Joints. These include most of the joints in the body. The adjacent ends of the bones are covered with hyaline cartilage and are surrounded by a fibrous *articular* capsule, which is strengthened by ligaments and lined with synovial membrane except over the articular cartilage. Tendons of muscles pass over these joints and play an important part in stabilizing the joint. The hyaline cartilage provides a smooth surface for the opposing bones, lubricated by synovial fluid (see page 74). Freely movable joints are classified by the kind of motion permitted.

1. GLIDING JOINTS (ARTICULATIO PLANA). These joints permit gliding movement only, as in the joints between the carpal bones of the wrist, between the tarsal

bones of the ankle, and between the articular processes of the vertebrae. The articular surfaces are nearly flat, or one may be slightly convex and the other slightly concave.

2. HINGE JOINTS (GINGLYMUS). Hinge joints allow angular movement in one direction, like a door on its hinges. The articular surfaces are of such shape as to permit motion in the forward-and-backward plane. These movements are called flexion and extension, as may be seen in the joint between the humerus and ulna, in the knee and ankle joints, and in the articulations of the phalanges. Strong collateral ligaments form their chief bond of union. The knee joint is described as a hinge joint, but it is actually much more complicated. There is the articulation between each condyle of the femur and the corresponding meniscus and the condyle of the tibia, and the articulation between the femur and the patella. The bones are held together by the articular capsule and 10 ligaments. (See Figure 5–7.) There are two menisci, the medial and the lateral; they are crescent-shaped structures of fibrocartilage; the surfaces are smooth and covered with synovial membrane. The upper surfaces are concave and in contact with the femur; the lower surfaces are flat and rest on the head of the tibia. The peripheral border of each one is thick and convex and attached to the inside of the joint capsule. The inner border is thin, concave, and unattached. The *function* of the menisci is to *deepen* the surfaces of the head of the tibia for articulation with the condyles of the femur.

3. CONDYLOID JOINTS (ARTICULATIO ELLIPSOIDEA). These permit an angular movement in two directions, as when an ovoid articular surface or condyle of bone is received into an elliptical cavity, e.g., the wrist joint. Movements permitted in this form of articulation include flexion, exten-

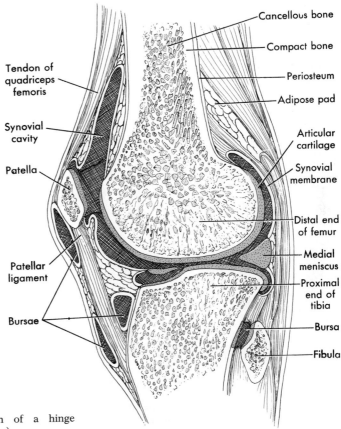

Figure 5–4. Longitudinal section of a hinge joint—the knee. (Modified from Ham.)

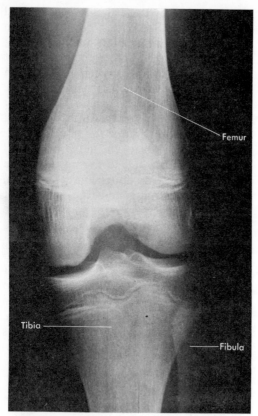

Figure 5-5. X-ray of the knee. (Courtesy of Radiology Department, The New York Hospital, New York, N.Y.)

Figure 5-6. X-ray of elbow. *A.* Elbow flexed, lateral view. *B.* Elbow extended. (Courtesy of Radiology Department, The New York Hospital, New York, N.Y.)

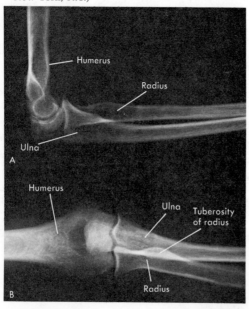

sion, adduction, abduction, and circumduction but no axial rotation.

4. SADDLE JOINTS (ARTICULATIO SELLARIS). In this type of joint the articular surface of each of the articular bones is concave in one direction and convex in another. The metacarpal bone of the thumb is articulated with the trapezium bone of the carpus by a saddle joint. The movements at these joints are the same as in condyloid joints.

5. PIVOT JOINTS (TROCHOID). These are joints with a rotary movement in one axis. In this form a ring rotates around a pivot or a pivotlike process rotates within a ring being formed of bone and cartilage. In the articulation of the axis and atlas the front of the ring is formed by the arter. r arch of the atlas and the back by the transverse ligament. The odontoid process of the axis forms a pivot, and around this pivot the ring rotates, carrying the head with it. In the proximal articulation of the radius and ulna, the head of the radius rotates within the ring formed by the radial notch of the ulna and the annular ligament. The hand is attached to the lower end of the radius, and the radius, in rotating, carries the hand with it; thus, the palm of the hand is alternately turned forward and backward. When the palm is turned forward or upward, the attitude is called *supination*; when backward or downward, *pronation*.

6. BALL-AND-SOCKET JOINTS (SPHEROIDEA ENARTHROSIS). These have an angular movement in all directions and a pivot movement. In this form of joint a more or less rounded head lies in a cuplike cavity, such as the head of the femur in the acetabulum and the head of the humerus in the glenoid cavity of the scapula. The shoulder joint is the most freely movable joint in the body.

Movement. Bones at their joints are capable of different kinds of movement, which rarely occur singly but usually in combination, thus allowing great variety.

Gliding Movement. This is the simplest kind of motion that can take place in a joint, one surface moving over another without any angular or rotatory movement. The costovertebral articulations permit a slight gliding of the heads and tubercles of the ribs on the bodies and transverse processes of the vertebrae.

Angular Movement. This occurs only between long bones, and by it the angle between two bones is either increased or diminished. It includes flexion, extension, abduction, and adduction.

1. FLEXION. A limb is flexed when it is bent, e.g., bending the arm at the elbow. The angle between the two bones is *decreased.*

2. EXTENSION. This is the reverse of flexion. The angle between the two bones is *increased.*

3. ABDUCTION. This term means drawn away from the middle line of the body, e.g., lifting the arm away from or at right angles to the body.

4. ADDUCTION. This term means brought to, or nearer, the middle line of the body, e.g., bringing the arm to the side of the body.

Both abduction and adduction have a different meaning when used with reference to the fingers and toes. In the hand, abduction and adduction refer to an imaginary line drawn through the middle finger, and in the foot, to an imaginary line drawn through the second toe.

Circumduction. This means that form of motion which takes place between the head of a bone and its articular cavity, when the bone is made to circumscribe a

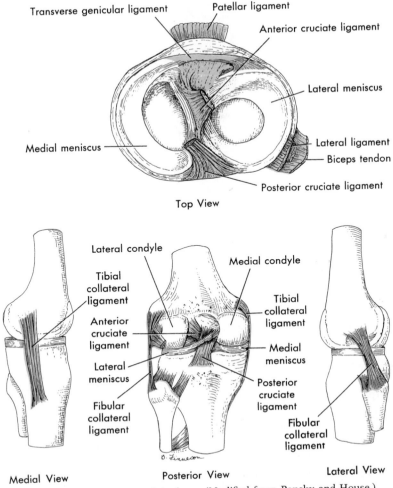

Top View

Medial View   Posterior View   Lateral View

Figure 5–7. Ligaments of the knee. (Modified from Pansky and House.)

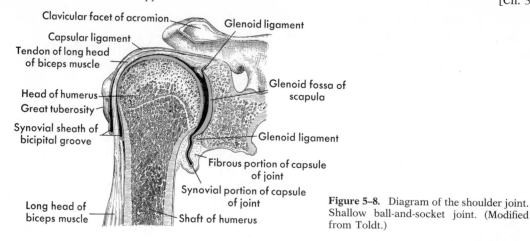

Clavicular facet of acromion

Glenoid ligament

Capsular ligament

Tendon of long head of biceps muscle

Head of humerus

Great tuberosity

Synovial sheath of bicipital groove

Glenoid fossa of scapula

Glenoid ligament

Fibrous portion of capsule of joint

Synovial portion of capsule of joint

Long head of biceps muscle

Shaft of humerus

Figure 5–8. Diagram of the shoulder joint. Shallow ball-and-socket joint. (Modified from Toldt.)

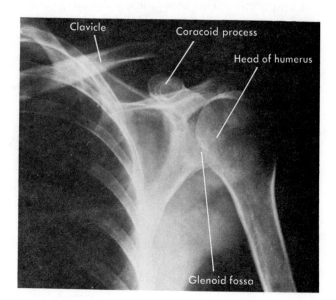

Clavicle

Coracoid process

Head of humerus

Glenoid fossa

Figure 5–9. X-ray of shoulder joint. (Courtesy of Radiology Department, The New York Hospital, New York, N.Y.)

conical space by rotation around an imaginary axis, e.g., swinging the arms or legs.

Rotation. This means a form of movement in which a bone moves around a central axis without undergoing any displacement from this axis, e.g., rotation of the atlas around the odontoid process of the axis.

Complete rotation, as of a wheel, is not possible in any joints of the body.

Inversion. In movement at the ankle the sole of the foot turns inward.

Eversion. In movement at the ankle the sole of the foot turns outward.

Frequent Disorders of Joints. *Sprain.* A wrenching or twisting of a joint accompa-

nied by a stretching or tearing of the ligaments or tendons is called a sprain.

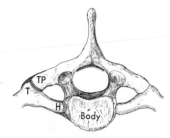

Figure 5–10. Diagram of gliding joints between the head of a rib and the body of a vertebra, and also between the tubercle of a rib and the transverse process of a vertebra. *H,* Head of rib; *T,* tubercle of rib; *TP,* transverse process.

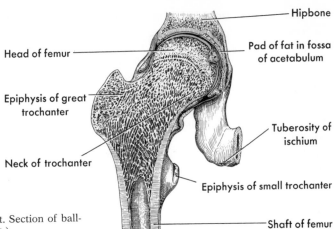

Head of femur

Hipbone

Pad of fat in fossa of acetabulum

Epiphysis of great trochanter

Tuberosity of ischium

Neck of trochanter

Epiphysis of small trochanter

Shaft of femur

Figure 5–11. Diagram of the hip joint. Section of ball-and-socket joint. (Modified from Toldt.)

Dislocation. If, in addition to a sprain, a bone of a joint is displaced, the injury is called a dislocation.

Ankylosis. This is immobility and consolidation of a joint.

Bursitis. Bursitis is an acute inflammation of the synovial bursa. It may be caused by excess tension or from some systemic or local inflammatory process. There may be deposits of calcium that

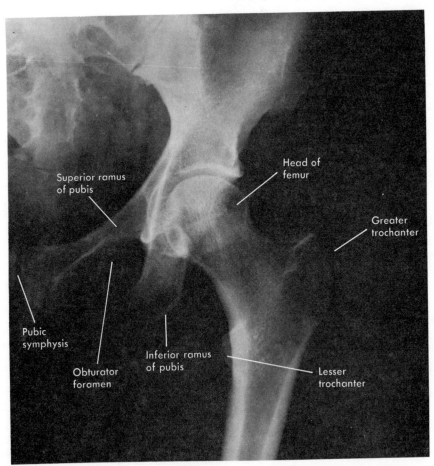

Superior ramus of pubis

Head of femur

Greater trochanter

Pubic symphysis

Obturator foramen

Inferior ramus of pubis

Lesser trochanter

Figure 5–12. X-ray of hip joint. (Courtesy of Radiology Department, The New York Hospital, New York, N.Y.)

interfere with motion. If bursitis becomes chronic, the joint may become stiff even though the joint itself is not involved. Joints most often involved include the shoulder, elbow, and knee joints.

Arthritis. This is an inflammation of joints that is common and painful. There are many varieties. The most common types are:

1. RHEUMATOID ARTHRITIS. This is a systemic disease with widespread involvement of connective tissues. The synovial membrane is thickened and ankylosis results.

2. OSTEOARTHRITIS. This is a degenerative condition that occurs in joints that are subject to a great deal of wear and tear. It usually occurs in individuals after the age of 45 years. The spine and joints of the lower extremity are most frequently involved.

There are softening of the cartilage, separation of fibers, and eventually disintegration of the cartilage. As the cartilage thins, the perichondrium and periosteum are irritated, which in turn stimulates cartilaginous and bony proliferation at the joint margins.

3. GOUTY ARTHRITIS. This is a metabolic disorder due to a disturbance of purine metabolism. Uric acid is elevated in the blood and crystals may be formed in the joints. The joint most frequently involved is the metatarsophalangeal of the great toe.

4. RHEUMATIC FEVER. This is a disease involving the synovial tissues, tendons, and other connective tissues around joints. Cartilage and bone are not involved. It is an acute inflammatory process that does not show residual changes in the articular system, but may leave permanent damage to the valves of the heart, which becomes evident in later life.

Questions for Discussion

1. How are joints classified? Give examples of each.
2. After a knee injury, there may be considerable swelling. Explain. What structures of the knee are affected?
3. A frequent knee injury to football players involves ligaments and the menisci. Explain the menisci and give their location and structure.
4. What are the causes of osteoarthritis and gouty arthritis?

Summary

Joints or articulations—connections between bones

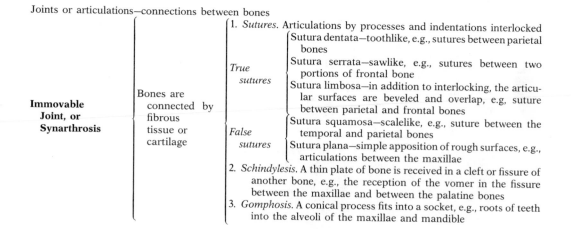

Immovable Joint, or Synarthrosis — Bones are connected by fibrous tissue or cartilage

1. *Sutures.* Articulations by processes and indentations interlocked

True sutures
- Sutura dentata—toothlike, e.g., sutures between parietal bones
- Sutura serrata—sawlike, e.g., sutures between two portions of frontal bone
- Sutura limbosa—in addition to interlocking, the articular surfaces are beveled and overlap, e.g, suture between parietal and frontal bones

False sutures
- Sutura squamosa—scalelike, e.g., suture between the temporal and parietal bones
- Sutura plana—simple apposition of rough surfaces, e.g., articulations between the maxillae

2. *Schindylesis.* A thin plate of bone is received in a cleft or fissure of another bone, e.g., the reception of the vomer in the fissure between the maxillae and between the palatine bones

3. *Gomphosis.* A conical process fits into a socket, e.g., roots of teeth into the alveoli of the maxillae and mandible

Slightly Movable Joint, or Amphiarthrosis	Bones are connected by disks of cartilage or interosseous ligaments	1. *Symphysis.* The bones are united by a plate or disk of fibrocartilage of considerable thickness 2. *Synchondrosis.* Temporary form of joint as the cartilage is changed to bone before adult life, as found between the epiphysis and bodies of long bone
Movable Joint, or Diarthrosis	1. Hyaline cartilage covering adjacent ends of the bones 2. Fibrous capsule strengthened by ligaments 3. Synovial membrane lining fibrous capsule	1. *Articulatio plana.* Gliding joint; articulates by surfaces that glide upon each other 2. *Ginglymus.* Hinge joint; moves backward and forward in one plane 3. *Articulatio ellipsoidea.* Condyloid joint; ovoid head received into elliptic cavity 4. *Articulatio sellars.* Saddle joint; articular surfaces are concavo-convex 5. *Trochoid.* Pivot joint; articulates by a process turning within a ring or by a ring turning around a pivot 6. *Spheroidea enarthrosis.* Ball-and-socket joint; articulates by a globular head in a cuplike cavity
Movement	1. Gliding movement 2. Angular 3. Circumduction 4. Rotation 5. Inversion 6. Eversion	Flexion Adduction Extension Abduction

Additional Readings

BASMAJIAN, J. V.: *Primary Anatomy*, 7th ed. Williams & Wilkins Co., Baltimore, 1976, pp. 75–112.

GARDNER, E.; GRAY, D. J.; and O'RAHILLY, R.: *Anatomy—A Regional Study of Human Structure*, 4th ed. W. B. Saunders Co., Philadelphia, 1975, Chapter 2.

JAFFE, H. E.: *Metabolic, Degenerative and Inflammatory Diseases of Bones and Joints.* Lea & Febiger, Philadelphia, 1972.

ROMANES, G. J. (ed.): *Cunningham's Textbook of Anatomy.* 11th ed. Oxford University Press, New York, 1972, pp. 207–55.

WARWICK, R., and WILLIAMS, P. (eds.): *Gray's Anatomy*, 35th Brit. ed. W. B. Saunders Co., 1973, pp. 388–471.

CHAPTER  6

Skeletal Muscle

Chapter Outline

PROPERTIES AND GENERAL FEATURES OF
    SKELETAL MUSCLE
CONNECTIVE TISSUE AND MUSCLE
    TENDONS AND THE ARRANGEMENT OF MUSCLE
      FIBERS
NERVE SUPPLY OF MUSCLES
ULTRASTRUCTURE OF MUSCLE: THE
    BASIS OF MUSCLE CONTRACTION
EXCITATION-CONTRACTION COUPLING
MECHANICS OF MUSCLE CONTRACTION
    THE STIMULUS
    THE TWITCH RESPONSE

TYPES OF CONTRACTION
    Summation of Contractions
    Tonus
LENGTH-TENSION RELATIONSHIPS IN
    MUSCLE
HEAT FORMATION
SOURCES OF ENERGY FOR MUSCLE
    CONTRACTION
    EXERCISE AND FATIGUE
MUSCLES AND THE BONY LEVERS
ACTIONS OF SKELETAL MUSCLES

Motion is an important activity of the body that is made possible by the special function of contractility in muscle tissue. All motion in the body is due to three types of muscle: skeletal, cardiac, and smooth. Whereas skeletal muscle is under voluntary control, the contraction of cardiac and smooth muscle is involuntary. The contraction of skeletal muscle allows for the movement of the entire body or parts of the body from place to place and also gives rise to the movements associated with breathing. Smooth muscle is responsible for the characteristic movements associated with the alimentary canal, and it controls and adjusts the diameter of blood vessels and the ducts of glandular tissue. The beating (repetitive contraction and relaxation) of the heart is responsible for moving blood to all the tissues of the body.

The motion, therefore, produced by the contraction and subsequent relaxation of muscle is intimately related to all physiologic activities. In this chapter, as well as in Chapters 7 and 8, we will direct our attention to the action and function of skeletal muscle. The activity of cardiac muscle will be discussed with the heart (Chapter 17). The varied effects caused by the activity of smooth muscle will each be considered with its appropriate system, e.g., circulatory (Chapter 16), respiratory (Chapter 19), digestive (Chapter 20), reproductive (Chapter 25).

Properties and General Features
of Skeletal Muscle

In the human, muscular tissue constitutes 40 to 50 per cent of the body weight.

Special characteristics of muscle tissue are irritability (excitability), contractility, extensibility, and elasticity. *Irritability,* or *excitability,* is the property of receiving stimuli and responding to them. All cells possess this property. The response of any tissue to stimulation is to perform its special function, which, in the case of muscular tissue, is contraction. *Contractility* is the property that enables muscles to change their shape and become shorter and thicker. This property is characteristic of all protoplasm, but is more highly developed in muscular tissue than in any other. *Extensibility* of a living muscle cell means that it can be stretched, or extended, and *elasticity* means that it readily returns to its original form when the stretching force is removed.

Skeletal muscles have the greatest diversity in shape, size, power, speed of contraction, and the means by which they are attached to bones or other tissues. The numerous designs of muscles and tendons are the foundation of all voluntary movements. Muscles have been named for such features as shape (trapezius), size (teres major), situation (pectoralis), attachments (sternocleidomastoid), direction (obliquus), and function (extensor carpi ulnaris).

Connective Tissue and Muscle

Muscles are attached to bones by dense connective tissue called *tendon.* Tendons are usually cordlike structures. When they appear as flattened sheets, as in the abdominal muscles, they are referred to as an *aponeurosis.* A looser type of connective tissue extends into and surrounds all the muscle fibers making up the whole muscle. There is a specific distribution and arrangement of this connective tissue, and it is referred to by special terms. The connective tissue that surrounds the whole

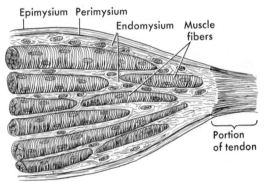

Figure 6–2. Longitudinal section of muscle illustrating connective tissue of muscle fibers attaching to tendon.

muscle is called *epimysium.* Strands of connective tissue from the epimysium pierce the muscle and surround groups or bundles of muscle fibers, this is called the *perimysium.* These bundles of muscle fibers are referred to as *fascicles.* Again, strands of connective tissue from the perimysium dip into a muscle fascicle and surround each individual muscle fiber. This connective tissue is called the *endomysium* (Figure 6–1).

Muscle fibers vary from 0.01 mm to 0.1 mm in thickness and from 3 mm to more than 40 cm in length. Some fibers may extend from the point of origin to insertion of the whole muscle, but *all* the fibers *do not.* It is by means of the connective tissue that all the fascicles of muscle are attached to the tendons (Figure 6–2). Hence,

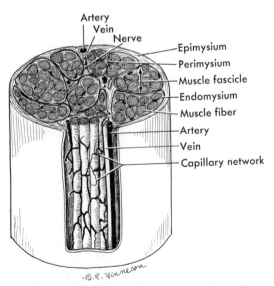

Figure 6–1. Diagrammatic representation of a muscle showing arrangement of connective tissue and blood vessels.

when individual muscle fibers contract, whether or not they extend the full length of the muscle, their shortening can be manifested via the connective tissue. The connective tissue associated with the muscle fibers is referred to as the *parallel elastic component* of muscle, whereas the tendons are called the *series elastic component*. Each of these components has a role in the contraction of muscle.

The connective tissue not only binds together the individual fibers and groups of fibers, but also it is the tissue in which blood vessels and nerves can ramify to reach all the individual muscle fibers. All muscles have a rich supply of blood vessels and nerves.

TENDONS AND THE ARRANGEMENT OF MUSCLE FIBERS

The arrangement of tendons and muscle fibers varies. Often tendons form long, rounded cords on either end of a fleshy muscle belly. This is referred to as a *fusiform*, or spindle-shaped, muscle (e.g., biceps brachii). A tendon may extend into the muscle, the muscle fibers and their accompanying connective tissue arising from it on each side like the barbs from the axis of a feather—hence the term *bipennate*, or bipenniform muscle (e.g., rectus femoris). In some muscles a tendinous core extends through one side of the muscle, with all the fascicles of muscle reaching the tendon from one side only. This appearance is much like half a feather—hence the term *unipennate* muscles (e.g., flexor pollicis longus). Several tendons within a muscle give rise to a *multipennate* type of muscle (e.g., deltoid). Thus, tendons add the needed area for muscle fiber attachment, as bone by itself would not provide the surface required. The different types of tendon and fiber arrangement also allow for the varying actions.

Nerve Supply of Muscles

In order for a muscle to contract it must be innervated. Muscles are supplied with both sensory and motor nerves. Sensory nerves receive and transmit to the appropriate areas in the brain the condition of the muscle: i.e., is it relaxed, contracted, what is the degree of contraction? Motor nerves in a sense tell the muscle to contract.

Many motor nerves supply a given muscle. A single motor nerve and the bundle of muscle fibers it supplies is called a *motor unit*. When a few muscle fibers make up a motor unit, that muscle is capable of rapid, fine, precise movements (e.g., muscles of eye, approximately a dozen fibers compose the unit). This is in contrast to a motor unit made up of many muscle fibers, in which the muscle action is much coarser and slower (e.g., postural muscles, several hundred individual muscle fibers comprise the motor unit). The area where the nerve ending meets the muscle is called the *motor end plate, neuromuscular junction*, or *myoneural junction* (Figure 6–3).

The structural components and details of the neuromuscular junction have been greatly clarified by use of electron microscopy. The nerve endings appear as small oval to rounded structures called boutons, terminal feet, or end feet. These rest in specialized, highly enfolded regions on the muscle membrane called *subneural clefts, synaptic folds*, or the *subneural apparatus*. The nerve terminals contain numerous small vesicles (400 to 600 Å in diameter) called *synaptic vesicles*, mitochondria, and a few filaments and microtubules. The synaptic vesicles are generally thought to be the sites of storage of the neurotransmitter substance *acetylcholine*. The currently held concept is that when an impulse or action potential passes down the

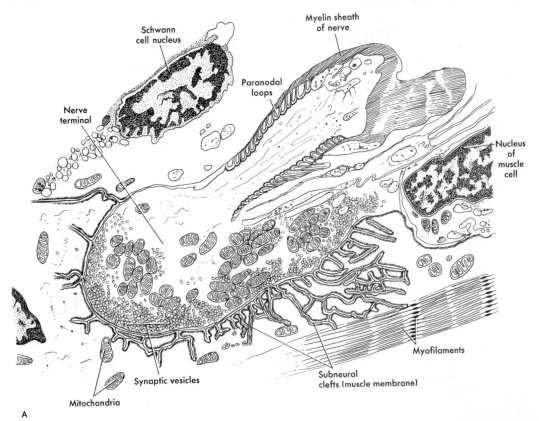

A

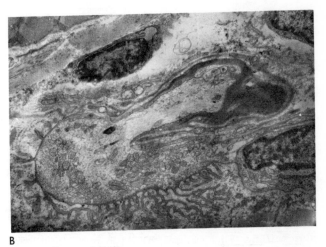

B

Figure 6–3. *A.* Drawing of mammalian neuromuscular junction. *B.* Electron micrograph of neuromuscular junction.

nerve fiber and reaches the terminal arborizations of the nerve at the motor end plate, acetylcholine is released. The acetylcholine diffuses across the synaptic space and activates the sarcolemma of the muscle, thereby increasing the permeability to specific ions. This increase in ion flow causes the muscle membrane to become excited, or depolarize, one of the first events leading to muscle contraction. Acetylcholine is rapidly destroyed by the enzyme *acetylcholinesterase,* which is located at the end plate region. This allows the muscle membrane to return to its resting state, ready to receive another wave of excitation.

Ultrastructure of Muscle: The Basis of Muscle Contraction

Each individual muscle fiber is made up of many smaller *myofibrils* (1 to 3 mμ in diameter). The myofibrils, in turn, are made up of many *myofilaments*. Myofilaments are of two types: (1) thick filaments, which are composed of the protein *myosin*; and (2) thin filaments, which are composed of the protein *actin*. The arrangement of the myofilaments gives rise to the banding pattern seen in "striated" muscle. When striated muscle is viewed in longitudinal sections, the following arrangement of myofilaments is evident (Figure 6–4). The parallel arrays of thick myosin filaments are the principal elements of the region called the *A-band*. The thick myosin filament is thicker in the center and tapers at its ends. Except for its central portion, the myosin filament has many short projections. These projections appear to form cross bridges between thick and thin filaments. The thinner actin filaments extend from either end of a thick, transversely oriented dense area, the *Z-line*. A central Z-line with actin filaments on either side makes up the *I-band*. The small area where actin and myosin filaments do not overlap, and consisting only of portions of myosin, is called the *H-band*. The region of myofilaments between two Z-lines is called the *sarcomere*. It is the functional unit of skeletal muscle.

The relative lengths of the bands are different when the muscle shortens, is stretched, and is at rest. This observation proved to be the basis for the sliding-filament hypothesis of muscle contraction originally proposed by Hansen and Huxley. In contraction, the thick and thin filaments retain the same length, but slide

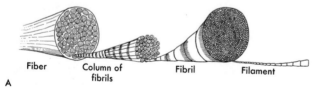

Fiber   Column of fibrils   Fibril   Filament

A

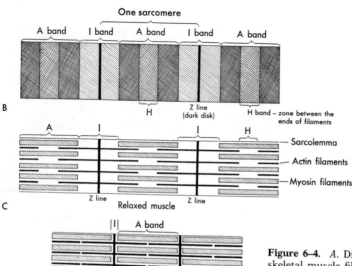

Figure 6–4. *A.* Diagram of the organization of one skeletal muscle fiber. *B.* Representation of banding pattern evident in longitudinal section. *C.* Representation of myofilaments in a relaxed muscle. *D.* Myofilaments in a contracted muscle.

Figure 6–5. Electron micrograph of the fine structural detail of striated muscle fibrils in longitudinal section. Letters correspond to those of Figure 6–4. (Courtesy of Dr. F. DeCaro, PMC Colleges.)

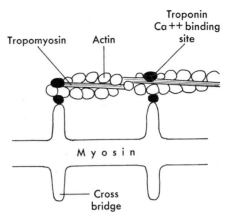

Figure 6–6. Schematic representation of actin and myosin.

past each other. The ends of the actin filaments thus extend farther into the A-band, decreasing and ultimately obliterating the H-band. Each of the sarcomeres shorten, and there is an overall shortening of the myofibril. The projections, or cross bridges of the myosin filament, engage the actin filaments and apparently create a pulling force. The chemical events involved in the longitudinal displacement of one set of fibrils in relation to the other also appear to involve the cross bridges. Evidence to date indicates that the force of contraction is produced by a cyclic process occurring at cross bridges and that each cross bridge may attach to a succession of sites along its neighboring actin filament, thereby causing the filaments to move past one another. The enzyme ATPase is localized on a portion of

the myosin's cross bridges, and this activates the breakdown of ATP, yielding the energy necessary in energizing this sliding process.

In order for actin to combine with myosin, *calcium* must be present. Calcium has a specific receptive protein. This protein is *troponin*, which is found associated with another protein, *tropomyosin* (Figure 6–6). Both troponin and tropomyosin, referred to as regulator or modulating proteins, are present in the actin filaments. In the absence of calcium, troponin in collaboration with tropomyosin inhibits the binding of actin and myosin. When calcium is present, this inhibitory action is removed and the contractile mechanism is activated.

Excitation-Contraction Coupling

We have seen that excitation of the muscle membrane will lead to contraction. It was not, however, until recently that we understood how the wave of excitation proceeded from the muscle membrane into the fiber to reach all of the contractile elements. The discovery by electron microscopists of the *sarcoplasmic reticulum* and the *transverse tubular system*, or *T-system*, has given us a partial answer to link the events of excitation and contraction. The sarcoplasmic reticulum forms a

sleevelike structure that surrounds each of the myofibrils and has no connection with the surface membrane. It consists of flattened longitudinal tubules that at regular intervals (in mammals at the A-I junction) enlarge to form *lateral sacs*. Situated between two lateral sacs is the smaller-diameter tubule of the T-system. The complex of two lateral sacs with a central T-tubule is called a *triad*. The T-tubules interconnect with each other and reach the surface of the muscle membrane. Thus, the

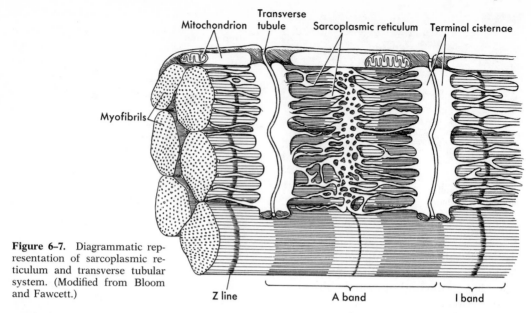

Mitochondrion   Transverse tubule   Sarcoplasmic reticulum   Terminal cisternae

Myofibrils

Z line   A band   I band

Figure 6–7. Diagrammatic representation of sarcoplasmic reticulum and transverse tubular system. (Modified from Bloom and Fawcett.)

lumens of tubules of the T-system not only are open to the extracellular medium but also act as a network of channels conveying excitation inwardly to reach all the contractile elements.

Calcium, which is essential for muscle contraction, is stored in the longitudinal elements of the sarcoplasmic reticulum. We do not as yet know how, but during excitation calcium is released. Following contraction calcium is pumped back into the sarcoplasmic reticulum, using energy derived from the splitting of ATP.

The following is a summary of the sequence of events in muscle contraction that have been discussed thus far:

1. Nerve impulse, or action potential, travels down the nerve fiber, reaches terminal arborizations, and causes release of acetylcholine.

2. Acetylcholine diffuses across the synaptic space and excites or depolarizes the muscle membrane.

3. Excitation of muscle membrane is transmitted via the T-system to all contractile elements.

4. There is release of calcium ions from sarcoplasmic reticulum surrounding the myofibrils.

5. Calcium ions activate actin and myosin binding at cross bridges of myosin by inhibition of inhibitory proteins troponin and tropomyosin.

6. ATPase of myosin splits ATP, releasing the energy needed for sliding of actin over myosin.

7. Contraction of muscle.

8. Recapture of calcium ions by sarcoplasmic reticulum.

9. Inhibitory effects of troponin and tropomyosin in action, actin and myosin interactions blocked.

10. Relaxation of muscle.

Mechanics of Muscle Contraction

The mechanics of muscle contraction have for the most part been defined and studied by the use of isolated nerve muscle preparations. In these preparations the parameters of stimulation (duration and intensity) can be controlled and moni-

tored, and the response of the muscle to stimulation can be displayed on a recording device. Both experimentally and in the body, the magnitude or strength of muscle contraction is directly affected by (1) the strength of the stimulus, (2) the speed of application of the stimulus, (3) the duration of the stimulus, (4) the weight of the load imposed on the muscle, (5) the initial length of the muscle fibers, and (6) the temperature, which in man is optimal at 37° C (98.6° F).

THE STIMULUS

A muscle is excitable because the muscle fibers composing it are excitable. All protoplasm possesses the property of *excitability*. Any force that affects this excitability is called a *stimulus*. Physiologically, a stimulus represents a change in the environment of the muscle cells. Protoplasm also possesses the property of *conductivity*, and when stimulated at one point the response may travel through the cell. The response is the specialized one that is characteristic for the tissue stimulated; in muscles it is contraction. The muscles are stimulated by impulses conveyed to them by nerve fibers.

The weakest stimulus possible to excite *some* motor units, thereby causing contraction of muscle, is called a *minimal stimulus*. Any stimulus weaker than this is known as *subminimal* (subliminal, liminal). Two subminimal stimuli, each in itself too weak to cause contraction, may, when applied in rapid succession, by their additive forces be equivalent to a minimal stimulus and result in contraction. A stimulus that excites *all* the motor units of the muscle is called *maximal*. The magnitude of a muscle contraction due to a maximal stimulus is greater than that due to a minimal stimulus.

Immediately following a stimulus there is a very brief interval of time known as the *absolute refractory period*, during which a second stimulus, no matter how strong, will not cause a contraction. This is followed by the *relative refractory period*, often called the period of depressed excitability, during which the muscle regains its irritability.

It has been found that if the stimulus applied to a *single muscle fiber* is strong enough to produce a response, it will give a contraction that is maximal, no matter

what the strength of the stimulus. This is the *all-or-none* law. This means that each muscle fiber gives a maximal response or none at all. Fatigue and varying conditions of nutrition may alter the cell's response, but increasing the strength of the stimulus will not change the response.

THE TWITCH RESPONSE

Following a single maximal stimulus to either the nerve supplying the muscle, or the muscle itself, a brief contraction occurs called a *twitch*. There is a brief interval of time from the application of the stimulus to the start of contraction called the *latent period*. It is followed by a *period of contraction* and, in turn, by a period of *relaxation* (Figure 6–8). The contraction time and the total twitch duration vary a great deal in different types of muscle and in different species of animals. There are "fast muscles" in which the twitch duration may be as short as 7.5 msec. Fast muscles are primarily concerned with fine, rapid, precise movement (e.g., the medial rectus of the eye). "Slow muscles" have twitch durations up to 100 msec in duration and are principally involved in strong, sustained movements (e.g., the muscles that maintain the upright posture). Most of the muscles of the body are mixed, containing both fast and slow fibers.

TYPES OF CONTRACTION

Contraction refers to a series of internal events that are observable externally by either *shortening* or *tension*. A muscle contraction (twitch) that results in shortening of the whole muscle is referred to as *isotonic* (the length changes, but the tension remains the same). This type of contraction is seen in lifting a load, in the movements associated with walking and running, and when work is performed by the muscle. The weight of the load lifted in an isotonic contraction will determine the velocity of shortening and the duration of the contraction. Namely, a light load can be moved much more quickly and over a

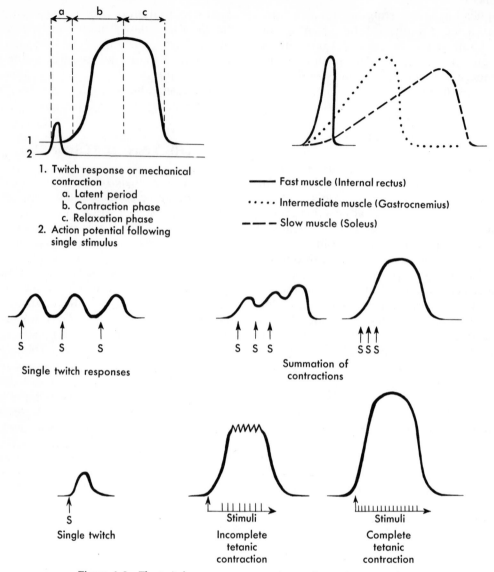

Figure 6–8. The twitch response, summation, and tetanic contraction.

greater distance than a heavier load. Some load is necessary to obtain a contraction, but increase of load beyond the optimum decreases the height of contraction.

When the muscle length does not change but tension develops, the contraction is called *isometric* (same length). Such contractions produce tension rather than shortening and work. The tension developed during an isometric contraction is utilized to oppose other forces such as gravity, in supporting a weight in a fixed position, and to maintain posture.

Muscles of the body can contract both isometrically and isotonically, and usually most contractions are a mixture of the two. In the act of running the mixture of these two types of contraction is seen in the leg muscles in that isometric contraction keeps the limbs stiff when the legs hit the ground, and isotonic contractions move the limbs.

Summation of Contractions. When two or three stimuli are applied in rapid succession, so that the mechanical contraction of the first stimulus is not completed, a fusion or *summation* of contractions occurs. The magnitude of tension, or shortening of the summated contraction, is greater than that which was produced by the individual single stimulus (Figure 6–8).

A *tetanic contraction*, or *tetanus*, is the mechanical response of a muscle resulting from a repetitive stimulation, at a frequency that is sufficiently rapid to result in a sustained maximal summation. If a high rate of stimulation is used, the contraction is continuous, and a smooth contraction or complete tetanus is obtained. If the stimulus rate is low, the contraction is not smoothly maintained (incomplete tetanus), but fluctuates at the frequency of the stimulating pulses. During a complete tetanic contraction the greatest magnitude of response can be produced by the muscle, and is usually three to four times greater than the twitch response (Figure 6–8).

In both summation and tetanus there is a fusion of the mechanical contractions. It must be emphasized, however, that the nerve impulses or action potentials remain separate. The time factor between nerve impulses is a critical factor. If nerve impulses were to arrive at a frequency such that the action potential from one stimulus arrived during the refractory period of the preceding action potential, tetanus would not result.

In the living animal, muscle fibers are usually not activated at the high frequencies necessary for a complete tetanic contraction. Instead, incomplete tetanic contractions occur. The whole muscle, however, exhibits a smooth, sustained contraction, since impulses are normally delivered to different motor units at different times.

Tonus. Muscle is always under some degree of contraction, and this is referred to as *tonus* or *tone*. The tone of skeletal muscles gives a certain firmness and maintains a slight, steady pull upon their attachments; it also functions in the maintenance of a certain pressure upon the contents of the abdominal cavity. By means of tonic contractions in skeletal muscle, posture is maintained for long periods with little or no evidence of fatigue. Absence of fatigue is brought about mainly by means of different groups of muscle fibers contracting in relays, giving alternating periods of rest and activity for given muscle groups. In man the antigravity muscles (e.g., extensors of neck and back) exhibit the highest degree of tonus. During sleep, the tonus of muscles is at a minimum.

Length-Tension Relationships in Muscle

The tension developed in a muscle is dependent upon the initial length of the muscle, i.e., its length at the time of stimulation. The relationship of length and tension developed can be experimentally studied in whole muscle and in single muscle fibers. The length of a relaxed muscle, free of its bony attachments, is called the *equilibrium length*. A muscle at equilibrium length will develop very little tension when it is stimulated and contracts. As the muscle is stretched or in-creased in length, the tension developed increases in a linear fashion until a maximum is reached. The length of the muscle at which the developed tension is maximal is called the *resting length*. Stretching the muscle beyond its resting length will not increase tension and will result in rupture of the fibers. The muscles of the body are attached to bones under some stretch. Specifically, they are at their resting length and, therefore, capable of producing a maximal tension or force on contraction.

COVERED PARTS IN CLASS.

Heat Formation

Muscles form the major source of body heat. The chemical changes that occur in muscle cells form mechanical energy used for movement and also release heat energy that is used to maintain body temperature. Heat is liberated by the chemical processes that change the muscle from a relaxed to an active state, regardless of whether the muscle shortens or not. The heat formed during stimulation and the events of the contractile process is called *initial heat*. It is followed by *relaxation heat*. The *recovery* heat is associated with the metabolic processes involved in the resynthesis of ATP.

Sources of Energy for Muscle Contraction

The source of energy for muscle contraction is ATP. There are three means of providing a continual supply of ATP: (1) creatine phosphate, (2) glycolysis, and (3) oxidative phosphorylation (citric acid or Krebs cycle).

Resting muscle contains stores of ATP and creatine phosphate. This stored ATP is immediately used up during the first few seconds of muscle contraction and ADP levels increase. *Creatine phosphate* is able to *rapidly* supply ATP to contracting muscle. This molecule contains both energy and phosphate, which are transferred to a molecule of ADP to form ATP and creatine. If the contraction is short, the muscle can derive all of its ATP from creatine phosphate. If a continued mild form of muscular activity is to be maintained, other sources of ATP must be available.

Glycolysis is the metabolic process by which glycogen (stored form of glucose in muscle) or glucose from the blood is broken down to form pyruvic acid. The process results in the synthesis of ATP. If oxygen is readily available to the muscle, pyruvic acid is further broken down to carbon dioxide and water by *oxidative phosphorylation* in the citric acid cycle to yield more ATP.

The breakdown of ATP, the combination of creatine phosphate with ADP, and glycolysis can occur both aerobically (in presence of oxygen) and anaerobically (in absence of oxygen). Oxidative phosphorylation is an aerobic process; i.e., it requires the presence of oxygen. If oxygen supplies to the muscle become exhausted, as in extensive muscular activity, the pyruvic acid formed in glycolysis is converted to lactic acid. As lactic acid accumulates in the muscle, fatigue becomes evident and

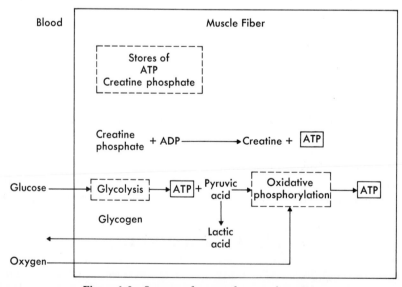

Figure 6–9. Sources of energy for muscle activity.

the muscle is in *oxygen debt*. In oxygen debt lactic acid diffuses out into the blood and is carried to the liver, where it is reoxidized to pyruvic acid. In the liver pyruvic acid can enter the citric acid cycle and be broken down to carbon dioxide and water, or it is reconverted to glycogen.

When oxygen supplies are adequate, the production of lactic acid is inhibited and pyruvic acid enters the citric acid cycle— the common pathway for terminal oxidation of carbohydrates, fats, and amino acids (page 525)—with the production of carbon dioxide, water, and energy in the form of ATP. The metabolic reactions in the presence of oxygen make available from 10 to 12 times as much energy as do anaerobic reactions. Muscles operating under anaerobic conditions use six to eight times as much glycogen to do the same amount of work as could be accomplished under aerobic conditions.

EXERCISE AND FATIGUE

In muscles undergoing contraction, the first effect of the formation of carbon dioxide and lactic acid is to increase irritability; however, if a muscle is continuously stimulated, the strength of contraction becomes progressively less until the muscle refuses to respond. This is true fatigue and is caused, in part at least, by hypoxia and the toxic effects of metabolites (carbon dioxide, acid phosphate, lactic acid) that accumulate during exercise. The loss of nutritive materials may also be a factor in fatigue, but recent conceptions stress the accumulation of metabolites.

In moderate exercise the system is able to eliminate these substances readily. After prolonged contractions a period of rest may be necessary to furnish opportunity for the blood to carry the fatigue substances to the excretory organs and nutritive materials and oxygen to the muscle. Probably it is chiefly lactic acid that brings on fatigue by disturbing the hydrogen ion concentration of the cell fluids, thus inhibiting the enzyme action that is responsible for further breakdown of glycogen. It has been demonstrated that injection of the blood of a fatigued animal into a rested one will promptly bring on signs of fatigue.

Exercise stimulates circulation and thereby brings about a change in conditions for cells in all locations throughout the body. This great stirring-up effect of exercise brings fresh blood via the arterioles, and the local pressure as well as the fluid environment of all cells is changed. Exercise has been shown to increase the size, strength, and tone of the muscle fibers. Massage and passive exercise may, if necessary, be used as a partial substitute for exercise. Although physical recreation is desirable for aiding metabolic processes, continued use of fatigued muscles is injurious if, during such conditions, the muscles exhaust their glycogen supply and utilize the protein of their own cells. Under normal conditions it is the sensation of fatigue that protects us from such extremes.

Muscles and the Bony Levers

Direct muscular contraction alone is not entirely responsible for bodily motions. Intermediate action of bony levers is also essential. In the body, cooperative functioning of bones and muscles forms levers. A knowledge of levers gives a basis for understanding the principles underlying good posture and the movements of the body.

A *simple lever* is a rigid rod that is free to move about on some fixed point or support called the *fulcrum*. It is acted upon at two different points by (1) the *resistance* (*weight*), which may be thought of as something to be overcome or balanced, and (2) the *force* (*effort*) that is exerted to overcome the resistance. In the body bones of varying shapes are levers, and the

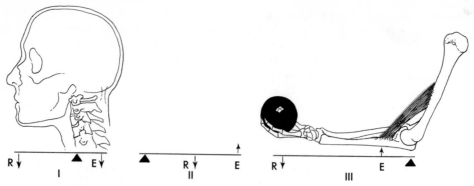

Figure 6–10. Diagram of simple levers. Note insertion of muscle in relation to fulcrum and resistance. The effort is applied at the place of insertion of muscle to the bone. The ▲ represents the fulcrum; *R*, the resistance; *E*, the effort; *arrows*, the direction of motion. There are no levers of the second class in the body. See discussion in text.

resistance may be a part of the body to be moved or some object to be lifted or both of these. The muscular effort is applied to the bone at the insertion of the muscle and brings about the motion or work.

For example, when the forearm is raised, the elbow is the fulcrum, the weight of the forearm is the resistance, and the pull due to contraction of the biceps muscle is the effort.

Levers act according to a law that may be stated thus: When the lever is in equilibrium, the effort times the effort arm equals the resistance times the resistance arm ($E \times EA = R \times RA$).

The "resistance arm" is the perpendicular distance from the fulcrum to the line of action of the resistance (weight). The "effort arm" is the perpendicular distance from the fulcrum to the line of action of the effort (force.)

For example, if the distance from the effort to the fulcrum is the same as the distance from the resistance to the fulcrum, an effort of 5 lb will balance a resistance of 5 lb.

Levers may be divided into three classes according to the relative position of the fulcrum, the effort, and the resistance. In levers of the *first class* the fulcrum lies between the effort and the resistance, as in a set of scales. In this type of lever the resistance is moved in the opposite direc-

tion to that in which the effort is applied. When the head is raised, the facial portion of the skull is the resistance, moving upon the atlanto-occipital joint as a fulcrum, while the muscles of the back produce the effort.

In levers of the *second class* the resistance lies between the fulcrum and the effort and moves, therefore, in the same direction as that in which the effort is applied, as in the raising of a wheelbarrow. There are no levers of the second class in the body.

In levers of the *third class* the effort is exerted between the fulcrum and the resistance. Levers in which the resistance arm is thus longer than the effort arm produce rapid delicate movements wherein the effort used must be greater than the resistance. The flexing of the forearm is a lever of this type, as are most of the levers of the body. The *law of levers* applies in the maintenance of correct posture. The head held erect in correct standing posture rests on the atlas as a fulcrum, with little or no muscular effort being exerted to maintain this position. The head in this position is in the "line of gravity," which passes through the hip joints, knee joints, and the balls of the feet (Figure 4–22, page 109). When the shoulders are stooped and the head is bent forward, constant muscular effort is exerted against the pull of gravity on the head.

Actions of Skeletal Muscles

The attachment of a muscle to bone at one end is the *origin* and at the other the *insertion*. Usually the origin is the more fixed point, and the insertion the mobile point, but the roles may be reversed in different actions. A knowledge of the origin and insertion of a muscle often reveals the action of the muscle.

Skeletal muscles are arranged in groups with specific functions to perform: flexion and extension, external and internal rotation, abduction and adduction. For example, in flexing of the elbow, several muscle groups are involved in varying degrees. The *agonists*, or prime movers, give power for flexion; the opposing group, the *antagonists*, contribute to smooth movements by their power to maintain tone yet relax and give way to movement of the flexor group. Other groups of muscles act to hold the arm and shoulder in a suitable position for action and are called *fixation* muscles. The *synergists* are muscles that assist the agonists (prime movers) and reduce undesired action or unnecessary movement. Activity of these opposing muscle groups is coordinated in relation to degree of tension exerted. When tension of the flexor muscles is increased, the tone of the extensor muscles is decreased; movement is controlled and position is maintained against varying degrees of pressures or pulls.

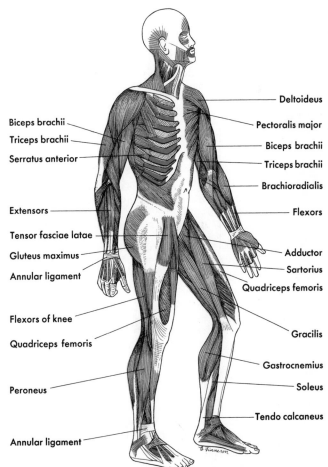

Figure 6–11. The human body, showing muscles. (Courtesy of Willian Wood & Co.)

Questions for Discussion

1. Define the following terms:
   Epimysium   Multipennate
   Perimysium   Tendon
   Fascicle   Aponeurosis
2. Describe the ultrastructure of the neuromuscular junction, and list the functions or advantages of synaptic vesicles and the highly enfolded subneural clefts.
3. What is the sarcomere? Describe the sliding-filament hypothesis.
4. What is the role of calcium in muscle contraction?
5. Define each of the following:
   Minimal stimulus   All-or-none law
   Maximal stimulus   Absolute refractory period
6. What is the twitch response? How would you characterize fast and slow muscles?

7. Give a description of an isotonic, isometric contraction and tetanic contraction.
8. Discuss and give examples of energy sources for muscle activity.
9. What are the causes of muscle fatigue following prolonged exercise?
10. Name the three classes of levers and give examples, where applicable, of their locations in the human body.
11. Define the following functional classification of muscles:
    Synergists
    Agonists
    Antagonists
12. What is meant by the terms *origin* and *insertion* of a muscle?

Summary

Skeletal muscle has the properties of irritability, contractility, extensibility, and elasticity. It makes up the muscular system and constitutes 40 to 50 per cent of body weight.

Connective Tissue and Muscle

- Muscle attached to bones by dense connective tissue called tendon
- Arrangement of tendon and muscle
  - Fusiform   Unipennate
  - Bipennate   Multipennate
- Looser type of connective tissue extends into and surrounds all muscle fibers making up whole muscle; blood vessels and nerves ramify through connective tissue
- Terms
  - *Epimysium*—surrounds whole muscle
  - *Perimysium*—surround bundles of muscle fibers; these bundles called fascicles
  - *Endomysium*—Surrounds individual muscle fibers

Nerve Supply of Muscles

- Muscles supplied with sensory and motor nerves
- *Sensory nerves* convey to the central nervous system the state of contraction of the muscle
- *Motor nerves* convey impulses from the central nervous system to muscles and control their contraction
- *Motor unit* is the single motor nerve and the bundle of muscle fibers it supplies
- Area where motor nerve ending meets muscle called neuromuscular junction (motor end plate, myoneural junction)
- Fine structural details of neuromuscular junction
  1. Nerve terminal oval to round structure; rests in specialized region of muscle membrane that is highly enfolded, called subneural clefts (synaptic folds, subneural apparatus)
  2. Nerve terminal contains mitochondria, few filaments and microtubules, and many synaptic vesicles that contain acetylcholine
  3. Acetylcholine released on stimulation of nerve; this causes muscle membrane to become excited, thereby initiating contraction
  4. Action of acetylcholine blocked by enzyme acetylcholinesterase, which is located at end plate region

Ultrastructure of Muscle

1. Each individual muscle fiber made up of many smaller myofibrils
2. Myofibrils made up of many myofilaments
3. Two type of myofilaments: *actin* and *myosin*
4. Arrangement of myofilaments gives rise to banding pattern seen in "striated" muscle
   - Regions seen in longitudinal sections
     - A-band   I-band
     - H-band   Z-line
     - *Sarcomere*—region between two Z-lines; the functional unit of skeletal muscle

Muscle Contraction		1. Sliding-filament hypothesis—change in relative length of I- and H-band and sarcomere when muscle, at rest, shortens, and is stretched 2. Interaction of actin and myosin at cross bridges of myosin filaments; process that requires ATP 3. Actin-myosin interaction requires calcium 4. Troponin and tropomyosin—regulator proteins present in actin filament 5. Troponin is receptive protein for calcium 6. Calcium stored in longitudinal elements of sarcoplasmic reticulum
Mechanics of Muscle Contraction	**The Stimulus**	Minimal stimulus—weakest stimulus to excite some motor units Subminimal stimulus—a stimulus with a value just below that of a minimal stimulus Maximal stimulus—excites all motor units Absolute refractory period—brief interval of time immediately following a contraction when no stimulus will cause contraction Relative refractory period—interval of depressed excitability All-or-none law—stimulus to a *single* muscle fiber will cause a maximal response or none at all
	Twitch response—contraction of muscle seen following a single maximal stimulus to either nerve or muscle—latent period; period of contraction; period of relaxation *Isotonic (twitch) contraction*—length changes, tension remains the same *Isometric (twitch) contraction*—same length, but tension changes *Summation*—a greater magnitude of tension or shortening due to rapid succession of 2 to 3 stimuli *Tetanus*—greatest magnitude of response, usually 3 to 4 times greater than twitch response; results from repetitive stimulation at a rate that maintains sustained maximal summation *Tonus*—a continual level of some degree of contraction of all muscles that gives a certain firmness and maintains a slight, steady pull on bony attachments	
Energy Sources for Muscle Contraction		Source of energy is ATP Means of providing continual supply of ATP: Creatine phosphate / Glycolysis / Oxidative phosphorylation (citric acid or Krebs cycle) Muscle fatigue due to accumulation of lactic acid, muscle in oxygen debt
Muscles and the Bony Levers		*Lever*—a rigid rod free to move about on some fixed point, the *fulcrum* Lever acted upon by: *Resistance* (weight)—something to be overcome or balanced / *Force* (effort)—that which is exerted to overcome the resistance Three types of levers: 1. Fulcrum lies between the effort and the resistance — Example, raising head 2. Resistance lies between fulcrum and effort 3. Effort exerted between fulcrum and resistance — Example, flexing forearm

Additional Readings

BLOOM, W., and FAWCETT, D. W.: *A Textbook of Histology,* 10th ed. W. B. Saunders Co., Philadelphia, 1975, pp. 295–315.

GREEP, R. O., and WEISS, L: *Histology,* 3rd ed. McGraw-Hill Book Co., New York, 1973, Chapter 7.

GUYTON, A. C.: *Basic Human Physiology: Normal Function and Mechanisms of Disease.* W. B. Saunders Co., Philadelphia, 1971, Chapters 6 and 7.

HOYLE, G.: How is muscle turned on and off? *Sci. Amer.,* **222:**84–93, April, 1970.

LESTER, H. A.: The response to acetylcholine. *Sci. Amer.,* **236:**106–18, February, 1977.

MURRAY, J. M., and WEBER, A.: The cooperative action of muscle proteins. *Sci. Amer.,* **230:**58–71, February, 1974.

VANDER, A. J.: SHERMAN, J. H.; and LUCIANO, D. S.: *Human Physiology: The Mechanisms of Body Function,* 2nd ed. McGraw-Hill Book Co., New York, 1975, Chapter 8.

Muscles of the Head, Neck, and Trunk: Blood Supply and Innervation

Chapter Outline

MUSCLES OF FACIAL EXPRESSION	MUSCLES OF ABDOMINAL WALL
MUSCLES OF MASTICATION	MOVEMENT OF THE VERTEBRAL COLUMN
EXTRINSIC MUSCLES OF THE TONGUE	MUSCLES OF RESPIRATION
MOVEMENT OF THE HEAD	BLOOD CIRCULATION TO THE MUSCLES OF
BLOOD CIRCULATION TO MUSCLES OF HEAD, FACE, AND NECK	RESPIRATION—MAIN BLOOD VESSELS

The muscles of the head, neck, and trunk have been placed in chart form for ease in learning the relationships of nerve and blood supply to muscle function. * Indicates muscles to be emphasized for first-level learning.

Muscles of Facial Expression

Expression	Muscle	Origin	Insertion	Nerve Supply	Function
Surprise	*Epicranius (occipito frontalis)	Occipital bone and mastoid portion of temporal bone	Galea aponeurotica (epicranial part)	Posterior auricular branches of facial nerve	Occipital portion draws scalp backward
	Frontal portion	Epicranial aponeurosis (galea aponeurotica)	Skin above supraorbital line	Temporal branch of facial nerve	Most powerful in raising eyebrows and wrinkling forehead
Smiling or laughing	Zygomaticus minor	Zygomatic bone and descends obliquely to its insertion	Upper lip between greater alar cartilage and orbicularis oris muscle	Buccal branches of facial nerve	Draws upper lip upward and outward

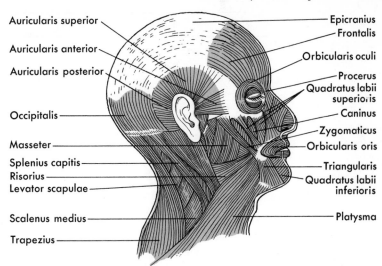

Auricularis superior
Auricularis anterior
Auricularis posterior
Occipitalis
Masseter
Splenius capitis
Risorius
Levator scapulae
Scalenus medius
Trapezius

Epicranius
Frontalis
Orbicularis oculi
Procerus
Quadratus labii superioris
Caninus
Zygomaticus
Orbicularis oris
Triangularis
Quadratus labii inferioris
Platysma

Figure 7–1. The superficial muscles of the head and neck.

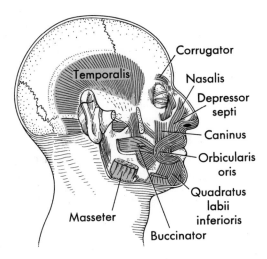

Corrugator
Nasalis
Depressor septi
Caninus
Orbicularis oris
Quadratus labii inferioris
Buccinator
Temporalis
Masseter

Figure 7–2. The temporal and deep muscles about the mouth.

Expression	Muscle	Origin	Insertion	Nerve Supply	Function
Sadness	Levator labii superioris	Below infraorbital foramen of maxilla	Orbicularis oris	Buccal branches of facial nerve	Elevates upper lip
Irony	Depressor labii inferioris	Mandible between symphysis and mental foramen	Integument, lower lip	Mandibular branch of facial nerve	Draws lower lip downward
	*Buccinator	Alveolar processes of the maxilla and mandible	Orbicularis oris at angle of mouth	Motor fibers Buccal branches from facial nerve	Principal muscle of cheek; compresses cheek; important in mastication
Laughing or smiling	Zygomaticus major	Zygomatic bone	Orbicularis oris at angle of mouth	Branches of facial nerve	Pulls angle of mouth upward and backward

Expression	Muscle	Origin	Insertion	Nerve Supply	Function
Doubt, disdain, contempt	Mentalis	Incisor fossa of the mandible	Skin of chin	Mandibular and buccal branches of facial nerve	Raises and protrudes lower lip
	*Orbicularis oris (ring-shaped muscle of the mouth)	Layers of muscle fibers surrounding the opening of the mouth	Angle of mouth	Buccal branches of facial nerve	Compresses and closes lips
	Depressor anguli oris (triangularis)	Oblique line of mandible	Orbicularis oris muscle	Mandibular, branches of facial nerve	Depresses angle of mouth
	Levator anguli oris (caninus)	Canine fossa below infraorbital foramen	Angle of mouth	Buccal branches of facial nerve	Furrow is deepened into expression of disdain
Horror	*Platysma (broad sheet muscle)	Skin and fascia covering pectoral and deltoid muscles	Mandible and muscles about angle of mouth	Cervical branch of facial nerve	Draws outer part of lower lip inferiorly and posteriorly, widens aperture at corner of mouth
Strain and tenseness	Risorius	Fascia over masseter muscle	Skin at angle of mouth	Fibers of mandibular and buccal branches of facial nerve	Retracts angle of mouth
Frowning, suffering	Corrugator supercilii	Medial end of superciliary arch (fibers pass upward and lateralward)	Deep surface of skin above supraorbital arch	Branches of temporal and zygomatic branches of facial nerve	Frowning and principal muscle in expressions of suffering
Muscles of the nose	Nasalis, depressor septi, and procerus	Small muscles of nose cover nasal bones	In skin, lower part of forehead, and nose	Buccal branches of facial nerve	Constrict and enlarge apertures of the nares

Muscles of Mastication

Name	Origin	Insertion	Nerve Supply	Function
Masseter	Zygomatic process of maxilla and zygomatic arch	Superior half of ramus and lateral surface of the mandible	Masseteric nerve from mandibular division of trigeminal nerve	Closes jaws
Temporalis	Temporal fossa of skull and from deep surface of temporal fascia	Medial surface, apex, and coronoid process of mandible	Deep temporal nerves from mandibular division of trigeminal nerve	Raises mandible and closes mouth, draws mandible backward

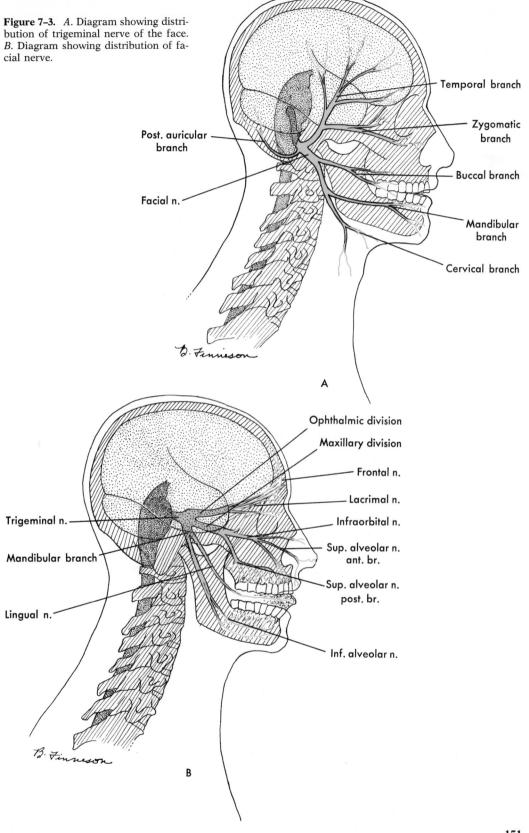

Figure 7–3. *A.* Diagram showing distribution of trigeminal nerve of the face. *B.* Diagram showing distribution of facial nerve.

Temporal branch

Zygomatic branch

Post. auricular branch

Buccal branch

Facial n.

Mandibular branch

Cervical branch

A

Ophthalmic division

Maxillary division

Frontal n.

Lacrimal n.

Trigeminal n.

Infraorbital n.

Sup. alveolar n. ant. br.

Mandibular branch

Sup. alveolar n. post. br.

Lingual n.

Inf. alveolar n.

B

Name	Origin	Insertion	Nerve Supply	Function
Pterygoideus medialis (internal pterygoid)	Medial surface, lateral pterygoid plate; pyramidal process of palatine bone; tuberosity of maxilla	Ramus of mandible Inferior and posterior part of ramus and mandibular foramen	Medial pterygoid, branch of mandibular division of trigeminal nerve	Raises mandible and closes mouth
Pterygoideus lateralis (external pterygoid)	Two heads: upper head from great wing of sphenoid; lower head from lateral surface of pterygoid plate	Upper head: condyle of mandible Lower head: articular disk of temporal mandibular articulation	Lateral pterygoid nerve from mandibular division of trigeminal nerve	Opens jaw; protrudes mandible; moves mandible from side to side

Extrinsic Muscles of the Tongue

These muscles are concerned with speaking, mastication, and swallowing.

Name	Origin	Insertion	Nerve Supply	Function
Genioglossus	Superior mental spine of mandible	Entire length of undersurface of tongue and, by a thin aponeurosis, upper part of body of hyoid bone	Hypoglossal	Thrusts tongue forward and depresses it
Styloglossus	Styloid process of temporal bone	Whole length of side and under part of tongue	Hypoglossal	Draws tongue upward and backward
Hyoglossus	Side and body of greater cornu of hyoid bone	Fibers pass upward and enter side of tongue between other muscles	Hypoglossal	Depresses tongue and draws down its sides
Stylohyoid (stylohyoideus)	Posterior and lateral surface of styloid process	Body of hyoid bone	Branch of facial nerve	Draws hyoid bone superiorly and posteriorly

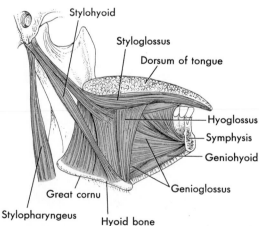

Figure 7–4. The muscles of the tongue viewed from the right side.

Name	Origin	Insertion	Nerve Supply	Function
Geniohyoid (geniohyoideus)	Inferior mental spine, inner surface	Anterior surface, body of hyoid bone	Branch of first cervical nerve through hypoglossal	Draws tongue and hyoid bone anteriorly

Movement of the Head

Muscles	Origin	Insertion	Nerve Supply	Function
*Sternocleidomastoideus (two heads)	Upper sternum and inner border of the clavicle	Inserted by a strong tendon into the lateral surface of the mastoid process	Spinal part of the accessory nerve and anterior branches of the second and third cervicals	One side flexes cervical vertebral column laterally and rotates it; both muscles acting together flex cervical vertebral column bringing head ventrally, at the same time elevating chin
Splenius capitis	Lower half ligamentum nuchae and spinous processes of the seven cervical vertebrae and upper four thoracic vertebrae	Outer part of occipital bone and mastoid process of temporal bone	Lateral branches of dorsal division of the middle and lower cervical nerves	When both muscles act together, head is pulled backward; when they act alone, head is rotated to same side
Semispinalis capitis	Series of tendons from transverse processes of first, six, or seven thoracic, and seventh cervical vertebrae	Occipital bone	Branches of dorsal primary divisions of middle and lower cervical nerves	Extends head and rotates it to opposite side

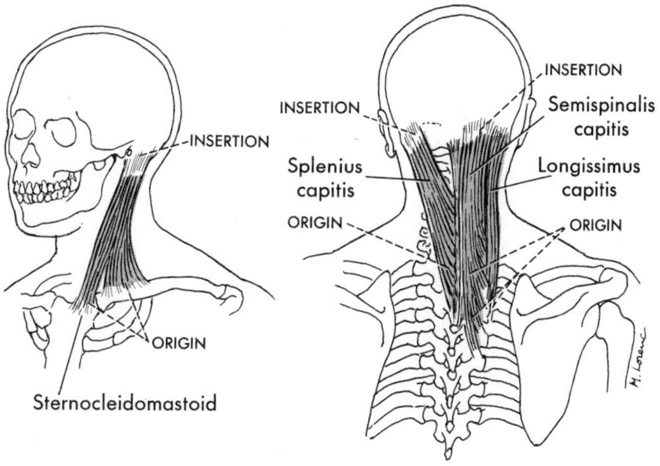

Figure 7–5. The muscles for flexion and extension of the head.

Muscles	Origin	Insertion	Nerve Supply	Function
Longus capitis	Transverse processes of upper four thoracic vertebrae	Posterior margin of mastoid process	Branches of dorsal division of middle and lower cervical nerves	When both muscles contract, head is extended; when they act alone, head is bent to same side

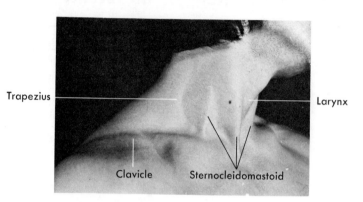

Figure 7–6. Surface anatomy of the neck.

BLOOD CIRCULATION TO MUSCLES OF HEAD, FACE, AND NECK

Arteries

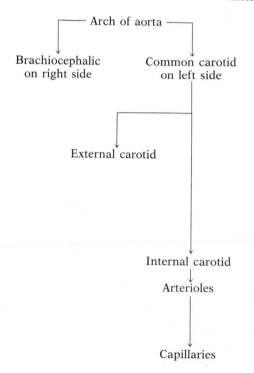

The left common carotid is an inch or two longer than the right. The right common carotid is a division of the brachiocephalic artery. They ascend obliquely on either side of the neck until on a level with the upper border of the thyroid cartilage where they divide into two great branches: (1) the external carotid, and (2) the internal carotid. These subdivide into many branches.

This artery is more superficial. Each side has eight main branches, which continue to divide into smaller branches. Each one is usually named in relation to the part supplied. These arteries supply blood to all the muscles of the face, scalp, and tongue and to the thyroid and parathyroid gland.

Each internal carotid has many branches. Important branches are the cerebral, distributed to the brain, and the ophthalmic, which enters the orbit through the optic foramen and distributes branches to the orbit, the muscles, and bulb of the eye.

See Figure 11–19 (page 254).

Veins

VENOUS RETURN

Venules

↓

Veins

↓

External jugular

↓

Subclavian vein

↓

Superior vena cava

↓

Right atrium

Tributaries to the external jugular veins are usually named in terms of structures from which they receive blood.

The external jugular veins are the chief superficial veins of the neck. They are formed in the substance of the parotid glands by the union of the posterior facial and the posterior auricular veins of each side of the face on a level with the angle of the mandible. Each vein terminates in the subclavian vein. The external jugular receives most of the blood from the greater part of the scalp and deep parts of the face. Blood from the tongue and all other deep structures of the face and neck enters the internal jugular vein → brachiocephalic → superior vena cava → right atrium.

Muscles of Abdominal Wall

KNOW NAMES, POS+ ACTIONS.

Muscle	Origin	Insertion	Nerve Supply	Function
*Obliquus externus (external oblique)	External surfaces and inferior border of lower eight ribs	Anterior half of outer lip of iliac crest; anterior rectus sheath to linea alba	Iliohypogastric, ilioinguinal nerves; branches from eighth through twelfth intercostal nerves	Compresses abdominal contents in forced expiration; one side alone bends vertebral column laterally

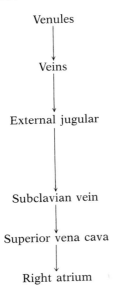

Pectoralis major

Serratus anterior

Rectus abdominis

External oblique

Internal oblique

External intercostals

Transversus abdominis

Posterior layer of rectus sheath

Figure 7–7. The abdominal muscles.

Muscle	Origin	Insertion	Nerve Supply	Function
*Obliquus internus (internal oblique)	Iliac crest; lumbodorsal fascia and inguinal ligament	Costal cartilages of last three or four ribs	Iliohypogastric, ilioinguinal nerves; branches of eighth through twelfth intercostals	Compresses abdominal contents in forced expiration; one side acting alone bends vertebral column laterally
*Transversus abdominis (transversalis)	Lateral third of inguinal ligament and anterior three fourths of inner lip of iliac crest, lumbodorsal fascia, inner surface of cartilages of lower six ribs	Xiphoid process Linea alba and pubis	Iliohypogastric Ilioinguinal nerve; branches of seventh to twelfth intercostals	Constricts abdomen; in forced expiration

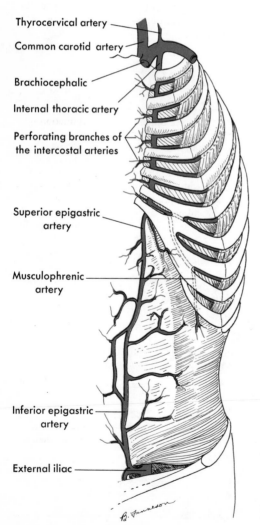

Thyrocervical artery

Common carotid artery

Brachiocephalic

Internal thoracic artery

Perforating branches of the intercostal arteries

Superior epigastric artery

Musculophrenic artery

Inferior epigastric artery

External iliac

Figure 7–8. Diagram showing blood supply to intercostal muscles and abdominal wall.

Muscle	Origin	Insertion	Nerve Supply	Function
*Rectus abdominis	Crest of pubis and ligaments; ventral surface of symphysis pubis	Cartilages of fifth, sixth, and seventh ribs and xiphoid process	Branches of seventh through twelfth intercostal nerves	Flexes vertebral column (lumbar portion); tenses anterior abdominal wall; compresses abdominal contents
*Levator ani *Note: not muscle of abdominal wall; creates floor of abdominopelvic cavity*	Posterior surface of body of pubic bone, spine of ischium and obturator fascia	Side of coccyx and fibrous band, which extends between coccyx and anus	Branches of pudendal plexes; fibers from fourth and sometimes third and fifth sacral nerves	Supports and slightly raises pelvic floor; resists increased intra-abdominal pressure

For arterial blood supply to these muscles and venous return see Figure 7–8.

Weak places in abdominal walls are the inguinal canal (fascia of transverse muscle, between superior spine of ilium and symphysis pubis), femoral ring, and umbilicus.

Movement of the Vertebral Column

Forward and backward movement of the spine is limited in the thoracic region, but movement is free in the lumbar region, particularly between the fourth and fifth lumbar vertebrae.

Muscle	Origin	Insertion	Nerve Supply	Function
*Quadratus lumborum (rectangular muscle)	Iliac crest and iliolumbar ligament	Inferior border of last rib and first four lumbar vertebrae	Branches of twelfth thoracic and first lumbar nerves	Flexes vertebral column laterally; fixes last two ribs in forced expiration
*Sacrospinalis	Anterior surface of a broad, thick tendon that is attached to the middle crest of sacrum; spinous processes of lumbar and eleventh and twelfth thoracic vertebrae	Fibers form a large mass of muscular tissue, which splits in the upper lumbar region into three columns: 1. Lateral, iliocostalis 2. Intermediate, longissimus 3. Medial, spinalis These are all attached to the ribs and vertebrae at various levels; see Figure 7–10	Lateral branches and posterior divisions of spinal nerves	Maintains vertebral column in erect posture against gravity

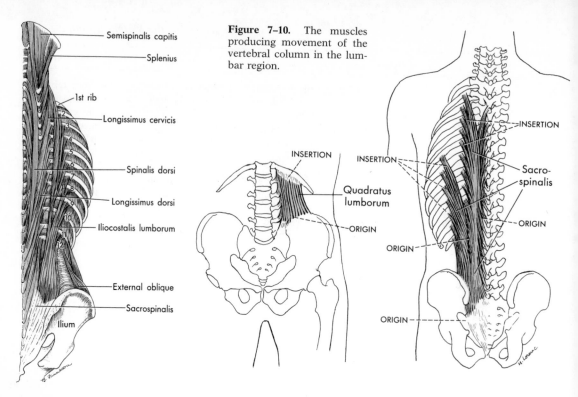

Figure 7–10. The muscles producing movement of the vertebral column in the lumbar region.

Semispinalis capitis

Splenius

1st rib

Longissimus cervicis

Spinalis dorsi

Longissimus dorsi

Iliocostalis lumborum

External oblique

Sacrospinalis

Ilium

INSERTION

INSERTION

INSERTION

Quadratus lumborum

Sacrospinalis

ORIGIN

ORIGIN

ORIGIN

ORIGIN

Figure 7–9. Diagram showing location of the deep muscles of the back.

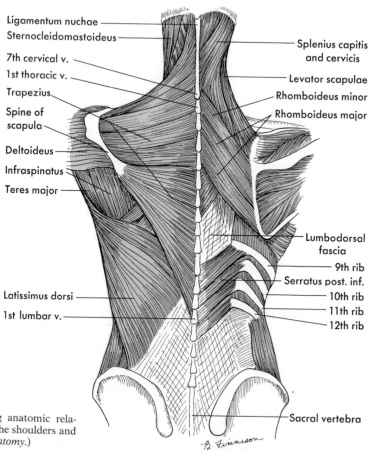

Ligamentum nuchae

Sternocleidomastoideus

7th cervical v.

1st thoracic v.

Trapezius

Spine of scapula

Deltoideus

Infraspinatus

Teres major

Latissimus dorsi

1st lumbar v.

Splenius capitis and cervicis

Levator scapulae

Rhomboideus minor

Rhomboideus major

Lumbodorsal fascia

9th rib

Serratus post. inf.

10th rib

11th rib

12th rib

Sacral vertebra

Figure 7–11. Diagram showing anatomic relations of some of the muscles of the shoulders and back. (Modified from *Gray's Anatomy*.)

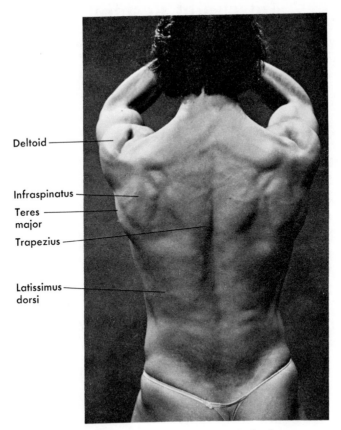

Figure 7–12. Surface anatomy of the back.

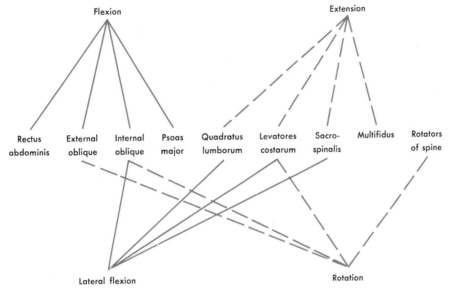

Figure 7–13. Summary of muscle action of the vertebrae.

Muscles of Respiration

Name	Origin	Insertion	Nerve Supply	Function
*Diaphragm—principal openings: Aortic Vena caval Esophageal	Sternal part: from dorsal side of xiphoid process Costal part: from costal cartilages and bone of ribs 7–12 Lumbar part: from lumbocostal arches and crura; see Figure 7–14	Central tendon—muscle fibers converge and become tendinous near central part of diaphragm	Phrenic nerve (derived from third, fourth, and fifth cervical nerves)	Principal muscle of respiration; during inspiration, central tendon pulled downward; vertical diameter (volume) of thoracic cavity increased; pressure in thoracic cavity decreased; air "pulled" into the lungs; pressure in abdominal cavity increased
*External intercostals (11 on either side)	Lower border of rib (caudal aspect); fibers run downward and forward	Upper border of rib below origin (cranial aspect)	Intercostal nerves	During inspiration, lifts ribs, increases volume of thoracic cavity
*Levatores costarum (12 on either side)	Ends of transverse processes of seventh cervical and upper II thoracic vertebrae	Outer surface of rib immediately caudal (below) to vertebrae from which it arises	Branches of intercostal nerves	Raises the ribs, increasing thoracic cavity

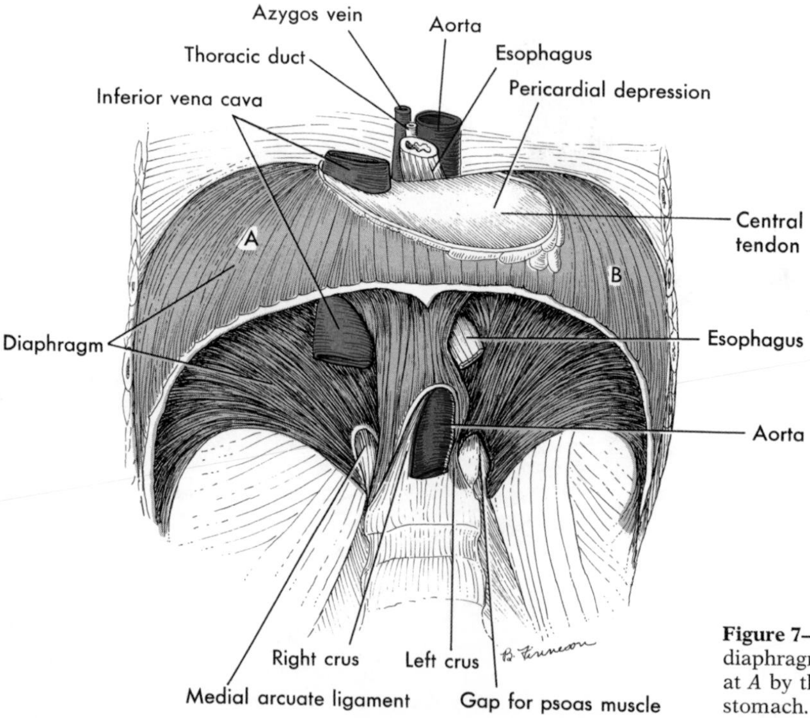

Figure 7–14. Frontal view of the diaphragm. It is slightly elevated at *A* by the liver and at *B* by the stomach.

Name	Origin	Insertion	Nerve Supply	Function
*Internal intercostals (11 on either side)	Ridge on inner surface of rib and corresponding costal cartilage; fibers run downward and backward	Cranial border of rib below origin	Intercostal nerves	In expiration, draws adjacent ribs together; decreases volume of thoracic cavity
The scaleni, anterior, medial, posterior	Transverse process of cervical and upper second or third thoracic vertebrae	Inserted into first and second ribs	Fourth, fifth, sixth, seventh, and eighth cervical nerves	In forced inspiration, raises first two ribs, increasing thoracic cavity
Serratus posterior superior	Caudal part of ligamentum nuchae and spines of seventh and first two thoracic vertebrae	Cranial borders of second, third, fourth, and fifth ribs	Branches of ventral divisions of first four thoracic nerves	Raises ribs, increasing thoracic cavity
Serratus posterior inferior	Spinous processes of last two thoracic and first two or three lumbar vertebrae	Inferior borders of last four ribs	Branches of ventral divisions of ninth to twelfth thoracic nerves	Draws these ribs outward and downward, counteracting inward pull of diaphragm
Transverse thoracic	Inner surface body of sternum, dorsal surface, xiphoid process, sternal ends of costal cartilages of last three or four true ribs	Caudal borders and inner surfaces of costal cartilages of second, third, fourth, fifth, and sixth ribs	Branches of intercostal nerves	Draws ventral part of ribs downward

Also see muscles of neck for action in respiration

BLOOD CIRCULATION TO THE MUSCLES OF RESPIRATION—MAIN BLOOD VESSELS

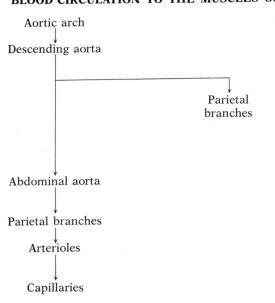

Aortic arch

↓

Descending aorta

Parietal branches

Abdominal aorta

↓

Parietal branches

↓

Arterioles

↓

Capillaries

The descending aorta extends from the lower border of the fourth to the twelfth thoracic vertebrae where it passes through the aortic hiatus in the diaphragm. There are *visceral* branches and *parietal* branches.

These include the *posterior intercostals, subcostals,* and *superior phrenic* arteries, which supply blood to the muscles of the thoracic wall, which include the diaphragm and other muscles of respiration.

This is a continuation of the thoracic aorta. There are both visceral and parietal branches.

These include the *inferior phrenic, lumbar,* and middle *sacral* arteries. They supply blood to the diaphragm, intercostal, and other body wall muscles. The diaphragm also receives blood from the musculophrenic branch of the internal thoracic artery.

Venous Return

Intercostal veins

Azygos veins

Inferior vena cava

Right atrium

The intercostal veins accompany the intercostal arteries and unite on the left side with the *azygos vein*, and on the right side with the *hemiazygos* and *accessory hemiazygos veins*. (See Figure 16–13, page 370.)

The azygos and hemiazygos veins receive the ascending lumbar veins from the abdominal region. In case of obstruction of the vena cava, the azygos veins form a channel by which blood from the lower part of the body can enter the superior vena cava.

Questions for Discussion

1. a. What is the principal muscle of respiration?
   b. Where are its origin and insertion?
   c. Which spinal nerves form the phrenic nerve?
2. What important anatomic structures pass through the diaphragm? What is their function?
3. Discuss the importance of the muscles of facial expression for the nurse.
4. Which muscles are used in forced respiration?

5. An individual has had a cerebral accident and has lost function of the left side of his face. What nerve or nerves are involved?
6. Discuss the nerve supply to the following structures:
   a. Tongue   d. Abdominal muscles
   b. Neck   e. Intercostal muscles
   c. Forehead   f. Face and head
7. Discuss in detail the blood supply to the head and face.
8. What is the blood supply to the intercostal muscles?

Additional Readings

BASMAJIAN, J. V.: *Grant's Method of Anatomy*, 9th ed. Williams & Wilkins Co., Baltimore, 1975, Sections 3, 6, and 7.

GARDNER, E.; GRAY, D. J.; and O'RAHILLY, R.: *Anatomy—A Regional Study of Human Structure*, 4th ed. W. B. Saunders Co., Philadelphia, 1975, Parts 4, 5, and 8.

HOLLINSHEAD, H. W.: *Textbook of Anatomy*, 3rd ed. Harper & Row, New York, 1974, Chapters 14, 19, 20, 23, 24, and 25.

PANSKY, B., and HOUSE, E. L.: *Review of Gross Anatomy*, 3rd ed. Macmillan Publishing Co., Inc., New York, 1975, Units 1, 2, 4, and 5.

ROMANES, G. J. (ed.): *Cunningham's Textbook of Anatomy*, 11th ed. Oxford University Press, New York, 1972, pp. 259–301.

WOODBURNE, R. T.: *Essentials of Human Anatomy*, 5th ed. Oxford University Press, New York, 1973, Sections III, V, and VI.

Muscles of the Extremities;
Blood Supply and Innervation

Chapter Outline

The muscles of the extremities have been placed in chart form for ease in learning the relationships of nerve and blood supply to muscle function. * Indicates muscles to be emphasized for first-level learning.

Movement of the Shoulder Girdle

Name	Origin	Insertion	Nerve Supply	Function
Flexion Levator scapulae	Upper four or five cervical vertebrae	Vertebral border of scapula	Dorsal scapular branches of third and fourth cervical nerves	Raises scapula, draws it medially, and rotates it to lower the lateral angle
Adduction *Rhomboideus major	Spines of first four or five thoracic vertebrae	Vertebral border of scapula between root of spine and inferior angle of scapula	Dorsal scapular nerve derived from fifth cervical nerve	Moves scapula backward and upward, producing slight rotation

Name	Origin	Insertion	Nerve Supply	Function
Rhomboideus minor	Inferior part of ligamentum nuchae and spinus processes of seventh cervical and first thoracic vertebrae	Vertebral border of scapula at root of spine	Dorsal scapular derived from fifth cervical nerve	Adducts scapula (backward and upward movement) and assists in adduction of arm
*Trapezius—flat, triangular muscle	Occipital bone, ligamentum nuchae, spinous processes of seventh cervical and spinous processes of all thoracic vertebrae	Acromial process of clavicle and spine of scapula	Spinal accessory and branches of third and fourth cervical nerves	All fibers rotate scapula; lower fibers draw scapula downward; upper fibers of one side draw head toward same side
*Pectoralis minor	Upper margins and outer surfaces of third, fourth, and fifth ribs and aponeuroses covering intercostals	Medial border and superior surface of coracoid process of scapula	Medial pectoral nerve (derived from eighth cervical and first thoracic nerves)	Draws scapula downward and rotates it to lower the lateral angle; raises third, fourth, and fifth ribs in forced inspiration when scapula is fixed
Subclavius	First rib and its cartilage	Groove on inferior surface of the clavicle	A special nerve from lateral trunk of brachial plexus (derived from fifth and sixth cervical nerves)	Draws shoulder forward and downward
*Serratus anterior	Outer surfaces and superior borders of first eight or nine ribs and from intercostals between ribs	Ventral surface of medial angle of vertebral border and inferior angle of scapula	Long thoracic nerve (derived from fifth, sixth, and seventh cervical nerves)	Rotates scapula; moves scapula forward as in act of pushing; assists in abduction of arm

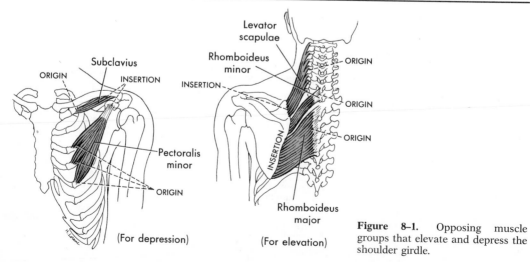

Figure 8–1. Opposing muscle groups that elevate and depress the shoulder girdle.

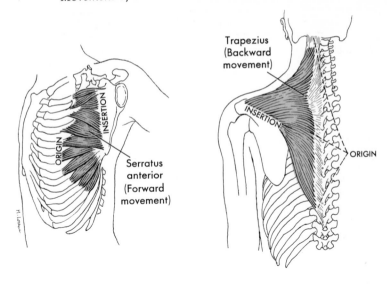

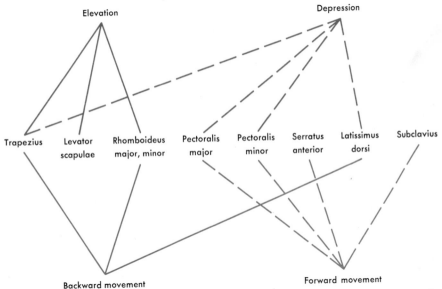

Figure 8–2. Opposing muscles that abduct and adduct the shoulders.

Figure 8–3. Summary of muscle action at the shoulder girdle, illustrating multiple functions of muscles.

Movement of Arm at Shoulder (Ball-and-Socket Joint)

Head of humerus fits into shallow glenoid cavity of scapula.

Name	Origin	Insertion	Nerve Supply	Function
Flexion *Coracobrachialis	Coracoid process of scapula	Middle of humerus at medial aspect	Musculocutaneous nerve (derived from sixth and seventh cervical nerves)	Carries arm forward (flexion); assists in adduction of arm

Name	Origin	Insertion	Nerve Supply	Function
Extension				
Teres major	Dorsal side, axillary border of scapula	Crest of lesser tubercle of humerus	Branches of lower subscapular nerve (derived from fifth and sixth cervical nerves)	Extends humerus and draws it down; helps to adduct and rotate arm medially
Abduction				
*Deltoideus, thick, triangular muscle, covers shoulder joint	Clavicle, acromion process, and posterior border of spine of scapula	Lateral surface of body of humerus	Branches of axillary nerve (derived from fifth and sixth cervical nerves)	Abduction (raises arm from side)
Supraspinatous	Fossa superior to spine of scapula, passes over shoulder joint	Inserted into highest facet of greater tubercle of humerus	Branches of suprascapular nerve (derived from fifth cervical nerve)	Assists deltoid in abduction of arm
*Pectoralis major, large, fan-shaped muscle	Anterior surface of sternal half of clavicle, ventral surface of sternum, and costal cartilages of ribs two to six	Fibers converge and form a thick mass that is inserted by a flat tendon into crest of greater tubercle of humerus	Lateral and medial pectoral nerves (derived from fifth, sixth, seventh, and eighth cervical, and first thoracic nerves)	Flexes, adducts, and rotates arm medially
Rotation				
Infraspinatus	Infraspinous fossa on posterior aspect of scapula	Middle facet of greater tubercle of humerus	Suprascapular nerve (derived from fifth and sixth cervical nerves)	Rotates arm outward
Teres minor, long, narrow muscle	Axillary border of scapula	Lowest facet of greater tubercle of humerus	Branches of axillary nerve (derived from fifth cervical nerve)	Functions with infraspinatus to rotate humerus laterally; weakly adducts it

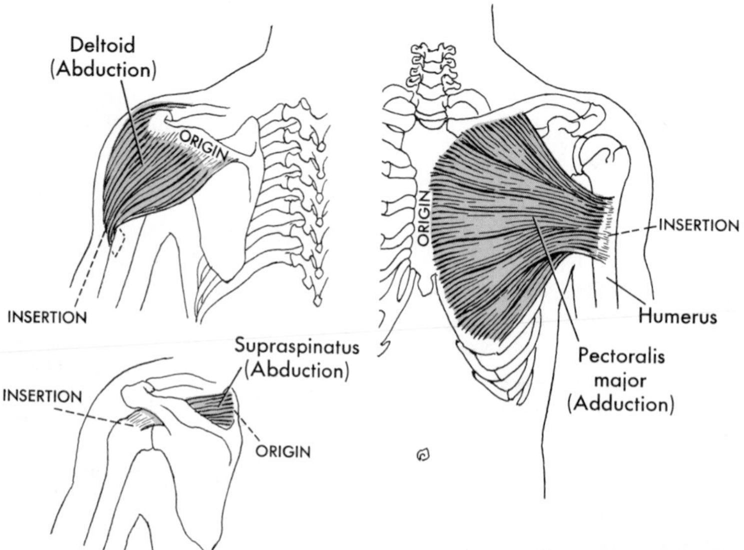

Figure 8–4. Muscles having opposing action for abduction and adduction of the humerus.

Name	Origin	Insertion	Nerve Supply	Function
*Latissimus dorsi, large, triangular muscle	Broad aponeurosis, which is attached to spinous processes of lower six thoracic vertebrae, spinous processes of lumbar vertebrae, crests of sacrum, and posterior part of crest of ilium and outer surface of lower four ribs	Fibers converge to form a flat tendon, which is inserted into bottom of intertubercular groove of humerus	Thoracodorsal nerve (derived from sixth, seventh, and eighth cervical nerves)	Rotates arm medially; extends and adducts humerus, draws shoulder downward and backward

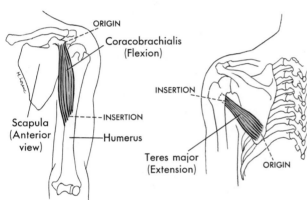

Figure 8–5. Opposing muscles that flex and extend the humerus.

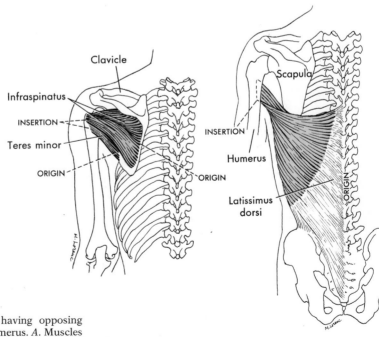

Figure 8–6. The muscles having opposing action for rotation of the humerus. *A.* Muscles for external rotation. *B.* Muscles for internal rotation.

A

B

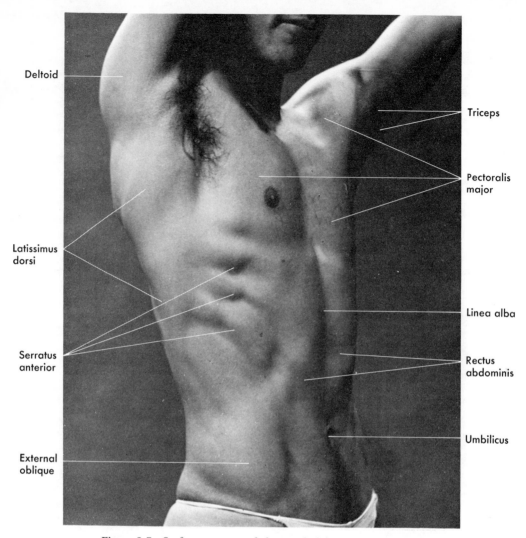

Deltoid

Triceps

Pectoralis
major

Latissimus
dorsi

Serratus
anterior

Linea alba

Rectus
abdominis

Umbilicus

External
oblique

Figure 8–7. Surface anatomy of chest and abdomen. Anterior view.

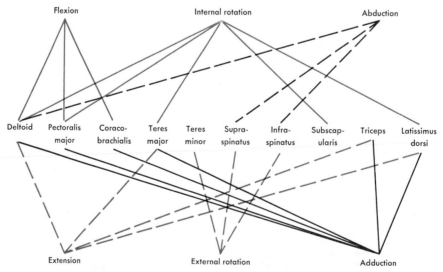

Flexion — Internal rotation — Abduction

Deltoid — Pectoralis major — Coraco-brachialis — Teres major — Teres minor — Supra-spinatus — Infra-spinatus — Subscap-ularis — Triceps — Latissimus dorsi

Extension — External rotation — Adduction

Figure 8–8. Summary of muscle action of the humerus.

Movement Between Humerus and Ulna (Elbow)

Name	Origin	Insertion	Nerve Supply	Function
*Brachialis	Distal half of anterior aspect of humerus	Tuberosity of ulna and rough depression, anterior surface of coronoid process	Branches of musculocutaneous nerve (derived from fifth and sixth cervical nerves)	Strong flexor of forearm
Triceps brachii (arises by three heads):		All fibers terminate in two aponeurotic laminae, which unite above the elbow and are inserted into proximal surface of olecranon of ulna	Branches of radial nerve (derived from seventh and eighth cervical nerves)	Great extensor of forearm; direct antagonist of brachialis; long head extends and adducts arm
Long head	Infraglenoid tuberosity of scapula			
Lateral head	Lateral side and posterior surface of body of humerus above radial groove			
Medial head	Posterior surface of body of humerus below radial groove			
*Biceps brachii				
Long head	Supraglenoid tuberosity	Common tendon inserts on rough posterior portion, tuberosity of radius	Branches of musculocutaneous nerve (derived from the fifth and sixth cervical nerves)	Flexes arm, flexes forearm, supinates hand; long head draws humerus toward glenoid fossa
Short head	Thick, flattened tendon from apex coracoid process	As above	As above	As above
Brachioradialis	Lower two thirds, ridge of humerus; lateral supracondyle ridge of humerus	Lateral side, base of styloid process of radius	Branches of radial nerve (derived from fifth and sixth cervical nerves)	Flexes forearm

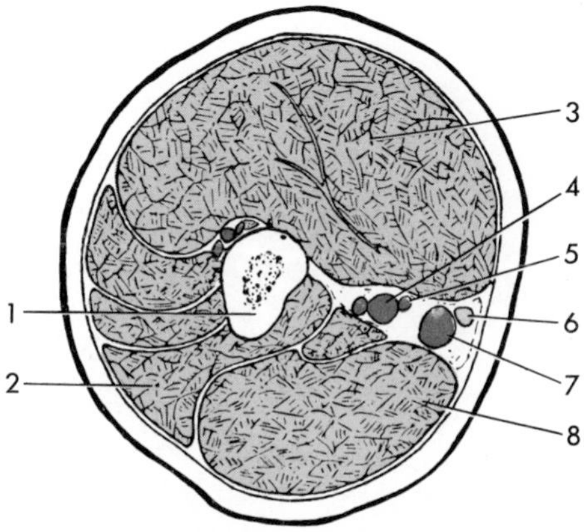

Figure 8–9. Transverse section through the middle of the right upper arm as seen from above. *1*, Humerus; *2*, brachialis muscle; *3*, triceps muscle; *4*, brachial artery; *5*, companion vein; *6*, ulnar nerve; *7*, basilic vein; *8*, biceps muscle.

Name	Origin	Insertion	Nerve Supply	Function
*Supinator	Lateral epicondyle of humerus and ridge of ulna below medial notch	Anterior and lateral margin of tuberosity and oblique line of radius	Branches of deep radial nerve (derived from fifth, sixth, and seventh cervical nerves)	Supination of hand
*Pronator teres, has two heads: humeral head, ulnar head	Medial epicondyle of humerus and coronoid process of ulna	Middle of lateral surface of body of radius	Branches of median nerve (derived from sixth and seventh cervical nerves)	Pronation of hand
Pronator quadratus	Distal part of volar surface of body of ulna	Volar surface of radius	Anterior interosseous branch of median nerve (derived from eighth cervical and first thoracic nerves)	Pronation of hand

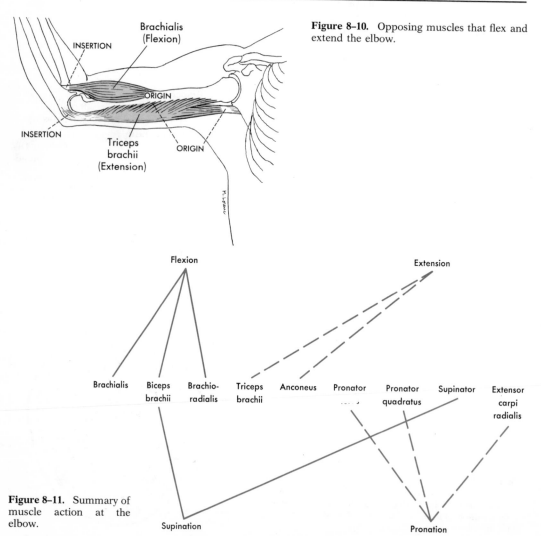

Figure 8–10. Opposing muscles that flex and extend the elbow.

Figure 8–11. Summary of muscle action at the elbow.

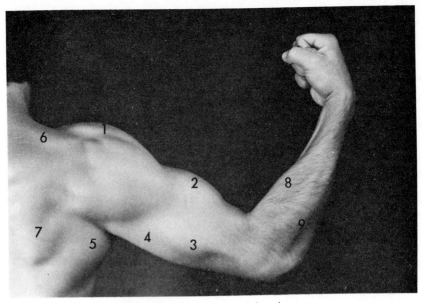

Figure 8–12. Surface anatomy of upper extremity, posterior view.

1. Deltoid
2. Biceps
3. Triceps, lateral head
4. Triceps, long head
5. Latissimus dorsi

6. Trapezius
7. Infraspinatus
8. Brachioradialis
9. Extensors

(From Drakontides, A. B.; Miller, M. A.; and Leavell, L. C.: *Anatomy and Physiology: Workbook and Laboratory Manual,* 2nd ed. Macmillan Publishing Co., Inc., New York, 1977.)

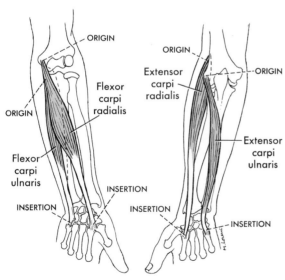

Figure 8–13. Opposing groups of muscles that flex and extend the wrist (left hand).

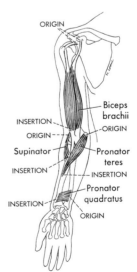

Figure 8–14. Opposing muscles that supinate and pronate the hand (right arm).

Movement of the Wrist

Name	Origin	Insertion	Nerve Supply	Function
*Flexor carpi radialis	Medial epicondyle of humerus	Base of second and third metacarpal bones	Branches of median nerve (derived from sixth and seventh cervical nerves)	Flexion of hand, assists in abduction
*Flexor carpi ulnaris	Medial epicondyle of humerus and upper two thirds of dorsal border of ulna	Pisiform, hamate, and fifth metacarpal bones	Branches of ulnar nerve (derived from eighth cervical nerve and first thoracic nerve)	Flexion of hand, assists in adduction of hand
*Extensor carpi radialis longus	Lower third lateral supracondylar ridge of humerus	Dorsal side of base of second metacarpal bone	Branch of radial nerve (derived from sixth and seventh cervical nerves)	Extends and abducts hand
Extensor carpi ulnaris	Lateral epicondyle of humerus and dorsal border of ulna	Ulnar side of base of fifth metacarpal bone	Branch of deep radial nerve (derived from sixth, seventh, and eighth cervical nerves)	Extends and adducts hand
*Extensor carpi radialis brevis	Lateral epicondyle of humerus	Dorsal surface of base of third metacarpal bone, radial side	Branches of radial nerve	Extends and may abduct hand

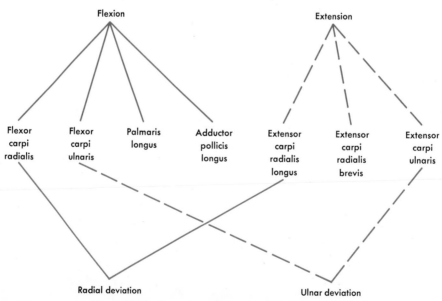

Figure 8–15. Summary of muscle action at the wrist.

Movement of the Fingers

Name	Origin	Insertion	Nerve Supply	Function
*Flexor digitorum profundus	Volar and medial surfaces of body of ulna and interosseous membrane	Bases of last phalanges	Branches of ulnar and median nerves (derived from eighth cervical and first thoracic nerves)	Flexes terminal phalanges of each finger and by continued action flexes other phalanges
Flexor digitorum superficialis	Medial epicondyle of humerus and medial side of coronoid process and oblique line of radius	Second phalanges of four fingers	Branches of median nerve (derived from seventh and eighth cervical and first thoracic nerves)	Flexes second phalanx of each finger; flexes hand
Extensor digitorum	Lateral epicondyle of humerus	Second and third phalanges of fingers	Branch of deep radial nerve (derived from sixth, seventh, and eighth cervical nerves)	Extends phalanges, then wrist
*Extensor indicis	Dorsal surface of body of ulna	Into tendon of extensor digitorum to index finger	Branch of deep radial nerve (derived from sixth, seventh, and eighth cervical nerves)	Extends index finger
Extensor digiti minimi	With extensor digitorum	First phalanx of little finger	Branch of deep radial nerve (derived from sixth, seventh, and eighth cervical nerves)	Extends little finger

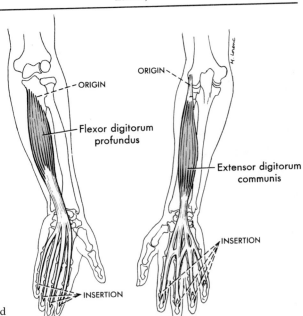

Figure 8–16. Opposing muscles that flex and extend the fingers (left hand).

Movement of the Thumb

Name	Origin	Insertion	Nerve Supply	Function
*Flexor pollicis longus	Volar surface of body of radius	Base of distal phalanx of thumb	Branch of palmar interosseous from median nerve (derived from eighth cervical and first thoracic nerves)	Flexes second phalanx of thumb and by continued action flexes first phalanx
Flexor pollicis brevis	Trapezium bone	Radial sides of and medial base of proximal phalanx of thumb	Branches of median and ulnar nerves (derived from eighth cervical and first thoracic nerves)	Flexes and adducts thumb
*Extensor pollicis longus	Lateral side of dorsal surface of body of ulna	Base of last phalanx of thumb	Branch of deep radial nerve (derived from sixth, seventh, and eighth cervical nerves)	Extends second phalanx of thumb
Extensor pollicis brevis	Dorsal surface of body of radius and interosseous membrane	Base of first phalanx of thumb	Branch of deep radial nerve (derived from sixth and seventh cervical nerves)	Extends first phalanx of thumb
Abductor pollicis longus	Lateral part of dorsal surface of body of ulna and interosseous membrane	Radial side of base of first metacarpal bone	Branch of deep radial nerve (derived from sixth and seventh cervical nerves)	Abducts thumb and, by continued action, the wrist

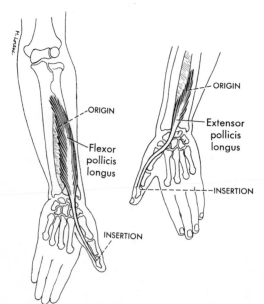

Figure 8–17. Opposing muscles that flex and extend the thumbs (left hand).

Figure 8–18. The muscles concerned with adduction and abduction of the thumb (left hand).

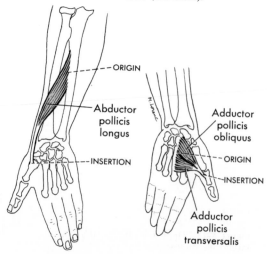

Name	Origin	Insertion	Nerve Supply	Function
Abductor pollicis brevis	Transverse carpal ligament and tuberosity of scaphoid and trapezium	Radial side of first phalanx of thumb	Branch of median nerves (derived from sixth and seventh cervical nerves)	Abducts thumb
Adductor pollicis	Distal two thirds of palmar surface of third metacarpal bone; capitate bone	Ulnar side of base of proximal phalanx of thumb	Branch of the deep palmar branch of ulnar nerve (derived from eighth cervical and first thoracic nerves)	Adducts thumb, brings thumb toward palm

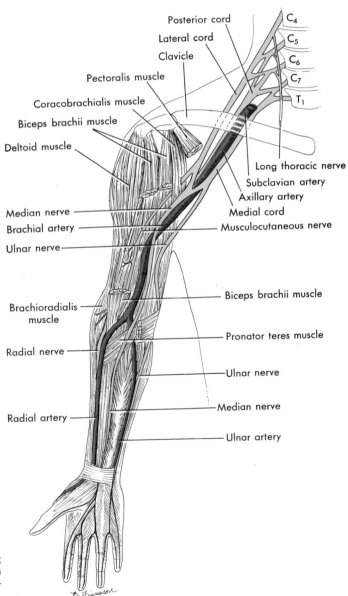

Figure 8–19. Diagram illustrating the brachial plexus and distribution of nerves. Note their relation to arteries and to muscles (right arm).

Name	Origin	Insertion	Nerve Supply	Function
*Opponens pollicis	Ridge of trapezium	Length of metacarpal bone of thumb	Branch of median nerve derived from sixth and seventh cervical nerves	Abducts, flexes, and rotates metacarpal bone of thumb

BLOOD CIRCULATION TO THE UPPER EXTREMITY—MAIN BRANCHES

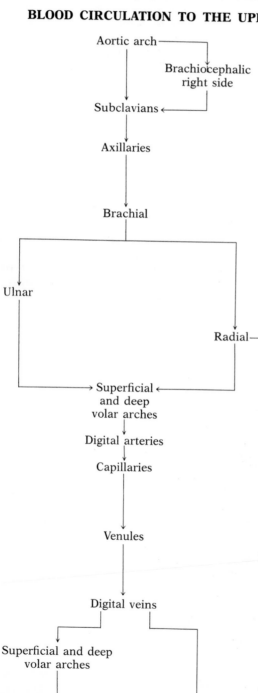

Aortic arch

Brachiocephalic right side

Subclavians

Axillaries

Brachial

Ulnar

Radial

Superficial and deep volar arches

Digital arteries

Capillaries

Venules

Digital veins

Superficial and deep volar arches

Branches to muscles of scapular region, shoulder, chest, and mammary glands.

These extend from outer border of first rib to lower border of tendon of teres major muscles where they become the brachial. Distribute branches to chest, shoulder, and arm.

Lies in the depression of the inner border of the biceps muscles and divides into two main branches. It distributes many branches to the muscles of the upper arm, both flexors and extensors.

Ulnar extends along ulnar border of forearm into palm of the hand. Supplies flexor and extensor muscles of wrist and hand. Extends along radial side of forearm. Several branches to muscles of forearm and comes to the surface at wrist—a pulse point.

The ulnar and radial arteries anastomose, forming arches that supply the hand with blood.

These go to each finger.

These function as a diffusion membrane so that substances needed by muscle cells and substances formed in them may enter or leave the bloodstream.

Venous Return
In all of the muscle cells, bones, joints, and other structures venules unite to form veins. Veins coalesce and form larger veins. Study Figure 8–22.

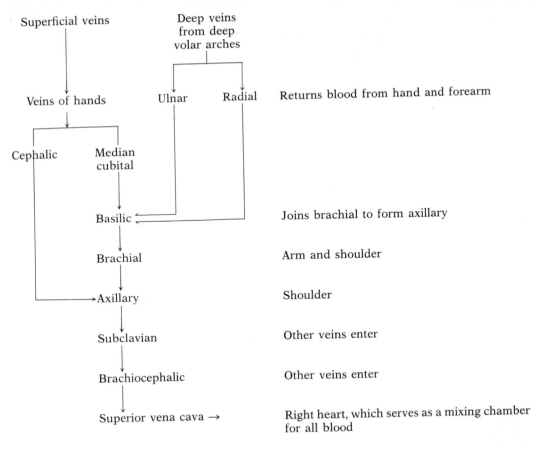

Superficial veins → Veins of hands

Deep veins from deep volar arches → Ulnar, Radial → Returns blood from hand and forearm

Cephalic

Median cubital

Basilic ← Joins brachial to form axillary

Brachial → Arm and shoulder

Axillary → Shoulder

Subclavian → Other veins enter

Brachiocephalic → Other veins enter

Superior vena cava → Right heart, which serves as a mixing chamber for all blood

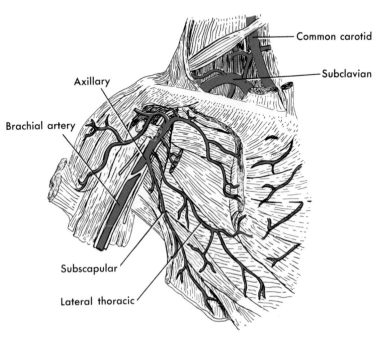

Common carotid

Subclavian

Axillary

Brachial artery

Subscapular

Lateral thoracic

Figure 8–20. Subclavian and axillary arteries.

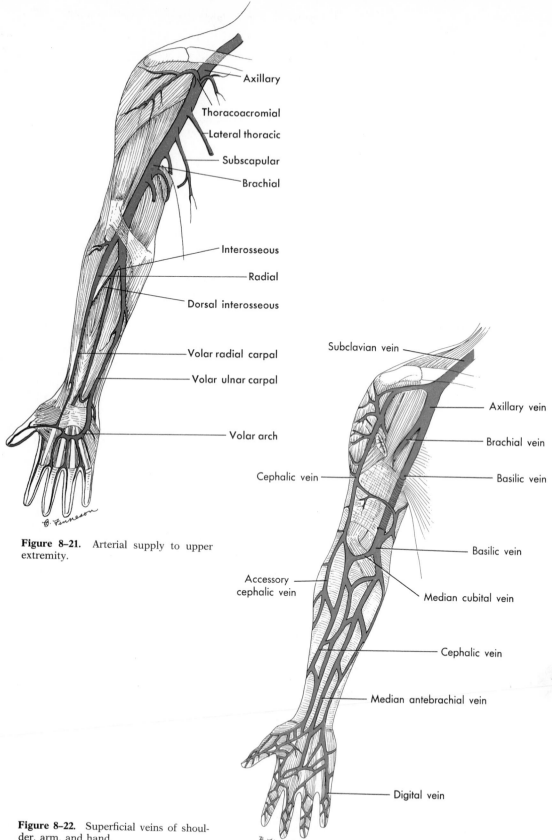

Figure 8–21. Arterial supply to upper extremity.

Axillary

Thoracoacromial

Lateral thoracic

Subscapular

Brachial

Interosseous

Radial

Dorsal interosseous

Volar radial carpal

Volar ulnar carpal

Volar arch

Subclavian vein

Axillary vein

Brachial vein

Cephalic vein

Basilic vein

Basilic vein

Accessory cephalic vein

Median cubital vein

Cephalic vein

Median antebrachial vein

Digital vein

Figure 8–22. Superficial veins of shoulder, arm, and hand.

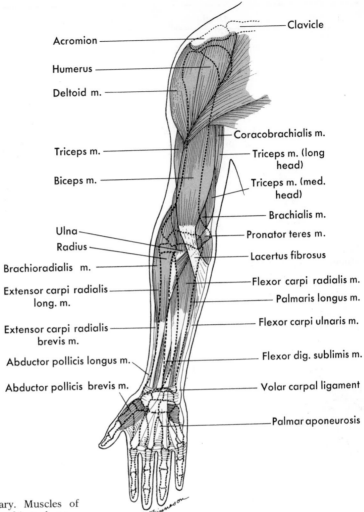

Figure 8–23. Diagram summary. Muscles of upper extremity, showing relationship to bone.

Movement at the Hip Joint (Ball-and-Socket Joint)

Name	Origin	Insertion	Nerve Supply	Function
Flexion				
*Psoas major (Psoas magnus)	Bodies and transverse processes of last thoracic and all lumbar vertebrae	Lesser trochanter of femur and iliac fossa	Lumbar nerves two and three	Flexes thigh on pelvis and helps to rotate thigh medially
Psoas minor (Psoas parvus)	Sides of bodies of T_{12} and L_1	Iliac fascia and body of femur	Branches of femoral nerve (fibers from first, second, and third lumbar nerves)	Acts with psoas major

Name	Origin	Insertion	Nerve Supply	Function
Extension				
*Gluteus max-imus	Iliac crest, sacrum, side of coccyx, and aponeuroses of sacrospinalis	Fascia lata and gluteal ridge of femur	Inferior gluteal (fibers from fifth lumbar and first and second sacral)	Extends femur and rotates it outward
Abduction				
Gluteus medius	Outer surface of ilium and gluteal aponeurosis covering it	Lateral surface of greater trochanter	Superior gluteal nerve (fibers from fourth and fifth lumbar and first sacral nerves)	Abduction of thigh and medial rotation
*Tensor fasciae latae	Anterior crest and spine of ilium	Iliotibial band of fascia lata	Superior gluteal nerve	Tenses fascia lata; flexes and abducts thigh
Adduction				
*Adductor longus	Anterior pubic bone			
Adductor brevis	Outer surface of inferior ramus of pubis	Linea aspera of femur	Anterior and posterior branches of obturator nerve (fibers from third and fourth lumbar nerves)	Adducts, flexes, and rotates thigh medially; lower portion of magnus is a powerful extensor
Adductor magnus	Inferior ramus of pubis and inferior ramus of ischium and ischial tuberosity			
Lateral Rotation				
Piriformis	Anterior surface of sacrum	Upper border of great trochanter	Branches from first and second sacral	Rotates thigh laterally abducts it
Quadratus femoris	Tuberosity of ischium	Upper part of linea quadrata	Special branch from sacral plexus (fibers from fourth and fifth lumbar and first sacral nerves)	Lateral rotation of thigh
Obturator				
Internus	Internal surface, rami of pubis and ischium	Forepart of great trochanter	Obturator nerve from lumbo-sacral plexus	Lateral rotation of thigh
Externus	Side obturator membrane and ischiopubic rami	Trochanteric fossa of femur	Branch of obturator nerve	Lateral rotation of thigh
Medial Rotation				
Gluteus medius (anterior part)	Outer surace of ilium	Lateral surface of great trochanter	Branches of superior gluteal (fibers from fourth and fifth lumbar and first sacral nerves)	Abducts thigh and rotates it medially
Gluteus mini-mus	Outer surface of ilium	Anterior border of great trochanter	Branch of superior gluteal (fibers from fourth and fifth lumbar and first sacral nerves)	Medial rotation of thigh and abduction

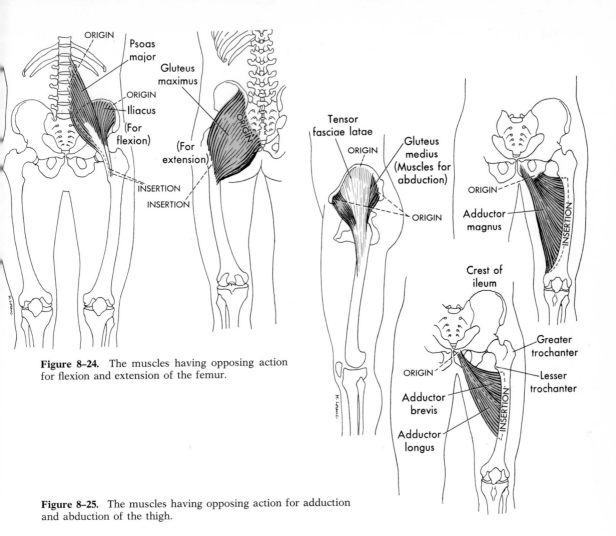

Figure 8–24. The muscles having opposing action for flexion and extension of the femur.

Figure 8–25. The muscles having opposing action for adduction and abduction of the thigh.

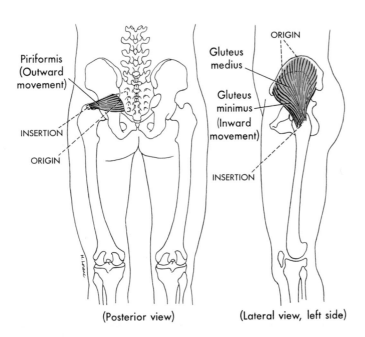

(Posterior view)   (Lateral view, left side)

Figure 8–26. Opposing muscles for rotation of the thigh. Other outward rotators are not shown.

Figure 8–27. Summary of muscle action at the hip.

Movement at the Knee Joint (Hinge Joint)

Name	Origin	Insertion	Nerve Supply	Function
Flexion				
*Biceps femoris				
Long head	Tuberosity of ischium	Lateral side, head of fibula	Tibial nerve (fibers from first three sacral nerves)	Flexes leg and extends thigh
Short head	Lateral side, linea aspera of femur	Lateral condyle of tibia	Peroneal (fibers from fifth lumbar and first two sacral nerves)	
*Semitendinosus	Tuberosity of ischium	Upper part of body of tibia	Tibial (fibers from fifth lumbar and first and second sacral nerves)	Flexes leg and extends thigh
*Semi-membranosus	Tuberosity of ischium	Medial condyle of tibia	Tibial (fibers from fifth lumbar and first and second sacral nerves)	Flexes leg and extends thigh
Popliteus	Lateral condyle of femur	Posterior surface of body of tibia	Tibial (fibers from fifth lumbar and first and second sacral nerves)	Flexes leg

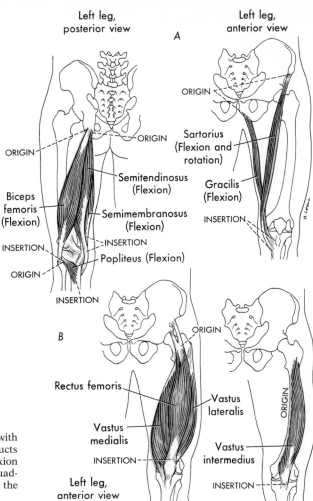

Left leg, posterior view

Left leg, anterior view

A

ORIGIN

ORIGIN

Sartorius (Flexion and rotation)

ORIGIN

Biceps femoris (Flexion)

Semitendinosus (Flexion)

Gracilis (Flexion)

Semimembranosus (Flexion)

INSERTION

INSERTION

INSERTION

Popliteus (Flexion)

ORIGIN

INSERTION

B

ORIGIN

Rectus femoris

Vastus lateralis

Vastus medialis

INSERTION

Vastus intermedius

ORIGIN

Left leg, anterior view

INSERTION

Figure 8–28. *A.* The muscles concerned with flexion of the knee. The gracilis also adducts the thigh, and the sartorius assists in flexion and lateral rotation of the thigh. *B.* The quadriceps femoris muscle for extension of the knee.

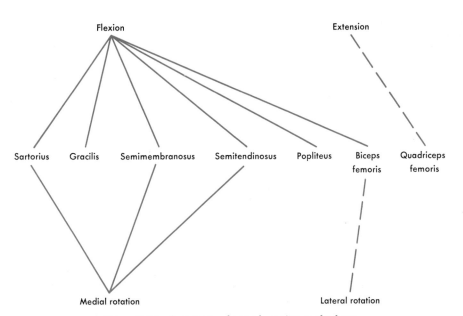

Flexion

Extension

Sartorius   Gracilis   Semimembranosus   Semitendinosus   Popliteus   Biceps femoris   Quadriceps femoris

Medial rotation

Lateral rotation

Figure 8–29. Summary of muscle action at the knee.

Name	Origin	Insertion	Nerve Supply	Function
*Gracilis	Symphysis pubis and pubic arch	Medial surface of tibia, below condyle	Obturator	Adducts thigh and flexes leg
*Sartorius	Anterior superior spine of ilium	Medial surface of body of tibia	Femoral nerve	Flexes leg, flexes thigh and rotates it laterally
Extension				
*Quadriceps femoris (four heads)				
1. Rectus femoris	Two tendons; one, anterior inferior iliac spine; the other, brim of acetabulum	The four tendons unite at distal part of thigh to form a strong tendon, which is inserted into base of patella and tuberosity of tibia	Branches of femoral nerve (fibers from second, third, and fourth lumbar nerves)	Entire quadriceps extends leg; rectus division also flexes thigh
2. Vastus lateralis	Great trochanter and linea aspera of femur			
3. Vastus medialis	Medial lip of linea aspera			
4. Vastus intermedius	Ventral and lateral surfaces of body of femur			

Movement at the Ankle Joint

Name	Origin	Insertion	Nerve Supply	Function
Plantar Flexion				
*Gastrocnemius	Two heads from medial and lateral condyles of femur and adjacent part of capsule of knee	Through tendocalcaneus to calcaneus	Tibial (fibers from first and second sacral nerves)	Plantar flexes foot (points toes downward)
*Soleus	Posterior surface, head of fibula and medial border of tibia	Through tendocalcaneus to calcaneus	Tibial (fibers from first and second sacral nerves)	Plantar flexes foot
Plantaris	Linea aspera of femur and oblique popliteal ligament of knee	Posterior part of calcaneus via tendocalcaneus	Tibial (fibers from fourth and fifth lumbar and first sacral nerves)	Plantar flexes foot
Peroneus longus	Head and lateral surface of body of fibula and lateral condyle of tibia	Lateral side, base of first metatarsal bone, and lateral side of first cuneiform bone	Branches of peroneal nerve (fibers from fourth and fifth lumbar and first sacral nerves)	Plantar flexes and everts foot
Peroneus brevis	Lateral surface, distal two thirds of body of fibula	Tuberosity at base of fifth metatarsal bone, lateral side	Branch of superficial peroneal nerve (fibers as above)	Plantar flexes and pronates (everts and abducts) foot

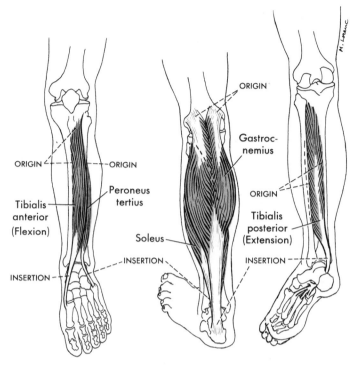

Figure 8–30. The muscles having opposing action for flexion and extension of the ankle.

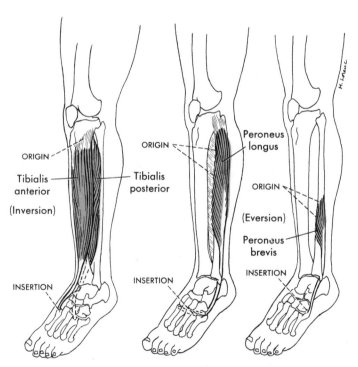

Figure 8–31. Opposing muscles for inversion (supination) and eversion (pronation) of the foot.

Name	Origin	Insertion	Nerve Supply	Function
Dorsiflexion				
*Tibialis anterior	Lateral condyle and upper portion of body of tibia and interosseous membrane	Undersurface of first cuneiform and base of first metatarsal	Branches of deep peroneal (fibers from fourth and fifth lumbar and first sacral nerves)	Dorsally flexes and supinates foot (adducts and inverts)
Peroneus tertius	Lower, medial surface of fibula and interosseous membrane	Medial and plantar surface of first cuneiform and base of fifth metatarsal bone	Branches of deep peroneal (fibers as above)	Dorsally flexes and pronates foot
*Posterior tibial (tibialis posterior)	Whole posterior surface of interosseous membrane, lateral side of posterior surface of body of tibia	Tuberosity of navicular bone, gives extension to third cuneiform, cuboid, and base of second, third, and fourth metatarsal bones	Branch of tibial nerve (derived from fifth lumbar and first sacral nerves)	Supinates (adducts and inverts foot) and plantar flexes foot

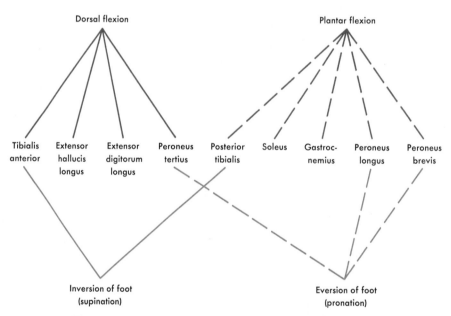

Figure 8–32. Summary of muscle action at the ankle and foot.

Deep Muscles of the Toes

Name	Origin	Insertion	Nerve Supply	Function
Flexor digitorum brevis	Medial process, tuberosity of calcaneus	Sides of second phalanx of lateral four toes	Branches of medial plantar nerve (fibers of fourth and fifth lumbar nerves)	Flexes second phalanges of the four small toes

Name	Origin	Insertion	Nerve Supply	Function
Abductor hallucis	Medial process, tuberosity of calcaneus	Tibial side of base of first phalanx of great toe	Medial plantar nerve (fibers of fourth and fifth lumbar nerves)	Abducts great toe
Abductor digiti minimi	Lateral process, tuberosity of calcaneus	Fibular side of base of first phalanx of little toe	Branches of lateral plantar nerve (fibers of first and second sacral)	Abducts small toe
Quadratus plantae	Two heads on either side of calcaneus	Tendons of flexor digitorum longus	Branches of lateral plantar nerve	Flexes terminal phalanges of the four small toes

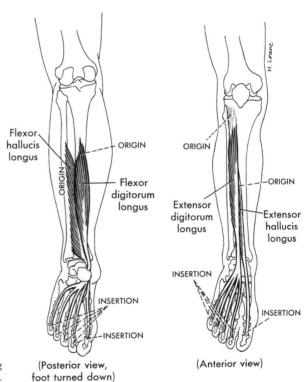

Figure 8–33. The muscles having opposing action for flexion and extension of the toes.

(Posterior view, foot turned down)

(Anterior view)

Movement of the Toes

Name	Origin	Insertion	Nerve Supply	Function
Flexion				
Flexor hallucis longus	Inferior two thirds of posterior surface of fibula	Base of terminal phalanx of great toe	Tibial nerve (fibers of fifth lumbar and first and second sacral nerves)	Flexes great toe, plantar flexes and inverts foot

Name	Origin	Insertion	Nerve Supply	Function
Flexor digitorum longus	Posterior surface of body of tibia	By four tendons into last phalanges of four outer toes	Tibial nerve	Flexes terminal phalanges of the four toes; plantar flexes and inverts foot
Extension				
Extensor hallucis longus	Inferior two thirds of posterior surface of body of fibula	Base of terminal phalanx of great toe	Branch of deep peroneal nerve (fibers of fourth and fifth lumbar and first sacral nerves)	Extends second phalanx of great toe; dorsally flexes and everts foot
Extensor digitorum longus	Lateral condyle of tibia and anterior surface of body of fibula	Second and third phalanges of the four lesser toes	Branches of deep peroneal (fibers of fourth and fifth lumbar and first sacral nerves)	Extends proximal phalanges of the four lesser toes; dorsally flexes and everts foot
Extensor digitorum brevis	Distal and superior lateral surfaces of calcaneus	Lateral sides of tendons of extensor digitorum longus of medial four toes	Branches of deep peroneal (fibers of fifth lumbar and first sacral nerves)	Extends proximal phalanges of the great and the adjacent three small toes

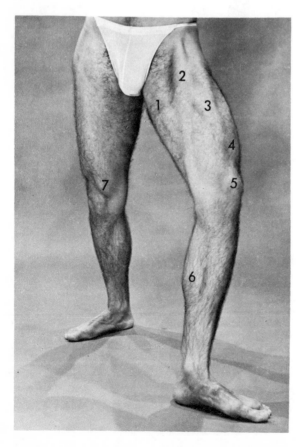

Figure 8–34. Surface anatomy of lower extremities, anterior aspect.
1. Adductor muscles
2. Sartorius
3. Rectus femoris
4. Vastus lateralis
5. Patella
6. Gastrocnemius
7. Vastus medialis
(From Drakontides, A. B.; Miller, M. A.; and Leavell, L. C.: *Anatomy and Physiology: Workbook and Laboratory Manual,* 2nd ed. Macmillan Publishing Co., Inc., New York, 1977.)

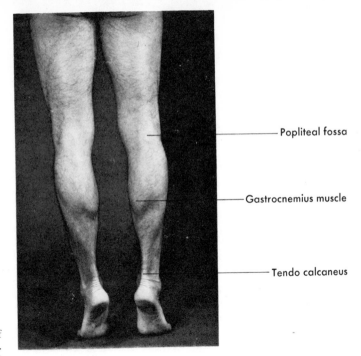

— Popliteal fossa

— Gastrocnemius muscle

— Tendo calcaneus

Figure 8–35. Surface anatomy of lower extremities. Posterior aspect.

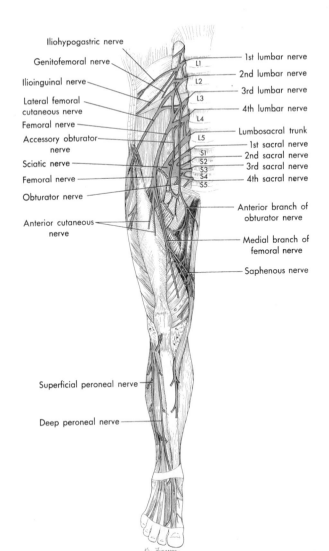

Iliohypogastric nerve

Genitofemoral nerve

Ilioinguinal nerve

Lateral femoral cutaneous nerve

Femoral nerve

Accessory obturator nerve

Sciatic nerve

Femoral nerve

Obturator nerve

Anterior cutaneous nerve

Superficial peroneal nerve

Deep peroneal nerve

L1 — 1st lumbar nerve
L2 — 2nd lumbar nerve
L3 — 3rd lumbar nerve
L4 — 4th lumbar nerve
— Lumbosacral trunk
L5 — 1st sacral nerve
S1 — 2nd sacral nerve
S2 — 3rd sacral nerve
S3 — 4th sacral nerve
S4
S5

— Anterior branch of obturator nerve

— Medial branch of femoral nerve

— Saphenous nerve

Figure 8–36. Diagram illustrating the lumbosacral plexus and distribution of nerves. Note their relation to muscles of the leg (right).

Figure 8–37. Nerve supply to right lower extremity. Posterior view.

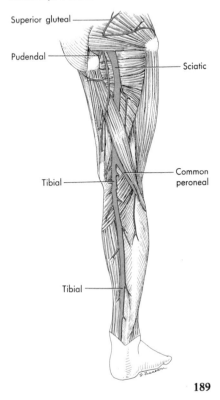

Superior gluteal —

Pudendal —

— Sciatic

Tibial —

— Common peroneal

Tibial —

189

BLOOD CIRCULATION TO THE LOWER EXTREMITY

Name of Artery	Location and Distribution

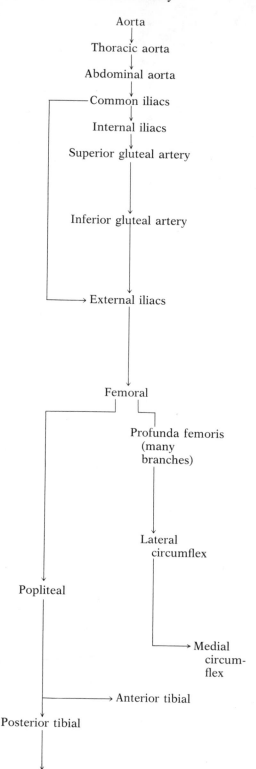

Parietal and visceral branches.

Superior branch anastomoses with the inferior gluteal and lateral sacral arteries. *Deep branch* anastomoses with the deep iliac and lateral femoral arteries.

Many branches—anastomose with branches of the perforating arteries (so named because they perforate the adductor muscle). These arteries surround the gluteal region and posterior thigh. See Figure 8–38.

Passes obliquely, distally, and laterally along the medial border of the psoas major muscle—passes under the inguinal ligament, midway between the anterior superior spine of the ilium and the symphysis pubis, where it enters the thigh and becomes the femoral.

The femoral artery extends from the inguinal (Poupart's) ligament to an opening in the adductor magnus muscle, where it becomes the popliteal. In the first part of its course the artery lies along the middle of the depression on the inner aspect of the thigh, known as the femoral (Scarpa's) triangle. Here pulsation may be felt. The femoral artery supplies blood to the muscles and fascia of the thigh. Branches also extend to the abdominal walls.

The *lateral femoral circumflex artery* anastomoses with branches of the popliteal to form the *circumpatellar* anastomosis, which surrounds the knee joints. The popliteal sends branches to the knee joint and all the posterior femoral muscles, the gastrocnemius and soleus muscles, and skin on back of the leg.

The medial branch supplies the adductors, gracilis, and external obturator muscles.

Distributes blood to the front of the leg.

Posterior tibial lies at the back of the leg and distributes blood to muscles of the back of the leg and nutrient vessels to tibia and fibula.

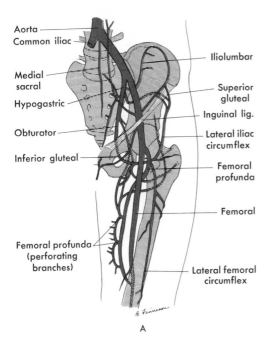

Aorta
Common iliac
Iliolumbar
Medial sacral
Superior gluteal
Hypogastric
Inguinal lig.
Obturator
Lateral iliac circumflex
Inferior gluteal
Femoral profunda
Femoral
Femoral profunda (perforating branches)
Lateral femoral circumflex

A

Figure 8–38. Arterial blood supply to the lower extremity.

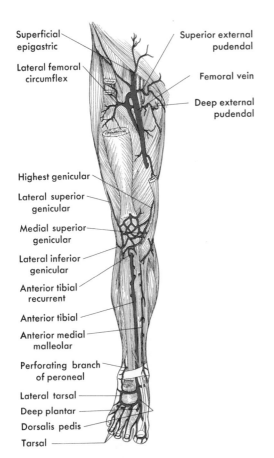

Superficial epigastric
Superior external pudendal
Lateral femoral circumflex
Femoral vein
Deep external pudendal
Highest genicular
Lateral superior genicular
Medial superior genicular
Lateral inferior genicular
Anterior tibial recurrent
Anterior tibial
Anterior medial malleolar
Perforating branch of peroneal
Lateral tarsal
Deep plantar
Dorsalis pedis
Tarsal

B

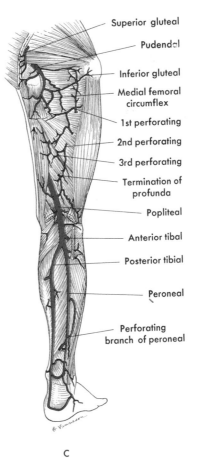

Superior gluteal
Pudendal
Inferior gluteal
Medial femoral circumflex
1st perforating
2nd perforating
3rd perforating
Termination of profunda
Popliteal
Anterior tibal
Posterior tibial
Peroneal
Perforating branch of peroneal

C

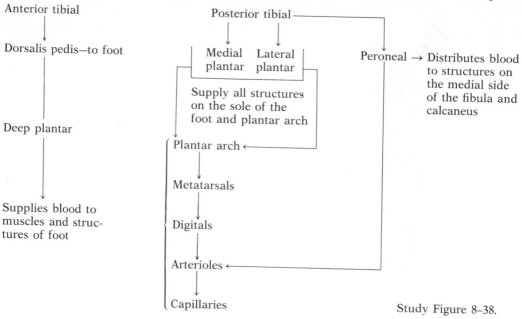

Anterior tibial

Dorsalis pedis—to foot

Deep plantar

Supplies blood to
muscles and struc-
tures of foot

Posterior tibial

Medial | Lateral
plantar | plantar

Supply all structures
on the sole of the
foot and plantar arch

Peroneal → Distributes blood
to structures on
the medial side
of the fibula and
calcaneus

Plantar arch ←

Metatarsals

Digitals

Arterioles ←

Capillaries

Study Figure 8–38.

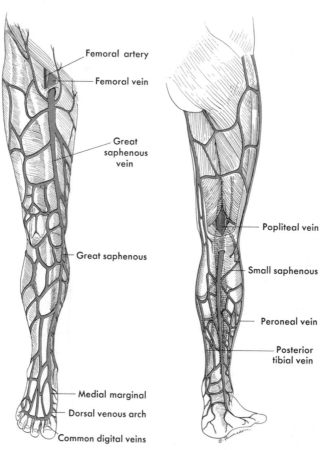

Femoral artery

Femoral vein

Great
saphenous
vein

Great saphenous

Medial marginal

Dorsal venous arch

Common digital veins

Popliteal vein

Small saphenous

Peroneal vein

Posterior
tibial vein

Figure 8–39. Venous return from the lower extremity.

Veins

Blood from the lower extremity is returned by a superficial and deep set of veins. The deep veins accompany the arteries and the superficial veins are beneath the skin between the layers of superficial fascia. All veins from the lower extremity possess numerous valves.

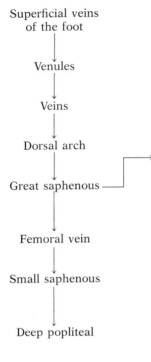

Superficial veins
of the foot

Venules

Veins

Dorsal arch

Great saphenous — Begins in the medial marginal vein of the dorsum of the foot and extends upward on the medial side of the leg and thigh and terminates in the femoral vein (about 4 cm below the inguinal ligament). In the leg it anastomoses with the small saphenous vein. It receives many tributaries. Veins from the posterior and medial aspects of the thigh frequently unite to form an accessory saphenous vein, which joins the great saphenous.

Femoral vein

Small saphenous — Begins behind the lateral malleolus as a continuation of the lateral marginal vein, passes up the back of the leg, and terminates in the deep popliteal vein. It receives many branches from the deep veins on the dorsum of the foot and from the back of the leg. Before it joins the popliteal it gives off a branch that runs upward and forward and joins the great saphenous. The small saphenous has about 10 to 12 branches, one of which is located near its termination in the popliteal vein.

Deep popliteal

The deep veins possess many valves and accompany the arteries and their branches.

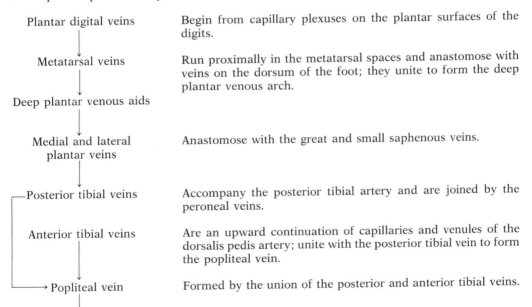

Plantar digital veins — Begin from capillary plexuses on the plantar surfaces of the digits.

Metatarsal veins — Run proximally in the metatarsal spaces and anastomose with veins on the dorsum of the foot; they unite to form the deep plantar venous arch.

Deep plantar venous aids

Medial and lateral plantar veins — Anastomose with the great and small saphenous veins.

Posterior tibial veins — Accompany the posterior tibial artery and are joined by the peroneal veins.

Anterior tibial veins — Are an upward continuation of capillaries and venules of the dorsalis pedis artery; unite with the posterior tibial vein to form the popliteal vein.

Popliteal vein — Formed by the union of the posterior and anterior tibial veins.

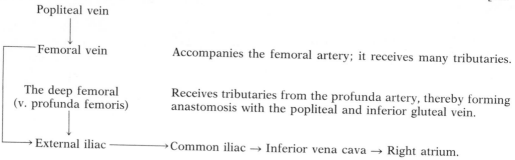

Popliteal vein

Femoral vein — Accompanies the femoral artery; it receives many tributaries.

The deep femoral (v. profunda femoris) — Receives tributaries from the profunda artery, thereby forming anastomosis with the popliteal and inferior gluteal vein.

→ External iliac ———→ Common iliac → Inferior vena cava → Right atrium.

Spinal Nerves

Spinal Nerves		Plexuses Formed	Some Main Nerves	Major Distribution
Cervical	1			To muscles of occipital triangle
	2	Cervical C2–C4	Branches from plexus	Skin and muscles of cervical region and neck; trapezius, etc.
	3		Phrenic nerve (chiefly C4)	Motor to diaphragm
	4		Branches from plexus	Deltoid Supraspinatus / Pectoralis Infraspinatus / Rhomboides Biceps
	5	**Brachial** C5–T1	Median and ulnar nerve	Flexor carpi radialis Flexor carpi ulnaris / Flexor digitorum sublimis Flexor digitorum profondus / Flexor pollicis longus Flexor pollicis brevis / Pronators
	6			
	7		Radial nerve	Triceps Extensor carpi radialis / Brachialis Extensor carpi ulnaris / Brachioradialis Extensor policis / Supinator Extensor indicis proprius / Extensor digitorum
	8			
Thoracic	1			
Thoracic	1-12		Intercostal nerves	Levators costarum Back muscles / Intercostal muscles / Abdominal muscles
Lumbar	1		Femoral nerve	Iliopsoas Quadriceps femoris / Sartorius Knee
	2			
	3		Ventral branches of L5–S2	External rotators of thigh
	4		Obturator nerve	Gracilis, adductor muscles
	5		Gluteal nerves	Gluteal muscles
Sacral	1	Lumbosacral T12–S3	Sciatic nerve	Biceps femoris (long head) / Semitendinosus / Semimembranosus
	2		Tibial	Posterior tibial Flexor digitorum / Gastrocnemius Flexor hallucis longus / Soleus Small muscles of foot / Plantaris
	3			
	4		Tibial, superficial, and deep peroneal nerves	Biceps femoris (short head) / Anterior tibialis / Extensor digitorum longus and brevis / Extensor hallucis longus / Peroneus longus, brevis, tertius
	5			
Coccygeal	1			Muscles over coccyx

Questions for Discussion

1. What groups of muscles are used in the following activities?
   a. Combing the hair
   b. Climbing stairs
   c. Turning to look over the right shoulder
   d. Standing on tiptoe
2. Why is it important for you to know the location and function of the larger muscles and the relationship of nerves and blood vessels to them?
3. What would be the result of injury to the following nerves?
   a. The radial   b. The median

   c. The ulnar   e. The tibial
   d. The femoral   f. The peroneal
4. Discuss in detail the blood supply to:
   a. The arms   b. The lower extremities
   c. What is the venous return from these structures?
5. Review the main functions and nerve supply of the following muscles:
   a. Biceps brachii   d. Gluteus maximus
   b. Deltoid   e. Gastrocnemius
   c. Iliopsoas   f. Quadriceps
6. Which muscle or muscle groups oppose the muscles in 5, a, e, f?

Additional Readings

BASMAJIAN, J. V.: *Grant's Method of Anatomy,* 9th ed. Williams & Wilkins Co., Baltimore, 1975. Sections 2 and 5

GARDNER, E.; GRAY, D. J.; and O'RAHILLY, R.: *Anatomy–A Regional Study of Human Structure,* 4th ed. W. B. Saunders Co., Philadelphia, 1975, Parts 11–17, 18–25.

HOLLINSHEAD, H. W.: *Textbook of Anatomy,* 3rd ed. Harper & Row, New York, 1974, Chapters 11, 12, 13, 16, 17, and 18.

PANSKY, B., and HOUSE, E. L.: *Review of Gross Anatomy,* 3rd ed. Macmillan Publishing Co., Inc., New York, 1975, Units 3 and 6.

ROMANES, G. J. (ed.): *Cunningham's Textbook of Anatomy,* 11th ed. Oxford University Press, New York, 1972, pp. 302–94.

WOODBURNE, R. T.: *Essentials of Human Anatomy,* 5th ed. Oxford University Press, New York, 1973, Sections II and IX.

UNIT

Awareness and Response to the Environment

Interpretation of sensations, reason and understanding, and initiation of thought are readily apparent activities of the nervous system. Yet the nervous system, along with the endocrine system, is also the controller of unconscious activity, making the necessary adjustments for the functional needs of the body. The structure and function of the nervous and endocrine systems are discussed in this unit.

CHAPTER 9

Fundamental Aspects of the Nervous System

Chapter Outline

The nervous system, in conjunction with the endocrine system, controls the functions of the human body. It is because of the nervous system that the individual is made aware of changes in both his external and internal environments, and can respond in the appropriate manner. Further, in higher organisms, the nervous system is capable of processing and storing information for future use.

General Organization

The nervous system is divided into the (1) *central nervous system*, which includes the brain and spinal cord, and the (2) *peripheral nervous system*, which includes the cranial nerves, spinal nerves, and the autonomic nervous system.

The organization of the nervous system is complex but highly ordered. The central nervous system, as the name implies, is located in the central axis of the body; the brain, in the cranial cavity and the spinal cord, in the vertebral cavity. Projecting from the brain and the spinal cord are the nerves of the peripheral nervous system, which are distributed to all the areas of the body. Within the brain and the spinal cord are specific centers or areas that have very precise functions in integrating the activity of the nervous system. Basically, the nervous system may be described as a means

199

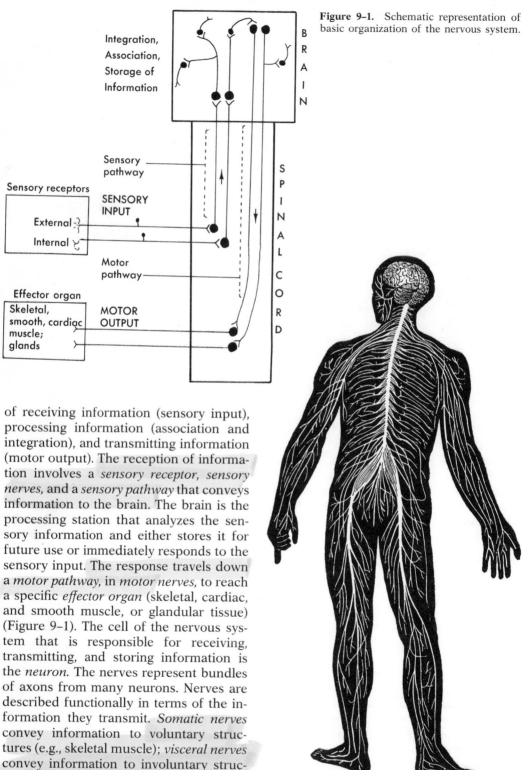

Figure 9–1.   Schematic representation of basic organization of the nervous system.

Integration,
Association,
Storage of
Information

B R A I N

Sensory pathway

Sensory receptors

SENSORY INPUT

External

Internal

S P I N A L   C O R D

Motor pathway

Effector organ

Skeletal, smooth, cardiac muscle; glands

MOTOR OUTPUT

of receiving information (sensory input), processing information (association and integration), and transmitting information (motor output). The reception of information involves a *sensory receptor, sensory nerves,* and a *sensory pathway* that conveys information to the brain. The brain is the processing station that analyzes the sensory information and either stores it for future use or immediately responds to the sensory input. The response travels down a *motor pathway,* in *motor nerves,* to reach a specific *effector organ* (skeletal, cardiac, and smooth muscle, or glandular tissue) (Figure 9–1). The cell of the nervous system that is responsible for receiving, transmitting, and storing information is the *neuron.* The nerves represent bundles of axons from many neurons. Nerves are described functionally in terms of the information they transmit. *Somatic nerves* convey information to voluntary structures (e.g., skeletal muscle); *visceral nerves* convey information to involuntary structures (e.g., smooth and cardiac muscle and glands). The direction of information is defined by the terms *sensory* or *afferent,*

Figure 9–2.   Diagram illustrating the brain, spinal cord, and spinal nerves.

meaning from a sensory receptor to the central nervous system, and *motor* or *efferent*, meaning from the central nervous system to an effector organ. Therefore, to fully describe the function of a nerve, we speak of somatic afferent or somatic efferent and visceral afferent and efferent nerves.

Neurons are both anatomically and functionally related by their processes to (1) other nerve cells, (2) muscular tissue, and (3) glandular cells. The area of contact between neurons is the *synapse;* between nerve and skeletal muscle, the *neuromuscular junction.*

In addition to neurons, the nervous system is composed of specialized supportive cells called *neuroglial cells.* These cells cannot transmit information, and act as the connective tissue of the nervous system.

The Neuron

The neuron is a cell that has a *cell body, perikaryon,* or *soma* and, projecting from it, several short radiating processes called *dendrites* and one long process called the *axon* (Figure 9–3). Functionally the dendrites convey information to the cell body and the axon transmits information away from the cell body. The axon, which extends over great lengths, often has branches or *axon collaterals* along its course and at its end exhibits fine ramifications.

The cell body of the neuron is much like any other cell. It has a nucleus, which is usually centrally located, and a prominent nucleolus. The cytoplasm contains neurofibrils, Golgi apparatus, mitochondria, and *Nissl bodies* or substance. Nissl bodies are found in great abundance throughout the cytoplasm and the dendrites. They are absent, however, in the area of the cell body from which the axon originates, the *axon hillock.* The Nissl bodies when observed with the electron microscope appear as rough-surfaced endoplasmic reticulum, made up of membranes, tubules, and accompanying ribosomes. Since endoplasmic reticulum represents sites of protein synthesis, the neuron is capable of synthesizing protein. Recently it has been shown in biochemical and autoradiographic studies that protein synthesized in the cell body moves or flows down the axon to reach the axonal endings. The functional significance of axoplasmic flow of protein has not as yet been definitively established. It has been implicated, however, in

trophic functions, namely, maintaining the functional integrity of other neurons and particularly effector organs such as skeletal muscle. Other theories relate axoplasmic flow with the movement of substances involved in synaptic transmission. Nissl bodies change their appearance—i.e., distribution, shape, and size—under different physiologic conditions, such as rest and fatigue, and in certain pathologic states.

Types of Neurons. The size and shape of the cell bodies vary, as do the number and means of branching of the dendrites. Therefore, many morphologically distinguishable types of neurons exist. There is a correlation with the morphologic characteristics and the function of the neuron. Three general types of neurons are usually described: unipolar, bipolar, and multipolar (Figure 9–4).

Unipolar neurons are sensory, are found in the dorsal root ganglia and other sensory ganglia, and have a round- to pear-shaped cell body and one process called the axon that branches into a central and a peripheral process. Although the process is called an axon, functionally the peripheral end acts as a dendrite while the centrally directed process acts as an axon. Embryologically unipolar cells begin with one process as the dendrite and another as the axon. During development, however, there is a fusion of these two processes. Because of this development these cells are often referred to as pseudounipolar.

Bipolar neurons are also sensory and

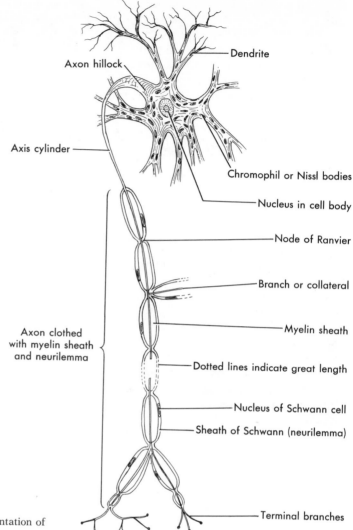

Figure 9–3. Diagrammatic representation of the neuron.

have two processes, the dendrite and the axon projecting from each end of the perikaryon. They are found in the retina and the olfactory epithelium.

By far the greatest majority of neurons are the *multipolar* type, and they are usually motor in function. The dendritic branching of these cells is quite varied and extensive. Multipolar cells are found in the ventral horn of the spinal cord; they are present in the cerebral cortex and are called pyramidal cells, and in the cerebellum these types of cells are called Purkinje cells.

Location of Cell Bodies. The cell bodies of neurons are localized in areas both inside and outside the central nervous system. In the central nervous system, groups of cell bodies, usually having a specific function, are called *nuclei* (in this instance the term *nucleus* has a different meaning from the nucleus of the cell). Groups of cell bodies located outside the central nervous system are known as *ganglia* (singular, *ganglion*). Ganglia are seen as swellings on sensory cranial nerves and on the dorsal roots of spinal nerves. The autonomic nervous system has many ganglia. These include the

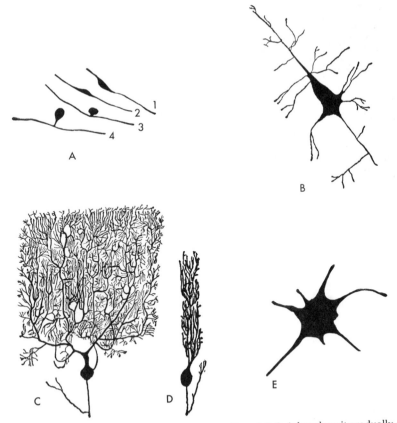

Figure 9–4. Types of neurons. *A.* Cell of dorsal root ganglion; *1, 2, 3, 4* show how it gradually develops into a unipolar cell. *B.* Pyramidal cell of cerebral cortex. *C* and *D.* Purkinje cells of cerebellum, *D* in profile view. *E.* Anterior horn cell of spinal cord. *B* through *E* are examples of multipolar cells.

sympathetic chain of ganglia that run parallel to the vertebral column from the thoracic to the lumbar level, the ganglia in the celiac and mesenteric plexuses, and less discrete ganglia found in the wall of the intestinal tract, the heart, and other visceral organs.

Nerves

A single nerve fiber consists of an axon surrounded by sheaths, called the myelin sheath and the neurilemma. In peripheral nerves, the cell responsible for these sheaths around the axon is the *Schwann cell.* Many Schwann cells are arranged sequentially along the length of the axon. The point where two successive Schwann cells abut is called the *node of Ranvier.* The nodes, then, constitute microscopic intervals before the beginning of the next Schwann cell, at which the axon is naked, a point that will be seen to have functional importance in the conduction of nerve impulses. During development each Schwann cell wraps itself *one* or *more* times around the axon. Many, repeated wrappings form a thick *myelin sheath.* The myelin sheath is composed of alternating layers of lipids and proteins that represent successive layers of the plasma membrane of the Schwann cell wrapped spirally around the axon (Figure 9–5). The body of the Schwann cell remains on the outside

Axon   Schwann cell

Figure 9–5. Diagram to show coiling of a Schwann cell around an axon. Myelin will form in the clear areas between the layers of Schwann cell cytoplasm. (Modified from Crosby, Humphrey, and Lauer.)

of the myelin sheath and is referred to as the *neurilemma* or *sheath of Schwann*. Nerve fibers that have extensive or thick myelin sheaths are called myelinated; those without extensive wrapping of the Schwann cell are called unmyelinated or naked fibers (Figure 9–6).

Many nerve fibers are grouped together as *fascicles* and are held together by connective tissue to form nerve trunks (e.g., the sciatic and median nerves). The outer layer of connective tissue of a nerve trunk, holding together many fascicles, is called the *epineurium*. Each individual fascicle is

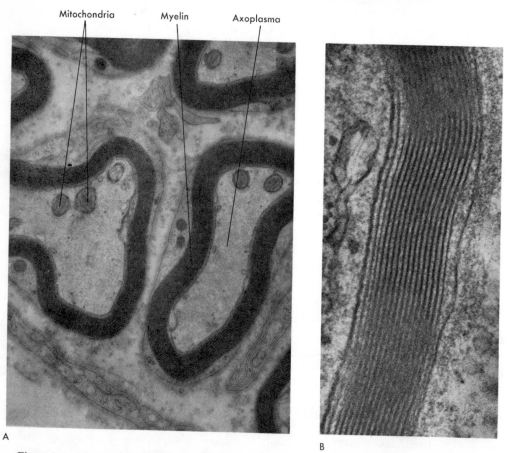

Mitochondria   Myelin   Axoplasma

A

B

Figure 9–6. Electron micrographs of myelinated nerve fibers. *A.* Portion of two nerve fibers and their myelin sheaths. *B.* Higher magnification showing laminated appearance of myelin. (Compare with Figure 3–29, *C* and *D*, page 73.)

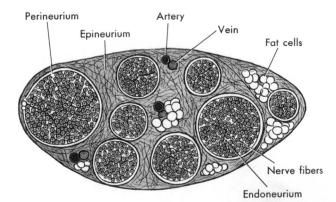

Figure 9–7. Transverse section of the sciatic nerve of a cat. This nerve consists of eight bundles (funiculi) of nerve fibers. Each bundle has its own wrappings (perineurium); and all the bundles are embedded in connective tissue (epineurium) in which arteries, veins, and fat cells can be seen. See Figure 9–3 for the structure of a nerve fiber.

encased by connective tissue called *perineurium,* and each individual nerve fiber is surrounded by *endoneurium.* (This arrangement of connective tissue is quite similar to that seen in skeletal muscle.) Blood vessels traverse both the epineurium and perineurium (Figure 9-7).

Types of Nerve Fibers. On the basis of the degree of myelination and certain functional properties, nerve fibers are classified as A, B, and C types. A-type fibers are heavily myelinated and are of large diameter, 5 to 20 mμ, and they transmit impulses at conduction velocities of 30 to 120 m per second. A-type fibers are both sensory (afferent) and motor (efferent) and are somatic. B-type fibers have smaller diameters, less than 3 mμ; transmit impulses at conduction velocities of 3 to 15 m per second; and functionally are efferent, preganglionic[1] axons found in autonomic nerves. Those of the C group are unmyelinated and, therefore, are the smallest-diameter fibers, 0.3 to 1.2 mμ, and have the slowest conduction speeds, 0.5 to 2.0 m per second. C-type fibers are efferent postganglionic[2] sympathetic fibers, and afferent fibers that are found in peripheral nerves and dorsal roots.

Neuroglia

Neuroglia represent supportive cells of the nervous system, which do not have the ability of transmitting impulses. The number of cells comprising neuroglia exceeds that of neurons (about 5 to 1). The term *neuroglia* is applied to the *ependymal cells* that line the ventricles of the brain and spinal cord, to the *neuroglial cells* of the central nervous system, to the *satellite* or capsular cells of peripheral ganglia, and to the *Schwann cells* of peripheral nerves.

The cells comprising neuroglia in the central nervous system include (1) the *astrocytes,* which are of the fibrous or protoplasmic type and are found associated with neurons and blood vessels by means of their processes (these cells because of their anatomic relationship to neurons and blood vessels are thought to act as a means of transferring nutrients and/or waste materials); (2) *oligodendrocytes,* which are the cells responsible for forming myelin in the nerve fibers of the central nervous system; and (3) *microglia,* which act as scavengers or macrophages of the central nervous system, and can multiply and migrate to a site of damage. Whereas all of the other cells comprising neuroglia are of ectodermal origin, the microglia are of mesodermal origin.

---

[1] Preganglionic: defines a nerve fiber before it makes synaptic contact with another neuron in a ganglion.
[2] Postganglionic: defines the nerve fiber of the second neuron, after a synapse in the ganglion.

Sensory Receptors

Sensory receptors are the terminal ends of sensory nerve fibers that pick up all sensory information. Structurally they are either free nerve endings or they are specialized into morphologically distinct structures. Sensory receptors are categorized in reference to their location, namely: (1) *exteroceptors*, which receive stimuli from the body surface; (2) *interoceptors*, which receive stimuli from internal organs; and (3) *proprioceptors*, which receive stimuli from muscles, tendons, and joints.

Sensory receptors are also classified in reference to their function.

Mechanoreceptors. These are receptors that are stimulated by differences of pressure. The sensation and the receptors involved are:

1. *Pressure and touch:* free nerve endings in all tissues, pacinian corpuscles and Meissner's corpuscles in the skin.
2. *Position of joints and of extremities* (*kinesthesia*): Golgi tendon organs, muscle spindles, and free nerve endings.
3. *Equilibrium and balance:* receptors in the vestibular apparatus of the inner ear.
4. *Pressure- and stretch-sensitive receptors in visceral organs such as in blood vessels:* the baroreceptors found in the arch of the aorta and the carotid sinus in the wall of the internal carotid arteries; in the lungs receptors probably as free nerve endings and pacinian corpuscles.
5. *Hearing:* hair cells in the basilar membrane of the inner ear, which respond to sound wave vibrations.

Thermoreceptors. These are usually free nerve endings and respond to variations in temperature. *Ruffini's end bulbs* are thought to be the receptors of heat stimuli, and *Krause's end bulbs* are thought to respond to cold stimuli.

Photosensitive Receptors. These are found in the retina of the eye and are the *rods* and *cones*. Both of these receptors respond to light, the rods being most sensitive and adapted for night vision, whereas the cones are responsible for vision in bright light and for color vision.

Chemoreceptors. These include the taste buds of the tongue and pharynx, responding to different chemicals in food; olfactory cells of the nose, responding to chemicals in the air; and receptors of cells in the carotid and aortic bodies, responding to pH and to oxygen and carbon dioxide levels of the blood (nonconscious sensations).

Specificity of Receptors. In order for nerve impulses to be initiated by receptors, the stimulus must be *adequate* and at least *minimal*. By an adequate stimulus is meant one whose physical energy is appropriate for the receptor, e.g., light for the eye or sound for the ear. A minimal, or threshold, stimulus refers to the least change in stimulus strength that can excite the receptor.

Intensity of stimulus determines the degree of change imposed on the receptor and is "coded" in the frequency (number per second) of the impulses initiated. With increase in intensity of stimulus, the frequency of discharge by the receptor increases. Furthermore, most receptor systems in their frequency of discharge code information related not only to the absolute strength of the stimulus, but also to the *rate of change* of stimulus strength. The more rapid the increase in intensity of a stimulus, the more rapidly the receptor discharges impulses to the nerve fiber.

Adaptation. When an adequate and minimal stimulus is applied to a receptor, the receptor is excited, and nerve impulses are generated and conducted to the central nervous system by the afferent nerve fiber. The number of impulses resulting from a given stimulus depends on two things: (1) the strength, rate of change of strength, and duration of the stimulus and (2) the inherent properties of the receptors. Some receptors continue to elicit action potentials for as long as the stimulus endures. Other receptors discharge only at the time the stimulus is applied and soon cease discharging, even though the stimulus continues. The former are classed as *slowly adapting* receptors; the latter are *rapidly adapting*. Slowly adapting receptors provide neural information to the central nervous system about steady-state conditions. Rapidly adapting receptors report to the nervous system primarily about changes in conditions. For example, receptors at the base of hairs discharge at the moment the hair is bent but promptly cease discharging even though the hair remains in the bent position. Only movement of the hair is signaled to the central nervous system. In contrast, muscle spindles are slowly adapting. If a muscle is stretched, its muscle spindles discharge at the increased rate for the entire duration of the stretch. Hence, the muscle spindle is said to be slowly adapting. It reports to the nervous system on the static length of a muscle as well as change in length.

The Nerve Impulse

The Resting State. All cells possess the property of excitability; however, this property is most extensively developed in nerve and muscle cells. In addition to excitability, nerve and muscle cells have the property of conductivity. Thus, these cells can respond to a stimulus and can transmit the stimulus.

The nerve membrane that is not actively transmitting an impulse, or is in the resting state, is characterized by a number of features. Due to the process of *diffusion* and *active transport*, the concentration and type of ions are different on the inside and outside of the membrane.

Ion	Concentration (mEq/liter water)	
	Inside Cell	Outside Cell
Sodium (Na⁺)	15	150
Potassium (K⁺)	150	5.5
Chloride (Cl⁻)	9	125
Anions (A⁻)	150	5

The potassium and sodium ions have a primary role in excitable tissue. The resting membrane is 50 to 100 times more permeable to potassium than it is to sodium. Potassium, therefore, diffuses through the membrane with relative ease, but sodium diffuses only with difficulty.

Inside the nerve membrane, there are a large number of anions, such as phosphate ions, organic sulfate ions, and protein ions, that cannot diffuse through the nerve membrane at all or that diffuse very poorly. Whatever sodium does gain entry into the cell is removed by an active transport mechanism known as the *"sodium pump."* Although the membrane is relatively permeable to potassium, there is also a "potassium pump," which actively pumps potassium to the inside of the cell.

Due to the combined processes of diffusion and active transport, a condition is created in which more negatively charged ions are on the inside of the membrane (the anions are impermeable and therefore remain inside) and more positively charged ions are on the outside (Na⁺, which is relatively impermeable, in addition to some K⁺). Therefore, there is electronegativity on the inside and electropositivity on the outside (Figure 9–8, *A*). This condition results in a *membrane potential*, which in the resting state can be measured, and the voltage is found to be about 70 to 90 millivolts (mv) negative to the outside.

The Active State: The Action Potential. A sudden increase in the permeability of the nerve membrane to sodium ions will result in a sequence of very rapid changes in the membrane potential. This can be caused

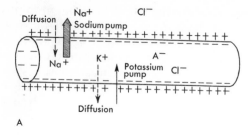

A

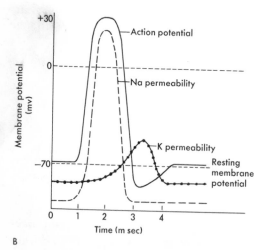

B

Figure 9–8. *A.* Schematic representation of the resting membrane of a nerve fiber. *B.* The action potential and related permeability changes.

by *electrical stimulation* (experimentally applying an electric current), by *chemicals* (physiologically by neurotransmitters such as acetylcholine), and by *mechanical forces* (pressure change in pacinian corpuscles, in hair cells of the ear), all of which are referred to as the stimulus. The stimulus must be of adequate strength. The sequence of changes that occur is termed the *action potential.*

As the permeability of the membrane to sodium ions suddenly increases, a great number of sodium ions begin to enter the cell and positive charges build up on the *inside* of the membrane. Indeed, the initial entry of Na+ ions cancels the polarity of the membrane and for a brief instant the membrane potential has a value of zero. This is called *depolarization.* As Na+ ions continue to enter, the inside now becomes *positive* and the outside *negative;* hence a reverse (opposite) polarity is established and can be measured at about 30 mv posi-

tive to the outside. Almost immediately after these changes have occurred, the pores of the membrane again become impermeable to sodium ions. However, the permeability to K+ now increases; therefore, K+ ions move out of the cell and positive charges again build up on the outside of the cell, returning the membrane to its resting state. This is called *repolarization.* A simultaneous increase in the permeability of sodium and potassium would result in an energetically wasteful interchange of Na+ and K+ and would not cause the membrane potential to vary in a purposeful manner (Figure 9–8, *B*).

Propagation of the Action Potential. An action potential at any one point on an excitable membrane excites adjacent portions of the membrane and results in propagation of the action potential. At the initial site of depolarization a *local circuit* of current flows inward through the depolarized membrane and outward through the resting membrane, thereby completing a circuit. Current flow through the resting portion of the membrane in some way increases the membrane's permeability to sodium, and the process of membrane activation occurs in each successive portion of the membrane. The transmission of this depolarization process along either nerve or muscle fiber is the nerve or muscle impulse, or the propagated action potential (Figure 9–9, *A*).

The action potential is an all-or-none response. As long as there is an adequate stimulus, the response seen will be maximal. An increase in the strength of the stimulus will not further increase the response. However, weak stimuli occurring at a rapid rate may summate their effects and cause an action potential.

Conduction in Nerve Fibers. In those nerve fibers that are unmyelinated every portion of the membrane must be depolarized, so that the action potential is propagated over the entire length of the nerve. In myelinated nerve fibers the myelin sheath acts as an insulator, since ions cannot flow through the thick myelin sheath. However, at the nodes of Ranvier the membrane is free of myelin. In myeli-

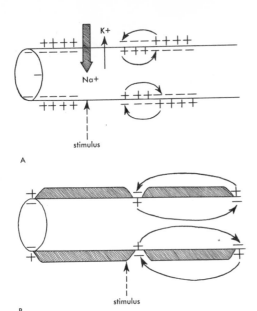

A

B

Figure 9–9. *A.* Propagation of nerve impulse in unmyelinated fiber. *B.* Saltatory conduction in myelinated nerve fiber.

nated fibers, impulses are conducted from node to node, instead of successively along the entire fiber, as in unmyelinated fibers. This process in which the impulse "jumps"

from node to node is termed *saltatory conduction* (the term *saltatory* is from the Latin word *saltare*, meaning "to jump"). Since only the nodal regions are depolarized, both energy and, more significantly, time are conserved. Hence, this mechanism explains the high rates of conduction velocity seen in myelinated nerve fibers (Figure 9–9, *B*).

Excitability Periods. The excitability of the nerve fiber has three phases: the *absolute refractory period*, the *relative refractory period*, and the *normal period*. During the absolute refractory period, which represents the interval of the action potential, no stimulus no matter how strong will excite the nerve. In other words, the membrane is fully excited and cannot respond any more. The relative refractory period represents that interval when the membrane is repolarizing and returning to a resting state. During the relative refractory period a stimulus stronger than normal is needed to cause excitation. Finally the normal period simply represents the resting state.

The Synapse

The synapse is the point of continuity that permits transmission of activity from one neuron to another in one direction only. At the synapse the neurons are closely apposed but do not touch each other; they are separated by a small space called the *synaptic cleft*. Since the neuron is a cell that has a number of anatomically distinct parts, different means of contact or synapses can occur. Hence, if the axon of one neuron contacts the cell body of another neuron, there is an *axosomatic* synapse (soma, referring to the cell body); if the axon of one neuron contacts the dendrite of another neuron, there is an *axodendritic* synapse; and if the axon of one neuron contacts the axon of another neuron, there is an *axoaxonic* synapse (Figure 9–10). Most recently a dendro-dendritic type of synapse has been described.

The terms *presynaptic* and *postsynaptic* are used to refer to the junction between the processes of two neurons or a neuron and an effector organ. For example, the axonal endings of a motor neuron on a muscle are presynaptic and the muscle membrane is the postsynaptic site. Transmission of activity from one neuron to another or to an effector organ (muscle, glands) involves a *chemical transmitter substance*. Substances known to act as chemical transmitters include acetylcholine, norepinephrine, 5-hydroxytryptamine (serotonin), and gamma aminobutyric acid (GABA). At the nerve terminal endings the chemical transmitter is found localized in small membrane-bound structures called *synaptic vesicles* (Figure 9–11). In addition to synaptic vesicles, mitochondria and a high content of enzymes necessary for the synthesis of the neurotransmitter are also

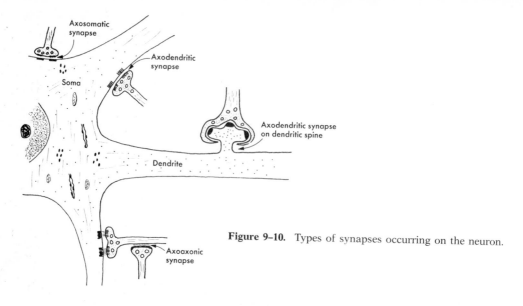

Figure 9–10. Types of synapses occurring on the neuron.

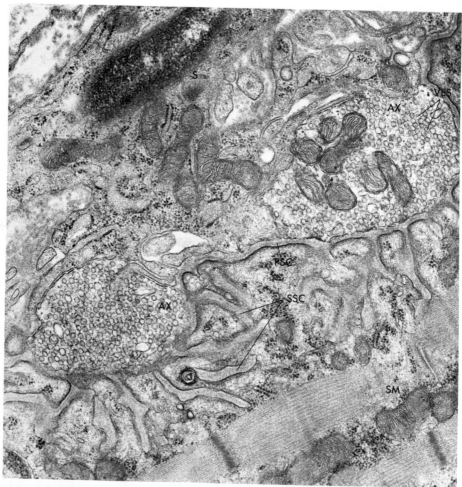

Figure 9–11. Electron micrograph of motor nerve terminals. *S*, portion of Schwann cell; *AX*, axonal ending; *VES*, vesicles; *PSC*, primary synaptic cleft; *SSC*, secondary synaptic cleft; *SM*, skeletal muscle. (See also Figure 6–3, page 135.)

present. In the process of transmission, synaptic vesicles in the presynaptic site release their transmitter substance, and this diffuses across the synaptic cleft to reach the postsynaptic membrane. The complexity of the nervous system in both its structure and function is continually in evidence. Within the central nervous system, it has been estimated that literally hundreds of presynaptic terminals lie on the surface of each cell body and its processes (Figure 9–12). The network of communication of information, therefore, is almost infinite.

Properties of Synapses. Conduction at synapses differs from that along nerve fibers. The chief differences are:

1. Conduction at the synapse is slower than conduction along a nerve fiber. This reflects the time for diffusion of the chemical transmitter across the synaptic cleft to reach the postsynaptic membrane.

2. There are excitatory and inhibitory synapses. This is dependent on the transmitter that is released and whether it excites or inhibits the second neuron. One transmitter thought to exert an inhibitory action is gamma aminobutyric acid.

3. The synapses are more readily susceptible to fatigue and are more easily affected by anesthetics and drugs.

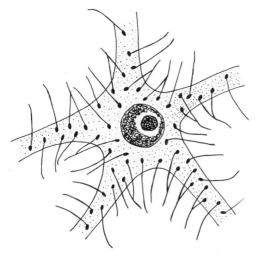

Figure 9–12. Representation of presynaptic terminals on neuron cell body and its processes.

4. Synaptic transmission is always one way.

Facilitation. You will recall that each neuron has many presynaptic contacts, which represent all the neurons making contact with the postsynaptic neuron. If only one or two of these presynaptic neurons are activated, this is not enough to excite the postsynaptic neuron. The presynaptic terminals of several neurons must be activated in order to excite the postsynaptic neuron. A single presynaptic impulse, however, makes the second neuron more excitable and therefore more responsive to excitatory impulses from other neurons. This excited neuron, which does not discharge an impulse, is said to be *facilitated.*

Inhibition. One tends to think of activation of a neuron and production of the impulse as resulting in action, e.g., stimulation of a second neuron, stimulating muscle to contract or glands to secrete. This is not always the case; some neurons are inhibitory and their activity decreases the likelihood of the second neuron discharging. Inhibition, like excitation, is related to changes in ionic membrane permeability. When an inhibitory impulse reaches the synapse, the membrane permeability to potassium of the second neuron is increased, and the cell, instead of being depolarized, is hyperpolarized; the inside becomes more negatively charged with respect to the outside and is therefore less excitable (or inhibited). Inhibitory neurons of the somatic nervous system are located in the central nervous system; those of the autonomic nervous system are in the periphery. The spinal cord has many fibers that descend from inhibitory cells of the brain. When a motor neuron affecting skeletal muscle action receives an impulse from these fibers, the result is inhibition, or decreased excitability of the motor neuron, and a decrease or stopping of its discharge. A good example is relaxation of flexor muscles of the arm when extensor muscles are contracted—one set of motor neurons is inhibited when the other set is stimulated.

The Reflex Arc and Response

The functional unit of the nervous system is the *reflex arc* (Figure 9–13). It consists of five essential parts:

1. A sensory receptor
2. A sensory (afferent) neuron
3. One or more synapses within the central nervous system
4. A motor (efferent) neuron
5. An effector organ

The sensory receptor responds to changes in the external and internal environment and transmits this information along sensory (afferent) fibers to the central nervous system, the spinal cord, or brain. Within the central nervous system, if only one synapse is involved, (i.e., between the sensory and motor neuron), the reflex is termed *monosynaptic*. If there is more than one synapse involved, the reflex is termed *polysynaptic*. The neurons between the sensory and motor neurons in a polysynaptic reflex are called *internuncial* or *interneurons*. The fibers of the motor neuron will convey information to an effector organ. If the effector organ is skeletal muscle, we speak of a *somatic* reflex arc. When smooth muscle, cardiac muscle, or gland tissue is the effector organ, we speak of *autonomic* reflex arcs. The motor neuron or anterior horn cell represents the final site at which the response can be modified (e.g., either enhanced or inhibited via synaptic connections from higher centers). Because of this, the term *final common pathway* has been applied to the anterior horn cell.

Reflex arcs represent the physiologic mechanism by which the organism responds adaptively to its external and internal environment. Thus, the hand is quickly withdrawn from a harmful source, such as fire; the tension and tone of skeletal muscle are continually adjusted, thereby maintaining the appropriate position in space and allowing for purposeful movements (e.g., walking, running, holding objects); the blood pressure, heart rate, respiratory rate, to mention but a few, are continuously being adjusted to meet a particular physical or physiologic state. Indeed, the list of examples in which the reflex arc and its response are involved encompasses all the parts and functions of the body.

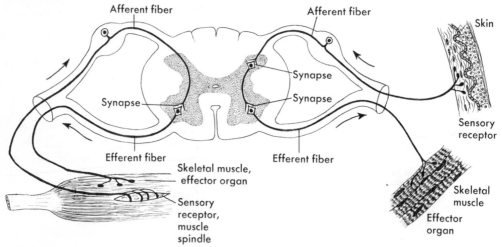

Figure 9–13. Schematic representation of the reflex arc. The left side of diagram depicts a monosynaptic reflex, and the right side illustrates a polysynaptic reflex.

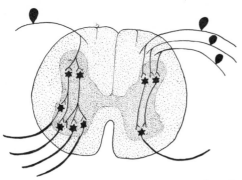

Figure 9–14. A cross section of the spinal cord, showing on the right a converging reflex, on the left a spreading reflex.

The response observed following activation of the reflex arc is the *reflex*. A reflex represents activity that is involuntary, even when skeletal muscle is involved (somatic reflex). All reflexes can be inhibited as well as enhanced. Spreading of impulses via axon collaterals can involve greater numbers of motor neurons, and convergence can limit the number of motor neurons that will respond (Figure 9–14). Reflexes are stereotyped. For example, stimulation of a given sensory receptor will always cause the leg to jerk (the "knee jerk"); a foreign object in contact with the cornea will cause the corneal reflex (wink or blink reflex). If any part of the reflex arc is damaged, the reflex will not occur. Thus, the testing of reflexes is a means of diagnosis of the functional integrity of the nervous system.

Questions for Discussion

1. Name and discuss the structure and function of each part of the neuron.
2. Where are the cell bodies of neurons located?
3. Define the following:
   Schwann cell   Epineurium
   Myelin sheath   Endoneurium
   Node of Ranvier   Fascicle
4. Name the cells that comprise the neuroglia of the central nervous system and give the function of each cell.
5. Describe what is meant by the terms *exteroceptors, interoceptors,* and *proprioceptors* and give examples for each.
6. Discuss the events involved in the transmission of a nerve impulse.
7. What is meant by saltatory conduction?
8. What is the synapse? What are some of the properties of synapses?
9. Name the parts of the reflex arc.
10. What is meant by a somatic reflex arc? Give an example of a somatic reflex arc.
11. What is meant by the term *final common pathway?*
12. List some examples of reflexes by name, indicating what stimuli and responses are involved.

Summary

Basic Plan of the Nervous System	**Sensory Input**	Sensory receptors
		Sensory nerves
		Sensory pathways
	Association and Integration	Processing information within the central nervous system
	Motor Output	Motor pathways
		Motor nerves
		Effector organs

Divisions of the Nervous System

- **Central Nervous System**
  - Brain
  - Spinal cord
- **Peripheral Nervous System**
  - Cranial nerves
  - Spinal nerves
  - Autonomic nerves, parasympathetic and sympathetic divisions

The Neuron

1. Unit of structure and function of the nervous system
2. Capable of receiving and transmitting information
3. Neuron has a cell body (perikaryon, or soma); projecting from cell body one long process, the *axon*, and usually many short processes, the *dendrites*

Types of Neurons
- Unipolar
- Bipolar
- Multipolar

Location of Cell Bodies
1. Brain and spinal cord
2. In central nervous system groups of cell bodies with specific function called *nuclei*
3. Groups of cell bodies located outside the central nervous system called *ganglia*, e.g., dorsal root ganglia, ganglia of the autonomic nervous system

Nerves

1. Single nerve fiber consists of axon surrounded by sheaths: *myelin sheath* and *neurilemma* (sheath of Schwann)
2. In peripheral nerves *Schwann cell* responsible for sheaths
3. *Node of Ranvier*—point where two successive Schwann cells abut on nerve fiber
4. Nerve fibers are described as:
    - *Myelinated*—these have extensive layer of myelin
    - *Unmyelinated*—these fibers have no myelin or one that is sparse
5. Extent and thickness of myelin sheath and function of nerve fiber are the basis of classification of A, B, and C types of nerve fibers
6. Many nerve fibers grouped together as *fascicles* and held together by connective tissue to form nerve trunks
    - *Epineurium*—outer layers of connective tissue enclosing many fascicles
    - *Perineurium*—connective tissue encasing individual fascicles
    - *Endoneurium*—connective tissue around individual nerve fiber

Functional Classification of Nerves
- *Somatic nerves*—convey information to voluntary structures, e.g., skeletal muscles
- *Visceral nerves*—convey information to involuntary structures, e.g., smooth and cardiac muscle and glands
- *Sensory or afferent*—from a sensory receptor to the central nervous system
- *Motor or efferent*—from central nervous system to an effector organ

Neuroglia

Neuroglia represent supportive cells of the nervous system; these cells do not possess the property of transmission of impulses

Types
- *Ependymal cells*—line ventricles of brain and spinal canal
- *Satellite cells*—surround cell bodies in ganglia
- *Schwann cells*—surround peripheral nerves, responsible for myelin sheath
- *Neuroglial cells* of the central nervous system include:
    - *Astrocytes*—found in association with blood vessels and neurons
    - *Oligodendrocytes*—cells responsible for forming myelin in nerve fibers of the central nervous system
    - *Microglia*—macrophages of the central nervous system

Sensory Receptors

Characterized in reference to their location
1. *Exteroceptors*—receive stimuli from the body surface
2. *Interoceptors*—receive stimuli from internal organs
3. *Proprioceptors*—receive stimuli from muscles, tendons, and joints

Classified in reference to function
1. *Mechanoreceptors*
    - Pressure, touch: free endings, pacinian and Meissner's corpuscles
    - Kinesthesia: free endings and pacinian corpuscles on tendons, joints
    - Hearing: receptors in inner ear
    - Equilibrium: vestibular apparatus in inner ear
    - Muscle and tendon stretch: muscle spindles, Golgi tendon apparatus
2. *Thermoreceptors*: free nerve endings, Ruffini, Krause endings
3. *Chemoreceptors*: taste buds, olfactory cells, receptors of carotid and aortic bodies
4. *Photosensitive receptors*: retina of eye
5. *Baroreceptors*: in aortic arch and carotid sinus

Specificity of Receptors

Receptor stimuli

1. *Stimuli* are physical or chemical changes in the immediate environment
2. Changes in environment are converted into a rhythmical succession of nervous impulses
3. *Adequate* stimulus, the *kind* of physical or chemical change to which the receptor is sensitive
4. *Minimal* stimulus, the *least* change that can excite a receptor
5. *Subminimal* stimulus, a stimulus below the threshold level that fails to excite
6. *Intensity* of stimulus, probably related to the degree of change at the receptor and reflected in the frequency of impulses initiated

Adaptation

Adaptation of end organs varies in relation to the speed with which they reach approximate equilibrium with their environment
1. *Slow* adaptation, long trains (sequences) of nerve impulses issue from end organs in response to prolonged stimulation
2. *Rapid* adaptation, short trains of impulses issue from end organs even though stimulus is prolonged

Nerve Impulse

Resting Membrane

1. Due to process of diffusion and active transport, concentration and type of ions are different on inside and outside of membrane
2. In resting state more positively charged ions on outside of membrane and more negatively charged ions on inside

Active State: the Action Potential

Stimulus causes a sudden increase in permeability of nerve membrane to sodium ions, which causes rapid changes in membrane potential; sequence of changes that occur termed action potential

Events

1. Initial entry of Na^+ ions cancels polarity of membrane—*depolarization*
2. Na^+ ions inside of nerve membrane change polarity; inside becomes positive, outside negative
3. Return of membrane to resting state—*repolarization*

Conduction

Action potential is propagated along entire length of nerve in nerve fibers that are unmyelinated
In myelinated fibers impulses are conducted from one node of Ranvier to next; process referred to as *saltatory conduction*

Excitability Periods

Excitability of nerve fiber shows three periods:
1. Normal period or phase,
2. Absolute refractory period, a short period during which the nerve fiber is inexcitable to any stimulus regardless of strength
3. Relative refractory period, period of less excitability than normal phase—stronger-than-normal stimulus required to excite

The Synapse

Synapse, point of continuity that permits transmission of activity from one neuron to another

Types
axosomatic
axodendritic
axoaxonic
dendrodendritic

Transmission of activity from one neuron to another involves a *chemical neurotransmittor*

Properties of Synapses

1. Conduction at synapse slower than along nerve fiber; reflects time for diffusion of transmittor across synaptic cleft
2. Excitatory and inhibitory types of synapses
3. Fatigue occurs more readily at synapses; more easily affected by drugs
4. Transmission in one direction only

The Reflex Arc

Functional unit of the nervous system. Represents physiologic mechanism by which organism responds adaptively to its external and internal environment

Parts
Sensory receptor
Sensory (afferent) neuron
One or more synapses
Motor (efferent) neuron
Effector organ, e.g., voluntary and involuntary muscles, glands

Types, Anatomic
Monosynaptic reflex arc, e.g., "knee jerk"
Polysynaptic reflex arc, e.g., withdrawal of limb in response to noxious stimulus

Types, Functional
Somatic reflex arc—effector organ skeletal muscle
Autonomic reflex arc—effector organ smooth muscle, cardiac muscle, or gland tissue

Additional Readings

DeRobertis, E.: Molecular biology of synaptic receptors. *Science,* **171:**963, March 12, 1971.

Guyton, A. C.: *Basic Human Physiology: Normal Function and Mechanisms of Disease.* W. B. Saunders Co., Philadelphia, 1971, Chapters 5, 6, and 31.

Jacobson, M., and Hunt, R. K.: The origins of nerve-cell specificity. *Sci. Amer.,* **228:**26–35, February, 1973.

Kandel, E. R.: Nerve cells and behavior. *Sci. Amer.,* **223:**57–70, July, 1970.

Vander, A. J.: Sherman, J. H.; and Luciano, D. S.: *Human Physiology: The Mechanisms of Body Function,* 2nd ed. McGraw-Hill Book Co., New York, 1975, Chapter 6.

Whittaker, V. P.: Membranes in synaptic function. *Hosp. Prac.,* **9:**111–19, April, 1974.

Williams, P. L., and Warwick, R.: *Functional Neuroanatomy of Man.* W. B. Saunders Co., Philadelphia, 1975, pp. 766–804.

CHAPTER 10

The Spinal Cord and Spinal Nerves

Chapter Outline

The spinal cord is the path by which sensory information from the body can reach conscious and unconscious centers in the brain, and instructions from the brain can reach the body. The spinal cord also serves as a center for immediate responses to receptor stimulation through reflexes that occur within the cord.

General Organization

The spinal cord has an average length of 45 cm and originates at the level of the foramen magnum, where it is continuous with the medulla and extends downward to end between the first and second lumbar vertebrae. The spinal cord is cylindrical in shape and has two enlargements, at the cervical region and the lumbar region. It tapers at its inferior end into the *conus medullaris,* from which projects the *filum terminale.* The filum terminale extends and attaches to the posterior aspect of the coccyx.

The spinal cord is protected in three ways: (1) it lies within the spinal canal of the *bony* vertebral column; (2) it is surrounded by three membranes—an outer *dura mater,* an inner *arachnoid mater,* and the *pia mater,* which immediately surrounds the cord; and (3) it is suspended in a fluid medium, the *cerebrospinal fluid,* which acts as a shock cushion against mechanical damage.

Projecting from the whole length of the spinal cord are 31 pairs of spinal nerves, 8 of which are cervical, 12 thoracic, 5 lumbar, 5 sacral, and 1 coccygeal. Each spinal nerve has two roots, a ventral root that

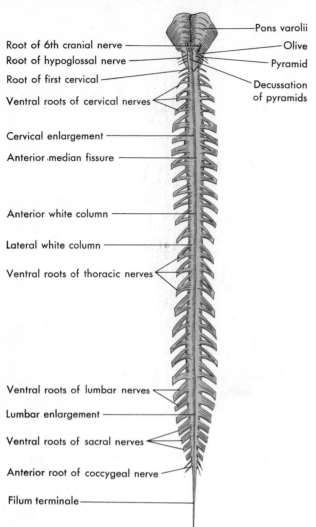

Root of 6th cranial nerve

Root of hypoglossal nerve

Root of first cervical

Ventral roots of cervical nerves

Cervical enlargement

Anterior median fissure

Anterior white column

Lateral white column

Ventral roots of thoracic nerves

Ventral roots of lumbar nerves

Lumbar enlargement

Ventral roots of sacral nerves

Anterior root of coccygeal nerve

Filum terminale

Pons varolii

Olive

Pyramid

Decussation of pyramids

Figure 10–1. Ventral view of the spinal cord. (Modified from Toldt.)

contains motor (efferent) fibers and a dorsal root that contains sensory (afferent) fibers. The two roots merge to form the spinal nerve. These nerves exit from the vertebral column via an intervertebral foramen, at the appropriate vertebral level (e.g., thoracic nerves will exit at thoracic vertebral levels). Since the spinal cord is shorter than the vertebral column, the lumbar and sacral nerves have very long roots extending from the cord to the proper intervertebral foramina, where dorsal and ventral roots are joined to form the spinal nerve. These roots descend in a bundle beyond the conus medullaris and,

as they resemble a horse's tail, the formation is known as the *cauda equina.*

When the spinal cord is freed from its meninges, the surface exhibits a number of longitudinal grooves. The two that are the most obvious are, on the anterior side, the deep *anterior median fissure* (ventral fissure) and, on the posterior side, the shallow *posterior median sulcus.*

A transverse section of the cord reveals a characteristic picture that can be seen with the naked eye. There is a central region, roughly in the shape of a butterfly or the letter H, appearing grayish in color, that is surrounded by a white-appearing

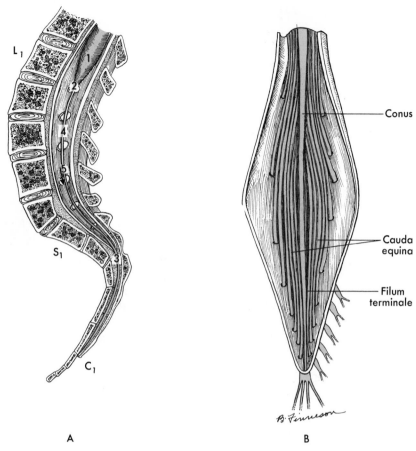

Figure 10–2. *A.* Longitudinal section of the vertebral column showing the end of the spinal cord. *1,* Beginning of conus; *2,* end of conus; *3,* filum punctures dura; *4,* spinal canal; *5,* foramen for exit of spinal nerve; C_1, first coccygeal vertebra; L_1, first lumbar vertebra; S_1, first sacral vertebra. *B.* The conus filum terminal and cauda equina. (Modified from Toldt.)

substance. Both in the spinal cord and in the brain, *gray matter* is composed principally of cell bodies of neurons, neuroglial cells, and unmyelinated nerve fibers, and *white matter* is composed primarily of nerve fibers. Since many of these fibers are myelinated, and myelin is composed of lipid material, this gives rise to the white appearance. The gray matter has features that are present at all levels of the cord. In the center is a small *central canal,* which opens into the fourth ventricle of the brain at its upper end and terminates in the filum terminale. The region of the gray matter surrounding the central canal (the bar of the letter H) is called the *gray commissure.* On each side of the gray commissure, or the vertical bars of the H, are the *posterior* (dorsal) *column* or *horn* and the *anterior* (ventral) *column* or *horn,* respectively. The region between the anterior and posterior columns is called the *lateral* or *intermediolateral column* or *horn.* This area appears most prominently in the thoracic and lumbar regions. In like manner the white matter is identified by regions, namely a *posterior funiculus,*[1] a *lateral funiculus,* and an *anterior funiculus.* Each funiculus, in turn, is divided into smaller segments or *fasciculi* (Figure 10–4).

It is important to know the anatomic divisions of the gray and white matter in the spinal cord, since each of these areas is associated with specific functional groups of neurons, either the cell bodies or their axons. Basically the following functional

[1] The word *funiculus* is Latin for "cord."

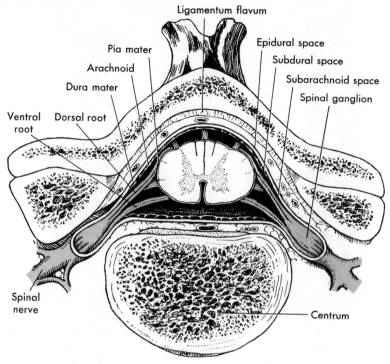

Figure 10–3. Transverse section of spinal cord and vertebra to show their relative positions. Note dorsal root ganglia lying in the intervertebral foramina and dorsal and ventral roots in the spinal canal. The spinal nerve is shown outside the vertebra, with branches to the dorsal body wall, ventral body wall, and viscera.

elements are found in gray and white matter. As you read this list keep in mind the parts of the reflex arc.

1. The dorsal (posterior) gray column contains sensory (afferent) fibers and internuncial neurons passing on sensory in-

formation. The cell bodies of the sensory fibers are located in the dorsal root ganglia, on the outside of the spinal cord proper. These sensory fibers (a) can pass through the dorsal gray columns and synapse with motor neurons in ventral gray columns of the same segment of cord (this

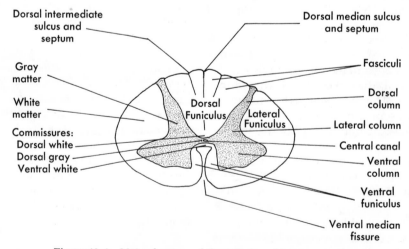

Figure 10–4. Major features of the spinal cord in cross section.

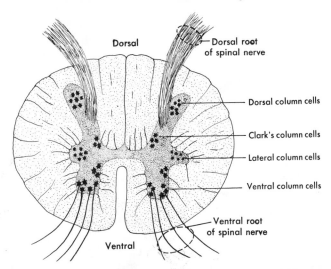

Figure 10–5. Cross section of the spinal cord to show some of the groups of nerve cells.

would represent a monosynaptic reflex arc) or (b) can synapse with internuncial neurons in dorsal gray columns. The fibers of these internuncials can extend to the ventral gray column of the same segment of cord; extend either to higher or lower segments of cord; extend to areas in the brain; and extend from one side of the cord to another. The internuncial neurons are relay neurons, conveying sensory information to a final motor neuron. Since more than one neuron is involved, the reflex arc is termed *polysynaptic.*

2. The ventral (anterior) gray column contains cell bodies from which the efferent (motor) fibers of the spinal nerves arise, as well as internuncial neurons.

3. The lateral gray column contains the cell bodies of sympathetic preganglionic neurons.

4. The posterior, lateral, and anterior funiculi of the white matter contain bundles of axons having particular functions. Within the nervous system, a bundle of axons with the same function connecting one part of the central nervous system with another is called a *tract* or *pathway.* If

the bundle of axons carries sensory (afferent) information, it is called an *ascending pathway* (spinal cord to brain). These are pathways to the brain for information entering the cord in afferent fibers of spinal nerves. If the information is motor (efferent), it is called a descending pathway (brain to spinal cord). These pathways transfer information from the brain to the motor neurons of the spinal nerves.

The relative amounts of gray and white matter vary at the particular level of the cord (cervical, thoracic, lumbar, and sacral). Gray matter is greater in proportion in the cervical and lumbar enlargements. This reflects the large increase in the number of neurons for the control of the upper and lower extremities. The white matter increases in amount from the lower to upper parts of the cord. The greater amounts of white matter in cervical regions reflect the presence of all the pathways connecting the spinal cord and brain, whereas only those necessary for a given level and below are present at lower segments.

The Major Pathways of the Spinal Cord[2]

ASCENDING PATHWAYS

Information from peripheral sensory receptors is conveyed through the nervous system by a series of neurons that make up ascending pathways. These ascending pathways are usually made up of a series of three neurons, each with a long axon: (1) a first-order neuron, the cell body of which is in the dorsal root ganglion or a sensory ganglion of a cranial nerve; (2) a second-order neuron, whose cell body is in the central nervous system; and (3) a third-order neuron, the cell body being in the thalamus.

Posterior Funiculus. Ascending pathways are present in the posterior (dorsal) funiculus. This region is divided into two smaller fasciculi—the *fasciculus gracilis* and the *fasciculus cuneatus.* Both of these pathways convey sensory information of general kinesthetic or proprioceptive na-

[2]The name given a spinal pathway usually gives information as to whether it is ascending or descending and where the pathway begins and terminates. For example, the spinothalamic tract is ascending, begins in the spinal cord (spino), and ends in the thalamus (thalamic); the rubrospinal tract is descending, begins in the red nucleus (rubro) of the midbrain, and ends in the spinal cord.

ture (i.e., sense of position and movement, touch and pressure sensations). The fasciculus gracilis relays these sensations from the lower extremity and the lower part of the trunk, and the fasciculus cuneatus carries these sensations from the upper extremity, trunk, and neck.

The fibers of fasciculus gracilis are medially placed in the dorsal funiculus and are made up of long ascending fibers from the sacral, lumbar, and lower thoracic dorsal root ganglia. The fibers of fasciculus cuneatus are more laterally placed and are made up of long ascending fibers from upper thoracic and cervical dorsal root ganglia. The fibers of the fasciculus gracilis and cuneatus synapse in the lower portion of the medulla in the nucleus gracilis and cuneatus, respectively. Fibers of the cells of these two nuclei cross to the opposite side in the *decussation of the medial lemniscus,* ascend as the *medial lemniscus,* synapse in the thalamus, and from the thalamus cells send fibers to the postcentral gyrus of the cortex (sensory area).

Lesions that damage these conscious proprioceptive pathways result in defects in muscle sense and in stereognosis. This causes marked disturbances of gait, with the individual stumbling, staggering, and falling.

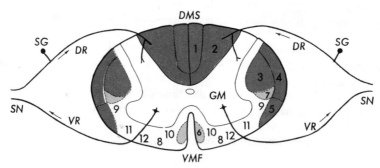

Figure 10–6. Diagram to show general location of some of the conduction paths as seen in a transverse section of the spinal cord. *DMS,* dorsal median sulcus; *DR,* dorsal root; *GM,* gray matter; *SG,* spinal ganglion; *SN,* spinal nerve; *VMF,* ventral median fissure; *VR,* ventral root. *1,* Fasciculus gracilis (tract of Goll); *2,* fasciculus cuneatus (tract of Burdach); *3,* lateral cerebrospinal fasciculus (crossed pyramidal tract); *4,* dorsal spinocerebellar fasciculus; *5,* ventrolateral spinocerebellar fasciculus (Gowers' tract); *6,* ventral cerebrospinal fasciculus (direct pyramidal tract); *7,* rubrospinal tract; *8,* ventral spinothalamic tracts; *9,* lateral spinothalamic tracts; *10,* vestibulospinal tracts; *11,* tectospinal tracts; *12,* olivospinal tracts.

Spinothalamic Pathways. These pathways arise from large cells in the dorsal gray column of the cord, which have received sensory information from peripheral nerves. Most of the fibers in these pathways cross in the cord and ascend in the white matter of the opposite side as the lateral and ventral spinothalamic tracts (Figure 10–6, areas 8 and 9). These fibers eventually terminate in the thalamus. The lateral spinothalamic tract conveys information of pain and temperature; the ventral spinothalamic tract conveys information of touch and pressure.

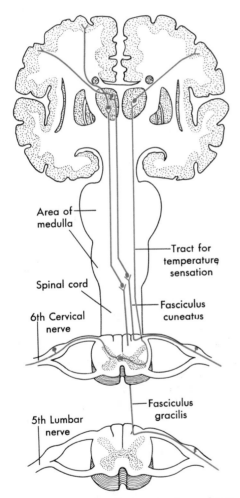

Figure 10–7. Ascending tracts, three shown. Fasciculus (tract) gracilis, concerned with proprioceptor impulses from lower part of body; fasciculus cuneatus, concerned with proprioceptor impulses from upper part of body; and spinothalamic fasciculus, concerned with pain and temperature sensations.

Labels in figure:
Area of medulla
Tract for temperature sensation
Spinal cord
Fasciculus cuneatus
6th Cervical nerve
Fasciculus gracilis
5th Lumbar nerve

Dorsal Spinocerebellar Pathway. This pathway begins from cells in the medial gray of the cord (Clarke's column) that pass to the white matter of the cord on the same side and project directly to the cerebellum via the inferior cerebellar peduncle (restiform body). This pathway conveys unconscious proprioceptive information from muscles and tendons, primarily from the lower extremities and trunk (Figure 10–6, area 4).

Ventral Spinocerebellar Pathway. This arises from cells in the intermediate gray of the cord. Most of the fibers cross to the opposite side of the cord and reach the cerebellum via the superior cerebellar peduncle (brachium conjunctivum). This pathway conveys unconscious proprioceptive information from all parts of the body (Figure 10–6, area 5).

Both spinocerebellar pathways are composed of only two neurons, the first-order neuron being the dorsal root ganglion cells and the second-order neuron the cell bodies in the spinal cord gray matter. Since these spinocerebellar pathways convey impulses from sensory receptors in muscles, tendons, and joints to the cerebellum, this allows the cerebellum to exert a control in regulating voluntary and reflex muscular activity.

DESCENDING PATHWAYS

These pathways convey information from the brain to the spinal cord, or from the brain to lower regions within the brain. The neurons in the brain that form descending motor pathways are known as *upper motor neurons.* This is in contrast to the motor neurons in the anterior gray columns of the spinal cord, which are known as *lower motor neurons.*

Corticospinal Pathways. The corticospinal tracts (cerebrospinal, pyramidal) are long pathways that originate in the cerebral cortex, descend through the internal capsule, midbrain, and pons, and pass through the pyramids of the medulla to end in the spinal cord at the anterior horn

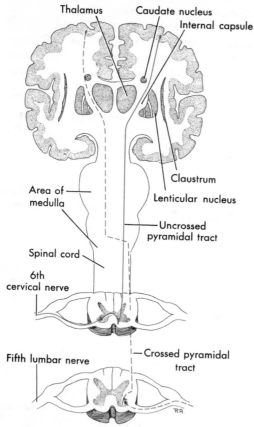

Thalamus   Caudate nucleus
Internal capsule

Claustrum

Area of medulla

Lenticular nucleus

Uncrossed pyramidal tract

Spinal cord

6th cervical nerve

Fifth lumbar nerve

Crossed pyramidal tract

Figure 10–8. Descending tracts, two shown. The lateral corticospinal tract (crossed pyramidal) and the ventral corticospinal tract (direct pyramidal). These tracts contain motor fibers to skeletal muscles.

cell (lower motor neuron). As this tract descends, collateral fibers are given off to the motor nuclei of the cranial nerves. Approximately 85 per cent of the fibers of the corticospinal pathway cross to the opposite side as the pyramidal decussation to make up the *lateral corticospinal tract* of the spinal cord (Figure 10–6, area 3). The remaining fibers remain uncrossed to make up the *anterior (ventral) corticospinal tract* (Figure 10–6, area 6). The anterior corticospinal tract terminates primarily in the cervical segments, whereas the lateral corticospinal tract terminates at all levels of the spinal cord. Both corticospinal tracts convey impulses that bring about volitional movements, especially those concerned with fine movements that are essential for developing motor skills. The

fibers of these tracts are large and heavily myelinated. Myelination begins before birth and is not complete until about the third year. (Compare and contrast the motor movements of the infant and the developing child.)

Vestibulospinal Pathway. This pathway originates from cells in the vestibular nucleus of the medulla and descends on the same side in the cord to terminate at the motor neurons in the ventral gray matter of the cord. Since the vestibular nucleus receives fibers from the vestibular portion of the eighth cranial nerve and from the cerebellum, this tract conveys impulses from the middle ear and cerebellum that exert a modulating influence on the muscles of the extremities and trunk, thereby helping to maintain equilibrium and posture (Figure 10–6, area 10).

Rubrospinal Pathway. This tract originates from cells in the red nucleus (rubro) of the midbrain. The fibers cross and descend in the cord and terminate around cells in the dorsal part of the ventral gray of the cord of the thoracic region. The red nucleus relays impulses from the cerebellum and vestibular apparatus to the motor nuclei of the brain stem and spinal cord and in this way participates in the coordination of reflex postural adjustments.

Tectospinal Pathway. This tract originates from cells in the colliculi of the midbrain. The fibers cross and descend in the cord to terminate around cells in the ventral gray. The fibers convey impulses that mediate reflex activity of the muscles of the head and neck in response to optic stimuli and perhaps auditory stimuli (Figure 10–6, area 11).

Olivospinal Pathway. This tract originates from cells in the olivary nucleus of the medulla and perhaps other higher centers of the brain and terminates at the motor neuron cells in the ventral gray matter of the cord. The olivary nucleus receives and sends impulses to the cerebellum, thereby interacting with this center of equilibrium. The olivospinal tract is probably con-

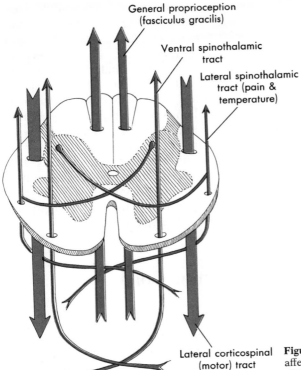

General proprioception
(fasciculus gracilis)

Ventral spinothalamic
tract

Lateral spinothalamic
tract (pain &
temperature)

Lateral corticospinal
(motor) tract

Figure 10–9. Diagram showing location of major afferent and efferent spinal tracts and the direction of impulse.

Figure 10–10. Motor (*red*) and sensory (*blue*) conduction paths and reflex arcs of the spinal cord. (Modified from Toldt.)

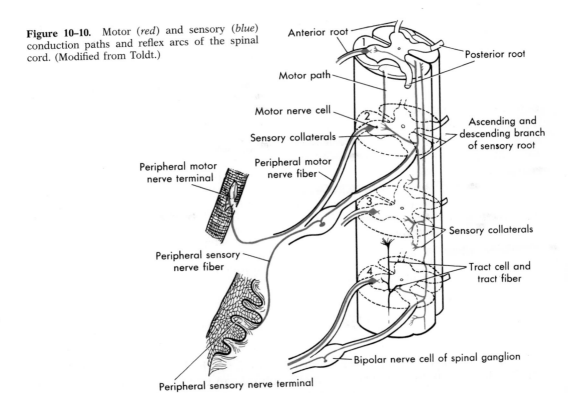

Anterior root

Posterior root

Motor path

Motor nerve cell

Sensory collaterals

Ascending and
descending branch
of sensory root

Peripheral motor
nerve terminal

Peripheral motor
nerve fiber

Peripheral sensory
nerve fiber

Sensory collaterals

Tract cell and
tract fiber

Peripheral sensory
nerve terminal

Bipolar nerve cell of spinal ganglion

cerned with the coordination of movements of the head and arms (Figure 10–6, area 12).

All these descending pathways terminate ultimately on the motor neurons of the ventral gray of the cord in either excitatory or inhibitory synapses. The combined effect of these descending impulses determines the frequency of discharge of the motor neurons, hence the strength of skeletal muscle contraction in reflex or volitional movement.

Other Descending Pathways. There are also descending fibers in the cord that terminate on the autonomic preganglionic motor neurons innervating smooth muscle, cardiac muscle, and glandular epithelium. The hypothalamus is the chief coordinating center of the autonomic system. Activities of the hypothalamus, in turn, are coordinated with those of the thalamus and cortex. The fibers descend from the hypothalamus in a rather diffuse manner and terminate on autonomic cells in the ventrolateral gray of the cord.

In addition to long descending fiber tracts, there are short fiber tracts connecting neighboring segments of the cord, forming part of the intrinsic, or segmental, reflexes of the spinal cord.

Functions of the Spinal Cord

Two major functions are apparent from the foregoing discussion: reflexes and transmission of information. Afferent impulses are forwarded upward; efferent impulses are transmitted downward. The motor neurons in the anterior column at each level of the spinal cord are bombarded with impulses from many pathways. Some are inhibitory, others facilitatory. Thus impulse discharge by this motor neuron is in response to a "summation" of these influences, and for this reason it is called the "final common pathway."

Reflexes Involving the Spinal Cord. Some of the reflexes that involve the spinal cord are (1) those concerned with withdrawal from harmful stimuli, called *flexion reflexes;* walking involves the flexion reflexes; (2) the *extensor* or *stretch reflex,* such as the knee jerk and those concerned with posture and muscle tone maintenance; (3) the *scratch reflex,* or responses to local irritation.

Some spinal cord reflexes are very complex and involve many segments of the spinal cord. In some instances cord reflexes involve the viscera, as in the reflex that empties the bladder. Afferent nerve impulses from receptors in the bladder wall enter the sacral and lumbar regions of the cord, synapse on internuncial neurons, which in turn excite motor neurons that contract and empty the bladder. The same sensory impulses travel to higher brain levels and provide the conscious sensation of a full bladder.

Reflexes whose adjusting mechanism is in the spinal cord may be inhibited by centers in the cerebrum. Micturition is an example. Micturition is initially brought about as a spinal reflex, the stimulus starting from receptors in the bladder itself, so that when the bladder is filled, it automatically empties itself. By training in early infancy, this reflex may be inhibited from the cerebrum, so that micturition takes place only under voluntary control. The same is true of defecation. These are examples of "modulated" primitive reflexes. If the conducting paths to the cerebrum are interrupted, but the spinal reflex path is intact, involuntary micturition and defecation may still be possible.

The Spinal Nerves

There are 31 pairs of spinal nerves, arranged in the following groups, and named for the region of the vertebral column from which they emerge:

Cervical	8 pairs
Thoracic	12 pairs
Lumbar	5 pairs
Sacral	5 pairs
Coccygeal	1 pair

The first cervical nerve arises from the medulla oblongata and leaves the spinal canal between the occipital bone and the atlas. The other cervical spinal nerves arise from the spinal cord, and each leaves the

spinal canal through an intervertebral foramen *above* the vertebra whose number it bears; e.g., the seventh thoracic nerve emerges through the foramen between the sixth and seventh vertebrae. The eighth spinal nerve emerges from the vertebral column below the seventh cervical vertebra. All the other spinal nerves emerge from the cord below the vertebra whose number it bears. The coccygeal nerve passes from the lower extremity of the canal.

Mixed Nerves. The spinal nerves consist mainly of myelinated nerve fibers and are called mixed nerves because they contain both motor and sensory fibers. Each spinal nerve has two roots, a ventral root and a dorsal root. The fibers of the ventral root *arise from nerve cells comprising the gray matter* in the ventral column and convey motor impulses from the spinal cord to the periphery.

The fibers of the dorsal root arise from the cells composing the enlargement, or *ganglion,* of the dorsal root situated in the openings between the arches of the vertebrae. These cells are unipolar, giving off a single fiber that divides into two processes. One extends to a sensory end organ of the skin or of a muscle, tendon, or joint, and the other extends into the spinal cord, forming the dorsal root of the spinal nerve, and conveys sensory information from the periphery to the spinal cord.

Branches and Distribution of Spinal Nerves. After leaving the spinal column, each spinal nerve divides into two major branches or rami (ramus means branch), the *dorsal branch* and the *ventral branch,* and a third smaller branch, the *meningeal* or *recurrent* (distributed to the meninges). The dorsal branches supply the muscles and skin of the back of the head, neck, and trunk. The ventral branches supply the extremities and parts of the body wall in front of the spine. There is a fourth, or *visceral,* branch, *present only in nerves from the first thoracic to the third lumbar.* These connect with the sympathetic ganglia by means of fibers that pass from the nerve to the ganglia (the white and gray

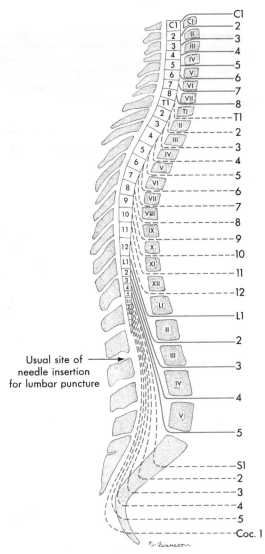

Usual site of needle insertion for lumbar puncture

Figure 10–11. Diagram showing segmental relationships of the spinal cord to the vertebral column. Note the site of origin of each spinal nerve in the cord and its point of exit from the vertebral column.

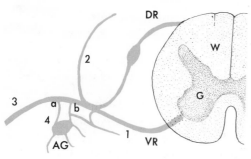

Figure 10–12. The four branches of a typical thoracic spinal nerve. *1*, Meningeal, or recurrent, branch; *2*, dorsal branch; *3*, ventral branch; *4*, visceral branch; *a*, gray ramus; *b*, white ramus; *AG*, autonomic ganglion; *DR*, dorsal root of spinal nerve; *G*, gray matter; *VR*, ventral root of spinal nerve; *W*, white matter.

rami, see pages 300–302). Extending from the sympathetic ganglia to their final distribution are the autonomic nerves. These nerves form plexuses called the cardiac, the celiac or solar, the hypogastric, the pelvic, and the enteric. Some nerve fibers from the sympathetic ganglia return to and are distributed with the spinal nerve to supply sweat glands, arrector pili muscles, and the smooth muscle of all blood vessels.

Nerve Plexuses. With the exception of the thoracic spinal nerves, the ventral rami of the other spinal nerves merge together in a complex intertwining and interchanging of fibers that are known as plexuses. From plexuses specific peripheral nerves are formed. In this way the nerve has fibers that represent more than one spinal cord segment. For example, the radial nerve is made up of fibers from the fifth through eighth cervical spinal cord segments and the first thoracic spinal cord segment. In effect, therefore, damage of the sixth spinal nerve would not totally destroy the functional capacity of the radial nerve.

In the thoracic region a plexus is not formed, but the fibers pass as intercostal nerves out into the intercostal spaces to supply the intercostal muscles, the upper abdominal muscles, and the skin of the abdomen and chest.

The Cervical Plexus. This plexus is formed by the ventral branches of the first four cervical nerves. The second, third, and fourth nerves divide into an upper and a lower branch; these in turn unite to form three loops from which peripheral nerves are distributed. Some fibers of these nerves join the hypoglossal, vagus,

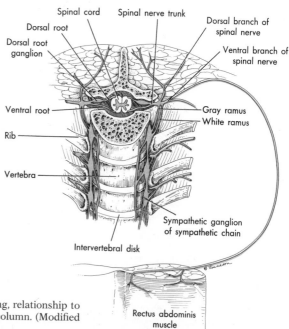

Figure 10–13. Diagram of spinal nerve branching, relationship to spinal cord, sympathetic ganglia, and vertebral column. (Modified from Pansky and House.)

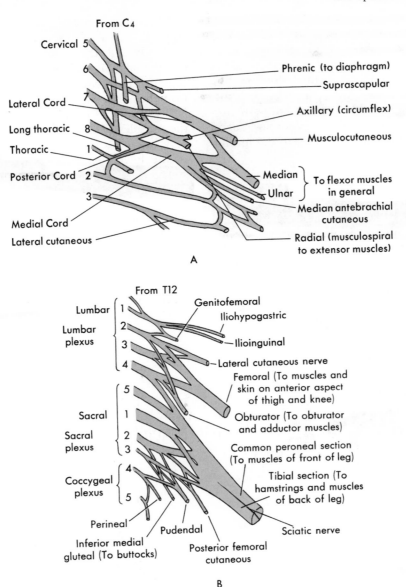

From C₄
Cervical 5
6
7
Lateral Cord
Long thoracic 8
Thoracic 1
Posterior Cord 2
3
Medial Cord
Lateral cutaneous

Phrenic (to diaphragm)
Suprascapular
Axillary (circumflex)
Musculocutaneous
Median
Ulnar
} To flexor muscles in general
Median antebrachial cutaneous
Radial (musculospiral to extensor muscles)

A

From T12
Lumbar 1
Lumbar plexus 2
3
4
Sacral 5
Sacral 1
plexus 2
3
Coccygeal 4
plexus 5
Perineal
Inferior medial gluteal (To buttocks)
Pudendal
Posterior femoral cutaneous

Genitofemoral
Iliohypogastric
Ilioinguinal
Lateral cutaneous nerve
Femoral (To muscles and skin on anterior aspect of thigh and knee)
Obturator (To obturator and adductor muscles)
Common peroneal section (To muscles of front of leg)
Tibial section (To hamstrings and muscles of back of leg)
Sciatic nerve

B

Figure 10–14. *A.* Diagram of brachial plexus, showing distribution of the larger nerves. See page 194 for distribution of the nerves. *B.* Diagram of lumbosacral plexus and distribution of some of the larger nerves. See page 194 for distribution of the nerves.

and accessory cranial nerves in their course.

The Brachial Plexus. This plexus is formed by the union of the ventral branches of the last four cervical and the first thoracic nerves. See Figure 10–14 for the fiber composition and nerves formed. Three major cords are formed, the lateral, medial, and posterior, from which the five major nerves supplying the upper extremities arise, namely the median, ulnar,

radial, axillary, and musculocutaneous nerves.

The Lumbar Plexus. This plexus is formed by a few fibers from the twelfth thoracic and the anterior primary divisions of the first four lumbar nerves. Figure 10–14 illustrates the peripheral distribution of the fibers to the muscles. The largest nerves formed are the femoral and obturator nerves.

The Sacral Plexus. This plexus is

formed by a few fibers from the fourth lumbar nerve, all of the fifth, and the first, second, and third sacral nerves. Figure 10–14 illustrates the joining of these fibers and the great nerves formed. The largest is the great sciatic, which supplies the muscles of the lower extremity. Many neuroanatomists group the lumbar and sacral plexuses together and refer to the lumbosacral plexus.

Names of Peripheral Nerves. Many of the larger branches given off from the spinal nerves bear the same name as the artery they accompany or the part they supply. Thus, the radial nerve passes down the radial side of the forearm in company with the radial artery; the intercostal nerves pass between the ribs in company with the intercostal arteries. Exceptions to this are the two sciatic nerves, which pass down from the sacral plexus, one on each side of the body near the center of each buttock and the back of each thigh, to the popliteal region, where each divides into two large branches that supply the legs and feet. Motor branches from these nerves pass to the muscles of the legs and feet, and sensory branches are distributed to the skin of the lower extremities. A chart of the spinal nerves and their distribution appears on page 194.

Degeneration and Regeneration of Nerves. Since the cell body is essential for the nutrition of the whole cell, it follows that if the processes of a neuron are cut off, they will suffer from malnutrition and die. If, for instance, a spinal nerve is cut, all the peripheral part will degenerate, because the fibers have been cut off from their cell bodies. The cut connective

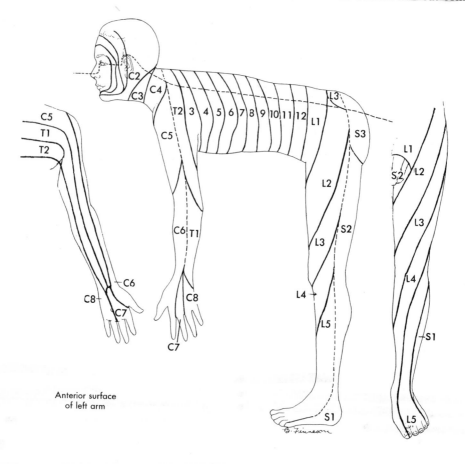

Figure 10–15.   Distribution of spinal nerves. (Modified from Pansky and House.)

tissue framework reunites, but the cut ends of the fibers cannot unite, for the peripheral or severed portion of the nerve fiber begins to degenerate. Its medullary sheath breaks up into a mass of fatty molecules and is gradually absorbed, and finally the central fiber also disappears. In regeneration, many new fibers start to grow out from the central end of the severed axon. If one of these "growth sprouts" makes a successful penetration into the peripheral end of the neurilemma, all other growth sprouts wither and die. The successful sprout continues to grow toward the end organ and is destined to become the new functional fiber. In time it will become surrounded with a myelin sheath. The Schwann cells of peripheral fibers play an important part in both the degeneration and regeneration of the cut fiber. Restoration of function in the nerve may not occur for several months, during which time it is presumed the new nerve fibers are slowly finding their way along the course of those that have been destroyed. Since the nerve fibers of the brain and cord have little or no neurilemma, regeneration after injury does not occur.

Questions for Discussion

1. List the names of the membranes that cover the spinal cord in correct order, beginning from the outermost covering.
2. What is the cauda equina?
3. Which two ascending pathways are located in the posterior funiculus? What is the type of sensory information conveyed in these pathways?
4. Which ascending pathway is responsible for conveying information of pain and temperature?
5. What is meant by the terms *upper* and *lower motor neurons?*
6. Differentiate between the symptoms of an upper and a lower motor neuron lesion.
7. Trace the corticospinal tracts (both anterior and lateral) from their origins to terminations.
8. Differentiate between the symptoms caused by a crush injury to the fourth lumbar vertebra and an injury that involved the tenth thoracic vertebra.
9. An injury to the shoulder that pulled excessively on the arm and dislocated the shoulder might injure what nerve fibers?
10. What is the distribution of the terminal branches of a spinal nerve?

Summary

Spinal Cord Location and Shape
1. Located in spinal canal
2. Average length, 45 cm
3. Originates at level of foramen magnum and extends downward to end between first and second lumbar vertebrae
4. Cylindrical in shape; two enlargements, one at the cervical region and another at the lumbar region
5. Tapers at its inferior end into conus medullaris, from which projects filum terminale

Spinal Cord Protected by

Membranes
Dura mater—outer membrane
Arachnoid mater—inner membrane
Pia mater—closely invests spinal cord

Bony Vertebral Column
Fluid Medium—the cerebrospinal fluid

Appearance in Transverse Section
1. Central region of gray matter in shape of letter H surrounded by white matter
2. In center of gray matter is central canal
3. Region of gray matter around central canal is *gray commissure*
4. Gray matter divided into columns or horns
   Posterior
   Anterior
   Intermediolateral
5. White matter divided into funiculi
   Posterior
   Anterior
   Lateral

Pathways of Spinal Cord	**Major Ascending Pathways**—convey sensory information to brain	Fasciculus gracilis / Fasciculus cuneatus	Convey information of general kinesthesia, touch, and pressure
		Spinothalamic pathways	*Lateral spinothalamic*—pain and temperature / *Ventral spinothalamic*—touch and pressure
		Spinocerebellar pathways	*Dorsal spinocerebellar*—unconscious proprioceptive information from muscles and tendons, primarily from lower extremities and trunk / *Ventral spinocerebellar*—unconscious proprioceptive information from all parts of the body
	Major Descending Pathways—convey information from brain to spinal cord or from higher centers to lower centers within brain	Corticospinal pathways	*Lateral and anterior*—both convey impulses that bring about volitional movements involved with fine movements
		Vestibulospinal pathways	Impulses from inner ear and cerebellum, influence on muscles of extremities and trunk, thereby maintaining equilibrium and posture
		Rubrospinal pathway	Participates in coordination of reflex postural adjustments
		Tectospinal pathway	Participates in reflex activity of muscles of head and neck in response to optic and auditory stimuli
		Olivospinal pathway	Involved in coordination of movements of head and arms
Spinal Nerves	**Number**	Cervical / Thoracic / Lumbar / Sacral / Coccygeal	8 pairs / 12 pairs / 5 pairs / 5 pairs / 1 pair _____ 31 pairs
	Variety	Myelinated / Mixed { Sensory / Motor }	
	Origin—two roots	Ventral, or motor, in gray matter of cord / Dorsal, or sensory, in spinal ganglia	
	Distribution— four main branches	Ventral supplies extremities and parts of body in front of spine / Dorsal supplies muscles and parts of the body in back of the spine / Visceral extends to lateral chain of ganglia and viscera / Meningeal, or recurrent, branch returns to meninges	
Nerve Plexuses	*Major Plexuses*	*Ventral Rami of Spinal Nerves*	
	Cervical / Brachial / Lumbar / Sacral	C_1 through C_4 / C_5 through T_1 / T_{12} through L_4 / L_4 through S_3	

Additional Readings

BARR, M. L.: *The Human Nervous System,* 2nd ed. Harper & Row, Publishers, New York, 1974, Chapter 5.

GUYTON, A. C.: *Basic Human Physiology: Normal Function and Mechanisms of Disease.* W. B. Saunders Co., Philadelphia, 1971, Chapter 37.

NOBACK, C. R., and DEMAREST, R. J.: *The Human Nervous System: Basic Principles of Neurobiology,* 2nd ed. McGraw-Hill Book Co., New York, 1975, Chapter 5.

PEARSON, K.: The control of walking. *Sci. Amer.,* 235: 72–86, December, 1976.

ROMANES, G. J. (ed.): *Cunningham's Textbook of Anatomy,* 11th ed. Oxford University Press, New York, 1972, pp. 734–74.

VANDER, A. J.; SHERMAN, J. H.; and LUCIANO, D. S.: *Human Physiology: The Mechanisms of Body Function,* 2nd ed. McGraw-Hill Book Co., New York, 1975, Chapter 17.

WILLIAMS, P. L., and WARWICK, R.: *Functional Neuroanatomy of Man.* W. B. Saunders Co., Philadelphia, 1975, pp. 806–40.

The Brain and Cranial Nerves

Chapter Outline

The brain is the center that receives, integrates, stores, and responds to all sensory information. It is the brain that controls and modulates all the activities of the body.

The Brain

The brain lies in the cranial cavity, a bony cavity that offers protection, as do the meninges covering the brain and the cerebrospinal fluid. The average weight of the adult brain is about 1400 gm (3 lb), which is approximately 2 per cent of the total body weight. The brain is divided into three major areas: the *forebrain*, the *midbrain*, and the *hindbrain* (Figure 11–1). These areas, in turn, are further subdivided. Each area of the brain is associated with a particular part of the ventricular system of the brain (cavities within the brain containing cerebrospinal fluid).

Forebrain (prosencephalon)

A. Cerebral hemispheres (telencephalon)
  1. Cerebral cortex ⎫ Lateral
  2. Basal ganglia (nuclei) ⎭ ventricles

B. Diencephalon
  1. Thalamus ⎫ Third
  2. Hypothalamus ⎭ ventricle

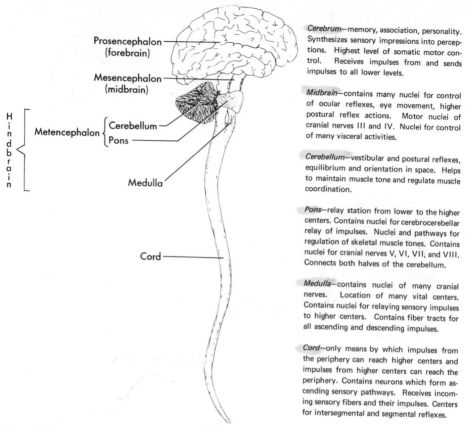

Prosencephalon (forebrain)

Mesencephalon (midbrain)

Metencephalon { Cerebellum / Pons

Hindbrain

Medulla

Cord

Cerebrum—memory, association, personality. Synthesizes sensory impressions into perceptions. Highest level of somatic motor control. Receives impulses from and sends impulses to all lower levels.

Midbrain—contains many nuclei for control of ocular reflexes, eye movement, higher postural reflex actions. Motor nuclei of cranial nerves III and IV. Nuclei for control of many visceral activities.

Cerebellum—vestibular and postural reflexes, equilibrium and orientation in space. Helps to maintain muscle tone and regulate muscle coordination.

Pons—relay station from lower to the higher centers. Contains nuclei for cerebrocerebellar relay of impulses. Nuclei and pathways for regulation of skeletal muscle tones. Contains nuclei for cranial nerves V, VI, VII, and VIII. Connects both halves of the cerebellum.

Medulla—contains nuclei of many cranial nerves. Location of many vital centers. Contains nuclei for relaying sensory impulses to higher centers. Contains fiber tracts for all ascending and descending impulses.

Cord—only means by which impulses from the periphery can reach higher centers and impulses from higher centers can reach the periphery. Contains neurons which form ascending sensory pathways. Receives incoming sensory fibers and their impulses. Centers for intersegmental and segmental reflexes.

Figure 11–1. Diagram of the major parts of the brain. The cerebellum has been displaced to show midbrain.

Midbrain (mesencephalon)
  A. Cerebral peduncles   } Cerebral
  B. Corpora quadrigemina } aqueduct
Hindbrain (rhombencephalon)
  A. Cerebellum } (meten-
  B. Pons   } cephalon) } Fourth
  C. Medulla   (myelen- } ventricle
           cephalon)

The brain stem includes the diencephalon, midbrain, pons, and medulla. It is most clearly evident when the cerebral hemispheres and cerebellum are removed.

DEVELOPMENT AND GROWTH

The anatomic organization of the brain is complex. A key to an understanding of the organization of the adult brain is the study of the development of the brain. Along with the cardiovascular system, the nervous system is one of the first systems to function during embryonic life. The nervous system is derived from ectoderm and develops from a region called the neural plate. The neural plate elongates and its lateral edges are thereby raised, forming neural folds. These folds meet to form a neural tube. The anterior portion of the neural tube develops into the brain, and the lower portion becomes the spinal cord. The cavity of the neural tube persists in the adult as the *ventricles* of the brain and the *central canal* of the spinal cord. As the nervous system develops, the cephalic end of the neural tube enlarges into three areas called the "primary brain vesicles." These three areas are the forerunners of the three divisions of the brain—namely, the prosencephalon, or *forebrain;* the mesencephalon, or *midbrain;* and the rhombencephalon, or *hindbrain* (Figure 11–2).

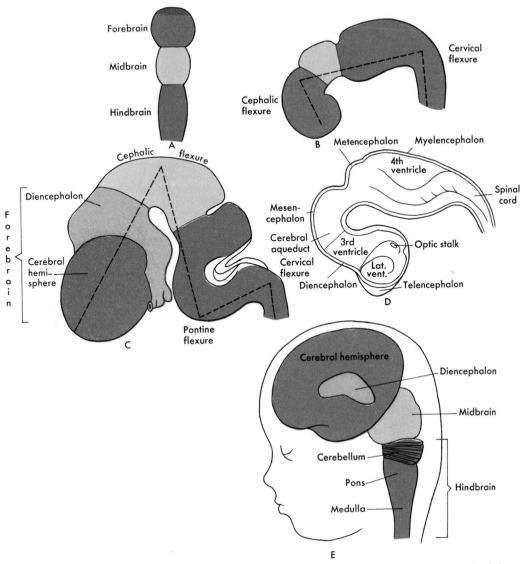

Figure 11–2. Stages in the development of the human brain. *A.* Early three-vesicle stage. *B* and *C.* Flexures of brain. *D.* Hemisection in early stage of development showing cavities of brain. *E.* The brain at three months.

The brain begins as a simple tube; however, the enlargements of the primary brain vesicles increase the dimensions of the brain so that it exceeds the space of the cranial cavity. A very simple means is employed to fit this enlarged brain into the cranial cavity, and this is folding. The development of a "folded" brain from a tubelike structure is the result of the complex consolidation of the following processes: (1) three bends known as flexures; (2) the growth of portions of the cerebral hemispheres *over* the diencephalon, midbrain, and cerebellum; and (3) the formation of sulci and gyri in the cerebral cortex and the cerebellum. The three flexures are the cephalic, cervical, and pontine. The cephalic flexure remains the most obvious and is evident in the final structure of the brain. Due to this flexure the forebrain is placed in a horizontal position over the midbrain and hindbrain. In man there is an enormous development of the cerebral hemispheres of the forebrain. This creates

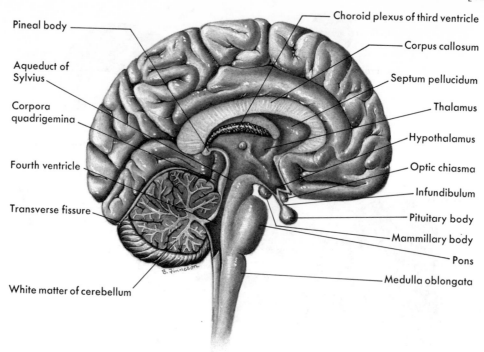

Figure 11–3. Midsagittal section of the brain. The cerebral hemisphere covers the brain stem and the cerebellum.

DIVISIONS OF THE BRAIN AND FUNCTION

FOREBRAIN

The two major divisions of the forebrain are the cerebral hemispheres and the diencephalon. The cerebral hemispheres are subdivided into the cerebral cortex and the basal ganglia (nuclei). These structures surround the lateral ventricles of the brain.

Cerebral Hemispheres. The two cerebral hemispheres connected by the *corpus callosum* are referred to as the *cerebrum*. The cerebrum is the largest part of the brain and fills the whole upper portion of the skull. The entire surface, both upper and lower, is composed of layers of gray matter and is called the *cortex*. Beneath the gray matter is white matter. The white matter of the cerebrum consists of three types of fibers. These are projection fibers, association fibers, and commissural fibers. *Projection fibers* include (1) descending pathways, which originate in the cortex (axons of pyramidal cells) and project to other areas of the brain and spinal cord, and (2) ascending pathways, primarily from the thalamus, which project and terminate in the cortex. *Association fibers* are the axons of pyramidal cells that extend to other areas of the cortex within the same cerebral hemispheres. *Commissural fibers* are the axons of pyramidal cells that interconnect one area of one cerebral hemisphere with its counterpart in the other cerebral hemisphere. An example of commissural fibers is the large extensive tract of white matter seen connecting the two hemispheres, the corpus callosum. These three types of fibers of the white matter are the means of linkage of all the areas of the brain and the brain and spinal cord (Figure 11–5).

a situation in which the cerebral hemispheres cover the rest of the brain (this can be seen in the sagittal section of the brain in Figure 11–3). The formation of sulci and gyri is another expression of growth and accommodation to a given space.

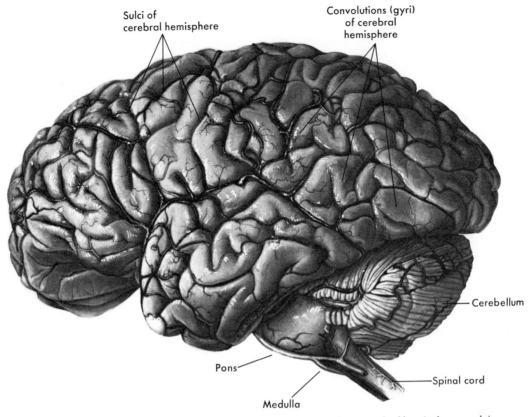

Sulci of
cerebral hemisphere

Convolutions (gyri)
of cerebral
hemisphere

Cerebellum

Pons

Spinal cord

Medulla

Figure 11–4. Lateral view of the brain. Note the blood vessels, the large cerebral hemisphere overlying the cerebellum, pons, and medulla.

Fissures, Convolutions, and Lobes of the Cerebral Hemispheres. In the early stages of embryonic development the surface of the cerebrum is comparatively smooth. As fetal development progresses, the surface becomes covered with furrows, which vary in depth. The deeper furrows are called *fissures,* the shallow ones *sulci,* and the ridges between sulci are called *gyri* or *convolutions.* The patterns formed on the surface of the brain by the sulci and gyri are variable, and no two brains have precisely the same pattern. Certain fissures and sulci, however, are constant and represent landmarks that localize specific functional areas and divide each hemisphere into lobes. Each cerebral hemisphere is divided into the frontal, parietal, occipital, and temporal lobes and the insula (island of Reil[1]). The frontal, pari-

etal, occipital, and temporal lobes are evident on gross observation of the brain and correspond in location to the bones of the skull. The insula can only be seen on

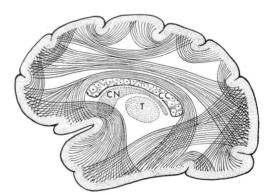

Figure 11–5. Section of brain, showing association fibers connecting the gyri and commissural fibers connecting the two sides of the brain. The commissural fibers are cut across and appear as dots. The myelin sheaths surrounding these fibers appear as white areas. *CC,* corpus callosum; *CN,* caudate nucleus; *T,* thalamus.

[1] Johann Christian Reil, Dutch physiologist (1759-1813).

dissection, as it is covered by the frontal and parietal lobes.

The *longitudinal cerebral fissure* is beneath the sagittal suture of the skull. This fissure separates the two cerebral hemispheres. The separation is complete anteriorly and posteriorly, but the middle region is made up of the wide band of commissural fibers, the *corpus callosum*, which unites the two hemispheres.

The *transverse fissure* is located between the cerebrum and the cerebellum. Whereas the longitudinal fissure divides the cerebrum into two hemispheres, and the transverse fissure divides the cerebrum from the cerebellum, the other important fissures and sulci divide each hemisphere into lobes.

The *central sulcus* or *fissure of Rolando* is situated approximately beneath the coronal suture of the skull. This sulcus is the boundary between the *frontal* and *parietal* lobes (Figure 11–6). The *lateral cerebral fissure* or *fissure of Sylvius* is found on the lateral surface of the hemispheres and separates the *frontal* and *temporal* lobes (Figure 11–6). The *parieto-occipital fissure* is the least obvious. There is not a marked separation of the *occipital lobe* from the *parietal* and *temporal lobes;* however, the parieto-occipital fissure serves as a boundary (Figure 11–6).

The fissures and lobes of the cerebrum that have thus far been discussed are all evident on the surface of the hemispheres. The fifth lobe, the insula (island of Reil), is not seen when the surface of the hemisphere is examined, because it lies within the lateral cerebral fissure, and the overlying convolutions of the parietal and frontal lobes must be lifted up before the insula may be seen.

Localized Areas of Function. In higher vertebrates the cerebrum constitutes a larger proportion of the brain than in lower forms. The cerebrum contains areas that govern all mental activities, such as reason, intelligence, will, and memory. It is the discriminating area of consciousness, the interpreter of sensations (correlation), and the instigator and coordinator of voluntary acts; it exerts strong control (both facilitatory and inhibitory) over many reflex acts, such as laughing, weeping, micturition, and defecation.

Consciousness and memory are two areas of cerebral activity that encompass much or all of its other more specific activities. The conscious brain is kept aware of environmental changes by way of afferent nerve impulses and responds appropriately. The unconscious brain fails to respond to these changes; only very basic physiologic activities and reflexes persist during unconsciousness. For example, the cardiovascular and respiratory systems continue to function.

As the result of numerous experiments on animals and close observation of the effects of electrical stimulation of the cerebral cortex on human individuals and clinical results of cerebral disease, physiologists have been able to localize certain areas in the brain that control motor, sensory, and other activities (Figure 11–7).

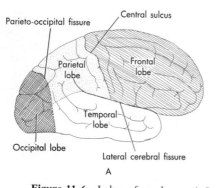

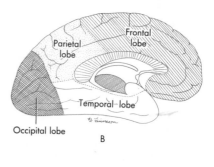

Figure 11–6. Lobes of cerebrum. *A.* Lateral view. *B.* Medial view, brain stem removed.

Some knowledge has been gained concerning the areas in the cerebrum that are concerned with higher mental activities. In no case, however, is the control of a function limited to a single center, for practically all mental processes involve the discharge of nervous energy from one center to another. All parts of the cerebrum interconnect. Change in the nervous activity of any one part alters the excitability of the whole. Any activity, therefore, is the result of all the changes throughout the whole of the cortex.

The portions of the cerebrum that govern muscular movement are known as *motor areas,* those controlling sensation as the *sensory areas,* and those connected with the higher faculties, such as reason and will, as *association areas.*

Motor Areas. The surface of the brain involved in the function of movement is the precentral gyrus of the frontal lobe, i.e., the gray matter immediately in front of the central sulcus. The large pyramidal cells whose fibers form corticospinal pathways are located in the precentral gyrus and arranged so that motor cells for toe movement are located in the lowest area on the medial side of the cortex and motor

cells for face movement are located near the lateral cerebral fissure. Figure 11–8 illustrates the spatial arrangement of both motor and sensory areas.

Sensory Areas. The somatic sensory area occupies the part of the cortex behind the central sulcus and can be divided into regions like those of the motor area just in front of the sulcus. The *visual area* is situated in the posterior part of the occipital lobe and the *auditory area,* in the superior part of the temporal lobe.

The area for the sense of *taste* has been located deep in the fissure of Sylvius near the island of Reil. See Figure 11–7 for *alimentary system* areas. The area concerned with interpretation of the sense of *smell* is located on the medial aspect of the temporal lobe (uncus).

The *motor speech* area is located in the frontal lobe, anterior to the laryngeal area in the premotor region of the motor cortex. It is known as Broca's area. The temporal speech region is believed to be concerned with choice of thoughts to be expressed, and the parietal region, with choice of words used in the expression of thoughts. In right-handed persons the area is more fully developed in the left hemi-

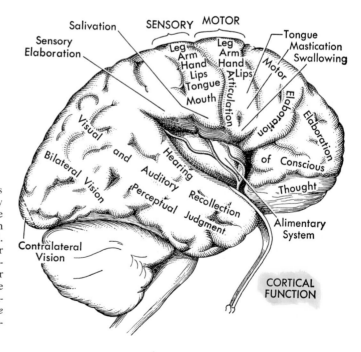

Figure 11–7. Cortical function. This illustration will serve as a summary restatement of conclusions, some hypothetical (e.g., the elaboration zones), others firmly established. The suggestion that the anterior portion of the occipital cortex is related to both fields of vision rather than to one alone is derived from the results of stimulation. (From Penfield, W., and Rasmussen, T.: *The Cerebral Cortex of Man.* The Macmillan Co., New York, 1950.)

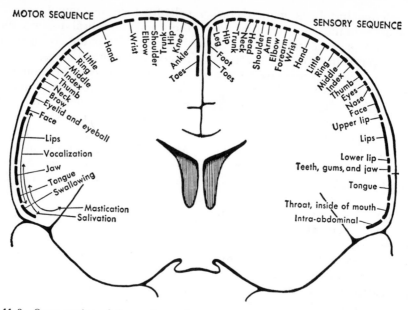

Figure 11–8. Cross section of the cerebrum through the sensorimotor region with the motor and sensory sequences indicated. The lengths of the solid bars represent an estimate of the average relative cortical areas from which the corresponding responses were elicited. (From Penfield, W., and Rasmussen, T.: *The Cerebral Cortex of Man.* The Macmillan Co., New York, 1950.)

sphere; and in left-handed persons, in the right hemisphere. The basis of language is a series of memory pictures. The mind must know and recall the names of things in order to mention them; it must have seen or heard things in order to describe them and to have learned the words to express these ideas. Even this is not enough. All these factors must work together under the influence of the center for articulate speech, which is in close connection with those for the larynx, tongue, and the muscles of the face. Injury to these centers results in some form of inability to speak (aphasia), to write (agraphia), or to understand spoken words (word deafness) or written words (word blindness). It is customary, therefore, to distinguish two types of aphasia, i.e., motor and sensory. By *motor aphasia* is meant the condition of those who are unable to speak although there is no paralysis of the muscles of articulation. By *sensory aphasia* is meant the condition of those who are unable to understand written, printed, or spoken symbols of words, although the sense of vision and that of hearing are unimpaired. These centers are really memory centers, and

aphasia is due to loss of memory of words, meaning of words seen or heard, or formation of letters.

Association Areas. These are cortical areas that surround the motor and sensory areas. They are made up of association fibers that connect motor and sensory areas. Association areas are essential to general sensations, such as the recognition and comprehension of weight, shape, texture, and form, and for the recognition of the written and spoken word. Animals that are capable of acquiring conditioned reflexes have a greater development of these areas. It is thought that the association areas are plastic and register the effects of individual experience.

After removal of the cerebrum, any animal becomes a simple reflex animal. In other words, all its actions are then removed from volition and consciousness. All responses that depend upon memory of acquired experience are lost.

Rhinencephalon—Limbic System. The term *rhinencephalon* means olfactory brain. It is anatomically represented by the olfactory bulb and tract and those areas of the cerebral hemispheres that

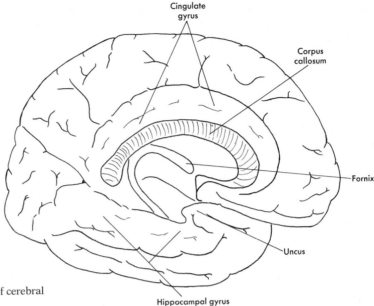

Figure 11–9. Medial surface of cerebral hemisphere.

receive and integrate olfactory impulses. The areas of the cerebral hemispheres involved with olfaction are present on the medial surfaces of the hemispheres and include the cingulate gyrus and hippocampus (Figure 11–9). In addition to the reception and integration of olfactory information, however, the rhinencephalon is involved in behavioral and emotional expressions. In order to separate these functions from olfaction, the term *limbic system* is used. The limbic system has connections with the amygdala (one of the basal ganglia), thalamus, hypothalamus,

and midbrain. The regions of the brain comprising the limbic system are represented in Figure 11–10. The connections of the limbic system with the hypothalamus, which is the major center regulating the autonomic nervous system, play a major role in the autonomic reactions that are associated with emotional and behavioral patterns. The limbic system is often referred to as the visceral brain. Within this system both "pleasure" and "punishment" centers have been identified in animals by means of localized electrical stimulation. Continual stimulation of a "pleasure"

Figure 11–10. Diagrammatic representation of the limbic system.

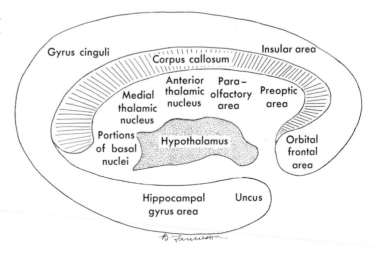

center, for example, will cause the animal to suffer great pain in order to achieve continual electrical stimulation of such a center. Stimulation of a "punishment" center causes the animal to avoid further stimulation.

Basal Ganglia (Nuclei): Location and Function. The term *basal ganglia* (nuclei) refers to masses of gray matter situated deep within the cerebral hemispheres. These basal ganglia are in close relationship to parts of the diencephalon (thalamus, hypothalamus) and are functionally integrated to motor activities. Included under the basal ganglia are the *caudate nucleus;* the *lentiform* or *lenticular nucleus*, which is subdivided into the *putamen* and the *globus pallidus;* the *amygdaloid nuclear complex;* and the *claustrum.* The caudate nucleus is composed of a head, body, and tail. Its relationship to the thalamus, the lenticular nucleus, and the amygdaloid nucleus is shown in Figure 11–11. The claustrum is a thin plate of gray matter that lies between the lenticular nucleus and the cortex of the insula.

Related both anatomically in location and functionally to the basal ganglia is a region of white matter called the *internal capsule.* The internal capsule contains corticospinal, corticobulbar, and sensory fibers. In appropriate sections of the brain the internal capsule and the lentiform nucleus appear striated and, therefore, are collectively referred to as the *corpus striatum.*

The basal ganglia, in addition to nuclei of other regions of the brain, are collectively referred to as the *extrapyramidal system.* This system is involved in somatic motor functions, as is the pyramidal (corticospinal) system. The functions of some of the basal ganglia and associated brain stem nuclei are indicated in Table 11–1.

The basal ganglia are involved in the control of movement and posture. This is well known since lesions in these areas cause disturbances of motor activity, such as tremor, involuntary movements, and muscular rigidity. Parkinsonism (paralysis agitans) is an example of a disease that affects the extrapyramidal system (probably the basal ganglia). Although clinical data reveal the involvement of these areas in movement, we cannot as yet visualize the total integrative role of the basal ganglia in the normal function of motor activity.

Diencephalon. The two major divisions of the diencephalon are the *thalamus* and the *hypothalamus.* These structures surround the third ventricle of the brain.

Thalamus. The thalamus is a large bilateral, oval structure located above the midbrain. It receives all sensory impulses either directly or indirectly from all parts of the body, with the exception of olfactory sensations. It also receives impulses from the cerebellum, cerebral cortex, and other areas of the brain. The thalamus is made up of many nuclei, each concerned with a specific function. Each nucleus is named—e.g., the medial geniculate body and the lateral geniculate body, which relay auditory and visual information, respectively. Some examples of thalamic nuclei are listed in Table 11–2.

Some of the impulses received by the thalamus are associated, and some are simply relayed to specific areas of the cerebral cortex. In this way it is evident that the thalamus functions as a sensory integrating center of great physiologic importance. Bodily well-being or malaise is believed to be interpreted by the thalamus rather than by the cortex. Appreciation of part of temperature, crude touch, and pain is possible even though the sensory cortex is destroyed. Through connections with the hypothalamus, influence

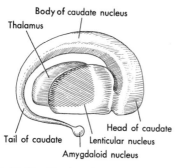

Figure 11–11. The basal ganglia in relation to the thalamus.

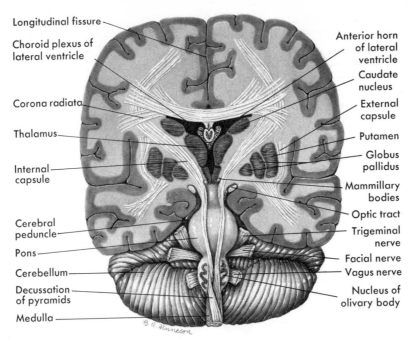

Longitudinal fissure
Choroid plexus of lateral ventricle
Corona radiata
Thalamus
Internal capsule
Cerebral peduncle
Pons
Cerebellum
Decussation of pyramids
Medulla

Anterior horn of lateral ventricle
Caudate nucleus
External capsule
Putamen
Globus pallidus
Mammillary bodies
Optic tract
Trigeminal nerve
Facial nerve
Vagus nerve
Nucleus of olivary body

Figure 11-12. Section of brain to show basal ganglia, internal capsule, pons, cerebellum, and medulla.

is exerted on both visceral and somatic activities.

Hypothalamus. The hypothalamus lies below the thalamus and forms part of the lateral walls and floor of the third ventricle. It is a control center for visceral activities by means of (1) neural connections with the posterior pituitary gland, the thalamus, and the midbrain, and (2)

the blood supply to the anterior pituitary gland, through which "releasing factors" synthesized in the hypothalamus reach the pituitary and regulate hormonal secretion.

The hypothalamus is made up of nuclei (like the thalamus), each of which are involved with particular functions. The mammillary bodies appear as small, round nuclear masses that are evident on the

TABLE 11-1
FUNCTIONS OF BASAL GANGLIA AND ASSOCIATED BRAIN STEM NUCLEI

The Caudate Nucleus and Putamen	Regulates gross intentional movements of the body that are performed *unconsciously* through pathways into the globus pallidus and thalamus to the cerebral cortex and finally downward into the spinal cord via the corticospinal and extrapyramidal spinal pathways. The motor cortex controls *conscious* specific fine movements
Globus Pallidus	Is concerned with regulation of muscle tone essential for intended, specific movements of the body. It is believed that it can also excite the cerebral cortex
The Subthalamic Nuclei	Are concerned with various rhythmical motions such as walking and running
The Red Nucleus	Receives impulses from the dentate nucleus of the cerebellum and from the caudate nucleus and putamen. It sends fibers down into the spinal cord. It is concerned with movements of the head and upper trunk and is necessary for reflexes concerned with righting oneself in space
The Substantia Nigra	Receives many fibers from other basal nuclei. It is believed to function in controlling associative movements

TABLE 11–2
SOME OF THE THALAMIC NUCLEI CONNECTIONS

Impulses from	
Touch receptors Pressure receptors Joint receptors Temperature receptors Pain receptors	Are received by cells in the posteroventral nucleus of the thalamus. With the possible exception of the pain impulse, these impulses are relayed to the postcentral gyrus in parietal lobe cortex for interpretation
Taste fibers	Are projected to the thalamus from the geniculate ganglion and are relayed to the postcentral cortex close to the face area
Auditory fibers	Are projected from the cochlear nuclei and inferior colliculi to the medial geniculate bodies of the thalamus and from there relayed to the temporal cortex
Visual fibers	Are projected to the lateral geniculate bodies of the thalamus and from there relayed to the calcarine cortex

basal surface of the brain. They lie below the floor of the third ventricle and behind the optic chiasma. The mammillary bodies receive fibers from the olfactory areas of the brain and from ascending pathways and send fibers to the thalamus and to other brain nuclei. They form relay stations for olfactory fibers and are concerned with olfactory reflexes.

Functions of the hypothalamus will be referred to in detail in appropriate areas of the text; however, they may be summarized here:

1. AUTONOMIC NERVOUS CONTROL. The main subcortical control center for regulation of parasympathetic and sympathetic activities is the hypothalamus. In general the anterior and medial portions are related to the parasympathetic system; the posterior and lateral portions regulate the sympathetic system.

2. CARDIOVASCULAR REGULATION. Stimulation of the posterior hypothalamus causes a rise in blood pressure and an increase in heart rate. Stimulation of the preoptic area has the opposite effect. These effects are produced via impulses to the cardiovascular centers of the medulla.

3. TEMPERATURE REGULATION. The temperature of the blood flowing through the preoptic and anterior hypothalamic areas causes heat loss through vasodilation in the skin, sweating, and increased respiration if the temperature is above normal (37° C, 98.6° F). If the blood temperature is

below normal, there are constriction of skin blood vessels, cessation of sweating, increased activity of the adrenal medulla, and perhaps shivering. Thus the external temperature of air has little effect on the cells and tissue fluid of the body.

4. FOOD INTAKE. Two centers in the hypothalamus regulate the amount of food ingested. The "feeding center" in the lateral hypothalamus is presumably stimulated by hunger sensations and the overall need for food (such as accompanies starvation) since electrical stimulation of this area causes eating of whatever food is available to the animal (appetizing or not). The "satiety center" in the medial hypothalamus is stimulated when the animal has taken in enough food. Electrical stimulation of the satiety center inhibits the feeding center, and the animal abruptly stops eating. Destruction of this nucleus, on the other hand, produces animals who eat until they are too obese to move. They always "feel hungry" because there is no satiety center to inhibit the feeding center.

5. WATER BALANCE. Cells in the supraoptic and paraventricular nuclei respond to the osmotic pressure of the blood—the "osmoreceptors." When there is an increase in the osmotic pressure due to water lack, the antidiuretic hormone (ADH) is secreted, reaching the posterior pituitary by traveling along the nerve fiber from this area to the gland. The action of ADH when secreted in the bloodstream is

to cause the tubular cells of the kidney to conserve water, thus permitting the blood to become less concentrated. A "thirst" area is found near the satiety center, which, when stimulated, causes the animal to seek water. In the normal individual the thirst area is stimulated when the osmolality of blood is elevated.

6. GASTROINTESTINAL ACTIVITY. The dorsomedial nucleus causes increased peristalsis and secretion from intestinal glands, when stimulated.

7. SLEEPING-WAKING ACTIVITY. The upper portion of the reticular activating system lies in the central lower hypothalamus. This activity will be discussed below under Reticular Formation.

8. EMOTIONS. Experimental stimulation of the inferior lateral portion of the hypothalamus in animals results in extreme excitation with symptoms of rage, such as hissing and arching of the back in cats and elevated blood pressure and cardiac rate. This evidence together with the knowledge that in man many of the "symptoms" of various emotions are those controlled by the autonomic nervous system indicates that the hypothalamus may be a center for emotional responses.

MIDBRAIN

The midbrain (mesencephalon) is a short, constricted portion that connects the pons and cerebellum with the hemispheres of the cerebrum. It is directed upward and forward and consists of (1) a pair of cylindrical bodies called the *cerebral peduncles*, which are made up largely of the descending and ascending fiber tracts from the cerebrum above, the cerebellum, medulla, and spinal cord below; (2) four rounded eminences, called the *corpora quadrigemina* (Figure 11–3, page 236), which contain important correlation centers and also nuclei concerned with motor coordination (the superior colliculi for optic reflexes, the inferior for auditory reflexes); and (3) an intervening passage or tunnel, the cerebral aqueduct (aqueduct of Sylvius), which serves as a communication between the third and fourth ventricles.

HINDBRAIN

Cerebellum and Function. The cerebellum occupies the lower and posterior part of the skull cavity. It is below the posterior portion of the cerebrum, from which it is separated by the *tentorium cerebelli*, a fold of the dura mater, and behind the pons and the upper part of the medulla. It is oval, constricted in the center, and flattened from above downward. The constricted central portion is called the *vermis*, and the lateral expanded portions are called the *hemispheres*.

The surface of the cerebellum consists of gray matter that is folded into numerous parallel, long gyri called *folia*. The gray matter contains cells from which fibers pass to form synapses in other areas of the brain and cells with which fibers entering the cerebellum from other parts of the brain synapse. The cerebellum is connected with the cerebrum by the *superior peduncles*, with the pons by the *middle*

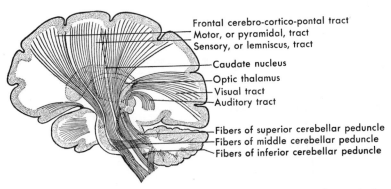

Figure 11–13. Dorsoventral section of brain, showing some of the fiber tracts from spinal cord to the cerebral cortex. Fiber tracts to cerebellum are also shown.

peduncles, and with the medulla oblongata by the *inferior peduncles* (Figure 11–13). These peduncles are bundles of fibers. Impulses from the motor centers in the cerebrum, from the semicircular canals of the inner ear, and from the muscles enter the cerebellum by way of these bundles. Outgoing impulses are transmitted to the motor centers in the cerebrum, down the cord, and thence to the muscles.

The cerebellum receives tactile, kinesthetic, auditory, visual, cortical, and pontine impulses. It sends nerve impulses into all the motor centers innervating the body wall and helps to maintain posture and equilibrium and the tone of the voluntary muscles. It modifies the stretch reflexes. Movements elicited by spinal reflexes are also modified. The *dentate nucleus* is large and receives most of the fibers from Purkinje cells and relays the impulses to the thalamus. From there impulses are sent on to the frontal motor cortex (Figure 11–14).

There are other smaller nuclei that relay impulses to the reticular formation of the midbrain and to the red nucleus. None of the activities of the cerebellum comes into consciousness. In man, injury to the cerebellum results in muscular weakness, loss of tone, and inability to accurately control the movements of the skeletal muscles. There may be difficulty in walking due to

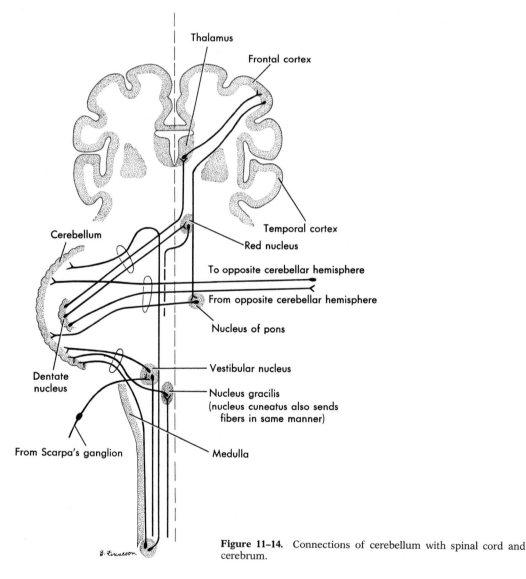

Figure 11–14. Connections of cerebellum with spinal cord and cerebrum.

poor coordination of the muscles of the legs or difficulty in talking due to lack of coordination of the muscles moving the tongue and jaw. The area of the body affected is determined by the location and extent of the injury to the cerebellum. If both sides of the cerebellum are injured, the lack of muscle tone and coordination may be so great that the person is helpless.

Pons and Function. The pons is situated in the front of the cerebellum between the midbrain and the medulla oblongata. It consists of interlaced transverse and longitudinal white fibers intermixed with gray matter. The transverse fibers are those derived from the middle peduncles of the cerebellum and serve to join its two halves. The longitudinal fibers connect the medulla with the cerebrum. In it also are the nuclei of all or a part of the fibers of the fifth, sixth, seventh, and eighth cranial nerves.

The pons is a bridge of union between the two halves of the cerebellum and a bridge between the medulla and the midbrain. The fifth (trigeminal) nerve emerges from the side of the pons near its upper border. The sixth (abducens), seventh (facial), and eighth (vestibulocochlear) nerves emerge in the superficial furrow that separates the pons from the medulla (Figure 11–15). There is a pneumotaxic center in the pons that participates in the regulation of respiration.

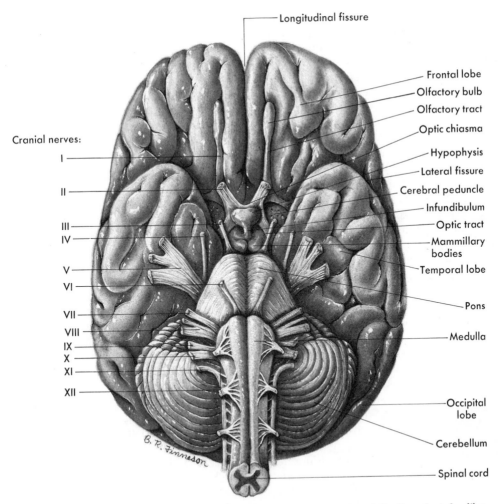

Figure 11–15. Base of brain showing cerebrum, cerebellum, pons, and medulla. Note the infundibulum to which the pituitary, or hypophysis, is attached. The numerals indicate the cranial nerves.

Medulla Oblongata. The medulla oblongata (spinal bulb) is continuous with the spinal cord, which, on passing into the cranial cavity through the foramen magnum, widens into a pyramid-shaped mass that extends to the lower margin of the pons. Externally, the medulla resembles the upper part of the spinal cord, but the internal structure is different. All the afferent and efferent tracts of the spinal cord are represented in the medulla, and many of them decussate, or cross, from one side to the other, whereas others terminate in the medulla. The nerve cells of the medulla are grouped to form *nuclei,* some of which are centers in which the cranial nerves arise. The motor fibers of the glossopharyngeal and of the vagus nerves, also the cranial portion of the accessory nerves, arise in the *nucleus ambiguus.* The hypoglossal nerve arises in the *hypoglossal nucleus.* Some of the nuclei are relay stations of sensory tracts to the brain, e.g., the *nucleus gracilis* and *nucleus cuneatus.* Some serve as centers for the control of bodily functions, e.g., the *cardiac, vasoconstrictor,* and *respiratory centers.*

Function. The medulla serves as an organ of conduction for the passage of impulses between the cord and the brain. It contains (1) the cardiac, (2) the vasoconstrictor, and (3) the respiratory centers and controls many reflex activities.

THE CARDIAC INHIBITORY CENTER. This consists of a bilateral group of cells lying in the medulla at the level of the nucleus of the *vagus nerve.* The vagal fibers from this center go to the heart and unite with the cardiac branches from the thoracolumbar nerves to form the *cardiac plexus* (Figure 17–11, page 396), which envelops the arch and ascending portion of the aorta. From the cardiac plexus the heart receives these *inhibitory fibers.* This center constantly discharges impulses that tend to hold the heart to a slower rate than it would assume if this check did not exist. The activity of the heart is also influenced by impulses from the *cardiac sympathetic nerves,* which increase the rate of the heartbeat and are called accelerator nerves. The inhibitory and accelerator fibers are true antagonists, having opposing effects on the heart.

THE VASOCONSTRICTOR CENTER. This consists of a bilateral group of cells in the medulla. Fibers from these cells descend in the cord and at various levels form synapses with autonomic spinal neurons in the lateral columns of gray matter. These spinal neurons are preganglionic vasoconstrictor fibers that terminate on postganglionic neurons in the sympathetic ganglion. The postganglionic neurons innervate the smooth muscle of the arteriole walls. The center is in a constant state of activity, which is increased or decreased reflexly by excitatory or inhibitory impulses from peripheral somatic and visceral receptors. One must conceive of different cells in this center being connected by definite vasoconstrictor paths with different parts of the body, e.g., the intestines or the skin. Cells that, when stimulated, cause peripheral vasoconstriction and a rise of arterial pressure are called the *pressor center.* Cells that cause an opposite effect, i.e., a decrease in peripheral vasoconstriction and a fall in arterial pressure, are called the *depressor center.*

THE RESPIRATORY CENTER. This consists of a bilateral group of cells located in the medulla. The center is automatic, possessing an inherently rhythmical activity. However, it is very responsive to impulses from sensory fibers of most cranial and spinal nerves and from regions of the brain. The effect of sensory nerves on activity of the respiratory center is to alter the rate and depth of respiration. Sensory fibers that alter the activity of the cardiac and vasoconstrictor centers may also affect the respiratory center, as do alterations of oxygen and carbon dioxide tensions of the blood.

In addition to the control of respiration and circulation, many other reflex activities are effected through the medulla by means of the vagus and other cranial nerves, which originate in this region. Such reflex activities are sneezing, coughing, vomiting, winking, and the movements and secretions of the alimentary canal.

Reticular Formation. The reticular formation is composed of large and small nerve cells and an intricate system of interlacing fibers that run in all directions. It begins in the lower part of the medulla and extends through the brain stem, where it terminates in the diencephalon. The reticular cells receive collaterals from all the great ascending pathways and nuclei. Efferent impulses are sent to both higher and lower brain centers. The reticular area is believed to be essential for control of cortical activities such as initiation and maintenance of alert wakefulness; hence, it has been called the *activating* or *arousal system*. It exerts some sort of a regulatory action on the brain and cord by either excitation or inhibition of certain activities, such as spinal cord reflexes, or voluntary movements. Muscle tone and smooth, coordinated muscle activity are dependent upon this system. Its function is not to relay a specific impulse or message but simply to arouse the brain for action. Any stimulus reaching the reticular formation alerts the cortex to a state of wakefulness so that, when a specific stimulus reaches a specific cortical

center, it may be identified and action may result, as, for example, being wakened by a loud noise or other unusual environmental condition. Impulses through the reticular system can be blocked by hypnotic drugs; thus, emotional response to environmental stress is reduced. Injury or disease to the system results in permanent unconsciousness.

INTEGRATED FUNCTION OF THE BRAIN

Although we list various functions for each area of the brain and have accumulated a great deal of information about interconnections within the brain, there is still a great amount of knowledge to be gained. Most of our concepts of brain function have come from pathologic studies. Correlations are made of damage (lesions due to hemorrhage or tumor) of specific areas of the brain and functional changes. More recently the concept of brain chemistry has proven to be of great importance in an understanding of brain function. The disease parkinsonism, which

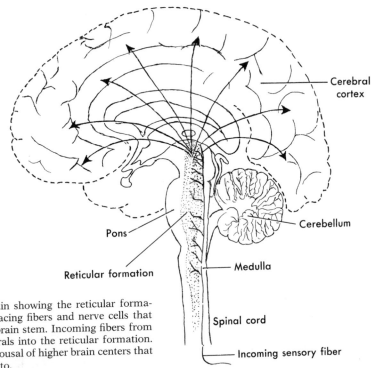

Figure 11–16. Section of brain showing the reticular formation. It is composed of interlacing fibers and nerve cells that form the central core of the brain stem. Incoming fibers from the spinal tracts send collaterals into the reticular formation. *Arrows* indicate the general arousal of higher brain centers that the reticular system projects to.

affects the basal ganglia, can be corrected by administration of drugs that increase the concentration of dopamine, a substance normally found in the region of the basal ganglia. The many drugs that affect mood and behavior have as their mechanism of action an effect on the chemical environment of the brain.

On the basis of behavior we can define consciousness; however, there seems to be no specific control center in the brain for consciousness, although the hypothalamus and the reticular formation are known to play important roles. Sedatives, which produce sleep, act by interfering with the transmission of impulses to the cortex from those deeper brain areas. Sleep is similar to unconsciousness in that both conditions involve unawareness of surroundings; however, one can be aroused from sleep but not from unconsciousness. Furthermore, the electrical activity that can be recorded as an electroencephalogram (EEG) is different during sleep and coma.

Prolonged lack of sleep results in progressive malfunction of the brain to the point of psychotic manifestations if forced wakefulness is long enough. The inescapable conclusion is that sleep is necessary for normal brain functioning.

At certain stages of sleep (varying from relatively light, easily arousable sleep to very deep sleep) dreaming occurs. It is possible to recognize the occurrence of dreaming by rapid eye movements (REM sleep) and other physiologic changes. When sleep is interrupted during REM sleep, the subject reports that he was dreaming. If the subject is awakened every time he starts to dream for a succession of nights, and then is permitted to sleep undisturbed, the amount of sleeping time devoted to dreaming is greatly increased—an indication that dreaming may be a necessary part of sleep, perhaps as a release of psychic energy.

Memory is the ability to recall a thought more than once. Short-term memory represents the recollection of thoughts, such as a telephone number or a license number, for a brief interval of time, perhaps an hour or days, whereas long-term memory lasts for weeks, months, or years. The ability of the nervous system to store memories is the basis of learning. Memory and learning are activities that involve higher centers of the brain, such as the cortex; however, other areas are involved. For example, sensory impulses from the eye or ear are transmitted to the appropriate cortical areas, so that at a later date they may be recalled—for pleasure or interpretation and thought.

The anatomic and physiologic bases of memory and learning are yet to be fully clarified. Many proposals have been offered. For example, it has been suggested that as a particular task is repeated the area of the cortex associated with the task becomes thicker. This apparently represents changes in presynaptic terminals, possibly the number of synaptic contacts, the size of terminals, or the chemical composition of terminals. Conversely, in animals that have lost their eyesight there is an apparent thinning of the visual cortex. Another theory proposes that nucleic acids (DNA and RNA) are involved in memory and learning. Needless to say, research continues in an attempt to define and clarify further the many integrated and complex functions of the brain and nervous system.

The Meninges

The brain and spinal cord are enclosed within *three* membranes. These are named from without inward: the dura mater, arachnoid mater, and pia mater (Figure 11–17).

Dura Mater. The dura mater is a dense membrane of fibrous connective tissue containing a great many blood vessels. The cranial and spinal portions of the dura mater differ and are described separately,

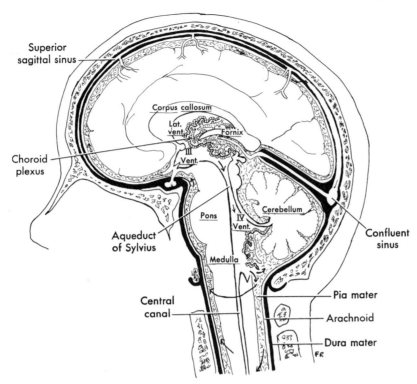

Figure 11–17. Diagram showing relationship of brain and cord to meninges. The choroid plexus forms the cerebrospinal fluid. (Modified from Rasmussen.)

but they form one complete membrane. The *cranial dura mater* is arranged in *two* layers, which are closely connected except where they separate to form sinuses for the passage of venous blood. The outer, or endosteal, layer is adherent to the bones of the skull and forms their internal periosteum. The inner, or meningeal, layer covers the brain and sends numerous prolongations inward for the support and protection of the different lobes of the brain. These projections also form sinuses that return the blood from the brain, and sheaths for the nerves that pass out of the skull.

A process of the dura mater extends down into the longitudinal cerebral fissure and separates the two cerebral hemispheres. It is called the *falx cerebri*, because it is narrow in front and broader behind, thus resembling a sickle in shape. It contains important venous sinuses between its two layers. A second process of the dura mater extends into the transverse fissure of the cerebrum and covers the upper surface of the cerebellum and the undersurface of the cerebrum. It is called the *tentorium cerebelli* and also contains important venous sinuses.

The *spinal dura mater* forms a loose sheath around the spinal cord and consists of only the *inner layer* of the dura mater; the outer layer ceases at the foramen magnum, and its place is taken by the periosteum lining the vertebral canal. Between the spinal dura mater and the arachnoid mater is a potential cavity, the *subdural cavity*, which contains only enough fluid to moisten their contiguous surfaces.

Arachnoid Mater. The arachnoid mater is a delicate fibrous membrane placed between the dura mater and the pia mater. The cranial portion invests the brain loosely and, with the exception of the longitudinal fissure, it passes over the various convolutions and sulci and does not dip down into them. The spinal portion is tubular and surrounds the cord loosely. The *subarachnoid space*, between the arachnoid mater and the pia mater, contains a spongy connective tissue forming trabeculae and is filled with cerebrospinal fluid.

Pia Mater. The pia mater is a vascular membrane consisting of a plexus of blood vessels held together by fine areolar connective tissue. The cranial portion invests the surface of the brain and dips down between the convolutions. The spinal portion is thicker and less vascular than the cranial. It is closely adherent to the entire surface of the spinal cord and sends a process into the ventral fissure.

Ventricles of the Brain

The brain contains cavities called *ventricles* (Figure 11–18) that are lined with cells called ependymal and filled with cerebrospinal fluid. Each cerebral hemisphere contains a *lateral ventricle*. The lateral ventricle is divided into four parts: an anterior horn located in the frontal lobe, a body located in the parietal lobe, an inferior or temporal horn in the temporal lobe, and an occipital horn in the occipital lobe. The cavity of the lateral ventricles is large and may become overdistended with cerebrospinal fluid in certain pathologic conditions.

The *third ventricle* is behind the lateral ventricles but connected with each one by means of small openings called the foramina of Monro.[2] The *fourth ventricle* is in front of the cerebellum, behind the pons and the medulla. The third communicates with the fourth by means of a slender canal called the *cerebral aqueduct (aqueduct of Sylvius[3])*. In the roof of the fourth ventricle there is an opening called the foramen of Magendie.[4] In the lateral wall there are two openings called the foramina of Luschka.[5] By means of these three openings, the ventricles communicate with the subarachnoid space, and the cerebrospinal fluid can circulate from one to the other.

Each ventricle contains a choroid plexus, which is a rich network of blood vessels of the pia mater that are in contact with the ependymal cells lining the ventricle. The choroid plexuses have a role in the formation of cerebrospinal fluid.

[3] François Sylvius, French anatomist (1614–1672).
[4] François Magendie, French physiologist (1783–1855).
[5] Hubert von Luschka, German anatomist (1820–1875).

[2] Alexander Monro (Primus), Scottish anatomist (1697–1767).

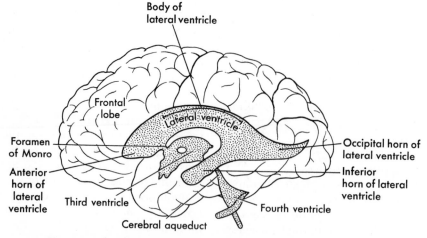

Figure 11–18. Diagram showing ventricles of the brain. Lateral view.

The Cerebrospinal Fluid

The meningeal membranes and the spaces between filled with fluid form a pad enclosing the brain and cord on all sides. Cerebrospinal fluid is secreted and diffused from the blood by the ependymal cells that cover the *choroid plexuses* of the ventricles. The choroid plexuses are highly vascular folds or processes of the pia mater that are found in the ventricles. The capillary network is intricate and resembles other cerebral capillaries. These differ greatly in their selective permeability from those in other parts of the body. For this reason drugs in the bloodstream often do not penetrate brain tissue and infections of the brain are difficult to cure. The term *blood-brain barrier* is sometimes used to refer to this phenomenon. The choroid plexus capillaries are covered by thick, highly differentiated cells that have the structure needed for *active* transport. After filling the lateral ventricles, the cerebrospinal fluid escapes by the foramen of Monro into the third ventricle and thence by the aqueduct into the fourth ventricle. From the fourth ventricle the fluid is poured through the medial foramen of Magendie and the two lateral foramina of Luschka into the subarachnoid spaces and reaches the cisterna magna. From the cisterna magna the cerebrospinal fluid may pass down the spinal canal within the subarachnoid space where it circulates around and upward and finally enters the venous circulation. From the cisterna magna this fluid also bathes all parts of the brain. From the subarachnoid spaces it is absorbed through the villi of the arachnoid mater, which project into the dural venous sinuses; a small amount passes into the perineural lymphatics of the cranial and spinal nerves. Experimentally, it has been found that dyes added to the cerebrospinal fluid travel along the course of certain cranial nerves, especially the olfactory. This loophole affords an opportunity for the entry of infection from the nasal cavities to the cerebral cavity.

The cerebrospinal fluid is highly variable in quantity, which is usually given as from 80 to 200 ml. It is colorless, alkaline, and has a specific gravity of 1.004 to 1.008. It consists of water with traces of protein, some glucose, and electrolytes, as in blood plasma, a few lymphocytes, and some pituitary hormones. Since the cerebrospinal fluid bathes the neural tissue of the cord and brain, its composition sets the composition of the extracellular fluid. It is now known that metabolic energy is continually expended to maintain the constant composition of cerebrospinal fluid in spite of fluctuations in arterial blood. If the membranes of the brain or cord are inflamed, there is usually a change in the normal characteristics of the fluid.

Infection and inflammation of the meninges of the brain will quickly spread to those of the cord. Such inflammation results in increased secretion, which, as it collects in a confined bony cavity, gives rise to symptoms of pressure, such as headache, slow pulse, slow respirations, and partial or complete unconsciousness. Cerebrospinal fluid may be removed by lumbar puncture. The needle by which the fluid is withdrawn is usually inserted between the third and fourth lumbar vertebrae (Figure 10–11, page 227) (thus below the end of the cord) into the subarachnoid space, the patient usually lying on his left side, with knees drawn up in order to arch the back, so as to slightly separate the vertebrae. In this position the pressure of the fluid is normally 70 to 180 mm of water. This pressure is raised when pressure in the brain is elevated, as in cerebral edema or inflammatory swelling. The fluid, or exudate, will contain the products of the inflammatory process and the organisms causing it. Lumbar puncture is used for (1) diagnosis of meningitis, syphilis, increased intracranial pressure, cerebral hemorrhage, and intracranial tumors; and (2) therapeutic effect: (a) to relieve pressure in meningitis and hydrocephalus; and (b) rarely for the introduction of sera, such as antimeningitis serum, or drugs.

Blood Supply

The internal carotid and vertebral arteries are the source of blood to the brain. The vertebral arteries arise from the subclavian arteries and ascend on either side to the level of the sixth cervical vertebra, where they enter the foramina of the transverse processes and continue upward in the foramina of the upper six cervical vertebrae (Figure 11–19). They wind behind the atlas, enter the skull through the foramen magnum, and unite to form the basilar artery.

Circle of Willis.[6] This is an arterial anas-

[6]Thomas Willis, English anatomist (1621–1675).

tomosis at the base of the brain (Figure 11–20). It is formed by the union of the *anterior cerebral arteries,* which are branches of the internal carotid, and the *posterior cerebral arteries,* which are branches of the basilar. The circle extends from the upper border of the pons and encircles the optic chiasma. It ends by dividing into the two posterior cerebral arteries. These two arteries are connected on either side with the internal carotid by the posterior communicating arteries. In front, the anterior cerebral arteries are connected by the anterior communicating artery. These arteries form a complete circle. This arrangement (1) equalizes the

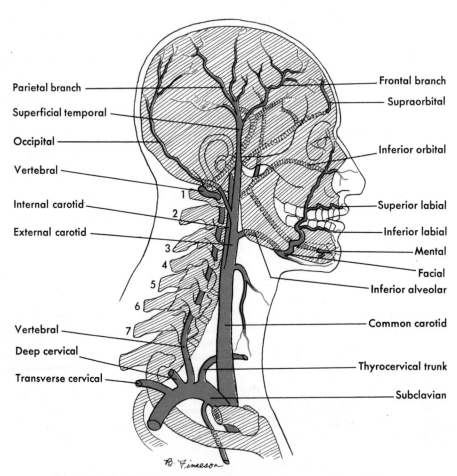

Figure 11–19. Diagram of arterial blood supply to head, face, and brain.

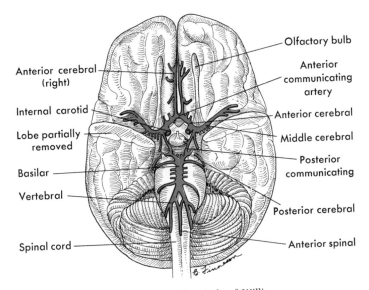

Olfactory bulb

Anterior cerebral (right)

Anterior communicating artery

Internal carotid

Anterior cerebral

Lobe partially removed

Middle cerebral

Posterior communicating

Basilar

Vertebral

Posterior cerebral

Spinal cord

Anterior spinal

Figure 11–20. The circle of Willis.

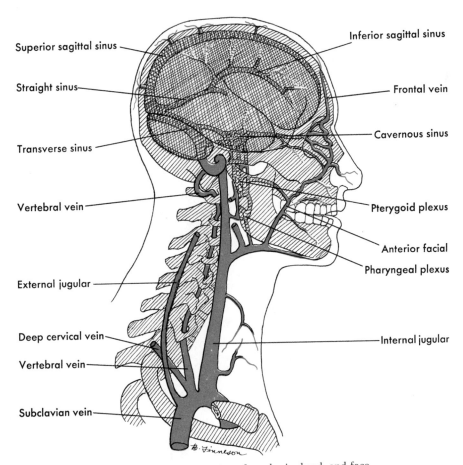

Superior sagittal sinus

Inferior sagittal sinus

Straight sinus

Frontal vein

Transverse sinus

Cavernous sinus

Vertebral vein

Pterygoid plexus

External jugular

Anterior facial

Pharyngeal plexus

Deep cervical vein

Internal jugular

Vertebral vein

Subclavian vein

Figure 11–21. Venous return from brain, head, and face.

circulation of the blood in the brain and (2) in case of destruction of one of the arteries, provides for the blood reaching the brain through other vessels. Venous drainage from the brain is through the large sinuses, shown in Figure 11–21, and into the internal jugular veins at either side of the neck.

Cranial Nerves:
Structure, Location, and Function

Twelve pairs of cranial nerves emerge from the undersurface of the brain and pass through the foramina in the base of the cranium. They are classified as motor, sensory, and mixed nerves (Figure 11–15, page 247).

The origin of the cranial nerves is comparable to that of the spinal nerves. The motor fibers of the spinal nerves arise from cell bodies in the ventral columns of the cord, and the sensory fibers arise from cell bodies in the ganglia outside the cord. The motor cranial nerves arise from cell bodies within the brain, which constitute their *nuclei of origin*. The sensory cranial nerves arise from groups of nerve cells outside the brain. These cells may form ganglia on the trunks of the nerves, or they may be located in peripheral sensory organs, such as the nose and eyes. The central processes of the sensory nerves run into the brain and end by arborizing on nerve cells that form their *nuclei of termination*. The nuclei of origin of the motor nerves and the nuclei of termination of the sensory nerves are connected with the cerebral cortex.

Numbers and Names. The cranial nerves are named according to the order in which they arise from the brain, and also by names that describe their nature, function, or distribution.

I.	Olfactory	Sensory
II.	Optic	Sensory
III.	Oculomotor	Motor
IV.	Trochlear	Motor
V.	Trigeminal	Mixed
VI.	Abducens	Motor
VII.	Facial	Mixed
VIII.	Vestibulocochlear	Sensory
IX.	Glossopharyngeal	Mixed
X.	Vagus	Mixed
XI.	Accessory	Motor
XII.	Hypoglossal	Motor

Table 11–3 indicates the functions and principal connections in the brain. Figure 11–22, *A* and *B*, shows the brain-stem nuclei of nerves III through XII. Nerves I and II are in reality tracts, rather than true nerves, since the primary receptor neuron terminates in a ganglion near the specialized receptors, the olfactory bulb in the case of nerve I, and the retina in the case of nerve II.

The central connections of the optic nerve are shown in Figure 11–23. Axons from ganglion cells of the retina make up the optic nerve. The optic fibers from each retina pass backward through the optic foramen; shortly after leaving the orbit, the two nerves come together, and the fibers from the nasal portions of the retinas cross. This is called the *optic chiasma* and is an incomplete crossing of fibers because the fibers from the temporal retinas do not cross.

The *optic tracts* are the continuation of optic nerves after the optic chiasma. Nerve fibers of the optic tract synapse in the *lateral geniculate body* of the thalamus. Cell bodies in the geniculate body form the optic radiation and terminate in the visual cortex of the occipital lobe. A branch of the optic nerve fibers passes to the superior colliculi. Cell bodies in the superior colliculi send collateral fibers to the nuclei of the oculomotor, trochlear, and abducens nucleus. The main fiber bundle from the superior colliculus descends into

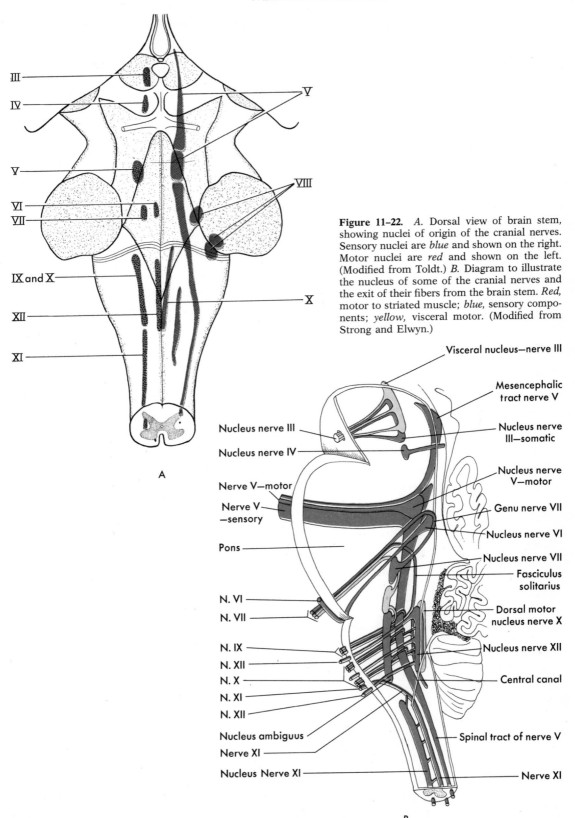

Figure 11–22. *A.* Dorsal view of brain stem, showing nuclei of origin of the cranial nerves. Sensory nuclei are *blue* and shown on the right. Motor nuclei are *red* and shown on the left. (Modified from Toldt.) *B.* Diagram to illustrate the nucleus of some of the cranial nerves and the exit of their fibers from the brain stem. *Red,* motor to striated muscle; *blue,* sensory components; *yellow,* visceral motor. (Modified from Strong and Elwyn.)

TABLE 11–3

CRANIAL NERVES

No. Name	Sensory (S) Motor (M)	Function	Central Connections
I Olfactory	S	Smell: from receptors in nasal mucosa, via olfactory bulb to olfactory cortex (hippocampus)	Mammillary bodies, nuclei of cranial nerves for reflexes of swallowing, secretion, and motility of digestive tract
II Optic	S	Vision: from receptors in retina to lateral geniculate body of thalamus	From thalamus to cortex; nuclei of nerves III, IV, VI for visual reflexes and to spinal cord tracts for body reflexes, e.g., turning head from light
III Oculomotor	M	Movement of eyeball up, down, and inward Raising the lid Constriction of pupil	Nuclei of nerves III, IV, VI have reflex connection with their nuclei of the opposite side of brain stem, with each other for visual reflexes; from the cochlear nuclei for reflexes related to sound; from the vestibular nuclei for correlating sight with balance and equilibrium; from the cortex for voluntary eye movement
	(s) *	Kinesthetic sensation from muscles innervated	
IV Trochlear	M	Movement of eyeball up and out	Reflex connections as for nerve III
	(s)	Kinesthetic sensation from muscle innervated	
V Trigeminal 3 branches:		Sensation from all 3 branches enters the semilunar ganglion and nucleus	To thalamus, then postcentral gyrus of cortex Reflex connections of trigeminal sensory and motor nuclei with each other, those of opposite side, and those of nerves VII, IX
1. Ophthalmic	S	Sensations of pain, touch, temperature from conjunctiva, skin of nose, upper lid, forehead, nasal mucosa	
2. Maxillary	S	Sensations of pain, touch, temperature from cheek, upper lip, nasal mucosa, hard palate, upper jaw and teeth, maxillary sinuses Kinesthetic sensation from muscles of mastication	From cortex for voluntary movement of muscles

*Indicates presence of a few fibers, central connections unknown.

TABLE 11–3 (*Continued*)

No.	Name	Sensory (S) Motor (M)	Function	Central Connections
		M	Contraction of muscles of mastication	
	3. Mandibular	S	Sensations of pain, temperature, touch from lower jaw, teeth and lips, mouth, anterior tongue, external ear, and meatus	
			Kinesthetic sensation from muscles innervated	
VI	Abducens	M	Movement of eyeball laterally	Reflex connections as with nerve III
		(s)	Kinesthesia of muscle innervated	
VII	Facial	M	Contraction of superficial face and scalp muscles and of stapedius muscle of middle ear	Reflex connections between motor and sensory nuclei of nerve VII, from opposite sides; with extrapyramidal and tectospinal tracts to spinal nerves; with nuclei of nerve V
			Secretion of glands of mucous membrane of nose and mouth; of submaxillary, sublingual, and lacrimal glands	
		S	Taste from anterior two thirds of tongue: from taste buds to nucleus solitarius in medulla	From nucleus to thalamus and postcentral gyrus
			Kinesthesia from muscles innervated	
VIII	Vestibulocochlear (2 *separate* parts) Cochlear (auditory)	S	Hearing: from receptors in cochlea of inner ear to medial geniculate body of thalamus	From thalamus to auditory cortex in temporal lobe. Reflex connections between nuclei of nerves III, IV, VI; with tectospinal tract to spinal nerves, for reflexes to sound
	Vestibular	S	Position sense and equilibrium: from receptors in semicircular canals of inner ear to vestibular nucleus	From vestibular nucleus to cerebellum, then to cortex. Reflex connections with opposite side, with vestibulospinal tract for postural reflexes; with nuclei of nerves III, IV, VI for reflex eye movement
IX	Glossopharyngeal	M	Tongue movement, swallowing	Reflex connections between sensory and

TABLE 11–3 (*Continued*)

No.	Name	Sensory (S) Motor (M)	Function	Central Connections
IX	Glossopharyngeal (*cont.*)	S	General sensation from pharynx, soft palate, posterior tongue, eustachian tube, and tympanic membrane Taste from posterior third of tongue: from taste buds to nucleus solitarius Chemoreceptor sensation of blood levels of O_2, CO_2, pH in carotid body; from these receptors to nucleus solitarius	motor nuclei of nerve IX on both sides, with nuclei of nerve X, with extrapyramidal and tectospinal tracts for gag and swallowing reflexes From nucleus to thalamus and postcentral gyrus of cortex Reflex connections with cardiac center of medulla, with sympathetic centers for blood pressure regulation, with respiratory center for adjusting rate and depth of breathing
X	Vagus	M	Swallowing: contraction of muscles of soft palate, pharynx Speaking: contraction of intrinsic muscles of larynx Constriction of smooth muscle of bronchi Increased peristalsis: contraction of smooth muscle of abdominal viscera Decreased rate of cardiac muscle contraction Secretion of glands of stomach and intestines	Reflex connections with extrapyramidal and tectospinal tracts, with thalamus and nuclei of nerves V, VII, IX Reflex connections between motor and sensory nuclei of vagus nerve
		S	Pain, touch, temperature sensation from auditory meatus, meninges, general sensation from pharynx, larynx, trachea, esophagus; sensations of nausea, abdominal distention, lung stretch	From sensory nucleus to thalamus and postcentral gyrus
XI	Accessory	M	Speaking: contraction of intrinsic muscles of larynx Shoulder and head movement: contraction of trapezius and sternocleidomastoid muscles	Reflex connections with corticospinal tract for voluntary movement; with tectospinal and vestibulospinal tracts for postural reflexes

XII Hypoglossal	M	Speaking, mastication, swallowing: contraction of tongue muscles	Reflex connections with cortex, extrapyramidal and tectospinal tracts; with sensory nuclei of nerves V, IX, X

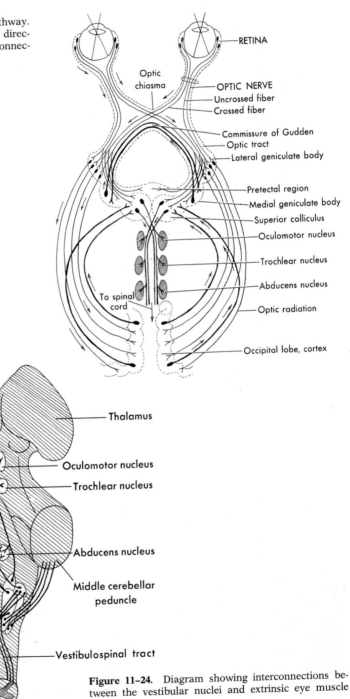

Figure 11–23. Diagram of visual pathway. Note the arrows as they indicate the direction in which impulses travel, and connections between nuclei.

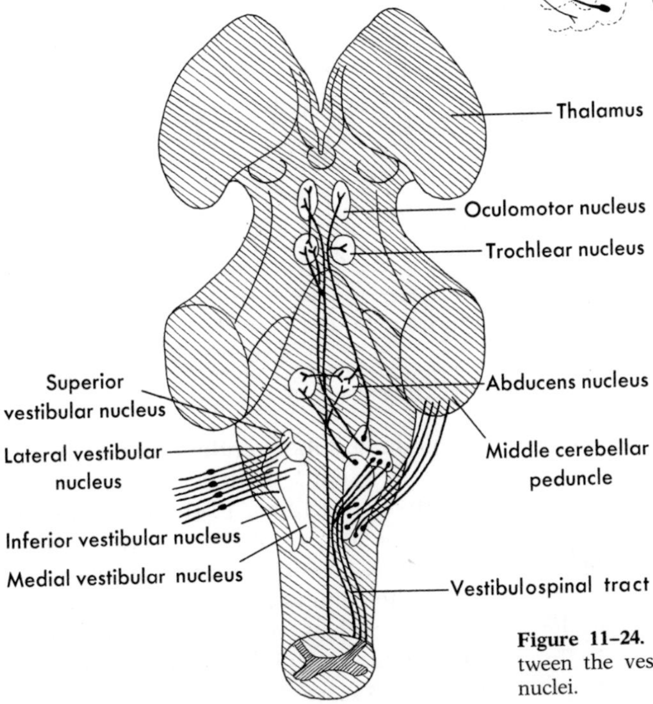

Figure 11–24. Diagram showing interconnections between the vestibular nuclei and extrinsic eye muscle nuclei.

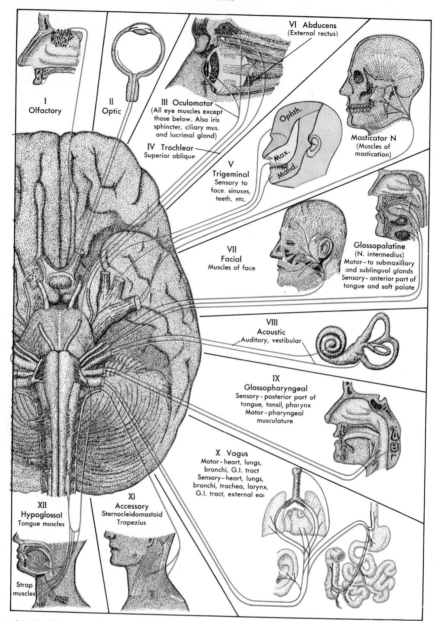

Figure 11–25. Diagram showing base of brain, the emergence of the cranial nerves, their distribution to the structures, and the functions with which they are concerned. *Blue,* sensory; *red,* motor. (Reproduced from full-color illustration in *Ciba Collection of Medical Illustrations,* Vol. 1, "Nervous System." Courtesy of Ciba Pharmaceutical Products, Inc., Summit, N.J.)

the upper part of the spinal cord (tectospinal tract) and terminates on the large motor cells of the ventral gray column.

The *vestibular center,* at the junction of the medulla and pons, shown in Figure 11–24, is the site of termination of most of the vestibular fibers in the vestibular nerve. Some pass without synapsing to the portion of the cerebellum lying closest to the medulla. Fibers ending in the vestibular nuclei synapse with second-order neurons that send fibers to the same area of the cerebellum, to other areas of the brain stem, as well as to the vestibulospinal tract and to the centers for control of extraocular eye muscles. Thus eye movement is coordinated reflexly with movement of the body.

Questions for Discussion

1. a. What is the blood supply of the brain?
   b. Describe the circle of Willis.
2. Name and discuss five functions of the hypothalamus.
3. Which part of the brain receives practically all the incoming sensory impulses? Explain what happens to these impulses.
4. Explain the arousal mechanism of the brain.
5. What part of the brain contains the nuclei for the following cranial nerves? What are their functions?

   a. Oculomotor
   b. Vagus
   c. Facial
6. List some of the functions of the medulla.
7. An individual has had a cerebrovascular accident (hemorrhage) that results in pressure on the internal capsule on the right side of the brain. Where will there be loss of motor function? Explain.

Summary

Divisions of the Brain

- **Forebrain**
  - Cerebral hemispheres
    - Cerebral cortex
    - Basal ganglia (nuclei)
  - Diencephalon
    - Thalamus
    - Hypothalamus
- **Midbrain**
  - Cerebral peduncles
  - Corpora quadrigemina
- **Hindbrain**
  - Cerebellum
  - Pons
  - Medulla

Development and Growth

1. Derived from ectoderm and develops from a region called the neural plate
2. Neural plate develops into neural tube
3. Anterior portion of neural tube develops into brain, posterior portion becomes the spinal cord
4. Cavity of neural tube persists as ventricles of the brain and central canal of the spinal cord
5. At cephalic end of neural tube, formation of three primary brain vesicles—forerunners of forebrain, midbrain, and hindbrain
6. Via process of flexures; growth of cerebral hemispheres over diencephalon, midbrain, and cerebellum; formation of sulci and gyri, brain assuming its final form

Cerebral Hemispheres

- **Description**
  1. Two cerebral hemispheres connected by corpus callosum, referred to as cerebrum
  2. Cerebrum, largest part of brain, fills upper portion of skull
  3. Surface composed of layer of gray matter called cortex
  4. Beneath gray matter is white matter

  - **Types of fibers in white matter**
    - Projection
    - Association
    - Commissural
  - **Surface of cerebral hemispheres**
    - Fissures
    - Sulci
    - Convolutions
  - **Fissures**
    - Longitudinal
    - Transverse
    - Central sulcus, or fissure of Rolando
    - Lateral cerebral, or fissure of Sylvius
    - Parieto-occipital

Cerebral Hemispheres *(cont.)*

Functions — Governs all mental activities
- Organ of associative memory
- Reason
- Intelligence
- Will
- Seat of consciousness
- Interpreter of sensations
- Instigator of voluntary acts
- Exerts a controlling force on reflex acts
- Sleep—necessary for recovery of neurons

Areas

- **Lobes**
  - Frontal
  - Parietal
  - Occipital
  - Temporal
  - Insula, or island of Rei

- **Motor area** — In front of central sulcus

- **Sensory areas** — Behind the central sulcus
  - Visual—occipital lobe
  - Auditory—superior part of the temporal lobe
  - Olfactory
  - Gustatory — anterior part of temporal lobe

- **Association areas**—cerebral tissue surrounding motor and sensory areas; interconnect and integrate motor and sensory areas.

Rhinencephalon-Limbic System
- Rhinencephalon means olfactory brain; anatomically represented by olfactory bulb and tract, and areas of cerebral hemispheres involved with olfactory impulses
- Rhinencephalon and connections with amygdala, thalamus, hypothalamus, and midbrain also involved with behavioral and emotional expression; term *limbic system* employed to define these functions

Basal Ganglia (Nuclei)
- Masses of gray matter located deep within cerebral hemispheres, associated with lateral ventricles
- **Parts**
  - Caudate nucleus
  - Lentiform nucleus
  - Putamen
  - Globus pallidus
  - Amygdaloid nuclear complex
  - Claustrum
- Related both anatomically in location and functionally with the internal capsule
- Function—involved in somatic motor functions

Thalamus
1. Bilateral oval structures above midbrain, associated with third ventricle
2. Made up of many nuclei, each concerned with a specific function
3. Receives all sensory information with exception of olfaction

Hypothalamus
1. Associated with third ventricle
2. Composed of many nuclei, each with specific function
- **Functions**
  - Autonomic nervous control
  - Cardiovascular regulation
  - Temperature regulation
  - Food intake
  - Water balance
  - Gastrointestinal activity
  - Sleep-waking activity
  - Emotions

Midbrain — **Description**
- Short, constricted portion connects pons and cerebellum with the hemispheres of the cerebrum
- Consists of
  - Pair of cerebral peduncles
  - The corpora quadrigemina
  - The cerebral aqueduct
- Contains nuclei of the III and IV cranial nerves

Cerebellum — **Description**
- Oval, constricted in center
- Central portion called vermis
- Lateral portions called hemispheres
- Gray matter on exterior
- White matter in interior
- Connected with cerebrum by superior peduncles
- Connected with pons by middle peduncles
- Connected with medulla by inferior peduncles

Cerebellum (*cont.*)	Function	Participates in the coordination and integration of posture and all voluntary movements
Pons	Description	Situated between the midbrain and the medulla oblongata. Consists of interlaced transverse and longitudinal white fibers mixed with gray matter
	Function	Connects two halves of cerebellum and also medulla with cerebrum Contains nuclei of trigeminal, abducens, facial, and vestibulocochlear nerves Participates in the regulation of respiration

Medulla Oblongata	Description	Pyramid-shaped mass, upward continuation of cord. Sensory and motor tracts of spinal cord represented. Many of them cross from one side to the other in the medulla; some end in medulla Gray matter forms nuclei
		Nuclei serve as — Centers in which cranial nerves arise, centers for control of bodily functions / Relay stations of sensory tracts to brain
	Function	Vital centers — Cardiac center / Vasoconstrictor center / Respiratory center
		Controls such reflex activities as — Sneezing / Coughing / Vomiting / Winking / Movement and secretion of digestive tract

Reticular Formation	Description	Network of interlacing cells and fibers Extends from upper spinal cord to diencephalon
	Function	Alerts cortex to wakefulness Sends efferent impulses to higher and lower centers

The Meninges (Coverings of the Brain)	Dura mater Arachnoid mater Pia mater

Blood Supply	Source	Internal carotid Vertebral arteries

Cranial Nerves	12 pairs classified as motor, sensory, and mixed nerves

Additional Readings

Altered States of Awareness. Readings from *Scientific American*. W. H. Freeman & Co., San Francisco, 1972.

BARR, M. L.: *The Human Nervous System*. Harper & Row, Publishers, New York, 1974.

EVARTS, E. V.: Brain mechanism in movement. *Sci. Amer.*, **229**:96–103, July, 1973.

GESCHWIND, N.: Organization of language and the brain. *Science*, **170**:940, November 27, 1970.

GUILLEMIN, R., and BURGUS, R.: The hormones of the hypothalamus. *Sci. Amer.*, **227**:24–33, November, 1972.

HEIMER, L.: Pathways in the brain. *Sci. Amer.*, **225**:48–60, July, 1971.

LLINAS, R. R.: The cortex of the cerebellum. *Sci. Amer.*, **232**:56–71, January, 1975.

LURIA, A. R.: The functional organization of the brain. *Sci. Amer.*, **222**:66–78, March, 1970.

MYERS, R. D., and VEALE, W. L.: Body temperature: Possible ionic mechanism in the hypothalamus controlling the set point. *Science*, **170**:95, October 2, 1970.

The Nature and Nurture of Behavior, Developmental Psychobiology. Readings from *Scientific American*. W. H. Freeman & Co., San Francisco, 1972.

NOBACK, C. R., and DEMAREST, R. J.: *The Human Nervous System: Basic Principles of Neurobiology*, 2nd ed. McGraw-Hill Book Co., New York, 1975.

ROSENZIVEIG, M. R.; BENNETT, E. L.; and DIAMOND, M. C.: Brain changes in response to experience. *Sci. Amer.*, **226**:22–29, February, 1972.

VANDER, A. J.; SHERMAN, J. H.; and LUCIANO, D. S.: *Human Physiology: The Mechanisms of Body Function*, 2nd ed. McGraw-Hill Book Co., New York, 1975, Chapter 18.

CHAPTER 12

Sensations

Chapter Outline

Through the integrity of his sense organs man is able to perceive his environment.

All sensory information from the external and internal environment reaches the brain. How the brain employs this sensory input is the basis of sensations. Thus, *sensations* are the conscious awareness of processes that take place within the brain as a consequence of sensory input. Sensations can give rise to a motor response or they may be stored as memory concepts that may be called into play at any time. Although sensations represent interpretation within the brain, they are recognized as occurring in the periphery. In reality, individuals see and hear in the brain and the eye and ear merely serve as the end organs responding to the stimulus.

With the exception of the sense of smell (olfaction), *all* sensory information is projected to the *thalamus*. From the thalamus each sensory input is conveyed to a specific sensory projection area in the cerebral cortex. By the processes of facilitation and inhibition at this relay station in the thalamus, sensory information to the cortex may be either directed or dissipated. Thus, when an individual is concentrating on a particular task, his awareness of distracting sounds is diminished. The brain, therefore, makes the decision as to whether or not the sensory information is pertinent to the welfare or activity of the individual at that particular time.

Classification of Sensations

One means of classifying sensations is based on the part of the body to which the sensation is projected, namely, *external* (*exteroceptive*) and *internal* (*interoceptive* and *proprioceptive*). External sensations include sight, hearing, taste, smell, touch, pressure, and temperature. Internal sensations include pain, position sense, vestibular sense, hunger, thirst, nausea, and other less well-defined sensations from the viscera.

The sensations of taste, smell, hearing, balance, and vision are classified as *special senses.* Each of these senses has receptors that are structurally specialized and are located in specific areas.

Sensations are categorized by the quality of the sensation itself as *epicritic* and *protopathic*. Epicritic sensations are highly critical and discriminatory, such as differentiating between two points placed close together on the skin—vibratory and touch sensations—and differentiating between temperature variations. Protopathic sensations are poorly localized and less specific. They include crude touch, burning types of pain, and deep muscle aches and pains.

General Sensations

TACTILE SENSATION

At least four kinds of receptors—free nerve endings, Meissner's[1] corpuscles, pacinian[2] corpuscles, and nerve fibers at the base of each hair—give rise to touch sensations. The hair nerve ending is stimulated by change of position of the hair, producing a brief burst of impulses followed by rapid adaptation. Meissner's and pacinian corpuscles are also rapidly adapting. These can sense a quick touch or deformation of tissue, but not a sustained touch. Free nerve endings slowly adapt to continuous stimulation, as we all know from experience with particles "in the eye," which we continue to feel until they are removed. Although the cornea of the eye has only free nerve endings, it gives rise to all qualities of sensation.

Pressure sensations differ from touch sensations in that they are more enduring. We can discriminate between a light pressure of a finger on the arm and a hard pressure. Free nerve endings in the superficial and deep layers of skin and other tissue are receptors of pressure, as well as the pacinian corpuscles.

KINESTHESIA, OR POSITION SENSE

Kinesthesia is the recognition of location and rate of movement of the parts of

---

[1] Georg Meissner, German physiologist (1829–1905).
[2] Filippo Pacini, Italian anatomist (1812–1883).

the body in relation to other parts. Ruffini's[3] end organs as well as the Golgi[4] endings in joint capsules and ligaments around joints give rise to position sense. Some of these receptors respond to flexion of the joint, others to extension. Thus we are conscious of the degree and direction to which the elbow is flexed or extended. The receptors that are stimulated by joint movement signal rate, as well as direction and extent of movement.

Degree of muscle stretch and tension are detected by special muscle receptors. However, these sensory impulses are transmitted to the cerebellum rather than to the cortex; so they produce no conscious sensation.

[3]Angelo Ruffini, Italian anatomist (1864–1929).
[4]Camillo Golgi, Italian histologist (1843–1926).

TEMPERATURE

Degrees of temperature from 12° to 50° C are perceived and discriminated by humans. Some regions of the body, such as the fingertips and face, have morphologically specific structures that are attuned to particular temperature ranges. Krause[5] end bulbs discharge at all steady-state skin temperatures between 12° and 35° C and paradoxically at 50° C. These receptors also discharge whenever the skin is rapidly cooled from any temperature. Most receptors are paralyzed or inactivated by cold if maintained for any length of time at 12° C or less. Warm re-

[5]Wilhelm Johann Friedrich Krause, German anatomist (1833–1910).

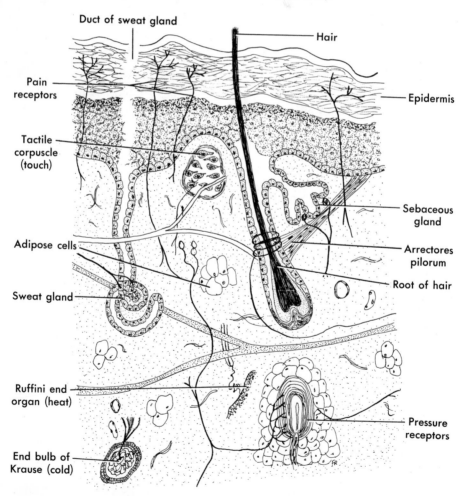

Figure 12–1. Diagram of skin showing receptors for pain, pressure, touch, heat, and cold.

ceptors, the Ruffini end organs, are found at greater skin depths, are more sparse, and have lower discharge frequencies than cold receptors. They discharge at skin temperatures between 25° and 45° C and increase their firing rate whenever the skin is warmed. In skin regions lacking morphologically specific receptors, temperature sensation is less acute and is subserved by free nerve endings. Hot and burning sensations result from the combined sensory inputs from cold, warm, and free nerve endings.

These receptors respond strongly to *change* in temperature. Because of this, and owing to adaptation, one feels colder in a cooling environment than one would at the same temperature if it were steady. Total adaptation does not occur so that it is possible to feel cold and warmth, even though the intensity is less than on initial exposure.

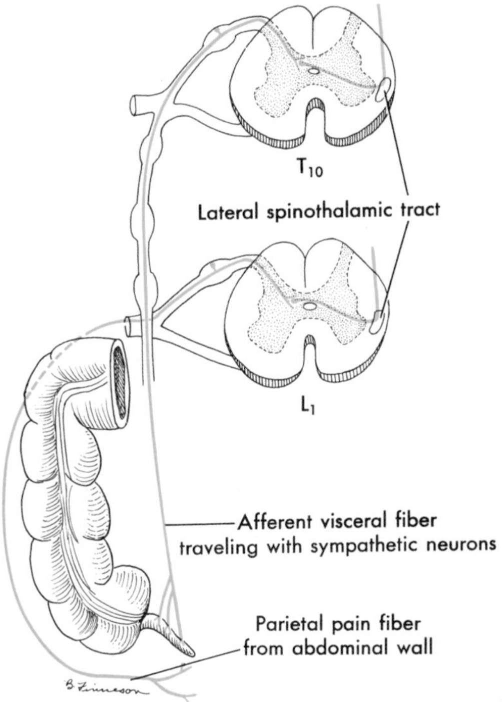

Figure 12–2. Neural pathway for visceral pain.

PAIN

Pain perception is a protective mechanism that occurs whenever tissue is damaged. Receptors for pain are the free nerve endings which are found in the skin, muscles, joints, tendons, dura mater, periosteum, and arterial walls. Other stimuli for pain include excessive heat or cold and blocking of the blood flow to a tissue. Spasm of muscle of the digestive tract, or distention of the intestine, common bile duct, and ureter, all cause pain—probably by interfering with the blood supply as the tissue is stretched or squeezed, and ischemia results.

Impulses for pain are transmitted to the thalamus by way of the lateral spinothalamic tract. Awareness of pain takes place in the thalamus, but localization and recognition of the kind and intensity of pain take place in the postcentral convolution of the cerebral cortex. Pain impulses are also relayed to other thalamic nuclei and to the hypothalamus.

All individuals have about the same

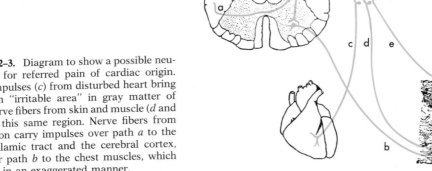

Figure 12–3. Diagram to show a possible neural path for referred pain of cardiac origin. Nerve impulses (*c*) from disturbed heart bring about an "irritable area" in gray matter of cord. Nerve fibers from skin and muscle (*d* and *e*) enter this same region. Nerve fibers from this region carry impulses over path *a* to the spinothalamic tract and the cerebral cortex, and over path *b* to the chest muscles, which contract in an exaggerated manner.

threshold for pain, but the reaction to pain varies widely between individuals, depending upon such factors as ethnocultural background, childhood experiences, and emotional status. Because of its emotional overlay, pain is very different from other sensations.

Pain may be described as sharp, dull, boring, piercing, pricking, aching, throbbing, stabbing, burning, constant, and intermittent. Aching pain is usually a deep pain with varying degrees of intensity and may be either diffuse or localized. Burning pain results from diffuse stimulation of all pain receptors in an area, e.g., burns or other types of tissue injury. As the intensity of the pain stimulus increases and more nerve fibers are involved, the intensity of the pain also increases. Deep pain is frequently associated with nausea and a fall in blood pressure, whereas superficial pain will quicken the pulse and raise blood pressure.

Headache is usually due to external causes rather than causes inside the head. Tension of muscles at the back of the head, straining eye muscles to focus for long periods, or inflammation of the sinuses will result in aching of the head in the related areas. Pain from within the head occurs when the meninges are stretched, as with increased spinal fluid pressure, or loss of spinal fluid, and when blood vessels are dilated. The latter is thought to be the cause of migraine headaches.

Referred Pain. In some instances visceral pain is not localized specifically in the organ but is felt on the surface of the body. Pain of this kind is spoken of as *referred pain*. It has been shown that the different visceral organs have a more or less definite relation to certain areas of the skin. Pain arising from stimuli in the intes-

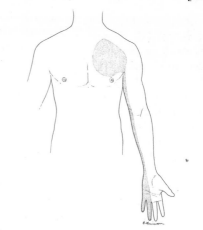

Figure 12–4. Referred pain. Pain from cardiac region is referred to the left side of the chest and down the inside of the left arm.

tines is located in the skin of the back, loins, and abdomen, in the area supplied by the ninth, tenth, and eleventh thoracic nerves. Pain from irritation in the stomach is referred to the skin over the xiphoid process, that from the heart to the scapular region. An explanation for referred pain is that the pain is referred to the skin region that is supplied from the spinal segment from which the organ in question receives its sensory fibers. The misreference results from excitation of a secondary neuron in the spinal cord that also normally is excited by neurons that supply that particular skin area. Examples of referred pain are: in appendicitis, the abdominal pains are often remote from the usual position of the appendix; in some pneumonia cases, abdominal pain is the prominent symptom; in angina pectoris, the pain radiates to the left shoulder and down the left arm (Figures 12–3 and 12–4). In this instance pain fibers from the heart are carried in the first, second, and third thoracic roots along with afferent fibers from the chest wall and arm.

Visceral Sensations

HUNGER

The feeling that is commonly designated as hunger occurs normally at a certain time before meals and is usually projected to the region of the stomach. It is presumably due to contractions of the empty stomach, which stimulate the receptors distributed to the mucous membrane. If food is not taken, hunger increases in intensity for a time and is likely to cause fatigue and headache. Professional fasters state that after a few days the pangs of hunger diminish and sometimes disappear. In illness hunger contractions may not occur at all, even when the food taken is not sufficient. Probably this results from a lack of muscular tone in the stomach. On the other hand, hunger contractions may be frequent and severe even if an abundance of food is taken regularly, as in diabetes, or following a period of starvation.

Appetite. Appetite is similar to hunger but is less related to physiologic activity, such as stomach contractions. The desire for a specific food is related to appetite, whereas when one is hungry, any one of a variety of foods may satisfy. Cultural and social factors influence one's appetite so that certain foods satisfying and desirable to natives of the United States may be repulsive to natives of Africa.

Food intake is regulated by the hypothalamus (see page 244), and many centers in the brain stem and spinal cord are involved in the actual process of eating, e.g., salivation, chewing, swallowing.

THIRST

This sensation may be defined as a conscious desire for water, which is to be distinguished from the desire to moisten the mouth and tongue. Thirst, as mentioned in the previous chapter, arises from the body's need for water, when tissues are dehydrated, and when the blood osmolality is increased. The "thirst" center in the hypothalamus, in conjunction with the osmoreceptors in the nearby area that secrete a hormone (antidiuretic) to cause kidney conservation of water, thus aids in the maintenance of total body water.

Special Senses

TASTE

The adequate stimulus for taste receptors is a substance in solution. In the case of dry substances saliva serves as the solvent. It is also necessary that the surface of the organs of taste be moist. The substances that excite the special sensation of *taste* act by producing a permeability change in the taste buds, and this change initiates the nerve impulses.

Taste Buds. Taste buds are ovoid bodies with an external layer of supporting cells, and contain in the interior a number of elongated cells, which end in hairlike processes that project through the central taste pore. These cells are the sense cells, and the hairlike processes probably are the parts stimulated by the dissolved substances. The taste buds are found chiefly in the surface of the tongue, though some are scattered over the soft palate, fauces, and epiglottis.

The Tongue. The tongue is a freely movable muscular organ consisting of two distinct halves united in the center. The root of the tongue is directed backward and is attached to the hyoid bone by several muscles. It is connected with the epiglottis by three folds of mucous membrane, and with the soft palate by means of the glossopalatine arches.

Papillae of the Tongue. The tongue is covered with mucous membrane, and the upper surface is studded with papillae. The papillae are projections of connective tissue covered with stratified squamous epithelium and contain loops of capillaries, among which nerve fibers are distributed. The papillae give the tongue its characteristic rough appearance. There are four varieties of these papillae:

Vallate (circumvallate) papillae are the largest, are circular, and form a V-shaped row near the root of the tongue. They contain *taste buds.*

Fungiform papillae, so named because

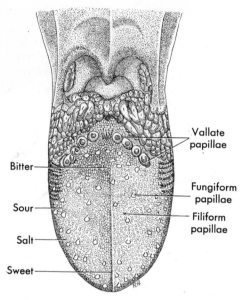

Vallate papillae

Bitter

Fungiform papillae

Sour

Filiform papillae

Salt

Sweet

Figure 12–5. The upper surface of the tongue, showing kinds of papillae and areas for taste.

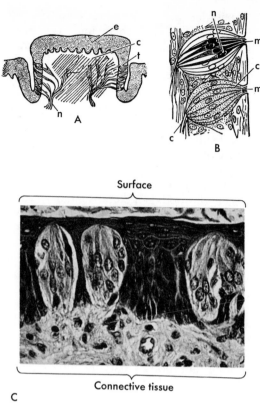

Surface

Connective tissue

C

Figure 12–6. *A.* A vallate papilla cut lengthwise, showing (*c*) corium, (*e*) epidermis, (*n*) nerve fibers, and (*t*) taste buds. *B.* The two taste buds at *t* more highly magnified—the lower as seen from the outside, showing (*c*) the outer or supporting cells; the upper as seen in section, showing (*n*) four inner cells with processes (*m*) projecting at the mouth of the bud. *C.* Histologic appearance of taste buds.

they resemble fungi in shape, are found principally on the tip and sides of the tongue.

Filiform papillae cover the anterior two thirds of the tongue and bear delicate brushlike processes that seem to be specially connected with the sense of touch.

Simple papillae similar to those of the skin cover the larger papillae and the whole of the mucous membrane of the dorsum of the tongue. All papillae contain taste buds.

The sense of touch is very highly developed in the tongue, as are the senses of temperature and pain. Tactile and muscular feedback depends on these to a great extent, i.e., the accuracy of the tongue in many of its important uses—speech, mastication, deglutition, sucking.

Classes of Taste. Taste sensations are very numerous, but four fundamental, or primary, sensations are recognized, namely, salty, bitter, acid, and sweet. Quantitative appreciation of taste is not great. All other taste sensations are combinations of these or combinations of one or more of them with sensations of odor or with sensations derived from stimulation of other receptors in the tongue. The seemingly great variety of taste sensations is due to the fact that they are confused or combined with simultaneous odor sensations. Thus the flavors in fruits are designated as tastes because they are experienced at the time these objects are eaten. If the nasal cavities are closed, as by holding the nose, the so-called taste often disappears in large measure. Very disagreeable tastes are usually due to unpleasant odor sensations, hence the practice of holding the nose when swallowing a nauseous drug. On the other hand, some volatile substances that enter the mouth through the nostrils and stimulate the taste buds are interpreted as odors. The odor of chloroform is largely due to stimulation of the sweet taste buds of the tongue.

Taste on the posterior third of the tongue is mediated by the glossopharyngeal nerve. Taste on the anterior two thirds of the tongue is mediated by the facial nerve. The vagus nerve mediates taste sensations from around the epiglottis. These fibers convey taste impulses to the medulla and pons where they terminate on cell bodies of the secondary fibers. These secondary fibers cross in the medulla and then ascend to the thalamus. Tertiary neurons project to the primary receiving area of the sensory cortex, on the parietal lobe near the sylvian fissure.

SMELL

The sensory endings for the sense of smell are located in the olfactory membrane over the surface of the superior nasal conchae and the upper part of the septum. These sensory nerve endings are the least specialized of the special senses.

Olfactory Nerves. The olfactory sensory endings are modified epithelial cells scattered freely among the columnar epithelium of the mucous membrane. These sensory cells are called *olfactory cells*, and the other epithelial cells, supporting cells.

Olfactory cells are bipolar and slender; peripheral hairlike processes known as olfactory hairs extend beyond the surface of the epithelial membrane. The central or deep process passes through the basement membrane and joins adjacent processes to form bundles of unmyelinated fibers of the *olfactory nerve*. These bundles of nerves form a plexus in the submucosa and eventually form about 20 or more nerves that pierce the cribriform plate of the ethmoid bone and end in a mass of gray matter called the olfactory bulb. In the olfactory bulb these fibers form synapses with the dendrites of the mitral cells. Impulses from the mitral cells are conducted to their various terminations in the olfactory lobe, of either the same or the opposite side.

The cells of the bulb lie in synaptic relation to cells whose processes form the olfactory tract and finally terminate in the

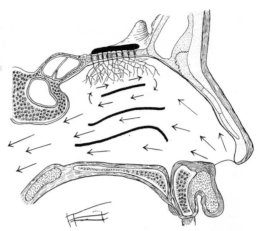

Figure 12–7. Diagram of lateral wall of left nasal cavity. The three black lines represent the region of the inferior, middle, and superior conchae. *Arrows* indicate the direction of air flow. The olfactory lobe is shown with nerve fibers extending through orifices in the cribriform plate of the ethmoid bone. Olfactory nerve fibers are distributed in the mucosa above the superior conchae, the cell bodies lying in the nasal mucosa.

gray of the cortex in the parolfactory area and hippocampal gyrus of the temporal lobe. Some of the fibers terminate in other nuclei of the brain, and through many synaptic connections may reach the thalamus.

The nerve fibers that ramify over the lower part of the lining membrane of the nasal cavity are branches of the fifth, or trigeminal, nerve. These fibers furnish the tactile sense and enable one to perceive, by the nose, the sensations of cold, heat, tickling, pain, and tension, or pressure. It is these nerve fibers that are excited by strong irritants (ammonia or pepper).

Odoriferous Substances. Such substances emit particles that usually are in gaseous form. In order to stimulate, these particles must penetrate into the upper part of the nasal chamber. In the olfactory area the cells are always bathed in fluid from special glands, so that the diffusing particles dissolve. The fluid acts chemically upon the sensitive hairs of the olfactory cells, initiating impulses that travel to the olfactory lobe and give rise to the sensation of smell. Few odors are detected in a dry, hot environment.

To smell anything particularly well, air is sniffed into the higher nasal chambers, thus bringing the odoriferous particles in greater numbers into contact with the olfactory hairs. Odors can also reach the nose by way of the mouth.

Each substance smelled causes its own particular sensation, and one is able not only to recognize a multitude of distinct odors but also to distinguish individual odors in a mixed odor. These odors are difficult to classify, i.e., it is not possible to pick out what might be called the fundamental odor sensations.

The sensation of smell develops quickly after the contact of the odoriferous stimulus and may last a long time. When the stimulus is repeated, the sensation very soon dies out, and the end organs of the sensory cells quickly become adapted. This accounts for the fact that one may easily become accustomed to unpleasant odors, an advantage when these odors have to be endured. On the other hand, it emphasizes the importance of acting on the first sensation of a disagreeable odor, so as not to become accustomed to it.

The olfactory center in the uncus and the hippocampus of the brain is widely connected with other areas of the cerebrum. Olfactory memories may be vivid.

The sense of smell is widely and closely connected with the other senses and with many psychic activities.

HEARING

The auditory apparatus consists of the external ear; the middle ear, or tympanic cavity; the internal ear, or labyrinth; and the acoustic nerve and acoustic center.

External Ear. This consists of an expanded portion, named the pinna or auricula, and the external acoustic meatus, or auditory canal.

Pinna. The pinna projects from the side of the head. It consists of a framework of cartilage, containing some adipose tissue and muscles; in the lobe, the cartilage is replaced by soft connective tissues. The pinna is covered with skin and joined to the surrounding parts by ligaments and muscles. It is very irregular in shape. The pinna serves to some extent to collect sound waves and direct them toward the external acoustic meatus.

External Acoustic Meatus (External Auditory Canal). This is a tubular passage, about 2.5 cm in length, which leads from the concha to the tympanic membrane. It

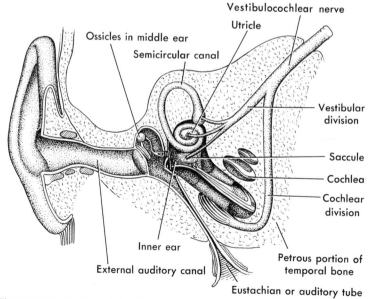

Figure 12–8. Section of the right ear showing middle and inner ear structures.

forms an S-shaped curve and is directed inward, forward, and upward, then inward and backward. Lifting the pinna upward and backward tends to straighten the canal, but in children it is best straightened by drawing the pinna downward and backward. The external portion of this canal consists of cartilage, which is continuous with that of the pinna; the internal portion is hollowed out of the temporal bone. It is lined by a prolongation of the skin, which in the outer half of the canal is very thick and not at all sensitive, and in the inner half is thin and highly sensitive. Near the orifice the skin is furnished with a few hairs and further inward with modified sweat glands, and the ceruminous glands, which secrete the yellow, pasty cerumen, or earwax. The hairs and the cerumen protect the ear from the entrance of foreign substances.

The *tympanic membrane* (membrana tympani) separates the auditory canal from the tympanic cavity. It consists of a thin layer of fibrous tissue covered externally with skin and internally with mucous membrane. It is ovoid and extends obliquely downward and inward and forms an angle with the floor of the meatus. It is chiefly innervated by a branch of the mandibular nerve (branch of fifth, the trigeminal, nerve).

Tympanic Cavity, or Middle Ear. This is a small, irregular bony cavity, situated in the petrous portion of the temporal bone. This air cavity is so small that probably five or six drops of water would fill it. It is separated from the external auditory canal by the tympanic membrane, and from the internal ear by a very thin bony wall ($1/24$ in.) in which there are two small openings: the *fenestra vestibuli* (*oval*) and the *fenestra cochleae* (*round*). In the posterior, or mastoid, wall there is an opening into the mastoid antrum and mastoid cells; and because of this, infection of the middle ear may extend into the mastoid cells and cause mastoiditis. The temporal bone at this point is very porous, and any suppurative process is exceedingly dangerous, for the infection may travel inward and invade the brain. In the anterior, or carotid, wall is an opening into the eustachian tube, a small canal that leads to the nasopharynx. Thus, there are five openings in the middle ear; namely, the opening between it and the auditory canal; the fenestra vestibuli and the fenestra cochleae, which connect with the internal ear; the opening into the mastoid cells; and the opening into the eustachian tube. The walls of the tympanic cavity are lined with mucous membrane, which is continuous anteriorly with the mucous membrane of the auditory tube and posteriorly with that of the mastoid antrum and mastoid cells.

Ossicles. Stretching across the cavity of the middle ear from the tympanic membrane to the fenestra vestibuli are three tiny, movable bones, named, because of their shapes, the *malleus*, or hammer, the *incus*, or anvil, and the *stapes*, or stirrup. The handle of the malleus is attached to the base of the incus. The long process of the incus is attached to the stapes, and the footpiece of the stapes occupies the fenestra vestibuli. These little bones are held in position, attached to each other, to the tympanic membrane, and to the edge of the fenestra vestibuli, by minute ligaments and muscles. They are set in motion with every movement of the tympanic membrane. Vibrations of the membrane are communicated to the malleus, received by the incus, and transmitted to the stapes, which rocks in the fenestra vestibuli and is therefore capable of transmitting to the fluid in the cavity of the labyrinth the impulses it receives. These bones form a se-

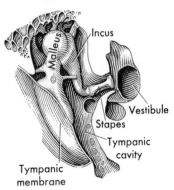

Figure 12–9. Chain of ossicles and their ligaments, seen from the front.

ries of levers, the effect of which is to magnify the force of the vibrations received at the tympanum about 10 times that at the oval window.

The auditory, or *eustachian, tube* connects the cavity of the middle ear with the pharynx. It is about 36 mm long and about 3 mm ($\frac{1}{8}$ in.) in diameter at its narrowest part and lined with mucous membrane. By means of this tube the pressure of the air on both sides of the tympanic membrane is equalized. In inflammatory conditions, the auditory tube may become occluded and may prevent this equalization. Under such conditions, hearing is much impaired until the tube is opened. The pharyngeal opening of the tube is closed except when swallowing, yawning, or sneezing.

Internal Ear, or Labyrinth. This receives the ultimate terminations of the vestibulo-cochlear nerve. It consists of an *osseous labyrinth,* which is composed of a series of peculiarly shaped cavities, hollowed out of the petrous portion of the temporal bone and named from their shape:

$$\text{Osseous labyrinth} \begin{cases} \text{1. The vestibule} \\ \text{2. The cochlea (snail shell)} \\ \text{3. The semicircular canals} \end{cases}$$

Within the osseous labyrinth is a *membranous labyrinth,* having the same general form as the cavities in which it is contained, though considerably smaller,

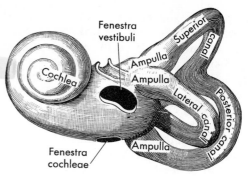

Fenestra vestibuli

Cochlea

Superior canal

Ampulla

Ampulla

Lateral canal

Posterior canal

Fenestra cochleae

Ampulla

Figure 12–10. Left osseous labyrinth, viewed from lateral side. (Modified from Cunningham.)

being separated from the bony walls by a quantity of fluid called the *perilymph.* It does not float loosely in this liquid but is attached to the bone by fibrous bands. The cavity of the membranous labyrinth contains fluid—the *endolymph*—and on its walls the ramifications of the acoustic nerve are distributed.

The *vestibule* is the central cavity of the osseous labyrinth; it is situated behind the cochlea and in front of the semicircular canals. It communicates with the middle ear by means of the fenestra vestibuli in its lateral or tympanic wall. The membranous labyrinth of the vestibule does not conform to the shape of the bony cavity but consists of two small sacs, called respectively the *saccule* and the *utricle.* The saccule is the smaller of the two and is situated near the opening of the scala vestibuli of the cochlea; the utricle is larger and

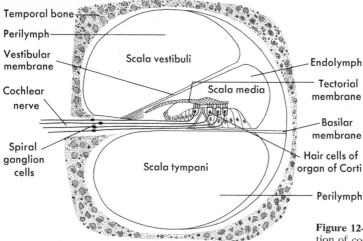

Temporal bone

Perilymph

Vestibular membrane

Scala vestibuli

Cochlear nerve

Scala media

Spiral ganglion cells

Scala tympani

Endolymph

Tectorial membrane

Basilar membrane

Hair cells of organ of Corti

Perilymph

Figure 12–11. Diagram showing cross section of cochlea.

occupies the upper and back part of the vestibule. These sacs are not directly connected with each other. From the posterior wall of the saccule, a canal, the *ductus endolymphaticus*, is given off. This duct is joined by a duct from the utricle and ends in a blind pouch on the posterior surface of the petrous portion of the temporal bone. The utricle, saccule, and ducts contain endolymph and are surrounded by perilymph. The inner wall of the saccule and utricle consists of two kinds of modified columnar cells on a basement membrane. One is a specialized nerve cell provided with stiff hairs, which project into the endolymph. Between the nerve cells are supporting cells, which are not ciliated and are not connected with nerve endings. The hair cells serve as end organs for fibers of the vestibular branch of the vestibulocochlear nerve, which arborize around the base of each hair cell. Small crystals of calcium carbonate, called *otoliths*, are located over the hair cells of the utricle. The otoliths give weight to the hair cells and make them more sensitive to change in position. Impulses are set up when the otoliths pull or push on the hairs.

The *cochlea* forms the anterior part of the bony labyrinth and is placed almost horizontally in front of the vestibule. It resembles a snail shell and consists of a spiral canal of $2\frac{3}{4}$ turns around a hollow, conical central pillar called the *modiolus,* from which a thin *lamina* of bone projects like a spiral shelf about halfway toward the outer wall of the canal. Within the bony cochlea is a *membranous cochlea,* which begins at the fenestra ovalis and duplicates the bony structure.

The *basilar membrane* stretches from the free border of the lamina to the outer wall of the bony cochlea and completely divides its cavities into two passages, or *scalae*, which, however, communicate with each other at the apex of the modiolus by a small opening. The upper passage is the scala vestibuli, which terminates at the fenestra vestibuli; and the lower is the scala tympani, which terminates at the fenestra cochleae.

From the free border of the lamina, a

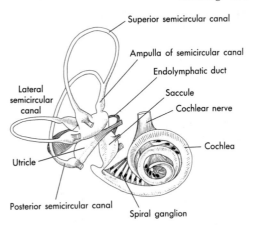

Figure 12–12. The bony labyrinth has been removed to show the membranous labyrinth that is filled with endolymph and from which all the nerves of the internal ear arise.

second membrane, called the *vestibular membrane* (Reissner[6]), extends to the outer wall of the cochlea and is attached some distance above the basilar membrane. A triangular canal, called the *ductus cochlearis* or *scala media*, is thus formed between the scala vestibuli above and the scala tympani below.

On the basilar membrane the sound-sensitive epithelium of the *organ of Corti*[7] is located. This consists of a large number of *rod-shaped cells* and *hair cells*, extending into the endolymph of the scala media. The *tectorial membrane* projects from the spiral lamina over these cells of the organs of Corti. The hairs of the hair cells are attached to the tectorial membrane so that as the basilar membrane vibrates, the hair cells are alternately stretched and relaxed. The fibers of the cochlear branch of the vestibulocochlear nerve arise in the nerve cells of the *spiral ganglion*, which is situated in the modiolus. These cells are bipolar and send fibers toward the brain in the vestibulocochlear nerve and the other fibers to end in terminal arborizations around the hair cells of the organ of Corti.

The *semicircular canals* are three bony canals lying above and behind the vestibule and communicating with it by five openings, in one of which two tubes join.

[6]Ernst Reissner, German anatomist (1824–1878).
[7]Alfonso Corti, Italian anatomist (1822–1888).

They are known as the *superior, posterior,* and *lateral* canals, and their position is such that each one is at right angles to the other two. One end of each tube is enlarged and forms what is known as the *ampulla.*

The *semicircular ducts* (membranous semicircular canals) are similar to the bony canals in number, shape, and general form, but their diameter is less. They open by five orifices into the utricle, one opening being common to the medial end of the superior and the upper end of the posterior duct. In that ampullae the membranous canal is attached to the bony canal, and the epithelium is thrown into a ridge (the *crista ampullaris*) of cells with hairlike processes, which project into the endolymph. The hair cells are covered with a gelatinous substance that contains the otoliths. These are minute particles of calcium carbonate that pull and push on the hairs when the position of the head is changed. Hair movement initiates reflexes that maintain balance and equilibrium. Some of the peripheral terminations of the vestibular branch of the vestibulocochlear nerve are distributed to these cells. Between the hair cells are supporting cells.

Acoustic Center. The primary acoustic center is in the *temporal lobe* of the cerebrum. Removal of both temporal lobes produces complete deafness. Removal of one temporal lobe impairs hearing. This suggests that some fibers from each ear cross at some point in their afferent pathways and terminate in the opposite cortex. This is similar to the partial decussation of visual fibers occurring in the optic chiasma.

The vestibulocochlear nerve (VIII) is sensory and contains at least two sets of fibers, which differ in their origin, destination, and function. One set of fibers is known as the cochlear division and the other as the vestibular.

The cochlear nerve arises from bipolar cells in the spiral ganglion of the cochlea. The peripheral fibers pass to the cells of the organ of Corti, at which point the sound waves initiate the nerve impulses. The central fibers, forming part of the cochlear branch of VIII, pass into the lat-

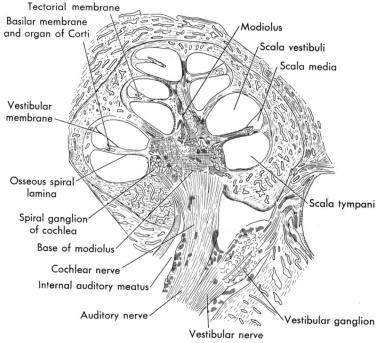

Figure 12–13. Section of the cochlea, showing the scalae. *Red* indicates blood supply. (Modified from Toldt.)

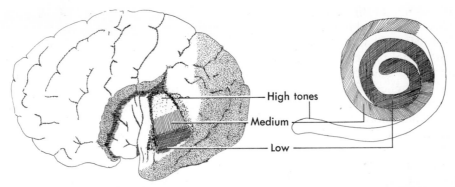

Figure 12–14. Diagram showing relationship between cochlea and acoustic area of cortex and interpretations of sound in the brain.

eral border of the medulla, terminating in the dorsal and ventral cochlear nuclei. From these nuclei the path is continued by secondary neurons to the auditory centers in the medial geniculate bodies of the thalamus. Some of the fibers cross and some do not. The cell bodies of the thalamus form the third cell in the primary auditory pathway, the processes of which terminate in the temporal lobe of the cortex.

The vestibular nerve arises from bipolar cells in the *vestibular ganglion* (ganglion of Scarpa), situated in the internal acoustic meatus. The peripheral fibers divide into three branches, which are distributed to the hair cells of the saccule, the utricle, and the ampullae of the semicircular canals. The central fibers, forming part of the vestibular branch, terminate in the vestibular nuclei in the medulla (Figure 11–24, page 261). From these nuclei some fibers project to the cerebellum, and others pass down the spinal cord as the vestibular spinal tract to form connections with motor centers of the spinal nerves. Connections between vestibular and ocular nuclei provide the neural circuitry for reflex eye movements that accompany changes in the position of the body in space.

Process of Hearing. All bodies that produce sound are in a state of vibration and communicate their vibrations to the air with which they are in contact. The range of air vibrations for sound is from 20 to 20,000 vibrations per second.

When these air waves, set in motion by sonorous bodies, enter the external auditory canal, they set the tympanic membrane vibrating. The vibrations of the tympanic membrane are then communicated by means of the auditory ossicles stretched across the middle ear to the perilymph and then to the endolymph of the inner ear. The movements of the fluids, in rhythm with the air, stimulate the nerve endings in the organ of Corti; and from these, impulses are conveyed to the center of hearing in the temporal cortex. Characteristics of the sensation of sound are *loudness*, which varies with the *amplitude* of vibrations; *pitch*, which varies with the frequency of vibrations; and *timbre*, which is due to the pattern the complex of vibrations makes.

A unit for measuring the loudness of sound is the *bel* (which measures air pressure changes); for convenience a tenth of a bel, or a decibel, is used. Zero decibels is the threshold of hearing, and normal conversation is about 65 decibels.

The *resonance* theory of Helmholtz, postulates that the cochlea is the analyzer of sound. The theory makes use of the elastic nature of the basilar membrane. This structure is relatively massive at its distal end, has a relatively low stiffness, and requires the movement of a relatively large mass of perilymph to move it because of its distance from the stapes. Low frequencies of vibration are thought to act here. Just inside the oval window and stapes the basilar membrane is lighter, narrower, and more rigid; at this site there is less fluid to move. High frequencies tend to produce a maximal effect here, i.e., the amplitude of oscillation is greater at this site than further on in the cochlea. Man is thought to be able to distinguish more than 10,000 pitches of tone. There are also more than 15,000 hair cells on the basilar membrane and more than 15,000 fibers in the cochlear nerve. This almost constitutes an adequate physical mechanism for discrimination in auditory sensation. Abnormal conditions in any part of the auditory mechanism may excite the auditory nerve and give rise to noises that are described as rushing, roaring, humming, and ringing.

Since the cochlea is embedded in a bony cavity, vibrations from any or all skull bones can cause fluid vibrations of the cochlea itself. Thus, in deafness due to calcification and immobility of ossicles, it is possible to hear by bone conduction. Some hearing-aid devices amplify air waves into vibrations that will be readily transmitted through the mastoid bone. If disease has destroyed the receptor hair cells or the afferent nerve fibers, hearing aids will be of no use. Prolonged and excessive noise causes deafness, presumably through damage to the fibers in the organ of Corti.

EQUILIBRIUM

Among the various means (such as sight, touch, and muscular sense) whereby one is enabled to maintain equilibrium, coordinate movements, and become aware of position in space, one of the most important is the vestibular apparatus. It contains mechanoreceptors specialized to detect changes in both the motion and position of the head. The vestibular apparatus consists of the three membranous *semicircular canals*, the connecting *utricle*, and the *saccule*, which connects to the utricle and the membranous cochlear duct (Figure 12–12). It is located in the bony channels of the inner ear, in the temporal bone of the skull. This system of canals and ducts is filled with endolymph. Basically the vestibular apparatus contains specialized epithelium consisting of sensory receptor hair cells that are innervated by afferent neurons, and supporting cells.

On each side of the head the three semicircular canals are arranged at right angles to one another. An expanded end of each canal, the ampulla, contains a crista ampullaris which is the sensory receptor. The crista contains sensory hair cells, supporting cells, and a gelatinous mass, the cupula, overriding the hair cells. There are two types of hair cells: type I cells are flask-shaped and their inferior ends are enclosed by a chalice-like nerve terminal; type II cells are columnar in shape and innervated by numerous small synaptic endings. When observed with the electron microscope, both type I and type II hair cells are characterized by the presence of a single cilium commonly called a *kinocilium,* and 50 to 100 straight hairs which are highly specialized microvilli called stereocilia.

The function of the semicircular canals is the detection of angular acceleration and deceleration of the head. When the head rotates, the inertia of the endolymph within the semicircular canals causes a relative movement of the endolymph in a direction opposite to that of the head. The movement of the endolymph causes the cupula to bend in the direction of its flow, thereby bending and thus stimulating the sensory hair cells. If rotation of the head continues, the friction between the endolymph and the walls of the semicircular canal will eventually cause the endolymph to attain the same rate and direction of motion as the head, the cupula will not be bent, distortion of hair cells will subside, and stimulation will cease. When head rotation stops, the reverse sequence of events will occur. If the head is turned to the left, the sensory hair cells of the left semicircular canal are stimulated, but in the right semicircular canal the movement of the sensory hairs is in the opposite direction and no sensory stimulation occurs, with the result that the sensation of rotation to the left is perceived.

The utricle and possibly the saccule contain the receptors of the vestibular system which detect changes in head position relative to the direction of the forces of gravity. The sensory epithelium of both the utricle and saccule is the same as that of the crista having both type I and type II cells and supporting cells. It is covered by a gelatinous membrane in which are embedded numerous calcium carbonate crystals, the otoliths. Bending of the head causes movement of the otoliths, which in turn stimulates the receptor hair cells.

The sensory hair cells of both the semicircular canals and the utricle and saccule transmit impulses via the sensory fibers of the vestibular nerve, whose cell bodies make up the vestibular ganglion, which is located in the internal auditory meatus. Fibers of the vestibular nerve pass through a number of relay stations to the brain stem and the cerebellum. Information from the vestibular apparatus controls the muscles which move the eyes, thus in spite of changes in the position of the head, the eyes remain fixed on the same point. In addition, vestibular information is utilized (vestibulospinal pathways) in reflex mechanisms for maintaining upright posture. The cerebellum links the impulses that arise from stimulation of the sensory nerves of the vestibular apparatus to the motor centers of the cerebrum and spinal cord, thus acting as a coordination center that maintains posture and equilibrium.

VISION

The visual apparatus consists of the eyeball, the optic nerve, and the visual center in the brain. In addition to these essential organs, there are accessory organs that are necessary for the protection and functioning of the eyeball.

Accessory Organs of Eye. Under this heading are grouped eyebrows, eyelids, conjunctiva, lacrimal apparatus, muscles of the eyeball, and the fascia bulbi.

The eyebrow is a thickened ridge of skin, covered with short hairs. It is situated on the upper border of the orbit and protects the eye from too vivid light, perspiration, etc.

The eyelids (palpebrae) are two movable folds placed in front of the eye. They are covered externally with skin and internally with a mucous membrane, the conjunctiva, which is reflected from them over the bulb of the eye. They are composed of muscle fibers and dense fibrous tissue known as the *tarsal plates*. The upper lid is attached to a small muscle called the elevator of the upper lid (*levator palpebrae superioris*). Arranged as a sphincter around both lids is the *orbicularis oculi* muscle, which closes the eyelids.

The slit between the edges of the lids is called the palpebral fissure. It is the size of this fissure that causes the appearance of large and small eyes, as the size of the eyeball itself varies but little. The eyelids provide protection for the eye—movable shades that cover the eye during sleep, protect the eye from bright light and foreign objects, and spread the lubricating secretions of the eye over the surface of the eyeball.

Eyelashes and Sebaceous Glands. From the margin of each eyelid, a row of short, thick hairs—the eyelashes—project. The follicles of the eyelashes receive a lubricating fluid from the sebaceous glands which open into them. If these glands become infected, a sty results. A *sty*, therefore, is comparable to a pimple or *furuncle* resulting from the infection of retained sebaceous fluid in other regions of the skin.

Lying between the conjunctiva and the tarsal cartilage of each eyelid is a row of elongated sebaceous glands—the tarsal, or meibomian, glands—the ducts of which open on the edge of the eyelid. The secretion of these glands lubricates their edges and prevents adhesion of the eyelids. Distention of the gland is termed a *chalazion*.

Conjunctiva. The mucous membrane that lines the eyelids and is reflected over the forepart of the eyeball is called the conjunctiva. It is continuous with the lining membrane of the ducts of the tarsal glands, the lacrimal ducts, lacrimal sac, nasolacrimal duct, and nose.

Lacrimal Apparatus. This apparatus consists of the lacrimal gland, the lacrimal ducts, the lacrimal sac, and the nasolacrimal duct.

The *lacrimal gland* is a compound gland

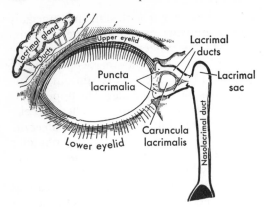

Figure 12–15. The lacrimal apparatus. (Modified from *Gray's Anatomy*.)

and is lodged in a depression of the frontal bone at the upper and outer angle of the orbit. It is about the size and shape of an almond and consists of two portions, a superior and inferior, which are partially separated by a fibrous septum.

Six to twelve minute *ducts* lead from the gland to the surface of the conjunctiva of the upper lid. The secretion (tears) is usually just enough to keep the eye moist and, after passing over the surface of the eyeball, flows through the *puncta* into two tiny *lacrimal ducts* and is conveyed into the *lacrimal sac* at the inner angle of the eye. The *lacrimal sac* is the expanded upper end of the *nasolacrimal* duct, a small canal that opens into the nose. It is oval and measures from 12 to 15 mm in length. The *caruncula lacrimalis* (caruncle) is a small reddish body situated at the medial commissure. It contains sebaceous and sudo-riferous glands and forms the whitish secretion that collects in this region.

The lacrimal gland secretes tears. This secretion is a dilute solution of various salts in water, which also contains small quantities of mucin. The ducts leading from the lacrimal gland carry tears to the eyeball, and the lids spread it over the surface. Ordinarily this secretion is evaporated, or carried away by the nasolacrimal duct, as fast as formed; but under certain circumstances, as when the conjunctiva is irritated or when painful emotions arise, the secretion of the lacrimal gland exceeds the drainage power of the nasolacrimal duct, and the fluid, accumulating between the lids, at length overflows and runs down the cheeks. The purpose of the lacrimal secretion is to keep the surface of the eyes moist and to help remove microorganisms and dust.

Inflammation from the nose may spread to the nasolacrimal ducts, blocking them and thus cause a slow dropping of tears from the inner angle of the eye. The lacrimal glands do not develop sufficiently to secrete tears until about the fourth month of life, hence, the need for protecting a baby's eyes from bright light and dust.

Fascia Bulbi. Between the pad of fat and the eyeball is a thin membrane—the fascia bulbi—which envelops the eyeball from the optic nerve to the ciliary region and forms a socket in which the eyeball rotates.

Muscles. Muscles of the eye are listed in Table 12–1.

TABLE 12–1
MUSCLES OF THE EYE AND LIDS

Muscle	Origin	Insertion	Nerve Supply	Function
I. Extrinsic				
Superior rectus	Apex of orbital cavity	Upper and central portion of eyeball	Oculomotor	Rolls eyeball upward
Inferior rectus	Apex of orbital cavity	Lower and central portion of eyeball	Oculomotor	Rolls eyeball downward
Lateral rectus	Apex of orbital cavity	Midway on outer side of eyeball	Abducens	Rolls eyeball laterally

TABLE 12–1 (*Continued*)

Muscle	Origin	Insertion	Nerve Supply	Function
Medial rectus	Apex of orbital cavity	Midway on inner side of eyeball	Oculomotor	Rolls eyeball medially
Superior oblique	Apex of orbital cavity	Eyeball between superior and lateral recti	Trochlear	Rolls eyeball on its axis, directs cornea downward and laterally
Inferior oblique	Orbital plate of maxilla	Eyeball between superior and lateral recti	Oculomotor	Rolls eyeball on its axis, directs cornea upward and laterally
II. Intrinsic Sphincter pupillae		Circular muscle attached at circumference to sclera, cornea, and ciliary processes	Parasympathetic fibers, oculomotor	Constriction of pupil
Dilator pupillae		Radiating fibers from pupil outward, attached at circumference to sclera, cornea, and ciliary processes	Sympathetic fibers, from ciliary ganglion and trigeminal nerve	Dilation of pupil
III. Muscles of Lid Orbicularis oculi	Nasal part of frontal bone; frontal process of maxilla; medial palpebral ligament	Palpebral portion is inserted into lateral palpebral raphe; orbital portion surrounds orbit; upper fibers blend with frontalis and corrugator muscle	Temporal and zygomatic branches of facial nerve	Palpebral portion closes lids gently; orbital portion, stronger closing
Levator palpebrae superioris	Inferior surface of small wing of sphenoid	Tarsus and orbicularis oculi and skin of eyelid	Fibers from oculomotor nerve	Raises upper eyelid, antagonist to orbicularis oculi

The muscles of the eye and eyelids receive their blood supply from both the internal and external carotid arteries (Fig. 11–19, page 254). See Figure 11–21, page 255, for venous return.

The Orbits. The orbits are the bony cavities in which the eyeballs are contained. Each orbit is shaped like a funnel; the large end, directed outward and forward, forms a strong bony edge that protects the eyeball. The small end is directed backward and inward and is pierced by a large opening—the optic foramen—through which the optic nerve passes into the cranial cavity, and the ophthalmic artery passes from the cranial cavity to the eye. A larger opening to the outer side of the foramen—the superior orbital fissure— provides a passage for the orbital branches of the middle meningeal artery and the nerves that carry impulses to and from the muscles, i.e., the oculomotor, the trochlear, the abducens, and the ophthalmic. The inner portion of the eyeball contains a pad of fat which serves to cushion the eyeball and diminishes when starvation occurs. Then the eyeballs sink into the orbit.

Bulb of Eye, or Eyeball. The bulb of the eye is spherical, but its transverse diameter is less than the anteroposterior so that it projects anteriorly and looks as if a section of a smaller sphere had been engrafted on the front of it.

The bulb of the eye is composed of three coats, or tunics. From the outside of the eyeball inward toward its center these are:

Fibrous: (1) sclera, (2) cornea
Vascular: (1) choroid, (2) ciliary body, (3) iris
Nervous: retina

Fibrous Tunic. This is formed by the sclera and cornea.

1. *The sclera,* or *white of the eye,* covers the posterior five sixths of the eyeball. It is composed of a firm, unyielding, fibrous membrane, thicker behind than in front, and serves to maintain the shape of the eyeball and to protect the delicate structures contained within it. It is opaque, white, and smooth externally; behind, it is pierced by the optic nerve. Internally it is brown in color and is separated from the choroid by a fluid space. It is supplied with few blood vessels. A *venous sinus*—the canal of Schlemm[8]—encircles the cornea at the corneoscleral junction. Its nerves are derived from the ciliary.

2. *The cornea* covers the anterior sixth of the eyeball. It is directly continuous with the sclera, which, however, overlaps it slightly above and below. The cornea, like the sclera, is composed of fibrous tissue, which is firm and unyielding, but, unlike the sclera, it has no color and is perfectly transparent; it has been aptly

[8] Friedrich Schlemm, German anatomist (1795–1858).

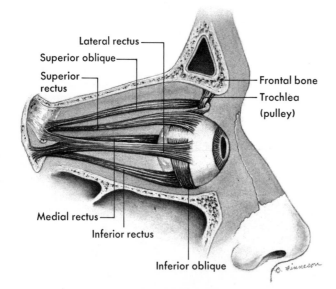

Lateral rectus
Superior oblique
Superior rectus
Frontal bone
Trochlea (pulley)
Medial rectus
Inferior rectus
Inferior oblique

Figure 12–16. The extrinsic muscles of the eyeball in the right orbit. Note tendinous insertions of superior and inferior oblique muscles between the superior and lateral recti. Lateral view.

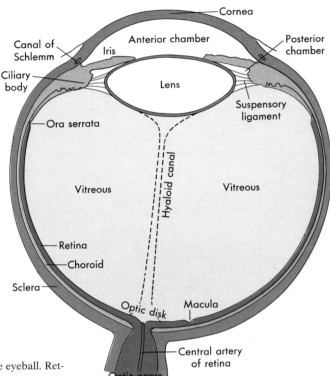

Figure 12–17. Horizontal section of the eyeball. Retina, *red;* choroid, *yellow;* sclera, *blue.*

termed the "window of the eye." The cornea is well supplied with nerves (derived from the ciliary) and lymph spaces that surround the nerves but is destitute of blood vessels, so that it is dependent on the lymph for nutrients. Injury to the cornea causes scarring and impairs vision.

Vascular Tunic (Uvea, or Uveal Tract). This consists, from behind forward, of the choroid, the ciliary body, and the iris.

1. *The choroid* is a thin, dark-brown membrane lining the inner surface of the posterior five sixths of the sclera. It is pierced behind by the optic nerve. The inner surface is attached to the pigmented layer of the retina and extends anteriorly to the ora serrata. It consists of a dense capillary plexus and small arteries and veins carrying blood to and from the plexus. Between these vessels are pigment cells which with other cells form a network, or stroma. The blood vessels and pigment cells render this membrane dark and opaque, so that it darkens the chamber of the eye by preventing the reflection of light. It extends to the ciliary body.

2. *The ciliary body* includes the orbicu-

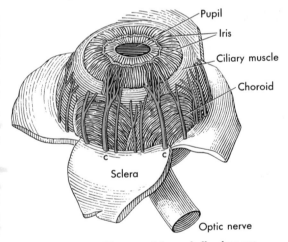

Figure 12–18. Diagram of the eyeball, sclera cut and turned back. Note pupil, iris, ciliary muscle, choroid, ciliary nerves (*c*), and optic nerve.

laris ciliaris, the ciliary processes, and the ciliaris muscle. The orbicularis ciliaris is a zone about 4 mm in width, which is directly continuous with the anterior part of the choroid.

Just behind the edge of the cornea, the choroid is folded inward and arranged in radiating folds, like a plaited ruffle, around the margin of the lens. There are from 60 to 80 of these folds, and they constitute the ciliary processes. They are well supplied with nerves and blood vessels and also support a muscle, the ciliaris (ciliary) muscle. The fibers of this muscle arise from the sclera near the cornea and, extending backward, are inserted into the outer surface of the ciliary processes and the choroid. This muscle is the chief agent in accommodation. When it contracts, it draws forward the ciliary processes, relaxes the suspensory ligament of the lens, and allows the elastic lens to resume a more convex form.

3. *The iris* is a circular, colored disk suspended in the aqueous humor in front of the lens and behind the cornea. It is attached at its circumference to the ciliary processes, with which it is practically continuous, and is also connected to the sclera and cornea at the point where they join one another. Except for this attachment at its circumference, it hangs free in the interior of the eyeball. In the middle of the iris is a circular hole, the *pupil*, through which light is admitted into the eye chamber. The iris is composed of connective tissue containing branched cells, numerous blood vessels, and nerves. The color of the eye is related to the number and size of pigment-bearing cells in the iris. If there is no pigment or very little, the eye is blue; with increasing amounts of pigment the eye is gray, brown, or black. It also contains two sets of antagonistic muscles described in the chart on page 283.

The posterior surface of the iris is covered by layers of pigmented epithelium designed to prevent the entrance of light.

FUNCTION OF THE IRIS. The function of the iris is to regulate the amount of light entering the eye and thus assist in obtaining clear images. This regulation is accomplished by the action of the muscles described above, as their contraction or relaxation determines the size of the pupil. When the eye is accommodated for a near object or stimulated by a bright light, the sphincter muscle contracts and diminishes the size of the pupil. When, on the other hand, the eye is accommodated for a dis-

tant object or the light is dim, the dilator muscle contracts and increases the size of the pupil.

Retina. The retina, the innermost coat of the eyeball, is a delicate membrane of tissue that receives the images of external objects and transfers the impressions evoked by them to the center of sight in the cortex of the cerebrum. It occupies the space between the choroid coat and the hyaloid membrane of the vitreous body and extends forward almost to the posterior margin of the ciliary body, where it terminates in a jagged margin known as the *ora serrata*. The retina is made up of ten layers; from external to internal these are:

1. Pigment epithelium
2. Layer of rods and cones
3. External limiting membrane
4. Outer nuclear layer
5. Outer plexiform layer
6. Inner nuclear layer
7. Inner plexiform layer
8. Ganglion cell layer
9. Optic nerve fiber layer
10. Internal limiting membrane

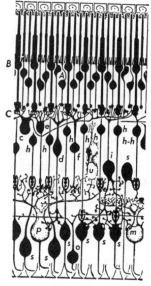

Figure 12–19. Scheme of the structure of the primate retina, as shown by Polyak, 1941. *A,* rod; *B,* cone; *C,* horizontal cell; *d,* diffuse ganglionic cell. This diagram illustrates the very complex arrangement of retinal cells. (Courtesy of University of Chicago Press.)

The second layer, called the layer of rods and cones, is the light-sensitive layer where light energy is converted to nerve impulses that are transmitted to the brain by the optic nerve. To reach the light-sensitive receptors lying in this second retinal layer, light must pass through the cornea, aqueous humor, lens, vitreous humor, and each of the retinal layers before striking the rods and cones.

In the center of the receptor layer of the retina, there is a small special region, the fovea, in which there are only cones. Moving from the fovea to the periphery of the retina the concentration of cones diminishes and the concentration of rods increases until at the most peripheral edges of the retina one finds only rods.

The rods and cones are two different types of visual receptors. They differ both structurally and functionally, each subserving important roles in vision.

The rods are particularly important for vision in dim illumination because it takes very little light to stimulate them. Although the rods are stimulated by most of the wavelengths in the visible spectrum, they cannot produce a color sensation. They permit light and dark discrimination and form and movement perception but provide poor visual acuity. Rhodopsin is the photosensitive pigment within the rods. This material is bleached by light, and it is slowly regenerated in the dark. It takes almost an hour in total darkness for complete resynthesis of rhodopsin. As the rhodopsin is regenerated, the sensitivity of the rods progressively increases. This gradual increase in retinal sensitivity is called dark adaptation. Vitamin A and nicotinamide are essential for resynthesis of rhodopsin and any vitamin A deficiency will delay or reduce rhodopsin regeneration. The resulting condition is known as night blindness (nyctalopia); the rods are less capable of increasing their sensitivity in dim illumination.

In contrast the cones are particularly important for color vision in bright illumination. The cones need more intense light to be excited than do the rods, and the cones maximally alter their sensitivity to decreased illumination within 10 to 15 minutes. Their range of sensitivity is small compared to the 100 million sensitivity range over which the rods can function. The cones are responsible for visual acuity. In the fovea, where only cones exist, there is only one cone per nerve fiber. It is as if each receptor has its own "private line" to the brain. The cones also contain a photosensitive pigment, which has not been isolated in man, but is thought to be similar to rhodopsin. Three types of cones exist, containing different types of pigment that make them sensitive to red, green, or blue color selectively.

BLIND SPOT. The optic nerve pierces the eyeball not exactly at its most posterior point but a little to the inner side, called the blind spot. Because it contains no receptor cells, this point is insensitive to light. The central artery of the retina, a branch of the ophthalmic artery, and its vein pass into the retina along with the optic nerve.

MACULA LUTEA. One point of the retina is of great importance—the macula lutea, or yellow spot. It is situated about 2.08 mm ($\frac{1}{12}$ in.) to the outer side of the exit of the optic nerve and is the exact center of the retina. In its center is a tiny pit—*fovea centralis*—which is the center of direct vision. At this point there is an absence of rods but a great increase in the number of cones. This is the region of greatest visual acuity. In reading, the eyes move so as to bring the rays of light from word after word into the center of the fovea.

Refracting Media. The eye contains four refracting media. These are cornea, aqueous humor, crystalline lens and capsule, and vitreous body. The cornea and the aqueous humor form the first refracting media. The *aqueous humor* fills the forward chamber; the latter is the space bounded by the cornea in front and by the lens, suspensory ligament, and ciliary body behind. This space is partially divided by the iris into an anterior and a posterior chamber. The aqueous humor is a clear, watery solution containing minute amounts of salts, mainly sodium chloride.

It is derived mainly from the capillaries by diffusion and it drains away through the veins and through the spaces of Fontana[9] into the venous canal of Schlemm and then on into the larger veins of the eyeball.

The *crystalline lens* enclosed in its capsule is a transparent, refractive body, with convex anterior and posterior surfaces. It is placed directly behind the pupil, where it is retained in position by the counterbalancing pressure of the aqueous humor in front and the vitreous body behind, and by its own suspensory ligament, formed in part by the hyaloid membrane and in part by fibers derived from the ciliary processes. The posterior surface is considerably more curved than the anterior, and the curvature of each varies with the period of life. In infancy, the lens is almost spherical; in the adult, of medium convexity; and in the aged, considerably flattened. The capsule surrounding the lens is elastic, and with age it loses its original elasticity. Its refractive power is much greater than that of the aqueous or vitreous body. In cataract the lens or its capsule becomes less transparent and blurs, causing loss of vision. Cataracts are treated by removing the opaque lens and compensating with an artificial lens, i.e., eyeglasses.

The *vitreous body*, a semifluid albuminous tissue enclosed in a thin membrane, the hyaloid membrane, fills the posterior four fifths of the bulb of the eye. The vitreous body distends the greater part of the sclera, supports the retina, which lies upon its surface, and preserves the spheroidal shape of the eyeball. Its refractive power, though slightly greater than that of the aqueous humor, does not differ much from that of water.

In glaucoma, intraocular pressure increases, cupping the optic disk and interfering with the proper distribution of blood to all the inner tissues of the eye. This increased pressure may be due to increased blood pressure in the larger blood vessels of the eye, to altered osmotic conditions of blood and eye fluids, to rigidity of the eyeball, or to improper functioning of intrinsic muscles of the eye. It may or may not cause pain. If the increased pressure persists, it can lead to irreversible damage to the visual cells.

Perception of Light and Color. Electromagnetic vibrations from the sun or other light source occur in waves of varying length. Vibrations from 400 mμ to 800 mμ long are called the visible spectrum (light and color waves) (Table 12–2). Those shorter than this are known as ultraviolet rays; those much longer are electrical waves. When light waves enter the eye, they give rise to impulses that are carried by the optic nerve to the occipital cortex, where the visual sensation results.

Color blindness refers to the inability to discriminate colors properly. About 9 per cent of normal healthy males are color blind to some degree.

Color-blind individuals are classifiable in relation to normal subjects who are called *trichromats*. Individuals lacking one type of cone pigment are called *dichromats*. The most usual type of dichromat is the *protanope*, who lacks sensitivity to wavelengths in the red end of the spectrum. Another type of dichromat is the *deuteranope*, whose sensitivity is weak or lacking in the green region of the spectrum. Though the deuteranopes do not have a red deficiency, they confuse red and green. The *tritanope* is a rare type of dichromat who lacks blue-type cones. These individuals confuse blue and green. Some individuals have all three types of cone pigments, but one type may be deficient rather than absent, and they are called *anomalous trichromats*. These individuals rarely misname colors, but they confuse colors. These confusions provide the basis for some of the standard "hidden-figure" type of color tests.

Refraction. The central components of light waves enter the eyes perpendicularly, and the sides obliquely. For clear vision

---

[9]Felice Fontana, Italian physiologist (1720–1805).

TABLE 12–2

WAVE LENGTHS AND COLOR

723 mμ – 647 mμ	= red
647 mμ – 585 mμ	= orange
585 mμ – 575 mμ	= yellow
575 mμ – 492 mμ	= green
492 mμ – 455 mμ	= blue
455 mμ – 424 mμ	= indigo
424 mμ – 397 mμ	= violet

the oblique rays must converge and come to a focus with the central rays on the retina. The cornea, aqueous humor, crystalline lens, and vitreous humor form a system of refractory devices. Rays of light are bent, or undergo refraction, chiefly on entering the cornea from the air, on entering the lens from the aqueous humor, and on leaving the lens and entering the vitreous fluid.

Physiology of Vision. Visible objects reflect light rays that fall upon them. These reflected rays are brought to a focus on the rods and cones of the retina, and the resulting nerve impulses are transmitted via the optic nerve and thence through various relay stations to the centers of vision in the occipital lobes of the cerebrum. From here it is believed the impulses are transmitted to the association areas, where they awaken memories that enable one to interpret their meaning. The cones of the fovea centralis are the place of most acute vision and the part on which the light rays are focused when the eyes are accommodated for near objects. In a bright light the object is focused directly on the fovea, and the reflexes controlling accommodation help to bring this about. In a dim light the tendency is to diverge the eyes and thus bring the image into the peripheral and sensitive part of the retina. The visual field includes the entire expanse of space seen in a given instant without moving the eyes.

The temporal portion of the retina receives light waves from the nasal field of vision. The nasal portion of the retina receives light waves from the temporal field of vision. These fields may be tested to

Left eye

Right eye

Right occipital cortex

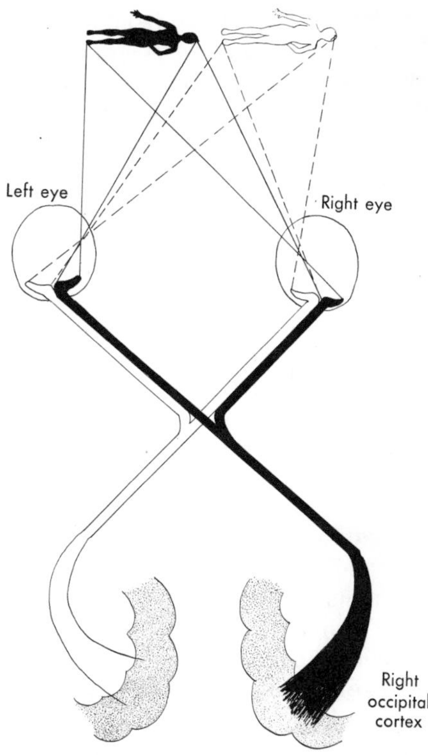

Figure 12–20. Diagram showing crossing of optic fibers. The fibers from the *nasal* half of the left retina and the fibers from the lateral half of the right retina are projected to the right occipital cortex and vice versa for the opposite side. See Figure 11–22 for nerve fibers in optic pathway.

determine the specific areas of retinal (or optic tract) damage.

Binocular Vision. The value of two eyes instead of one is that true binocular vision is possible. This is distinctive in that it is stereoscopic. A stereoscopic picture consists of two views taken from slightly different angles. In stereoscopic vision, two

Figure 12–21. *Left:* Diagrams illustrating rays of light converging in (*A*) normal eye, (*B*) myopic eye, and (*C*) hypermetropic eye. The parallel lines indicate light rays entering the eye; *X* is the point of convergence, or focus. In *A* the rays are brought to a focus (*X*) on the retina. In *C* they would come to a focus behind the retina. *Right:* Diagrams illustrating the convergence of light rays in a normal eye (*A*) and the effects of concave lens (*B*) and convex lens (*C*) on rays of light.

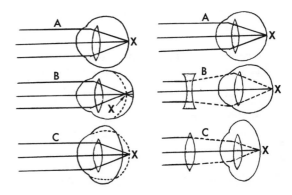

optical images are made from slightly different angles. This gives the impression of distance and depth and is equivalent to adding a third dimension to the visual field. The processes necessary for binocular vision are convergence, or turning the eyes inward; change in the size of the pupil; accommodation; and refraction. In binocular vision it is necessary to turn the eyes inward, in order that the two images of a given object may lie upon what are called corresponding points of the two retinas. Excitation of two corresponding points causes only one sensation, which is the reason why binocular vision is not ordinarily double vision. Convergence of the eyes is brought about by innervation of the medial rectus muscles. The correspondence of the two retinas and of the movements of the eyeballs is produced by a close connection of the nerve centers controlling the contraction of eye muscles.

Change in Size of Pupil. When one looks at a near object in a bright light, the pupil contracts, so that the entering rays are directed to the central part of the lens, i.e., the part where the convexity and the consequent refractive power are greatest, and to the fovea centralis. In a dim light the pupil is dilated, causing a diffusion of the rays to the peripheral parts of the retina where the concentration of rhodopsin is high. The constriction of the pupil is brought about by the reflex contraction of the circular muscle of the iris by impulses from parasympathetic fibers in the oculomotor nerve in response to strong light stimulating the retina. In dim light the stimulation of the retina is lessened and the pupil dilates. In excitement, fear, etc., its dilation is due to stimulation of autonomic nerve fibers that are distributed by the ophthalmic branch of the trigeminal nerve.

Accommodation. Accommodation is the adjustment of the eye to focus on objects at different distances, because for an image to be sharply focused it must fall on the macula. Accommodation involves three coordinated responses: (1) the convergence of the eyes, (2) pupillary constriction, and (3) alteration in the refractiveness of the lens. The first two reflexes were discussed in the preceding paragraphs. Alteration in lens refractiveness is also a reflex response. A blurred image on the retina initiates afferent impulses in the optic nerve that signal the motor center of the ciliary muscle, the chief effector in accommodation. When the eye is at rest or fixed upon distant objects, the suspensory ligament, which extends from the ciliary processes to the capsule of the lens, exerts a tension upon the capsule of the lens, which keeps the lens flattened, particularly the anterior surface to which it is attached. When the eye fixates on near objects, as in reading or sewing, the ciliary muscle contracts and draws forward the choroid coat, which in turn releases the tension of the suspensory ligament upon the capsule of the elastic lens and allows the anterior surface to become more convex. The accommodation for near objects is an active process and is always more or less fatiguing. On the contrary, the accommodation for distant objects is a passive process; consequently the eye rests for an indefinite time upon remote objects without fatigue.

Inversion of Images. Owing to refraction, light rays as they enter the eye cause the image of external objects on the retina to be *inverted*. The question then arises, "Why is it that objects do not appear to be upside down?" This question is answered if it is remembered that actual visual sensations take place in the brain and that the projection of these sensations to the exterior is a secondary act that has been learned from experience.

Abnormal Conditions That Interfere with Refraction. The normal eye is one in which at a distance of about 20 ft parallel rays of light focus on the retina when the eye is at rest. Such an eye is designated as emmetropic, or normal. Any abnormality in the refractive surfaces or the shape of the eyeball prevents this focusing of parallel rays and makes the eye ametropic, or abnormal.

The most common refractive conditions are myopia, hypermetropia, presbyopia, and astigmatism (Figure 12–21).

MYOPIA. Myopia, or nearsightedness, is a condition in which rays of light converge too soon and are brought to a focus before reaching the retina. This is the opposite of hypermetropia and is caused by a cornea or lens that is too convex or an eyeball of too great depth. This condition is remedied by wearing concave lenses, which cause parallel rays of light to diverge before they converge and focus on the retina.

HYPERMETROPIA. Hypermetropia, or farsightedness, is a condition in which rays of light from near objects do not converge soon enough and are brought to a focus behind the retina.

A hypermetropic eye must accommodate slightly for distant objects and overaccommodate for near objects. Hypermetropia is usually caused by a flattened condition of the lens or cornea, or an eyeball that is too shallow; and convex lenses are used to concentrate and focus the rays in a shorter distance.

PRESBYOPIA. Presbyopia is a defective condition of accommodation in which distant objects are seen distinctly but near objects are indistinct. It occurs as an aging process and is caused by a loss of the elasticity of the lens and lack of tone of the ciliary muscle.

ASTIGMATISM. Astigmatism means that the curvature of the refracting surfaces is unequal; e.g., the cornea is more curved vertically than it is in a horizontal direction or vice versa.

The commonest form is that in which the vertical curvature is greater than the horizontal and is described as regular astigmatism.

Questions for Discussion

1. Define the terms "epicritic" and "protopathic sensations."
2. Explain how receptors function as a protective mechanism.
3. Explain why injury to the right occipital lobe (visual area) causes partial loss of vision in both eyes.
4. When an individual has a severe "cold in the head," sense of taste for most foods is lost. Explain.
5. How does receptor adaptation affect sensation?

6. Explain the phenomenon of referred pain.
7. Trace the pathway for hearing from the external ear to the cerebral cortex.
8. Trace the nerve fibers forming the optic pathway from the retina to the visual area of the occipital lobe.
9. Explain how an individual perceives color.
10. Explain how the eye accommodates to near and to far vision.

Summary

Sensation—Interpreted in brain, may be modified or ignored

Tactile Sense
- Touch receptors—free nerve endings, hair nerve endings, Meissner's, pacinian corpuscles; all rapidly adapting except possibly free nerve endings
- Pressure different from touch; wide range of degree of pressure sensed

Kinesthesia
- Identifies location of parts of body, movement of parts
- Receptors located in joints, ligaments, joint capsules, pacinian, Ruffini end organs, Golgi end organs

Temperature
- Warmth—Ruffini end organs; cold—Krause end organ
- Range of temperature: 12° through 50° C; above this or below, pain results
- Adaptation occurs but incompletely

Pain
- Receptor is free nerve ending
- Stimulus is tissue damage, excessive heat or cold, inadequate blood supply, spasm of muscle, stretch of tubes
- Headache due to external head muscle spasm, stretching of meninges, dilation of brain blood vessels
- Referred pain is visceral pain that is referred to skin area supplied with nerve fibers from same spinal segment
- Function of pain is protective

Hunger
- Normal gastric hunger due to contractions of empty stomach, acting on nerves distributed to mucous membrane, and to hypothalamic centers
- Hunger contractions may be frequent and severe, even when food is taken regularly, as in diabetes

Appetite
- Aroused in part through sensory nerves of taste and smell, associated with previous experiences
- Thought of food associated with appetite induces flow of saliva and gastric fluid

Thirst
- Center in hypothalamus responds to dehydration and concentration of blood
- Aids in maintaining body water balance

Taste
- Sensory apparatus
  1. Taste buds are end organs
  2. Nerve fibers of trigeminal, facial, and glossopharyngeal, vagus nerves
  3. Center in sensory cortex of parietal lobe
- Solution of savory substances must come in contact with taste buds
- Taste buds are distributed over
  - Surface of tongue
  - Soft palate and fauces
  - Tonsils and pharynx

Tongue
- Freely movable muscular organ
- Attached to hyoid bone, epiglottis, and the glossopalatine arches
- Surface covered by papillae containing capillaries and nerves
  - Vallate
  - Fungiform
  - Filiform
  - Simple
- **Nerves**
  - Sensory
    - Lingual branch of trigeminal
    - Chorda tympani, branch of the facial
    - Glossopharyngeal
  - Motor—hypoglossal
- **Sense of**
  1. Taste
  2. Temperature
  3. Pressure
  4. Pain

  are all well developed

Classification of Taste Sensations
- **Four primary sensations**
  - Salty, bitter, acid, sweet
- **All others are**
  - Combinations of primary sensations
  - Combinations of one or more plus odor

Smell
- Sensory apparatus
  - Olfactory nerve endings
  - Olfactory nerve fibers spread out in fine network over surface of superior nasal conchae and upper third of septum
  - Olfactory bulb and center in hippocampus and paraolfactory area
- Odors
  - Minute particles usually in gaseous form
    - Must be capable of solution in mucus
- Olfactory center in the brain is widely connected with other areas of the cerebrum
- Branches of trigeminal nerve found in lining of lower part of nose (pressure)

Hearing
- Auditory apparatus
  - External ear
  - Middle ear, or tympanic cavity
  - Internal ear, or labyrinth
  - Vestibulocochlear nerve
  - Acoustic center in temporal lobe

Ear

External ear
- Pinna, or auricle
  - **Structure**—cartilaginous framework, some fatty and muscular tissue, covered with skin
  - **Function**—collects sound waves
- External acoustic meatus
  - 2.5 cm long, partly cartilage, partly bone
  - Leads from the concha to the tympanic membrane
  - Near orifice skin is furnished with hairs and ceruminous glands
  - Ceruminous glands secrete a yellow, pasty substance (wax)

Middle ear
- An irregular cavity in the temporal bone
- Five or six drops of water will fill it
- Bones
  - Malleus (hammer)
  - Incus (anvil)
  - Stapes (stirrup)
- Five openings
  - Opening between it and external auditory canal, covered by tympanic membrane
  - Fenestra vestibuli at end of scala vestibuli
  - Fenestra cochleae at end of scala tympani

    Connect with internal ear
  - Opening into mastoid antrum and mastoid cells
  - Eustachian (auditory) tube—connects with the nasopharynx; equalizes pressures

Ear (cont.)

Internal ear

Osseous labyrinth
- Vestibule behind the cochlea, in front of the semicircular canals
- Semicircular canals
  - Three in number
  - Open into vestibule
- Vestibular branch of vestibulocochlear nerve distributed to vestibule and semicircular canals
- Cochlea
  - A spiral canal 2¾ turns around modiolus
  - Cochlear branch of the vestibulocochlear nerve

Membranous labyrinth
- Surrounded by perilymph
- Contains endolymph
- In the vestibule forms the
  - Saccule
  - Utricle
- Lines the semicircular canals
- Lines the cochlea
  - *Basilar membrane* extends from free border of lamina to outer wall of cochlea and separates the scala vestibuli and the scala tympani. Supports organ of Corti
  - *Vestibular membrane* extends from free border of lamina to outer wall of cochlea and is attached above basilar membrane, forms *scala media*

Vestibulocochlear nerve
- Cochlear arises from bipolar cells in spiral ganglion
  - Peripheral fibers from cells terminate in and around the cells of the organ of Corti
  - Central fibers pass into the medulla and terminate in two nuclei
- Vestibular arises from bipolar cells in vestibular ganglion
  - Peripheral fibers terminate in hair cells of saccule, utricle, and ampullae of the semicircular canal
  - Central fibers pass into the medulla and terminate in two nuclei

Process of Hearing
- Air waves enter external auditory canal, set tympanic membrane vibrating, vibrations communicated to ossicles, transmitted through fenestra vestibuli to perilymph, stimulate nerve endings in organ of Corti, impulses carried to center of hearing in brain

Unit of measure of sound intensity is the *bel*. A tenth of a bel, or decibel, is used for measurement of hearing. Zero decibels is the threshold of hearing. Sixty-five decibels—normal conversation

Sense of Equilibrium
- Function of the vestibule and semicircular canals
- Lining membrane supplied with sensory hairs and otoliths that connect with vestibular nerve
- Movement of the endolymph stimulates the sensory hairs; this is transmitted to the vestibular branch of the VIIIth nerve, thence to cerebellum

Visual Apparatus
- Bulb of the eye
- Optic nerve
- Center in brain
- Accessory organs
  - Eyebrows
  - Eyelids
  - Conjunctiva
  - Lacrimal apparatus
  - Muscles of the eyeball
  - Fascia bulbi

Accessory Organs

Eyebrows
- Thickened ridges of skin furnished with short, thick hairs
- Protect eyes from vivid light, perspiration, etc.

Eyelids
- Folds of connective tissue covered with skin, lined with mucous membrane, conjunctiva, which is also reflected over the eyeball
- Provided with lashes
- Upper lid raised by levator palpebrae superioris
- Both lids closed by orbicularis oculi muscle
- Slit between lids called palpebral fissure
- **Function**
  1. Cover the eyes
  2. Protect eyes from bright light and foreign objects
  3. Spread lubricating secretions over surface of eyeball

Eyelashes and sebaceous glands
- Margin of each lid, a row of short hairs project
- Sebaceous glands connected with lashes
- Meibomian glands between conjunctiva and tarsal cartilage of each lid
- Secretion lubricates edges, prevents adhesion of lids

Conjunctiva
- Mucous membrane, lines eyelids and is reflected over eyeball. Continuous with mucous membrane of lacrimal ducts and nose

Accessory Organs (*cont.*)

Lacrimal apparatus
- Lacrimal gland—in the upper and outer part of the orbit. Secretes tears
- Lacrimal ducts begin at puncta and open into lacrimal sac
- Lacrimal sac, expansion of upper end of nasolacrimal duct. Between lateral ducts is the lacrimal caruncle
- Nasolacrimal canal—extends from lacrimal sac to nose

Tears
- Secretion constant, carried off by nasal duct
- Dilute solution of various salts in water, also mucin
- Function
  - Keep surface of eyes moist
  - Help to remove microorganisms, dust, etc.

Muscles

Extrinsic
- Superior rectus
- Inferior rectus
- Medial, or internal, rectus
- Lateral, or external, rectus
- Superior oblique
- Inferior oblique

Intrinsic
- Ciliary muscle — Determines position and shape of lens
- Muscles of iris
  - Contractor of pupil
  - Dilator of pupil

Nerves of Eye
1. Optic nerve concerned with vision only
2. Oculomotor controls
   - Medial rectus muscle
   - Superior rectus muscle
   - Inferior rectus muscle
   - Inferior oblique muscle
   - Ciliary muscle
   - Circular muscle of iris
3. Trochlear controls superior oblique muscle
4. Abducens controls lateral rectus muscle
5. Ophthalmic supplies general sensation, such as pressure, muscle sense, and pain and sympathetic fibers to dilator pupillae

Orbit
- A bony cavity formed by seven bones — Frontal, malar, maxilla, palatine, ethmoid, sphenoid, lacrimal
- Contains eyeball, muscles, nerves, vessels, lacrimal glands, fat, fascia bulbi, and fascia holding structures in place
- Lined by fibrous tissue
- Pad of fat—supports eyeball
- Fascia bulbi is a serous sac that envelops eyeball from optic nerve to ciliary region
- Shaped like funnel
  - Large end directed outward and forward
  - Small end directed backward and inward
- **Optic foramen**—opening for passage of optic nerve and ophthalmic artery
- **Superior orbital fissure**—opening for passage of orbital branches of middle meningeal artery and oculomotor, trochlear, abducens, and ophthalmic nerves

Bulb of the Eye
- Spherical, but it projects anteriorly
- **Coats**
  1. Fibrous—sclera and cornea
  2. Vascular—choroid, ciliary body, and iris
  3. Nervous—retina
- **Refracting media**
  1. Cornea and aqueous humor
  2. Crystalline lens and capsule
  3. Vitreous body

Fibrous Tunics
- **Sclera**
  - Tough, fibrous, sclera
  - Covers posterior five sixths of eyeball
  - Opaque, white and smooth externally, brown internally
- **Cornea**
  - Fibrous, transparent—covers one sixth of eyeball
  - Well supplied with nerve fibers

Vascular Tunic
- **Choroid**
  - Composed of dense capillary network and stroma of cells, some of which are pigmented, lines the sclera
- **Ciliary body**
  - Includes the orbicularis ciliaris, the ciliary processes, and the ciliaris muscle. The orbicularis ciliaris is a zone about 4 mm in width that is continuous with anterior part of choroid
  - Ciliary processes 60–80 radiating folds, arranged like a plaited ruffle around margin of lens
  - Support ciliaris muscle—action of this muscle determines shape or refractiveness of lens

Vascular Tunic (*cont.*)	**Iris**	A circular colored disk suspended in front of lens and behind cornea. Hangs free except for attachment at circumference to the ciliary processes and choroid. Central perforation—pupil
		Pupil contracted by circular, or sphincter, muscle
		Pupil dilated by radial, or dilator, muscle
		Composed of connective tissue, containing numerous blood vessels and nerves. Contains pigment cells
		Function—regulates size of pupil and thereby amount of light entering eye

Nervous Tunic, or Retina

Nervous layer—contains elements essential for reception of rays of light. Situated between the choroid coat and hyaloid membrane of the vitreous humor, extends forward and terminates in the *ora serrata*

The retina is composed of ten layers. The layer containing the rods and cones is the light-sensitive layer

1. Pigment epithelium
2. Layer of rods and cones
3. External limiting membrane
4. Outer nuclear layer
5. Outer plexiform layer
6. Inner nuclear layer
7. Inner plexiform layer
8. Ganglion cell layer
9. Optic nerve fiber layer
10. Internal limiting membrane

Blind spot	Entrance of optic nerve and central artery and vein of the retina
	There are no rods and cones
	Totally insensitive to light
Macula lutea	2 mm lateral side of blind spot
	Central pit—fovea centralis—is the center of direct vision

Refracting Media

Cornea	Transparent, refractive structure covering the anterior one sixth of the eye bulb
Aqueous humor	Aqueous chamber is between cornea in front and lens, suspensory ligament, and ciliary body behind. Aqueous humor is a watery solution containing minute amounts of salts
	Dialyzed from capillaries, drains away through canal of Schlemm
Crystalline lens	Transparent, refractive body enclosed in an elastic capsule
	Double convex in shape. Situated behind the pupil
	Held in position by counterbalancing of aqueous humor, vitreous body, and suspensory ligament
Vitreous body	Semifluid, albuminous tissue enclosed in hyaloid membrane
	Fills posterior four fifths of bulb of the eye, distends sclera, and supports retina

Perception of Light and Color	Waves vary in length	Waves between 400 and 800 mμ in length are called light and color waves—the visual spectrum

Refraction—bending or deviation in the course of rays of light, in passing obliquely from one transparent medium into another of different density

Vision

Visible objects reflect light waves that fall upon them

These reflected rays are brought to focus on receptors (rods and cones) of retina, where a chemical change in rhodopsin initiates nerve impulses, which are transmitted to optic nerve, and thence to centers of vision in occipital lobe of cerebrum, from here to association areas

Color blindness due to abnormal cones

Processes Necessary for Binocular Vision

1. Convergence, or turning the eyes inward, in order to place the image on corresponding points of the two retinae
2. Change in size of pupil—contracts in a bright light—dilates in a dim light
3. Accommodation—capacity of the eyes and lenses to adjust so that objects at varying distances can be seen clearly
4. Refraction—bending of light rays entering the pupil so they come to a focus on the retina

Abnormal Conditions	Myopia	{ Nearsightedness **Cause**—rays of light converge too soon
	Hypermetropia	{ Farsightedness **Cause**—rays of light do not converge soon enough
	Presbyopia	{ Defective condition of accommodation in which distant objects are seen distinctly but near objects are indistinct
	Astigmatism	{ Condition in which the curvature of the refracting surfaces is defective

Additional Readings

FAVREAU, O. E., and CORBALLIS, M. C.: Negative aftereffects in visual perception. *Sci. Amer.*, **235**:42–48, December, 1976.

GUYTON, A. C.: *Basic Human Physiology: Normal Function and Mechanisms of Disease.* W. B. Saunders Co., Philadelphia, 1971, Chapter 32, 33, 34, and 35.

HABER, R. N.: How we remember what we see. *Sci. Amer.*, **222**:104–12, May, 1970.

HARPEN, R.: *Human Senses in Action.* Williams & Wilkins Co., Baltimore, 1972.

JOHANSSON, J.: Visual motion perception. *Sci. Amer.*, **232**:76–87, June, 1975.

JULESZ, B.: Experiments in visual perception of texture. *Sci. Amer.*, **232**:34–43, April, 1975.

OSTER, G.: Auditory beats in the brain. *Sci. Amer.*, **229**:94–102, April, 1973.

PETTIGREW, J. D.: The neurophysiology of binocular vision. *Sci. Amer.*, **227**:84–95, August, 1972.

ROSS, J.: The resources of binocular perception. *Sci. Amer.*, **234**:80–86, March, 1976.

SNYDER, S. H.: Opiate receptors and internal opiates. *Sci. Amer.*, **236**:44–56, March, 1976.

VANDER, A. J.; SHERMAN, J. H.; and LUCIANO, D. S.: *Human Physiology: The Mechanisms of Body Function,* 2nd ed. McGraw-Hill Book Co., New York, 1975, Chapter 16.

The Autonomic Nervous System

Chapter Outline

The autonomic nervous system controls the internal environment—e.g., heart action, adjustment of circulation to meet body needs, secretion of digestive juices, and peristaltic activity. There is also a personal, nonintellectual response to one's environment, such as an emotional reaction to a given situation, which also involves the autonomic nervous system.

General Organization

The autonomic nervous system is divided into the sympathetic and parasympathetic divisions. It is concerned with visceral or involuntary functions. This visceral system is both afferent and efferent in function.

Most of the centers controlling these processes are located within the central nervous system; but there is coordination of cellular activities even within the walls of the viscera, and this intraorgan integration helps to control such activities as gastrointestinal motility and secretion, cardiac output, sweating, urinary output, arterial blood pressure, and many other

physiologic processes. Some autonomic functions are almost completely controlled by the central nervous system; others are only partly controlled by the central nervous system. Some visceral functions can be performed quite independently of any nervous control. The heart muscle contracts automatically; some of the glands are excited to secrete by chemical substances in the blood such as the secretion of pancreatic fluid in response to the hormone *secretin*. Even though such activities are not directly dependent on the nervous system, they may be regulated or modulated by the nervous

297

system, because in all visceral functions the nonnervous and the nervous cooperate in a most intimate way.

Visceral Afferent Fibers. The afferent fibers of many receptors in the viscera carry impulses from receptors in the organs to the spinal cord and brain. The afferent fibers of the vagus, with cell bodies located in the nodosal ganglion, innervate sensory endings in the heart, lungs, and other viscera of the thoracic and abdominal cavity. The pelvic nerve carries afferent fibers from receptors in the mesentery and in the colon wall. Visceral reflexes are mediated through the spinal cord and brain stem. Visceral pain is carried by fibers in the sympathetic nerves.

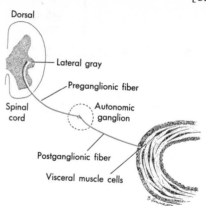

Figure 13–1. Diagram showing the relationship of the preganglionic and postganglionic neurons of the autonomic nervous system.

Motor, vasomotor, and secretory reflexes are mediated over autonomic neural arcs

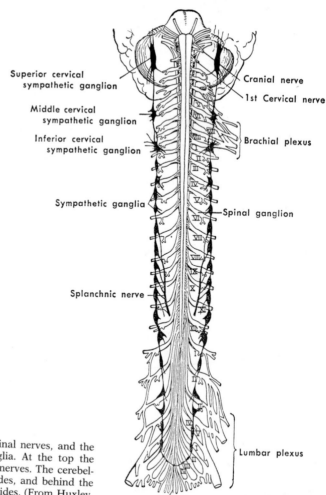

Figure 13–2. Diagram of spinal cord, spinal nerves, and the right and left chains of autonomic ganglia. At the top the medulla is seen with some of the cranial nerves. The cerebellum is seen behind the medulla at the sides, and behind the cerebellum the cerebrum is shown at the sides. (From Huxley, after Allen Thomson.)

and do not reach consciousness. Sensations of bladder and colon distention are mediated over the afferent fibers of the pelvic nerve.

Visceral Efferent Fibers. In the autonomic system two neurons connect the central nervous system and the end organ. The fiber of a neuron lying in the *central nervous system* extends to an *autonomic ganglion* and synapses on the dendrites or cell body of an autonomic neuron. The fiber of the second neuron passes from the ganglion to the effector to be innervated. The fiber of the first neuron is called the *preganglionic fiber;* the fiber of the second neuron is called the *postganglionic fiber.*

In the sympathetic division both preganglionic and postganglionic fibers are of approximately the same length. In the parasympathetic division the preganglionic fibers are much longer and the postganglionic fibers are short, as the cell bodies are usually in the organ being inrٍervated or in a ganglion quite close to the organ.

Craniosacral, or Parasympathetic, System

This system includes all the fibers that arise from the midbrain (tectal autonomics), from the medulla and pons (bulbar autonomics), and from the sacral region of the cord (sacral autonomics) (Figure 13–3).

The *tectal autonomics* arise from nuclei in the midbrain, send preganglionic fibers with the oculomotor nerve into the orbit, and pass to the ciliary ganglion, where they terminate by forming synapses with motor neurons whose axons (postganglionic fibers) proceed as the short ciliary nerves to the ciliary muscle of the eye and to the pupillary sphincters.

The *bulbar autonomics* arise from nuclei in the medulla and pons and emerge in the seventh, ninth, and tenth cranial nerves. Fibers from the seventh nerve are distributed to lacrimal, nasal, submaxillary, and sublingual glands. Fibers from the ninth nerve are distributed to the parotid gland. Fibers from the tenth nerve are distributed to the heart, lungs, esophagus, stomach, the small intestine, proximal half of the colon, gallbladder, liver, and pancreas.

Some of the fibers of the vagus nerve are distributed to the skeletal muscles of the larynx and pharynx from the nucleus ambiguus. The vagus also carries important afferent nerve fibers from pressor receptors in arteries and stretch receptors of the lungs to the medulla.

The *sacral autonomics* include autonomic fibers that emerge from the spinal cord. Neurons of the second, third, and fourth, and sometimes the first sacral spinal nerves send fibers to the pelvis, where they form the pelvic nerve, which sends fibers to the pelvic plexus, from which postganglionic fibers are distributed to the pelvic viscera. Motor fibers pass to the smooth muscle of the descending colon, rectum, anus, bladder, and reproductive organs.

Thoracolumbar, or Sympathetic, System

This includes (1) small neurons in the gray lateral columns of the thoracic and lumbar regions of the cord giving rise to preganglionic fibers; (2) the sympathetic ganglia and their postganglionic fibers—the lateral chain of the sympathetic trunk; and (3) the great prevertebral plexuses. Postganglionic fibers may arise either from a ganglion in the lateral chain or from a ganglion in one of the great plexuses (Figure 13–4).

The *sympathetic centers of the spinal cord* are composed of groups of cells lying in the lateral columns of the gray matter of

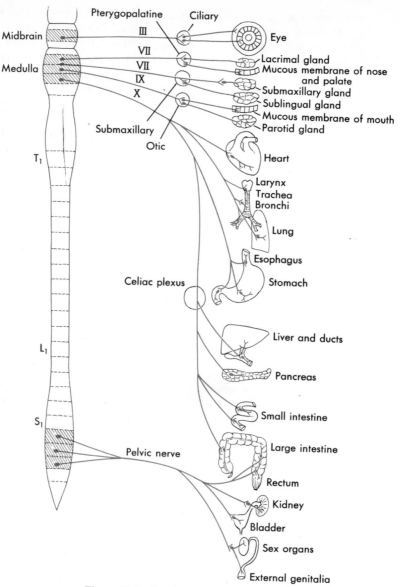

Figure 13–3. Craniosacral autonomic system.

the cord from the first thoracic to lumbar two or three. They give rise to preganglionic fibers that terminate in one of the sympathetic ganglia.

The sympathetic ganglia (Figure 13–2) consist of paired chains of ganglia that lie along the ventrolateral aspects of the vertebral column, extending from the base of the skull to the coccyx. They are grouped as cervical, thoracic, lumbar, and sacral, and, except in the neck, they correspond in number to the vertebrae against which they lie:

Cervical	Thoracic	Lumbar	Sacral
3 pairs	10–12 pairs	4 pairs	4–5 pairs

The sympathetic ganglia are connected with each other by nerve fibers called gangliated cords, and with the spinal nerves by branches called *rami communicantes.* In the thoracic and lumbar regions these communications consist of two rami, a white ramus and a gray ramus (Figure 13–5).

The white rami fibers are *myelinated fibers* passing from the central nervous

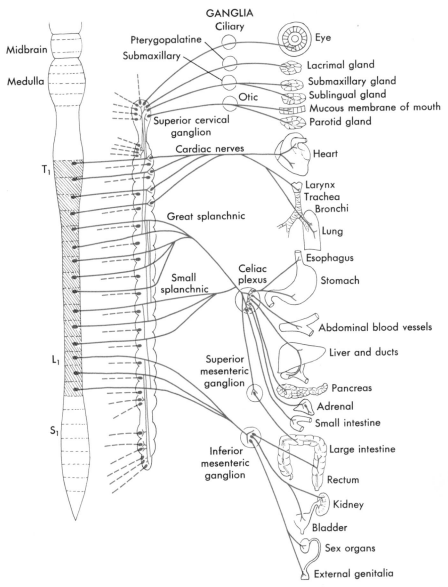

GANGLIA

Midbrain

Medulla

Ciliary
Pterygopalatine
Submaxillary
Otic
Superior cervical ganglion
Cardiac nerves

T_1

Great splanchnic

Small splanchnic

Celiac plexus

Superior mesenteric ganglion

L_1

S_1

Inferior mesenteric ganglion

Eye
Lacrimal gland
Submaxillary gland
Sublingual gland
Mucous membrane of mouth
Parotid gland
Heart
Larynx
Trachea
Bronchi
Lung
Esophagus
Stomach
Abdominal blood vessels
Liver and ducts
Pancreas
Adrenal
Small intestine
Large intestine
Rectum
Kidney
Bladder
Sex organs
External genitalia

Figure 13–4. Diagram of the thoracolumbar nervous system. *Dashed red lines* represent fibers to blood vessels (vasomotor), arrector pili muscles (pilomotor), and sweat glands (secretory).

system to the sympathetic ganglia. The gray rami fibers are *nonmyelinated postganglionic* fibers that are the axons of cells in the sympathetic ganglia and are distributed chiefly with the peripheral branches of all the spinal nerves to the periphery.

White Rami Communicantes. The cells of origin of these fibers lie in the lateral gray of the spinal cord from the first thoracic (T_1) to the second or third lumbar segments. The fibers leave by way of the an-

terior root with the somatic efferent axons. They leave the spinal nerve by way of the white rami and enter the sympathetic ganglia, where they may terminate on a sympathetic postganglionic neuron or may pass up or down in the sympathetic chain for some distance before ending on a sympathetic neuron. These are the fibers that form the gangliated cord of the sympathetic chain.

The fibers from T_1 to T_5 (preganglionic) emerge from the cord and synapse in the

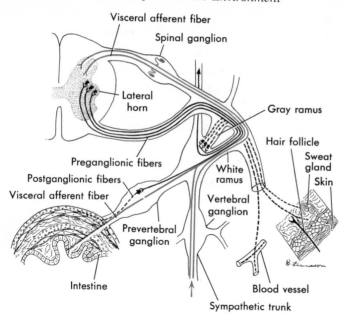

Visceral afferent fiber

Spinal ganglion

Lateral horn

Gray ramus

Hair follicle

Sweat gland

Skin

Preganglionic fibers

White ramus

Postganglionic fibers

Visceral afferent fiber

Vertebral ganglion

Prevertebral ganglion

Intestine

Blood vessel

Sympathetic trunk

Figure 13-5. Diagram showing origin of sympathetic preganglionic and postganglionic fibers and their distribution.

sympathetic ganglia. The fibers of the sympathetic cells in the ganglia (postganglionic fibers) are distributed to the heart, lungs, and blood vessels.

The fibers from T_6 to T_{12} form the splanchnic nerves. These preganglionic fibers pass through the sympathetic ganglia and terminate in the celiac ganglia (solar plexus). The postganglionic fibers are distributed to the esophagus, stomach, and intestine as far as the proximal colon, liver, and gallbladder.

The fibers from L_1 to L_3 form the *preganglionic* fibers that terminate in the inferior mesenteric ganglia. The *postganglionic* fibers are distributed to the distal part of the colon, rectum, and genitourinary organs.

Gray Rami Communicantes. These have their cells of origin in the sympathetic ganglia and are distributed via the spinal nerves to arteries, arterioles (vasoconstrictors), veins, venules, sweat glands, and pilomotor muscles (Figure 13-5).

The fibers to the blood vessels, glands, and walls of the viscera are distributed by various sympathetic ganglia. For the head region the fibers, after entering the sympathetic chain, pass upward and end in the *superior cervical ganglion;* from this

ganglion postganglionic fibers emerge from the various plexuses that arise from this ganglion. (See Figure 13-4.)

The Great Plexuses of the Thoracolumbar System. These consist of ganglia and fibers derived from the lateral chain ganglia and the spinal cord. They are situated in the thoracic, abdominal, and pelvic cavities and are named the cardiac, celiac, mesenteric, lumbar, and sacral plexuses.

Cardiac Plexus. This is situated at the base of the heart, lying on the arch and the ascending portion of the aorta.

Celiac Plexus (Solar Plexus). This is situated behind the stomach, between the suprarenal glands. It surrounds the celiac artery and the root of the superior mesenteric artery. It consists of two large ganglia and a dense network of nerve fibers uniting them. It receives the greater and lesser splanchnic nerves of both sides and some fibers from the vagi and gives off numerous secondary plexuses along the neighboring arteries. The names of the secondary plexuses indicate the arteries they accompany and the organs to which they distribute branches:

Aortic

Phrenic

Suprarenal

Renal

Hepatic   Superior mesenteric
Splenic   Inferior mesenteric
Superior gastric   Spermatic

These nerves form intricate networks, and any one organ may receive branches from several nerves. This increases the number of pathways and connections between the organs.

Mesenteric Plexus. This is situated in front of the last lumbar vertebra and the promontory of the sacrum. It is formed by the union of numerous filaments that descend on either side from the aortic plexus and from the lumbar ganglia; below, it divides into the lumbar and sacral plexuses.

Enteric System. This system includes the myenteric (Auerbach's[1]) plexus and the submucous (Meissner's[2]) plexus in the wall of the digestive canal. They extend from the upper level of the esophagus to the anal canal. The myenteric plexus is situated between the longitudinal and circular muscular coats in the wall of the gastrointestinal tract. This plexus is responsible for coordinating the timing and strength of contractions and secretions of the intestinal wall. The submucous plexus

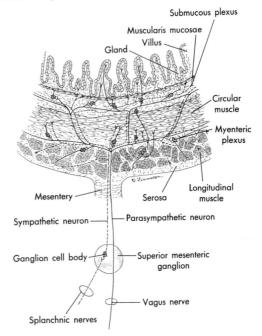

Figure 13–6. Distribution of autonomic neurons to the enteric system.

lies in the submucosa. These plexuses are intimately connected with each other (Figure 13–6).

Functions of the Autonomic Nervous System

The autonomic system innervates (1) all smooth muscular tissue in the body, (2) the heart, and (3) the glands. The ganglia serve as relay stations for many of the impulses passing from the midbrain, pons, and medulla, or spinal cord, or they may act independently of these influences.

In general, most organs have a double autonomic innervation, one from the thoracolumbar system and one from either the

[1] Leopold Auerbach, German anatomist (1828–1897).
[2] Georg Meissner, German anatomist and physiologist (1829–1905).

cranial or the sacral autonomic system. The functions of these two systems are usually antagonistic. With the exception of nicotine, which paralyzes all autonomic ganglia, most drugs that act on the autonomic system affect principally either the craniosacral system, as do atropine, pilocarpine, and physostigmine, or the thoracolumbar system, as do epinephrine and ergotoxine. There are new drugs that block the transmission of impulses from the preganglionic neurons to the postganglionic neurons. All of these interfere with the transmission of impulses in both the parasympathetic and sympathetic systems to varying degrees. These drugs are useful in the treatment of high blood pressure and intestinal hyperactivity.

Neural Transmission

As in the somatic nervous system, the action of the autonomic impulses causes release of a chemical transmitter substance. Autonomic fibers are classed as cholinergic and adrenergic fibers on the basis of the transmitter substance. All preganglionic fibers release from the vesicles in their nerve endings acetylcholine, which diffuses across the synaptic cleft and produces permeability changes in the membrane of the postganglionic neuron. The transmitter is rapidly hydrolyzed by the enzyme acetylcholinesterase. Since all preganglionic fibers of both parasympathetic and sympathetic systems release

acetylcholine they are called "cholinergic." All parasympathetic postganglionic fibers are also cholinergic. Most sympathetic postganglionic fibers release norepinephrine and are called "adrenergic." Monoamine oxidase is the enzyme that rapidly destroys norepinephrine (see Figure 13–8).

Cells stimulated by adrenergic fibers have been subdivided into two groups (α and β), based on their response to certain drugs. Stimulation of β fibers results in increased rate and strength of cardiac contraction, vasodilation, and bronchial relaxation.

Interdependence of the Craniosacral and Thoracolumbar Systems

Marked stimulation of one system, or even part of one system, is likely to stimulate some part of the other system, thus checking excessive stimulation with the untoward results that might follow. Stimulation of the part of the vagus that supplies the bronchial tubes may cause such marked constriction of the tubes that interference with breathing, pain, and distress may result; this in turn stimulates the thoracolumbar system to lessen the contraction of the tubes. For another example, the afferent branch of the vagus connected with the aorta is stimulated when the blood pressure within the vessel rises. These afferent impulses initiate efferent impulses that are (1) inhibitory to the heart, thus slowing its action, and (2) inhibitory to the vasoconstrictor center, thus lessening vasoconstriction.

Antagonistic actions of the craniosacral and thoracolumbar systems are listed in Table 13–1 (page 309) so that the results following stimulation of these two systems may be compared.

The chief subcortical center for regulation of both parasympathetic and sympathetic activities lies in the hypothalamus. The anterior and medial areas of the hypothalamus control parasympathetic activities. When this region is stimulated, there are slowing of the heart rate, increased motility and tone of the alimentary musculature, and vasodilation of peripheral blood vessels. This area is also concerned with maintaining water balance. Diabetes insipidus results if this area is destroyed.

The posterior and lateral hypothalamic regions are concerned with control of sympathetic activities. When these areas are stimulated, the prompt sympathetic responses are dilation of the pupil, increased heart rate, vasoconstriction causing an elevation of blood pressure, and inhibition of the digestive organs and bladder. These centers complement each other in regulation of body processes. For instance, if the body temperature falls, the "heat conservation" center in the caudal hypothalamus initiates (1) shivering, which promptly increases heat production, and (2) marked vasoconstriction of cutaneous blood vessels, which reduces heat loss through the skin. If body temperature increases, a "heat loss" center in the anterior hypothalamus responds by initiating (1) sweating, (2) dilatation of cutaneous blood vessels, and (3) constriction of splanchnic blood vessels, which shunts blood to the skin's surface where heat is removed from the body by radiation and conduction. Sweating cools the skin if the external environmental conditions are conducive to rapid evaporation. If the posterior hypothalamus is destroyed, a state of almost total lethargy results. The hypothalamus sends fibers down to the preganglionic autonomic centers of the brain stem and to the lateral gray of the spinal cord. By these connections there are pathways through which impulses from receptors, responding to changes in the environment, are transmitted to the thalamus, to the cortex, to the hypothalamus, and finally to the viscera. It is through these and other pathways that "fleeting thoughts" or other emotional crises affect the heart rate, vascular beds, and other autonomic physiologic processes.

Cortical centers that regulate autonomic activity are located in prefrontal lobes and temporal regions. Stimulation of these areas during emotional states arouses autonomic areas of the hypothalamic centers. There are also regulating centers in the thalamus. Both conscious and unconscious areas of the cortex can cause autonomic response. The action of the sympathetic nervous system is augmented by the hormones of the adrenal medulla. (See Chapter 14 for discussion.)

The thoracolumbar system is strongly stimulated by pain and unpleasant excitement such as anger, fear, or insecurity. The animal responses to anger and fear are fight and flight, and the conditions brought about by stimulating the thoracolumbar system are such as to favor these responses; i.e., the bronchial tubes are relaxed and rapid breathing is rendered easier; the constriction of the blood vessels in the stomach and intestines and the increased heart action deliver more blood to the skeletal muscles and thus provide them with the extra oxygen and nutrients needed for increased muscular activity; the supply of glucose from the liver is also increased, thus providing for greater production of energy. Increased activity of sweat glands produces perspiration. If the environmental conditions permit the evaporation of this excess perspiration, body temperature will be maintained more nearly constant. All these responses are closely connected with the adrenal glands, which secrete epinephrine and norepinephrine. The amount of secretion is increased when the thoracolumbar system is stimulated.

In acute stress situations the physiologic response is to prepare the body for fight and flight. However, it must be remembered that in certain stressful situations physiologic response may be mainly either sympathetic or parasympathetic or a combination of both. Excessive response to environmental conditions of portions of either of these divisions of the autonomic nervous system may predispose the individual to physiologic disorders classified as psychosomatic or psychovisceral diseases. Such changes may include a great variety of physiologic disorders, a few of which are hypertension, peptic ulcers, colitis, and headache.

Fibers from the various ganglia are distributed to organ listed in center column

Parasympathetic—Blue Fibers			Thoracolumbar—Red Fibers	
Nucleus of Origin of Preganglionic Cell	Postganglionic Cell Bodies, Ganglia	Name of Part	Postganglionic Cell Bodies, Peripheral Ganglia	Nucleus of Origin of Preganglionic Cell Body
Edinger-Westphal nucleus, midbrain	Ciliary ganglia	Eye, iris, ciliary muscle	Superior cervical sympathetic, no fibers to ciliary muscle	Lateral gray of cord, T_1–T_2 or T_3
Superior salivatory nucleus in pons	Pterygopalatine ganglia	Lacrimal glands	Superior and middle cervical sympathetic ganglia	Lateral gray of cord, T_1–T_2
Superior salivatory nucleus in pons	Submandibular ganglia	Submaxillary, submandibulary glands	Superior and middle cervical sympathetic ganglia	Lateral gray of cord T_1–T_3 or T_4
Inferior salivatory nucleus in medulla	Otic ganglia	Parotid glands	Superior and middle cervical sympathetic ganglia	Lateral gray of cord, T_1–T_3 or T_4
Dorsal motor nucleus of vagus	Ganglia of pulmonary plexus	Lungs and bronchi	Inferior cervical and T_1–T_5 sympathetic ganglia	Lateral gray of cord, T_1–T_5
Dorsal motor nucleus of vagus	Intracardiac ganglia of the atria	Heart	Superior, middle, and inferior cervical sympathetic ganglia and T_1–T_6 sympathetic ganglia	Lateral gray of cord, T_1–T_6
Dorsal motor nucleus of vagus	Myenteric and submucous plexuses	Esophagus	Sympathetic ganglia T_1–T_3	Lateral gray of cord, T_1–T_6
Dorsal motor nucleus of vagus	Myenteric and submucous plexuses	Stomach, small intestine, and transverse colon	Celiac and superior mesenteric ganglia	Lateral gray of cord, T_5–L_{11}
Autonomic nucleus of the lateral gray of cord, S_2–S_4	Ganglia of myenteric, submucous, and hemorrhoidal plexuses	Descending colon, rectum, and internal sphincter	Lumbar and inferior mesenteric sympathetic ganglia	Lateral gray of cord, T_{12}–L_3
Autonomic nucleus of the lateral gray in cord, S_2–S_4	Ganglia of vesical branches of internal iliac artery	Urinary bladder and internal urethral sphincter	Lumbar and inferior mesenteric sympathetic ganglia	Intermediolateral gray of cord, T_{12}–L_2
Autonomic nucleus of the lateral gray in cord, S_2–S_4	Ganglia along branches of aorta and internal iliac arteries	Reproductive organs	Lumbar, sacral, and inferior sympathetic ganglia	Intermediolateral gray of cord, T_{10}–L_2

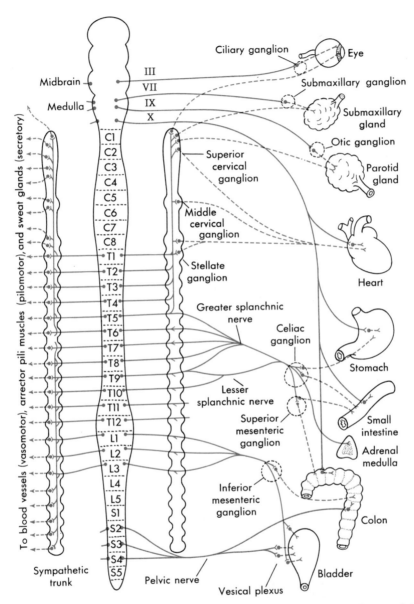

Figure 13–7. Diagrammatic representation of some of the chief conduction pathways of the autonomic nervous system. For clarity, the nerves to blood vessels, arrector pili muscles, and sweat glands are shown on the left side of the figure and the pathways to other visceral structures only on the right side. The sympathetic division is shown in red, the parasympathetic in blue. *Solid lines* represent preganglionic fibers; *broken lines* represent postganglionic fibers. (Modified from Bailey's *Textbook of Histology*, 13th ed, revised by P. E. Smith and W. M. Copenhaver. Courtesy of Williams & Wilkins Co.)

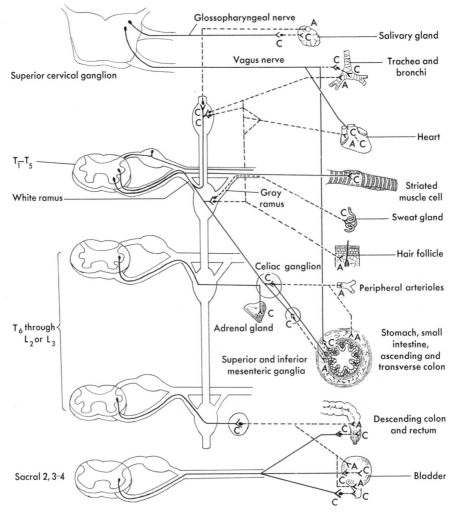

Figure 13–8. Diagram of autonomic nervous system showing cholinergic (*C*) and adrenergic (*A*) nerve endings.

Questions for Discussion

1. Discuss the physiologic means by which stress raises blood pressure.
2. An individual has been under stress and complains of increased peristalsis with severe cramps. The doctor orders an anticholinergic drug. Explain the reasons for the symptoms and why the drug was ordered.
3. Analyze and explain the physiologic response to stress.
4. Compare the physiologic responses of the sympathetic and parasympathetic systems on the heart and on the gastrointestinal wall.
5. What is the connection between the spinal cord and the sympathetic ganglia?
6. What is the enteric system?
7. What part of the brain is concerned with sympathetic activities?
8. Discuss neural transmission at the preganglionic and the postganglionic synapses.

TABLE 13–1
RESPONSES OF AUTONOMIC INNERVATION

Name of Part	Effect of Craniosacral (Parasympathetic) Stimulation— Cholinergic	Effect of Thoracolumbar (Sympathetic) Stimulation— Adrenergic
Eye—Iris	Constricts the pupil (miosis)	Dilates the pupil (mydriasis)
Ciliary muscle	Contracts ciliary muscle; accommodation of the lens for near vision	No effect
Lacrimal glands	Stimulates secretion	Little or no effect
Lungs—Bronchi	Constricts bronchial tubes	Dilates bronchial tubes
Heart—Muscle	Slows heart rate	Accelerates heart rate and strengthens ventricular contraction
Arterioles in viscera		
Lungs	No innervation	Very mildly constricts vessels
Coronary arteries	Constricts arteries	Vasodilation
Abdominal	No innervation	Vasoconstriction
Arterioles in somatic tissue		
Muscle	Vasodilation	Vasoconstriction; vasodilation
Skin	Vasodilation	Vasoconstriction
Glands—Sweat	No innervation	Marked sweating
Salivary	Increased secretion; thin, watery, containing many enzymes	Vasoconstriction; decrease in amount of saliva; becomes viscid in character
Gastric	⎰ Increased secretion	Secretion inhibited
Intestinal	⎱ Increased tension in walls	Walls of gut relaxed
Liver	No effect	Glucose released
Gallbladder and ducts	Stimulates bile flow	Inhibits bile flow
Kidney	No effect	Vasoconstriction, which leads to decreased urine flow
Bladder	Muscle wall contracted; internal sphincter relaxed	Muscle wall relaxed; internal sphincter constricted
Intestinal organs—		
Motility	Increased peristalsis and tone of wall increased	Decreased peristalsis and muscle tone; wall relaxed
Sphincters	Internal sphincter relaxed	Sphincter constricted
Adrenal gland		
Cortex	No effect	Increased secretion
Medulla	Little or no effect	Increased secretion
Basal metabolism	No effect	Metabolism markedly increased
Blood sugar	No effect	Increased; liver releases glycogen
Blood coagulation	No effect	Increased coagulation
Mental activity	No effect	Increased activity
Sex organs	Vasodilation and erection	Contraction of uterine musculature, ductus deferens, seminal vesicle, vasoconstriction
Piloerector muscles	No effect	Excited; hair stands on end

Summary

Autonomic or Efferent Visceral System

Craniosacral, or parasympathetic

- **Tectal autonomics** — Tectal autonomics—neurons that arise from roots in midbrain, pass to ciliary ganglia—terminate by forming synapses with motor neurons, whose axons proceed as ciliary nerves to the intrinsic muscles of the eye
- **Bulbar autonomics** — Bulbar autonomics—neurons arising from roots in pons and medulla, which emerge in seventh, ninth, and tenth cranial nerves
- **Sacral autonomics** — Sacral autonomics—neurons of the second, third, and fourth sacral spinal nerves send preganglionic fibers to pelvis—form pelvic nerve, which proceeds to pelvic plexus

Thoracolumbar, or sympathetic

Chain of ganglia that lie along ventrolateral aspects of vertebral column

- **Grouped as**
  - Cervical   3 pairs
  - Thoracic   10–12 pairs
  - Lumbar   4 pairs
  - Sacral   4–5 pairs
- **Connected**
  1. With each other by gangliated cords
  2. With spinal nerves by rami communicantes } White and gray
- **Distribution**
  1. In spinal nerves to blood vessels, glands, walls of viscera, and skin areas of body
  2. Fibers from fifth to tenth or eleventh thoracic ganglia converge to form { Great splanchnic / Small splanchnic
- **Three great plexuses** — Consist of masses of gray matter in thoracic and abdominal cavities. Form { Cardiac plexus / Celiac plexus / Hypogastric plexus } Connect with many others embedded in thoracic and abdominal viscera
- **Enteric** — Myenteric plexus—situated between the circular and longitudinal coats of digestive tube. Submucous plexus—lies in the submucosa

Function—neural homeostatic regulation. Innervates all smooth muscular tissue, the heart, and the glands. Most important factor is reflex stimulation. Ganglia serve as relay stations

Craniosacral and Thoracolumbar Systems

Many of the viscera are supplied with nerves from both craniosacral and thoracolumbar systems—functions of these two sets are often antagonistic

These two systems are interdependent, stimulation of one system or part of one system likely to stimulate some part of the other system

Thoracolumbar system is stimulated by intense excitement; craniosacral system is not

Nicotine paralyzes all autonomic ganglia. Most drugs affect either the craniosacral or the thoracolumbar, not both

Norepinephrine is the chemical mediator for postganglionic sympathetic nerve endings, except fibers to sweat glands,

Acetylcholine is the chemical mediator for all parasympathetic nerve endings and preganglionic sympathetic fibers

Subcortical Areas for Regulation of Parasympathetic Activities

Anterior and medial areas of hypothalamus—control parasympathetic activities

This area also concerned with water balance and body temperature regulation

Regulation of Sympathetic Activities

Posterior and lateral hypothalamic areas—control sympathetic activities

The parasympathetic and sympathetic nervous systems complement each other in regulation of body processes

Augmented by hormones of adrenal medulla

Cortical Control—Located in prefrontal lobes and temporal regions

Additional Readings

BURN, J. H.: *The Autonomic Nervous System for Students of Physiology and Pharmacology,* 5th ed. Blackwell Scientific Publications, Oxford, 1975.

CARRIER, O., Jr.: *Pharmacology of the Peripheral Autonomic Nervous System.* Year Book Medical Publishers, Inc., Chicago, 1972.

DiCARA, L. V.: Learning in the autonomic nervous system. *Sci. Amer.,* **222**:30–39, January, 1970.

GUYTON, A. C.: *Basic Human Physiology: Normal Function and Mechanisms of Disease.* W. B. Saunders Co., Philadelphia, 1971, Chapter 40.

NOBACK, C. R., and DEMAREST, R. J.: *The Human Nervous System: Basic Principles of Neurobiology,* 2nd ed. McGraw-Hill Book Co., New York, 1975, Chapter 6.

WILLIAMS, P. L., and WARWICK, R.: *Functional Neuroanatomy of Man.* W. B. Saunders Co., Philadelphia, 1975, pp. 1065–83.

CHAPTER **14**

The Endocrine System

Chapter Outline

Physiologic organization of body functions is brought about by the nervous system and the endocrine system. In general, the nervous system regulates the rapidly changing activities, such as skeletal movements, smooth muscle contraction, and many glandular secretions. The endocrine system regulates the metabolic functions of the body and the varying rates of chemical reactions within the cell.

Definitions. As explained in Chapter 3 (page 48), glands are of two types: exocrine and endocrine. Endocrine glands secrete their hormones directly into the bloodstream, and the hormone enters the general circulation, to reach all cells of the body. Only certain cells respond to the hormone, however; these responsive cells are termed *target cells.* A hormone thus may be defined as a chemical substance that is produced by a specific organ or tissue, is carried to other cells by the bloodstream, and exerts its action on *other* cells. Hormones have been grouped in two classes: local hormones and general hormones. *Local* hormones include the *pros-*

312

taglandins (p. 329) and *secretin, gastrin,* and *cholecystokinin,* secreted by cells of the digestive tract and discussed in Chapter 21. Local hormones have their action locally, hence their name. *General* hormones are discussed in this chapter and include the hormones of the following glands: pituitary, thyroid, parathyroid, adrenal, ovary, testis, and pancreas. (See Figure 14–1; see also Plate III, inserted between pages 536–37.)

Mechanism of Hormonal Action. The function of a hormone, in general, is to control the activity level of the target cells. How this control is exerted depends on the molecular structure of the hormone. Structurally hormones are of two types: *peptides*

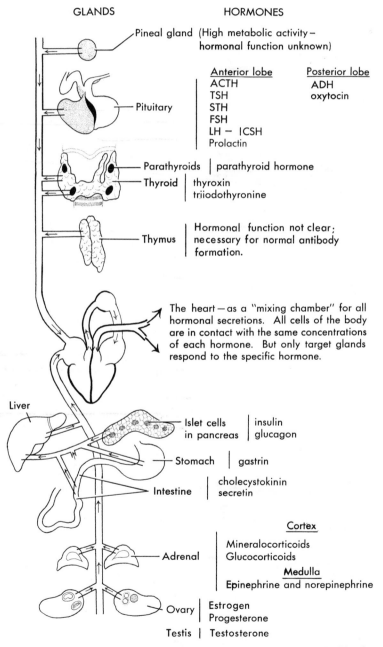

GLANDS   HORMONES

Pineal gland (High metabolic activity – hormonal function unknown)

Pituitary

Anterior lobe	Posterior lobe
ACTH	ADH
TSH	oxytocin
STH	
FSH	
LH — ICSH	
Prolactin	

Parathyroids | parathyroid hormone

Thyroid | thyroxin
triiodothyronine

Thymus | Hormonal function not clear; necessary for normal antibody formation.

The heart — as a "mixing chamber" for all hormonal secretions. All cells of the body are in contact with the same concentrations of each hormone. But only target glands respond to the specific hormone.

Liver

Islet cells in pancreas | insulin
glucagon

Stomach | gastrin

Intestine | cholecystokinin
secretin

Adrenal |

Cortex
Mineralocorticoids
Glucocorticoids
Medulla
Epinephrine and norepinephrine

Ovary | Estrogen
Progesterone

Testis | Testosterone

Figure 14–1. Diagram showing relationship of endocrine glands to the bloodstream.

(smaller portions of protein) and *steroids* (ringlike arrangements similar to cholesterol). The steroid hormones are those of the adrenal cortex, ovary, and testis. All other hormones are peptides. Two general mechanisms of action are known:

(1) formation of cyclic adenosine 3',5'-monophosphate (cyclic AMP) and
(2) activation of genes to form proteins.

1. The peptide hormones act by attaching to a specific receptor site *on the cell membrane*, activating an enzyme, adenyl cyclase, present in the membrance; this, in turn, promotes formation of cyclic AMP from ATP in the cytoplasm. Cyclic AMP then alters the activity level of enzymes within the cell to cause the specific effects associated with the hormone. Cyclic AMP has been called the "second messenger," the first being the hormone that prompts its formation. Since many hormones cause formation of cyclic AMP, its effect depends on the cell's ability to respond, i.e., the variety of enzymes within the cell. As an example, enzymes within the liver cells promote formation of glycogen from glucose, whereas those within bone cells promote the formation of bony tissue.

How two hormones, both causing formation of cyclic AMP, can have different effects in the same cell is a question that has not been answered as yet. It has been suggested that there are compartments of cyclic AMP associated with different receptor sites on the cell membrane, and that these compartments have access to different areas within the cell and, therefore, would influence different enzyme systems.

2. Hormones with steroid structure are small enough to enter through cell membranes. In the cytoplasm, they attach to receptor proteins; the combined protein-hormone then is transported into the nucleus. The hormone combination initiates formation of messenger RNA (page 37), which promotes formation of specific protein (enzymes) that will exert the cell response characteristic of the hormone.

Other mechanisms of action are probable, and not understood at the present time. Insulin, for example, alters the cell permeability to glucose, and growth hormone increases cell permeability to amino acids.

Control of Hormonal Secretion. Hormone secretion is regulated by negative feedback. In general, each gland tends to oversecrete its hormone. When the effects of the hormone are attained, the gland is inhibited from further secretion. If the gland undersecretes, the physiologic effects also decrease, and this feedback promotes the gland to secrete more hormone. Specific feedback controls are discussed below.

The posterior pituitary and the adrenal medulla are stimulated directly by the nervous system. Feedback to the nervous system may be afferent neural impulses or metabolites secreted by the glands.

Hormone Destruction. Hormones, once formed, do not continuously circulate in the blood. They are removed by the liver, primarily, where breakdown to inactive forms occurs. The residue is excreted in urine. The rate of inactivation may be altered by liver damage.

Hypothalamic Releasing Factors. Special neurons in the hypothalamus synthesize and secrete hormones called *hypothalamic releasing* and *inhibitory factors*. These neurons send fibers into the *median eminence* of the hypothalamus (Figure 14–3), the lowermost portion of the hypothalamus; the special nerve endings of these fibers secrete the hormones into the pituitary portal capillaries and they are carried to the adenohypophysis. For most of the anterior pituitary hormones, it is the releasing factor that is significant; prolactin secretion is controlled by an inhibitory factor. The hypothalamic factors that are of major importance are:

1. Corticotropin releasing factor (CRF), which causes release of corticotropin.
2. Growth hormone releasing factor (GRF), which causes release of growth hormone, or somatotropin.
3. Thyrotropin releasing factor (TRF),

which causes release of thyroid-stimulating hormone.

4. Follicle-stimulating hormone releasing factor (FRF), which causes release of the follicle-stimulating hormone.

5. Luteinizing hormone releasing factor (LRF), which causes release of the luteinizing hormone.

6. Prolactin inhibitory factor (PIF), which *inhibits* the secretion of prolactin.

Pituitary Gland (Hypophysis)

The pituitary gland is located in the sella turcica of the sphenoid bone (Figure 14–2). The *infundibular stem* attaches the gland below to the hypothalamus above. The hypophysis consists of two lobes: the larger anterior lobe, the *adenohypophysis*, which is derived embryologically from primitive pharyngeal epithelium; and the posterior lobe, the *neurohypophysis*, which is derived from neural ectoderm. An *intermediate lobe* is found between these two main lobes and is prominent in animals but not in humans.

Blood Supply. The hypophysis is highly vascular and receives its blood from the superior and inferior hypophyseal arteries (Figure 14–3). These are branches of the internal carotid artery and of the posterior communicating artery.

The superior hypophyseal artery extends medially to the upper part of the hypophyseal stalk and divides into the anterior and posterior branches. These branches anastomose and supply the upper part of

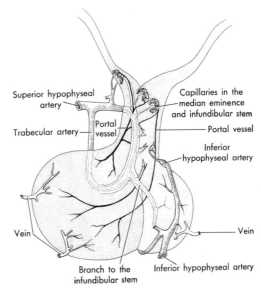

Figure showing: Superior hypophyseal artery, Trabecular artery, Portal vessel, Capillaries in the median eminence and infundibular stem, Portal vessel, Inferior hypophyseal artery, Vein, Vein, Branch to the infundibular stem, Inferior hypophyseal artery

Figure 14–3. Blood supply to the pituitary gland. (Modified from Crosby, Humphrey, and Lauer.)

the infundibular stem. Another branch supplies the superior surface and connective tissue of the pars distalis. This artery

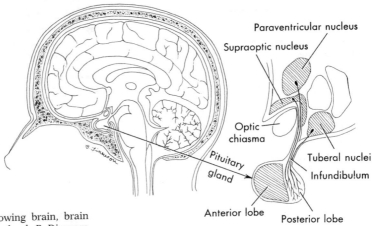

Figure labels: Paraventricular nucleus, Supraoptic nucleus, Optic chiasma, Pituitary gland, Tuberal nuclei, Infundibulum, Anterior lobe, Posterior lobe

Figure 14–2. *A.* Diagram showing brain, brain case, and location of pituitary gland. *B.* Diagram to show fibers from hypothalamus to the neurohypophysis.

A   B

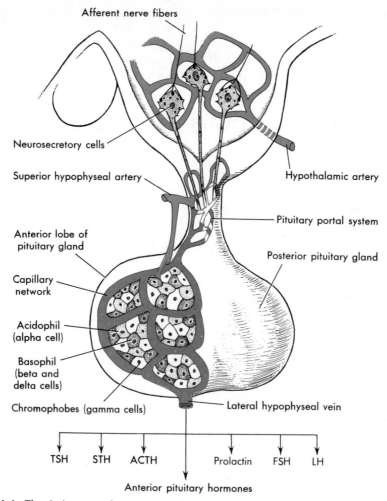

Afferent nerve fibers

Neurosecretory cells

Superior hypophyseal artery

Hypothalamic artery

Pituitary portal system

Anterior lobe of pituitary gland

Posterior pituitary gland

Capillary network

Acidophil (alpha cell)

Basophil (beta and delta cells)

Chromophobes (gamma cells)

Lateral hypophyseal vein

TSH   STH   ACTH   Prolactin   FSH   LH

Anterior pituitary hormones

Figure 14–4. The pituitary portal system. Note the relationship of capillaries in the hypothalamus and adenohypophysis.

does not supply the glandular tissue. Branches of the inferior hypophyseal artery divide into the medial and lateral branches and supply the *posterior lobe.* The blood supply to the *anterior lobe* is mainly through a portal system of veins (Figure 14–4). Branches of the internal carotid arteries break up into capillaries in the median eminences of the hypothalamus and in the lower infundibular stalk. The capillaries form portal vessels, which then terminate in blood capillaries and sinusoids in the pars distalis (anterior lobe) of the pituitary gland. The capillaries and sinusoids finally empty into the hypophyseal veins, which enter the circular and cavernous sinuses. The portal circula-

tion thus provides a means of conveying the releasing factors of the hypothalamus to the anterior pituitary. The neurohypophysis is well supplied with nerve fibers that descend in the hypophyseal tract; there is no nerve supply to the pars distalis. The nerve supply from the sympathetic plexus terminates around the internal carotid artery and is vasomotor in its function.

ADENOHYPOPHYSIS

The adenohypophysis consists of cells that give characteristic staining (chromophobic, indifferent to dyes; acidophilic, taking red, stain called alpha cells; and basophilic cells [beta cells], taking blue stain). The tissue is composed of irregular branching cords of epithelial cells, supported by fine reticular fibers. Between the cords of cells are many blood sinuses, which form a plexus. The adenohypophysis, influenced by hypothalamic releasing factors (page 314), forms hormones that have their primary site of action on other endocrine glands. These are *tropic* hormones and increase the activity of the target gland.

Corticotropin. Corticotropin (adreno-corticotropin, ACTH) is secreted by the chromophobic cells and has influence on the integrity of the adrenal cortex. Its greatest effect is to stimulate the secretion of the glucocorticoid hormones.

Somatotropic Hormone (STH), or Growth Hormone (GH). Somatotropic hormone is secreted by the acidophils of the gland. It increases the growth of tissues, especially of bone, muscle, and viscera. It is essential for growth and development and promotes both increased mitosis and increased sizes of cells. STH has effect on metabolic processes; these include increased rate of protein synthesis in tissue, decreased rate of carbohydrate utilization in striated muscle and adipose tissue, increased mobilization of stored fat, and increased use of fats for energy.

Excess of the growth hormone in early life results in giantism; the result of hyperactivity of the gland in later life is acromegaly, in which the jaws, bones, hands, and feet show overdevelopment and the features become enlarged and coarse. *Hypoproduction* of the growth hormone results in dwarfism. Growth of the body and sexual development are arrested. In the adult, the result is emaciation, muscular debility, loss of sexual function, and general apathy.

Thyrotropin, or Thyroid-Stimulating Hormone (TSH). Thyrotropin has several actions on the thyroid gland: it increases the size and number of the cells of the gland; it increases its secretory activity; and it increases the rate of iodine "trapping" in the glandular cells. Various emotional states can increase secretion of TSH, as does exposure to cold.

Gonadotropic Hormones (Follicle-Stimulating Hormone, FSH; Luteinizing Hormone, LH; Prolactin; Interstitial Cell–Stimulating Hormone, ICSH). These hormones increase the activity of the gonads, the ovaries and testes; the development of the breasts in females during pregnancy; and the secretion of milk. (They are discussed more fully in Chapter 25.)

FSH causes proliferation and maturation of the follicle cells of the ovary and stimulates the secretion of estrogen by these cells. In the male, FSH influences development of the sperm in the seminiferous tubules of the testis.

In the female, *LH* stimulates the corpus luteum to produce progesterone in the later stage of ovulation. In the male, LH is termed interstitial cell–stimulating hormone (ICSH) because it acts on these cells and stimulates the production of testosterone.

Prolactin aids in the development of the ovum in the later stage of the menstrual cycle, but its major function is its stimulating effect on the breast development during pregnancy and specifically to stimulate development of the milk-secreting cells. The stimulus of an infant sucking at the breast causes afferent impulses to be transmitted to the hypothalamus that inhibit PIF release, thus resulting in milk synthesis. Although prolactin is secreted in both sexes, its function in the male is not known.

INTERMEDIATE LOBE

The intermediate lobe produces melanocyte-stimulating hormone (MSH). MSH stimulates the melanocytes of the skin, cells that contain the black pigment melanin. Its chemical structure is similar to that of ACTH, and both cause darkening of the skin. In humans the secretion of MSH is poorly understood, and secretion is probably in small amounts.

NEUROHYPOPHYSIS

This consists of the infundibular process and stem and the median eminence. The infundibular process (posterior pituitary) is composed of numerous nerve fibers and glial-like cells known as pituicytes. Nerve fibers arise from cells located in the paraventricular and supraoptic nuclei of the hypothalamus and descend in the hypothalamohypophyseal tract and terminate on capillaries in the posterior pituitary (Figure 14–3). The hormones of the posterior pituitary are formed in these hypothalamic nuclei and move down the nerve fibers via axoplasmic flow into the posterior pituitary, where they enter the bloodstream directly, or are temporarily retained at the nerve ending for later secretion. (Aggregated hormonal material is observable with the light microscope and is known as Hering bodies.)

Two hormones are stored in the posterior lobe: *antidiuretic hormone* and *oxytocin*. Both hormones are polypeptides and have been synthesized in the laboratory.

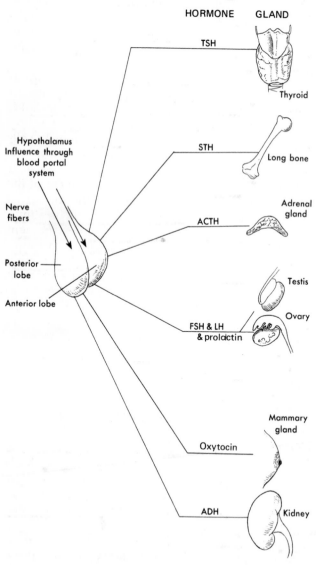

HORMONE   GLAND

TSH — Thyroid

STH — Long bone

ACTH — Adrenal gland

FSH & LH & prolactin — Testis / Ovary

Oxytocin — Mammary gland

ADH — Kidney

Hypothalamus Influence through blood portal system

Nerve fibers

Posterior lobe

Anterior lobe

TSH (thyroid-stimulating hormone) influences structure and secretory activity of the thyroid. The thyroid influences metabolic rate. Hypothalamic releasing factor is *thyrotropin-releasing* factor (TRF). Feedback-level of thyroxin in the bloodstream.

STH (somatotrophic hormone) influences growth of long bones, muscles, and viscera. In some way it is related to protein metabolism. It acts directly on all tissues. Hypothalamic releasing factor is *somatotropin-releasing* factor (SRF).

ACTH influences the activities of the *adrenal cortex*. It supports secretion of *mineralocorticoids* and controls secretion of *glucocorticoids*. Hypothalamic releasing factor is *corticotropin-releasing* factor (CRF). Feedback—serum cortisol level. Mineralocorticoid secretion is controlled by Na^+, K^+ serum levels, and renin-angiotensin system.

FSH (follicle-stimulating hormone) influences ripening of follicles, production of estrogen, and activity of the seminiferous tubules. Feedback is level of estrogen in the bloodstream.

LH (luteinizing hormone) influences secreting cells of the ovaries and testes and maintains their normal activity. In the male, stimulates secretion of testosterone. The hypothalamic releasing factors, *gonadotropin-releasing* factors, control the secretion of FSH and LH but not prolactin. Feedback—level of progesterone or testosterone in bloodstream. Prolactin stimulates development of milk-secreting cells of breast. PIF release inhibited by nerve impulses due to infant suckling.

Oxytocin is secreted during parturition—causes uterine contraction and active ejection of milk.

ADH (antidiuretic hormone) controls water reabsorption in the kidney tubules. Secreted by cells in the hypothalamus, trickles down nerve fibers to posterior pituitary where it is stored. Feedback mechanisms—osmolarity of bloodstream on osmoreceptors in hypothalamus.

Figure 14–5. Diagram showing the major functions of the pituitary hormones on physiologic activities.

Antidiuretic Hormone (ADH), or Vasopressin. Antidiuretic hormone increases the permeability of the cells of the distal tubule and collecting ducts in the kidney, hence decreases urine formation. In the absence of ADH large amounts of urine with a very low specific gravity are eliminated (polyuria).

Secretion of ADH is regulated by the osmolality of the blood. Cells in the supraoptic nuclei (Figure 14–2) function as osmoreceptors that are sensitive to the concentration of solutes in plasma. A rise in osmotic pressure increases the secretion of ADH and increases water reabsorption. (See Chapter 24.) In other words, concentrated body fluids stimulate the osmoreceptors and increase secretion of ADH, whereas dilute concentrations of body fluids inhibit ADH secretion.

Diabetes insipidus, a disease in which the urinary output is greatly increased, was formerly thought to result from hyposecretion of the posterior lobe; but it has been shown that a similar polyuria results from injury to the hypothalamic region (supraoptic nuclei) of the brain. It is therefore probable that involvement of either the posterior lobe of the pituitary or the hypothalamus will cause the disease.

Oxytocin. Oxytocin has a stimulating effect upon smooth muscles of the pregnant uterus; physiologic concentrations have no effect upon the nonpregnant uterus. Oxytocin is secreted in increased amounts during parturition, thereby increasing the contraction of the uterus. Oxytocin is also secreted during the process of lactation and causes ejection of milk from the alveoli into the ducts so that the infant can obtain it by suckling. The suckling stimulus increases the secretion of oxytocin. Figure 14–5 summarizes the actions of the pituitary gland on physiologic activities.

Thyroid Gland

The thyroid gland (Figure 14–6) consists of two lobes, situated at the sides of the trachea and thyroid cartilage. These lobes are connected by strands of thyroid tissue called the isthmus, ventral to the trachea. The external layer of the thyroid is connective tissue that extends inward as trabeculae and divides the gland into closed follicles of irregular size. Centrally, each of these follicles contains a colloid or jellylike substance that is secreted by the columnar epithelial cells that line the follicle. This substance is called *thyroglobulin.*

An abundant blood supply is derived from the external carotids and the subclavian arteries and is returned via the superior, middle, and inferior thyroid veins to the jugular and left brachiocephalic veins. It has been estimated that about 4 to 5 liters of blood pour through the gland per hour.

The nerves are derived from the second to fifth thoracic spinal nerves through the superior and middle cervical ganglia of the thoracolumbar system and from the vagus and glossopharyngeal nerves of the craniosacral system.

The hormones of the thyroid are thyroxin and triiodothyronine. The latter is more active than thyroxin, but its action is not as sustained. The thyrotropic hormone of the pituitary stimulates the growth and activity of the follicular cells of the thyroid, and in this way the formation of thyroid hormones is controlled.

Ingested iodides are absorbed from the intestine into the blood and are selectively removed from the blood or "trapped" by the thyroid gland. The iodides are combined with the amino acid tyrosine to form diiodothyronine, triiodothyronine, and finally *thyroxin.* This process occurs within the thyroglobulin molecule, to which the tyrosine molecules are attached. There are about nine thyroxin molecules for each

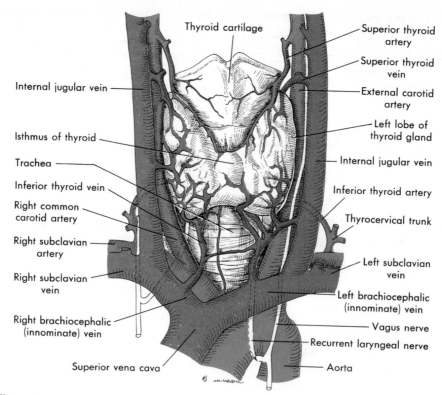

Figure 14–6. The thyroid gland and related blood vessels. (Modified from Pansky and House.)

triiodothyronine molecule. Storage of these hormones goes on until, under the influence of TSH, they are secreted. the negative feedback for TSH is the circulating level of thyroxin and triiodothyronine. When these hormones are released into the blood, they adsorb to plasma globulin and albumin and are then slowly released to the cells of the body.

Functions of Thyroxin

1. Thyroxin regulates the metabolic and oxidative rates in tissue cells, hence it is involved in maintaining body temperature. It increases the rate of glucose absorption from the intestine and increases the rate of glucose utilization by cells. These effects are believed to be through enzyme systems in mitochondria.

2. Thyroxin stimulates the growth and differentiation of tissues in young persons.

3. Thyroxin influences conversion of glycogen from noncarbohydrate sources and conversion of glycogen to glucose, thus raising blood sugar.

4. Thyroxin increases quantities of certain oxidative enzymes in the mitochondria.

5. Thyroxin influences the rate of metabolism of lipids, proteins, carbohydrates, water, vitamins, and minerals.

6. Thyroxin increases protein synthesis as indicated by (a) increased turnover of RNA, and (b) increased incorporation of amino acids into protein by the mitochondria.

7. In the young person both physical and mental development are influenced and in the adult thyroxin stimulates mental processes.

One milligram of thyroxin increases the metabolic rate about 2.5 per cent. Carbohydrates, proteins, and fats are all oxidized in increased amounts. The thyroid hormones regulate cell metabolism primarily by controlling the activities of oxidizing enzymes within the mitochondria. This in turn increases the activity of carbohydrases, amidases, transferases, and proteolytic enzymes.

When thyroxin is given experimentally to animals or human beings, nitrogen loss exceeds intake. Glucose tolerance is decreased and stored glycogen is also decreased, although blood glucose may be normal or below.

Cases of hypoactivity of the gland may show as much as 50 per cent decrease in the metabolic rate; cases of hyperactivity may show an increase up to 60 to 100 per cent and more. Injection or feeding of thyroid tissue results in increased basal metabolism, loss of weight, increase in elimination of nitrogen, increased heart rate—which results at times in an abnormality called tachycardia—and nervous excitability.

The size of the thyroid varies with age, sex, and general nutrition. It is relatively larger in the young, in women, and in the well nourished. Removal of the gland does not cause death but, unless the hormone is replaced, brings about marked changes, such as lowered basal metabolism and general malnutrition. Disturbances in the secretion of thyroxin are classed under two headings: (1) hypothyroidism, or decreased secretion, and (2) hyperthyroidism, or excess secretion. The liver removes excess thyroxin from the bloodstream.

Goiter is an enlargement of the gland. It may result from increased functional activity due to a decrease in the iodine content of the gland. This in turn is usually due to a decrease or lack of iodine in water and food. Goiter occurs frequently in adolescent girls, but its incidence is greatly reduced if iodine is given.

A goiter may also be due to the presence of a tumor or increased thyroxin secretion.

Hypothyroidism. In man certain pathologic conditions are caused by hypothyroidism, i.e., cretinism and myxedema.

Cretinism is caused by congenital defects of the thyroid or by atrophy in early life. The growth of the skeleton ceases, although the bones may become thicker than normal and there is marked arrest of mental development. Children so afflicted are called cretins. They are not only dwarfed but ill proportioned, having large heads, protruding abdomens, weak muscles, and slow speech.

Myxedema is a condition that results from atrophy or removal of the thyroid in adult life. The most marked symptoms of this condition are slowness of both body and mind, usually associated with tremors and twitchings. The skin becomes rough and dry, owing to lack of cutaneous secretions, and assumes a yellow, waxlike appearance. There is an overgrowth of the subcutaneous tissues, which in time is replaced by fat; the hair grows coarse and falls out; the face and hands are swollen and puffy; the metabolic rate is low and the mental activities apathetic. Cretinism and myxedema are both caused by insufficient secretion of thyroxin, which may be supplied by synthetic thyroid hormone. The treatment must be kept up throughout the patient's life.

Hyperthyroidism. Overactivity of the thyroid gland, i.e., increase in the amount of internal secretion, produces a condition called Graves' disease or exophthalmic goiter. It is characterized by protruding eyeballs, quickened and sometimes irregular heart action, elevated temperature, nervousness, and insomnia. The appetite may be excessive, but this is accompanied by loss of weight due to increased metabolism and digestive disturbances. This condition is sometimes remedied by removing part of the gland.

Parathyroid Glands

The parathyroids, usually four in number and arranged in pairs, are independent of the thyroid both in origin and in function but are usually located on its dorsal surface. These small reddish glands are about 6 to 7 mm long and 2 to 3 mm thick. Accessory nodules are sometimes found surrounding the glands or embedded in connective tissue. The glands consist of closely packed epithelial cells richly supplied with capillaries from branches of the inferior and superior thyroid arteries. The nerve supply is from the vagus and glossopharyngeal nerves of the craniosacral system and from the cervical autonomics of the thoracolumbar system.

The parathyroids secrete a hormone, parathyroid hormone, or *parathormone*, protein in nature, which plays an important role in the maintenance of the normal calcium level of the blood. It also regulates phosphorus metabolism and increases the rate of calcium reabsorption in the renal tubules. *Calcitonin*, produced by cells in the thyroid gland, is another hormone controlling calcium ion concentration in the body. Its action is to prevent excessive calcium levels in the blood.

Parathormone influences the reabsorption of phosphate in the renal tubules and in the presence of vitamin D stimulates the absorption of calcium in the intestine. Parathormone exerts action on bone and stimulates osteoclastic activity. This may be a phagocytic action whereby bony particles are digested with a final release of calcium into the blood.

Acute symptoms of *hypoparathyroidism*, known clinically as tetany, may result from removal of the parathyroids or may possibly occur spontaneously; the concentration of blood calcium falls and increased nerve irritability causes muscle spasms that may result in death if untreated. Symptoms are relieved by giving solutions of calcium.

In *hyperparathyroidism* there are muscular weakness, pain in the bones, and an increase in the calcium in blood and urine. The bones show decalcification and deformity, and spontaneous fractures may occur. These symptoms are often caused by a parathyroid tumor, the removal of which brings about a reduction in the blood calcium, and considerable resolidification of the bones occurs. Deposits of calcium may occur, especially in the kidney, and nitrogenous wastes may become excessive in blood and lymph.

Adrenal Glands

The adrenal glands (Figure 14–7) are two small bodies that lie at the superior pole of each kidney. The right adrenal gland is somewhat triangular in shape and the left one more semilunar. They vary in size, and the average weight of each is about 5 to 9 gm. Each gland is surrounded by a thin capsule and consists of two parts known as the cortex, or external tissue, and the medulla, or chromophil tissue. These parts differ in origin and function.

Embryologically the adrenal cortex develops from mesoderm; the medulla has the same embryologic origin as the thoracolumbar division of the autonomic nerv-

ous system, namely, from an outgrowth from the neural ectoderm. It is also functionally related to the sympathetic nervous system. EPINEPHRINE

At birth the adrenal glands weigh about 8 gm, which is proportionately 20 times greater than in the adult. They gradually decrease in size, and by one year of age the two adrenals weigh 4.5 gm. The more mature the infant, the more adequately the adrenal cortex responds to stress.

Blood Supply. The arteries supplying this highly vascular gland are derived from the aorta, the inferior phrenics, and the renal

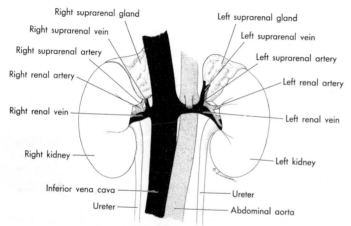

Right suprarenal gland
Right suprarenal vein
Right suprarenal artery
Right renal artery
Right renal vein
Right kidney
Inferior vena cava
Ureter

Left suprarenal gland
Left suprarenal vein
Left suprarenal artery
Left renal artery
Left renal vein
Left kidney
Ureter
Abdominal aorta

Figure 14–7. Diagram showing position of adrenal glands and kidneys.

arteries. Blood is returned via the suprarenal veins. It has been estimated that a quantity of blood equal to about six times each gland's weight passes through it per minute.

Nerves. The nerve fibers are derived from the celiac and renal plexuses (splanchnic nerves). Removal of the gland has long been known to be followed by prostration, muscular weakness, and lowered vascular tone, with subsequent death in a few days. These symptoms are caused by the removal of the adrenal cortex. Removal of the medulla causes no serious disturbance if the individual is not subjected to stress.

ADRENAL CORTEX

The adrenal cortex secretes a group of hormones called corticosteroids. These include the *mineralocorticoids* and the *glucocorticoids*. It also secretes small amounts of *androgenic* hormones and minute quantities of *estrogenic* hormones.

The hormones of the adrenal cortex include *cortisol, corticosterone, aldosterone, androgens, estrogens,* and *progesterone.* Cortisol and corticosterone are called glucocorticoids because they have a specific effect on glucose metabolism. *Aldosterone* has been called a *mineralocorticoid* because its chief action is to promote sodium retention and potassium excretion.

Mineralocorticoids. The salt or mineral hormone *aldosterone* functions at the renal tubule and stimulates the reabsorption of Na^+, which then attracts Cl^- and causes its reabsorption into the blood. In this way the Na^+ and Cl^- content of the extracellular fluids is maintained. Water is reabsorbed with the salt as the result of increased osmotic pressure. The reabsorption of K^+ is depressed and hence the Na^+/K^+ ratio is controlled in body fluids. Fluid balance is also controlled through regulation of these electrolytes. Aldosterone secretion is increased by

1. An increase in extracellular potassium
2. A decrease in extracellular sodium
3. An increase in angiotensin formation
4. Secretion of ACTH

Potassium level is by far the most important influence on aldosterone secretion (which acts to return the high serum level of potassium to normal), and low sodium level is probably the next most significant control. ACTH is necessary for maintenance of the cells producing aldosterone but does not directly regulate aldosterone secretion. Angiotensin (p. 548) was thought to increase aldosterone secretion; however, recent studies have cast some doubt on the exact effect. Whereas high potassium and low sodium levels increase the secretion of aldosterone, the opposite conditions (low potassium and high sodium) result in decreased secretion of aldosterone.

Glucocorticoids. The actions of the glucocorticoids are chiefly from cortisol. The glucocorticoids are hormones whose predominant action is the regulation of metabolism of carbohydrate, protein, and fat.

Effects on Carbohydrate Metabolism. These hormones stimulate gluconeogenesis by the liver. The liver forms glycogen from noncarbohydrate sources and decreases the rate of glucose utilization by the cell.

Effects on Protein Metabolism. The glucocorticoids reduce protein stores by decreased anabolism of protein, a decreased rate of protein synthesis, and mobilization of amino acids from the tissues. Liver protein is increased, and in turn plasma proteins are released into the blood. Glucocorticoids increase permeability of the liver cells to amino acids and decrease the permeability of muscle cells to amino acids.

Effects on Fat Metabolism. As need arises, the glucocorticoids either promote fat mobilization from fat depots (adipokinesis) or increase the rate of storage of fat deposition as adipose tissue (lipogenesis).

Effects on Lymph and Blood. The corticosteroids influence the activity of lymphoid tissue and the number of eosinophils

INFECTION ↑

in circulating blood. Administration of adrenocortical hormones decreases the number of circulating eosinophils and the size of lymphoid tissue. Large doses of the glucocorticoids reduce or may completely block antibody formation in lymphoid tissue.

Stress Situations. The adrenal cortex in some way aids the body to cope more effectively with adverse environmental situations, or what is commonly known as stress situations. See discussion and Table 14–1.

Gonadal Hormones. Several hormones have been isolated from the adrenal cortex that have influence on the sex organs, for example, androgen, estrogen, and progestins.

When the adrenal cortex, ovaries, or testes are removed, certain physiologic changes take place. Removal of the ovaries or testes results in an enlargement of the adrenal cortex. If the adrenal glands fail to function and blood level of cortisol falls, corticotropin, secreted by the pituitary gland, is greatly increased. This is believed to explain why there is decreased production of the gonadal hormones. On the other hand, sex hormones can affect adrenocortical function. Estrogens from the ovary inhibit the formation of FSH by the pituitary; however, at the same time corticotropin is increased.

Cortical hypofunction is recognized as the cause of Addison's[1] disease, which, if untreated, is fatal within from one to three years. Symptoms of Addison's disease are muscular weakness and general apathy, gastrointestinal disturbances, pigmentation of the skin and mucous membranes, loss of weight, and depressed sexual function. The pigmentation is the outstanding symptom and is due to excessive deposition of the normal cutaneous pigment, melanin. Treatment with cortical extract gives favorable results.

Cortical hyperfunction appears to be associated with cortical tumors. It results in the young in precocious sexual development with profuse growth of hair, and in the case of adult females, in the development of secondary male characteristics.

Stress. In any type of severe injury to tissues or cells, excessive exercise, starvation, infections, or emotional stress, there is a definite pattern of physiologic response. Selye[2] found that no matter what the stressor was, including physical restraint, certain changes always occured in the animal. These were atrophy of the thymus gland, marked enlargement of the adrenal cortex, and hemorrhage into the gastric mucosa. The reasons for all these changes have not been clarified. It is known that in stress situations ACTH is released, and the response is an increased production of adrenocortical hormones. There is also increased sympathetic nervous system activity with a release of epinephrine, which in turn stimulates production of ACTH as well as bringing about the usual body adjustments for emergency physiology. Table 14–1 indicates many of the physiologic adjustments necessary for survival of the individual.

On the other hand, the helpful properties of stress should be remembered. For instance, normal bone needs the stress of weight bearing if it is to retain its mineral content. In fact, all tissues of the body need moderate stress to maintain normal functioning. The aim should not be to eliminate stress from life situations, but effort should be made to understand and moderate stress before it becomes excessive.

ADRENAL MEDULLA

The adrenal medulla consists of large, granular cells arranged in a network. It secretes two hormones, amine in nature, called epinephrine (Adrenalin) and norepinephrine (noradrenalin).

Epinephrine affects all structures of the body innervated by the sympathetic nervous system and thereby reinforces its action.

The hormones *epinephrine* and *norepinephrine* belong to a group of chemicals known as *catecholamines* and are referred to as *sympathomimetic amines* because they have the same physiologic effect as the sympathetic nervous system. Through the glycogenolytic effect of *catecholamines*, these hormones provide the body with a means of sparing carbohydrates for the brain during periods of hypoglycemia. They also have ability to stimulate the re-

[1] Thomas Addison, English physician and teacher (1793–1860).

[2] Hans Selye, Canadian endocrinologist (1907–).

TABLE 14–1
RESPONSE OF BODY TO STRESS

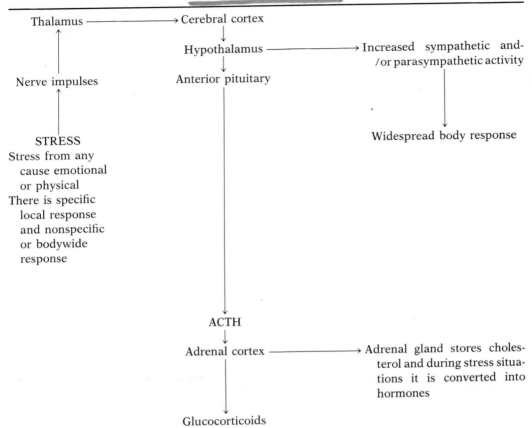

Thalamus ————→ Cerebral cortex
↓
Hypothalamus ————————→ Increased sympathetic and-
/or parasympathetic activity
↑ ↓
Nerve impulses Anterior pituitary
↑
STRESS Widespread body response
Stress from any
cause emotional
or physical
There is specific
local response
and nonspecific
or bodywide
response

ACTH
↓
Adrenal cortex ————————→ Adrenal gland stores cholesterol and during stress situa-
tions it is converted into
hormones
↓
Glucocorticoids

Carbohydrate metabolism altered—blood glucose increased
Gluconeogenesis increased—decreased use of glucose by tissues
Fat *mobilization* enhanced
Protein metabolism altered—proteins are *mobilized* from tissue
  cells
Glucocorticoids—suppress inflammatory processes and block
  inflammatory response to allergic reactions, but do not affect
  basic allergic reactions between antibody and antigen
Permeability of capillary membranes increased—rapid transfer
  of substances between cells and fluids
Lysis of eosinophils and atrophy of lymphoid tissues
Increased production of red cells

lease of free fatty acids from adipose tissue. These, in turn, are used by muscle tissue of the body in place of glucose, thereby providing another physiologic means of conserving circulating glucose. On the whole, norepinephrine is more effective in vasoconstriction and epinephrine has a more pronounced effect on carbohydrate metabolism. (See Table 14–2.)

The differences in effect of epinephrine and norepinephrine are believed to be due to type of receptor on the effector cell, one class having alpha receptors, the other beta receptors.

Examples of the use of epinephrine in medicine are to bring about vasoconstriction, which prolongs the action of local anesthetics and

TABLE 14–2
COMPARISON OF EFFECTS OF EPINEPHRINE AND NOREPINEPHRINE

Epinephrine	Norepinephrine
1. Cardiac acceleration	Little effect
2. Increased blood flow in muscle	Vasoconstriction in muscle
3. Increased myocardial strength	About the same
4. Increased cardiac output and venous return	Slight effect
5. Systolic pressure raised	Pressor effects marked, raises both systolic and diastolic pressure
6. Hyperglycemic effects marked	Much less effect
7. Marked increase in basal metabolic rate	Much less effect
8. Lipolytic effects, liberates nonesterified fatty acids from fat depots	Slightly greater effects
9. Excitation of central nervous system	No effect
10. Myometrial relaxation	Pilomotor contraction
11. Dilation of iris	No effect
12. Bronchial relaxation	No effect

reduces the absorption of the substance producing the anesthesia; to relax the bronchioles in asthma; to constrict arterioles of the mucous membranes and of the skin, thus reducing the loss of blood in minor operations.

Cannon[3] and his associates noted that the supply of epinephrine to the blood by the adrenal medulla is increased under emotional stress. This increase results in a more rapid and forceful heartbeat; a greater flow of blood to the muscles, central nervous system, and heart; an increased output of glucose from the liver; inhibition of the intestinal muscle coat; and closure of the sphincters. The muscle coat of the bronchi is relaxed and there is contraction of the splenic capsule, which gives more blood to circulation. By means of these reactions the stable environment of cells is maintained during periods of great functional demands. Whether these reactions are in direct response to the increased secretion of the medulla or are brought about by the action of the sympathetic nervous system alone is undetermined. Nevertheless, it may be said that the medulla in normal physiologic activities carries on certain emergency functions such as constriction of skin and splanchnic blood vessels, vasodilation in skeletal and cardiac muscle, a rise in general blood pressure and increased output of the heart, the liberation of glucose supplies from the liver, the relaxation of gastrointestinal smooth muscles, decrease in the coagulation time of the blood, discharge of red blood cells from the spleen, and an increase in the depth and rate of respiration. These and other accompanying reactions, such as emotional responses, result in equipping the individual to meet the emergency at hand.

Ovary

The ovary produces two hormones, whose actions are understood. (1) Follicular hormone, or *estrogen*, is present in the blood of females from puberty to the menopause, reaching its highest concentration just before ovulation. Its actions are on the reproductive tract (uterus, fallopian tubes, vagina) and the breasts; it is responsible for the secondary sex charac-

[3] Walter B. Cannon, American physiologist (1871–1945).

teristics of the female. In addition, its cyclic rise during the menstrual cycle is responsible for typical changes in the lining of the uterus preceding menstruation (see page 587). It is present in large amounts in the blood during pregnancy. It is also present in the urine of males and has been found in the tissues of growing plants and animals. Several different kinds of estrogens have been isolated from human blood plasma. Three are present in significant amounts: β-estradiol, estrone, and estriol. β-Estradiol is the most important and most potent. (2) The corpus luteum hormone, or *progesterone*, supplements the action of estrogen, promoting further development of the uterine mucosa in preparation for implantation of the developing ovum. It is essential for the growth of the mammary glands and for the complete development of the maternal portion of the placenta and of the mammary glands during pregnancy.

In the nonpregnant female most of the progesterone is secreted during the latter half of each ovarian cycle. It is secreted in larger quantities than is estrogen. Progesterone secretion is under the influence of LH.

Table 14–3 summarizes the effects of estrogen and progesterone.

TABLE 14–3
EFFECTS OF ESTROGEN AND PROGESTERONE

Name	Site of Action	Effect
Estrogen	Fallopian tubes and endometrium	Causes glandular cells to proliferate Increases number of ciliated epithelial cells Activity of cilia enhanced
	Breasts	Initiates growth of breast Causes fat deposition and development of stromal tissues. Lobules and alveoli develop to slight extent
	Skeleton	Causes osteoblastic activity at puberty. There is a rapid growth rate, which causes early uniting of the epiphyses with the shafts of long bone
	Pelvis	Broadens pelvis and size of vagina
	Calcium and phosphate in blood stream	Causes retention to promote bone growth
	Metabolism	Apparently has no effect on metabolic rate but does cause greater deposition of fat and increases total body proteins
	Skin	Causes skin to develop a soft, smooth texture and increases its vascularity
	Electrolyte balance	Causes Na^+ and Cl^- and H_2O retention by the kidney
Progesterone	Uterus	Promotes secretory changes and storage of nutrients in the endometrium, thus prepares it for implantation of the fertilized ovum Inhibits contractility of the myometrium
	Fallopian tubes	Promotes secretions of mucosa
	Breasts	Promotes final development of lobules and alveoli; causes proliferation of cells and causes cells to become secretory. Enlarges breasts
	Electrolyte balance	Enhances Na^+, Cl^-, and H_2O reabsorption from distal tubules of kidney
	Protein	Has a mild catabolic effect

Testis

The testis produces the male sex hormone, *testosterone*. This hormone promotes the development of male secondary sex characteristics, leading to increase in growth of tissues and muscularity, in addition to development of the sex organs. The term *androgen* is used to refer to any hormone that, like testosterone, promotes the development of male characteristics. Androgens other than testosterone are produced in the testis (i.e., androstenedione) but they are not as potent. Additionally, the adrenal cortex of both males and females produces androgens (and estrogens). The breakdown products of androgens, 17-ketosteroids, are present in urine and provide a measure of the secretion of testosterone or other androgens. Testosterone production is under the control of ICSH (LH).

Pancreas

Deep within the pancreas are special groups of cells, the *islets of Langerhans*[4] (Figure 14–8), which secrete two hormones, *insulin* and *glucagon*. Two main types of cells are identifiable in the islets and are called alpha and beta; the alpha cells produce glucagon, the beta cells produce insulin.

Insulin promotes the diffusion of glucose across the cell membrane of most cells, with the exception of the brain; its effect is dramatic and results in prompt fall of blood sugar level. The glucose concentration in the blood is the stimulus for secretion of insulin. When the glucose level is high, insulin is promptly secreted until the glucose resumes a normal level; when the glucose level is low, insulin secretion stops. Glucose uptake by the cell facilitates glycogen synthesis in liver and skeletal muscle. In addition, insulin stimulates the enzyme that catalyzes the conversion of glucose to glycogen. Other effects include inhibition of triglyceride breakdown, promotion of protein synthesis, and influence on many liver enzymes other than those involved in glycogen formation. The lack of insulin results in opposite effects: high blood sugar level, breakdown of triglycerides and adipose tissue stores, and decreased protein synthesis. (See also page 520.)

Glucagon has the opposite effect on blood glucose; it raises the blood glucose level by stimulating liver gluconeogenesis and glycogenolysis. Glucagon is secreted in response to *low* blood sugar level.

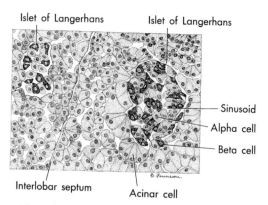

Islet of Langerhans   Islet of Langerhans

Sinusoid

Alpha cell

Beta cell

Interlobar septum   Acinar cell

Figure 14–8. Diagram of microscopic view of pancreas showing islets of Langerhans. Alpha cells light, beta cells dark.

[4]Paul Langerhans, German pathologist (1847–1888).

Other Possible Endocrine Glands

Two other glands have been proposed, although without definitive proof: the *pineal body* and the *thymus*. Experimental data support the idea that the pineal body is a gland, although no hormone has been identified. The thymus, like the pineal body, may be an endocrine gland, with production of a hormone; however, definitive evidence is lacking.

Local Hormones

In addition to the gastrointestinal hormones mentioned at the beginning of the chapter, another important group must be mentioned: the *prostaglandins* (so named because they were first identified in the prostate gland). These are a group of chemicals derived from the fatty acid arachidonic acid and are found in a wide variety of tissues. These include male and female reproductive tracts, heart, lung, thymus, brain, intestine, liver, kidney, and pancreas. The type of prostaglandin varies slightly according to tissue and species. Prostaglandins have an important effect on lipolysis of fat depots and blood level of fatty acids; on activity of smooth muscle of the digestive tract, cardiovascular system, and respiratory system; and on organs of the reproductive system. It is too early yet to know the exact action of each of the prostaglandins; however, their wide distribution signifies their metabolic importance.

Questions for Discussion

1. Name the hormones of the anterior pituitary and explain the function of each.
2. Explain the anatomic relationship between the hypothalamus and the hypophysis.
3. What is the hypothalamic-portal system?
4. List the hormones of the pancreas and explain their function.
5. Explain the feedback mechanism for control of the secretion of one hormone.
6. Which of the adrenal hormones are essential for life? Explain.
7. Differentiate between the actions of epinephrine and norepinephrine.
8. Explain briefly the meaning of "negative feedback."

Summary

General Hormones	{ Secreted into circulation; act on target cells { Have effect on distant organs
Local Hormones	{ Function at point of origin
Endocrine Glands	{ Pituitary, thyroid, parathyroid, adrenal, ovary, testis, pancreas
Mechanism of Action	{ Based on structure of hormone { Peptides—initiate formation of cyclic adenosine 3',5'-monophosphate (cyclic AMP) { Steroids—initiate formation of messenger RNA, with production of cytoplasmic enzymes

Control of Hormonal Secretion	Negative feedback—high level of target cell secretion or response inhibits release of stimulating substance, and vice versa
Hormone Destruction	Occurs in liver, breakdown products excreted by kidney
Hypothalamic Releasing Factors	Corticotropin releasing factor (CRF) Growth hormone releasing factor (GRF) Thyrotropin releasing factor (TRF) Follicle-stimulating hormone releasing factor (FRF) Luteinizing hormone releasing factor (LRF)
Hypothalamic Inhibitory Factor	Prolactin inhibitory factor (PIF)

SUMMARY OF ENDOCRINE GLANDS

Name of Gland	Location	Secretion	Probable Function	Diseases Associated with It
Pituitary, or Hypophysis Cerebri Consists of an anterior and posterior lobe and infundibular stem	Lodged in the sella turcica of the sphenoid bone	Posterior lobe *Antidiuretic hormone (ADH)*—formed in special areas of hypothalamus	Antidiuretic—controls water reabsorption in the distal kidney tubule Pressor effects unimportant physiologically	**Hypofunction** (or hypothalamic lesion) results in *diabetes insipidus*
		Oxytocin—formed in special areas of hypothalamus	Oxytocic effect—muscles of pregnant uterus contract	
		Anterior lobe	ACTH influences adrenal cortex secretion STH influences growth of tissues TSH influences structure and secretory activity of thyroid gland Gonadotropic hormones FSH, prolactin, LH, ISCH influence gonadal functions	**Hypofunction** may result in dwarfism, polyuria, depressed sexual function, and scanty growth of hair **Hyperfunction** may result in gigantism (early life), acromegaly (adult life), and glycosuria
Thyroid Weighs about 1 oz Consists of two lobes connected by an isthmus	In front of trachea and beside thyroid cartilage	*Thyroxin*—contains 65% of the body iodine *Triiodothyronine*	Influence the general rate of oxidation in the body, also growth and development in the young Influence through enzyme systems	**Hypothyroidism** may result in cretinism (early life) and myxedema (adult life) **Hyperthyroidism** may result in Graves' disease (exophthalmic goiter)

SUMMARY OF ENDOCRINE GLANDS (*Continued*)

Name of Gland	Location	Secretion	Probable Function	Diseases Associated with It
Parathyroids Four small glands	Between the posterior borders of the lobes of the thyroid gland and its capsule	*Parathyroid hormone (parathormone)*	Regulates the blood-calcium level and the irritability of the nervous system and muscles Increases excretion of inorganic phosphate in the renal tubule	When the parathyroids are removed, tetany develops **Hyperparathyroidism** may result in muscular weakness and high blood calcium
Adrenal Glands Two small glands, each surrounded by a fibrous capsule and consisting of two parts {cortex {medulla *Cortex* consists of epithelial cells arranged in columns. These cells are derived from the part of the mesoderm that gives rise to the kidneys	Placed above and in front of the upper end of each kidney	*Several hormones* (glucocorticoids, mineralocorticoids)	Regulate electrolyte and water balance. Influences fat, carbohydrate, and protein metabolism; activity of lymphoid tissue and sexual organs. Aids body to cope more effectively with stress situations.	Removal causes death **Hypofunction** results in Addison's disease and poor response to stress **Hyperfunction** may cause precocious sexual development (Cushing's syndrome) and virilism in the adult female
Medulla consists of a network of large granular cells that when treated with chromic acid give a yellow or brown reaction. Derived from neural crest of ectoderm		Medulla— *Epinephrine*	Constitutes a reserve mechanism that comes into action at times of stress. Epinephrine augments the response of sympathetic nerves, increases the heartbeat, increases blood supply to muscles, nervous system, and heart, and increases output of glucose from liver. Raises blood pressure.	Removal of the medulla causes no serious physiologic disturbance

Summary of Endocrine Glands (*Continued*)

Name of Gland	Location	Secretion	Probable Function	Diseases Associated with It
Medulla (*cont.*)		*Norepinephrine*	Vasoconstriction, elevates blood pressure	
Gonads *Ovaries* Two almond-shaped bodies which weigh 2–3.5 gm	One on each side of the uterus, attached to the broad ligament and below the uterine tubes	*Estrogen*	Seems to act by maintaining nutrition and mature size of female reproductive organs	Excessive ovarian function produces precocious puberty Diminished ovarian function characterized by late onset of menstruation, faulty development of genital organs, delayed menstruation
		Progesterone	Sensitizes the mucous membrane of the uterus so that it responds to the contact of the developing ovum and assists in implantation	
Testes Two glandular organs which weigh 10.5–14 gm	In the scrotum	*Testosterone*	Influences the development of secondary sex characteristics in the male	The tendency to become obese after castration
Pancreas A compound gland which weighs between 2 and 3 oz	In front of the first and second lumbar vertebrae behind the stomach	Islets of Langerhans furnish two hormones *Insulin* and	1. Facilitates movement of glucose across cell membrane, exception is brain 2. Accelerates the synthesis of sugar to glycogen and the storage of glycogen 3. Restricts production of glucose in liver from protein and fat	
		Glucagon	Aids in the conversion of glycogen to glucose in the liver	
Local Hormones	Various tissues	*Prostaglandins* (several types)	Not clearly understood at present	Unknown

Additional Readings

GILLIE, R. B.: Endemic goiter. *Sci. Amer.*, **224:**92–101, June, 1971.

GUILLEMIN, R., and BURGUS, R.: Hormones of the hypothalamus. *Sci. Amer.*, **225:**24–33, November, 1972.

LEVINE, S.: Stress and behavior. *Sci. Amer.*, **224:**26–31, January, 1971.

McEWEN, B.: Interactions between hormones and nerve tissue. *Sci. Amer.*, **235:**48–58, July, 1976.

O'MALLEY, B., and SCHRADER, W.: Receptors of steroid hormones. *Sci. Amer.*, **234:**32–43, February, 1976.

PASTEN, I.: Cyclic AMP. *Sci. Amer.*, **227:**97–105, August, 1972.

PIKE, J.: Prostaglandins. *Sci. Amer.*, **225:**84–92, November, 1971.

RASSMUSSEN, H., and PECHET, M.: Calcitonin. *Sci. Amer.*, **223:**42–50, October, 1970.

VANDER, A. J.; SHERMAN, J. H.; and LUCIANO, D. S.: *Human Physiology: The Mechanisms of Body Function*, 2nd ed. McGraw-Hill Book Co., New York, 1975, Chapter 7.

UNIT **IV**

Body Maintenance: Distribution of Energy Sources and Nutrients

For the body to be maintained in a state of health, the individual cells, tissues, and organs must have a continuous supply of oxygen and other nutrients. Conversely, there must also be removal of carbon dioxide and waste materials. The body system with the specialized function of transporting these materials to and fro is the circulatory system: the heart and blood vessels through which the blood is constantly moving.

CHAPTER 15

The Blood; Immunity

Chapter Outline

Blood is the transporting medium of the body in which materials move from one organ to another and one part of the body to another. It helps to maintain uniform temperature, to regulate acid-base balance, and to defend against infection. In short, it is a vital fluid.

Blood

Characteristics. Blood is bright red, approaching scarlet in the arteries, and somewhat darker in the veins. It is a viscous liquid with a specific gravity of 1.041 to 1.067. In the adult, blood volume is about one thirteenth of body weight, with little variation in the quantity. The major factor influencing blood volume variation is the degree of adiposity of the individual. Since adipose tissue is less vascular than other tissues, as the proportion of fatty tissue in the total body mass rises, the blood volume decreases. Both men and women who are lean have about 79 ml/kg of body weight, or about 5.5 L of blood for one whose weight is 70 kg.

Blood consists of cells, or *corpuscles*, suspended in an intercellular liquid, the *plasma*. The percentage of blood made up of red blood cells (the *hematocrit*) is 47 per

337

cent, ±7 for men and ±5 for women. Dehydration causes the blood to become more concentrated and the hematocrit therefore rises.

Erythrocytes (Red Blood Cells). Under the microscope erythrocytes, as seen in Figure 15–1, *A–D*, are circular disks, without nuclei and biconcave in profile. The

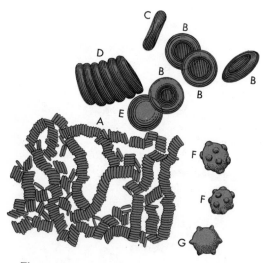

Figure 15–1. Erythrocytes of the blood, magnified. *A.* Moderately magnified; the erythrocytes are seen lying in rouleaux. *B.* Erythrocytes much more highly magnified, face view. *C.* In profile. *D.* In rouleau, more highly magnified. *F.* Erythrocytes puckered or crenated all over. *G.* Same at edge only.

average size is 7.7 μm in diameter. They have a distinct uniformity of size, shape, and color. Erythrocytes consist of a framework, or stroma, in which hemoglobin is deposited. The stroma includes the cell membrane and is composed of protein and lipid substances. The blood group antigens A and B and Rh are located in the stroma. Erythrocytes are soft, flexible, and elastic, so that they readily squeeze through passages narrower than their own diameters and immediately resume their normal shape.

Erythropoiesis. In the embryo the first blood cells arise from mesenchymal cells in the yolk sac. The next phase of development is chiefly in the liver and to some extent the spleen. Erythropoiesis takes place in bone marrow at about the fifth month and decreases in the liver as it increases in red bone marrow. After birth red bone marrow is the only tissue concerned with red cell formation. Red cells arise from primitive, undifferentiated cells in the *red marrow* of bone called hemocytoblasts. The hemocytoblast loses its nucleus and cytoplasmic granules, becomes smaller, accumulates hemoglobin, loses its nucleus and cytoplasmic organelles, enters the bloodstream, and assumes the shape of the erythrocyte.

For complete maturation of the red cells vitamin B_{12} is necessary. The parietal

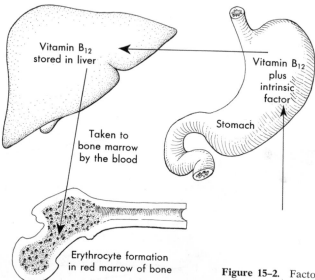

Vitamin B_{12} stored in liver

Vitamin B_{12} plus intrinsic factor

Stomach

Taken to bone marrow by the blood

Erythrocyte formation in red marrow of bone

Figure 15–2. Factors necessary for *maturation* of red blood cells.

cells of the stomach elaborate a substance called the *intrinsic factor*. The function of the intrinsic factor is to promote absorption of vitamin B_{12}, which is present in food. Absorbed vitamin B_{12} is stored in the liver and liberated as needed, carried in the blood to the bone marrow where it functions enzymatically to complete the maturation of the red blood cell. Proteins, several vitamins, folic acid, copper, cobalt, and iron are essential for red cell formation.

The life-span of erythrocytes is about 120 days. Disintegration occurs as the cells age, become more fragile, and rupture as they squeeze through small capillaries, particularly in the spleen. When the spleen is removed, the number of circulating old and abnormal cells is increased.

The average number of erythrocytes in a cubic millimeter of normal blood is given as 5,500,000 to 7,000,000 for men and 4,500,000 to 6,000,000 for women. When the blood is hypoxic, *erythropoietin*, a glycoprotein, appears in blood and stimulates the marrow to produce more red cells. Erythropoietin is thought to be formed from globulin (p. 344) under the influence of an enzyme produced in the kidney.

Hemoglobin. Hemoglobin is comprised of a complex protein molecule named *globin* and a nonprotein portion named *heme* (hematin), which contains iron. One red cell contains several million molecules of hemoglobin. Under normal conditions the adult body produces about 6.25 gm of hemoglobin per day. The destruction of hemoglobin occurs coincidentally with that of the erythrocyte by the reticuloendethelial cells (p. 380) of the spleen. Iron is separated from the heme molecule, enters the circulation, and is reused in formation of hemoglobin. The remainder of the heme molecule is transformed to bilirubin and excreted by the liver in bile. The entire function of hemoglobin depends upon its capacity to combine with oxygen in the lungs and then release it readily in the capillaries of the tissues. In the tissues it picks up carbon dioxide.

In adults 100 ml of normal blood contains on the average between 11.5 and 19 gm of hemoglobin—in males the average is 14 to 18 gm; in females, 11.5 to 16 gm. In children (4 to 13 years old) the average is 12 gm; at birth it is about 17.2 gm of hemoglobin per 100 ml of blood.

Hemolysis. Disruption of the erythrocyte membrane leads to the cell's hemoglobin content going into solution in the plasma. This process is hemolysis, or laking. The resulting colorless erythrocytes are referred to as "ghosts." Hemolysis in the body may result from hypotonic solutions given intravenously, antigen-antibody reactions between the red cell antigens and plasma agglutinins (page 347), or self-produced (autoimmune) antibodies aganist the red cells.

Functions of Erythrocytes. The functions of the erythrocytes include carrying oxygen to the tissues, carrying carbon dioxide from the tissues, and maintaining normal acid-base balance (pH).

In the capillaries of the lungs hemoglobin becomes fully saturated with oxygen, forming oxyhemoglobin. The erythrocytes carry the oxyhemoglobin to the capillaries of the tissues, where they give up the oxygen. Here part of the oxyhemoglobin becomes reduced hemoglobin and is ready to be carried to the lungs for a fresh supply of oxygen. The color of the blood is dependent upon the combination of the hemoglobin with oxygen; when the hemoglobin has its full complement of oxygen, the blood has a bright-red hue, and when the amount is decreased, it changes to a dark-crimson hue.

Polycythemia. The condition in which there is an increase of erythrocytes above the normal is called polycythemia. Conditions associated with cyanosis and residence in high altitudes are usually followed by polycythemia. It is thought that low atmospheric pressure existing in high altitudes decreases the ability of hemoglobin to combine with oxygen, and this reduction of oxygen tends to stimulate the formation of new cells. This result represents the chief benefit anemic people derive from residence in high altitudes.

In shock due to diffusion of plasma to tissues, after profuse perspiration, diarrhea, and loss of body fluids from other causes, there is an apparent (not real) increase in erythrocytes, due to a decreased amount of plasma.

Anemia. This term refers to a deficiency of erythrocytes or a deficiency of hemoglobin in the cells. A deficiency of erythrocytes results from (1) hemorrhage, (2) hemolysis, or (3) inability to produce new erythrocytes due usually to inadequate intake of iron.

Primary anemia, a pernicious anemia, results if the stomach fails to elaborate the intrinsic factor. Such anemias have been treated successfully by introducing purified vitamin B_{12} directly into the body by hypodermic injection.

Leukocytes (White Blood Cells). White cells are granular, translucent, ameboid cells, variable in size. They have been called the mobile units of the reticuloendothelial system. Some of them are formed in the red bone marrow and others are formed in lymphatic tissue.

The number of white cells in a cubic millimeter of blood is from 5000 to 9000 (about 1 white to 700 red). An increase in number is designated as *leukocytosis* and occurs in such infections as pneumonia, appendicitis, or other acute infection. A decrease in the number of leukocytes is designated as *leukopenia*.[1] Physiologic leukocytosis up to 10,000 occurs under normal conditions, such as digestion, exercise, pregnancy, and cold baths. More than 10,000 per cubic millimeter usually indicates pathologic leukocytosis.

White cells are classified according to structure, cytoplasmic granules, and their reaction to dyes. (See Figure 15-3.)

WHITE CELLS

Number	Kind	Size
Granulocytes, 60–70%	Neutrophils	9–12 μm
	Eosinophils	10–14 μm
	Basophils	8–10 μm
Agranulocytes Lymphocytes, 20–30%	Small	7–10 μm
	Large	Up to 20 μm
Monocytes 5–8%	Mononuclear	9–12 μm
	Transitional	

[1] This should not be confused with *leukemia*, a disease characterized by an increase in the white cells of the blood. *Temporary* increases to 20,000 or more after exercise, etc., are thought to be due to changes in circulation.

Granulocytes, or Granular Leukocytes. In the adult, granulocytes are formed in the red marrow of bones, developing from the myeloblast. Their life-span is short, less than 24 hours in many instances. They are destroyed as they become old and fragile, like the erythrocytes, or are killed in counteracting bacterial infection.

1. *Neutrophils,* or polymorphonuclear leukocytes ("polys"), have a nucleus that is lobulated and becomes more so with age, and cytoplasmic granules that stain with neutral dyes. They form from 55 to 65 per cent of the total leukocyte number. Their phagocytic property is extreme for most bacteria. They ingest bacteria (*phagocytosis*), which then are destroyed by the lysosomal enzymes within the leukocyte.

2. *Eosinophils* are similar in size and structure to the neutrophils, but the granules of the cytoplasm are larger and stain with acid dyes such as eosin. Normally they are present in small numbers (2 to 4 per cent), but increase in numbers in allergic conditions, worm infestation, and malaria. It has been suggested that eosinophils phagocytize the products of antigen-antibody reaction.

3. *Basophils* have a polymorphic nucleus, and the granules of the cytoplasm stain with basic dye. They are found in small numbers (½ per cent). Their function is phagocytosis. However, their number increases during the healing process of inflammation and during chronic inflammatory processes. Basophils contain relatively large amounts of histamine (about one half of the histamine content of normal blood).

Agranulocytes, or Agranular Leukocytes. 1. Several types of *lymphocytes* are found in the blood: B-lymphocytes, from bone marrow; T-lymphocytes from the thymus gland; and others derived from lymph nodes and spleen. Lymphocytes include small cells with a round nucleus that are metabolically rather inactive; larger cells with a similar nucleus, and with ribosomes and endoplasmic reticulum; and small lymphocytes which will migrate to connective tissue and become *plasma cells*, capable of protein synthesis.

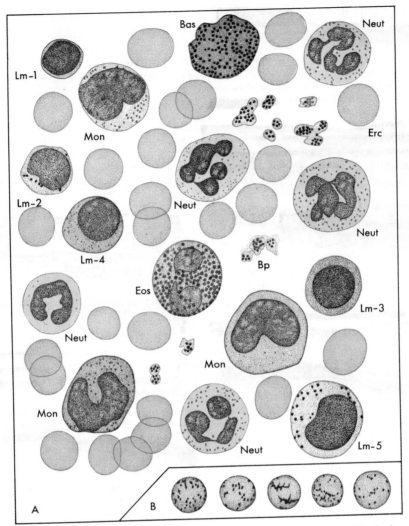

Figure 15–3. *A.* Cells from normal human blood (Wright's stain). *Bas,* basophilic leukocyte; *Bp,* aggregations of blood platelets; *Eos,* eosinophilic leukocyte; *Erc,* erythrocytes; *Lm 1–5,* lymphocytes (*1–3* are small and medium sizes and *4–5* are the less numerous larger forms); *Mon,* monocytes; *Neut,* neutrophilic leukocytes. *B.* Reticulocytes from normal human blood stained with dilute cresyl blue. (Modified from *Bailey's Textbook of Histology,* 13th ed., revised by P. E. Smith and W. M. Copenhaver. Courtesy of Williams & Wilkins Co.)

The number of lymphocytes is high in early life, decreasing from 50 per cent to about 35 per cent of the total leukocyte count at 10 years of age. Some lymphocytes live for only a few days, others live for many years. Their primary function is to assist in body defense through synthesis and release of antibody molecules (immune globulins). (See page 351).

2. *Monocytes* include the large mononuclear and transition types of leukocytes. They are large cells, each with an indented eccentric nucleus, and they can function effectively as phagocytes. Recent studies show that they have potential for tissue growth, development of enzyme systems, and synthesis of protein. The monocyte is formed chiefly in bone marrow, from which they enter the circulation. Many monocytes leave the bloodstream to become fixed to tissues. They are then called *tissue macrophages,* or *histiocytes.* During

the course of inflammation histiocytes divide and multiply, in many instances effectively walling off the infected area.

The proportion of the different classes of leukocytes varies during disease conditions, especially bacterial infections. *Differential counts* have a great practical value in diagnosis. Such a count requires the total number of leukocytes in 1 cu mm of blood and the number of each of the types. A high absolute count with a high neutrophil percentage indicates severe infection and good body resistance.

Leukocyte Movement. Leukocytes are capable of ameboid-type movement, hence the term *wandering* cells. They can squeeze through the walls of capillaries into surrounding tissues by a process called *diapedesis;* it occurs normally but is greatly increased in infection and inflammation. Diapedesis also refers to the migration of erythrocytes through the capillary wall. In microscopic examination of capillary blood flow, leukocytes are seen to hug the walls of the vessel, thus facilitating diapedesis. *Phagocytosis* refers to the engulfing of bacteria and is facilitated by the combination of globulins in the bloodstream, called *opsonins*, which combine with the foreign material and allow the leukocyte to adhere to the surface.

Functions of Leukocytes. In summary, the leukocytes are motile cells in the bloodstream and play a major role in the body defenses against infection and injury. This includes: (1) phagocytosis and destruction of bacteria, (2) production of immune globulins, (3) ingestion of cellular debris at the site of inflammation, and (4) a role as macrophages to become part of the reticuloendothelial system.

Inflammation. When tissues become inflamed as a result of bacterial invasion, or trauma such as a cut, chemicals, or heat, the substance *histamine* is released from most cells in the tissues. This increases the blood flow to the area, as well as increasing the permeability of capillaries. Thus, large amounts of fluid, including fibrinogen, enter the inflamed area. Walling off occurs then as a result of the coagulating effect of fibrinogen on tissue fluid, and macrophage activity.

Neutrophils are attracted to the area by chemotaxis, move by diapedesis through capillary walls, and begin phagocytosis of the injurious agents. Macrophages as well participate in this activity.

The classic symptoms of redness, heat, swelling, and pain are due to irritation by toxins of the bacteria, to increased supply of blood, to engorgement of blood vessels, and to the collection of fluid in the tissues (edema).

Secondary to the inflammatory process is increased production of neutrophils by bone marrow, possibly as a result of a *leukocyte-promoting factor* liberated from the inflamed tissue. The increased production of neutrophils by bone marrow provides large numbers of cells capable of phagocytosis.

As neutrophils and macrophages engulf bacteria in large numbers, the cells may die. After several days a cavity is formed in the center of the inflammation that contains necrotic tissue, dead bacteria and leukocytes, and other material exuded from blood vessels. Such a mixture is termed *pus;* the process is *suppuration. Resolution*, or recovery from inflammation, is concurrent with gradual decrease in size of the inflamed area, and complete healing occurs. If a large area of tissue damage has occurred, healing may require the production of scar tissue (page 82).

Thrombocytes (Blood Platelets). These are fragments of the cytoplasm of large megakaryocytes, a type of cell formed in red marrow. Thrombocytes are disk-shaped bodies about 2 to 4 μm in diameter. The average number is about 400,000 per cubic millimeter of blood.

Thrombocytes are initiators of the process that controls the cessation of bleeding when a blood vessel is injured, as in a cut. Contact with the collagen fibers of the vascular wall changes the form of the platelet: numerous irradiating processes protrude from the surface, and they become sticky so that they stick to the collagen fibers. They secrete adenosine di-

TABLE 15–1
CHARACTERISTICS AND FUNCTIONS OF THE WHITE BLOOD CELLS

Name	Number	Where Formed	Nucleus	Cytoplasm	Motility	Function
Agranular leukocytes						
Lymphocytes		Reticular tis-	Very large, single,	Stains pale blue.	Movement ac-	Form serum
3 mo to 5 yr	52–64%	sue of	generally spher-	Occasional scat-	tive in con-	globulin, both
3 yr to 5 yr	34–48%	lymph	ical, may be in-	tered reddish-	nective tissue.	beta and
5 yr to 15 yr	28–42%	glands and	dented. Sharply	violet granules.	Leave blood-	gamma. Form
Adults	20–30%	nodes;	defined. Stains	Cytoplasm	stream in	antibodies at
		spleen;	blue Paler nu-	abundant	large num-	site of inflam-
		bone mar-	clei		bers, espe-	mation. Mature
		row and			cially in fast-	cell is called
		other lymph			ing	plasma cell
		tissue				
Monocytes	6–8%	Lymph	Single, lobulated	Abundant—	Marked.	Phagocytic prop-
		glands,	or deeply in-	cytoplasm—	Migrate	erties for cell
		spleen	dented, or	stains a gray	readily	debris and for-
			horseshoe	blue	through capil-	eign material
			shaped. Stains		lary walls into	excellent
			blue		the connec-	Young cells most
					tive tissues	active
Granular leuko-	55–65%	In bone mar-	Lobulated—1 to 5	Fine neutrophilic	Marked.	Phagocytic prop-
cytes		row from	or more lobes.	granules in cyto-	Migrate from	erties extreme
Neutrophils		neutrophilic	Stains deep	plasm—pink	bloodstream.	for many bacte-
Filamented		myelocytes	blue. Number	cast	Believed to be	ria. Increased in
types have 2			of lobes is sig-		removed from	infections and
or more lobes			nificant in rela-		bloodstream	inflammatory
in nucleus;			tion to the rela-		by reticulo-	conditions. Rep-
nonfilamented			tive degree of		endothelial	resents degree
types, 1 lobe			maturity of		system	of infection.
			cells			They are
						proteolytic.
						By disintegra-
						tion and rup-
						ture of the cell
						itself become
						pus corpuscles.
Eosinophils	1–3%	In bone mar-	Shape irregular.	Sky-blue tinge	Less marked.	Little phagocytic
		row from	Stains blue, but	with many	Often extra-	action. In-
		eosinophilic	less deeply than	coarse, uniform,	vascular, es-	creased in in-
		myelocytes	neutrophils.	round or oval	pecially in	fection by para-
			Usually 2 lobes	bright-red gran-	fluids under	sites, especially
			in nucleus	ules	the linings of	worms. Number
					respiratory	decreased by
					and digestive	glucocorticoids
					tracts	
Basophils	0.25–0.7%	In bone mar-	Light purple, in-	Mauve color with	Least motile.	Phagocytosis
		row from	dented or	many large	Contain rela-	
		basophilic	slightly lobu-	deep-purple	tively large	
		myelocytes	lated, centrally	granules	amounts of	
			located. Quite		histamine	
			hidden by gran-			
			ules			

phosphate (ADP), which in turn acts on nearby platelets to cause them to become sticky. Thus, a cycle occurs producing a platelet plug for the injured wall, and ces-sation of bleeding. Following this the co-agulation process (page 346) solidifies the plug.

Blood Plasma. Blood plasma is a complex fluid of a clear amber color. It contains a great variety of substances, as might be inferred from its double relation to the cells, serving as it does as a source of nutrition and as a means of removing products of metabolism.

Water. About nine tenths of the plasma is water. This proportion is kept fairly constant by water intake and water output by the kidneys. There are also the continual exchanges of fluid that take place between the blood, intercellular tissue fluid, and the cells. Plasma has a specific gravity of 1.026, viscosity about five to six times greater than water, and pH of 7.4 with an average range of 7.32 to 7.41.

Blood Proteins. The hemoglobin of erythrocytes represents about two thirds of the blood proteins, and plasma proteins, about one third.

Plasma Proteins. There are three major types of proteins in the plasma of circulating blood: serum albumin, serum globulin, and fibrinogen. Albumin is concerned with maintaining oncotic pressure; antibodies of globulins; fibrinogen aids in blood clotting.

Practically all of the serum albumin and fibrinogen are formed in the liver. Globu- lins are formed by the liver and the lymphatic system.

The normal range of concentration of the plasma proteins is from 6.8 to 8.5 gm per 100 ml of plasma. These figures vary with age; in premature infants the concentration is low. There is a gradual increase with age. Normal levels are established at about 18 to 20 months.

Serum albumin forms about 53 per cent of the total plasma proteins. Globulins form about 43 per cent and fibrinogen about 4 per cent of the total plasma proteins. The liver forms plasma proteins rapidly, as much as 100 gm in 24 hours. However, the synthesis of plasma proteins by the liver depends on the concentration of amino acids in the blood. Figure 15–4 shows the electrophoretic pattern of the plasma proteins.

The blood proteins serve to maintain colloidal (oncotic) characteristics of blood and give it viscosity. All of the plasma proteins, but especially serum albumin, are concerned with the regulation of blood volume. They are responsible for the colloidal osmotic pressure, which provides the "pull pressure" of the plasma essential for holding and pulling water from the tissue fluid into the blood vessels. Plasma

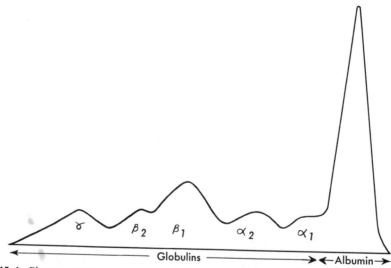

Figure 15–4. Plasma proteins may be separated by electrophoresis, yielding the tracing shown above. Globulins are of three varieties: alpha (α), beta (β), and gamma (γ). The higher the peak on the tracing, the greater the amount.

proteins serve as a source of nutrition for the tissues of the body and contribute to the solution and transport of lipids, fat-soluble vitamins, bile salts, and hormones in the blood by means of adsorption.

Antibodies are found in the serum globulin. Fibrinogen is essential for blood clotting. The plasma proteins aid in the regulation of acid-base balance.

Prothrombin is a plasma globulin, found in the blood in a concentration of about 15 mg per 100 ml. It is formed continually in the liver and is used by the body for the coagulation of blood. Vitamin K is essential for its synthesis.

Heparin is a conjugated polysaccharide that is secreted into the blood continuously by the *mast cells* found in the connective tissue surrounding capillary networks. It is a powerful anticoagulant. Further discussion of this role is found on page 346.

Nutrients. These are the end products resulting from the digestion of food—amino acids, glucose, and neutral fats. Under normal conditions amino acids are present in a small proportion; glucose concentration is from 80 to 120 mg per 100 ml of blood. Temporary increases in these amounts may follow the ingestion of a large quantity of food.

Cholesterol is found in all tissues and body fluids. Its source in the body is (1) from absorption in the intestinal tract from saturated fats (exogenous cholesterol) and (2) from formation in large quantities by the cells, especially liver cells (endogenous cholesterol). It is an essential component of all cells, especially nerve tissue and steroid hormones.

Cholesterol is used by the liver to form cholic acid, which in turn helps to form bile salts; by the adrenal gland to form cortical hormones; by the ovaries to form progesterone and perhaps estrogen; and by the testes to form testosterone. Large quantities of cholesterol are precipitated in the corneum of the skin. Cholesterol and other lipids in the skin help prevent water evaporation.

The *electrolytes* found in the blood are derived from food and from the chemical reactions going on in the body. The most abundant is sodium chloride. (See Chapter 24.)

Gases. Dissolved gases—oxygen, nitrogen, and carbon dioxide—are found in the blood. Oxygen is continuously entering the blood as it circulates through the lungs,

TABLE 15–2
SUMMARY—COMPOSITION OF BLOOD PLASMA

Electrolytes

Cations	mEq/L
Sodium	138–142
Potassium	4–5
Calcium	4.5–5
Magnesium	2
Anions	
Chloride	103
Phosphates	2
Sulfates	1
Bicarbonates	27
Organic acid	6
Protein	16

Other Minerals

Iron ⎫
Copper ⎬ Traces
Iodine ⎭

Enzymes ⎫
Vitamins ⎬ Variable concentrations
Hormones ⎭

Proteins	Gm %
Serum albumin	5–6
Serum globulin	3
Fibrinogen	0.4
Nonprotein Nitrogen	Mg %
Urea	26
Uric acid	3
Creatinine	1.0
Creatine	0.4
Ammonium salts	0.2
Nutrients	Gm %
Glucose	80–120
Lactic acid	7–8
Amino acids	5–6
Fatty acids	368–370
Cholesterol	150–182
Phospholipids	200
Cerebrosidin	15
Lecithin	10–15
Water	90% by volume

Gases

Nitrogen
Oxygen 2% (not very soluble)
Carbon dioxide 60–64% (soluble)

but rapidly attaches to hemoglobin and only a small proportion is found in solution. Similarly, carbon dioxide from the tissues is rapidly changed to carbonic acid and to sodium bicarbonate, so that little is in solution.

COAGULATION

Coagulation Process. Blood drawn from a living body is fluid. It soon becomes viscid and, if left undisturbed, forms a soft jelly. As the cells settle out of the plasma, a pale, straw-colored liquid begins to form on the surface, and finally the entire jelly separates into a firm mass, or *clot*, and a liquid called *blood serum*. If a portion of the clot is examined under a microscope, it is seen to consist of a network of fine needlelike fibers, in the meshes of which are entangled the red and some of the white cells. As the clot shrinks, the red cells are held more firmly by this network; but some of the white cells, owing to their power of ameboid movement, escape into the serum. The needle like fibers are composed of fibrin. The basic steps in the coagulation process are as follows:

1. Thromboplastin and serotonin are released from injured tissues.

2. Thromboplastin initiates a series of chemical reactions that convert prothrombin into thrombin.

3. Thrombin functions as an enzyme to convert fibrinogen into fibrin threads that enmesh platelets, red cells, and plasma to form the clot.

Blood contains the substances antithromboplastin (antithrombin) and antiprothrombin (heparin) concerned with *preventing* the clotting of blood in the blood vessels, and three substances concerned with the clotting of blood. These include (1) fibrinogen, (2) calcium ions, and (3) prothrombin (thrombogen). When blood clots, prothrombin and calcium ions form thrombin, and thrombin changes fibrinogen to fibrin, which is insoluble. The fibrin and the blood cells form the clot.

For the blood to clot, the two substances concerned with the prevention of clotting must be neutralized. These substances are neutralized by thromboplastin, which is set free by the crushed tissue cells, the platelets, or thrombocytes, and the blood corpuscles. This accounts for the fact that blood clots only when tissues are wounded.

Tables 15–3 and 15–4 present the *factors* concerned with the clotting process and where they function.

The time it takes for the blood of human beings to clot is usually about four to six minutes. Estimation of *clotting time* is important as a preliminary to surgery. The normal time varies with the type of test performed.

In normal individuals, when the ear lobe is punctured, the bleeding will stop spontaneously in a very few minutes. The time required for cessation of bleeding is called *bleeding time.* This process is controlled by vascular constriction and perhaps a platelet factor rather than by coagulation alone. Normal bleeding time is about one to four minutes.

Intravascular Clotting. It is well known that clots occasionally form within the blood vessels. The most frequent causes are:

1. Any foreign material, even air, that is introduced into the blood and not absorbed may stimulate the formation of thrombin and a clot.

2. When the internal coat of a blood vessel is injured, as for instance by a ligature or the bruising incidental to operations, the platelets become "sticky" and, together with the coagulation process, form a *thrombus* or clot.

The products of bacteria and other toxic substances may injure the lining of a blood vessel and produce the same result. Inflammation of the lining of a vein is called *phlebitis*.

THROMBUS AND EMBOLUS. A clot that forms inside a blood vessel is called a thrombus, and the condition is called *thrombosis*. A thrombus may be broken up and disappear, but the danger is that it may lodge in the lung or certain parts of the brain, where it blocks circulation and causes instant death. A thrombus that becomes dislodged from its place of formation is called an embolus. Such a condition is called *embolism*.

Hemorrhage. During hemorrhage blood pressure falls and the heart rate is accelerated in an effort to maintain cardiac output. The liver and spleen give up all possible blood to increase venous return. If hemorrhage is not controlled, there is further reduction in arterial

TABLE 15–3
BLOOD COAGULATION FACTORS

Source	Factors	Synonyms and Description
Liver	I	Fibrinogen; a globulin
Liver	II	Prothrombin; an albumin; vitamin K essential for synthesis
Injured tissues and blood during coagulation process	III	Thromboplastin (tissue); extrinsic prothrombin activator
Food	IV	Calcium
Liver	V	Proaccelerin; a labile factor in plasma accelerator globulin (AcG)
	(No six)	
Liver	VII	Proconvertin; a stable factor; a globulin Serum prothrombin conversion accelerator (SPCA)
Source unknown; gene controlling production of VIII is on the X chromosome	VIII	Antihemophilic factor (AHF); a globulin, thromboplastinogen; plasma thromboplastic factor (PTF); platelet cofactor I
Liver	IX	Plasma thromboplastic component (PTC); a globulin; Christmas factor (CF); platelet cofactor II
Liver	X	Stuart-Prower factor found in serum and plasma, not consumed during coagulation, is a globulin
Liver	XI	Plasma thromboplastin antecedent (PTA)
Liver	XII	Hageman factor; when substance is absent, blood does not coagulate in a normal period of time on contact with a glass surface
	XIII	Fibrin-stabilizing factor

blood pressure. Vasoconstriction is marked, the pulse is thready and rapid, the skin clammy and cold, the individual is restless, anxious, and air hungry. Blood flow to the tissues is decreased and the cell needs in relation to oxygen are not met.

The ischemic kidney initiates the secretion of renin, a vasoexcitatory material. Renin in the presence of enzymes converts angiotensin I to angiotensin II, which constricts arterioles and raises blood pressure. In this way blood supply to the kidney is increased. If hemorrhage has not been controlled and pressure continues to fall, the arterioles and precapillary sphincters relax and open and more blood moves into the capillaries. When this occurs, there is danger of irreversible shock due to hemorrhage.

Regeneration of the Blood After Hemorrhage. During hemorrhage it is probable that a healthy individual may recover from the loss of blood amounting to 3 per cent of the body weight. Experiments on animals show that the plasma of blood regains its normal volume within a few hours after a slight hemorrhage and within 24 to 48 hours if much blood has been lost. The number of red cells and hemoglobin is restored slowly, returning to normal after a number of days or even weeks.

BLOOD GROUPS

Red cells contain antigens that cause them to agglutinate in the presence of the specific antibody. These antigens (or agglutinogens) are genetically determined and are of two types: A and B. As shown in Table 15–5, the inheritance is as mendelian dominant and group O is recessive. It is possible to have either A or B antigens in the red cells, to have both (type AB), or to

TABLE 15–4
SUMMARY—STEPS IN THE COAGULATION OF BLOOD

Extrinsic System	Intrinsic System

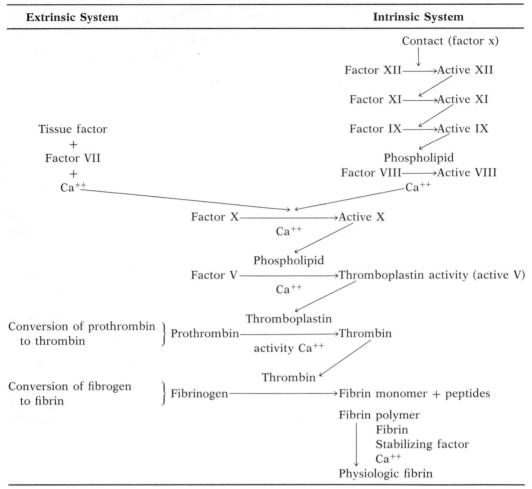

have neither antigen (type O). Agglutinins in the plasma are globulins and are of the opposite type: type A blood contains b agglutinins, type B blood contains a agglutinins, type AB has neither agglutinin, type O has both a and b agglutinins. (See Table 15–6.)

In the embryo agglutinogens are found in the red blood cells about the sixth week. At birth the concentration is about one fifth of the adult level. Normal concentrations are reached during adolescence. Agglutinins as a rule are not present in the blood of the newborn. Specific agglutinins are formed in blood plasma within two weeks and reach the highest concentration at about 10 years of age. Agglutinin concentration is variable in all individuals at all ages. Once established, blood groups do not change—that is, once a group B always a group B.

Blood types are determined by adding whole blood to serum of a known type. If the cells are agglutinated by serum a, agglutinogen A must be present; similarly, agglutination with serum b indicates blood of type B. As indicated in Table 15–7, in practice it is necessary only to use serum a and serum b to test for the four blood groups.

These are the major blood groups; there are many subgroups. The most important are A_1 and A_2. The recognized subgroups are A_1, A_2, A_1B, and A_2B. About 80 per cent of individuals in group A belong to subgroup A_1; 20 per cent belong to subgroup

TABLE 15–5
INHERITANCE OF BLOOD GROUPS*

Mother ♀   Offspring   Father ♂
Ib Ib (group A)   (no recessive gene)   Ia Ia (group B) (no recessive gene)
—Ib Ia—
—Ib Ia—

Possibilities are that all children will be group AB

Ia Ib (group AB)   Offspring   ii (group O) (recessive genes)
—Ib i—
—Ia i—

Possibilities are that children will be group A with recessive genes and group B with recessive genes

Ia i (group B)   Offspring   ii
ii
—Ia i—

Possibilities are that children will be group B with recessive gene and group O

Ib i (group A)   Offspring   ii
ii
—Ib i—

Possibilities are that children will be group A with recessive gene and group O

*Symbols: Ib Ib = group A
Ia Ia = group B
Ia Ib = group AB
ii = group O (recessive)

TABLE 15–6
ABO BLOOD GROUPS

Genotype	Name of Blood Group	Antigens in Red Cells	Antibodies in Serum	Incidence in Caucasians per cent
OO	O	—	Anti-A and anti-B	45
OA or AA	A	A	Anti-B (b)	41
OB or BB	B	B	Anti-A (a)	10
AB	AB	A and B	—	4

A_2; about 60 per cent of individuals in group AB belong to subgroup A_1B; and 40 per cent belong to A_2B.

Cross Matching. Before a blood transfusion is given, as a safety measure, cross matching is done to determine compatibilities. This means that a suspension of red cells from the donor and a small amount of defibrinated serum from the recipient are mixed together to determine whether or not agglutination occurs. A second test is done to cross-match the cells of the recipient to the serum of the donor. If no agglutination occurs, it can be assumed that the blood of the donor and the blood of the recipient are of the same type.

Rh Factor.[2] If the Rh antigen is present on the red blood cell, the individual will be Rh positive. If the antigen is *absent*, the individual will be Rh negative. About 83 per cent of American Caucasians and 93

[2]Rh factor is named for the fact that it was first found in the Rhesus monkey.

TABLE 15–7
BLOOD AGGLUTINATION

		Agglutinins in Serum (Antibodies)			
	Cells	ab	b	a	o
	O	−	−	−	−
Agglutinogens (Antigens)	A	+	−	+	−
	B	+	+	−	−
	AB	+	+	+	−

+ Denotes agglutination.
− Denotes absence of agglutination.
When antigen A meets antibody a, or when antigen B meets antibody b, agglutination takes place.

per cent of American Negros are Rh positive

In giving a blood transfusion, if an Rh-negative person receives Rh-positive blood, the recipient will develop an anti-Rh agglutinin, which may cause hemolysis of Rh-positive blood, if there is a subsequent transfusion of this type. Anti-Rh agglutinins are similar to the a and b agglutinins in their action, as they attach to Rh-positive red blood cells and cause them to agglutinate.

The major Rh types in blood include Rh_o, Rh′, and Rh″ factors. The Rh_o antigen is the one that is strongly antigenic.

According to the Fischer-Race concept, there are three sets of allelic genes: C and c, D and d, E and e; every person inherits a total of three genes from each parent, one gene from each pair. The resultant possible codes of genes gives cde, Cde, cdE, CdE, cDe, CDe, cDE, and CDE. Using all possible combinations, there are 36 different genotypes possible. Persons who have the D (Rh_o) antigen are considered Rh positive and those who do not have the D (Rh_o) antigen are considered Rh negative.

It is believed that most cases (about 90 per cent) of *erythroblastosis fetalis* are caused by the production of anti-Rh agglutinins in the mother's blood (if the mother is Rh negative and the father is Rh positive, the child may be Rh positive).

TABLE 15–8
INHERITANCE OF RH FACTORS

Possibilities—all children will type Rh positive but carry a recessive gene.

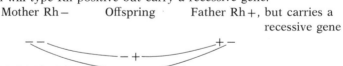

Possibilities—there is a 50–50 chance that children will carry recessive genes and a 50–50 chance that the children will carry a dominant gene for Rh+.

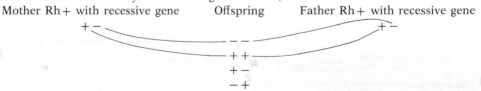

Possibilities—three children will type for Rh positive, but *two* will carry a recessive gene. One child will be Rh negative.

Leakage of agglutinogens through fetal circulation into mother's circulation causes formation of anti-Rh agglutinins, which in turn destroy the red cells of the fetus. However, not all children born of such parents develop hemolytic reactions. Ten per cent of erythroblastosis fetalis is due to ABO incompatibilities.

Immunity

Immune System. The immune system (lymph nodes, thymus, spleen, and lymphoid tissue in the tonsils and intestine) protects the individual from a variety of invading organisms to which the body does not have natural resistance. Lymphocytes are capable of recognizing foreign agents, such as bacteria and viruses, and to defend against their multiplication. The term *antigen* is used to refer to such an agent that can stimulate the production of antibody.

Types of Lymphocytes. In the embryo the lymphocytes are formed in the bone marrow from a stem cell and migrate to either the thymus gland or some other part of the body, possibly the liver or spleen. Those that are preprocessed in the thymus are called T-lymphocytes, or *sensitized* lymphocytes. They are responsible for *cellular immunity*, attaching to and destroying the foreign agent. B-lymphocytes (named for the fact that in the chicken they are preprocessed in the gastrointestinal tract, in the bursa of Fabricius) are responsible for *humoral immunity*, i.e., the production of antibodies which are released into the bloodstream against a foreign agent.

Following the preprocessing, both types of lymphocytes migrate to lymphoid tissue late in development of the embryo, just prior to birth or within the first few months of life. Here they remain. A characteristic of these lymphocytes is their great specificity in recognition of a particular antigen. Basic to this specificity is the concept that there are different populations (*clones*) of lymphocytes, which developed during the preprocessing stage. Each clone is responsive to a particular antigen, and to only that one. If the antigen is not present in the circulation, the lymphocyte clone remains dormant, perhaps throughout life.

Antibody Production. When an antigen enters the system, the B-lymphocyte clone specific to that antigen responds by increasing in size, and it is transformed eventually to the mature plasma cells that secrete the antibody. This antibody is secreted into the lymph and from there it enters the circulation. Some of the B-lymphocytes, however, of the same clone that divides and matures to produce antibody, do not go on to the stage of mature plasma cell but instead divide, increasing the number of B-lymphocytes with the memory of that specific antigen. Thus, on a second exposure to the antigen there is a much more rapid and efficient production of that type antibody (the reason for giving "booster shots" in immunization).

Antibodies belong to the family of proteins in the blood called gamma globulins, or immune globulins (Ig). There are several types of immune globulins, designated by the letters G, A, M, D, and E. IgG and IgM provide the bulk of specific antibodies against infection. IgA antibodies are produced by lymphoid tissue in the gastrointestinal tract and act locally. IgE is responsible for the physiologic manifestations of allergic reactions.

T-Lymphocyte Response. Upon exposure to a specific antigen, the appropriate clone of T-lymphocytes is sensitized to release a variety of chemicals injurious to the antigenic agent. As with the B-lymphocytes, some of the clone of T-lymphocytes reproduce to serve as a "memory bank" and on a second exposure to the specific antigen they greatly speed up the immune response. In contrast to the B-lymphocyte activity, which is located in lymphoid tissue removed from the invasion site and from which antibodies reach the antigenic agent via the circulation, T-lymphocytes travel to the invasion site to release their chemicals locally. A further difference between the B- and T-lymphocytes is in regard to the microorganisms that activate each type. B-lymphocytes are stimulated by bacteria primarily; T-lymphocytes are stimulated by viruses, fungi, and parasites (such as worms). In addition, T-lymphocytes are responsible for destruction of transplanted tissue and organs.

Current efforts to transplant tissue such as the kidney from one individual to another are hampered by the normal immune mechanism. Only grafts from a genetically closely related individual will be tolerated by the recipient. However, recent research indicates that it is possible to "type" individuals on the basis of antigens located on the leukocyte, thus making possible the matching of a recipient with a suitable donor.

Complement System. Antigen-antibody complexes may activate another "system" known as the complement system—a group of plasma proteins. Activated complement proteins enhance vasodilation, capillary permeability, and phagocytosis. They also facilitate neutrophil movement by acting as chemotactic agents. *Chemotaxis* is the attraction of neutrophils by a chemical agent. Thus, the complement system is a powerful aid to the inflammatory response (page 342).

Questions for Discussion

1. What are the functions of:
   a. Erythrocytes?
   b. Leukocytes?
   c. Platelets?
2. What are the functions of plasma proteins?
3. What is the advantage of transfusing whole blood rather than plasma?
4. What physiologic responses occur in a blood donor to replace the blood lost?
5. Why is it necessary to have a laboratory report on cross matching, blood groups, and Rh factors before a transfusion can be done?
6. What is cross matching?
7. What is believed to be the function of the thymus gland?
8. What are antibodies? Are they helpful or harmful? In what ways?

9. Mr. Green was admitted to the hospital with an infected toe. There was a red streak up the leg, associated with large painful masses in the groin. His foot and leg were edematous. He was taken to surgery for drainage of the toe infection and given antibiotics, and the leg was placed at rest.
   a. What was the cause of the red streak in the leg?
   b. The masses in the groin were enlarged lymph nodes. Explain why they were so large and painful.
   c. Where else in the body are lymph nodes found and what is their function?
   d. What were some of the probable causes of the edema in Mr. Green's leg and why was it placed at rest?

Summary

Blood	Description	Color	{ Bright red in arteries	
			Dark red in veins	
		Sticky fluid		
		Specific gravity varies between 1.041 and 1.067		
		Reaction varies from pH 7.35–7.45		
	Composition	Cells, about ½ of volume	Erythrocytes	
			Leukocytes	{ Agranular leukocytes
				Granular leukocytes
			Blood platelets, or thrombocytes	
		Plasma, about ½ of volume; hematocrit 47%		

Erythrocytes, or Red Cells

Description
- Biconcave disks about 7.7 μm in diameter
- Stroma containing protein, lipid substance, and cholesterol, blood group antigens A, B, and Rh antigen
- Contain hemoglobin
- Have no nuclei
- Soft, flexible, and elastic

Number
- Cubic millimeter of blood contains about
  - 5,000,000 for men
  - 4,500,000 for women
  - Varies even in health
- Polycythemia—increase above normal
- Anemia—deficiency of erythrocytes or deficiency of hemoglobin in the cells

Functions
- Contains oxygen and carbon dioxide
- Maintenance of viscosity, pH value, etc.

Hemolysis—loss of hemoglobin from the erythrocyte is called hemolysis

Life cycle
- Before birth—originate in liver, spleen, and red marrow
- After birth may originate in endothelial cells of blood capillaries of the red marrow of bones
- Lose their nuclei before being forced into the circulation, which suggests that their term of existence is short
- Disintegrate
  1. Undergo hemolysis in bloodstream
  2. Destroyed in spleen, lymph nodes, and liver
- Life-span 120 days

Hemoglobin
- Complex protein molecule
- Combines with oxygen to form oxyhemoglobin
- Body forms 6.25 gm per day
- **Function**—transports oxygen to tissue cells and carbon dioxide from tissue cells

Leukocytes, or White Cells

Description
- Nucleated cells
- Larger than erythrocytes

Number
- Cubic millimeter of blood
  - 5000 to 9000
  - Increase = leukocytosis
  - Decrease = leukopenia

Varieties
- Agranular leukocytes
  - Lymphocytes
    - a. Small
    - b. Large
  - Monocytes
    - a. Large mononuclear
    - b. Transitional
- Granular leukocytes
  - a. Polymorphonuclear, or neutrophils
  - b. Eosinophils, or acidophils
  - c. Basophils

Functions
- Different for different types
1. Protect the body from pathogenic bacteria
2. Promote tissue repair
3. Production of immune globulins

Life cycle
- Lymphocytes arise from the reticular tissue of the lymph tissue of the body
- Granular leukocytes arise from cells of bone marrow
- Decreased by
  1. Invading bacteria
  2. Hemorrhage
  3. Formation of granulation tissue or tissue regeneration
- Life-span short—few days

Inflammation

Process
1. Irritation from injury or bacteria
2. Increased blood supply
3. Migration of leukocytes, macrophages
4. Exudation of plasma
5. Walling off of inflamed area

Symptoms
- Redness
- Heat
- Swelling
- Pain
- Loss of function of inflamed part

Inflammation (*cont.*)	**Result**	a. Resolution—white cells destroy bacteria, clear up debris, and return to blood b. Suppuration—bacteria destroy white cells and form pus c. Pus consists of: Bacteria { Dead / Living }; Phagocytes; Disintegrated tissue cells; Exudate from blood vessels
Thrombocytes, or Blood Platelets	**Description**	Disk-shaped bodies Formed from megakaryocytes in bone marrow by fragmentation of the cytoplasm Assist in clotting of blood
Plasma	**Water**—about 99%	
	Blood proteins	Fibrinogen Serum globulin Serum albumin
	Nutrients	Amino acids Glucose Neutral fats Cholesterol
	Salts	Chlorides / Sulfates / Phosphates / Carbonates — of — Sodium / Calcium / Magnesium / Potassium
	Iron	In the form of transferrin
	Other organic substances	Purine bases Urea Uric acid Creatine Creatinine
	Gases	Oxygen Carbon dioxide Nitrogen
	Special substances	Internal secretions—hormones—antithrombin—antiprothrombin—prothrombin—heparin Enzymes
Coagulation (Clotting)	**Process**	Cellular elements of blood and tissues → tissue extract Thromboplastin neutralizes antithrombin and antiprothrombin Prothrombin + calcium ions + thromboplastic substance + platelet accelerator → thrombin Thrombin + serum activator → active thrombin Active thrombin + fibrinogen + platelet factor → insoluble fibrin Fibrin + cells of blood → *clot*
	Serum—blood minus clotting elements	
Bleeding Time	**Definition**—Time required for cessation of bleeding after an injury (1 to 4 min) **Process**—Vasoconstriction and a platelet factor	
Intravascular Clotting	**Theory to account for absence of clotting**	Absence of tissue extracts Presence of antithrombin and antiprothrombin
	Causes	Any foreign material introduced into blood and not absorbed will stimulate clotting Injury to internal coat of blood vessels
	Thrombus—name given to clot that forms inside vessel	
	Embolus	A thrombus that has become dislodged from place of formation
Regeneration of Blood After Hemorrhage		Plasma is regenerated rapidly, red cells within a few days or weeks
Immunity	**Immune System**	Lymph nodes, thymus, spleen, scattered lymphoid tissue
	Antigen	Foreign protein that stimulates production of antibodies
	Antibody	Globulins that destroy or inactivate antigen Usually protective
	Types of Lymphocytes	T-lymphocytes are preprocessed in embryo in thymus B-lymphocytes are preprocessed in bone marrow

| Immunity (*cont.*) | Immune Response | Bacterial antigen stimulates B-lymphocytes to produce globulins harmful to bacteria; these enter circulation
T-lymphocytes stimulated by viruses, fungi, and parasites to produce substances harmful to antigen; these are produced locally in tissue. T-lymphocytes destroy transplanted tissue
Lymphocytes have "memory" of antigen; second exposure results in rapid production of antibody |
| | Complement | Group of plasma proteins aiding in immunity may be used in the antigen-antibody combination; also cause vasodilation, aid phagocytosis |

Additional Readings

BURNET, M.: Mechanisms of immunity. *Sci. Amer.*, **204**:58–67, January, 1961.

HAM, A. W.: *Histology*, 7th ed. J. B. Lippincott Co., Philadelphia, 1974, Chapters 10, 11, and 12.

JERNE, N. The immune system. *Sci. Amer.*, **229**:52–60, July, 1973.

MAYER, M.: The complement system. *Sci. Amer.*, **229**:54–66, November, 1973.

Ross, R.: Wound healing. *Sci. Amer.*, **220**:40–50, June, 1969.

SELKURT, E. E.: *Basic Physiology for the Health Sciences.* Little, Brown & Co., Boston, 1975, Chapter 15.

VANDER, A. J.; SHERMAN, J. H.; and LUCIANO, D. S.: *Human Physiology: The Mechanisms of Body Function,* 2nd ed. McGraw-Hill Book Co., New York, 1975, pp. 228–29 and Chapter 15.

CHAPTER 16

The Circulatory System

Chapter Outline

The cells of the body are directly dependent on the constantly moving blood for supply of nutrients and for removal of cell secretions and waste products. It is at the capillary level that this physical movement of substances occurs. Transportation of blood to the capillaries is the work of the arteries; movement of blood to the heart is the work of the veins. In addition, the lymphatic system returns tissue fluid to the circulation.

The circulatory system includes the heart and blood vessels (arteries, capillaries, veins) and is a closed system through which blood continuously flows. The heart will be discussed in the next chapter; the *vascular system*, through which blood flows, and the *lymphatic* system, which transports *lymph*, are presented in this chapter.

Vascular System

Although there are similarities in structure of the blood vessels, certain structural variations are specific to the arteries, the capillaries, and the veins, and to their specific functions. Arteries contain blood under very high pressure and are therefore elastic, muscular, and thick walled. Capillaries must permit diffusion of material through their walls, and they have a single layer of cells in their walls. Veins contain blood under a low pressure, and their walls are much thinner than those of the arteries. Each of these three are discussed in greater detail in the following pages.

Arteries. Arteries are composed of three coats (Figure 16–1):

1. An inner coat (*tunica intima*) consists of three layers—a layer of endothelial cells, a layer of delicate connective tissue, which is found only in vessels of considerable size, and an elastic layer consisting of a membrane or network of elastic fibers.

2. A middle coat (*tunica media*) consists mainly of smooth muscle fibers with various amounts of elastic and collagenous tissue. In the larger arteries elastic fibers form layers that alternate with the layers of muscle fibers. In the largest arteries

Figure label:
Tunica media   Tunica intima

Tunica externa

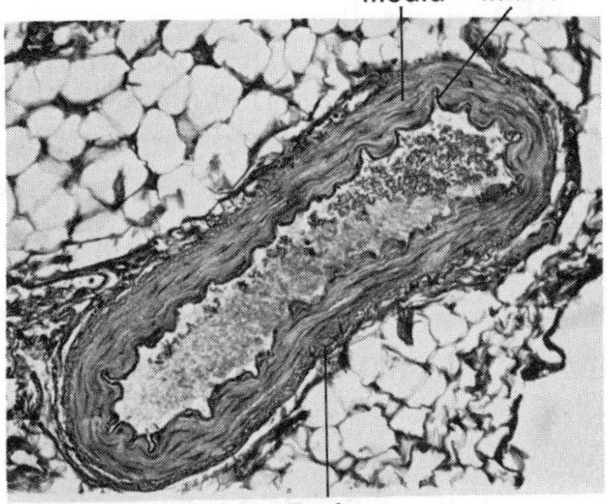

Figure 16–1. Cross section of elastic artery (photomicrograph).

connective tissue fibers have also been found in this coat.

3. The external coat (*tunica externa*, or *adventitia*) is composed of loose connective tissue in which there are scattered smooth muscle cells or bundles of cells arranged longitudinally. In all but the smallest arteries this coat contains some elastic tissue. The structure and relative thickness vary with the size of the artery.

The great extensibility of the arteries enables them to receive the additional amount of blood forced into them at each contraction of the heart. Elasticity of arteries serves as a buffer to the large volume of blood forced into the system by the heartbeat. If these vessels were rigid (as is true in arteriosclerosis), the systolic blood pressure would be markedly increased.

The strength of an artery depends largely on the external coat; it is far less easily cut or torn than the other coats and serves to resist undue expansion of the vessel.

The arteries do not collapse when empty, and when an artery is severed, the orifice remains open. The muscular coat, however, contracts somewhat in the region of the opening, and the elastic fibers cause the artery to retract a little within its sheath, so as to diminish its caliber and permit a blood clot to plug the orifice. This property of a severed artery is an important factor in the arrest of hemorrhage.

Most large arteries are accompanied by a *nerve* and *one or two veins*, all surrounded by a sheath of connective tissue, which helps to support and hold these structures in position (see Figure 16–5).

Classes of Arteries. Arteries may be divided into three classes: (1) elastic arteries, (2) muscular arteries, and (3) arterioles.

Elastic arteries include those large arteries leading directly from the heart (the aorta and pulmonary artery); they are also called *conducting arteries* because they conduct blood from the heart to the muscular arteries. These arteries are very large and may measure 3 cm in diameter. The middle coat contains a large amount of elastic tissue that permits the artery to expand as the blood is pumped in from the heart, and to passively contract when the cardiac contraction has ceased, and blood flow into the artery is momentarily interrupted.

Muscular arteries are of medium size, with a middle coat that is chiefly muscular. They are often called *distributing arteries*, for they distribute the blood to the various organs, and by constriction and dilation variously regulate the flow of blood according to the activity and need of the organ or tissue.

Arterioles are also muscular arteries but of a much smaller size, as the name implies. The relatively small lumen and thick, muscular wall facilitate regulation of the blood pressure of the body in general, and of the capillaries specifically. It is important that the arterial pressure be greatly reduced before the blood flows in to the capillary bed, because of the fragile walls of capillaries. This reduction of pressure is made possible by the structure and action of the arterioles.

Blood Supply of Arteries. The blood flowing through the arteries nourishes only the inner coat. Middle and external coats are supplied with small arteries, capillaries, and veins, the *vasa vasorum* (blood vessels of the blood vessels).

Division of Arteries. The way in which the arteries divide varies. (1) An artery may give off several branches in succession and still continue as a main trunk, e.g., the thoracic or abdominal portion of the aorta. (2) A short trunk may subdivide into several branches at the same point, e.g., the celiac artery. (3) An artery may divide into two branches of nearly equal size, e.g., the division of the aorta into the two common iliacs. Arteries usually occupy locations protected against accidental injury or the effects of local pressure. Arteries usually pursue a fairly straight course, but in some parts of the body they are *tortuous*. The external maxillary (facial) artery, both in the neck and on the face, and the arteries of the lips (inferior

and superior labial) are extremely tortuous and thereby accommodate themselves to the varied movements in speaking, laughing, and turning the head.

Anastomosis of Vessels. The distal ends of arteries unite at frequent intervals, when they are said to anastomose. Such anastomoses permit free communication between the currents of the blood, tend to obviate the effects of local interruption, and promote equality of distribution and of pressure. Anastomoses occur between the larger as well as the smaller arteries. Where great activity of the circulation is necessary, as in the brain, two branches of equal size unite; e.g., the two vertebral arteries unite to form the basilar (Figure 11–20, page 255). In the abdomen the intestinal arteries have frequent anastomoses between their larger branches. In the limbs, anastomoses are most numerous around the joints, the branches of the arteries above uniting with branches from the arteries below (Figure 16–2).

Anastomoses are of importance to the surgeon. By their enlargement, a collateral circulation is established after an artery is ligated. This means that subsidiary vascular channels, which are present in the circulatory network, form a secondary circulation through a part. The effectiveness with which these new channels transport blood varies.

A *plexus*, or network, is formed by the anastomosis of a number of arteries in a limited area. *Arteriovenous anastomoses* occur in some regions, e.g., lips, nail beds, and palm of the hand.

Vasomotor Nerves. The muscular tissue in the walls of the blood vessels is well supplied with nerve fibers, chiefly from the sympathetic portion of the autonomic system. These nerve fibers are called *vasomotor* and are divided into two sets: (1) vasoconstrictor and (2) vasodilator. A center in the medulla oblongata (vasoconstrictor center) is constantly sending impulses to the vessels, thus keeping them in a state of tone. The vasoconstrictor center is a *reflex center* and is connected with afferent fibers coming from all parts of the body. Vasoconstrictor fibers are *sympa-*

Figure 16-2. Anastomosis of arteries. *A.* Diagram showing types of branching. *B.* Scanning electron micrograph of inner view of anastomosis. (Courtesy of Dr. D. G. Gomez, Cornell University Medical College, New York, N.Y.)

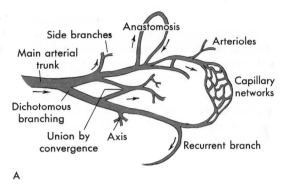

A

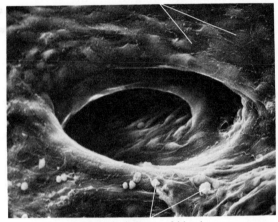

Endothelial cells

Formed elements of blood

B

thetic and are widely distributed to arteries and arterioles. They mediate constriction of vessels, and by tonic action speed of blood flow is controlled. Vasodilator nerve fibers have several origins and are found on the sympathetic, parasympathetic, and somatic sensory nerves. There is no direct evidence that they are tonically active, but they appear to "discharge selectively" when a local increase in blood flow is needed.

There is a diffuse network of sympathetic nerve fibers in the adventitia of all arteries, called the *periarterial plexus.* Nerve fibers are also present in the muscular coat. Arterioles are directly and completely under *nervous* control. Pressure from increased volume, exerted on the bloodstream in the muscular arteries, causes relaxation of the arterioles and more blood can move through to the cap-

illary bed. The exact function of vasodilator nerve fibers is not well understood. Sudden, widespread relaxation of arterioles lowers blood pressure by decreasing peripheral resistance, and shock may result.

Capillaries. The capillaries are exceedingly minute vessels that average about 7 to 9 μm in diameter. They connect the arterioles (smallest arteries) with the venules (smallest veins) (see Figures 16-3 and 16-4).

Structure. Capillaries are composed of endothelium which is continuous with the endothelium lining the arteries, veins, and heart. They vary somewhat in structure in different types of tissue, but there are three main types: *continuous* capillaries, *fenestrated* capillaries, and *sinusoids.*

In continuous capillaries, found in all types of muscle, in connective tissue and in the central nervous system, fluid leaves and enters between the borders of the endothelial cells at the junctional complex. In fenestrated capillaries, found in the renal glomeruli, endocrine glands, and the lamina propria of the intestines, movement of fluid occurs not only at the junctional complexes but also at special *fenestrae*, minute pores in the membrane. These pores are covered with an exceedingly thin layer of cytoplasm; therefore fluid and dissolved substances enter and leave more readily at these sites. Sinusoids are enlarged capillaries found in the liver, spleen, and bone marrow. Their shape is often not cylindrical but conforms to the arrangement of cells in the surrounding tissue, e.g., the cord arrangement of liver cells.

Capillaries have a thin basement membrane; in certain tissues special cells, *pericytes* (Rouget cells), surround the capillary. Their function is uncertain.

Distribution. The capillaries communicate freely with one another and form interlacing networks of variable form and size in the different tissues. All the tissues, with the exception of the cartilages, hair, nails, cuticle, and cornea of the eye, are traversed by networks of capillary vessels.

The capillary diameter is so small that the blood cells often must pass through them in single file, and very frequently the cell is larger than the caliber of the vessel and becomes distorted as it passes through. In many parts the capillaries lie so close together that a pin's point cannot be inserted between them. They are most abundant and form the finest networks in those organs where the blood is needed for purposes other than local nutrition, such as, for example, *secretion or absorption.*

Function. It is in the capillaries that the exchange of materials occurs, and the object of the vascular mechanism is to cause the blood to flow through these vessels in a steady stream. It has been estimated that there are about 7000 sq meters of blood capillaries in the adult body. This gives a large area for exchange of substances between the blood and tissue fluid. In the glandular organs the capillaries supply the substances requisite for secretion; in the ductless glands they also take up the products of secretion; in the alimentary canal they take up some of the digested food; in the lungs they absorb oxygen and give up carbon dioxide; in the kidneys they discharge the waste products collected from other parts; all of the time, everywhere in the body, through their walls an interchange is going on which is essential to the life of the body. The greater the metabolic activity of the tissue, the denser the capillary nets.

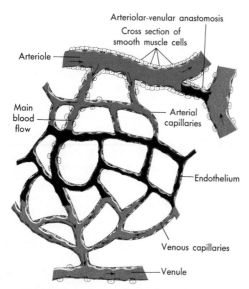

Figure 16–3. Diagram of capillary bed showing arteriole and venule. *Arrows* indicate direction of blood flow.

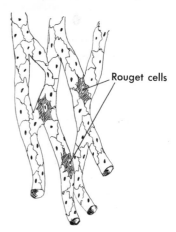

Figure 16–4. Capillary network magnified.

Figure 16–5. *A.* Cross section through an artery, vein, and nerve. *B.* Photomicrograph of artery and vein in cross section. Note the differences in the walls of the artery and vein.

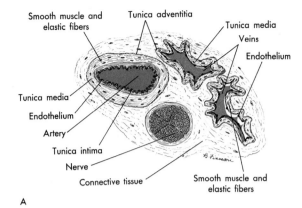

A

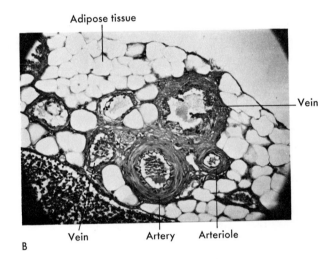

B

Veins. The veins return blood to the heart and are formed by the confluence of the capillaries. The structure of the veins is similar to that of the arteries. They have three coats: (1) an inner endothelial lining, (2) a thin muscular layer (which is absent in small veins), and (3) an external layer of areolar connective tissue. The main differences between the veins and arteries are: (1) the middle coat is not as well developed and not as elastic in the veins; (2) many of the veins are provided with valves; (3) the walls of veins are much thinner than those of arteries and hence tend to collapse when not filled with blood. In general, the total diameter of the veins returning blood from any organ is at least twice the diameter of the arteries carrying blood to the organ.

Valves. The valves are semilunar folds of the internal coat of the veins and usually consist of two flaps, rarely one or three.

The convex border is attached to the side of the vein, and the free edge points toward the heart. Their function is to prevent reflux of the blood and keep it flowing in the right direction, i.e., toward the heart.

If for any reason the blood in its onward course toward the heart is driven backward, the refluent blood, getting between the wall of the vein and the flaps of the valve, will press them inward until their edges meet in the middle of the channel and close it. The valves are most numerous in the veins where reflux is most likely to occur, i.e., the veins of the extremities. For the same reason a greater number are found in the lower than in the

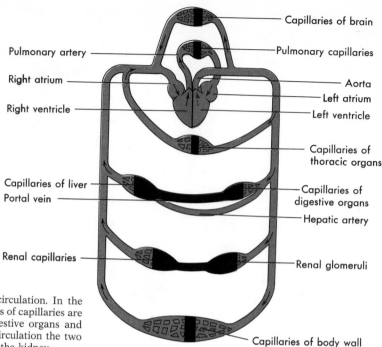

Figure 16–6. Diagram of circulation. In the *portal* circulation the two sets of capillaries are in different organs, the digestive organs and the liver, and in the *renal* circulation the two sets are in the same organ, the kidney.

upper limbs. They are absent in many of the small veins, in the large veins of the trunk, and in veins not subjected to muscular pressure. The veins, like the arteries, are supplied with blood vessels.

It must be remembered that, although the arteries, capillaries, and veins each have the distinctive structure described, it is difficult to draw the line between the arteriole and the large capillary and between the large capillary and the venule. The veins, on leaving the capillary networks, only gradually assume their several coats, and the arteries dispense with their coats in the same imperceptible way as they approach the capillaries.

Arterial System

Aorta. The aorta is the main trunk of the arterial system. In its course the aorta forms a continuous trunk, which gradually diminishes in size from its commencement to its termination. It gives off large and small branches along its course.

The aorta is called by different names throughout its length: (1) the ascending aorta, (2) the arch of the aorta, and (3) the descending aorta, which (a) above the diaphragm is referred to as the thoracic aorta and (b) below the diaphragm is called the abdominal aorta.

Ascending Aorta. This is short, about 5 cm (2 in.) in length, and is contained within the pericardium. The only branches of the ascending aorta are the right and left *coronary arteries,* which are described in the next chapter.

Aortic Arch. This extends from the ascending aorta upward, backward, and to the left in front of the trachea, then backward and downward on the left side of the body of the fourth thoracic vertebra, where it becomes continuous with the descending aorta. Three branches are given off from the arch of the aorta—the *brachiocephalic,* the *left common carotid,* and the *left subclavian* arteries. Branches of these arteries supply the head and the upper extremities.

The *brachiocephalic* artery arises from

the right upper surface of the arch, and ascends obliquely toward the right until, reaching a level with the upper margin of the clavicle, it divides into the right common carotid and right subclavian arteries.

Descending Aorta. This extends from the body of the fourth thoracic vertebra to the body of the fourth lumbar vertebra.

The *thoracic aorta* is comparatively straight and extends from the fourth thoracic vertebra on the left side to the aortic opening in the diaphragm in front of the last thoracic vertebra. Branches from the thoracic aorta supply the body wall of the thoracic cavity and its viscera.

The *abdominal aorta* commences at the aortic opening of the diaphragm in front of the lower border of the last thoracic vertebra and terminates below by dividing into the two common iliac arteries. The bifurcation usually occurs opposite the body of the fourth lumbar vertebra, which

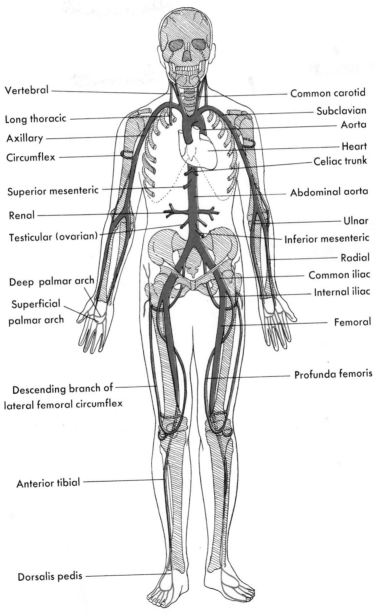

Vertebral — — Common carotid
Long thoracic — — Subclavian
— Aorta
Axillary —
Circumflex — — Heart
— Celiac trunk
Superior mesenteric —
— Abdominal aorta
Renal —
— Ulnar
Testicular (ovarian) — Inferior mesenteric
— Radial
— Common iliac
Deep palmar arch — Internal iliac
Superficial palmar arch — Femoral

— Profunda femoris
Descending branch of lateral femoral circumflex

Anterior tibial —

Dorsalis pedis —

Figure 16–7. Arterial system.

corresponds to a spot on the front of the abdomen slightly below and to the left of the umbilicus. Branches from the abdominal aorta supply the body wall of the abdominal cavity and the viscera it contains.

Arteries of the Head and Neck. The principal arteries are the two common carotids and the vertebral (Figure 11–19, page 254).

Arteries of the Thorax. The branches derived from the thoracic aorta are numerous but small, and the consequent decrease in size in the diameter of the aorta is not marked.

These branches may be divided into two sets: (a) the visceral, or those that supply the viscera, and (b) the parietal, or those that supply the walls of the chest cavity.

Visceral Group	Parietal Group
Pericardial arteries	Intercostal arteries
Bronchial arteries	Subcostal arteries
Esophageal arteries	Superior phrenic
Mediastinal arteries	arteries

Pericardial arteries are small and are distributed to the pericardium.

Bronchial arteries extend to the lungs. They vary in number, size, and origin. There are two left bronchial arteries, which arise from the thoracic aorta, and one right bronchial artery, which arises from the first aortic intercostal or from the upper left bronchial. Each vessel runs along the back part of the corresponding bronchus, dividing and subdividing along the bronchial tubes, supplying them and the cellular tissue of the lungs.

Esophageal arteries are four or five in number; they arise from the front of the aorta and form a chain of anastomoses along the esophagus. They anastomose with the esophageal branches of the thyroid arteries above and with ascending branches from the left gastric and the left inferior phrenic arteries below.

Mediastinal arteries are numerous small arteries that supply the nodes and areolar tissue in the posterior mediastinum.

Intercostal arteries are usually nine in number on each side; they arise from the back of the aorta and are distributed to the

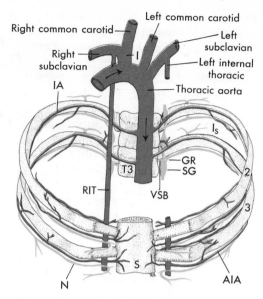

Figure 16–8. Internal thoracic artery and its branches to the intercostal muscles. *2* and *3*, Second and third ribs; *AIA*, anterior intercostal artery; *GR*, gray ramus; *I*, brachiocephalic artery; *IA*, intercostal artery; I_s, intercostal space; *N*, nerve; *RIT*, right internal thoracic artery; *S*, sternum; *SG*, sympathetic ganglion; *T3*, third thoracic vertebra; *VSB*, ventral somatic branch of spinal nerve.

lower nine intercostal spaces. Each intercostal artery is accompanied by a vein and a nerve, and each one gives off numerous branches to the muscle and skin (Figure 7–8, page 156) and to the vertebral column and its contents.

Subcostal arteries lie below the last ribs and are the lowest pair of branches derived from the thoracic aorta.

Superior phrenic arteries are small. They arise from the lower part of the thoracic aorta and are distributed to the posterior part of the upper surface of the diaphragm.

Arteries of the Pulmonary System. The blood vessels of the pulmonary system include (1) the *pulmonary artery* and all its branches, (2) the *capillaries* that connect these branches with the veins, and (3) the *pulmonary veins*. (See Chapter 19.)

Blood Vessels of the Systemic System. These consist of (1) the *aorta* and all the arteries that originate from it, including the terminal branches called arterioles;

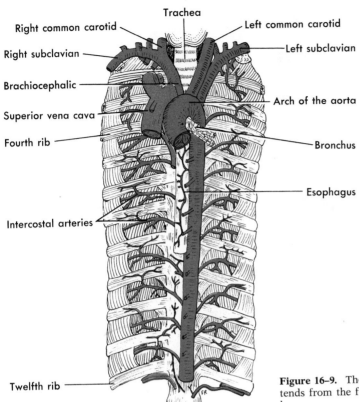

Right common carotid
Trachea
Left common carotid
Right subclavian
Left subclavian
Brachiocephalic
Superior vena cava
Arch of the aorta
Fourth rib
Bronchus
Esophagus
Intercostal arteries
Twelfth rib

Figure 16–9. Thoracic aorta. The thoracic aorta extends from the fourth to the twelfth thoracic vertebrae.

(2) the capillaries that connect the arterioles and venules, and (3) all the venules and veins of the body that empty into the superior and inferior venae cavae and then into the heart, as well as those that empty directly into the heart (coronary veins).

Arteries of the Abdomen. The branches derived from the abdominal aorta may be divided into two groups:

Visceral Branches	Parietal Branches
Celiac (celiac axis)	Inferior phrenics
Superior mesenteric (page 469)	Lumbars
Suprarenals (page 322)	Middle sacral
Renals (page 545)	
Internal spermatics (male) Ovarian (female) (page 586)	
Inferior mesenteric (page 472)	

The visceral branches are discussed with the organs supplied by them.

Inferior phrenics (two) arise from the aorta or celiac artery. They are distributed to the undersurface of the diaphragm.

Lumbar arteries, usually four on each side, are analogous to the intercostals. They arise from the back of the aorta opposite the bodies of the upper four lumbar vertebrae. These arteries distribute branches to the muscles and skin of the back. A spinal branch enters the spinal canal and is distributed to the spinal cord and its membranes.

The *middle sacral artery* arises from the posterior part of the abdominal aorta and descends in front of the fourth and fifth lumbar vertebrae and the sacrum and ends on the pelvic surface of the coccyx. It supplies the sacrum, sacral canal, coccyx, gluteus maximus, and the rectum.

Arteries of the Pelvis. At about the fourth lumbar vertebra the aorta divides into two

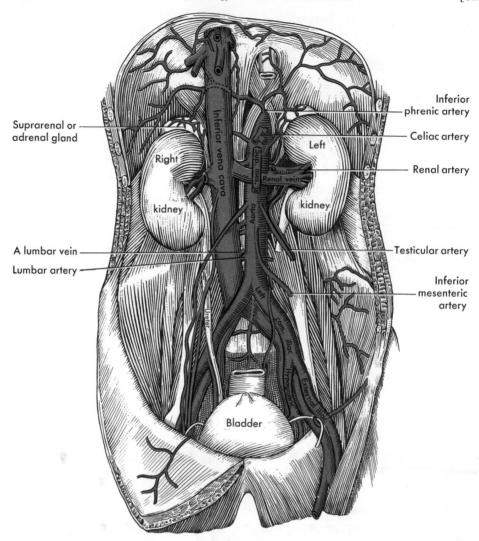

Figure 16–10. The abdominal aorta and inferior vena cava. The abdominal aorta bifurcates into the right and left common iliac arteries opposite the fourth lumbar vertebra.

common iliac arteries. These arteries divide into the hypogastric, or internal, which sends branches to the pelvic walls, pelvic viscera, external genitalia, the buttocks, and the medial side of each thigh. The external iliacs descend and supply the muscles of the thigh, and its branches supply the leg and foot. (See page 191.)

Arteries of the Upper Extremity. The subclavian artery is the first portion of a long trunk that forms the main artery of each upper limb. Different portions are named in terms of the regions through which they pass, viz., *subclavian, axillary,* and *brachial,* which divides into the *radial* and *ulnar* arteries. The right *subclavian* arises at the division of the *brachiocephalic* and the left *subclavian* from the arch of the aorta.

Branches of the subclavian go to the neck and thorax:

The *thyrocervical* sends branches to the thyroid, trachea, esophagus, and muscles of the neck and scapula.

The *internal thoracic* artery extends downward under the costal cartilages to the level of the sixth intercostal space, where it branches into the musculophrenic and superior epigastric arteries. It sends branches to the mammary glands, the diaphragm, areolar tissue and lymph nodes of the mediastinum, intercostal muscles, pericardium, and abdominal muscles.

The *costocervical* sends branches to the upper part of the back, the neck, and the spinal cord and its membranes.

Arteries of the Lower Extremity. The external iliac arteries form a large continuous trunk that extends downward in the lower limb and is named, in successive parts of its course, the *femoral*, the *popliteal*, and the *posterior tibial*. The *femoral artery* lies in the upper three fourths of the thigh, its limits being marked above by the *inguinal ligament* and below by the opening in the adductor magnus muscle. It then becomes the popliteal artery. In the first part of its course the artery lies along the depression on the inner aspect of the thigh, known as the femoral triangle (Scarpa's triangle).[1] Here the pulsation of

the artery may be felt and circulation through the vessel may be easily controlled by pressure (Figure 16–7, page 363). Branches from the femoral artery supply the abdominal walls and external genitalia, and a descending branch, the *lateral circumflex*, anastomoses with branches of the popliteal to form the *circumpatellar anastomosis*, which surrounds the knee joint. As it descends it becomes the *popliteal artery*, then the *posterior tibial artery*, and gives off other branches. At the ankle it divides into the *medial* and *lateral plantar* arteries, which supply the structures of the foot.

The *peroneal artery* is a large branch of the posterior tibial and supplies the structures of the medial side of the *fibula* and *calcaneus*. The *anterior tibial* is a small branch of the popliteal artery and extends along the front of the leg and at the ankle becomes the *dorsalis pedis* artery. The dorsalis pedis anastomoses with branches from the posterior tibial and supplies blood to the foot (Figure 16–7).

Venous System

Arteries begin as *large vessels* that gradually become *smaller* until they end in arterioles, which merge into capillaries. The *veins* begin as small branches, the *venules*, at first scarcely distinguishable from *capillaries*, which unite to form larger and larger vessels. They differ from arteries in their large size, greater number, thinner walls, and the presence of valves in many of them.

Veins of the Head and Neck. The *external jugulars*, one on each side, which return blood from the parotid glands, facial mus-

cles, exterior cranium, and other superficial structures, terminate in the subclavian veins.

The *internal jugular veins* are continuous with the lateral sinuses and begin in the jugular foramen at the base of the skull. They descend on either side of the neck and unite first with the external carotid, then with the common carotid, and join the subclavian at a right angle to form the brachiocephalic vein. They *receive* blood from the veins and sinuses of the cranial cavity, the superior part of the face, and the neck (See Figures 16–11 and 16–12.)

Veins of the Chest. The *brachiocephalic vein* is formed by the union of the *subclavian* and *internal jugular* veins. It receives blood from the head, neck, mammary

[1]The femoral triangle (Scarpa's triangle) corresponds to the depression just below the fold of the groin. Its apex is directed downward. It is bounded above by the inguinal ligament, and the sides are formed laterally by the sartorius muscle and medially by the adductor longus. Antonio Scarpa, Italian anatomist (1752–1832).

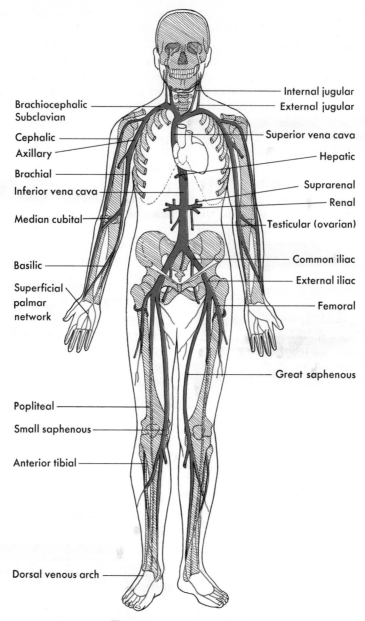

Brachiocephalic
Subclavian

Cephalic

Axillary

Brachial

Inferior vena cava

Median cubital

Basilic

Superficial
palmar
network

Popliteal

Small saphenous

Anterior tibial

Dorsal venous arch

Internal jugular

External jugular

Superior vena cava

Hepatic

Suprarenal

Renal

Testicular (ovarian)

Common iliac

External iliac

Femoral

Great saphenous

Figure 16–11. Venous system.

glands, and upper part of the thorax. It empties into the superior vena cava.

The *internal thoracic veins* receive tributaries corresponding to the branches of the artery. They unite to form a single trunk and end in the *brachiocephalic* vein.

The *superior vena cava* is formed by the union of the right and left brachiocephalic veins and opens into the *right atrium*, opposite the third right costal cartilage.

A supplementary channel between the inferior and superior venae cavae is formed by the *azygos veins*. There are three, and they lie along the front of the vertebral column. In case of obstruction these veins form a channel by which blood can be returned from the lower part of the body to the superior vena cava.

The *azygos vein* (right, or major, azygos) begins opposite the first or second lumbar

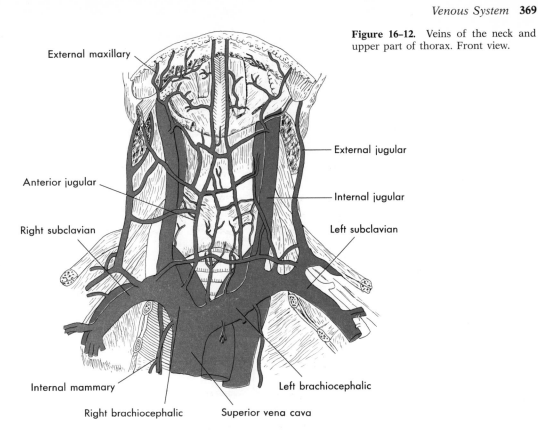

External maxillary

Anterior jugular

Right subclavian

Internal mammary

Right brachiocephalic

External jugular

Internal jugular

Left subclavian

Left brachiocephalic

Superior vena cava

Figure 16–12. Veins of the neck and upper part of thorax. Front view.

vertebra as the *right ascending lumbar vein* or by a branch of the *right renal vein* or from the *inferior vena cava.* (The lumbar veins empty into the inferior vena cava. They correspond to the lumbar arteries given off by the abdominal aorta and return the blood from the muscles and skin of the loins and walls of the abdomen.)

The azygos vein ascends on the right side of the vertebral column to the level of the fourth thoracic vertebra, where it arches over the root of the right lung and empties into the superior vena cava.

The hemiazygos vein (left lower, or minor, azygos) begins at the left lumbar or renal vein. It ascends on the left side of the vertebral column, and at about the level of the ninth thoracic vertebra it connects with the right azygos vein. It receives the lower four or five intercostal veins of the left side and some esophageal and mediastinal veins.

The *accessory hemiazygos vein* (left upper azygos) connects above with the

highest left intercostal vein and opens below into either the azygos or the hemiazygos. It receives veins from the three or four intercostal spaces between the highest left intercostal vein and highest tributary of the hemiazygos; the left bronchial vein sometimes opens into it.

The azygos veins return blood from the intercostal muscles to the superior vena cava. The internal thoracic veins are venae comitantes for the internal thoracic arteries and are tributaries of the right and left brachiocephalic veins.

Bronchial Veins. A bronchial vein is formed at the root of each lung and returns the blood from the larger bronchi and from the structures at the root of the lung; that of the right side opens into the azygos vein near its termination. (See Chapter 19.)

Veins of the Abdomen and Pelvis. *External iliac veins* are continuations of the

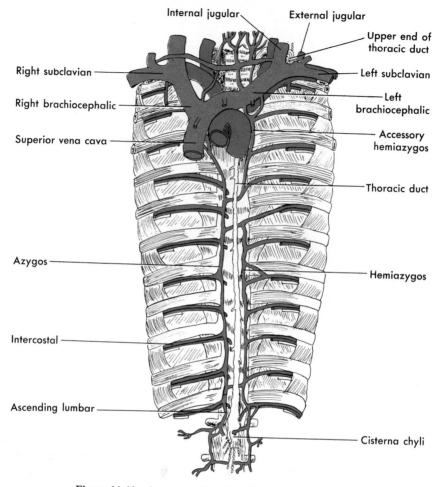

Figure 16–13. Azygos and intercostal veins; thoracic duct.

femoral veins. The iliac veins are formed by the union of veins corresponding to the branches of the hypogastric arteries.

The *internal* (hypogastric) and *external iliac* veins unite to form the *common iliacs*. The *common iliacs* extend from the base of the sacrum to the fifth lumbar vertebra and then unite to form the *inferior vena cava*. The *inferior vena cava* begins at the junction of the two common iliacs and ascends along the right side of the aorta, perforates the diaphragm, and terminates by entering the right atrium of the heart. It receives veins having the names of the parietal and visceral branches of the abdominal aorta. These veins are the lumbar, renal, suprarenal, inferior phrenic, hepatic, and right spermatic or ovarian.

There are a few exceptions:
1. The right suprarenal vein empties into the inferior vena cava; the left empties into the left renal or left inferior phrenic.
2. The right inferior phrenic empties into the inferior vena cava; the left often consists of two branches, one of which empties into the left renal or suprarenal vein and the other into the inferior vena cava.
3. The hepatic veins empty into the inferior vena cava, but they commence in the sinusoids of the liver.
4. The right spermatic vein empties into the inferior vena cava, but the left empties into the left renal vein. The ovarian veins end in the same way as the spermatic veins in the male.

The portal and accessory portal system includes blood returning from the spleen and digestive organs, through the liver, and on into the hepatic vein. These systems are discussed in detail with the digestive organs (Chapter 20).

The systemic veins return blood from all parts of the body to the right atrium of the heart. The systemic veins consist of three sets: *superficial, deep,* and *venous* sinuses. The *superficial veins* are found just beneath the skin in the superficial fascia. The superficial and deep veins frequently unite. The *deep veins* accompany the arteries and are usually enclosed in the same sheath. The deep veins accompany the smaller arteries and are called *venae comitantes,* or companion veins. The larger arteries have only one accompanying vein, called *vena comes.* The deep veins do not accompany the arteries in certain parts of the body, such as the skull and liver and the larger veins from the bones. *The venous sinuses* are canals found only in the interior of the skull. (See Chapter 11.)

The *systemic veins* are divided into three groups: (1) veins that empty into the heart, (2) veins that empty into the superior vena cava, and (3) veins that empty into the inferior vena cava. Veins emptying into the superior vena cava are the veins of the head, neck, upper extremities, and thorax, and the azygos. Veins emptying into the inferior vena cava include those from the lower extremities, abdomen, and pelvis, and the azygos vein.

Veins of the Upper Extremity. Blood from the upper extremity is returned by a *deep* and *superficial* set of veins. The *deep veins* are the venae comitantes of the hand, forearm, and arm and are called by the same name as the arteries. There are frequent anastomoses with one another and with the superficial veins. *The superficial veins* are larger than the deep veins and return more blood, especially from the distal part of the limb. They include the following:

The *cephalic vein* begins in the dorsal network of the hand and winds upward around the radial border of the forearm to below the bend of the elbow, where it joins the *accessory cephalic* vein to form the *cephalic* of the upper arm.

The *basilic vein* begins in the ulnar part of the dorsal network and extends along the posterior surface of the ulnar to below the elbow, where it is joined by the *median cubital.*

The *axillary vein* is a continuation of the basilic. It ends at the outer border of the first rib in the subclavian vein. It receives blood from the *brachial cephalic vein* and other veins.

The *subclavian vein* is a continuation of the axillary vein and unites with the internal jugular to form the brachiocephalic vein. At the junction with the *internal jugular* and *left subclavian* vein, it receives the thoracic duct and the *right subclavian* vein receives the right lymphatic duct.

Veins of the Lower Extremity. Blood from the lower extremity is returned by a *superficial* and a *deep* set of veins. The *superficial* veins are beneath the skin between the layers of fascia. The *deep veins* accompany the arteries. Both sets are provided with valves, which are more numerous in the deep veins. The superficial veins include the *great saphenous,* which begins in the marginal vein of the dorsum of the foot, extends up the medial side of the leg and thigh, and ends in the *femoral vein* just below the inguinal ligament. It receives many branches from the sole of the foot; in the leg it anastomoses with the small *saphenous vein* and receives many *cutaneous veins.* Branches from the posterior and medial aspects of the thigh frequently unite to form an *accessory saphenous* vein. The *small saphenous vein* (see Figure 16–11) begins behind the lateral malleolus as a continuation of the *lateral marginal vein* and extends up the back of the leg to end in the *deep popliteal* vein. It receives many branches from the deep veins on the dorsum of the foot and from the back of the leg. The *deep veins* accompany the arteries below the knee. They are in pairs and are called by the same names as the arteries. Veins from the foot empty into the *anterior* and *posterior* tibial veins, which unite to form the *popliteal vein,* which is continued as the *femoral* and becomes the *external iliac vein.* (See Figure 16–11.)

Blood Circulation in Upper Extremity

Blood leaves the left ventricle, traverses arteries, arterioles, capillaries, venules, and veins, and is returned to the right atrium via the superior vena cava

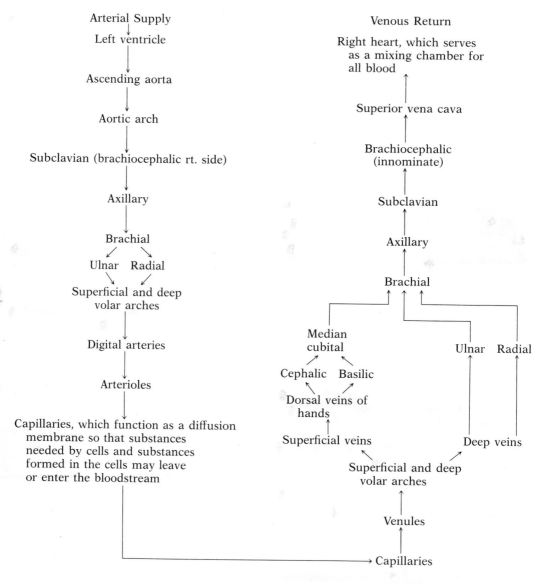

Arterial Supply

Left ventricle

Ascending aorta

Aortic arch

Subclavian (brachiocephalic rt. side)

Axillary

Brachial

Ulnar   Radial

Superficial and deep
volar arches

Digital arteries

Arterioles

Capillaries, which function as a diffusion
membrane so that substances
needed by cells and substances
formed in the cells may leave
or enter the bloodstream

Venous Return

Right heart, which serves
as a mixing chamber for
all blood

Superior vena cava

Brachiocephalic
(innominate)

Subclavian

Axillary

Brachial

Median
cubital   Ulnar   Radial

Cephalic   Basilic

Dorsal veins of
hands

Superficial veins   Deep veins

Superficial and deep
volar arches

Venules

Capillaries

Blood Circulation in Lower Extremity

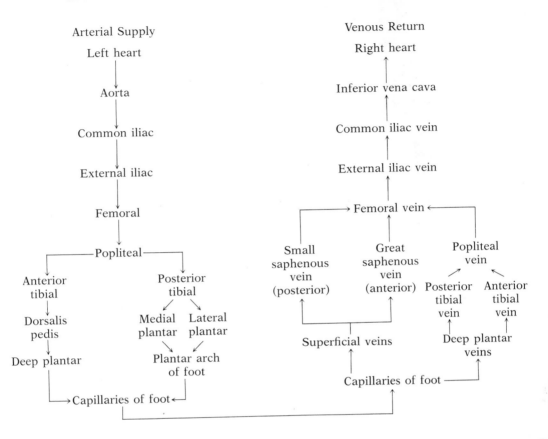

Lymphatic System

The lymphatic system is composed of the *lymph vessels, lymph nodes, spleen, thymus,* and scattered *lymphoid tissue* in the intestinal tract, including the *tonsils.* The fluid that flows throughout this system of vessels is called *lymph.*

Composition of Lymph. Lymph, tissue fluid, and plasma are similar in composition. Lymph consists of a fluid plasma containing a variable number of lymphocytes, a few granulocytes, no blood platelets (hence clots slowly), carbon dioxide, and *very* small quantities of oxygen. Other contained substances vary in kinds and amounts in relation to the location of lymphatic vessels. In the lymphatics of the intestine, fat content is high during digestion. Water, glucose, and salts are in about the same concentration as in blood plasma. Protein concentration is lower. Enzymes and antibodies are also present. Lymph has a specific gravity between 1.015 and 1.023.

Lymph Formation. Lymph is interstitial fluid that flows into the lymphatic capillaries. It is formed by the physical process of filtration. Colloidal substances from tissue fluid are returned to lymph capillaries rather than to the blood. Water, crystalloids, and other substances also enter the lymph capillaries. Since the process of tissue fluid formation is continuous, lymph formation is also continuous. The lymph system supplements the capillaries and veins in the return of the tissue fluid to the blood.

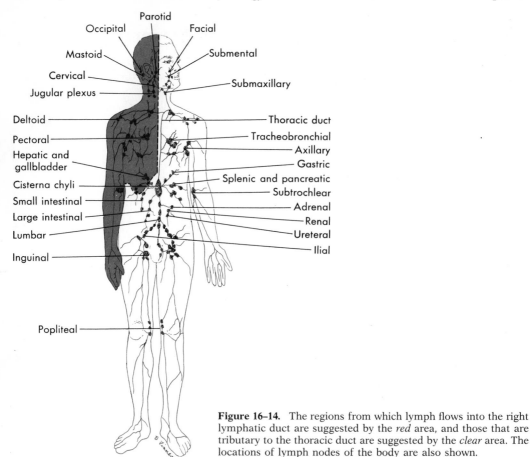

Parotid
Occipital
Facial
Mastoid
Submental
Cervical
Jugular plexus
Submaxillary
Deltoid
Thoracic duct
Pectoral
Tracheobronchial
Hepatic and gallbladder
Axillary
Gastric
Cisterna chyli
Splenic and pancreatic
Small intestinal
Subtrochlear
Large intestinal
Adrenal
Lumbar
Renal
Inguinal
Ureteral
Ilial
Popliteal

Figure 16–14. The regions from which lymph flows into the right lymphatic duct are suggested by the *red* area, and those that are tributary to the thoracic duct are suggested by the *clear* area. The locations of lymph nodes of the body are also shown.

Filtration pressure normally is highest in blood capillaries (as compared with tissue fluid pressure and lymph pressure) because of the beating heart and elastic arteries. Substances like the colloids, therefore, that are filtered out of the blood capillaries cannot enter them again but can enter the lymphatic capillaries.

Lymph Vessels. The plan upon which the lymphatic system is constructed is similar to that of the blood vascular system, if the heart and the arteries are omitted. In the tissues are located the closed *ends of minute microscopic vessels, called lymph capillaries*, which are comparable to, and often larger and more permeable than, the blood capillaries. The lymph capillaries are distributed in the same manner as the blood capillaries. Just as the blood capillaries unite to form veins, the lymph capillaries unite to form larger vessels called

lymphatics. The lymphatics continue to unite and form larger and larger vessels until finally they converge into two main channels, (1) the thoracic duct, and (2) the right lymphatic duct.

The thoracic duct, or *left lymphatic,* begins in the dilatation called the *cisterna chyli* (chyle cistern), located on the front of the body of the second lumbar vertebra. It ascends upward in front of the bodies of the vertebrae and enters the brachiocephalic vein at the angle of junction of the left internal jugular and left subclavian veins. It is from 38 to 45 cm long, about 4 to 6 mm in diameter, and has several valves. At its termination a pair of valves prevent the passage of venous blood into the duct. It receives the lymph from the left side of the head, neck, and chest, all of the abdomen, and both lower limbs, and also the chyle from the lacteals. Its dilatation, the cisterna chyli, receives the lymph from the lower extremities and from the walls and viscera of the pelvis and abdomen.

The right lymphatic duct is a short vessel, usually about 1.25 cm in length. It pours its contents into the brachiocephalic vein at the junction of the right internal jugular and subclavian veins. Its orifice is guarded by two semilunar valves.

The lymphatics from the right side of the head, the right arm, and the upper part of the trunk enter the right lymphatic duct.

Structure. The lymphatics resemble the veins in their structure as well as in their arrangement. The smallest consists of a single layer of endothelial cells, which have a dentated outline. The larger vessels have three coats similar to those of the veins, except that they are thinner and

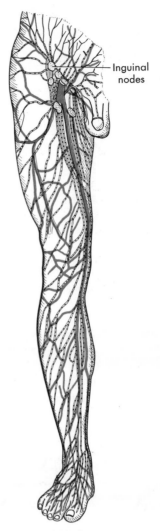

Figure 16–16. The lymph nodes and vessels of the lower extremity.

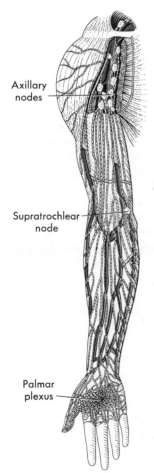

Figure 16–15. The lymph nodes and vessels of the upper extremity.

more transparent. Their valves are like those of the veins but are so close together that when distended they give the vessel a beaded or jointed appearance. They are usually absent in the smaller networks. The valves allow the passage of material from the smaller to the larger lymphatics and from these into the veins and prevent backflow.

Distribution. In general, the lymph vessels accompany and are closely parallel to the veins. Lymph vessels have been found in nearly every tissue and organ that contain blood vessels. The nails, cuticle, and hair are without them, but they permeate most other organs. No lymphatic *capillar-*

Figure 16-17. Diagram illustrating valves of lymphatics.

ies have been found in the central nervous system, the *internal* ear, cartilage, epidermis, spleen, or eyeball. Lymphatic vessels have not yet been demonstrated in the cornea, but lymph spaces are represented by the channels in which nerve fibers run. These channels are lined by an endothelium.

Classification. The lymph, like the blood in the veins, is returned from the limbs and viscera by a *superficial* and a *deep* set of vessels. The superficial lymph vessels are placed immediately beneath the skin and accompany the superficial veins. In certain regions they join the deep lymphatics by penetrating the deep fasciae. In the interior of the body they lie in the submucous tissue throughout the whole length of the gastropulmonary and genitourinary tracts and in the muscular tissue of the thoracic and abdominal walls. The deep lymphatics accompany the deep veins. They are few in number and are larger than the superficial lymphatics.

Lacteals. The lymphatics that have their origin in the villi of the small intestine are called *lacteals*. During the process of digestion they are filled with chyle, white in color from the fat particles suspended in the lymph. The lacteals enter the lymphatic vessels that run between the layers of the mesentery, pass through the mesenteric nodes, and finally terminate in the cisterna chyli.

Function. The function of the lymphatics is to convey tissue fluid from the tissues to the veins. Functionally, they may be considered supplementary to the capillaries and the veins, as they gather up a part of the fluid that exudes through the thin capillary walls and return it to the general circulation. Removal of protein from tissue fluid is a most important additional function. In the brachiocephalic veins lymph becomes mixed with the blood and enters the superior vena cava and then the right atrium of the heart. The function of the lacteals is to help in the absorption of digested food, especially fats.

Factors Controlling the Flow of Lymph. The flow of tissue fluid from the tissue spaces to the lymph capillaries and on to the veins is maintained chiefly by three factors.

Difference in Pressure. The tissue fluid is under greater pressure than the lymph in the lymph capillaries, and the pressure in the larger lymphatics near the ducts is much less than in the smaller vessels. Consequently the lymphatics form a system of vessels leading from a region of high pressure, the tissues, to a region of low pressure, the interior of the large veins of the neck.

Muscular Movements and Valves. Contractions of the skeletal muscles compress the lymph vessels and force the lymph on toward the larger ducts. The numerous valves prevent a return flow in the backward direction. The flow of lymph from resting muscles is small in quantity, but during muscular exercise and massage it is increased. The flow of chyle is greatly assisted by the peristaltic and rhythmical contractions of the muscular coats of the intestines. Pulsation waves moving over the enormous number of minute arteries existing everywhere act to "push ahead" the lymph from valve to valve, on to the larger lymphatics.

Respiratory Movements. During each inspiration the pressure on the thoracic duct is less than on the lymphatics outside the thorax, and lymph is accordingly

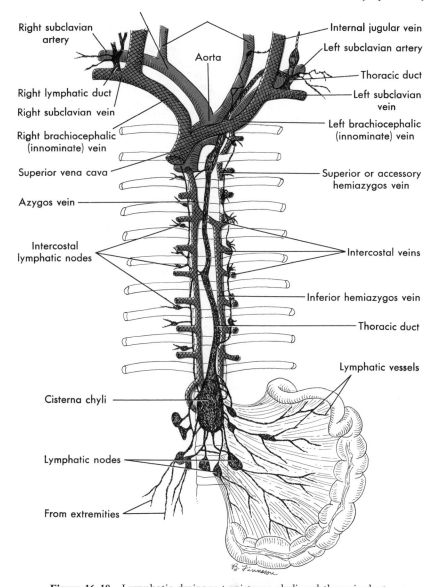

Right subclavian artery
Internal jugular vein
Aorta
Left subclavian artery
Thoracic duct
Right lymphatic duct
Right subclavian vein
Left subclavian vein
Right brachiocephalic (innominate) vein
Left brachiocephalic (innominate) vein
Superior vena cava
Superior or accessory hemiazygos vein
Azygos vein
Intercostal lymphatic nodes
Intercostal veins
Inferior hemiazygos vein
Thoracic duct
Lymphatic vessels
Cisterna chyli
Lymphatic nodes
From extremities

Figure 16–18. Lymphatic drainage to cisterna chyli and thoracic duct.

sucked into the duct. During the succeeding expiration the pressure on the thoracic duct is increased, and some of its contents, prevented by the valves from escaping below, are pressed out into the brachiocephalic veins.

Lymph Nodes. Lymph nodes (Figure 16–20) are small, oval or bean-shaped bodies, varying in size from that of a pinhead to that of an almond, and are located along the course of the lymphatics. They generally present a slight depression, called the

hilus, on one side. The blood vessels enter and leave through the hilus. The outer covering is a capsule of connective tissue containing a few smooth muscle fibers. The capsule sends fibrous bands called *trabeculae* into the substance of the node, dividing it into irregular spaces, which communicate freely with each other. The irregular spaces are occupied by a mass of lymphoid tissue, which, however, does not quite fill them as it never touches the capsule or trabeculae but leaves a narrow interval between itself and them. The

Figure 16–19. The lymph nodes and vessels of the dorsal body wall. (Modified from Toldt.)

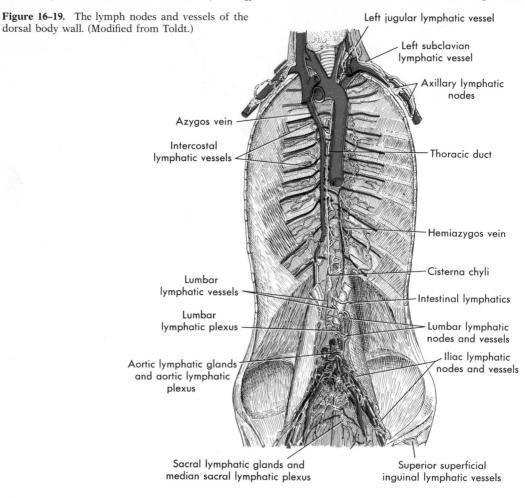

Left jugular lymphatic vessel

Left subclavian lymphatic vessel

Axillary lymphatic nodes

Azygos vein

Intercostal lymphatic vessels

Thoracic duct

Hemiazygos vein

Cisterna chyli

Lumbar lymphatic vessels

Intestinal lymphatics

Lumbar lymphatic plexus

Lumbar lymphatic nodes and vessels

Aortic lymphatic glands and aortic lymphatic plexus

Iliac lymphatic nodes and vessels

Sacral lymphatic glands and median sacral lymphatic plexus

Superior superficial inguinal lymphatic vessels

spaces thus left form channels for the passage of the lymph, which enters by several afferent vessels. After circulating through the node, the lymph is carried out by efferent vessels, which emerge from the hilus. The trabeculae support a free supply of blood vessels. It is said that no lymph on its way from the lymph capillaries ever reaches the bloodstream without passing through at least one node.

Location of Nodes. There are a superficial and a deep set of nodes just as there are a superficial and a deep set of lymphatics and veins. Occasionally, a node exists alone, but they are usually in groups or chains at the sides of the great blood vessels. Lymph nodes are found on the back of the head and neck, draining the scalp; around the sternomastoid muscle, draining the back of the tongue, the pharynx, nasal cavities, roof of the mouth, and

face; and under the rami of the mandible, draining the floor of the mouth.

In the upper extremities there are three groups—a small one at the bend of the elbow, which drains the hand and forearm; a larger group in the axillary space, into which the first group drains; and a still larger group under the pectoral muscles. The last-named drains the mammary gland and the skin and muscles of the thorax.

In the lower extremities there is usually a small node at the upper part of the anterior tibial vessels, and in the popliteal space back of the knee there are several; but the greater number are massed in the groin. These nodes drain the lower extremities and the lower part of the abdominal wall. The lymph nodes of the abdomen and pelvis are divided into a parietal and a visceral group. The parietal nodes

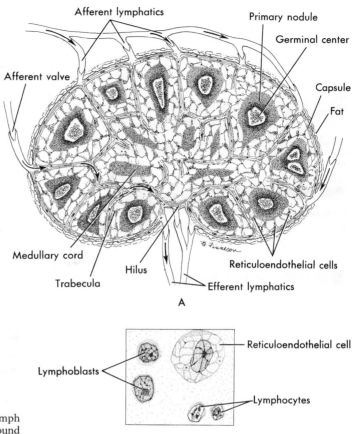

Figure 16–20. *A.* Diagram of a lymph node, highly magnified. *B.* Cells found in the lymph nodes.

are behind the peritoneum and in close association with the larger blood vessels. The visceral nodes are associated with the visceral arteries. The lymph nodes of the thorax are similarly divided into a parietal set, situated in the thoracic wall, and a visceral set associated with the heart, pericardium, trachea, lungs, pleura, thymus, and esophagus.

Function. The lymph nodes are credited with two important functions.

1. They produce lymphocytes and antibodies. As lymph passes through the nodes, fresh lymphocytes are added to the fluid. The lymphocytes are formed in the nodes by cell division.

2. The nodes are located in the course of the lymph vessels, and the lymph takes a tortuous course among the cells of the node. This suggests that they serve as filters and are a defense against the spread of infection. The lymph draining from an infected area carries the products of sup-

puration, and perhaps the infecting organisms themselves, to the first nodes in its pathway. Unless the infection is severe, the odds are against the organisms, and the lymph is more or less "disinfected" before it passes on. Nodes engaged in such a struggle are usually enlarged and tender, and if they are overpowered, they themselves may become the foci of infection.

Edema. Excessive amount of tissue fluid is termed *edema* and may be caused by:
1. Obstruction to flow of lymph, as in infection of the lymph node or a tumor pressing on the lymph vessel.
2. Elevation in capillary pressure, usually due to increased venous pressure.
3. Increased permeability of the capillary membrane, as in hives accompanying an allergic reaction.

Spleen (Lien). The spleen is a highly vascular, flat, oval organ situated directly beneath the diaphragm, behind and to the left of the stomach. It is covered by peritoneum and held in position by folds of this membrane. Beneath the serous coat is a connective tissue capsule from which trabeculae run inward, forming a framework, in the interstices of which is found the *splenic pulp*, made up of a network of fibrillae and blood cells—red corpuscles, the various forms of white cells, and large, rounded phagocytic cells (the macrophages of the reticuloendothelial system), which engulf fragmentary red corpuscles and invading organisms. Scattered throughout the pulp are masses of lymphoid tissue called malpighian follicles.[2] Smooth muscle fibers are found in both the outer capsule and the trabeculae.

The blood supply is brought by the splenic artery, a branch of the celiac artery; the splenic artery divides into six or more branches, which enter the concave side of the spleen at a depression called the hilum. The arrangement of the blood vessels is peculiar to this organ. After entering, the arteries divide into many branches and terminate in tufts of arterioles, which open freely into the splenic pulp. Each follicle lies in close relation to a small artery. The blood is collected by thin-walled veins, which unite to form the splenic vein. The splenic vein unites with the superior mesenteric to form the portal vein, which carries the blood to the liver.

Function. 1. The spleen can store large quantities of blood and release it on physiologic demand.

2. The spleen forms red blood cells in the embryo.

3. The spleen destroys and removes fragile or aged red blood cells.

4. The malpighian follicles of the spleen are a place of origin for lymphocytes.

5. Phagocytosis—the reticuloendothelial cells in the spleen pulp and venous sinuses function as a secondary line of defense for removing bacteria, debris, or other infectious agents that have entered the bloodstream.

Enlargement of the spleen occurs in certain pathologic conditions, such as anemias, and

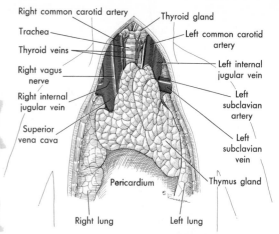

Figure 16–21. Thymus gland of newborn infant.

splenectomy gives favorable results. Enlargement also accompanies malaria, leukemia, and Hodgkin's disease, but removal in these cases is medically contraindicated.

Thymus. The thymus, as part of the lymphatic system, usually consists of two lobes, but they may unite to form a single lobe or may have an intermediate lobe between them. It is situated in the upper chest cavity along the trachea, overlapping the great blood vessels as they leave the heart. Each lobe has several lobules, each of which is composed of an outer *cortex* and *medulla.* The cortex is composed of closely arranged lymphocytes, which obscure the fewer number of reticular cells. In the medulla the reticulum is coarser and the lymphoid cells are fewer in number. There are many reticular cells. In the medulla are rounded nests of cells, 30 to 100 μm in diameter, called the corpuscles of Hassall.[3] The arteries are derived from the internal thoracic and the superior and inferior thyroids. The nerves are derived from the vagi of the craniosacral system and from the thoracolumbar system. At birth the thymus is large, and it gradually decreases in size after puberty, the corpuscles of Hassall disappearing more slowly than other portions. As indicated in the discussion of immunity in the preceding chapter, the thymus is an integral part of the immune system, preprocessing the T-lymphocytes prior to birth.

[2] Marcello Malpighi, a physician and professor of comparative anatomy at Bologna (1628–1694).

[3] Arthur Hill Hassall, English physician (1817–1894).

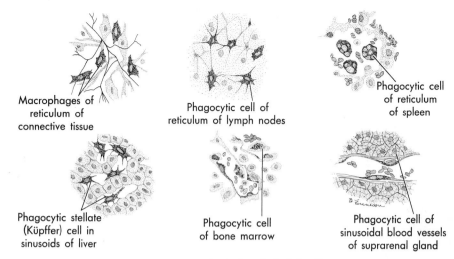

Macrophages of
reticulum of
connective tissue

Phagocytic cell of
reticulum of lymph nodes

Phagocytic cell
of reticulum
of spleen

Phagocytic stellate
(Küpffer) cell in
sinusoids of liver

Phagocytic cell
of bone marrow

Phagocytic cell of
sinusoidal blood vessels
of suprarenal gland

Figure 16–22. Reticuloendothelial cells.

Tonsillar Tissue. Such tissue is composed of a mass of lymphoid tissue embedded in mucous membrane. The epithelial lining dips between the lymphoid tissue forming crypts or glandlike pits. Many reticuloendothelial cells are found in the tonsils.

Reticuloendothelial Cells. The macrophages of a variety of tissues are called by some the *reticuloendothelial system;* their function is to phagocytize microorganisms or foreign particles. They are aggregated mainly in the spleen, bone marrow, lymphoid tissue, and scattered connective tissue. It is probable that macrophages participate in B-lymphocyte functioning prior to the actual formation of the antibody, although their role is not clear. Figure 16–22 illustrates the appearance of these cells.

Questions for Discussion

1. Where and how would you place pressure to stop bleeding from an artery? A vein?
2. How would you know whether it was a vein or artery bleeding?
3. Which veins are usually used for venipuncture? Why?
4. Where are the most accessible places to take the pulse? Name the arteries in each instance.
5. What are the differences in blood flow in arteries, veins, and capillaries?
6. How does the arterial wall receive its nutrients?
7. Discuss the structural differences between arteries, capillaries, and veins.

Summary

Arteries (characterized by elasticity)	Hollow vessels—carry blood from heart, break up into arteries and arterioles		
	Coats	1. Inner lining (intima)	Layer of endothelial cells Layer of connective tissue Layer of elastic tissue (fenestrated membrane)
		2. Middle coat (media)	Muscular and elastic tissue A few bundles of connective tissue
		3. External coat (adventitia)	Loose connective tissue with scattered smooth muscle cells

Arteries (cont.)
- Size
  - Aorta more than 1 in. in diameter
  - Arteries grow smaller as they subdivide
  - Smallest ones are microscopic and are called *arterioles*
- Usually deep seated for protection
- Division
  - Trunk gives off several branches
  - One branch that divides into several or two branches of nearly equal size
- Plexus—many anastomoses within limited area

Capillaries (characterized by multiplicity)
- Tiny vessels about 8μm in diameter. Connect arterioles to venules
- One layer of endothelial cells
- Communicate freely in networks

Veins (characterized by valves)
- Collapsible vessels, smallest ones called *venules*
- Begin where capillaries end
- Carry blood to the heart
- Three coats, same as arteries, but thinner
- Valves—semilunar pockets

Vasa vasorum—blood vessels that supply the wall of other blood vessels

Vasomotor—term applied to *nerve fibers* supplied to blood vessels
- Vasoconstrictor, well understood
- Vasodilator, not as well understood

Anastomosis—direct connection between artery and vein, bypassing capillary bed, or between distal ends of arteries

Division of the Vascular System

Pulmonary system
1. Pulmonary artery conveys venous blood from right ventricle to lungs
   - Right pulmonary artery—right lung
   - Left pulmonary artery—left lung
2. Capillaries connect arterioles to venules
3. Four pulmonary veins—two from each lung—convey oxygenated blood to left atrium

Systemic system
Provides for systemic circulation
1. Aorta and all its branches
2. Capillaries connect arterioles to venules
3. Veins empty into heart either directly or by means of inferior and superior venae cavae

Aorta

Ascending aorta
- About 5 cm long
- Branches
  - Right coronary
  - Left coronary

Arch of aorta
- Extends from ascending aorta to body of fourth thoracic vertebra
- Branches
  - Brachiocephalic
    - Right common carotid
    - Right subclavian
  - Left common carotid
  - Left subclavian

Descending aorta
- Thoracic aorta
  - Extends from lower border of fourth thoracic vertebra to aortic opening in diaphragm
  - Branches supply wall and viscera of thorax
- Abdominal aorta
  - Extends from aortic opening in diaphragm to body of fourth lumbar vertebra. Branches supply wall and viscera of abdominal cavity

Arteries of the Head and Face

- **Left common carotid** arises from arch of aorta
  - **External carotid**—Branches supply thyroid gland, tongue, throat, face, ear, and dura mater
  - **Internal carotid**—Branches supply brain, eye and its appendages, forehead, and nose
- **Right common carotid** arises at division of brachiocephalic
  - **External carotid**—Branches same as on left side
  - **Internal carotid**—Branches same as on left side

Arteries of the Chest

- **Visceral group**
  - **Pericardial**—to pericardium
  - **Bronchial**—nutrient vessels of lungs
  - **Esophageal**—to esophagus
  - **Mediastinal**—to nodes and areolar tissue in mediastinum
- **Parietal group**
  - **Intercostal**—to lower nine intercostal spaces
  - **Subcostal**—anastomose with superior epigastrics, lower intercostals, and lumbar arteries
  - **Superior phrenic**—upper surface of diaphragm

Arteries of the Abdomen, Visceral Branches

Celiac artery

- **Left gastric**–lesser curvature of stomach left to right
- **Hepatic artery** divides into right and left branches before entering liver
  - **Right gastric**–lesser curvature of stomach right to left
  - **Gastroduodenal**
    - *Superior pancreaticoduodenal* to duodenum and head of pancreas
    - *Right gastroepiploic*–greater curvature of stomach right to left
  - **Cystic**–gallbladder
- **Lienal**, or **splenic**
  - Branches to pancreas
  - **Left gastroepiploic**–greater curvature of stomach left to right

Superior mesenteric — Supplies small intestine except duodenum, cecum, ascending colon, half of transverse colon

Inferior mesenteric — Supplies left half transverse colon, whole of descending and sigmoid colon, continued as *superior hemorrhoidal artery* to rectum

Suprarenal — To suprarenal glands–anastomose with branches of phrenic and renal arteries

Renal — Each divides into four or five branches before entering kidney

Internal spermatic — Supply the testes

Ovarian — Supply ovaries, send small branches to the ureters and uterine tubes; one branch unites with uterine artery and assists in supplying uterus

Arteries of the Abdomen, Parietal Branches

Inferior phrenics–distributed to undersurface of diaphragm

Lumbar — Distribute branches to muscles and skin of back, to the spinal cord and its membranes, and to the lumbar vertebrae

Middle sacral–passes down to coccyx

Arteries of the Pelvis

Common iliac about 5 cm

- **Internal iliac** — Branches to pelvic walls, viscera, genitalia, buttocks, medial side of thighs
- **External iliac**
  - Branches to psoas major
  - Inferior epigastric
  - Deep iliac circumflex

Arteries of the Upper Extremity

Subclavian–forms main artery of each upper limb. Different portions named according to regions through which they pass

Right subclavian — Extends from the brachiocephalic to the first rib. Branches are { Vertebral, thyrocervical, Internal thoracic, Costocervical

Left subclavian — Extends from arch of aorta to first rib. Same branches as right subclavian

Axillary — In axillary regions. Branches to shoulders, chest, and arms

Brachial — Extend from axillary arteries to below bend of elbows, where they divide into ulnar and radial arteries

Ulnar — Extend along ulnar border of forearms to palms of hands

Radial — Extend along radial side of forearms to palms of hands

Form the superficial and deep *volar arches*

Arteries of the Lower Extremity

External iliacs form main arteries of lower limbs. Different portions named according to regions through which they pass

Femoral — Extend along inguinal ligaments to openings in adductor magnus muscles. Send branches to abdominal walls, external genitalia, muscles and fasciae of the thighs

Popliteal — Back of knees, send branches to knee joints, posterior leg muscles, gastrocnemius, soleus, skin of back of legs. Below knee joints divide into posterior tibials and anterior tibials

Posterior tibials — Back of legs, from bifurcation of popliteal to ankle. Send branches to back of legs and to tibiae and fibulae. Give off peroneal arteries about 1 in. below bifurcation of popliteals. *Peroneals* distribute blood to fibulae and calcaneus bones

Arteries of the Lower Extremity (*cont.*)	**Anterior tibials**	Front of legs from bifurcation of popliteal to ankle—then become the *dorsalis pedis arteries*
	Dorsalis pedis	Dorsum of each foot, anastomose with branches from posterior tibials and supply blood to feet
Veins		Begin small, grow larger
	Differ from arteries	Larger size / Greater number / Thinner walls / Valves / More frequent anastomoses
	Sets	Superficial, or cutaneous, beneath the skin
		Deep: Usually accompany the arteries. Exceptions are: Veins in skull and vertebral canal / Hepatic veins / Large coronary veins
		Venous sinuses—canals formed by separation of layers of dura mater
	Venae comitantes	Deep veins accompanying smaller arteries, such as brachial, radial, ulnar, peroneal, tibial, are in pairs. A single deep vein accompanying a larger artery, such as femoral, popliteal, axillary, and subclavian artery, is called a *vena comes*
	Veins to heart	**Coronary veins** from heart
		Superior vena cava — Veins of head, neck, thorax, and upper extremities empty into this vein
		Inferior vena cava — Veins of abdomen, pelvis, and lower extremities empty into this vein
Veins of the Head and Neck	**External jugular**	Formed in parotid glands, terminate in the subclavians. Receive blood from deep parts of the face and the exterior of the cranium
	Internal jugular	Continuous with the lateral sinuses, unite with subclavians to form the innominates. Receive blood from the veins and sinuses of the cranial cavity, superficial parts of face and neck
Veins of the Thorax	**Brachiocephalic**	Formed by union of internal jugular and subclavians. Receive internal thoracic veins / One on each side of body
	Superior vena cava	Formed by union of right and left brachiocephalic veins. 7.5 cm (3 in.) long / Opens into right atrium
	Supplementary channel	1. **Azygos vein** / 2. **Hemiazygos vein** / 3. **Accessory hemiazygos vein** — Connect inferior vena cava below with superior vena cava above
	Bronchial	Formed at the root of each lung / Return blood from larger bronchi and structures at root of lungs / Right bronchial vein empties into azygos / Left bronchial vein empties into highest left intercostal or the accessory hemiazygos
Veins of the Abdomen and Pelvis	**Hypogastric (internal iliac)**	Formed by union of veins corresponding to branches of hypogastric arteries
	Common iliac	Formed by union of external iliacs and hypogastrics. Extend from base of sacrum to the fifth lumbar vertebra
	Inferior vena cava	Formed by union of the common iliacs / Extends from fifth lumbar vertebra to the right atrium of the heart / Receives many tributaries corresponding to arteries given off from the aorta
Veins of the Upper Extremity	**Superficial veins**	Are larger, take a greater share in returning blood
		Cephalic — Begin in dorsal venous network, join accessory cephalics of arms below elbows, empty into axillaries
		Basilic — Begin in dorsal venous network, are joined by median basilics below elbows, are continued as axillaries

Veins of the Upper Extremity (*Cont.*)	**Deep veins**	Accompany arteries, are called by same names, i.e., metacarpals, radials, ulnars, brachials, axillaries, and subclavians
		Axillary — Are continuations of the basilics, end at outer border of first ribs, receive brachials, cephalics, and deep veins
		Subclavian — Are continuations of the axillaries, unite with internal jugulars to form brachincephalics
Veins of the Lower Extremity	**Superficial**	Are between the layers of superficial fasciae Provided with valves
		Great saphenous veins — Begin in medial marginal veins, extend upward, and end in femoral veins. Receive branches from soles of feet. Anastomose with small saphenous veins, receive cutaneous veins and accessory saphenous veins
		Small saphenous — Continuation of lateral marginal veins, end in deep popliteal veins Receive branches on dorsum of each foot and back of each leg
	Deep	Accompany the arteries and are called by same names. Provided with many valves
		Popliteal — Formed by union of anterior tibials and posterior tibials
		Femoral — Continuation of the popliteals and extend from opening in adductor magnus muscles to the inguinal ligaments 1. Receive blood from the superficial veins 2. Receive blood from deep veins of feet, legs, and thighs
	External iliac	Continuation of femoral veins. Extend from inguinal ligaments to the joints between sacral and iliac bones. They enter the inferior vena cava
Lymph	**Source**	Formed from tissue fluid by the physical process of filtration
	Description	Colorless or yellowish liquid Slightly alkaline reaction Salty taste. No odor Consists of blood plasma plus lymphocytes Specific gravity varies between 1.015 and 1.023 Contains a relatively low content of blood proteins Contains a relatively low content of nutrients Contains a relatively high content of products of metabolism
	Function	Carries products of metabolism from tissues to blood
Lymph Vascular System	**Lymph vessels**	Lymph capillaries Lymphatics Thoracic duct Right lymphatic duct Lacteals
	Lymph nodes	
Lymph Vessels	**Lymph capillaries**	Origin in tissues One coat of endothelium Start as blind ends of microscopic lymph capillaries, unite to form lymphatics Distribution comparable to that of blood capillaries
	Lymphatics—three coats—numerous valves	
	Thoracic duct	Begins in cisterna chyli, located on front of body of second lumbar vertebra 38–45 cm long; 4–6 mm in diameter Has three coats—numerous valves Receives lymph from side of head, neck, and thorax, left arm, all of abdomen, and both lower limbs. Receives chyle from lacteals Pours lymph and chyle into left brachiocephalic vein
	Right lymphatic duct	About 1.25 cm long Receives lymph from right side of hand, neck, and thorax, also right arm Pours lymph into right brachiocephalic vein

Lymph Vessels (*Cont.*)

Classification
- **Superficial**—beneath skin, accompany superficial veins
- **Deep**—accompany deep blood vessels

Lacteals
- Lymphatics of the intestines
- Many originate in villi of small intestine
- Contain
  - During digestion—chyle—lymph carrying emulsified fat
  - During period of fasting—lymph
- Absorb fatty substances

Functions—drain off lymph from all parts of the body and return it to the brachiocephalic veins

Lymph Nodes

Description
- Shape
  - Oval
  - Bean shaped
- Size varies from that of pinhead to that of almond
- Outer capsule—connective tissue with some muscle fibers
- Interior divided into irregular spaces like sponge
- Spaces partially filled with lymphoid tissue. Communicating channels for lymph, which enters by afferent, leaves by efferent vessels
- Are well supplied with blood
- Superficial and deep set
- Usually arranged in groups or chains at sides of great blood vessels

Location

Head
1. Back of head and neck draining scalp
2. Around sterno-cleidomastoid muscle draining
   - Back of tongue, pharynx, nasal cavities, roof of mouth, face
3. Under rami of mandible—draining floor of mouth

Upper extremities
1. Small group at bend of elbow drains
   - Hand
   - Forearm
2. Larger axillary group drains
   - First group
   - Axillary space
3. Larger group under pectoral muscles drains
   - Mammary gland
   - Skin and muscles of chest

Lower extremities
1. Node at upper part anterior tibial vessels
2. Several in popliteal space
3. Great number massed in groin
   — Drain lower extremities and lower abdominal wall

Abdomen and pelvis
1. Parietal group—behind peritoneum, in close association with large blood vessels
2. Visceral group—associated with the visceral arteries

Thorax
1. Parietal group situated in thoracic wall
2. Visceral group associated with
   - Heart, pericardium
   - Lungs, pleura
   - Thymus and esophagus

Functions
1. Formation of lymphocytes
2. Filtration—preventive and protective
3. Addition of serum globulins including antibodies

Factors Controlling Flow of Lymph
- Difference in hydrostatic and osmotic pressure
- Muscular movements and valves
- Respiration

Edema

Accumulation of lymph in the tissues

May be caused by
1. Excessive formation
2. Obstruction to flow of lymph from tissues
3. Pressure or obstruction to venous return

Spleen

Description
- Vascular, oval, flat lymph gland
- Beneath diaphragm, behind and to left of stomach
- Fibrous capsule surrounding network of trabeculae that contains splenic pulp
- Malpighian follicles—masses of lymphoid tissues scattered through the splenic pulp
- Blood supplied by lienal artery (branch of celiac), divides into six or more branches, which enter hilum

Spleen (*Cont.*)	Functions	1. Major place of destruction of red blood cells or place of preparation for their destruction by the liver 2. Reservoir of blood cells 3. Malpighian corpuscles give rise to lymphocytes 4. Formation of erythrocytes during fetal life and after birth if need arises
Thymus	Description	Two lateral lobes Temporary organ
	Location—upper part, chest cavity, along the trachea **Function**—preprocessing of T-lymphocytes before birth	
Tonsillar Tissue	Description	Interlacing network with phagocytic cells Composed of lymphoid tissue Many RE cells
Reticular Connective Tissue	Location	All loose connective tissue, lymph nodes, spleen, liver, sinusoids, bone marrow, adrenal glands, anterior lobe of pituitary, nervous system
	Function	Phagocytic properties paramount

Additional Readings

BLOOM, S.: Spontaneous rhythmic contraction of separated heart muscle cells. *Science,* **167:**1727–29, March 27, 1970.

GUYTON, A. C.: *Basic Human Physiology: Normal Functions and Mechanisms of Disease,* 2nd ed. W. B. Saunders Co., Philadelphia, 1977, Chapters 22, 23.

HAM, A. W.: *Histology,* 7th ed. J. B. Lippincott Co., Philadelphia, 1974, pp. 544–49, 581–90.

PANSKY, B., and HOUSE, E. L.: *Review of Gross Anatomy,* 3rd ed. Macmillan Publishing Co., Inc., New York, 1975, pp. 284–97.

VANDER, A. J.; SHERMAN, J. H.; and LUCIANO, D. S.: *Human Physiology: The Mechanisms of Body Function,* 2nd ed. McGraw-Hill Book Co., New York, 1975, pp. 231–45.

CHAPTER 17

The Heart

Chapter Outline

The pumping action of the heart is fundamental to adequate nutrition of cells and maintenance of internal environment. Without this pumping action cells would starve, waste products would build up, and life of the cell and of the individual would cease.

Location. The heart is a hollow, muscular organ, situated in the thorax between the lungs and above the central depression of the diaphragm. It is about the size of the closed fist, shaped like a blunt cone, and so suspended by the great vessels that the broader end, or *base*, is directed upward, backward, and to the right. The pointed end, or *apex*, points downward, forward, and to the left. As placed in the body, it has an oblique position, and the right side is almost in front of the left. The impact of the heart during contraction is felt against

the chest wall in the space between the fifth and sixth ribs, a little below the left nipple, and about 8 cm (3 in.) to the left of the median line. This impact is called the point of maximum impulse (PMI).

Heart Wall. The heart wall is composed of (1) an outer layer, the pericardium; (2) a middle layer, the *myocardium;* and (3) an inner layer, the *endocardium.* The epicardium is the serous membrane, or visceral pericardium.

Pericardium. The heart is covered by a

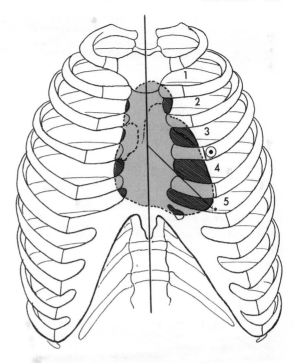

Figure 17–1. Heart in situ. *1, 2, 3, 4, 5.* Intercostal spaces; vertical line represents median line. Space outlined by the acute angle indicates the superficial cardiac region. ⊙ shows the location of the nipple, on the fourth rib. * indicates the area for placing the stethoscope to take the apex beat of the heart.

serous membrane called the pericardium. It consists of two parts: (1) an external fibrous portion and (2) an internal serous portion.

EXTERNAL FIBROUS PERICARDIUM. This is composed of fibrous tissue and is attached by its upper surface to the large blood vessels that emerge from the heart. It covers these vessels for about 3.8 cm (1½ in.) and blends with their sheaths. The lower border is adherent to the diaphragm, and the front surface is attached to the sternum.

INTERNAL, OR SEROUS, PERICARDIUM. This is a completely closed sac; it envelops the heart and lines the *fibrous* pericardium. The portion of the serous pericardium that lines it and is closely adherent to the heart is called the *visceral* portion (*viscus*, an organ); the remaining part of the serous pericardium, namely, that which lines the fibrous pericardium, is known as the *parietal* portion (*paries*, a wall). The visceral and parietal portions of this serous membrane are everywhere in contact. Between them is a small quantity of pericardial fluid preventing friction as their surfaces continually slide over each other with the constant beating of the heart. The pericardial fluid may aid in cushioning the heart, which is especially important with rapid bodily movements. Figure 17-2 shows the position of the heart within the pericardium.

Endocardium. The inner surface of the cavities of the heart is lined by a thin membrane called endocardium. It is com-

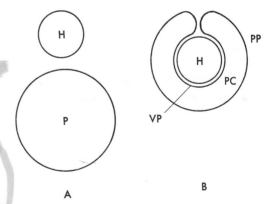

Figure 17–2. Diagram of the heart and serous pericardium. *A* shows heart (*H*) and pericardium (*P*) lying separately; *B* shows the pericardium invaginated by the heart; *PC*, pericardial cavity, which actually is a very narrow space filled with pericardial fluid; *PP*, parietal layer that lines the fibrous pericardium; *VP*, visceral layer that clings close to the heart muscle.

posed of endothelial cells and covers the valves, surrounds the chordae tendineae, and is continuous with the lining membrane of the large blood vessels. The endocardium contains many blood vessels, a few bundles of smooth muscle, and parts of the conducting system.

Myocardium. The main substance of the heart is cardiac muscle, called myocardium. This tissue includes the muscle bundles of (1) the atria, (2) the ventricles, and (3) the atrioventricular bundle (of His).[1]

1. The principal muscle bundles of the atria radiate from the area that surrounds the orifice of the superior vena cava. One, the interatrial bundle, connects the anterior surfaces of the two atria. The other atrial muscle bundles are confined to their respective atria, though they merge to some extent.

2. The muscle bundles of the ventricles begin in the atrioventricular fibrous rings. They form U-shaped bundles with the apex of the U toward the apex of the heart. There are many of these bundles, but for general description they may be divided into four groups. One group begins at the *left* atrioventricular ring, passes toward the right and the apex, where it forms whorls, and then ends either in the left ventricular wall, the papillary muscles, or the septum and the right ventricular wall (Figure 17–3).

A second bundle repeats this path except that it starts at the *right* ventricular ring, passes to the left in the anterior wall of the heart, and ends in the same structures as above.

These two groups form an outer layer that winds around both ventricles. Under these is a third group of muscle bundles that again wind around both ventricles. There is a fourth group of muscle bundles that wind around the left ventricle. Thus, the left ventricle has a much thicker wall than the right (Figure 17–4). During contraction the squeeze of the spirally arranged muscle bundles forces blood out of the ventricles.

3. The muscular tissue of the atria is not continuous with that of the ventricles. The walls are connected by fibrous tissue and the atrioventricular bundle of modified muscle cells. This bundle arises in connection with the atrioventricular (AV) node, which lies near the orifice of the coronary sinus in the right atrium. From this node the atrioventricular bundle passes forward to the membranous septum between the ventricles, where it divides into right and left bundles, one for each ventricle. In the muscular septum between the ventricles each bundle divides into numerous strands, which spread over the internal surface just under the endocardium. The greater part of the atrioventricular bundle consists of spindle-shaped muscle cells. The two bundles run about halfway down the septum until they come into contact with the Purkinje[2] fibers, specialized cardiac fibers that conduct the cardiac impulse much more rapidly than ordinary heart muscle (see Figure 17–14).

Cavities of the Heart. The heart is divided into right and left halves, frequently called the right heart and the left heart, by a muscular partition, the ventricular septum, which extends from the base of the ventricles to the apex of the heart. The atrial septum is inconspicuous. The two sides of the heart have no communication with each other after birth. The right side contains *venous* and the left side *arterial* blood. Each half is subdivided into two cavities: the upper, called the *atrium,* and the lower, the *ventricle.* The left ventricle ejects blood into the extensive systemic circulatory system under much higher pressure than is required of the right ventricle for ejecting blood into the relatively short pulmonary circulation. The structural arrangement of cardiac muscle fibers provides for the thicker muscle required in this pumping action. (See Figure 17–4.) Both the right and left sides of the heart contract and relax almost simultaneously.

Muscular columns, called the *trabeculae carneae (columnae carneae),* project from

---

[1] Wilhelm His, Jr., German physiologist (1863–1934).

[2] Johannes Purkinje, Bohemian physiologist (1787–1869).

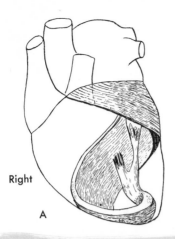
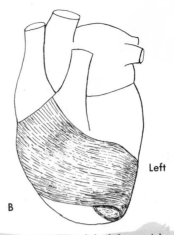

A, B. Note how the two groups of muscle fibers wind around both the right and the left ventricles on the outside.

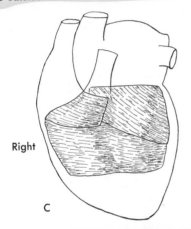
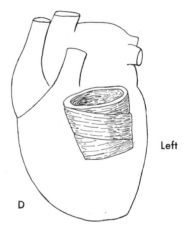

C. Note how the second layer of muscle fibers winds around both the right and left ventricles.
D. Note that the innermost layer winds around the left ventricle only.

Figure 17–3. Diagram showing arrangement of muscle fibers of the heart. When these muscles contract, what happens to the blood vessels? Position of heart more upright than normal. (Modified from Wiggers and Schaeffer.)

the inner surface of the ventricles. They are of three kinds: The first are attached along their entire length and form ridges,

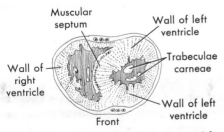

Figure 17–4. Cross section through ventricles, showing relative thickness of their walls and shape of cavities.

or columns. The second is a rounded bundle; the moderator band (*trabecula septomarginalis*) projects from the base of the anterior papillary muscle to the ventricular septum. It is formed largely of specialized fibers concerned with the conducting mechanisms of the heart. Third are the *papillary muscles*, which are continuous with the wall of each ventricle at its base. The apexes of the papillary muscles give rise to fibrous cords, called the *chordae tendineae*, which are attached to the cusps of the atrioventricular valves. These muscles contract when the ventricular walls contract.

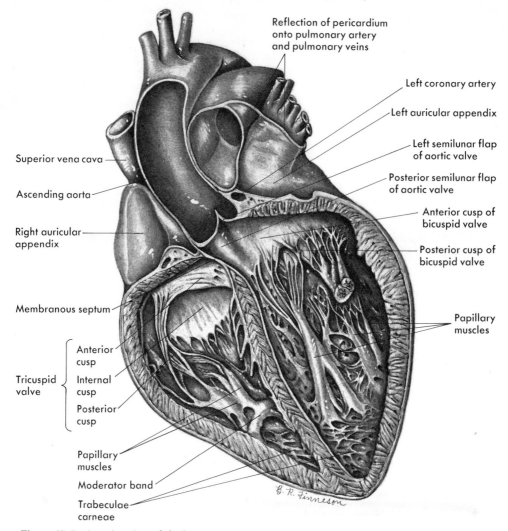

Reflection of pericardium onto pulmonary artery and pulmonary veins

Left coronary artery

Left auricular appendix

Left semilunar flap of aortic valve

Posterior semilunar flap of aortic valve

Anterior cusp of bicuspid valve

Posterior cusp of bicuspid valve

Papillary muscles

Superior vena cava

Ascending aorta

Right auricular appendix

Membranous septum

Tricuspid valve
- Anterior cusp
- Internal cusp
- Posterior cusp

Papillary muscles

Moderator band

Trabeculae carneae

Figure 17–5. Anterior view of the heart, sectioned through the aorta, pulmonary artery, and ventricles; atria are intact.

Orifices of the Heart. The orifices comprise the left and right atrioventricular orifices and the orifices of eight large blood vessels connected with the heart.

On the right side of the heart, the superior and inferior venae cavae and coronary sinus empty into the atrium, and the pulmonary artery leaves the ventricle.

On the left side of the heart, four pulmonary veins empty into the atrium, and the aorta leaves the ventricle. There are some smaller openings to receive blood directly from the heart substance, and before birth there is an opening between the right and left atria called the *foramen ovale.* Normally this closes soon after

birth. Its location is visible in Figure 26–12, page 614.

Valves of the Heart. Between each atrium and ventricle there is a somewhat constricted opening, the atrioventricular orifice, which is strengthened by fibrous rings and protected by valves. The openings into the aorta and pulmonary artery are also guarded by valves (Figures 17–6 and 17–7).

Tricuspid Valve. The right atrioventricular valve is composed of three irregular-shaped flaps, or cusps, and hence is named *tricuspid.* The flaps are formed mainly of fibrous tissue covered by endocardium. At their bases they are continu-

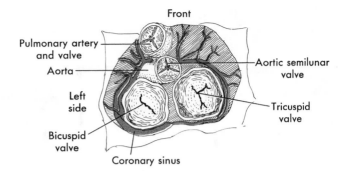

Figure 17–6. Valves of the heart as seen from above, atria removed.

ous with one another and form a ring-shaped membrane around the margin of the atrial openings; their pointed ends project into the ventricle and are attached by the chordae tendineae to small muscular pillars, the papillary muscles, in the interior of the ventricles.

Bicuspid Valve. The left atrioventricular valve consists of two flaps, or cusps, and is named the *bicuspid,* or *mitral,* valve. It is attached in the same manner as the tricuspid valve, which it closely resembles in structure except that it is much stronger and thicker in all its parts. Chordae tendineae are attached to the cusps and papillary muscles in the same way as on the right side; they are less numerous but thicker and stronger.

The tricuspid and bicuspid valves freely permit the flow of blood from the atria into the ventricles because the free edges of the flaps are pointed in the direction of the blood current; but any flow forced backward gets between the flaps and the walls of the ventricles and drives the flaps

upward until, meeting at their edges, they unite and form a complete transverse partition between the atria and ventricles. The valves remain open as long as the pressure of the blood is higher in the atria than in the ventricles. When the muscles of the ventricles begin to contract, the pressure in the ventricular chambers rises and the valves close. The valves are kept from everting into the atrial chambers by the chordae tendineae, which are kept taut by the papillary muscles.

Semilunar Valves. The orifice between the right ventricle and the pulmonary artery is guarded by the *pulmonary valve,* and the orifice between the left ventricle and the aorta is guarded by the *aortic valve.* These two valves are called *semilunar valves* and consist of three semilunar cusps, each cusp being attached by its convex margin to the inside of the artery where it joins the ventricle, while its free border projects into the lumen of the vessel. Between the cusps of the valve and the aortic wall are slight dilatations called the

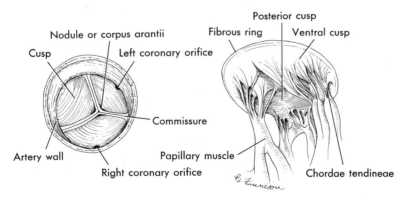

Figure 17–7. Valves of the left side of the heart. *A.* Aortic valve closed, seen from above. *B.* Mitral valve seen from below.

aortic sinuses or sinuses of Valsalva.[3] The coronary arteries have their origin from two of these sinuses.

The semilunar valves offer no resistance to the passage of blood from the heart into the arteries, as the free borders project into the arteries, but they form a complete barrier to the passage of blood in the opposite direction. In this case each pocket becomes filled with blood, and the free borders are floated out and distended so that they meet in the center of the vessel.

The orifices between the two caval veins and the right atrium and the orifices between the left atrium and the four pulmonary veins are not guarded by valves. The opening from the inferior vena cava is partly covered by a membrane known as the caval (eustachian) valve.

Lymph Vessels. The heart is richly supplied with lymph capillaries, which form a continuous network from the endocardium, through the muscle layers to the epicardium. These capillaries form larger vessels, which accompany the coronary blood vessels and finally enter the thoracic duct.

[3]Antonio Maria Valsalva, Italian anatomist (1666–1723).

Coronary Vessels. The blood vessels of the heart include the coronary arteries and coronary veins (Figures 17–8, 17–9, and 17–10).

Coronary Arteries. The *left coronary* artery has its origin in the left aortic sinus, runs under the left atrium, and divides into the anterior descending and circumflex branches. The anterior branch descends in the anterior ventricular sulcus to the apex, supplying branches to both ventricles. Occlusion of this artery is common. The circumflex runs in the left part of the coronary sulcus and curves around and nearly reaches the posterior sulcus. It supplies branches to the left atrium and ventricle.

The right coronary artery has its origin in the right aortic sinus and turns to the right under the right atrium to the coronary sulcus. There are two branches, the *posterior descending* and the *marginal branch*. Branches of the left and right coronaries anastomose and encircle the heart forming a crown, hence their name. They supply the heart muscle with blood. Blood within the cavities of the heart nourishes only the endocardium.

Coronary Veins. The coronary sinus receives most veins of the heart and terminates in the right atrium.

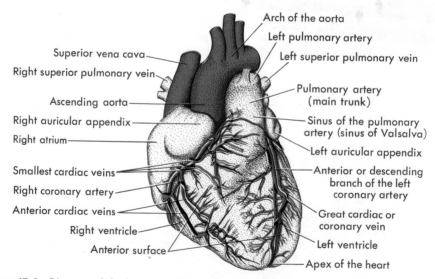

Figure 17–8. Diagram of the heart showing coronary arteries and veins. Anterior view. The great cardiac vein and artery indicate location of the septa. (Modified from Toldt.)

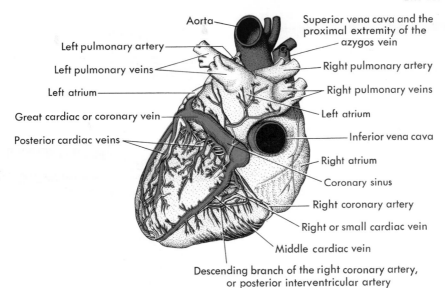

Figure 17–9. Diagram of the heart showing the coronary sinus and large veins on the dorsal wall of the heart. Which ventricle is to the left in the diagram? (Modified from Toldt.)

The great cardiac veins begin at the apex and ascend to empty into the coronary sinus.

There are smaller veins: the small cardiac, middle cardiac, and posterior veins, which begin at the apex and ascend to enter the coronary sinus.

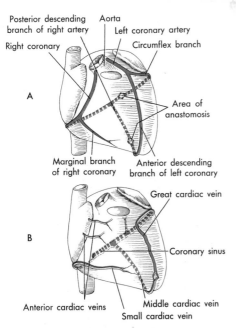

Figure 17–10. Diagram showing coronary arteries (*A*) and veins (*B*) and their connections. Dotted lines indicate vessels on posterior heart.

The mouths of the great and small cardiac veins have single cuspid valves, but are rarely efficient. Very small veins begin in the wall of the heart and open directly into the chambers of the heart (thebesian veins, or venae cordis minimae).

Coronary Circulation. The purpose of the coronary circulation is to distribute blood, containing oxygen, nutrients, and other substances, to the cardiac muscle cells and return to general circulation the products of metabolism.

The coronary arteries leave the aorta close to the heart. These arteries fill during myocardial relaxation and empty during contraction. Normally the rate of blood flow through these vessels is from 50 to 75 ml of blood per 100 gm of heart muscle per minute, depending upon the heart rate and heart volume. In other words, if the heart weighs 300 gm, from 150 to 225 ml of blood will flow through coronary vessels per minute. Since the output of the heart has been estimated to be about 3 to 4 liters per minute, this means that about 10 per cent of the heart output flows through the coronary arteries. This blood is returned to the heart via the coronary sinus, which opens directly into the right atrium. If this circulation is impaired (e.g., blockage of arteries by a thrombus or fatty deposit [atherosclerosis] in the arterial wall), damage to cardiac muscle cells may result with permanent scarring.

Collateral Circulation. The channels of communication between the artery-capillary-vein system of the heart are complex and numerous. There are direct channels between the coronary arteries and the chambers of the heart. At the apex of the heart the descending branches of both coronaries form an important anastomosis. In the myocardium of the posterior wall of the heart, branches of the circumflex artery anastomose with branches of the right coronary. Thus two crowns are formed around the heart. Since adequate valves are not found in the coronary vessels, it is possible that blood may backflow into the myocardium and enter the chambers of the heart.

Cardiac Cycle. The cardiac cycle consists of three phases: systole, diastole, and complete rest. Assuming a rate of 70 beats per minute, each cycle takes approximately 0.8 second, of which half (0.4 second) is complete rest (Figure 17–13, page 399). As the pulse rate increases, less time is spent at complete rest.

The Heart as a Pump. When the ventricles contract (*systole*), the muscle arrangement of the myocardium causes the size of the cavity within to decrease and at the same time the pressure rises. This causes the atrioventricular valves to close, and blood moves from the ventricle past the open semilunar valves into the pulmonary artery and the aorta. When the muscle relaxes (*diastole*), blood in these arteries causes the semilunar valves to float out so that they meet and securely prevent backflow of blood into the ventricular chambers. The decrease in pressure at this time also permits blood to move from the atria, past the open atrioventricular valves into the ventricle. Figure 17–11 shows these events during diastole and systole.

Cardiac Output. At each systole a volume of blood variously estimated at around 80 ml (man at rest) is forced from the left ventricle into the aorta. This is known as the *stroke volume*. A similar amount is forced from the right ventricle into the pulmonary artery. The total cardiac output per beat is, therefore, 160 ml.

With a pulse rate of 70, 5.6 liters (70 × 80 ml) of blood leave the left ventricle per minute. This is known as the *minute volume*. A similar amount leaves the right ventricle. With an increase or de-

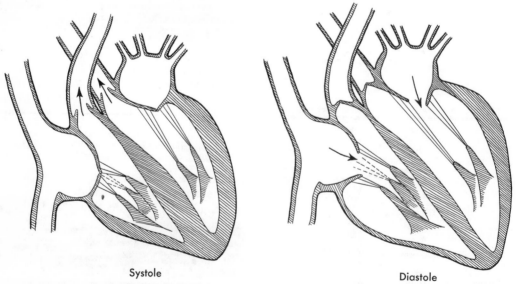

Systole   Diastole

Figure 17–11. Diagram of heart during systole and diastole. Note the change in thickness of the myocardium and overall size of heart.

crease in stroke volume, in pulse rate, or in both, the total output per minute would be increased or decreased. During exercise the total cardiac output may be doubled.

The heart muscle receives 10 per cent of cardiac output; the brain, 15 per cent; the liver, stomach, and intestines, 25 per cent; the kidneys, 20 per cent; and the soma, 30 per cent of the cardiac output. Blood supply to the brain is the most constantly maintained. In other organs the supply varies directly with activity. During diges-

tion the stomach and intestine receive far more blood than when at rest.

Sinoatrial (SA) Node and Atrioventricular (AV) Node. The SA and AV nodes are specialized cells found in the right atrial wall. The SA node is located beneath the opening of the superior vena cava. The specialized fibers are continuous over the atria.

The SA node is the causal factor of contraction. It initiates the electrical impulse that spreads out over both the atria (caus-

TABLE 17–1

MOVEMENT OF BLOOD THROUGH THE RIGHT HEART AND LUNGS

Venae cavae	During diastole of the atria, via the superior and the inferior venae cavae, coronary sinus, and other small vessels, blood enters and fills the right atrium and ventricle, which for the time may be thought of as a single chamber with the tricuspid valve open
Right atrium $\frac{5*}{0-2}$	The atrium contracts (atrial systole) and forces the blood over the open valve into the ventricle, which has been passively filled and now becomes well distended by the extra supply
Tricuspid valve	After a brief pause (0.1 second), rising muscle tension causes rapid rise in the pressure of the ventricle; when this pressure exceeds that of the atrium, the cusps approximate each other to close the valve. Chordae tendineae become taut and prevent the cusps from everting
Right ventricle Semilunar valve $\frac{16}{3}$	As soon as the rapidly rising pressure in the ventricle exceeds the pressure in the pulmonary artery, the pulmonary semilunar valve is forced open and blood moves on into the pulmonary artery. The valve closes rapidly
Pulmonary artery $\frac{20}{10}$ mean 15 Divides into two branches	The pulmonary artery divides into the right and left branches and transmits blood to the lungs
Lungs Capillaries 6	Here blood passes through innumerable capillaries that surround the alveoli of the lungs. Blood gives up carbon dioxide and the red cells are recharged with oxygen
Capillaries unite to form veins	The venules unite to form larger veins until finally two pulmonary veins from each lung are formed. These return the oxygenated blood to the heart and complete the pulmonary circulation
Pulmonary veins 4	
Left atrium $\frac{4-6}{1-2}$	

*Numbers indicate pressure, mm Hg.

ing them to contract) to the AV node. From here the impulse spreads down the entire bundle and finally reaches each cardiac muscle fiber of the ventricles, causing them to contract (Figure 17–12, *A*). The SA node has been called the pacemaker of the heart.

The AV node is located near the coronary sinus at the AV junction. The AV bundle descends in the septum, dividing into two bundle branches, which become continuous with the Purkinje fibers (Figure 17–13) about halfway down. The fibers run just beneath the endothelium toward the apex of the ventricle and then turn upward on the lateral walls. The Purkinje fibers supply the papillary muscles before the lateral walls and thus ensure that these muscles will take up the strain on the valve leaflets before the full force of contraction.

Wave of Contraction. If a stimulus is applied to one end of a muscle, a wave of contraction sweeps over the entire tissue. It is therefore easy to conceive how a wave of contraction can sweep over the muscular tissue of the atria, which is practically continuous. The question is—how is this wave transmitted to the muscular tissue of the ventricles, which, in man, is *not* continuous with that of the atria? The connecting pathway is furnished by the atrioventricular node (AV node), which transmits the nerve impulses by means of the AV bundle and causes the wave of contraction to spread from the atrioventricular openings over the ventricles to the mouths of the pulmonary artery and the aorta.

Cause of Cardiac Contraction. Contraction of cardiac muscle, like that of other types of muscle, is triggered by depolarization of the muscle membrane. However, cardiac muscle is autorhythmic—it is capable of self-excitation. The area with the fastest innate rhythm is the SA node; therefore, the electrical impulse for muscle membrane excitation originates here and spreads to adjacent areas of the atria. The branching arrangement of cardiac muscle cells and the tight fusion of cells at the sites of intercalated disks (Figure 3–25, p. 72) facilitate rapid and uniform spread

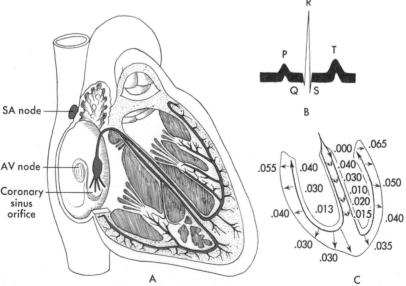

Figure 17–12. *A.* Diagram of the atrioventricular bundle of His. The atrioventricular (*AV*) node can be seen near the opening of the coronary sinus in the right atrium. Bundle indicated in red. Part of the sinoatrial node is seen in red between the base of the superior vena cava and the right auricular appendix. The tip of the left ventricular chamber is not shown. The figures in *C* indicate the time in seconds taken for the nerve impulses to reach the areas indicated. *B.* Electrocardiogram, showing excitation of different parts of the heart during one cycle. Wave P occurs during atrial excitation or systole; wave QRS occurs during ventricular systole; wave T occurs as ventricular excitation subsides.

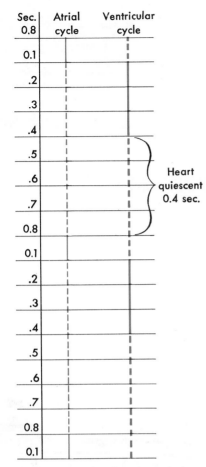

Sec. 0.8	Atrial cycle	Ventricular cycle
0.1		
.2		
.3		
.4		
.5		
.6		Heart quiescent 0.4 sec.
.7		
0.8		
0.1		
.2		
.3		
.4		
.5		
.6		
.7		
0.8		
0.1		

Figure 17–13. Atrial cycle and ventricular cycle, showing overlapping of diastole giving 0.4-second quiescent period of whole heart. Systole, *solid lines;* diastole, *dotted lines.*

Figure 17–14. Photomicrograph of Purkinje fibers.

of excitation, which continues through the AV node and bundle of His and over the ventricles.

The cause of this innate rhythmicity remains poorly understood; however, it is influenced by calcium, potassium, and sodium ions. Calcium has a stimulating effect on contraction, whereas potassium and sodium promote relaxation of the muscle. Alteration in the blood levels of these ions may result in inefficient contraction of the muscle, as well as in alteration of rhythm.

The heart is not in a continual state of contraction because of the *long refractory period of cardiac muscle.* From the time just before the contraction process begins, until some time after relaxation begins, the heart muscle is refractory to further excitation.

Electrocardiogram (ECG). Transmission of the depolarization wave, also called the *cardiac impulse,* generates electrical currents that are transmitted in small amounts to the surface of the skin. If electrodes are placed on opposite sides of the heart, the electrical impulses can be recorded—the *electrocardiogram.* The normal ECG is composed of a P wave, atrial depolarization, and a QRS complex occurring about 0.1 to 0.2 second later, ventricular *de*polarization. The T wave represents

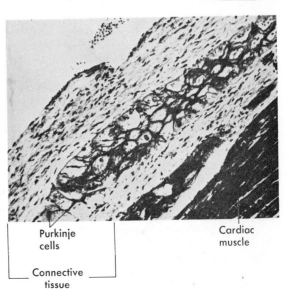

Purkinje cells

Cardiac muscle

Connective tissue

ventricular *re*polarization (Figure 17–12, *B*). (Atrial *re*polarization is masked by the QRS complex.) The impulse flow is extremely rapid (Figure 17–12, *C*), as it must be since the normal heart rate is about 70 times per minute when the individual is at rest.

Atrial Fibrillation. Fibrillation is irregular, weak, asynchronous contraction of muscle. In atrial fibrillation the impulses arriving at the SA node are weak and irregular so that it is stimulated irregularly and contraction of the ventricle may occur when it is not filled with the normal amount of blood. Thus, the pulse rate at the wrist may be of differing amplitude and irregular. At the same time all cardiac impulses that are heard with a stethoscope over the heart are not effective in causing movement of blood through the arterial system. This means that a *pulse deficit* occurs—the heart rate is greater than the pulse rate.

Ventricular Fibrillation. Spontaneous impulses originating in the ventricular myocardium in multiple areas cause ventricular fibrillation. Since the ventricles cannot forcefully empty, the condition cannot be tolerated for more than a very few minutes and it is life-threatening.

Heart Block. As a result of damage to the AV node, perhaps due to occlusion of a coronary artery branch supplying the septum, the node loses its ability to transmit impulses. The atria contract at a regular rate, but the ventricles contract at a slower rate, 40 to 50 beats per minute if the block is complete. At times the block is incomplete—some impulses are transmitted; e.g., in 2:1 block every second impulse is transmitted. It is possible now to treat this condition with a battery-operated *pacemaker*, which is inserted surgically under the skin in the fossa below the clavicle. Wires then lead directly through the chest wall to the heart, and the heart is stimulated electrically from the pacemaker.

Cardiac Arrest. When the heart ceases to beat, it is said to be *arrested*. Circulation may be temporarily maintained by forcefully compressing the heart by pressure on the lower third of the sternum. The depression of the sternum and ribs attached (which lie over the heart) causes mechanical compression of the heart (cardiac massage) and simulates ventricular contraction by forcing blood into the arteries by this means. Occasionally the heart will spontaneously resume its autorhythmicity. When accompanied by artificial respiration, cardiac massage is life-saving.

Heart Sounds. Listening to the heart with a stethoscope (*auscultation*) reveals two distinct sounds. The *first* heart sound (S_1) is low in pitch and relatively longer in duration than the *second* heart sound (S_2), which is relatively shorter and slightly higher in pitch. S_1 is due to the vibration of AV valves as they close and is heard best at the apex of the heart and in the fifth intercostal space near the sternum, on the left. S_2 is due to the vibration of the aortic and pulmonic valve closure; it is heard best at the second intercostal space close to the sternum. The abbreviations S_1 and S_2 derive from the fact that they are heard at the beginning and end of systole. Figure 17–15 illustrates these areas. Abnormal sounds called *murmurs* may be heard when valves are damaged as a result of inflammation. This causes turbulence of blood flow over the valves, which is audible.

Nerve Supply. The heart is supplied with two sets of motor nerve fibers. One set reaches the heart through the vagus nerves of the craniosacral system. Nerve impulses over these fibers have a tendency to slow or stop the heartbeat. They are *cholinergic* fibers (see page 304). The other set is sympathetic. The visceral branches of the first five thoracic nerves have their cells of origin in the lateral column of gray of the spinal cord. These fibers terminate in the sympathetic ganglia, forming the superior, middle, and inferior cardiac nerves. The postganglionic fibers pass to the heart where they quicken and augment the heartbeat; they are *β-adrenergic*. The vagus nerve has its origin in a nucleus in the medulla. The accelerator nerves also have connections in the medulla and through these centers either set may be stimulated.

In addition, the heart is supplied with afferent nerve fibers: one set from the aortic arch, called *depressor* fibers; the other set from the right side of the heart, called *pressor* fibers. Both sets of afferent fibers run within the sheath of the vagi to the cardiac center in the medulla. Impulses over the depressor fibers bring

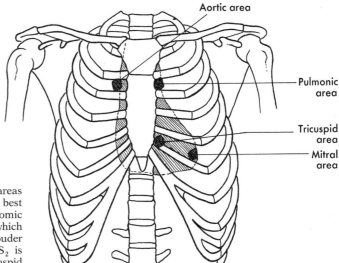

Aortic area

Pulmonic area

Tricuspid area

Mitral area

Figure 17–15. Diagram of the four areas on the chest where valve sounds can best be identified. These are not the anatomic areas of the valves, but the areas to which the sound is transmitted. S_1 is heard louder in the mitral and tricuspid areas; S_2 is heard louder in the aortic and tricuspid areas. Both S_1 and S_2 are heard in all areas.

about reflex inhibition of the heart—aortic reflex. Impulses over the pressor (sympathetic) fibers bring about reflex acceleration of the heart—right heart reflex.

Nervous Control of the Heart (see Figure 17–16). Although the heart contracts automatically and rhythmically, the continuously changing frequency and volume of the heart are controlled by the two sets of nerve fibers that enter the cardiac plexus. The sympathetic fibers follow along the coronary vessels and innervate all areas of both atria and ventricles. In general, stimulation of the sympathetic system increases the activity of the heart by *increasing* both force and rate of heartbeat, thereby increasing the effectiveness of the heart as a pump.

The vagus nerves chiefly innervate the atria and *decrease* the activity of the heart. At the SA node acetylcholine is secreted at the vagal endings, which decreases the rate and rhythm of the node. It also decreases excitability of the AV junctional fibers between the muscles of the atria and the Purkinje system, thereby slowing transmission of impulses.

This means that the heartbeat is controlled by two antagonistic influences, one tending to slow the heart action and the other to quicken it. If the inhibitory center is stimulated to greater activity, the heart

is slowed still further. If the activity of this center is depressed, the heart rate is increased, because the inhibitory action is removed. Stimulation of the accelerator nerves results in a quickened heartbeat.

Control of Heart Rate and Stroke Volume. The heart responds to changing body needs by alterations of rate and strength of contractions. This reflex response is due to three factors: (1) stretch "reflex," (2) pressure reflexes, and (3) chemoreceptor reflexes.

Stretch "Reflex." Cardiac muscle, like other muscle, increases its strength of contraction when it is stretched. When more blood enters the ventricles, as in exercise, the myocardium is stretched and therefore contracts more forcefully, thus increasing the stroke volume. This is sometimes called Starling's[4] law of the heart. During inspiration, with greater negative pressure in the thorax, more blood is drawn into the heart from the venae cavae, thus increasing the stroke volume slightly as compared with expiration.

Pressure Reflexes. There are receptors responsive to pressure in the large arteries close to the heart, in the large veins entering the heart, and in the right atrium.

[4] Ernest Starling, English physiologist (1866–1927).

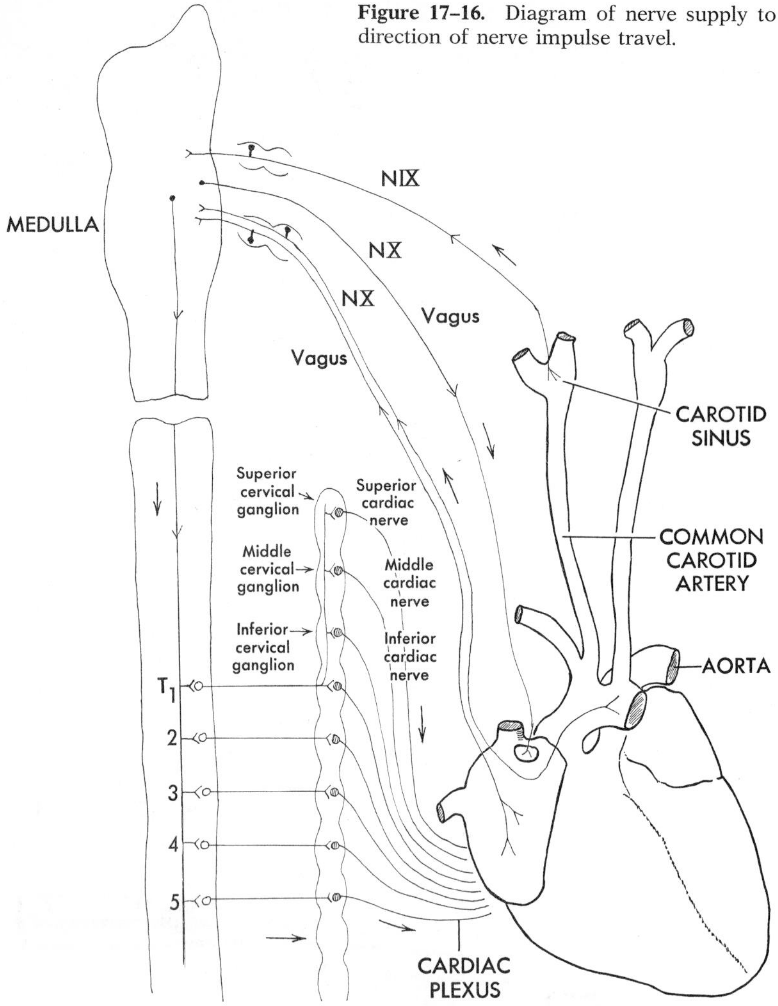

Figure 17–16. Diagram of nerve supply to heart. *Arrows* indicate the direction of nerve impulse travel.

These initiate the *aortic reflex,* the *carotid sinus reflex,* and the *right heart reflex* (Figure 17–17).

AORTIC REFLEX. Pressoreceptors in the aortic arch (aortic sinus) are sensitive to an *increase* in blood pressure. Impulses are conveyed over the afferent nerve fibers of cranial nerves IX and X to the cardiac centers of the medulla; the result is a *decrease* in heart rate. This may be considered a protective mechanism to prevent a dangerously high blood pressure.

CAROTID SINUS REFLEX. The carotid sinus is a slightly dilated area of the internal carotid artery at the point of bifurcation into the internal and external carotid arteries. Receptors in this sinus also respond to an *increase* in blood pressure, initiating impulses conveyed to the cardiac

centers over the cardiac branch of the ninth cranial nerve; the result is a decrease in heart rate.

RIGHT HEART REFLEX. There are receptors in the large veins entering the right heart and right atrium that are sensitive to changes in venous pressure. Increased pressure will accelerate heart action. Venous return and cardiac output are increased, which prevents pooling of blood in the venous system. Impulses are conveyed over afferent vagal fibers to the cardiac center in the medulla.

Pressoreceptors have been demonstrated on the pulmonary veins and in the wall of the left atrium. If pulmonary venous pressure increases, a reflex inhibition of the vasomotor center occurs that causes dilation of the vessels and reduction of

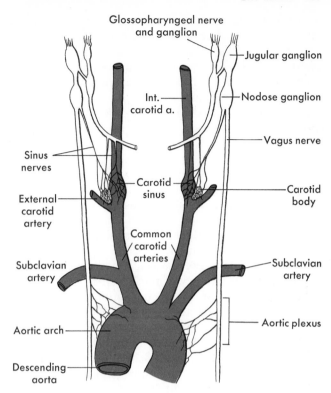

Glossopharyngeal nerve
and ganglion

Jugular ganglion

Int.
carotid a.

Nodose ganglion

Vagus nerve

Sinus
nerves

Carotid
sinus

Carotid
body

External
carotid
artery

Common
carotid
arteries

Subclavian
artery

Subclavian
artery

Aortic plexus

Aortic arch

Descending
aorta

Figure 17–17. Anterior view of aortic arch and carotid arteries, and the innervation related to aortic and carotid reflexes.

pressure in the lungs. This reflex is important because it protects against pulmonary edema resulting from excessive pulmonary pressure. At the same time there is reflex increase in stroke volume and rate to facilitate emptying of the ventricles.

Chemoreceptor Reflexes. At the bifurcation of the carotid arteries and in the arch of the aorta are small bodies, called, respectively, the *carotid body* and the *aortic body,* which are sensitive to oxygen *lack,* carbon dioxide excess, and hydrogen ion excess. Impulses generated traverse the glossopharyngeal and vagus nerves to the cardiac centers of the medulla, resulting in an increase in heart rate (and in respiratory rate through stimulation of the respiratory center).

Other Factors Influencing Cardiac Action. The frequency and strength of the heartbeat are affected by blood pressure; emotional excitement or keen interest; reflex influences that are of an unconscious character; the temperature of the blood; such characteristics of heart muscle as tone, irritability, contractility, and conductivity; physical factors such as size,

sex, age, posture, and muscular exercise; changes in the condition of the blood vessels; and certain hormones, e.g., epinephrine.

Under normal conditions the pulse rate is inversely related to the *arterial blood pressure;* that is, a rise in the arterial blood pressure causes a decrease in pulse rate, and a decrease in arterial blood pressure causes an increase in pulse rate.

The pulse rate is very susceptible to changing sensations. Especially is this true in *emotional excitement.* The heart also responds to reflex influences that are of an unconscious character, such as activity of the visceral organs. After meals the heart increases in rate and strength of beat.

Experimentally it has been demonstrated that abnormally high or low *temperatures* of the blood affect the frequency of the beat. If the heart is perfused with hot liquid, the rate is increased in proportion to the temperature until the maximum point, about 44° C (111.2° F), is reached. If the temperature is raised above this, the heart soon ceases to beat. In fever the increased rate of the heart action is thought to be due partly to the effect of the higher temperature of the blood on the heart muscle. On the other hand, if cold liquid is perfused through an animal heart, the rate is decreased, and the heart ceases to beat at about 17° C (62.6° F). Slowing of the heart by carefully inducing a fall in body temperature (hypothermia) permits cardiac surgery, which would not be possible with a rapidly moving heart at normal body temperature. At the same time this hypothermia slows cellular metabolism, thus decreasing the need for oxygen.

Conditions that affect the irritability, contractility, and conductivity of the heart muscle or reduce its normal *tone* are likely to change the frequency of the heartbeat, either accelerating or slowing the action. If the tone is decreased, the strength of the contractions is diminished.

In almost all warm-blooded animals the frequency of the heartbeat is in inverse proportion to the *size* of the body. An elephant's heart beats about 25 times per minute, a mouse's heart about 700 times per minute. Generally speaking, the smaller the animal, the more rapid is the consumption of oxygen in its tissues. The increased need for oxygen is met partly by a faster heart rate.

The heartbeat is somewhat more rapid in *women* than in *men.* The heart rate of a female fetus is generally 140 to 145 per minute, that of the male, 130 to 135.

Age has a marked influence. At birth the rate is about 140 per minute, at three years about 100, in youth about 90, in adult life about 75, in old age about 70, and in extreme old age 75 to 80.

The *posture* of the body influences the rate of the heartbeat. Typical figures are: standing, 80; sitting, 70; and recumbent, 66. If an individual remains in a recumbent position and keeps quiet, the work of the heart may be decreased considerably. This is the reason why patients with certain types of heart disease are kept in a recumbent position. On other occasions the physician may suggest the sitting position for his heart patient, if the patient is emotionally more relaxed in that position.

Muscular exercise increases the heart rate. This is due to (1) the activity of the cardioinhibitory center in the medulla being *depressed* by the motor impulses from the more anterior portions of the brain to the muscles, probably by means of collateral fibers to the cardiac center; (2) a stimulation of the cardioaccelerator center; (3) an increased secretion of epinephrine and other hormones that accelerate heart action; (4) increased temperature of the blood; and (5) the pressure of the contracting muscles (including respiratory movement) on the veins sending more blood to the heart, so that the right side is filled more rapidly. This increase of venous pressure reflexly accelerates the heartbeat.

In order to function effectively, the heart requires a certain amount of *resistance*, and normally this is offered by the blood vessels. The heart will beat more slowly and strongly in response to increased resistance, provided the resistance is not too great. In the latter case the heart is likely to dilate, and its action becomes frequent and weak. The most common cause of abnormally high resistance is abnormal constriction of the arterioles, and this in turn results in high systolic and diastolic pressure (hypertension). In the presence of arteriosclerosis the vessels do not stretch during contraction of the heart, which causes systolic blood pressure to rise (systolic hypertension). When the resistance is below normal, the heartbeats are frequent and weak. Lessened resistance is due either to a relaxed condition of the blood vessels or to the loss of much blood or of much fluid from the blood.

Certain internal secretions affect the frequency and strength of the heartbeat. *Thyroxin* produces a faster pulse. The partial removal of excessively active thyroid glands results in a slower heart rate. *Epinephrine* from the adrenal glands increases the frequency and force of the heartbeat.

Questions for Discussion

1. Why does edema occur when the individual has a constricted tricuspid valve?
2. Discuss the function of the right heart reflex in relation to exercise.
3. What is meant by coronary circulation? By collateral circulation?
4. Explain the value of an electrocardiogram.
5. Where does the cardiac impulse originate?
6. Discuss the location and function of the pressoreceptors.
7. Where are the chemoreceptors located? What substances are they sensitive to?
8. Which hormones affect the heart rate? Explain.

Summary

Cardiovascular System
- Heart
- Arteries—small arteries are named arterioles
- Capillaries
- Veins—small veins are named venules

Heart

Location
- Between lungs, within mediastinum
- Above diaphragm

Structure
- Outside covering—*pericardium*
  - Fibrous portion
  - Serous
    - Visceral
    - Parietal
- Muscle substance—*myocardium*
  - Muscle bundles of the atria
  - Muscle bundles of the ventricles
  - Atrioventricular bundle—bundle of muscular and nervous tissue located in septum between right and left heart, which connects the musculature of atria and ventricles
- Smooth lining on inside—*endocardium*

Cavities
- Right heart
  - Right atrium
    - Receives blood
    - Thin walls
  - Right ventricle
    - Expels blood into pulmonary artery
    - Thick walls
- Left heart
  - Left atrium
    - Receives blood
    - Thin walls
  - Left ventricle
    - Expels blood into aorta
    - Very thick walls

Orifices
- Right heart
  - Right atrium
    - Superior vena cava—returns blood from upper portion of body
    - Inferior vena cava—returns blood from lower portion of body
    - Coronary sinus—returns blood from heart muscle
  - Atrioventricular orifice between atrium and ventricle
  - Right ventricle
    - Pulmonary artery—carries blood from heart to lungs
- Left heart
  - Left atrium
    - Two right pulmonary veins
    - Two left pulmonary veins
    - Return blood from lungs
  - Atrioventricular orifice between atrium and ventricle
  - Left ventricle
    - Aorta—distributes blood to all parts of the body

Valves
- **Right atrioventricular or tricuspid valve**—composed of three cusps, situated in the right ventricle
- **Left atrioventricular or bicuspid, or mitral, valve**—composed of two strong, thick cusps, situated in the left ventricle
- **Function**—prevent flow of blood from ventricles into atria
- **Semilunar valves**
  - **Aortic**—composed of three half-moon-shaped pockets between aorta and left ventricle
  - **Pulmonary**—composed of three half-moon-shaped pockets between pulmonary artery and right ventricle
  - **Function**—prevent flow of blood from arteries into ventricles

Heart (*cont.*)

Lymph vessels { Heart well supplied with lymph capillaries

Coronary circulation
- Arteries { Right coronary / Left coronary } branches from aorta
- Veins {
  1. Cardiac veins empty into coronary sinus
  2. Three or four small veins empty into right atrium
  3. Veins of Thebesius empty into atrium and ventricles
- Function—distribute blood to cardiac muscle

Collateral circulation { Channels of communication—complex and numerous at apex of heart—descending branches of both coronaries anastomose

Pumping action { By rhythmical contractions blood is moved from veins through heart into arteries

Wave of contraction
- Starts at sinoatrial node, transmitted through the atrial muscle to the AV node, which in turn transmits it via the AV bundle to the ventricles
- *Heart block*—condition resulting from damage to atrioventricular bundle and consequent failure to transmit impulses from atria to ventricles. Rate of contraction of ventricles slower than that of atria.
- *Fibrillation*—rhythmical contractions interfered with. Atria undergo irregular twitching movements. Contractions of ventricles irregular and rapid. Difference between heart rate and pulse called *pulse deficit*

Cardiac cycle
- Coordinated contraction of cardiac muscle
1. **Systole**—contraction
2. **Diastole**—dilatation
3. **Rest**—quiescent
{ Cardiac cycles, 70–72 per minute / Occupies about 0.8 second / Systolic and rest period each about 0.4 second }

Heartbeat — Cause {
1. Unknown. Myogenic theory makes automatic contractility of muscle cells responsible
   - Stimulated by epinephrine and norepinephrine
   - Inhibited by acetylcholine
2. SA and AV nodes control frequency of electrical impulses over the heart from veins through heart to arteries
3. Rate and strength of beat meet needs of body; under chemical and nervous control

Heart sounds
- —turbulence of blood caused by closure of atrioventricular valves and contractions of the ventricles
- —turbulence of blood caused by closure of the semilunar valves
- Other sounds may be identified

Nerve supply
- Craniosacral—vagus—inhibitory to heart
- Thoracolumbar—visceral branches from first five thoracic nerves. Accelerators to heart; nerve endings are β-adrenergic

Cardiac nerves { Fibers from superior, middle, inferior ganglion { Form the superior, middle, and inferior cardiac nerves, accelerators to heart

Right heart reflex { Receptors in large veins of right heart / Increases heart rate and cardiac output

Aortic reflex { Receptors on arch of aorta / Sensitive to fluctuations in arterial pressure / A rise in pressure slows heartbeat

Carotid sinus reflex { Receptors respond to arterial pressure rise / Slows heart rate

Chemoreceptors { Located in carotid bodies and aortic bodies / Sensitive to lack of oxygen

Factors affecting
- Frequency of and strength of cardiac contraction
- Influenced by {
  - Emotional excitement
  - Unconscious reflex influences
  - Posture, muscular exercise, temperature of blood, sex
  - Epinephrine and other hormones

Additional Readings

GUYTON, A. C.: *Basic Human Physiology: Normal Functions and Mechanisms of Disease*, 2nd ed. W. B. Saunders Co., Philadelphia, 1977, Chapters 19, 20.

HAM, A. W.: *Histology*, 7th ed. J. B. Lippincott Co., Philadelphia, 1974, pp. 560–81.

PANSKY, B., and HOUSE, E. L.: *Review of Gross Anatomy*, 3rd ed. Macmillan Publishing Co., Inc., New York, 1975. Numerous diagrams; see specific arteries and veins.

CHAPTER 18

Dynamics of Circulation

Chapter Outline

DISTRIBUTION OF BLOOD
PHYSICAL ASPECTS OF BLOOD FLOW
  VISCOSITY
  LAMINAR FLOW
  CROSS-SECTIONAL AREA
INTERRELATIONSHIP OF PRESSURE AND
       BLOOD FLOW
PULSE PRESSURE
  HEART RATE
  PERIPHERAL RESISTANCE
  STROKE VOLUME
CONTROL OF ARTERIAL PRESSURE
  SYMPATHETIC NERVOUS SYSTEM
  METABOLIC FACTORS

CONTROL OF VENOUS PRESSURE
NORMAL VARIATIONS IN BLOOD
      PRESSURE
ABNORMAL VARIATIONS IN BLOOD
      PRESSURE
  HYPERTENSION
  HYPOTENSION
  SHOCK
PULSE
  POINTS TO NOTE IN FEELING A PULSE
MEASUREMENT OF ARTERIAL
      PRESSURE
MEASUREMENT OF VENOUS
      PRESSURE

The pumping of blood through the vascular system, by the heart, has as its sole purpose maintenance of an environment for the cell that is conducive to optimal functioning. Many factors influence blood flow; there is dynamic readjustment as cell needs change. Cell death follows impairment of the circulation.

Distribution of Blood. The greatest percentage of blood is located in the systemic circulation, and a lesser amount in the pulmonary circulation, as seen in Table 18–1. In the systemic circulation, the veins contain two thirds of the blood and only a small per cent is in the capillary bed.

TABLE 18–1
DISTRIBUTION OF BLOOD

Systemic circulation	84%	64% Veins
		15% Arteries
		5% Capillaries
Pulmonary circulation	9%	
Heart	7%	
	100%	

Blood ejected from the left ventricle into the aorta is distributed to tissues in varying amounts, as seen in Table 18–2. The amount of blood in a particular area is dependent on the basic vascularity of the tissue, how many capillaries it has (e.g., skeletal muscle as compared with bone), and the degree of activity of the tissue or organ. When the digestive organs are active, they need more blood than when they are inactive. If the skin is exposed to high temperatures, the arterioles that bring blood are dilated, and blood flow near the surface (the skin) is increased; this aids in heat radiation and control of body temperature. On the other hand, a cold environment causes constriction of the skin

407

TABLE 18–2
MOVEMENT OF BLOOD THROUGH ORGANS

Name of Part	Resting Tissue (amount of blood per 100 gm of tissue per unit of time)	Needs in General
Skeletal muscle	5 ml per minute	During exercise the amount of blood needed is proportional to metabolic activity; strenuous exertion, about 35 ml per 100 gm of tissue per minute
Bone	Receives a rich blood supply	During physical activity, there is increased blood flow
Heart and coronary arteries	Receive 10 per cent of cardiac output	The amount then varies directly with heart rate. Flow is slowest during systole
Lungs	Blood passing through the lungs per minute is directly related to heart rate and stroke volume; i.e., if stroke volume is 70 ml and pulse rate is 72, about 5040 ml of blood pass through the lungs in one minute	
Liver	200–300 ml per minute	Through a special mechanism located on the hepatic veins, the liver has the ability to either hold back or give blood to circulation as need arises
Stomach	25 ml per minute	When actively secreting digestive fluids and absorbing end products of digestion and fluids, the amount of blood through organs is tremendously increased
Intestines	65–70 ml per minute	
Kidneys	Receive about 1300 ml of the total resting output of the heart	Normally, each minute, the glomeruli filter about 120 ml of protein-free fluid, and at the tubule, about 119 ml are returned to the bloodstream
Thyroid gland	560 ml per minute	Metabolic activity is mediated through thyroxin; it is essential that blood flow and gland activity keep pace with metabolic needs
Spleen	40 ml per minute	The spleen serves as a reservoir for blood, and during excessive demands for blood by the body, the spleen releases blood
Human brain	Receives 15 per cent of cardiac output, or about 200 ml per minute	Not affected by physical or mental activity

arterioles with resulting paleness. The brain is an exception to this variation: blood supply is not reduced during sleep and not increased by mental activity.

This variation in blood flow is seen at the capillary level in Figure 18–1; as blood flow increases, the capillary bed is engorged, and as it decreases, some capillaries are closed off and are temporarily not active. It is possible to increase the flow to one area by decreasing flow to less active areas. In addition, blood that is "stored" in the blood reservoirs, the spleen and liver, is released into the general circulation.

Physical Aspects of Blood Flow. Blood flow may be simply defined as the amount of blood flowing past a particular point in

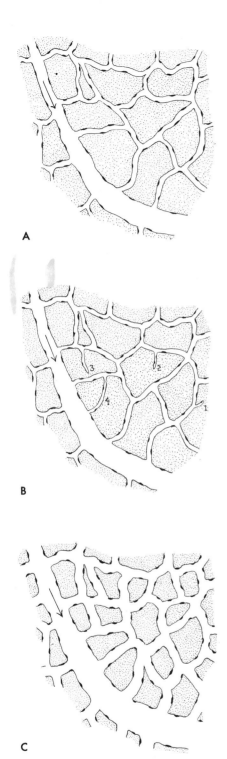

Figure 18–1. Diagrams showing opening of extra capillaries with increasing blood flow. *A* shows capillaries; *B* and *C* show progressive opening of an increasing number of capillaries. *1, 2, 3,* and *4* in *B* show the different degrees of the opening of capillaries. The *arrows* indicate the direction of blood flow.

the circulation in a given amount of time (as in milliliters per minute). Of the many factors that influence blood flow three are what might be considered as being of a physical nature: the viscosity of blood, the laminar flow of blood, and the cross-sectional area of the arterial system, the capillary bed, and the venous system.

Viscosity. Blood is viscous and, as with any fluid, friction develops at the point of contact with the inner wall of the containing vessel in which it flows. As viscosity increases (e.g., with dehydration), when the hematocrit is greatly elevated, the greater is the friction. Slowing of the blood flow in the arterial system is related in part to the friction.

Laminar Flow. When blood flows through a long, straight tube such as the aorta and large arteries and veins, it tends to flow in concentric layers, fastest in the center, and more slowly at the outermost layers, slowest of all in the layer adherent to the inner wall of the blood vessel. This type of concentric layer flow is termed *laminar* flow. Rapid movement in the center layers facilitates rapid movement of blood to distal branches; slow movement at the point of contact with the vessel wall is helpful for diffusion. Another important aspect of laminar flow is that a slight change in the diameter of the vessel has an immense effect on the blood flow. As the diameter increases, not only does the ability to transmit blood increase but also this increase is logarithmic: doubling of the diameter increases the blood flow not just double, but to the fourth power, or 16 times the amount.

Cross-Sectional Area. As the arterioles open, into the capillary bed, the total cross-sectional area increases dramatically, as seen in Figures 18–2 and 18–3. At the same time the speed of flow is decreasing dramatically. It must be apparent that slowly moving blood is conducive to diffusion of particles across the membrane.

Interrelationship of Pressure and Blood Flow. Blood flow to a particular area is determined by two factors: *resistance* to flow, primarily in the arterioles as the tone

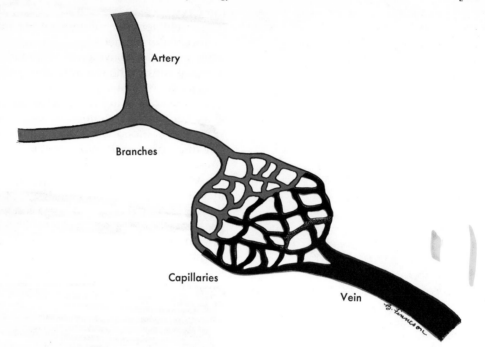

Figure 18–2. Diagram to illustrate cross-sectional areas. An artery divides into two branches. These are individually of smaller cross section than the main trunk, but united they exceed it. Linear velocity will be lower in the branches than in the parent artery. The sum of the cross-sectional areas of the capillaries is greater than that of the artery or vein.

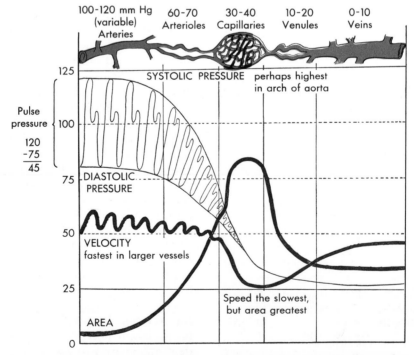

Figure 18–3. Diagram showing relationships between arterial, capillary, and venous blood pressures; area in various blood vessels; and speed of blood flow.

of the smooth muscle of the wall varies; and *pressure differences* at different points along the arterial branches, (i.e., blood flows from a vessel with higher pressure to one of lesser pressure).

Initially blood pressure is high, in the aorta. An average reading is $^{120}/_{80}$–120 mm Hg at the peak of systole, 80 mm Hg when diastole is most marked. The distensibility and elasticity of arterial walls act to propel blood in pulsations toward the capillaries. During this process the blood pressure is gradually decreasing until, at the arterial end of the capillary, it is only about 30 mm Hg and at the venous end about 10 mm Hg. The capillary walls are thin and weak, the capillaries are minute, and pressure is slight; however, it is high enough to facilitate movement of particles out of the capillary (page 38), and to promote onward movement of the blood. Pressure continues to fall (Table 18–3) in the venous system.

Veins are less distensible and elastic than arteries, and flow is not pulsatile (unless there is a blockage in venous return.) The speed of flow is rapid in the aorta, slowing progressively with arterial branching until in the capillaries the flow is relatively slow. As veins coalesce, flow increases again.

Pulse Pressure. The difference between the systolic and diastolic pressure is called the *pulse pressure;* e.g., a blood pressure of $^{120}/_{80}$ mm Hg equals a pulse pressure of 40 mm Hg. Pulse pressure is determined by *stroke volume, heart rate,* and *peripheral resistance.*

Heart Rate. When the heart rate is slow, there is a relatively long time for "runoff" of blood through the arterial system between beats. This time span influences diastolic pressure more than systolic, tending to decrease diastolic pressure and, therefore, to increase the pulse pressure. When the beat is rapid, the runoff time is shorter, diastolic pressure is higher, and the pulse pressure is decreased.

Peripheral Resistance. The resistance to blood flow has a direct influence on runoff. When the resistance is high, runoff will be slow, the aorta will remain filled with blood, and systolic pressure will be high.

Stroke Volume. An increase in *stroke volume* causes increased distention of the aorta and, therefore, a higher systolic pressure. If runoff time and peripheral resistance remain constant, an increase in stroke volume will increase pulse pressure; conversely, a decrease in stroke volume decreases pulse pressure.

Control of Arterial Pressure. Arterial blood pressure is determined by cardiac output and by peripheral resistance. Both output and resistance are under the control of the *sympathetic nervous system;* various *metabolic factors* influence the pressure as well.

Sympathetic Nervous System. Sympathetic control of cardiac output was discussed in the previous chapter (page 400); therefore, this section will focus on its effect on the arterioles. The arterioles have an abundant supply of nerve fibers, branches of the sympathetic system. An increased number of impulses flowing to the smooth muscle of the arterial wall increase the tone, thus increasing the arterial blood pressure. These impulses originate in the vasomotor centers of the medulla, which is tonically active at a low level: there is a constant outflow of impulses, which is responsible for the basic tone of the blood vessels. When the vasomotor center is stimulated—by emotion, cold, or impulses from the cardiac or respiratory centers or the hypothalamus—the outflow of impulses is increased and the arterial pressure rises. A negative feedback to the vasomotor centers occurs from the aortic and sinus node receptors (page 402); the result is a decrease in blood pressure.

Postural changes, such as changing from a sitting or lying-down position to a standing position, have a direct effect on these pressoreceptors. In the standing position blood fills the veins below the heart due to the effect of gravity; there is a short period of time when the return to the right side of the heart is less; stroke volume is less, and this activates the negative feedback effect of the sinus and aortic nodes.

TABLE 18–3

MOVEMENT OF BLOOD THROUGH THE LEFT HEART AND TO THE SOMATIC CAPILLARIES AND BACK TO THE LEFT HEART

Pulmonary veins
$\downarrow$ 4*

Left atrium

$\dfrac{6}{1-2}$

Bicuspid valve
$\downarrow$

Left ventricle

During diastole of the atria, oxygenated blood from the pulmonary veins enters the left atrium and fills it. The left ventricle relaxes and the pressure within it falls, the bicuspid valve opens and blood enters the ventricle and fills it. Systole of the atrium begins and ventricular filling is completed

$\dfrac{120}{7}$

Aortic semilunar valve
$\downarrow$

Aorta

Ventricular systole is initiated. Blood gets behind the cusps of the bicuspid valve and closes them. Intraventricular pressure rises, and when it exceeds the pressure in the aorta, the aortic semilunar valve opens and blood moves on into the aorta under high pressure. High pressure in the aorta causes the valve to close rapidly

$\dfrac{130}{80}$

Conducting arteries

Blood moves on into the conducting or elastic arteries (brachiocephalic, subclavian, common carotids, internal iliac, femoral, etc.)

$\dfrac{130}{80}$

Distributing arteries

gradually decreasing
$\downarrow$

Arterioles

Blood is forced onward through the elastic arteries into the distributing (muscular arteries) such as the axillary, radial, popliteal, tibial, and finally into the arterioles, where blood is moving in a steady stream, and then on into the capillaries, where the main work of the vascular bed is accomplished

$\dfrac{85}{60}$

Capillaries
$\mid$ 30

$\downarrow$

Venules
$\mid$ 10

$\downarrow$

The capillaries unite to form venules and these in turn unite to form veins, then larger veins, until blood finally reaches the right atrium and the circuit begins again

Veins

gradually decreasing

Right heart
$\mid$ 0–4

In the right heart blood contains its full complement of nutrients, hormones, and all other substances except oxygen needed by cells

$\downarrow$

Pulmonary artery

Blood moves from the right heart to the lungs

$\downarrow$

Lungs

Blood is forced on into the capillary network surrounding the air sacs

$\downarrow$

Lung capillaries

In these capillaries there is exchange of carbon dioxide and oxygen

$\downarrow$

Pulmonary veins

These veins take the oxygenated blood back to the left heart and the circuit begins again

*Figures are mm of mercury (Hg).

(Occasionally this decreased cardiac output is noticeable to the individual who jumps out of the bed to a standing position—the light-headed feeling is due to decreased blood pressure [*orthostatic hypotension*]; in the normal individual the feeling is of very short duration.)

Metabolic Factors. Certain hormones and electrolyte changes have an influence on the blood pressure.

HORMONES. Circulating *norepinephrine* and its secretion at sympathetic nerve endings have a stimulating effect on the sympathetic effectors, causing an increase in tone in the arterioles and an increase in arterial pressure. *Epinephrine* increases cardiac output and causes constriction of the vascular bed with the exception of skeletal muscle, which dilates. Physiologically this dilation of skeletal muscle vasculature may cancel out the pressure effects of increased cardiac output and increased peripheral resistance in other areas. Thus, there is usually no noticeable change in the arterial pressure.

ELECTROLYTES. Decreased oxygen level, increased carbon dioxide level, and decreased pH in the circulating blood activate the chemoreceptors in the aortic and carotid bodies, thus increasing heart rate, which tends to increase the arterial blood pressure. Locally, these same changes in oxygen, carbon dioxide, and pH have a vasodilating effect by causing relaxation of the smooth muscle of the arterioles. This increases the blood flow to that area but has little effect on arterial pressure because the sympathetic nervous system effect is more powerful. The local effect, however, is helpful when the tissue is active and needs increased blood supply.

Control of Venous Pressure. Strictly speaking, there is no control of venous pressure such as the control of arterial pressure. Venous pressure is always low (0 to 4 mm Hg in the venae cavae) and it is directly related to cardiac functioning. Venous pressure is elevated only when the ventricles fail to empty, in which case there is cardiac failure, of varying degrees. When this occurs, there is elevated pressure in the contributing veins, which may be seen to be engorged if they are superficial, e.g., the jugular veins.

Normal Variations in Blood Pressure. When one speaks of "blood pressure," it is the arterial pressure that is usually referred to, although, as has been mentioned, there is pressure in the capillaries and veins. The normal adult blood pressure range is 90 to 140 systolic and 60 to 90 diastolic.

At birth the average systolic pressure is 40 mm Hg. It increases rapidly during the first month to 80 mm. Then it increases slowly, and at the age of 12 years the average reaches 105 mm. At puberty a somewhat sudden increase occurs; the average is 120 mm. A steady, slow, but not marked increase in blood pressure occurs normally from adolescence to adulthood.

Blood pressure is increased by *muscular activity.* The amount of increase depends upon the amount of energy required for the activity and upon individual differences. Systolic pressure is raised slightly after *meals. Pain* and emotional factors, such as fear or worry, raise systolic pressure considerably. During quiet, restful *sleep* systolic pressure falls; the lowest point is reached during the first few hours. It rises slowly until the time of waking. Systolic pressure is about 8 to 10 mm lower in women than in men, until menopause occurs.

Abnormal Variations in Blood Pressure.

Hypertension. A pressure above $^{140}/_{90}$ mm Hg at rest is termed hypertension. Depending on the cause, hypertension is *secondary*, when it is caused by disease: pathology of the kidney, excessive secretion of epinephrine and norepinephrine as in a tumor of the adrenal gland, or perhaps overactivity of the thyroid gland. It is *primary*, or *essential*, when there is no known cause. Essential hypertension has a detrimental effect on the heart because the increased peripheral resistance is a force against which the heart must forcefully contract.

Blood pressure is raised above normal

when the distensibility of the arteries is reduced, as in arteriosclerosis; by various diseases of the heart, liver, and kidneys that interfere with the venous circulation; usually by fever; and by increased intracranial pressure, as in fracture of the skull. In negative acceleration, as when a plane dives, cardiac output increases and arterial pressure is increased. High pressure causes vessels in the head and brain to be overfilled with blood, and impairment of brain function may occur.

Hypotension. Continual blood pressures below $^{90}/_{60}$ mm Hg is termed hypotension. Its cause is not known, and if untreated it results in damage to the heart and kidney and increases the likelihood of cerebrovascular hemorrhage (stroke).

Shock. A severe drop in systolic pressure is termed shock. It may be of relatively short duration, as in primary shock, or relatively long, as in hemorrhagic shock.

Primary shock, also called *syncope*, or fainting, may result from psychologic or physical trauma. This transient loss of overall vasomotor tone results in pooling of blood in the lower extremity and decreased venous return, and the person usually falls to the floor. The flat position, particularly if the legs are elevated, facilitates venous return. The recovery of vasomotor tone is spontaneous and rapid.

Hemorrhagic shock is much more serious and may be life-threatening if untreated. Compensatory mechanisms, such as constriction of the vascular bed, increased heart rate, and redistribution of blood, permit loss of up to 1 liter of blood without a serious drop in systolic pressure. However, with larger losses of blood there is serious impairment of the circulation and a progressive drop in cardiac output as a result of the decreased blood volume. The *first stage* in compensation is formation of angiotensin (see page 548) and stimulation of the sympathetic nervous system. The result is vasoconstriction of all the nonvital areas, e.g., skin, skeletal muscle, the digestive tract, and kidney. At this time the low pressure at the arterial end of the capillary causes less fluid to be filtered out and more tissue fluid to be reabsorbed. The resultant dilution of the blood (seen as a decreased hematocrit) helps to maintain the total blood volume. Sympathetic stimulation of the heart increases the heart rate, and cardiac output is improved by the increased rate. At this time, the symptoms of shock (rapid, thready pulse, sweating, dilated pupils, and anxiety) are symptoms of sympathetic stimulation.

Unless treatment is initiated for the shock state, and if the individual's compensatory mechanisms are inadequate for the degree of shock, the second stage is reached, *progressive* shock. At this stage the blood volume is severely decreased (called *oligemia*) and tissue damage occurs: kidney cells fail to function and there is no urine output, liver cells fail to function, and eventually there is failure of electrolyte balance of the body. Usually the blood flow to the heart and brain is barely adequate, and as this too begins to fail, the last stage of shock occurs—*irreversible* shock. Death is inevitable.

The sequence of events described above indicates the vital importance of an adequate arterial pressure for normal functioning of cells.

Pulse. The alternate expansion and contraction of an artery constitute the pulse. The pulse *does not mark* the arrival of the ejected blood at the point felt, but represents the *pressure change* brought about by the ejection of blood from the heart into the already full aorta and propagated as a wave through the blood column and the arterial wall to the periphery.

All arteries have a pulse, but it is more readily counted wherever an artery approaches the surface of the body. These locations are as follows: the *radial* artery, at the wrist—the radial artery is usually employed for this purpose on account of its accessible situation; the *temporal* artery, above and to the outer side of the eye; the *external maxillary* (*facial*) artery, where it passes over the lower jaw bone, which is about on a line with the corners of the mouth; the *carotid* artery, on the side of the neck; the *brachial* artery, along the inner side of the biceps; the *femoral* artery, where it passes over the pelvic bone; the *popliteal* artery, behind the knee; the *dorsalis pedis,* over the instep of the foot.

Points to Note in Feeling a Pulse. In feeling a pulse, the following points should be noted:

1. The *frequency,* or *number of beats per minute,* should be normal for the individual concerned. The intervals between the beats should be of equal length. A pulse may be irregular in frequency and rhythm. When a pulsation is missed at regular or irregular intervals, the pulse is described as *intermittent.*

2. The *force,* or *strength,* of the pulsation. Each beat should be of equal strength. Irregularity of strength is due to lack of tone of the cardiac muscle or of the arteries. Occasionally the heartbeat appears to be divided, and two pulsations are felt, the second being weaker than the first. This is known as a *dicrotic* pulse.

3. The *tension,* or *resistance* offered by the artery to the finger, is an indication of the pressure of the blood within the vessels and the elasticity or inelasticity of the arterial walls. A pulse is described as *soft* when the tension is low and the wall of the artery is elastic. A pulse is described as *hard* when the tension is high and the wall of the artery is stiff, thick, and unyielding.

Measurement of Arterial Pressure. Arterial pressure may be measured directly by insertion of a tube into an artery; however, except experimentally, it is measured *indirectly* with the use of a sphygmomanometer. It consists of a scaled column of mercury (mercury manometer) marked in millimeters, which is connected by rubber tubing with an elastic air bag contained in a fabric cuff. The air bag is in turn connected with a small hand pump. Some instruments are constructed with a spring scale (aneroid manometer), shown in Figure 18–4, but the principle is the same. The air bag contained in the sleeve is wrapped snugly about the arm just above the elbow with the air bag centered over the brachial

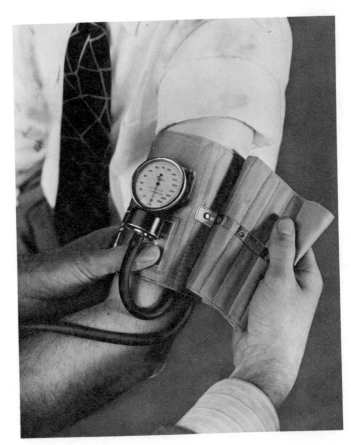

Figure 18–4. Method of using aneroid sphygmomanometer for measuring arterial blood pressure. (Courtesy of Taylor Instrument Co.)

artery. A finger is placed on the pulse at the wrist (as the bag is inflated), and a point is finally reached where the pulse disappears; then the bag is very slowly deflated until the pulse can just be felt. The pressure in the bag, therefore, against the artery from the outside, as indicated by the reading on the instrument, is approximately equal to the pressure that the blood exerts against the wall of the artery from the inside. This is the *systolic pressure* and is the greatest pressure that cardiac systole causes in the brachial artery. In the auscultation method of reading blood pressure, a stethoscope is placed over the brachial artery in the bend of the elbow. Blood pressure is then indicated by sounds heard through the stethoscope. The bag is inflated as before until all sounds cease. It is then slowly deflated until the pulse can just be heard. The reading on the manometer at this time indicates systolic pressure. The deflation

of the bag is then continued, and the reading on the manometer just before the last sound of the disappearing pulse indicates *diastolic pressure*, which is the lowest pressure that cardiac diastole causes in the brachial artery.

As pressure falls in the sphygmomanometer, the sounds heard change. First a clear, sharp, tapping sound is heard that corresponds to systolic pressure. The next sounds become softer, and as pressure continues to fall the sound gets louder again, then becomes muffled—this corresponds to diastolic pressure. The sound lasts during the next 4 to 6 mm of Hg fall, and then all sounds cease to be heard. Diastolic pressure is usually recorded when the muffled sound is heard and when the sound is completely lost $\frac{120}{80-75}\frac{\text{(systolic)}}{\text{(diastolic)}}$.

Measurement of Venous Pressure. Venous pressure is so low that it is measured in

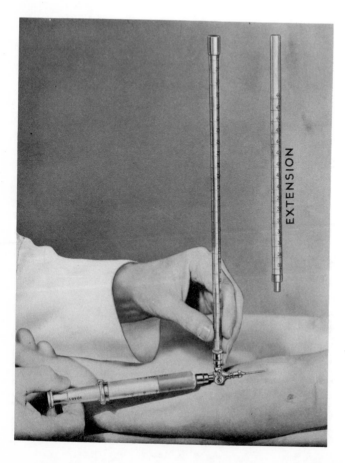

Figure 18–5. Use of water manometer for measuring venous blood pressure. (Courtesy of Taylor Instrument Co.)

terms of millimeters of water (actually a salt solution) instead of millimeters of mercury. It is measured directly by inserting a sterile flexible plastic tubing into the brachial vein, threading it back to the superior vena cava and right atrium. To this tubing is attached a manometer (Figure 18–5).

Venous pressure is usually 10 to 20 mm water. Although there is considerable variation from one individual to another, the venous pressure for the individual is fairly constant. Thus, it is the variation within the individual that is significant, rather than the absolute value in millimeters.

Questions for Discussion

1. How does the heart influence blood flow?
2. What changes in blood composition occur as it circulates through the following organs: right heart, lungs, liver, kidneys, adrenal glands?
3. What are the factors that maintain and modify arterial and venous pressure?
4. In taking a pulse what points are important to note and record?
5. Discuss the differences between arterial and venous pressures.

6. Assume that each time the heart beats, 80 ml of blood moves into the lungs and the heart rate is 72 beats per minute. How much blood moves through the heart to the lungs in one minute, one hour, one day, one week, one year, 10 years, 80 years?
7. How would exercise, rest, and sleep each affect the above findings?

Summary

Distribution of Blood	Pulmonary circulation, less than 20% Systemic circulation, about 85% Varies with activity of tissue		
Blood Flow	Definition—amount of blood moving past a particular point per unit of time Influenced by laminar flow, viscosity, cross-sectional area, resistance to flow, pressure differences at beginning and end of tube Flow is pulsatile in arteries, steady in veins		
Pulse Pressure	Difference between systolic and diastolic pressures Influenced by heart rate, stroke volume, peripheral resistance		
Arterial Blood Pressure	Determined by cardiac output and peripheral resistance		
	Control	**Nervous**	Vasomotor center in medulla Sympathetic nervous system—influences cardiac action, size of arterioles
		Metabolic	Norepinephrine, epinephrine Oxygen, carbon dioxide, pH levels
Venous Pressure	Always low, unless blockage to flow in veins Directly related to blood flow through heart and lungs		
Systolic Blood Pressure	Adult normal value: systolic 90–140 mm Hg; diastolic 60–90 mm Hg Normal variation: muscle activity, meals, emotion		
	Abnormal variation	Hypertension (essential, secondary) Hypotension (syncope, shock)	
Pulse	Alternate dilatation and contraction of artery, corresponding to heartbeat		
	Locations where pulse may be counted	Radial artery, temporal artery, external maxillary artery, carotid artery, brachial artery, femoral artery, popliteal artery, dorsalis pedis artery	
	Points to note	Frequency Force, or strength Tension, or resistance	Hard Soft

Pulse (*cont.*)	Pulse rate	Factors that influence heartbeat also influence pulse Average $\begin{cases} 65\text{–}70 \text{ in men} \\ 70\text{–}80 \text{ in women} \end{cases}$ Ratio of pulse to respiration is about 4 to 1
Measurement of Arterial Blood Pressure	Determined by use of sphygmomanometer	1. Two types $\begin{cases} \text{mercury} \\ \text{aneroid} \end{cases}$ 2. Types similar in principle; each consists of an air bag for attachment to arm, a hand pump for inflating the bag, and a scaled device for measurement of pressure in bag, which is equal to pressure of blood against wall of artery
Measurement of Venous Pressure		Directly determined by use of sterile catheter into venous system, attached to water manometer

Additional Readings

GUYTON, A. C.: *Basic Human Physiology: Normal Functions and Mechanisms of Disease*, 2nd ed. W. B. Saunders Co., Philadelphia, 1977, Chapters 5, 6, 7.

SELKURT, E. E.: *Basic Physiology for the Health Sciences.* Little, Brown & Co., Boston, 1975, Chapter 17.

VANDER, A. J.; SHERMAN, J. H.; and LUCIANO, D. S.: *Human Physiology: The Mechanisms of Body Function*, 2nd ed. McGraw-Hill Book Co., New York, 1975, pp. 246–82.

The Respiratory System

Chapter Outline

The normal sequence of chemical changes in tissue cells depends on oxygen, hence the need for a continuous supply. One of the chief end products of these chemical changes is carbon dioxide, hence the need for continuous elimination of carbon dioxide. In unicellular animals the intake of oxygen and the output of carbon dioxide occur at the surface by diffusion. As organisms increase in size and complexity, specialized structure is developed that functions to bring oxygen to the cells of the organism. The specialized structure for intake of oxygen is the respiratory system; in both unicellular and complex organisms, such as man, the process of taking in oxygen is termed *respiration*.

Respiratory Tract

Included in the respiratory tract are the larynx, trachea, bronchi, and lungs; the nose serves to take in air and transmit it to the pharynx and larynx. The diaphragm is the prime muscle involved in respiration; others are the intercostal muscles of the thorax and the abdominal muscles.

NOSE

The nose is the special organ of smell, but it has an important part in respiration, air intake. The nose consists of two parts: the external feature, the nose; and the internal cavities, the nasal fossae. It is composed of a triangular framework of bone and cartilage, covered by skin and lined by mucous membrane. On its undersurface are two oval openings, the nostrils (*anterior nares*), which are the external openings of the nasal cavities.

Nasal Cavities. These are two wedge-shaped cavities, separated from each other by a partition, or septum. The septum is formed in front by the crest of the nasal bones and the frontal spine and in the middle by the perpendicular plate of the ethmoid. The septum is usually bent more to one side than the other, a condition to be remembered when a tube must be inserted.

The conchae (Figure 19-1) and processes of the ethmoid, which are exceedingly light and spongy, project into the nasal cavities and divide them into three incomplete passages from before backward—the superior, middle, and inferior meatus. The palate and maxillae separate the nasal cavities from the mouth, and the horizontal plate of the ethmoid forms the partition between the cranial and nasal cavities.

The nasal cavities[1] communicate with the air in front by the anterior nares, and behind they open into the nasopharynx by the two *posterior nares*. The cavities are lined with mucous membrane. At the entrance each cavity or vestibule is lined with thick, stratified, squamous epithelium containing sebaceous glands and numerous coarse hairs. The middle, or respiratory, portion of the cavity is lined with pseudostratified ciliated epithelium with many goblet cells. The upper, or olfactory, portion is lined with neuroepithelium, which contains olfactory cells that are the receptors for smell. This membrane, which is highly vascular, is continuous externally with the skin and internally with the mucous membrane lining the sinuses and other structures connected with the nasal passages. Inflammatory conditions of the nasal mucous membrane may extend into the sinuses.

Under normal conditions breathing should take place through the nose. The arrangement of the conchae makes the upper part of the nasal passages very narrow; these passages are thickly lined and freely supplied with blood, which keeps the temperature relatively high and makes it possible to moisten and warm the air before it reaches the lungs. The hairs at the entrance to the nostrils and the cilia of the epithelium serve as filters to remove particles that may be carried in with the inspired air.

Nerves and Blood Vessels. The mucous membrane of the superior conchae and upper third of the septum contains the endings of the olfactory nerve fibers. The nerve fibers for the muscles of the nose are fibers of the facial (seventh cranial), and the skin receives fibers from the ophthalmic and maxillary nerves, which are branches of the trigeminal (fifth cranial). Blood is supplied to the external nose by

[1] Eleven bones enter into the formation of the nasal cavities: the floor is formed by the palatine (2) and part of the maxillae bones (2); the roof is formed chiefly by the horizontal plate of the ethmoid bone (1), the sphenoid (1), and the small nasal bones (2); in the outer walls we find, in addition to processes from other bones, the two conchae (2). The vomer (1) forms part of the septum.

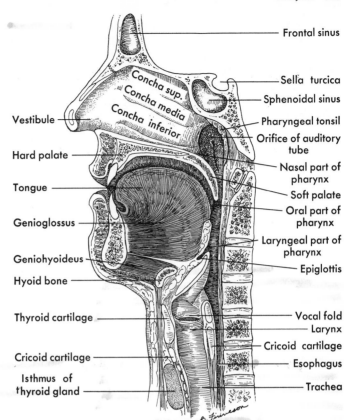

Figure 19–1. Sagittal section of nose, mouth, and pharynx. (Modified from *Gray's Anatomy*.)

branches from the external and internal maxillary arteries, which are derived from the external carotid. The lateral walls and the septum of the nasal cavities are supplied with nasal branches of the ethmoidal arteries, which are derived from the internal carotid.

The *mouth* serves as a passageway for the entrance of air, and the *pharynx* transmits the air from the nose or mouth to the larynx, but both are closely associated with digestion and will be described with the digestive organs.

LARYNX

The larynx, or organ of voice (Figure 19–2), is placed in the upper and front part of the neck between the root of the tongue and the trachea. Above and behind it lies the pharynx, which opens into the esophagus, and on either side of it lie the great vessels of the neck. The larynx is broad

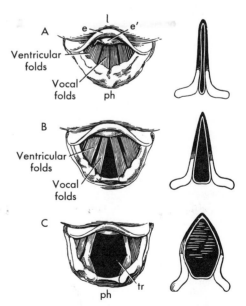

Figure 19–2. Glottis and vocal cords in different conditions: (*A*) while singing a high note, (*B*) in quiet breathing, and (*C*) during a deep inspiration. *l*, Base of tongue; *e*, upper free edge of epiglottis; *e'*, cushion of epiglottis; *ph*, part of anterior wall of pharynx; *tr*, trachea.

above and shaped somewhat like a triangular box, with flat sides and prominent ridge in front (the "Adam's apple"). Below it is narrow and rounded where it blends with the trachea. It is made up of nine fibrocartilages, united by extrinsic and intrinsic ligaments and moved by numerous muscles. These are listed in Table 19–1.

The larynx is lined throughout with mucous membrane, which is continuous above with that lining the pharynx and below with that lining the trachea.

The cavity of the larynx is divided into two parts by two folds of mucous membrane stretching from front to back but not quite meeting in the midline. They thus leave an elongated fissure called the *glottis*, which is the narrowest segment of the air passages. The glottis is protected by a lid of fibrocartilage called the *epiglottis* (Figure 19–4).

Vocal Folds. Embedded in the mucous membrane at the edges of the slit are fibrous and elastic ligaments, which strengthen the edges of the glottis and give them elasticity. These ligaments, covered with mucous membrane, are firmly attached at both ends to the cartilages of the larynx and are called the inferior or *true*

vocal folds, because they function in the production of the voice. Above the vocal folds are two ventricular folds, which do not function in the production of the voice but serve to keep the true vocal folds moist, in holding the breath, and in protecting the larynx during the swallowing of food. Figure 19–2 shows vocal folds in various positions.

The glottis varies in shape and size according to the action of muscles upon the laryngeal walls. When the larynx is at rest during quiet breathing, the glottis is V shaped; during a deep inspiration it becomes almost round, and during the production of a high note the edges of the folds approximate so closely as to leave scarcely any opening at all.

Muscles of the Larynx. Many of the muscles of the neck, face, lips, tongue, and diaphragm are concerned with speech. The muscles of the larynx are extrinsic and intrinsic.

The extrinsic muscles are listed in Table 19–2.

In prolonged inspiratory efforts, such as in singing, these muscles produce tension on the lower part of the cervical fascia and hence prevent the apices of the lungs and the large blood vessels from being compressed. In the act of swallowing, the larynx and hyoid bone are drawn up with the pharynx—these muscles depress them. In addition, they elevate and depress the thyroid cartilage.

The intrinsic muscles are confined entirely to the larynx; some are shown in Figure 19–3. The posterior cricoarytenoid muscles rotate the arytenoid cartilages outward, thereby separating the vocal folds. The lateral cricoarytenoid muscles

TABLE 19–1
CARTILAGES OF THE LARYNX

Single Cartilages

Thyroid, resembles a shield, rests on cricoid. Consists of two plates joined at acute angle in midline forming laryngeal prominence.

Cricoid, shaped like seal ring with signet part in back.

Epiglottis, shaped like leaf with stem inserted in union of two thyroid plates.

Paired Cartilages

Arytenoid, pyramid shaped, rest on upper border of cricoid cartilages on either side.

Corniculate, conical nodules of elastic cartilage, articulate with upper inner surface of arytenoid cartilages. They prolong the arytenoids backward and medially.

Cuneiform, elongated pieces of elastic cartilage on either side of the aryepiglottic fold.

TABLE 19–2
EXTRINSIC MUSCLES OF LARYNX

1. Infrahyoid Muscles	2. Suprahyoid Muscles (some of)
Omohyoid	Stylopharyngeus
Sternohyoid	Palatopharyngeus
Thyrohyoid	Inferior and middle constrictors of the pharynx
Sternothyroid	

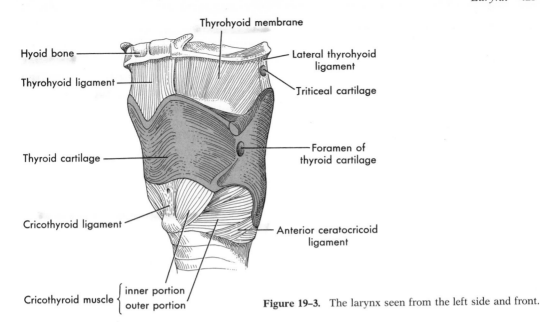

Hyoid bone

Thyrohyoid ligament

Thyroid cartilage

Cricothyroid ligament

Cricothyroid muscle { inner portion / outer portion

Thyrohyoid membrane

Lateral thyrohyoid ligament

Triticeal cartilage

Foramen of thyroid cartilage

Anterior ceratocricoid ligament

Figure 19-3. The larynx seen from the left side and front.

rotate the arytenoid cartilages inward, thereby approximating the vocal folds. The arytenoid muscles approximate the arytenoid cartilages, especially in the back. These muscles also open and close the glottis.

The cricothyroid muscles elevate the arch of the cricoid cartilage in front, causing the lamina to be depressed and thereby increasing the distance between the vocal processes and thyroid cartilages.

The thyroarytenoid muscles draw the arytenoid cartilages forward toward the thyroid, and in this way they shorten and relax the vocal folds. Working together, these muscles regulate the degree of tension on the vocal folds.

Nerves and Blood Vessels. The laryngeal nerves are derived from the internal and external branches of the superior laryngeal, branches of the vagi. Blood is supplied to the larynx by branches of the superior thyroid artery, which arises from the external carotid, and from the inferior thyroid, a branch of the thyroid axis, which arises from the subclavian artery.

Phonation. This term is applied to the production of vocal sounds. All the respiratory organs function in the production of vocal sounds, but the vocal folds, and the larynx are primarily concerned. The speech centers and parts of the brain that control all the movements of the tongue and jaw are of special importance. The organs of phonation in man are similar to those of many animals much lower in the scale of life; the association areas of the brain account for the greater variety of sounds that man can produce.

The vocal folds produce the *voice* by air movement in expiration which throws the two elastic folds into vibrations. These impart their vibrations to the column of air above them and so give rise to the sound that we call the voice. The pharynx, mouth, and nasal cavities above the glottis act as resonating cavities. The volume and force of the expired air and the amplitude of the vibrations of the vocal folds determine the loudness or intensity of the voice. The pitch of the voice depends upon the number of vibrations occurring in a given unit of time. This in turn is dependent on the length, thickness, and degree of elasticity of the vocal folds and the tension by which they are held. When the folds are tightly stretched and the glottis almost closed, the highest sounds are emitted.

Differences Between Male and Female Voice. The size of the larynx varies in different individuals, and this is one reason for differences in pitch. At the time of puberty, the growth of the larynx and the vocal folds is much more rapid and accentuated in the male than in the female. In the male the increase in the size of the larynx causes an increase in the length of the vocal folds and also gives rise to the laryngeal prominence. These changes in structure are accompanied by changes in the voice, which becomes deeper and lower. Before the characteristic adult voice is attained, there occurs what is described as a break in the voice, due to the inability of the individual to control the longer vocal folds.

TRACHEA

The trachea (Figure 19-4) is a membranous and cartilaginous tube, cylindrical, about 11.2 cm in length and about 2 to 2.5 cm from side to side. It lies in front of the esophagus and extends from the larynx on the level of the sixth cervical vertebra to the upper border of the fifth thoracic vertebra, where it divides into the two bronchi, one for each lung. The placement of the trachea in the thoracic cavity is such that the upper portion is more anterior than the lower. Thus, when one is lying prone, secretions in the bronchi and trachea tend to flow upward—facilitating drainage of fluid in lung infections.

The walls of the trachea are strengthened by C-shaped rings of hyaline cartilage placed so that the "open" portion is toward the esophagus. Like the larynx, it is lined with mucous membrane and has a ciliated epithelium on its inner surface; goblet cells are numerous. The mucous membrane, which extends into the bronchial tubes, keeps the internal surface of the air passages free from dust particles, the mucus entangles particles inhaled, and

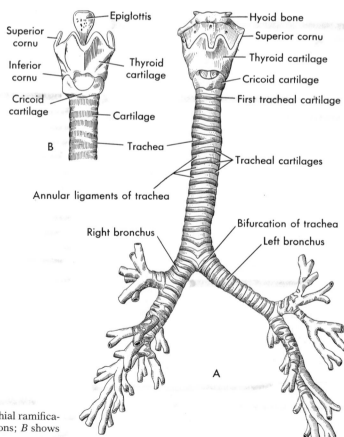

Figure 19–4. The trachea and bronchial ramification. Front view. *A* shows ramifications; *B* shows epiglottis. (Modified from Toldt.)

the movements of the cilia continually sweep this dust-laden mucus upward into the pharynx. The submucosa consists of loose connective tissue and contains mixed glands and adipose tissue.

BRONCHI

The two bronchi (Figure 19–5) into which the trachea divides differ slightly, the right bronchus being shorter, wider, and more vertical in direction than the left. They enter the right and left lung, respectively, and then break up into a great number of smaller branches, which are called the bronchial tubes and bronchioles. The two bronchi resemble the trachea in structure, but as the bronchial tubes divide and subdivide into the smaller bronchi, the incomplete rings of cartilage are replaced by cartilaginous plates, and as they further divide their walls become thinner, the small plates of cartilage cease, the fibrous tissue disappears, and the smallest tubes are composed of only a thin layer of smooth muscle and elastic tissue lined by ciliated epithelium. Each bronchiole terminates in an alveolar duct which in turn terminates in one or more elongated saccules called *alveolar sacs*. (See Figure 19–6.)

Nerves and Blood Supply to Trachea and Bronchi. The nerves are composed of fibers from the vagi, the recurrent nerves (which decrease the diameter of the bronchi), and the thoracolumbar system (which increase the diameter). Blood is supplied to the trachea by the inferior thyroid arteries.

LUNGS

The lungs (pulmones) are cone-shaped organs that fill the two lateral chambers of the thoracic cavity and are separated from each other by the heart and other contents of the mediastinum. Each lung presents an outer surface, which is convex, a base, which is concave to fit over the convex portion of the diaphragm, and an apex, which extends about 2.5 to 4 cm above the level of the sternal end of the first rib. Each lung is connected to the heart and trachea by the pulmonary artery, pulmonary vein, bronchial arteries and veins, the

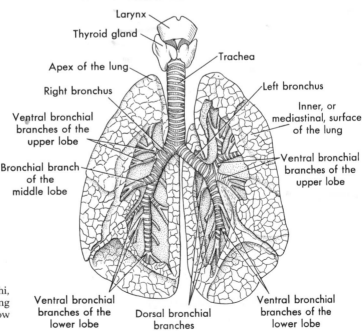

Figure 19–5. The trachea, bronchi, and bronchioles. Portions of lung tissue have been removed to show the branching.

Larynx

Thyroid gland

Trachea

Apex of the lung

Right bronchus

Left bronchus

Inner, or mediastinal, surface of the lung

Ventral bronchial branches of the upper lobe

Ventral bronchial branches of the upper lobe

Bronchial branch of the middle lobe

Ventral bronchial branches of the lower lobe

Dorsal bronchial branches

Ventral bronchial branches of the lower lobe

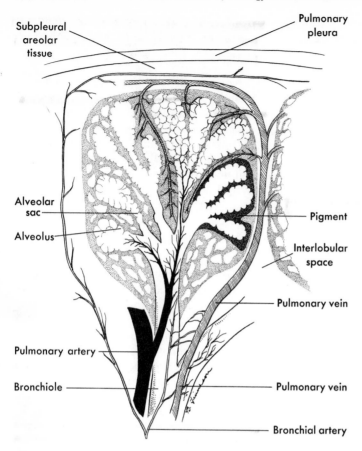

Subpleural areolar tissue

Pulmonary pleura

Alveolar sac

Alveolus

Pigment

Interlobular space

Pulmonary vein

Pulmonary artery

Bronchiole

Pulmonary vein

Bronchial artery

Figure 19–6. Diagram of a lobule of the lung showing a bronchiole dividing into two branches. Note the three alveolar sacs, upper left section, the alveoli of each opening into the common passageway. In the next group the first alveolar sac shows a pulmonary arteriole surrounding the opening of each alveolus. Around the group on the right is a deep deposit of pigment, such as occurs in old age and in persons who inhale coal dust, etc.

bronchus, plexuses of nerves, lymphatics, lymph nodes, and areolar tissue, which are covered by the pleura and constitute the *root* of the lung. On the inner surface is a vertical notch called the *hilum*, which gives passage to the structures that form the root of the lung. Below and in front of the hilum is a deep concavity, called the cardiac impression, where the heart lies; it is larger and deeper on the left than on the right lung, because the heart projects farther to the left side.

Right Lung. This is larger and broader than the left, due to the inclination of the heart to the left side; it is also shorter by 2.5 cm, as a result of the diaphragm's rising higher on the right side to accommodate the liver. It is divided by fissures into three lobes—superior, middle, and inferior.

Left Lung. This is smaller, narrower, and longer than the right and is divided into two lobes—superior and inferior. The projection of the fissures dividing these lobes is seen in Figure 19–7.

Composition of the Lung. The substance of the lungs is porous and spongy; owing to the presence of air, it crepitates when handled and floats in water. It consists of bronchioles and their terminal dilatations, numerous blood vessels, lymphatics, and nerves, and an abundance of elastic connective tissue. Each *lobe* of the lung is composed of many *lobules*, and into each lobule a terminal *bronchiole* enters and terminates in one or more alveolar sacs. Each sac presents on its surface numerous small pouches, or *alveoli*. It is estimated that there are 300,000 alveoli in the adult lung. Each alveolus is somewhat globular in form with a diameter of about 100 μm. Figure 19–8 shows their fragile structure.

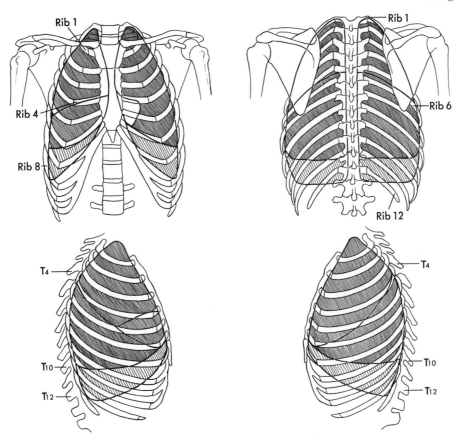

Figure 19–7. Surface projection of lung fissures.

Underlying the thin epithelial cells which make up the alveolar wall is a thin basement membrane approximating the epithelium and the capillary wall. The inner surface of the alveolus is covered with a lipid substance, called *surfactant,* which stabilizes the alveoli, preventing their collapse. The amount of surface exposed to the air and covered by the capillaries is enormous. It is estimated that the entire inner surface of the lungs amounts to about 90 sq m, more than 100 times the skin surface of the adult body. Of this lung area about 70 sq m are respiratory.

Nerves of the Lungs. The craniosacral nerve supply is made up of fibers in the vagus nerves. The thoracolumbar nerve supply is made up of fibers in the visceral branches of the first four thoracic spinal nerves.

Blood Vessels of the Lungs. Two sets of vessels are distributed to the lungs: (1) the pulmonary artery and its branches, the lung capillaries, and the four pulmonary veins and their branches; and (2) the bronchial arteries, capillaries, and veins.

The pulmonary artery conveys *venous* blood from the right ventricle to the lungs. The main trunk is a short, wide vessel about 5 cm in length and a little more than 3 cm in width. It arises from the right ventricle and passes upward, backward, and to the left. About the level of the intervertebral disk between the fifth and sixth thoracic vertebrae, it divides into two branches, the right and left pulmonary arteries, which pass to the right and left lungs. Before entering the lungs, each artery divides into two branches. The *right* pulmonary is longer and larger than the left. It runs horizontally to the right, *be-*

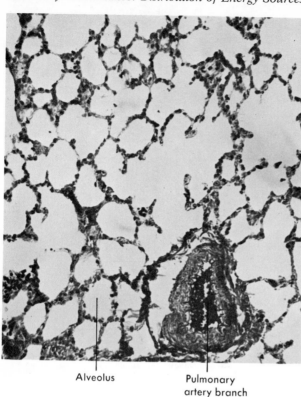

Figure 19–8. Appearance of lung tissue under the light microscope.

Alveolus   Pulmonary artery branch containing blood

hind the ascending aorta and superior vena cava to the root of the right lung, where it divides into two branches. The larger lower branch goes to the middle and lower lobes; the smaller upper branch goes to the upper lobe.

The *left* branch of the pulmonary artery is smaller and passes horizontally in *front* of the descending aorta and left bronchus to the root of the left lung, where it divides into two branches, one to each lobe. These branches divide and subdivide, grow smaller in size, and finally merge into capillaries, which form a network upon the walls of the air sacs (alveoli). These capillaries unite, grow larger in size, and gradually assume the characteristics of veins. The veins unite to form the pulmonary veins.

The air in the alveoli is separated from the blood in the capillaries only by the thin membranes forming their respective walls. The air sacs are surrounded by capillaries, which form a surface area of about 150 sq m for the exchange of oxygen and carbon

dioxide. Blood pressure in the pulmonary arteries is 20/10, and in the capillaries about 4 to 6 mm Hg. In other tissues the capillaries are surrounded by tissue fluid, which exerts pressure against the capillaries, but the capillaries in the lungs are not opposed by such pressures. This means that pressures must be lower to prevent disturbances of hydrostatic and osmotic forces of the blood in the lungs.

The pulmonary veins convey *oxygenated* blood from the lungs to the left atrium. these veins commence in the capillary network upon the air cells and unite to form one vein for each lobule. These further unite to form one vein for each lobe, two for the left lung and three for the right. The vein from the middle lobe of the right lung usually unites with that from the upper lobe, and finally two trunks from each lung are formed. They have no valves and open separately into the left atrium. (See Figures 19–9 and 19–10 for the relationship of these blood vessels to the heart.)

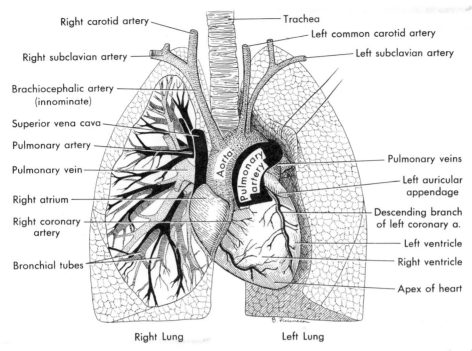

Right carotid artery

Right subclavian artery

Brachiocephalic artery (innominate)

Superior vena cava

Pulmonary artery

Pulmonary vein

Right atrium

Right coronary artery

Bronchial tubes

Trachea

Left common carotid artery

Left subclavian artery

Aorta

Pulmonary artery

Pulmonary veins

Left auricular appendage

Descending branch of left coronary a.

Left ventricle

Right ventricle

Apex of heart

Right Lung

Left Lung

Figure 19–9. The pulmonary artery and aorta. The dorsal part of the right lung has been removed, and the pulmonary vessels and the bronchi are thus exposed.

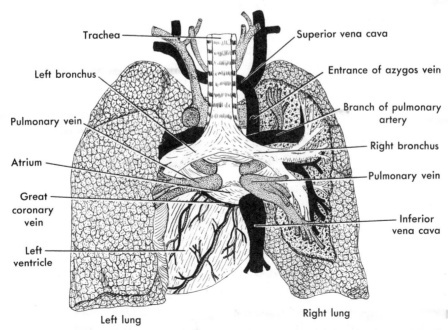

Trachea

Left bronchus

Pulmonary vein

Atrium

Great coronary vein

Left ventricle

Superior vena cava

Entrance of azygos vein

Branch of pulmonary artery

Right bronchus

Pulmonary vein

Inferior vena cava

Left lung

Right lung

Figure 19–10. The pulmonary artery and aorta. The front part of the right lung has been removed, and the pulmonary vessels and the bronchi are thus exposed.

The branches of the *bronchial arteries* supply blood to the lung substance—the bronchioles, coats of the blood vessels, the lymph nodes, and the pleura. The bronchial veins formed at the root of each lung receive veins that correspond to the branches of the bronchial arteries. Some of the blood supplied by the bronchial arteries passes into the pulmonary veins, but the greater amount is returned to the bronchial veins. The right bronchial vein ends in the azygos vein, the left in the highest intercostal or hemiazygos vein.

Pulmonary Blood Pressure. The pressure in the pulmonary artery is low ($^{20}/_{10}$ mm Hg) and capillary pressure is about 10 mm Hg. The higher oncotic pressure of the plasma proteins results in an inward pressure gradient, keeping the alveoli free of fluid.

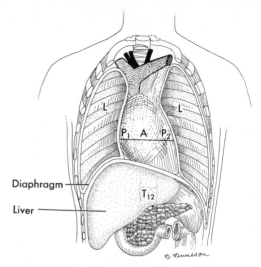

Figure 19–11. Diagram showing thoracic cavity. Line *A* indicates the mediastinum. P_1 and P_2 indicate the mediastinal pleura. *L* indicates lung spaces surrounded by pleura.

PLEURA

Each lung is enclosed in a serous sac, one layer of which is closely adherent to the inner chest wall and superior surface of the diaphragm—the parietal pleura; the other closely covers the lung—visceral or pulmonary pleura. The cavity thus formed is a potential space since the layers are in intimate contact. The pleura is a thin, transparent, moist membrane that forms serous fluid. The two layers move easily upon each other with respiratory movements of the thorax. If the surface of the pleura becomes inflamed (pleurisy), friction results, and the sounds produced by this rubbing can be heard through the stethoscope. Any collection of fluid, as with inflammation, in the pleural cavities will cause compression and possibly collapse of portions of the lung.

If a puncture occurs through the chest walls so that air enters between the two layers of the pleura (pneumothorax), the lung will collapse. If the puncture is closed, the air will be gradually absorbed, and the lung will resume its normal position. Certain types of tuberculosis are treated by artificial pneumothorax. This is accomplished by surgical removal of part of the chest wall or by injections of air into the pleural cavity.

MEDIASTINUM

The mediastinum, or interpleural space (Figure 19–11), lies between the right and left pleurae in the median plane of the thorax. It extends from the sternum to the spinal column and is entirely filled with the thoracic viscera, namely, the thymus, heart, aorta and its branches, pulmonary artery and veins, venae cavae, azygos vein, various veins, trachea, esophagus, thoracic duct, lymph nodes, and lymph vessels, all lying in connective tissue.

Respiration

The main purposes of respiration are to supply the cells of the body with oxygen and eliminate the excess carbon dioxide that results from oxidation. Respiration also helps to maintain the normal pH of body fluids and normal temperature of the body and eliminates about 350 ml of water each 24 hours.

It is common to discuss respiration under three headings:

1. The process of *ventilation* (*breathing*) may be subdivided into inspiration, or breathing in, and expiration, or breathing out.

2. *External respiration* includes external oxygen supply, or the passage of oxygen from the alveoli of the lungs to the blood, and external carbon dioxide elimination, or the passage of carbon dioxide from the blood to the alveoli.

3. *Internal respiration* includes internal oxygen supply, or the passage of oxygen from the blood to the tissue cells, and internal carbon dioxide elimination, or the passage of carbon dioxide from the tissue cells to the blood.

It is evident that external respiration is a process that takes place in the lungs and internal respiration is a process that takes place in the cells that make up the tissues.

Ventilation (Breathing). The thorax is a closed cavity that contains the lungs. The lungs may be thought of as elastic sacs, the interiors of which remain permanently open to the outside air by way of the bronchi, trachea, glottis, and nasopharynx; alveolar pressure, or the pressure against the lungs from inside, is therefore atmospheric pressure and is usually given as 760 mm Hg at sea level. The lungs are protected externally from atmospheric pressure by the walls of the thorax.

The primary muscle of respiration is the diaphragm—without which adequate respiration is impossible. Other muscles involved (see table, page 160) may be classed as accessory, even though some, such as the external intercostals (Figure 19–12), are involved in normal quiet breathing. The accessory muscles and those of the abdomen are noticeably involved during forced expiration and inspiration, as when there is oxygen lack.

During *inspiration* the upward-curving diaphragm contracts, pulling downward, and increasing the vertical length of the thoracic cage. The intercostal muscles pull the ribs upward and outward. Thus the

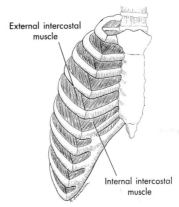

Figure 19–12. The intercostal muscles of the right thorax.

thoracic cage is increased in size laterally, dorsoventrally, and vertically. It is this *active* increase in size of the cage that causes lung expansion and the "pulling" in of air. Figure 19–13 illustrates the change in diaphragmatic position.

Expiration is passive; all the contracted muscles relax. Surface tension of fluid lining the alveoli causes a continual tendency of the alveoli to collapse, as does the elasticity of lung tissue. Therefore, with relaxation of respiratory muscles, the lung recoils and air is pushed out of the lungs. Forced expiration is made possible by action of accessory muscles to actively decrease the size of the thorax in all its dimensions. The diaphragm is *not* involved actively, but is pushed upward be-

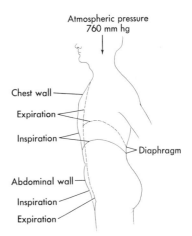

Figure 19–13. Changes in position of diaphragm and abdominal wall, and in size of thoracic cavity during respiration.

yond its normal position by contraction of abdominal muscles against the abdominal organs.

Deep ventilation aerates the lung more adequately than shallow ventilation because there is greater lung expansion, air is moved in and out in larger amounts, and more alveoli are involved.

The capacity of the lung to respond easily to changes in the size of the thoracic cage is termed *compliance*. If, through disease, lung tissue becomes fibrotic, its compliance decreases and the work of breathing increases.

Pressure Changes with Respiration (Figure 19–14). The intrapleural space is a potential space between the lung and the thoracic wall. The lungs fill the thoracic cavity, because the membranes constantly absorb any gas or fluids that may enter the space. However, the lungs have a continual tendency to collapse and to pull away from the thoracic wall. After the lungs have been *stretched on inspiration*, the tendency to collapse is increased to about −5 or −6 mm Hg, and on *expiration*, the collapse tendency is about −4 mm Hg in relation to atmospheric pressure of 760 mm.

The intrapleural pressure changes cause, in turn, a change in alveolar pressure. On *inspiration* pressures become slightly negative in relation to atmospheric pressure, about −3 mm Hg. Air is pulled inward through the respiratory airways. On *expiration* intra-alveolar pressure rises to about +3 mm Hg, which causes air to move *outward* through the respiratory airway.

Lung Capacities. After the lungs are once filled with air they are never completely emptied. In other words, no expiration ever completely empties the alveoli; neither are they completely filled. The quantity of air that a person can expel by a forcible expiration, after the deepest inspiration possible, is called the *vital capacity* and averages about 4000 to 4800 cc for an adult man. It is the sum of tidal, expiratory, and inspiratory reserve air. Figure 19–15 illustrates lung capacity.

Tidal air designates the amount of air that flows in and out of the lungs with each quiet respiratory movement. The average figure for an adult male is 500 cc.

Inspiratory reserve volume designates the amount of air that can be breathed in over and above the tidal air by the deepest

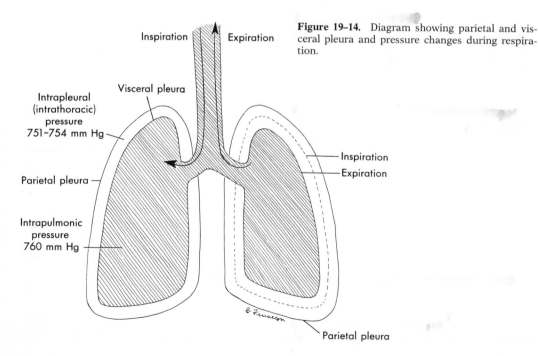

Figure 19–14. Diagram showing parietal and visceral pleura and pressure changes during respiration.

Inspiration   Expiration

Intrapleural (intrathoracic) pressure 751–754 mm Hg

Visceral pleura

Parietal pleura

Intrapulmonic pressure 760 mm Hg

Inspiration
Expiration

Parietal pleura

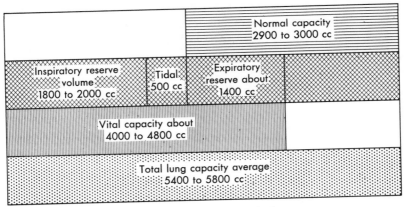

Figure 19–15. Capacities of the lung.

possible inspiration. It is estimated at about 1800 to 2000 cc.

Expiratory reserve volume is the amount of air that can be breathed out after a quiet expiration by the most forcible expiration. It is equal to about 1400 cc.

Residual air is the amount of air remaining in the air passages after the most powerful expiration. This has been estimated to be about 1200 cc and is air in the upper respiratory tract, trachea, bronchi, and bronchioles—the "dead space."

Reserve air is the residual air plus the supplemental air in the lungs which could be expelled with forced expiration, that is, about 3000 cc.

When the thorax is opened, the lungs collapse, driving out the supplemental and residual air; but before the alveoli are entirely emptied, the small bronchi leading to them collapse and entrap a little air in the alveoli. The small amount of air caught in this way is designated as *minimal air.*

Before birth the lungs contain no air. If, after birth, respirations are made, the lungs do not collapse completely on account of the capture of minimal air. Whether or not the lungs will float has constituted one of the facts used in medicolegal cases to determine if a child was stillborn.

Ventilatory Changes in Air. However dry the external air may be, the expired air is nearly, or quite, saturated with moisture. An average of about 500 ml of water is eliminated daily in the breath. Whatever the temperature of the external air, the expired air is nearly as warm as the blood, having a temperature between 36.7° and 37.8° C (98° and 100° F). In man, breathing is one of the subsidiary means by which the temperature and the water content of the body are regulated. The heat required to warm the expired air and vaporize the moisture is taken from the body and represents a daily loss of heat. It requires about 0.5 Cal to vaporize 1 gm of water.

Taking the respiratory rate as 18 per minute and respiratory depth at 500 cc, one breathes in and out 12,000 liters of air per day. Since inspired air contains about 20 per cent oxygen and expired air about 16 per cent, this difference of 4 per cent represents the oxygen retained by the body—some 480 liters per day—and used by the tissues. This amount is often stated as about 350 cc per minute.

External Respiration. This term is applied to the interchange of gases that takes place in the lungs. There is a continuous flow of blood through the capillaries, so that at least once or twice each minute all the blood in the body passes through the capillaries of the lungs. This means that the time during which any portion of blood is in a position for respiratory exchange is only a second or two. Yet during this time, the following changes take place: the blood loses carbon dioxide and moisture; it gains oxygen, which combines with the reduced hemoglobin of the red cells, or erythrocytes, forming oxyhemoglobin, and as a result of this the crimson color shifts to scarlet; and its temperature is slightly reduced.

TABLE 19–3
OXYGEN AND CARBON DIOXIDE VOLUMES
PER CENT IN BLOOD

Gas	Arterial Blood		Venous Blood	
	Total	in Solution	Total	in Solution
Oxygen	19–21	0.24	12–14	0.1
Carbon dioxide	48–50	2–2.5	56–58	3.0

Table 19–3 compares the average amounts of oxygen and carbon dioxide found in venous and in arterial blood. The actual amounts of oxygen and carbon dioxide in venous blood vary with the metabolic activity of the tissues and differ, therefore, in the various organs according to the state of activity of each organ and the volume of its blood supply per unit of time. The main result of the respiratory exchange is to keep the gas content of the arterial blood nearly constant at the figures given. Under normal conditions it is not possible to increase appreciably the amount of oxygen absorbed by the blood flowing through the lungs. It is possible to increase the amount of CO_2 in alveolar air (and decrease the blood level of CO_2) by increasing the depth of ventilation, as seen in Table 19–4.

From the pulmonary artery blood moves into the lung capillary networks, where the exchange of oxygen and carbon dioxide takes place. This is called *gas exchange* in the lungs. The blood contains a carbon dioxide concentration of about 56 to 58 per cent by volume and an oxygen concentration of about 12 to 14 per cent by volume, with some of each gas in solution. The rates of gaseous exchange are influenced by the following factors: (1) area of contact for the exchange, (2) length of time blood and air are in contact, (3) volume of blood passing through the alveolar network, (4) permeability of cells forming the capillary and alveolar membranes, (5) differences in concentrations of gases in alveolar air and the blood, and (6) rate at which chemical reaction takes place between the gases and the blood. Respiratory efficiency is also related to the number of red cells, hemoglobin content of the red cells, and size of the red cell.

In the alveoli the total area for exchange of gases has been estimated to be 25 to 50 times the total surface area of the body. The respiratory mechanism is delicately balanced, so that alveolar air remains constant at about 14 to 15 per cent by volume of oxygen and 5.5 per cent by volume of carbon dioxide. It is with this air that the blood is in contact for gaseous exchange. Blood in the alveolar capillaries is proportionate to the amount of physical activity. At rest, the amount of blood in alveolar capillaries is about half the amount present during exercise.

Translating the concepts of concentrations of oxygen and carbon dioxide expressed in volumes per cent into terms of partial pressures, they may be stated simply as follows (page 435):

TABLE 19–4
INCREASE IN RESPIRATORY DEPTH AND RATE ON INCREASING CO_2 IN INSPIRED AIR

% CO_2 in Inspired Air	Average Depth of Respiration, cc	Average Respiratory Rate	Minute Lung Volume, liters
Normal (0.04)	673	14	9.4
3.07	1216	15	18.2
5.14	1771	19	33.6

Atmospheric pressure at sea level exerts about 1 ton pressure per square foot, usually expressed in millimeters or inches. The total pressure of gases in the atmosphere is 760 mm Hg. Because oxygen forms 20.96 or about 21 per cent by volume of the atmosphere, its partial pressure is 21 per cent of 760, or 160 mm Hg (159.6). Carbon dioxide forms about 0.04 per cent by volume of the atmosphere, and its partial pressure is 0.04 per cent of 760, or 0.30 mm Hg. The other gases exert partial pressures in direct proportion to their volumes. Nitrogen exerts the highest partial pressure since its volume per cent is 79.

The tension of a gas in solution equals the partial pressure of that particular gas in the gas mixture with which the solution has established equilibrium. For example, in atmospheric air the partial pressure and tension of oxygen remain constant at 160 mm Hg and will change only if the relative concentration of oxygen in the total mixture is changed. The absolute amount or quantity of a gas in solution may vary, even if the partial pressure and tension remain constant. In applying the law of partial pressures, if the temperature remains constant the quantity of gas that goes into solution in any given liquid, e.g., plasma, is proportional to the partial pressure of the gas.

Table 19–5 illustrates the partial pressure of gas in inspired, expired, and alveolar air.

Analyzing the figures for oxygen in this table, it will be seen that oxygen pressure

TABLE 19–5
PARTIAL PRESSURES OF GASES IN THE ALVEOLUS

	Inspired Air (mm Hg)	Expired Air (mm Hg)	Alveolar Air (mm Hg)
Oxygen	160	116	103
Carbon dioxide	0.30	28	40
Nitrogen	594.70	569	570
Water vapor	5.00	47	47
	760.00	760	760

falls as atmospheric air moves into the alveolus; for carbon dioxide the decrease in pressure is in the reverse direction. That is, carbon dioxide pressure in the alveolar air is high, and in the inspired or atmospheric air the pressure is low. These inverse relationships between the pressures for oxygen and carbon dioxide in alveolar air promote the interchange of these gases in the lungs. The tensions of oxygen and carbon dioxide in alveolar air vary with depth and frequency of respiration. For example, in voluntary hyperventilation carbon dioxide tension falls and oxygen tension rises. If respiration is suspended for any reason, carbon dioxide tension rises and oxygen tension falls.

Nitrogen is an inert gas in relation to respiration; however, about 0.83 per cent by volume of the gas is dissolved in the plasma. This is neither used nor produced in the body.

Partial pressure[2] or tension of oxygen in blood varies in relation to venous and arterial blood. The tension of oxygen in arterial blood is about 100 mm Hg, and in venous blood, about 36 to 38 mm. The tension of carbon dioxide in arterial blood is about 40 to 45 mm, and in venous blood the tension varies directly in relation to muscular activity, but for mixed venous blood the average tension is about 48 to 50 mm.

In the lung the alveoli are separated from the capillaries by thin membranes that are permeable to the gases. Since the pressure gradient for oxygen in alveolar air is high and in the capillary tension low, there is rapid diffusion of oxygen from the alveolar air to the blood. The reverse is true for carbon dioxide: the pressure gradient is high in blood and low in the alveoli—hence equilibrium is rapidly and progressively established between oxygen and carbon dioxide in the blood and alveolar air as blood moves along in the capillary network.

Loss of carbon dioxide during external respiration is a significant factor in main-

[2]The partial pressure of a gas is indicated by the prefix "p", e.g., pO_2 100 mm.

taining the normal pH of body fluids, as will become apparent in the discussion of acid-base balance (page 563).

Internal Respiration. The exchange of oxygen and carbon dioxide in the tissues constitutes internal respiration and consists of the passage of oxygen from the blood into the tissue fluid and from the tissue fluid into the tissue cells and the passage of carbon dioxide from the cells into the tissue fluid and from the tissue fluid into the blood.

After the exchange of oxygen and carbon dioxide in the lungs, the aerated blood is returned to the heart and distributed to all parts of the body. As blood moves into the somatic and visceral capillaries, oxygen tension is high and carbon dioxide tension is low in the capillary network, and tension of oxygen in tissue fluid and cells is relatively low. Carbon dioxide tension is relatively high in the tissue fluid and cells and relatively low in the capillary. This means that the pressure gradients of the gases favor exchange between tissue fluid and blood. Changes in the pressure gradients cause a disequilibrium between blood plasma and oxyhemoglobin, favor the chemical changes taking place in blood between oxygen and carbon dioxide, and promote a steady, constant flow of oxygen from the blood to tissue fluid and cells and a constant flow of carbon dioxide from the cells and tissue fluid to blood. Complete removal of oxygen from the blood never occurs.

Carriage of Oxygen and Carbon Dioxide. As shown in Figure 19–16, oxygen reaches the cells in three steps: (1) environment to lungs, (2) lungs to blood, and (3) blood to cells.

As the blood flows through the lung capillaries, oxygen diffuses from the alveoli into the plasma and then into the red blood corpuscles and combines with the hemoglobin to form oxyhemoglobin. On leaving the lungs, practically all the hemoglobin exists as oxyhemoglobin, and the plasma is saturated with oxygen in solution. When the blood reaches the tissue capillaries, there is a continuous diffusion

of oxygen from erythrocytes to plasma, plasma to tissue fluid, tissue fluid to cells. The rate of this diffusion of oxygen depends upon the rate of use of oxygen by the cells.

The blood carries *from the lungs* about 20 volumes of oxygen per 100 volumes of blood, more than 19 per cent as oxyhemoglobin (stored as oxyhemoglobin) in the erythrocytes and less than 1 per cent (diffusible) in solution in the water of the blood, hence mainly in plasma. While the blood is flowing through the alveolar capillaries, oxygen diffuses from a place kept relatively high in diffusible oxygen (by breathing) to a place kept relatively low in diffusible oxygen (circulating blood).

The blood leaves in the tissues about 10 volumes of oxygen per 100 volumes of blood. While the blood is flowing through the tissue capillaries, oxygen diffuses from a place kept relatively high in diffusible oxygen (flowing blood) to a place kept relatively low in diffusible oxygen (used in cells).

Carbon dioxide is produced in living cells at varying rates, depending upon activity. Carbon dioxide is believed to enter the blood from the tissue cells by diffusion (Figure 19–17). Some of it remains in solution in the plasma, but most of it diffuses from plasma into the erythrocytes where the enzyme *carbonic anhydrase* catalyzes the formation of carbonic acid ($H_2O + CO_2 \rightarrow H_2CO_3$). This results in an increase in hydrogen ions and favors the decomposition of oxyhemoglobin ($HHbO_2 \rightarrow HHb + O_2$) and the escape of oxygen to tissue cells. Inasmuch as reduced hemoglobin is a weaker acid than oxyhemoglobin, this reaction decreases the hydrogen ion concentration within the erythrocyte. A further decrease results from the reaction of carbonic acid with the potassium salts of hemoglobin ($KHb + H_2CO_3 \rightarrow HHb + KHCO_3$). During these reactions the concentration of bicarbonate ions (HCO_3^-) within the red blood cells has increased; consequently these ions diffuse from erythrocytes into the plasma thus building up the sodium bicarbonate concentration in plasma. In compensation for the exit of these anions,

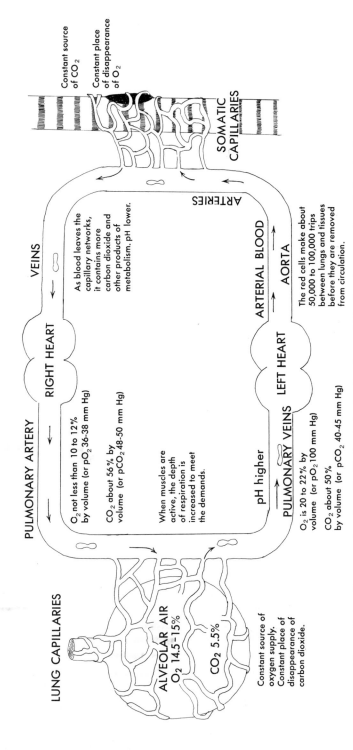

Figure 19–16. Oxygen and carbon dioxide levels in the alveolus and blood as it moves through the circulatory system.

LUNG CAPILLARIES

PULMONARY ARTERY

VEINS

RIGHT HEART

SOMATIC CAPILLARIES

Constant source of CO_2

Constant place of disappearance of O_2

As blood leaves the capillary networks, it contains more carbon dioxide and other products of metabolism. pH lower.

ARTERIES

ARTERIAL BLOOD

AORTA

The red cells make about 50,000 to 100,000 trips between lungs and tissues before they are removed from circulation.

O_2 not less than 10 to 12% by volume (or pO_2 36-38 mm Hg)

CO_2 about 56% by volume (or pCO_2 48-50 mm Hg)

When muscles are active, the depth of respiration is increased to meet the demands.

pH higher

LEFT HEART

PULMONARY VEINS

O_2 is 20 to 22% by volume (or pO_2 100 mm Hg)

CO_2 about 50% by volume (or pCO_2 40-45 mm Hg)

ALVEOLAR AIR
O_2 14.5–15%

CO_2 5.5%

Constant source of oxygen supply. Constant place of disappearance of carbon dioxide.

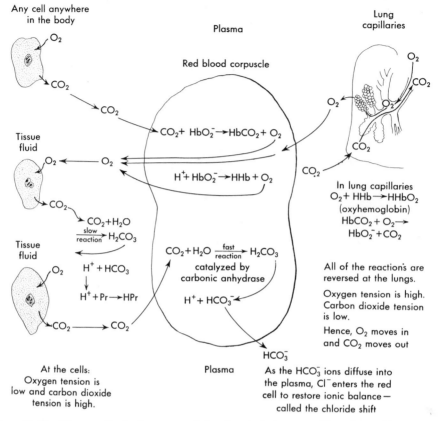

Figure 19–17. Diagram illustrating some of the chemical changes taking place with oxygen and carbon dioxide exchange at the cells and lungs. *Pr* refers to protein.

chloride ions (Cl^-) diffuse from plasma into the erythrocytes; this adjustment is known as the *chloride shift*. Part of the carbon dioxide that enters the erythrocytes combines directly with hemoglobin to form carbamino hemoglobin; this combination is limited, however, to the availability of free amino groups on the hemoglobin. Thus, various means are provided in erythrocytes and plasma for transportation of a relatively large quantity of carbon dioxide with remarkably little change in the pH of blood. When the blood reaches the lungs, oxygen is available and carbon dioxide can escape. With the entrance of oxygen, the sequence of reactions that took place at the tissue cells is reversed. The *concentration gradients* of the two substances concerned—oxygen and carbon dioxide—are reversed in the blood in the tissue capillaries and in the blood in the alveolar capillaries.

In summary, carbon dioxide is carried in venous blood:

1. As bicarbonate (HCO_3^-), about 90 per cent, one third in the red cell and two thirds in the plasma.
2. As carbamino hemoglobin.
3. As a small percentage dissolved physically in plasma.
4. In other chemical combinations.

Control of Respiration. The respiratory center is located in the medulla oblongata, and this center has connections in the pons. Both centers share in the control of respiration. The medulla is the center for nervous control of depth and frequency of respiration, that is, the quantity of air moved through the lungs during ventilation. In this center are inspiratory and expiratory neurons, which are intermixed to a great degree, so that the long-held

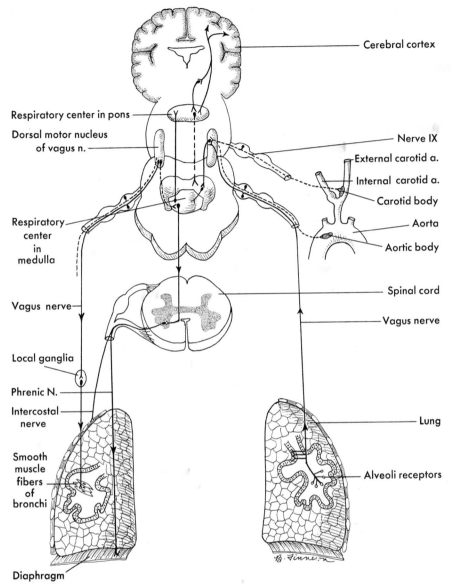

Figure 19–18. Nervous control of respiration, showing connections in the brain. *Arrows* indicate direction of nerve impulses.

idea of separate inspiratory and expiratory centers is debatable. There must be inhibition of expiratory neurons when inspiratory neurons discharge, since one cannot inhale and exhale simultaneously. Destruction of the respiratory area in the pons leads to serious interference with the rhythmical discharge of the medullary neurons.

The respiratory center receives impulses from the cerebral cortex (for speaking); from the periphery (for gasping, e.g., when cold water is applied to the skin); and from the vasomotor and cardiac centers. Perhaps the most significant incoming nerve impulses are those from the lung itself. The lung reflex (Hering–Breuer[3]) elicited by inflation of the lung results in deflating the lung, thus helping to maintain rhythmicity of respiration. As the lung is stretched on inspiration, impulses are

[3]Karl Hering, German physiologist (1834–1918); Josef Breuer, Austrian psychiatrist (1842–1925).

sent to inhibit the inspiration, thus preventing overstretch. Similarly, with expiration these neurons cease to discharge, permitting inspiration to begin again.

The respiratory center is sensitive to increased acidity, carbon dioxide level, anoxia, and increased temperature of blood, and to increased blood pressure. Of these, carbon dioxide and acidity are the most potent stimuli. Chemoreceptors in the carotid and aortic bodies are stimulated by low pH, high CO_2 level, and anoxia. As arterial pressure rises, receptors in the carotid sinus and aortic arch are stimulated and respiration is inhibited. Stimulation of the chemoreceptors increases the rate and depth of respiration.

From the respiratory center nerve impulses pass via the spinal cord and spinal nerves to the muscles of respiration, and respiration is adjusted to varying body needs. In Table 19–4 on page 434 it will be noted that the minute ventilation of lungs is multiplied more than three and a half times, mainly by increasing the completeness of contraction of the respiratory muscles rather than by causing them to contract a greater number of times.

Frequency of Respiration. The average rate of respiration for an adult is about 14 to 20 per minute. In health this rate may be increased by muscular exercise or emotion. Anything that affects the heartbeat will have a similar effect on the respirations. Age has a marked influence. The average rate during the first year of life is about 44 per minute, and at the age of five years, 26 per minute. It is reduced between the ages of 15 and 25 to the normal standard. Emotions have a distinct influence on respiratory activity, probably through the hypothalamus and pons.

Voluntary Control of Respiratory Rate. It is possible to increase or decrease the respiratory rate within certain limits, by voluntary effort, for a short time. If respirations are arrested or their frequency is diminished, the carbon dioxide concentration in the blood increases, and eventually the stimulus becomes too strong to be controlled. According to some observers, the "breaking point" is reached in 23 to 27 seconds. If, before the breath is held, sev-

eral breaths of pure oxygen are taken, the breaking point may be postponed; or, if the lungs are thoroughly aerated by forced breathing, so that the carbon dioxide is forced out and more oxygen breathed in, the breaking point may be postponed as long as eight minutes.

Cause of the First Respiration. The human fetus makes respiratory movements while in the uterus, possibly moving amniotic fluid in and out of the lungs. This may play a part in dilation of the future air passages. After birth and the interruption of the placental circulation, the first breath is taken. Three views are held regarding the immediate cause: that it is due to the increased amount of carbon dioxide in the blood, brought about by cutting the umbilical cord; that it is due to stimulation of the sensory nerves of the skin, due to cooler air, handling, drying, etc.; and that it is due to a combination of these causes.

If stimulation through the blood and stimulation through the nerves normally coincide, it may be that the essential cause is the increased tension of the carbon dioxide, and therefore the increased concentration in acidity, following the cutting of the cord.

During intrauterine life, the fetal blood is aerated so well by exchange with the maternal blood that there is not adequate carbon dioxide to act as a stimulus to the fetal respiratory center.

Respiratory Terms. BREATH SOUNDS. Each intake of air is accompanied by a low rustling sound, which can be heard if the ear is applied to the chest wall. It is thought that the dilation of the alveoli produces this sound, and absence of it indicates that the air is not entering the alveoli over which no sound is heard or that the lung is separated from the chest wall by effused fluid. The air passing in and out of the larynx, trachea, and bronchial tubes produces a louder sound, which is called a bronchial murmur. Normally this murmur is heard directly above or behind the tubes; but when the lung is consolidated as in pneumonia, it conducts sound more readily than usual, and the murmur is heard in other parts of the chest. In diseased conditions the normal sounds are modified in various ways and are then spoken of as *rales*.

EUPNEA. This term is applied to ordinary quiet respiration made without obvious effort.

DYSPNEA. Usually the term *dyspnea* is reserved for painful breathing, in which the expirations are active and forced. Dyspnea may be caused by (1) stimulation of the sensory nerves, particularly the pain nerves, (2) an increase in the hydrogen ion concentration of the blood, and (3) any condition that interferes with the normal rate of the respirations or of the heart action or prevents the passage of air in or out of the lungs.

HYPERPNEA. The word *hyperpnea* is applied to an increased rate and/or depth of respirations.

APNEA. The word means a lack of breathing. In physiologic literature, it is used to describe the cessation of breathing movements due to lack of stimulation of the respiratory center, brought about by rapid and prolonged ventilation of the lungs. In medical literature, the term is sometimes used as a synonym for asphyxia, or suffocation.

CHEYNE-STOKES RESPIRATIONS. This is a type of respiration that was first described by the two physicians[4] whose names it bears. It is an exaggeration of the type of respiration that is often seen during sleep in normal people. The respirations increase in force and frequency up to a certain point and then gradually decrease until they cease altogether; there is a short period of apnea, then the respirations recommence, and the cycle is repeated. Cheyne-Stokes respirations are associated with conditions that depress the respiratory center, especially in brain, heart, and kidney diseases.

EDEMATOUS RESPIRATION. When the air cells become infiltrated with fluid from the blood, the breathing becomes edematous and is recognized by the moist, rattling sounds, or rales, caused by the passage of the air through the fluid. It is a serious condition because it interferes with aeration of the blood and often results in asphyxia.

Cough. The cough reflex is a critically essential one, for it prevents obstruction of the airway. Foreign material, such as particles or irritating chemicals, stimulate nerve endings and impulses are transmitted to the respiratory center by the vagus nerves. This results in inspiration of a large volume of air; the epiglottis closes tightly; the abdominal muscles and other expiratory muscles contract forcefully. The result is a pressure rise in the lungs of about 100 mm Hg. When the epiglottis opens, there is rapid, strong expulsion of air from the lungs, which carries the foreign material outward.

[4]John Cheyne, Scottish physician (1777–1836). William Stokes, Irish physician (1804–1878).

Asphyxia. Asphyxia is produced by any condition that causes prolonged interference with the aeration of the blood, viz., obstruction to the entrance of air to the lungs, depression of the respiratory center, an insufficient supply of oxygen, or a lack of hemoglobin in the blood. The first stages are associated with dyspnea and convulsive movements; then the respirations become slow and shallow and are finally reduced to mere twitches. The skin is cyanosed, the pupils of the eyes dilate, the reflexes are abolished, and respirations cease. If the heart continues to beat, resuscitation is often accomplished by artificial respiration, even after breathing has ceased.

Artificial Respiration. Mouth-to-mouth resuscitation is the most efficient method of ventilating a person who has stopped breathing. This is possible because of the relatively high concentration of oxygen in expired air (see table, page 435). Other methods use manual or machine pressure to alter the size of the thoracic cage.

Hypoxia and Anoxia. Hypoxia is a general term that should be used for oxygen deficiency when arterial oxygen saturation is *not less* than 80 per cent. Anoxia is a marked deficiency or absence of oxygen reaching the tissue cells. It may be due to:

1. Decreased oxygen-carrying capacity of hemoglobin, due to presence of methemoglobin, or carbon monoxide hemoglobin, or low hemoglobin content of blood. This is called *anemic anoxia*.

2. *Anoxic anoxia*—the volume of oxygen in arterial blood is low due to low oxygen tension or inadequate pulmonary ventilation or pulmonary disease.

3. *Stagnant anoxia*—due to disturbances in circulation that slow down circulation in the capillaries, e.g., cardiac disease.

4. *Histotoxic anoxia*—due to interference with cellular respiration, as in cyanide poisoning.

Hypercapnia. Increased amounts of carbon dioxide are termed *hypercapnia;* the opposite is *hypocapnia*.

Oxygen Therapy. The signs of oxygen need are not always clearly defined. Cyanosis may occur. It results from an increase in reduced hemoglobin in the blood and consequently decreased oxyhemoglobin. Various studies indicate that even experienced observers may not detect this sign before the proportion of oxyhemoglobin has dropped from the normal 96 per cent to as low as 85 per cent. A person who is anemic and has a low hemoglobin level will not appear cyanotic although he may lack oxygen. In carbon monoxide poisoning, the skin becomes a cherry red rather than the blue of cyanosis. The earliest symptoms are frequently vague and related to decrease of oxygen to brain cells, which are very sensitive to oxygen shortage.

The purpose of oxygen therapy is to increase the oxygen tension of blood plasma and restore the oxyhemoglobin in the red blood cells to normal proportion, thus supplying oxygen to meet the needs of the tissue cells. This is done by providing oxygen in higher concentration than in air: by enriching the atmosphere in an oxygen tent, by introducing oxygen into the nasopharynx through a nasal tube, or by an oxygen mask. The concentration of oxygen is adjusted by regulating the rate of flow of the gas from a tank where it is stored under pressure. Concentrations of 40 to 60 per cent are usually used although even 100 per cent oxygen can be supplied through the mask technique. Ordinarily the gas is supplied at atmospheric pressure; the increased proportion of oxygen in the mixture increases the tension of oxygen in the alveoli and more oxygen will be absorbed across the alveolar-capillary membrane. Once the hemoglobin has been saturated with oxygen, little or no further advantage is obtained by increasing the oxygen concentration. In carbon monoxide poisoning, increased oxygen concentration in the blood favors release of carbon monoxide from its combination with hemoglobin and the hemoglobin thus released can combine with oxygen. When oxygen therapy is successful, the symptoms associated with hypoxia disappear.

Oxygen leaves the tank as a dry gas, free from water vapor, and as such would be very irritating to the membranes of the trachea, bronchi, and alveoli. Consequently the oxygen must be moistened by bubbling it through water; this is the function of the humidifier. It is also important that the air entering the lungs should be cool and, therefore, the temperature of the tent should be about 20° C (68° F). Increased oxygen concentration in air always increases the danger of fire. The temperature at which burning occurs becomes progressively lower as the oxygen concentration increases. Consequently, it is of utmost importance to safeguard the area where oxygen is in use against possible sources of ignition.

OXYGEN INTOXICATION. Increased oxygen tension in the blood does not significantly increase the amount of oxygen combined with hemoglobih, because the hemoglobin is already saturated; but it does continue to increase the amount of oxygen dissolved in the water of the blood. When this occurs to extreme limits, there is a change in the rates of many chemical reactions within tissue cells (metabolic rates).

The tissue most sensitive is the nervous system. Convulsions may occur or there may be actual brain cell destruction.

In the lungs, since the pulmonary membranes are directly exposed to high oxygen pressure, pulmonary edema may result. Oxygen intoxication will also interfere with the hemoglobin oxygen buffer system.

The alarming incidence of blindness in infants born prematurely has led to a better understanding of the relationship between oxygen and the development of blood vessels of the retina. In the human fetus, these vessels are developing during the fifth, sixth, and seventh months. Hence, they are in a sensitive developmental stage when the baby is born prematurely. Observations indicate that high concentrations of oxygen irritate and destroy the capillary network supplying blood to the retinal cells. The resulting lack of blood supply causes retrolental fibroplasia and blindness. For this reason, oxygen therapy is used very conservatively for the premature infant.

Low Barometric Pressure. Several changes that take place in the atmosphere as altitude increases affect man's reactions in high-altitude flying or space flight. These may be briefly stated as follows:

Oxygen concentration in air at 40,000 ft is about the same as at sea level, but barometric pressure is lower.

Oxygen begins to decrease at 80,000 ft, and above this height oxygen and nitrogen amounts decrease and the amounts of helium and hydrogen increase. Barometric pressure is also an important factor in high altitudes. At sea level the pressure is 760 mm Hg, while at 18,000 ft pressure is about half, or 380 mm Hg; and as altitude increases, barometric pressure continues to decrease. Another factor is temperature; the higher the altitude, the lower the temperature. With every 500 ft of rise in altitude there is about 1° C drop in temperature.

These atmospheric changes affect man considerably as ascent in an airplane is made. Oxygen lack causes anoxia; the barometric pressure changes and extremes of cold also directly affect and modify man's reactions. Above 15,000 ft symptoms of oxygen lack are present. The individual becomes quiet, and the lips, ear lobes, and nail beds become slightly bluish. As a higher altitude is reached, the blue color deepens, and weakness and dizziness occur. Since the brain cells are dependent upon oxygen, deprivation causes specific symptoms to become evident. The power of attention is diminished, muscular coordination is lessened, mental confusion is present, and there is interference with perception of objects. Breathing is embarrassed. At still higher altitudes these symptoms become more apparent; the individual may become irritable, the mental confusion increases, and speech is interfered with. Muscular coordination may be lost so that writing is impossible. There is difficulty in understanding written words. There may be drowsiness, headache, apathy, and loss of self-control. Vision and memory are impaired, judgments are unsound, pain sensations are dulled, and appreciation of passage of time is altered. Cyanosis becomes more apparent, dyspnea and vomiting may occur. Each sense is finally lost.

In airplane travel oxygen tension, pressures, and temperature are regulated by artificial means, thus preventing any physiologic changes that occur at high altitudes.

High Barometric Pressure. High barometric pressure increases the amount of gases that diffuse from the alveoli into the blood and eventually are dissolved in all body fluids. Increased oxygen absorption will damage tissues. Hemoglobin releases oxygen more rapidly, which results in cellular damage due to deranged cellular metabolism. The brain usually shows early effects, which result in twitching, convulsions, and coma. Such an increase in barometric pressure occurs in deep-sea exploration or, less severely, in scuba diving.

CARBON DIOXIDE. If, for example, a diver's helmet collects high concentrations of carbon dioxide, respiratory acidosis and coma can result.

NITROGEN. It is well known that high concentrations exert an anesthetic effect on the central nervous system. To avoid the problem, nitrogen is frequently replaced with helium, as it does not cause anesthetic effects and also leaves the body fluids rapidly. Nitrogen not only has anesthetic effects, but the gas bubbles in the body fluids when the diver surfaces. This condition is known as decompression sickness, the bends, caisson disease, and diver's paralysis. The most damage is done when bubbles develop in the brain and cord. Serious mental damage or paralysis may occur due to ruptured nerve fibers. Sudden gastrointestinal distention due to gas in the stomach and intestine may also occur. The condition can be prevented by slow ascent or by placing the diver or other worker in a decompression chamber.

In spite of the negative effects discussed above, high pressure chambers, called barometric chambers, have been developed for medical use. With controlled amounts of pressure and oxygen levels, increased amounts of oxygen will diffuse into the blood stream and be effective in treatment of infections caused by *anaerobic* bacteria.

Questions for Discussion

1. What is the relationship between lung compliance and ventilation?
2. What effect does increasing the oxygen content of air have on the blood?
3. Compare the pressure gradients of oxygen and carbon dioxide in the lungs and in the tissues.
4. List four factors that influence respiratory rate and depth; by what means is this influence mediated?
5. Discuss the nervous control of respiration.
6. Describe the effects of hyperventilation upon the respiratory center.

Summary

Respiration	All living organisms require continual supply of oxygen Chemical changes in tissue cells dependent upon it Carbon dioxide is one end product of chemical changes in cells, hence need for elimination of excess Exchange of these gases in lungs and cells constitutes respiration
Respiratory System	Air passes through nose or mouth to 1. Larynx 3. Bronchi 2. Trachea 4. Lungs

Nose

Function
- Special organ of the sense of smell
- Passageway for entrance of air to the respiratory organs
- Helps in phonation

External nose
- Framework of bone (nasal) and cartilage
- Covered with skin, lined with mucous membrane
- Nostrils are oval openings on undersurface, separated by a partition

Internal cavities or nasal fossae
- Two wedge-shaped cavities
- Extend from nostrils to pharynx
- Lined by mucous membrane, ciliated epithelium, well-vascularized
- Formed by
  - 2 palatine
  - 2 maxillae
  - 1 ethmoid
  - 1 sphenoid
  - 2 nasal
  - 2 conchae, and processes of the ethmoid
    - Superior meatus
    - Middle meatus
    - Inferior meatus
  - 1 vomer
  - 11 bones

Communicating sinuses
1. Frontal
2. Ethmoid
3. Maxillary, or antra of Highmore
4. Sphenoid

Nerves
1. Olfactory nerve—sense of smell
2. Facial nerve branches, the ophthalmic and maxillary

Arteries
- External maxillary } derived from the external carotid
- Internal maxillary }
- Ethmoidal arteries derived from internal carotid

Advantages of nasal breathing
- Air is
  - Warmed
  - Moistened
  - Filtered

Larynx
- Special organ of voice
- Triangular box made up of nine cartilages
- Muscles
  - Extrinsic
    - Infrahyoid group
    - Suprahyoid group
  - Intrinsic
- Situated between the tongue and trachea
- Contains vocal folds
- Slit or opening between cords called *glottis*, which is protected by leaf-shaped lid called *epiglottis*
- Connected with external air by
  - Mouth
  - Nose
- **Nerves**—derived from
  - Internal branches of superior laryngeal
  - External branches of superior laryngeal
- **Arteries**
  - Superior thyroid, branch of external carotid
  - Inferior thyroid

Phonation
- **Phonation**—production of vocal sounds
- **Organs of phonation**
  - Respiratory organs
  - Vocal folds
    - Lower pitch of male voice is due to greater length of vocal folds
  - Larynx, pharynx, mouth, nose, and tongue
  - Speech centers and parts of brain that control movements of the tongue and jaw, also association centers

Trachea
- Membranous and cartilaginous tube, 11.2 cm (4½ in.) long
- Strengthened by C-shaped rings of cartilage
  - Complete in front
  - Incomplete behind
- In front of esophagus—extends from larynx to upper border of fifth thoracic vertebra, where it divides into two bronchi
- **Nerves**
  - Branches of vagus
  - Recurrent laryngeal nerves
  - Autonomics
- **Arteries**—Inferior thyroid

Bronchi and Bronchioles
- Structure similar to trachea
- Right—shorter, wider, more vertical than left
- Divide into innumerable bronchial tubes, or bronchioles
- As tubes divide, their walls become thinner. Finer tubes consist of thin layer of muscular and elastic tissue lined by ciliated epithelium
- Each bronchiole terminates in elongated saccules called *alveolar sacs*
- Each alveolar sac terminates in small projections known as *alveoli*, or air sacs

Lungs
- **Location**—lateral chambers of thoracic cavity, separated by structures contained in mediastinum
  - **Cone-shaped organs**
    - Outer surface convex to fit in concave cavity
    - Base concave to fit over convex diaphragm
    - Apex about 2.5 to 4 cm (1 to 1½ in.) above the level of sternal end of first rib
    - Hilum, or depression on inner surface, gives passage to bronchi, blood vessels, lymphatics, and nerves
  - **Right**—larger, broader, shorter—three lobes
  - **Left**—smaller, narrower, longer—two lobes
  - **Anatomy**
    - Porous, spongy organs. Consist of bronchial tubes, atria, alveoli, also blood vessels, lymphatics, and nerves held together by connective tissues
  - **Blood vessels**
    - **Pulmonary artery**
      - Blood for aeration
      - Accompanies bronchial tubes
      - Plexus of capillaries around alveoli
      - Returned by pulmonary veins
    - **Bronchial arteries**—supply lung substance
  - **Blood pressure**
    - Very low; high oncotic pressure of plasma prevents fluid movement into alveolus

Pleura
- **Closed sac**—envelops lungs
- **Two layers**
  - Pulmonary, or visceral—next to lung
  - Parietal—outside of visceral
    - Normally in close contact—potential cavity
    - Moistened by serous fluid
- **Function**—to lessen friction by secretion of pleural fluid

Mediastinum
- Space between pleural sacs. Extends from sternum to spinal column. Contains the heart, large blood vessels connected with heart, trachea, esophagus, thoracic duct, various veins, lymph nodes, and nerves

Respiration
- **Function**
  - Increase the amount of oxygen
  - Decrease the amount of carbon dioxide
  - Help to maintain temperature
  - Help to eliminate water
- **Processes**
  - **Breathing**
    - **Inspiration**—process of taking air into lungs
    - **Expiration**—process of expelling air from lungs
    - **Normal rate**—16–18/minute
  - **External respiration**
    - External oxygen supply
    - External carbon dioxide elimination } Takes place in the lungs
  - **Internal respiration**
    - Internal oxygen supply
    - Internal carbon dioxide elimination } Takes place in the cells

Ventilation
- **Caused by change in cavity size**
  - Vertically
  - Dorsoventrally
  - Laterally
- **Inspiration**
  - Chest cavity enlarged
    - Elevation of ribs, dependent upon contraction of muscles of inspiration
    - Descent of diaphragm by contraction of diaphragmatic muscle
  - Enlargement of lungs—in proportion to enlargement of cavity—lungs in contact with thoracic walls
  - Air rushes in through trachea and bronchi
- **Expiration**
  - Chest cavity made smaller
    - Inspiratory muscles relax
    - Recoil of elastic thorax
    - Recoil of elastic lungs (compliance)
  - Air forced out through trachea
- Pressure changes with changes in size of thoracic cage

Inspired and Expired Air — **Changes effected**
1. Moisture increased. Expired air is saturated with moisture
2. Temperature increased. Expired air is as warm as blood
3. Heat to warm air and vaporize moisture taken from body
4. Oxygen decreased by 4.94%
5. Carbon dioxide increased by 4.34%

Capacity of Lungs
After lungs are once filled, they are never emptied during life
Vital capacity—quantity of air person can expel by forcible expiration after deepest inspiration possible—averages from 3500–4100 cc
Volumes
- Tidal
- Inspiratory reserve
- Expiratory reserve
- Residual
- Reserve

External Respiration
Takes place in alveoli
Blood
- Loses about 6% of carbon dioxide
- Gains about 8% of oxygen
  - Oxyhemoglobin
  - Scarlet color

Internal Respiration
Consists of
- Exchange of gases in the tissues
- Passage of oxygen from blood into tissue fluid and from tissue fluid into cells
- Passage of carbon dioxide from tissue cells into tissue fluid and from tissue fluid into blood

Important to remember blood does not give up all its oxygen to the tissues or all of its carbon dioxide in the lungs

Respiratory Center
Located in medulla oblongata and pons
Inspiratory, expiratory neurons
Efferent fibers from respiratory center travel down spinal cord and connect with fibers of vagi and sympathetic nerves distributed in the lung tissue
Afferent fibers lead to the respiratory center
Action
- Automatic, i.e., it is constantly sending impulses over afferent fibers
- Impulses from periphery may reach center
- **Rate and rhythm dependent on**
  - Vagus nerves, inhibit inspiration—Hering-Breuer reflex
  - Chemical condition of blood, i.e., hydrogen ion concentration of the blood and oxygen and carbon dioxide levels

Control of Respiration
Voluntary control for a short time
Breaking point reached in 23–27 seconds
If lungs are thoroughly aerated by forced breathing, breaking point may be postponed as long as 8 minutes

Cause of First Respiration
1. Increased amount of carbon dioxide in blood
2. Stimulation of sensory nerves of skin
3. Combination of these two causes

Respiratory Phenomena
Air passing into alveoli produces a fine, rustling sound
Air passing in and out of larynx, trachea, and bronchial tubes produces louder sound called bronchial murmur
In diseased conditions modified sounds are called *rales*
Eupnea—ordinary quiet respiration
Dyspnea—difficult breathing
Hyperpnea—excessive breathing
Apnea—lack of breathing
Cheyne-Stokes — Respirations increase in force and frequency, then gradually decrease and stop. Cycle repeated
Edematous—air cells filled with fluid, hence moist, rattling sounds
Asphyxia—oxygen starvation
Artificial respiration—mouth-to-mouth most efficient
Cough reflex—protective

High Barometric Pressure
Increases the amount of gases that diffuse from alveoli into the blood
Oxygen released rapidly
Altered cellular metabolism

Low Barometric Pressure	Decreases oxygen available for diffusion Many physiologic abnormalities
Oxygen Therapy	Hypoxia—oxygen lack, not lower than 80% Anoxia—oxygen deficiency or absence Cyanosis—bluish color due to oxygen need General purpose—to increase oxygen tension in blood plasma
Oxygen Intoxication	Nervous tissue most sensitive Pulmonary edema may occur High oxygen concentration irritates and destroys capillary networks in the retina of premature infants
Oxygen Lack	Causes tissue anoxia Disturbances of perception Diminished power of attention Individual becomes irritable Mental confusion Speech disturbed

Additional Readings

AVERY, M. E.: The lung of the newborn infant. *Sci. Amer.*, **228**:74–85, April, 1974.

GUYTON, A. C.: *Basic Human Physiology: Normal Function and Mechanisms of Disease*, 2nd ed., W. B. Saunders Co., Philadelphia, 1977. Chapters 27, 28, and 29.

HAM, A. W.: *Histology*, 7th ed. J. B. Lippincott Co., Philadelphia, 1974, Chapter 20.

PANSKY, B., and HOUSE, E. L.: *Review of Gross Anatomy*, 3rd ed. Macmillan Publishing Co., Inc., New York, 1975, pp. 122–33, 272–83, 364–65.

SELKURT, E. E.: *Basic Physiology for the Health Sciences*. Little, Brown & Co., Boston, 1975, pp. 349–454.

VANDER, A. J.; SHERMAN, J. H.; and LUCIANO, D. S.: *Human Physiology: The Mechanisms of Body Function*, 2nd ed. McGraw-Hill Book Co., New York, 1975, Chapter 10.

Body Maintenance: Processing and Utilization of Nutrients

The life of the cell requires the supply of essential materials for growth and for other structural activities, as well as a source of energy for these activities. Unit V is devoted to the kinds of materials needed by cells, how they enter the body, how they are used, and how the use is regulated.

The Digestive System

Chapter Outline

Within the digestive tract is carried on the necessary transformation of ingested complex food substances into the simpler substances that may pass into the bloodstream and be distributed to the cells of the body.

General Considerations

The means by which foods are transformed into simple substances are both physical and chemical and constitute the digestive processes; the organs that take part in them form the digestive system.

Processes concerned with such changes can be classified into two groups:

1. Those concerned with the breakdown

of food to particles small enough to pass through the wall of the alimentary tract into the body fluids.

Physical comminution brought about by mastication and by various types of muscular activity is described on following pages. A subsequent chemical comminution changes large molecules to molecules sufficiently small to pass into the blood. Normal motility of the alimentary tract and proper neuromuscular functioning, by which it is carried on, are essential to digestion.

2. Those concerned with moving foods along through the alimentary tract with optimum speed. This means slowly enough for all the necessary changes in each organ to be accomplished in preparation for those in the next organ and yet fast enough so that proper absorption

shall take place and bacterial decomposition or deleterious changes do not occur.

The structure of the digestive system is presented in this chapter; the changes occurring in food as it progresses through the digestive tract are presented in Chapter 21. Cellular use of nutrients is presented in Chapter 22.

The organs of the digestive system lie in the head, the thorax, the abdomen, and the pelvis. The greater portion is in the abdominal cavity (shown in Figure 20–1; see also Plate II, inserted between pages 536 and 537). The parts of the digestive system are the *alimentary canal* and the *accessory organs*. The alimentary canal is a continuous tube and as measured on the cadaver is about 9 m (30 ft) long; it extends from the mouth to the anus. In the living subject the length of the alimentary canal

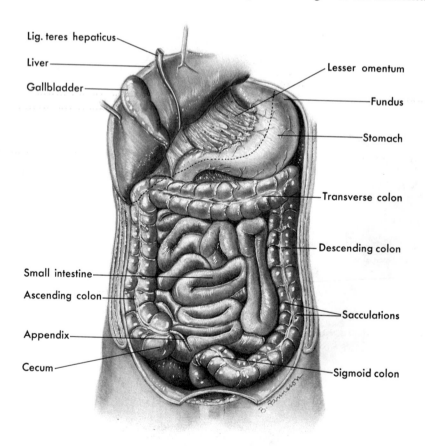

Figure 20–1. The stomach and intestines, front view, the great omentum having been removed and the liver turned up and to the right. The *dotted line* shows the normal position of the anterior border of the liver.

is much shorter, owing probably to shortening of its longitudinal coats. The accessory organs are the tongue, teeth, salivary glands, pancreas, liver, and gallbladder.

The Alimentary Canal

The divisions of the alimentary canal are:

Mouth cavity, containing tongue, orifices of ducts of salivary glands, and teeth
Pharynx
Esophagus
Stomach

Small intestine
{
Duodenum
Jejunum
Ileum

Large intestine
{
Cecum
Colon ———→
Rectum
Anal canal

{
Ascending
Transverse, or mesial
Descending
Sigmoid

The stomach, small intestine, and large intestine are often referred to as the gastrointestinal tract.

Mouth, Oral or Buccal Cavity (Figure 20–2). This is a cavity bounded laterally and in front by the cheeks and lips; behind, it communicates with the pharynx. The roof is formed by the hard and soft palate, and the greater part of the floor is formed by the tongue and sublingual region and lower jaw. The space bounded externally by the lips and cheeks and internally by the gums and teeth is called the *vestibule.* The cavity behind this is the *mouth cavity proper. The lips,* two musculomembranous folds, surround the orifice of the mouth and are important in speech.

The palate consists of a hard portion in front, formed by processes of the maxillary and palatine bones, which are covered by mucous membrane. Suspended from the posterior border is the soft palate, a movable fold of mucous membrane, enclosing muscle fibers, blood vessels, nerves, lymphatic tissue, and mucous glands. Hanging from the middle of its lower border is a conical process called the *uvula.*

The fauces is the name given to the aperture leading from the mouth into the pharynx, or throat cavity. At the base of

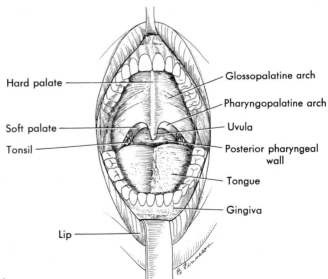

Hard palate
Soft palate
Tonsil
Lip

Glossopalatine arch
Pharyngopalatine arch
Uvula
Posterior pharyngeal wall
Tongue
Gingiva

Figure 20–2. Mouth and pharynx.

the uvula on either side is a curved fold of muscular tissue covered by mucous membrane, which shortly divides into two pillars; one runs downward, lateralward, and forward to the side of the base of the tongue; the other, downward, lateralward, and backward to the side of the pharynx. These arches are known respectively as the *glossopalatine arch* (anterior pillars of the fauces) and the *pharyngopalatine arch* (posterior pillars of the fauces).

The palatine tonsils are two masses of lymphoid tissue situated, one on either side, in the triangular space between the glossopalatine and the pharyngopalatine arches. The surface of the tonsils is marked by openings called crypts, which communicate with channels that course through the substance of the tissue. They are supplied with blood from the lingual and internal maxillary arteries, which are derived from the external carotid arteries, and receive nerve fibers from both divisions of the autonomic nervous system. Situated below the tongue are masses of lymphoid tissue called the *lingual tonsils;* however, the term *tonsil* as commonly used refers to the palatine tonsils.

The function of the tonsils is similar to that of other lymph nodes. They aid in the formation of lymphocytes and help protect the body from infection by acting as filters and preventing the entrance of microorganisms. If they are diseased, their protective function is reduced, and they may serve as foci of infection, which passes directly into the lymph and so into the blood. Inflammation of the palatine tonsils is called *tonsillitis.*

The palate, uvula, palatine arches, and tonsils are plainly seen if the mouth is widely opened and the tongue depressed.

Tongue. The tongue is the special organ of the sense of taste. It assists in mastication, deglutition, and digestion by movements that help to move the food and keep it between the teeth; the glands of the tongue secrete mucus, which (in addition to the salivary gland secretions) lubricates the food and makes swallowing easier; and stimulation of the end organs (taste buds) of the nerves of the sense of taste increases the secretion of saliva and starts

the first flow of gastric juices. The sense of taste is mediated over the sensory fibers of cranial nerve VII (anterior two thirds of the tongue) and cranial nerve IX (posterior third of the tongue). Probably more than half of the so-called tastes are due to stimulation of olfactory receptors rather than taste receptors. The tongue is essential for speech.

Salivary Glands (*Figure 20–3*). The mucous membrane lining the mouth contains many minute glands called *buccal glands,* which pass their secretion into the mouth. The chief secretion, however, is supplied by three pairs of compound saccular glands, the salivary glands, named *parotid,* submaxillary, and sublingual glands. Each *parotid* gland is placed just under and in front of the ear; its duct, the parotid (Stensen's[1]), opens upon the inner surface of the cheek opposite the second molar of the upper jaw. The *submandibular* and *sublingual* glands lie below the jaw and under the tongue, the submandibular being placed farther back than the sublingual. One duct (Wharton's[2]) from each submandibular and a number of small ducts from each sublingual open in the floor of the mouth beneath the tongue. The secretion of the salivary glands, mixed with that of the mucous glands of the mouth, the buccal secretion, is called *saliva.*

Nerves and Blood Vessels. The facial (VII) and glossopharyngeal (IX) nerves supply these glands. The fibers are both secretory and vasomotor and are derived from the craniosacral and thoracolumbar systems. Blood is supplied to the salivary glands by branches of the external carotid artery and is returned, after traveling through many branch arteries and capillaries, via the jugular veins.

Teeth (Dentes) (*Figure 20–4*). The alveolar processes of the maxillae and mandible contain *alveoli,* or sockets, for the teeth. Dense connective tissue covered by smooth mucous membrane—the gums, or gingivae—covers these processes and extends a little way into each socket. The sockets are lined with periosteum, which

---

[1] Nicolaus Stensen, Danish anatomist (1638–1686).
[2] Thomas Wharton, English anatomist (1610–1673).

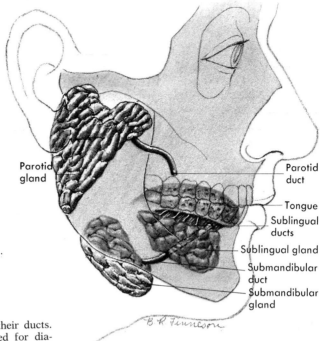

Figure 20–3. The salivary glands and their ducts. Sublingual ducts are somewhat enlarged for diagrammatic purpose.

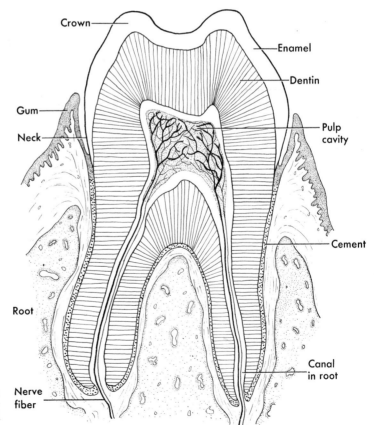

Figure 20–4. Section of human molar tooth. In the pulp cavity are located blood vessels and nerves.

connects with the gums and serves to attach the teeth to their sockets and as a source of nourishment.

Each tooth consists of three portions: the *root,* consisting of one to three divisions contained in the socket; the *crown,* which projects beyond the level of the gums; and the *neck,* or constricted portion between the root and the crown.

Each tooth is composed principally of *dentin*, which gives it shape and encloses a cavity, the pulp cavity. The dentin of the crown is capped by a dense layer of *enamel*. The dentin of the root is covered by *cement*. These three substances—enamel, dentin, and cement—are all harder than bone, enamel being the hardest substance found in the body. They are derived from epithelial tissue. The pulp cavity is just under the crown and is continuous with a canal that traverses the center of each root and opens by a small aperture at its extremity. It is filled with dental pulp, which consists of connective tissue holding a number of blood vessels and nerves, which enter by means of the canal from the root.

There are two sets of teeth developed during life: the first, deciduous, or milk, teeth; and the second, permanent.

DECIDUOUS TEETH. The deciduous set includes 20 teeth, 10 in each jaw: 4 incisors, 2 canines, and 4 molars. The cutting of these teeth usually begins at six months and ends at about the age of two years. In nearly all cases the teeth of the lower jaw appear before the corresponding ones of the upper jaw. Table 20–1 lists the deciduous teeth.

Another way of expressing the number of teeth is referred to as the "dentition formula." In such cases the formula is written as:

$$\frac{2:1:4:1:2}{2:1:4:1:2}$$

PERMANENT TEETH. During childhood the temporary teeth are replaced by the *permanent*. In the second set are 32 permanent teeth, 16 in each jaw. The first molar usually appears between five and seven years of age. Table 20–2 lists the permanent teeth.

The "dentition formula" for permanent teeth would be:

$$\frac{3:2:1:4:1:2:3}{3:2:1:4:1:2:3}$$

According to their shape and use the teeth are divided into incisors, canines, premolars, or bicuspids, and molars. *Incisors*, eight in number, form the four front teeth of each jaw. They have a sharp cutting edge and are especially adapted for biting food. *Canines* are four in number, two in each jaw. They have sharp, pointed edges, are longer than the incisors, and serve the same purpose in biting and tearing. *Premolars, or bicuspids,* are eight in number in the permanent set (none in the temporary set). There are four in each jaw, two placed just behind each of the canine teeth. They are broad, with two points or cusps on each crown, and have only one root, which is more or less completely divided into two. Their function is to grind food. *Molars* are 12 in number in the per-

TABLE 20–1
DECIDUOUS TEETH

	Molars	Canine	Incisors	Canine	Molars
Upper	2	1	4	1	2
Lower	2	1	4	1	2

The deciduous teeth are usually cut in the following order:

Lower central incisors	6–9 months
Upper incisors	8–10 months
Lower lateral incisors and first molars	15–21 months
Canines	16–20 months
Second molars	20–24 months

TABLE 20–2
PERMANENT TEETH

	Molars	Premolars	Canine	Incisors	Canine	Premolars	Molars
Upper	3	2	1	4	1	2	3
Lower	3	2	1	4	1	2	3

The permanent teeth appear at about the following periods:

First molars	6 years
Two central incisors	7 years
Two lateral incisors	8 years
First premolars	9 years
Second premolars	10 years
Canine	11–12 years
Second molars	12–13 years
Third molars	17–25 years

manent set (eight in the deciduous set). They have broad crowns with small, pointed projections, which make them well fitted for crushing food. Each upper molar has three roots, and each lower molar has two roots, which are grooved and indicate a tendency to division. The 12 molars do not all replace temporary teeth but are gradually added with the growth of the jaws. The hindmost molars are the last teeth to be added. They may not appear until 25 years of age, hence are called *late teeth* or "wisdom teeth."

Long before the teeth appear through the gums their formation and growth are in progress. The deciduous set begins to develop about the sixth week of intrauterine life; and the permanent set, with the exception of the second and third molars, begins to develop about the sixteenth week. About the third month after birth the second molars begin to grow, and about the third year, the third molars, or wisdom teeth, do likewise. Diseases such as rickets retard the eruption of the temporary teeth, and severe illness during childhood may interfere with the normal development of the permanent teeth so that they are marked with notches and ridges. Moreover cavities form in them readily. The diet of the mother during pregnancy and the diet of the child during the first years of life are important factors in determining the quality of the teeth and the development of caries. More recently the addition of fluorine salts to drinking water has been shown to retard caries formation.

The principal functions of the teeth are *biting* with the incisors and *chewing* or *mastication* with the molars. Mastication of the more solid foods is good for the teeth because the increased pressure has a massaging effect upon the gums, which tends to promote circulation in the pulp.

Pharynx. The pharynx is a musculomembranous tube shaped somewhat like a cone, with its broad end turned upward and its constricted end downward to end in the esophagus. It may be divided from above downward into three parts, nasal, oral, and laryngeal. The upper, or *nasopharynx,* lies behind the posterior nares and above the soft palate. The middle, or oral, part of the pharynx reaches from the soft palate to the level of the hyoid bone. The *laryngeal* part reaches from the hyoid bone to the esophagus. The pharynx communicates with the nose, ears, mouth, and larynx by seven apertures: two superior ones, leading into the back of the nose, the *posterior nares;* two on the lateral walls of the nasopharynx, leading into the auditory tubes, which communicate with the ears; one midway in front, the *fauces* connecting with the mouth in front; two below—one, the well-defined glottis, opening into the larynx, and the other, the poorly demarcated, opening into the esophagus.

The mucous membrane lining the pharynx is continuous with that lining the nasal cavities, the mouth, the auditory tubes, and the larynx. (See Figure 3–30, page 75.) It is well supplied with mucous glands. The walls of the pharynx are provided with sensory receptors, which are sensitive to mechanical stimulation and are important in the mechanisms of swallowing. When food or liquid stimulates these touch receptors, the complicated reflex of swallowing is initiated. If these sensory areas are anesthetized, as by swabbing the throat, swallowing becomes difficult. About the center of the posterior wall of the nasopharynx is a mass of lymphoid tissue, the pharyngeal tonsil. When abnormally large it is called *adenoids.*

Usually lymphoid tissue is larger in children than in adults and tends to become smaller with age. Owing to their position, adenoids may become infected or enlarged, block the auditory tubes, and interfere with the passage of air through the nose, as well as impair the hearing.

Nerves and Blood Vessels. Both divisions of the autonomic system supply nerve fibers to the pharynx. There are both sensory and motor fibers within the glossopharyngeal and vagus nerves. Blood is supplied by branches from the external carotid artery.

Functions. The function of the pharynx is to transmit air from the nose or mouth to the larynx and serve as a resonating cavity in the production of the voice. It also serves as a channel to transmit food from the mouth to the esophagus. Closure of the mouth and nasopharynx during deglutition, or swallowing, effectively shuts off the pharynx from the outside atmosphere. Contraction of pharyngeal muscles, combined with the thrust caused by other contracting muscles pushes the food downward and onward into the esophagus.

Esophagus. The esophagus is a muscular tube, about 23 to 25 cm long and 25 to 33 mm wide, which begins at the lower end of the pharynx, behind the trachea. It descends in the mediastinum in front of the vertebral column, passes through the diaphragm at the level of the tenth thoracic vertebra, and terminates in the upper, or cardiac, end of the stomach, about the level of the xiphoid process.

The walls of the esophagus are composed of four coats: (1) an external, or adventitious layer which merges with connective tissue of nearby structures, (2) a muscular, (3) a submucous, or areolar, and (4) an internal, or mucous, coat. The muscular coat consists of an external longitudinal and an internal circular layer. The muscles in the upper part of the esophagus are striated. These are gradually replaced by smooth muscle tissues. The lower third of the esophagus is completely smooth muscle. Contractions of the layers propel food to the stomach. The areolar coat serves to connect the muscular and mucous coats and to carry the larger blood and lymph vessels. The mucous membrane is arranged in longitudinal folds, which disappear when the esophagus is distended by the passage of food. It is studded with minute papillae and small glands, which secrete mucus to lubricate the canal.

Nerves and Blood Vessels. The nerve fibers from the vagus and the thoracolumbar nervous system form a plexus between the layers of the muscular coat and another in the submucous coat. Blood is supplied to the esophagus by arteries from the inferior thyroid branch of the thyrocervical trunk, which arises from the subclavian; from the thoracic aorta; from the left gastric branch of the celiac artery; and from the left inferior phrenic of the abdominal aorta. Blood is returned via the azygos, thyroid, and left gastric veins of the stomach.

Functions. The esophagus receives food from the pharynx and by a series of peristaltic contractions passes it on to the stomach. Muscle at the lower end of the esophagus acts as a sphincter, preventing reflux of material from the stomach during gastric peristalsis.

Before discussion of the digestive organs found in the abdominal cavity, consideration must be given to the general structure of the gastrointestinal tract, and to its outermost layer—the peritoneum.

General Plan of the Gastrointestinal Tract. The digestive tract is composed of four layers: mucous membrane (the innermost layer); submucosa; external muscle coat; and serosa.

Like mucous membrane elsewhere, the *gastrointestinal mucous membrane* is composed of a layer of columnar *epithelial cells*, its supporting *lamina propria*, and a thin layer of smooth muscle cells called the *muscularis mucosae*.

The *epithelial layer* varies somewhat, according to the particular function of different parts of the gastrointestinal tract; e.g., in some parts it is primarily absorptive and in some primarily secretive. The epithelial cells are all columnar (they are stratified squamous in the pharynx and esophagus). The mucous secretion of the columnar epithelium is supplemented by secretions from glands—some of which extend down into the lamina propria, some down into the muscularis mucosae, and some external to the tract (e.g., the salivary glands). The *lamina propria* is composed of loose areolar connective tissue. It has many interlacing collagen, reticular, and elastic fibers (in some places). Lymphocytes are found here, as well as blood and lymphatic capillaries. It forms the core of the villi of the small intestine (page 466). The *muscularis mucosae* is a thin layer of overlapping smooth muscle cells; these occasionally extend up into the villi.

The functions of mucous membrane, in general, are protection, secretion, and support (page 75). The gastrointestinal membrane in addition is absorptive, as will be described later.

The *submucosa* connects the mucous membrane to the muscularis externa, and consists of loose connective tissue containing larger blood vessels, and autonomic nerve plexuses (Figure 20–14 and 20–15, seen on page 468). It extends up into the circular folds of the small intestine. The *muscularis externa* is a double layer of smooth muscle fibers, an inner circular one and an outer longitudinal one. Its function is to propel the contents of the gastrointestinal tube downward (*peristalsis*). The outermost layer of the gastrointestinal tract (*serosa*) is a thin layer of loose connective tissue covered with mesothelial cells, which form the visceral peritoneum.

Peritoneum. The peritoneum is the largest serous membrane in the body, in the male consisting of a closed sac (in the female the fallopian tubes open into the peritoneal cavity). The *parietal* layer lines the walls of the abdominal cavity, and the *visceral* layer is reflected over most of the abdominal organs and it covers the upper surface of some of the pelvic organs. The space between the layers, the peritoneal cavity, is under normal conditions a potential cavity only, since the parietal and visceral layers are in lubricated contact. The arrangement of the peritoneum is very complex, for elongated sacs and double folds extend from it, to pass in between

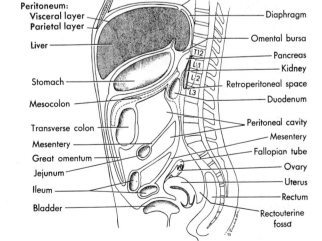

Figure 20–5. Diagram of the side view of the body, showing abdominal cavity, peritoneum, mesentery, and omentum. The *continuous lines* indicate the free surfaces of the peritoneum; the *dotted lines* indicate those parts of the peritoneum in which the free surfaces have disappeared.

Peritoneum:
Visceral layer
Parietal layer
Liver
Stomach
Mesocolon
Transverse colon
Mesentery
Great omentum
Jejunum
Ileum
Bladder

Diaphragm
Omental bursa
Pancreas
Kidney
Retroperitoneal space
Duodenum
Peritoneal cavity
Mesentery
Fallopian tube
Ovary
Uterus
Rectum
Rectouterine fossa

T12
L1
L2
L3

and either wholly or partially surround the viscera of the abdomen and pelvis.

Peritoneal Folds. Two important folds are the *greater omentum,* which hangs from the stomach in front of the intestines, and the *mesentery,* a continuation of the serous coat which attaches the small and much of the large intestine to the posterior abdominal wall. The mesentery supports the blood vessels and lymph vessels supplying the intestine, a portion of which is seen in Figure 20–12 (page 465). The mesentery is gathered in folds that attach to the dorsal abdominal wall along a short line of insertion, giving the mesentery the appearance of a ruffle, at the edge of which is the intestine.

The *omentum* is divided into greater and lesser parts. The *lesser omentum* extends from the lesser curvature of the stomach and upper duodenum to the liver. It is continuous with both layers of peritoneum. The *greater omentum,* a double layer of peritoneum, lies between the greater curve of the stomach and the spleen. From the stomach it hangs as a fat-laden apron down and in front of the small intestine. It is folded on itself so that four layers of peritoneum are in close contact. (See Figure 20–5.)

There are numerous lymph nodes in the omentum and other parts of the abdominal cavity, and many lymphatic vessels lead from the abdominal cavity into the bloodstream, so there is great rapidity with which fluids can leave the cavity. The nodes aid in protecting the peritoneal cavity against infections. (See Figure 16–18, page 377.)

Peritoneal Fossae (*Figure 20–6*). At several locations, pouches or *fossae* of peritoneum occur. Above the greater omentum are small fossae: duodenal fossae near the duodenal-jejunal juncture, and the cecal fossae; larger fossae are the pelvic fossae, and the paracolic gutters. Ordinarily there is a small amount of peritoneal fluid—enough for lubrication required by peristaltic movement of the intestine and compression of abdominal contents when the diaphragm contracts or body position changes. This fluid moves freely within the

confines of the peritoneal space. Excessive fluid formed in inflammation of the peritoneum (*peritonitis*) tends to collect in the fossae; position results in pooling of fluid in dependent areas—the pelvic fossae when the body is upright, and the posterior portion of the abdomen when the body is supine (Figure 20–6, *B*).

Stomach. In the abdominal cavity the esophagus ends in the stomach (gaster), which is a collapsible, saclike dilatation of the alimentary canal serving as a temporary receptacle for food. It lies obliquely in the epigastric, umbilical, and left hypochondriac regions of the abdomen, directly under the diaphragm. The shape and position of the stomach are modified by changes within itself and in the surrounding organs. These modifications are determined by the amount of the stomach contents, the stage of digestion that has been reached, the degree of development and power of the muscular walls, and the condition of the adjacent intestines. The stomach is never entirely empty, but always contains a little gastric fluid and mucin. When the stomach is contracted, its shape as seen from the front is comparable to that of a sickle. At an early stage of gastric digestion, the stomach usually consists of two segments, a large globular portion on the left and a narrow tubular portion on the right. When distended with food, it has the shape shown in Figure 20–7. The stomach presents two openings and two borders, or curvatures, the concave, or *lesser,* and the convex, or *greater* curvatures (Figures 20–7 and 20–8).

Component Parts (*Figures 20–7 and 20–8*). The *cardia* is the portion surrounding the esophageal opening. The upward turn of the stomach forms a J position. The *fundus* is the rounded end of the stomach, above the entrance of the esophagus. The opposite, or smaller, end is the *pyloric portion.* The central portion, between the fundus and the pyloric portion, is called the *body,* or corpus. The part of the stomach adjacent to the pyloric portion is the *antrum.* There are great differences in the position of the stomach. Much

APPENDICITIS

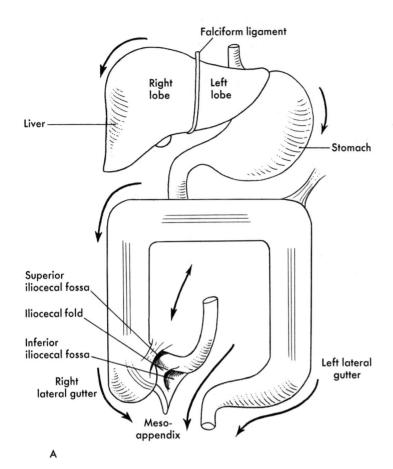

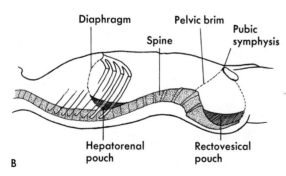

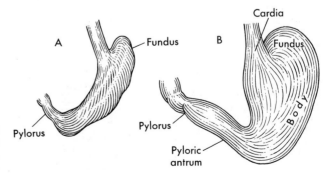

Figure 20–6. Diagram showing movement of fluid in abdomen. *A* shows abdominal gutters and peritoneal ileocecal junction. *Arrows* indicate movement of fluid in the upright position. *B.* In the supine position a large amount of fluid (e.g., hemorrhage) pools in the hollows formed by the posterior abdominal wall.

Figure 20–7. Form and outline of the stomach at different stages of digestion when seen from the front. *A.* Empty. *B.* Partially filled.

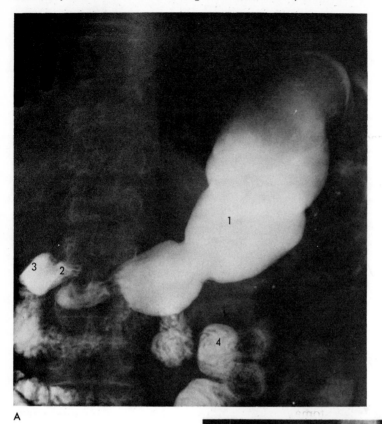

Figure 20–8. X-rays of upper gastrointestinal tract with barium contrast. *A.* Stomach and duodenum. *1,* Stomach (note contraction waves); *2,* pylorus; *3,* duodenum; *4,* jejunum. *B.* Small intestine. "Feathery" appearance of upper small intestine indicates plicae circulares; lower portion of small intestine is ileum. (Courtesy of the Radiology Department, The New York Hospital, New York, N.Y.)

depends upon body stature, position, respiratory movement, and content of the stomach.

Openings. The opening by which the esophagus communicates with the stomach is known as the *cardiac, or esophageal, orifice; the orifice that communicates with the duodenum is known as the *pyloric*. The pyloric aperture is guarded by a ringlike muscle, or sphincteric mechanism, which when contracted keeps the orifice closed. Although a distinct muscle is absent from the cardiac aperture, it is kept closed by the manner in which the muscles are arranged and the diaphragm is attached. Research shows that the pyloric antrum, pyloric sphincteric mechanism, and duodenal bulb function as a unit. The circular fibers of the pyloric sphincter serve to guard against backflow of intestinal contents into the stomach: normally the pyloric sphincter does *not* regulate stomach emptying. The movement of stomach contents into the small intestine is dependent upon the maintenance of a relatively small pressure gradient from the antrum to the pylorus. The food is kept in the stomach until such a pressure gradient is present. The relaxation of this aperture is related to the regular peristaltic waves moving over the stomach on to the duodenum.

Structure. The wall of the stomach consists of four coats: serous, muscular, submucous (or areolar), and mucous.

1. *The serosa* is covered by the visceral peritoneum. At the lesser curvature the two layers come together and are continued upward to the liver as the *lesser omentum*. At the greater curvature the two layers are continued downward as the apronlike *greater omentum,* which is suspended in front of the intestines.

2. *The muscular coat* of the stomach is beneath the serous coat and closely connected with it. It consists of three layers of smooth muscular tissue: an outer, longitudinal layer; a middle, or circular, layer; and an inner, less well-developed, oblique layer limited chiefly to the cardiac end of the stomach. This arrangement facilitates the muscular actions of the stomach by which it presses upon food and moves it back and forth.

3. *The submucous coat* consists of loose areolar connective tissue (lamina propria) connecting the muscular and mucous coats.

4. *The mucous coat* is thick, the thickness being mainly due to the fact that it is densely packed with small glands embedded in areolar connective tissue. It is covered with columnar epithelium and in its undistended condition is thrown into folds, or *rugae*. The surface is honeycombed by tiny, shallow pits, into which the ducts or mouths of the glands open. Figure 20–9 and its legend describe the coats and the tissues that compose them.

Gastric Glands (Figure 20–10). There are three varieties: cardiac, fundic, and pyloric.

Cardiac glands occur close to the cardiac orifice. They are of two kinds—simple tubular glands with short ducts, and compound racemose glands. *Fundic glands* are simple tubular glands that are found in the body and fundus of the stomach. These glands are lined with epithelial cells, of which there are two primary varieties: (1) cells lining the lumen of the tube, called *chief* cells, which secrete pepsinogen; (2) *parietal* cells, found behind the chief cells, which secrete H^+ and Cl^- ions, and the intrinsic factor. *Pyloric glands* are branched tubular glands found most plentifully about the pylorus. They secrete *pepsinogen* and *mucin,* but not acid.

Nerves and Blood Vessels. The stomach is supplied with thoracolumbar nerve fibers from the celiac plexus. Terminal branches of the right vagus are distributed to the posterior part of the organ; branches from the left vagus are distributed to the anterior part. Stimulation of the vagus fibers increases secretion and peristalsis. Stimulation of the thoracolumbar autonomic fibers has just the opposite effect, i.e., inhibits secretion and peristalsis. The blood vessels are derived from the celiac artery. The left gastric artery courses along the lesser curvature of the stomach from left to right, distributing branches to both surfaces. It anastomoses with the esophageal arteries at one end of its course and with the right gastric artery at the other. The right gastroepiploic artery courses from right to left

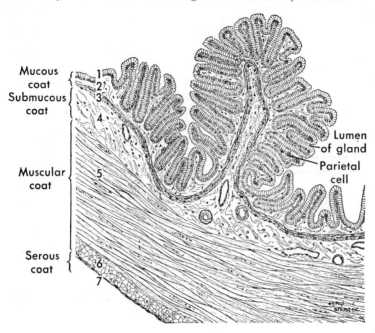

Mucous coat
Submucous coat

Muscular coat

Serous coat

Lumen of gland
Parietal cell

Figure 20–9. Diagram of stomach wall in cross section, highly magnified to show coats. One ruga covered with glands is shown. Parietal cells are shown communicating with the lumen of the glands by clefts between the chief cells that line the lumen. *1,* Columnar epithelium; *2,* areolar connective tissues; *3,* muscularis mucosae; *4,* areolar connective tissue; *5,* circular layer of smooth muscle; *6,* longitudinal layer of smooth muscle; *7,* areolar connective tissue and mesothelium.

along the greater curvature of the stomach and anastomoses with a branch of the splenic artery, the left gastroepiploic, which courses along the greater curvature from left to right. Blood is returned via the right gastroepiploic, which joins the superior mesenteric, the left gastroepiploic, and several short gastric veins that join the splenic and the left gastric. All of these eventually join the portal vein. A small quantity of blood is returned to the azygos vein instead of entering the portal vessel.

(See Figure 20–11; see also Figure 20–22, page 473.)

Functions. Probably the most important function of the stomach is to store food. Without a food reservoir it would be necessary to eat small amounts at frequent intervals. The digestive functions consist of chemical changes of the proteins of food under the action of the enzyme *pepsin*, formed from pepsinogen under the influence of HCl, and of maceration of the food bolus by the mechanics of contrac-

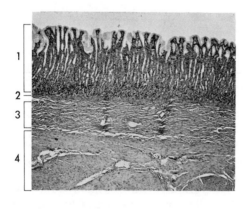

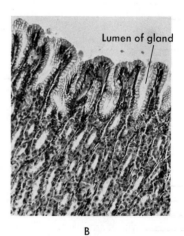

Lumen of gland

A

B

Figure 20–10. Section of stomach as seen with light microscope. *A.* Low-power view, showing layers (*1*) mucosa, (*2*) muscularis mucosae, (*3*) submucosa, and (*4*) smooth muscle. *B.* Gastric glands with higher magnification.

Figure 20–11. Celiac artery and its branches.

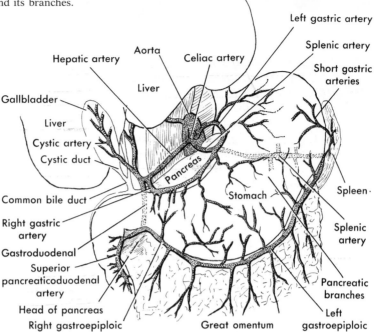

tions of the stomach musculature. The parietal cells of the fundus of the stomach elaborate a substance called the *intrinsic factor*, which is essential for absorption of vitamin B_{12}. (See page 339.)

Small Intestine (Figure 20–12). The small intestine extends from the pylorus to the colic valve. It is a coiled tube, which in the cadaver is about 7 m (23 ft) in length, and is contained in the central and lower part of the abdominal cavity.

At the beginning the diameter is about 2.8 cm but it gradually diminishes and is hardly 2.5 cm at the lower end. The small intestine is divided into three portions: the

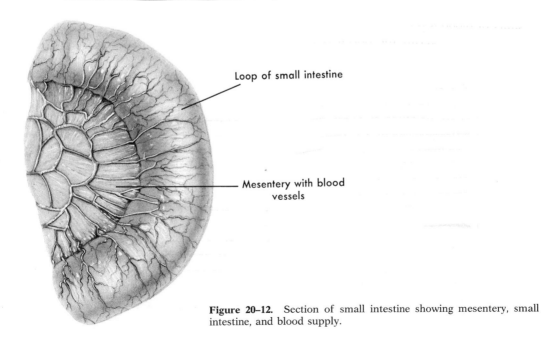

Loop of small intestine

Mesentery with blood vessels

Figure 20–12. Section of small intestine showing mesentery, small intestine, and blood supply.

duodenum, jejunum, and ileum. They are continuous, with only slight external variations in structure.

Duodenum. The duodenum is 25 cm long and is the shortest and broadest part of the small intestine. It extends from the pyloric end of the stomach to the jejunum. Beginning at the pylorus, the duodenum at first passes upward, backward, and to the right, beneath the liver. It then makes a sharp bend and passes downward in front of the right kidney; it makes a second bend, toward the left, and passes horizontally across the front of the vertebral column. On the left side, it ascends for about 2.5 cm and then ends in the jejunum opposite the second lumbar vertebra. Only the anterior surface of the duodenum is covered with peritoneum.

A musculofibrous band from tissue around the celiac artery and nearby diaphragm attaches to the junction of the duodenum and jejunum, acting as a suspensory ligament (ligament of Treitz[3]).

Jejunum. This constitutes about two fifths of the remainder, or 2.2 m, of the small intestine and extends from the duodenum to the ileum.

Ileum. This constitutes the remainder of the small intestine and extends from the jejunum to the large intestine, which it joins at a right angle. There is no definite point at which the jejunum ceases and the ileum begins, although the mucous membranes of the two divisions differ somewhat.

Coats of the Small Intestine. These are four in number and correspond in character and arrangement to those of the stomach. (1) The *serosa* with its covering of peritoneum forms an almost complete covering for the whole tube except for part of the duodenum. (2) The *muscular* coat of the small intestine has two layers: an outer, thinner layer with longitudinally arranged fibers and an inner, thicker layer with circularly arranged fibers. This arrangement aids the peristaltic action of the intestine. (3) The *submucous*, or loose connective tissue, coat connects the muscular

[3] Wenzel Treitz, Austrian anatomist (1819–1872).

and mucous coats. (4) The mucosa is thick, glandular, and very vascular. Loose areolar connective tissue, containing blood and lymphatic capillaries, supports the columnar epithelial cells and extends up into the villi. It is called *lamina propria.* The coats are illustrated in Figure 20–15 (page 468).

CIRCULAR FOLDS. About 3 or 4 cm beyond the pylorus the mucous and submucous coats of the small intestine are arranged in circular folds (valvulae conniventes, or plicae circulares), which project into the lumen of the tube (Figure 20–14, page 468). Some of these folds extend all the way around the circumference of the intestine; others extend part of the way. Unlike the rugae of the stomach, the circular folds do not disappear when the intestine is distended. About the middle of the jejunum they begin to decrease in size, and in the lower part of the ileum they almost entirely disappear. The major function of these folds is to present a greater surface area for secretion of digestive juices and absorption of digested food.

VILLI (Figures 20–13 and 20–15). Throughout the whole length of the small intestine the mucous membrane presents a velvety appearance due to minute, finger-like projections called *villi*, which number between 4,000,000 and 5,000,000 in man. Each villus consists of a central lymph channel called a *lacteal*, surrounded by a network of blood capillaries held together by lamina propria. This in turn is surrounded by a layer of columnar cells. After the food has been digested, it passes into the capillaries and lacteals of the villi. Owing to the large number of villi and the presence of microvilli on the absorbing surface of the columnar cells, the surface area of the small intestine is estimated to be about 10 sq m. The villi decrease somewhat in height and width along the small intestine, being tallest and widest in the duodenum and smallest in the terminal ileum.

Glands of the Small Intestine (*Figure 20–13*). The mucous membrane is thickly studded with secretory glands and lymph nodes. These are known as:

A Brunner's glands

B Lamina propria

Lacteal

C

D

Figure 20–13. Four sections of mucosa of the small intestine. *A*, *B*, and *C* as seen with the light microscope. *A*. Glands of duodenum. *B*. Plica circularis of jejunum; villi project outward from plica. *C*. Villus in longitudinal section. *D*. Villi as seen with the scanning microscope. (Courtesy of Mr. J. Patrikes, College of Physicians and Surgeons, Columbia University, New York, N.Y.)

Intestinal glands or crypts of Lieberkühn[4]

Duodenal or Brunner's[5] glands

Lymph nodules—(1) solitary lymph nodules, (2) aggregated lymph nodules

[4] Johann Nathanael Lieberkühn, German anatomist (1711–1756).
[5] Johann Conrad Brunner, Swiss anatomist (1653–1727).

Intestinal glands are found in the surface of the small intestine. They are simple tubular depressions in the mucous membrane, lined with columnar epithelium and opening upon the surface by circular apertures between villi, at their base.

Brunner's glands are located chiefly in the submucosa and pass their secretions by long ducts to the intestinal surface. These glands secrete mucus whereas the cells of

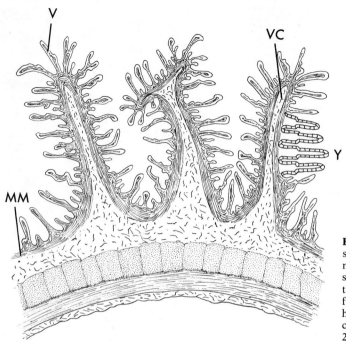

Figure 20–14. Longitudinal section of small intestine. Three valvulae conniventes (*VC*) (plicae circulares) are shown. Many villi (*V*) are shown on the valvulae and between them. At *Y* four villi with glands between them have been diagrammed. *MM* is muscularis mucosae. Compare with Figure 20–13, *B*.

other glands in the duodenum secrete an alkaline fluid, mucus, and enzymes.

Lymph Nodes. Along the length of the small intestine, in the lamina propria, are located numerous lymph nodes; they are most numerous in the ileum. Groups of lymph nodes gathered together are called Peyer's[6] patches. These occur most usually in the ileum, but may be found elsewhere.

[6] Johann Conrad Peyer, Swiss anatomist (1653–1712).

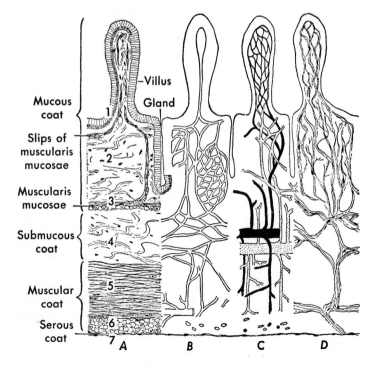

Figure 20–15. Diagram of a cross section of small intestine. *A* shows coats of intestinal wall and tissues of coats. *1*, Columnar epithelium; *2*, areolar connective tissue; *3*, muscularis mucosae; *4*, areolar connective tissue; *5*, circular layer of smooth muscle; *6*, longitudinal layer of smooth muscle; *7*, areolar connective tissue and mesothelium. *B* shows arrangement of central lacteal, lymph nodes, and lymph tubes. *C* shows blood supply; arteries and capillaries *black*, veins *stippled*. *D* shows nerve fibers, the submucous plexus lying in the submucosa, the myenteric plexus lying between the circular and longitudinal layers of the muscular coat. (*A* and *C* drawn from microscopic slide of injected specimen; *B* and *D* modified from Mall.)

One is seen in Figure 20–20 (page 472). The lymph nodes are part of the lymphatic system of the intestine, and lymph flows from the nodes into lymph vessels in the tissue below.

Nerve Supply. The vagus nerves supply secretory and motor fibers to the small intestine. Thoracolumbar nerve fibers are derived from the plexuses around the superior mesenteric artery. From this source they run to the myenteric plexus (Auerbach's plexus) of nerves and ganglia situated between the circular and longitudinal muscular fibers. Branches from this plexus are distributed to the muscular coats; and from these branches another plexus, the submucous (Meissner's) plexus, is derived (Figure 13–6, page 303). It sends fibers to the mucous membrane. The sensory fibers in the vagus nerve are concerned with intestinal reflexes, and the sensory fibers of the thoracolumbar nerves carry pain sensations. Thus, the pain from an ulcer in the duodenum is reduced after cutting of the thoracolumbar fibers, even though the ulcer is still active. Pain sensation from the intestine is thought to be caused by distention; there are no true pain receptors in the tissue.

Blood Supply (Figure 20–16). The superior mesenteric artery arising from the aorta, just below the celiac artery, supplies all of the small intestine except the duodenum, which is supplied by the gastroduodenal artery and its branch, the superior pancreaticoduodenal. These vessels distribute branches, which lie between the serous and muscular coats and form frequent anastomoses. Blood is returned by the superior mesenteric vein, which unites with the splenic to form the portal vein.

Functions. It is in the small intestine that the greatest amount of digestion and absorption takes place. It receives bile and pancreatic juice from the liver and the pancreas. The glands of the small intestine secrete succus entericus. The intestinal mucosa, containing glands and covered with villi, is arranged in folds, so that the surface areas for action of digestive juice

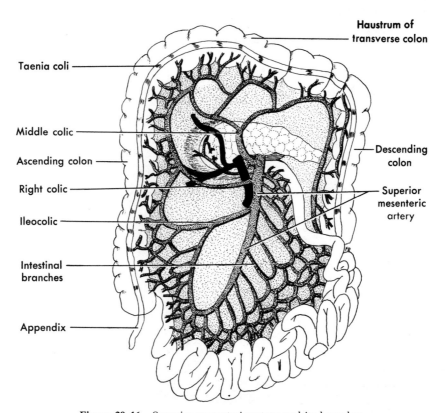

Figure 20–16. Superior mesenteric artery and its branches.

and absorption are greatly increased. Some of the cells of the mucous membrane (particularly in the duodenum) secrete the hormone *secretin*. Secretin is carried by the blood to the liver and pancreas, stimulating them to secretory activity.

Large Intestine. The large intestine is about 1.5 m long and is wider than the small intestine, particularly at its beginning, the cecum. It has four parts: cecum, with the appendix; colon; rectum; and anal canal. Figure 20–17 shows the large intestine following ingestion of a barium meal.

Cecum. The small intestine opens into the side wall of the large intestine about 6 cm above the commencement of the large intestine. This 6 cm of large intestine forms a blind pouch called the cecum. The opening from the ileum into the large intestine is provided with two large projecting lips of mucous membrane forming the colic, or ileocecal, valve, which allows the passage of material into the large intestine but prevents the passage of material in the opposite direction.

The appendix is a narrow tube attached to the end of the cecum. The length, diameter, direction, and relations of the appen-

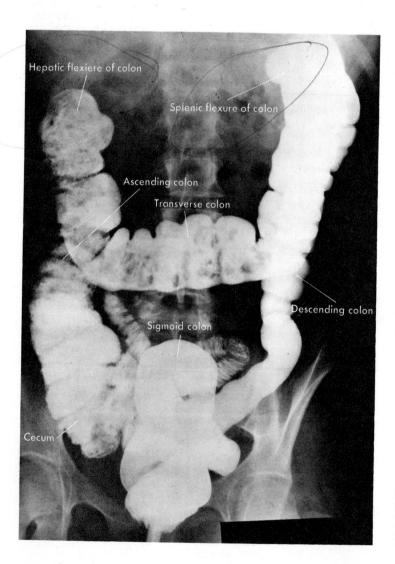

Figure 20–17. X-ray of large intestine with barium contrast. (Courtesy of Radiology Department, The New York Hospital, New York, N.Y.)

dix are very variable. The average length is about 7.5 cm.

possibly production of lymphocytes

The functions of the appendix are not known. It is most fully developed in the young adult and at this time is subject to inflammatory and gangrenous conditions commonly called *appendicitis*.

The reasons for this are that its structure does not allow for ready drainage, its blood supply is limited, and its circulation is easily interfered with because the vessels anastomose to a very limited extent.

Colon. The colon, although one continuous tube, is subdivided into the *ascending, transverse, descending,* and *sigmoid colon.* The ascending portion ascends on the right side of the abdomen until it reaches the undersurface of the liver, where it turns abruptly to the left (right colic or hepatic flexure) and is continued across the abdominal cavity as the transverse colon until, reaching the left side, it curves beneath the lower end of the spleen (left colic or splenic flexure) and passes downward as the descending colon. Reaching the left iliac region on a level with the margin of the crest of the ileum, it makes a curve like the letter S—hence its name of sigmoid—and finally ends in the rectum. Sigmoid colon, rectum, and anal canal are located in the pelvic cavity.

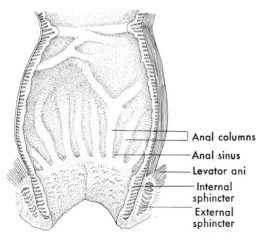

Figure 20–18. Longitudinal section of the anal canal. Shows anal columns, anal sinuses, and sphincter muscles.

Anal columns
Anal sinus
Levator ani
Internal sphincter
External sphincter

Rectum. The rectum is about 12 cm long and is continuous with the sigmoid colon and anal canal. From its origin at the third sacral vertebra it descends forward along the curve of the sacrum and coccyx and finally turns sharply backward to form the anal canal. In small children the rectum is much straighter than in adults.

Anal Canal (Figure 20–18). This is the terminal portion of the large intestine and is about 2.5 to 3.8 cm in length. The external aperture, called the *anus*, is guarded by an internal and external sphincter. It is kept closed except during defecation.

The condition known as *piles,* or *hemorrhoids*, is brought about by enlargement of the veins of the anal canal. They may be *external*, wherein enlargement is of the veins just outside the anal orifice, or *internal*, wherein the enlargement is of veins within the canal.

Coats of the Large Intestine. These are the usual four, except in some parts where the *serous* coat only partially covers it and in the anal canal, where the serous coat is lacking. The *muscular* coat consists of two layers of fibers, the external arranged longitudinally and the internal circularly. The longitudinal fibers form a thicker layer in some regions than in others. The thick areas form three separate bands, the *taeniae coli,* which extend from the cecum to the beginning of the rectum, where they spread out and form a longitudinal layer that encircles this portion. Because these bands (about 5 to 7 mm wide) are about one sixth shorter than the rest of the colon, the walls are puckered into numerous *sacculations.* The third coat consists of *submucous areolar tissue* and the fourth, or inner, coat consists of *mucous membrane.* The mucous coat possesses no villi and no circular folds. It contains intestinal glands and solitary lymph nodules, which closely resemble those of the small intestine. The coats are shown in Figure 20–19.

Nerves and Blood Vessels. Fibers from both divisions of the autonomic system reach the large intestine, nerves from the mesenteric and hypogastric plexuses being

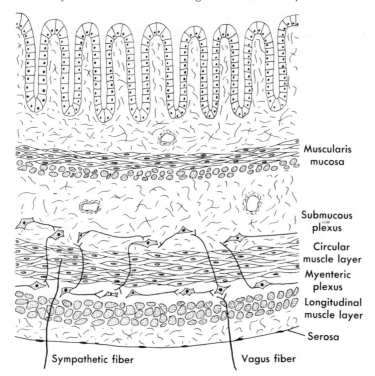

Muscularis mucosa

Submucous plexus

Circular muscle layer

Myenteric plexus

Longitudinal muscle layer

Serosa

Sympathetic fiber   Vagus fiber

Figure 20–19. Cross section of large intestine, showing muscle layers and autonomic plexuses.

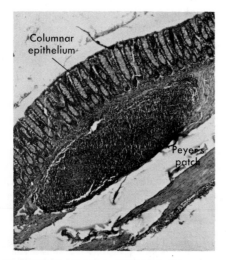

Columnar epithelium

Peyer's patch

Figure 20–20. Section of large intestine as seen with the light microscope. Glands and a Peyer's patch are visible.

Branches of the superior mesenteric artery supply the cecum, appendix, and the ascending and right half of the transverse colon. Branches of the inferior mesenteric artery supply the left half of the transverse colon, the descending colon, and the rectum. The rectum also receives branches from the internal iliac arteries. Blood from the large intestine is returned via the superior and inferior mesenteric veins; and blood from the rectum is returned via the superior and inferior mesenteric veins; and blood from the rectum is returned via the superior rectal, which joins the left colic vein, and the middle and inferior rectal, which join the internal iliac vein.

Functions. Nearly all the processes of food digestion and absorption are completed in the small intestine. Only the indigestible components remain to reach the colon. Perhaps the most important function of the colon is the reabsorption of water and electrolytes. By this process the liquid contents of the colon are dehydrated to form *feces.*

distributed in a way similar to that found in the small intestine.

The arteries are derived mainly from the superior and inferior mesenteric arteries.

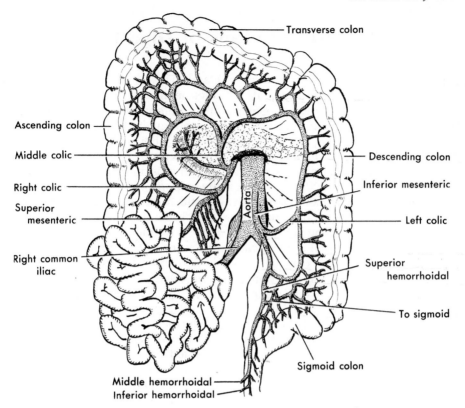

Figure 20–21. Inferior mesenteric artery and its branches.

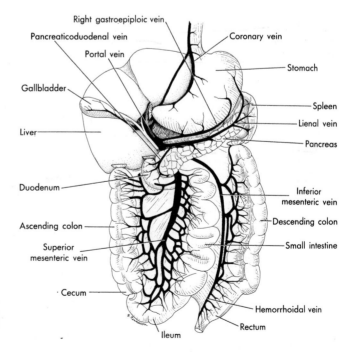

Figure 20–22. Veins of digestive tract. Liver has been pulled up and back. Portion of duodenum removed to show portal vein.

Accessory Organs of Digestion

The accessory organs of digestion are: the tongue, the teeth, the salivary glands, the pancreas, the liver, and the gallbladder. The first three have been described.

Pancreas (Figure 20–23). The pancreas is a soft, reddish- or yellowish-gray gland that lies in front of the first and second lumbar vertebrae and behind the stomach. In shape it is long and tapering and is divided into head, body, and tail. The right end, or head, is thicker and fills the curve of the duodenum, to which it is firmly attached. The left, free end is the tail and reaches to the spleen. The intervening portion is the body. Only the anterior surface is covered by peritoneum. Its average weight is between 60 and 90 gm; it is about 12.5 cm long and about 5 cm wide. The pancreas is a compound gland composed of lobules. Each lobule consists of one of the branches of the main duct, which terminates in a cluster of pouches, or alveoli. The lobules are joined together by areolar tissue to form lobes; and the lobes, united in the same manner, form the gland. The small ducts from each lobule open into one main duct about 3 mm in diameter, which runs transversely from the tail to the head through the substance of the gland. This is known as the pancreatic duct or duct of Wirsung.[7] The pancreatic and common bile ducts usually unite and pass obliquely through the wall of the duodenum about 7.5 cm below the pylorus. The short tube formed by the union of the two ducts is dilated into an ampulla, called the *ampulla of Vater*.[8] Sometimes the pancreatic duct and the common bile duct open separately into the duodenum and there is frequently an accessory duct (duct of Santorini[9]), which opens into the duodenum about 1 in. above the orifice of the main duct.

Islets of Langerhans. Between the alveoli small groups of cells are found, which are termed the islets of Langerhans[10] (interalveolar cell islets). The cells are surrounded by a rich capillary network and furnish the hormonal secretion of the pancreas (insulin and glucagon).

Functions. Two secretions are formed in the pancreas. (1) The pancreatic juice is an external secretion and is poured into the

[7] Johann Georg Wirsung, Bavarian anatomist (died 1643).
[8] Abraham Vater, German anatomist (1684–1751).
[9] Giovanni Domenico Santorini, Italian anatomist (1681–1739).
[10] Paul Langerhans, German anatomist (1847–1888).

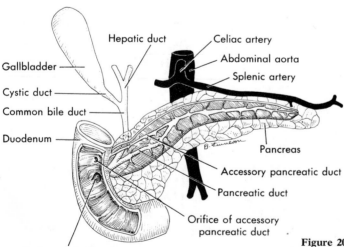

Figure 20–23. Diagram of pancreas showing its relationship to aorta, celiac artery, gallbladder, and hepatic and common bile duct.

duodenum during intestinal digestion; it contains digestive enzymes. (2) The secretion of the islets of Langerhans is internal (into the bloodstream): insulin and glucagon.

Liver. The liver (hepar) is the largest organ in the body, weighing ordinarily from 1.2 to 1.6 kg. It is located in the right hypochondriac and epigastric regions and frequently extends into the left hypochondriac region. The upper convex surface fits closely into the undersurface of the diaphragm. The under concave surface of the organ fits over the right kidney, the upper portion of the ascending colon, and the pyloric end of the stomach. The gallbladder is adherent to the undersurface of the liver (Figure 20–22).

The liver is divided into *right* and *left* lobes at the point of attachment of the falciform ligament (Figure 20–6); the *caudate* and *quadrate* lobes lie on the undersurface of the liver and are considered part of the right lobe. The liver is encapsulated by a fibrous capsule, which is reflected inward to divide the liver into its right and left lobes, and also enveloping the blood vessels and ducts that pass into and out of the liver. With the exception of a few small areas, the liver is enclosed in peritoneum.

Ligaments. The liver is connected to the undersurface of the diaphragm and the anterior walls of the abdomen by five ligaments, four of which—the falciform, the coronary, and the two lateral—are formed by folds of peritoneum. The fifth, or round, ligament is a fibrous cord resulting from the atrophy of the umbilical vein of intrauterine life.

Fossae. The liver has four fossae: the left sagittal; the portal, or transverse, which transmits the portal blood vessel, hepatic artery, nerves, hepatic duct, and lymphatics; the fossa for the gallbladder; and the fossa for the inferior vena cava.

The *porta hepatis* is on the inferior aspect of the liver and contains the right and left hepatic ducts (transmitting bile) and branches of the portal vein and hepatic artery. Lymph nodes, also located here, filter lymph drained from the liver and

gallbladder; efferent lymph vessels connect with the celiac nodes.

Nerve Supply. Hepatic nerve fibers are derived from the left vagus and the sympathetic celiac plexus.

Blood Supply. The liver receives blood from the hepatic artery, a branch of the celiac artery, and from the portal vein. The hepatic veins (right and left) drain directly into the inferior vena cava.

PORTAL SYSTEM. The veins that bring back the blood from the spleen, stomach, pancreas, and intestines are included in the portal system. Blood is collected from the spleen by veins that unite to form the *splenic*, or *lienal*, *vein*. This vein passes back of the pancreas from left to right and ends by uniting with the *superior mesenteric* to form the *portal vein*. Before this union takes place, the splenic receives *gastric veins*, *pancreatic veins*, and usually the *inferior mesenteric vein*, which returns the blood from the rectum, sigmoid, and descending colon. The *superior mesenteric vein* returns the blood from the small intestine, the cecum, and the ascending and transverse portions of the colon. (See Figure 20–22.)

The *portal vein*, formed at the level of the second lumbar vertebra by the union of the splenic and the superior mesenteric, passes upward and to the right to the transverse fissure of the liver. Here it divides into a right and a left branch, which accompany the right and left branches of the hepatic artery into the right and left lobes of the liver. Before entering the liver, the right branch usually receives the cystic vein, which returns blood from the gallbladder. The hepatic artery brings blood direct from the aorta, via the celiac artery to the liver. In the liver, blood from both sets of vessels enters into the interlobular vessels.

ACCESSORY PORTAL SYSTEM. Some of the veins that are tributaries to the portal vein have small branches whose blood reaches the heart via the superior and inferior venae cavae without going through the liver. For example, branches of the coronary vein of the stomach unite with the esophageal veins, which enter the azygos, on its way to the heart, thus bypassing the liver. The inferior mesenteric communicates with the hemorrhoidal veins, which empty into the hypogastric veins. There are also small communicating branches that unite the superior and inferior epigastric and internal thoracic veins and through the diaphragmatic veins with the azygos. These communications are called the accessory portal system and are important in returning blood to the superior and inferior venae cavae when there is interference with portal circulation.

Lymphatics. There are a superficial and a deep set of lymphatic vessels. They begin in irregular spaces in the lobules, form networks around the lobules, and run always from the center outward.

Liver Lobule. The liver is made up of many minute units called lobules. Each *lobule* is an irregular body composed of plates *of hepatic cells* held together by connective tissue. Between the plates of hepatic cells are capillaries, called *sinusoids*, which are formed from the portal vein and hepatic artery. Kupffer[11] cells are located along the sinusoids. Nerve fibers are also present. The plates are formed by two layers of cells with a bile canaliculus between them, which empties into a bile duct. The plates with their blood and lymph supply are the units of minute structure. Together they give an enormous area of contact between liver cells and capillaries for the volume of tissue concerned. Thus each lobule has all the following: (1) blood vessels in close connection with secretory cells, (2) cells that are capable of forming a secretion, and (3) ducts by which the secretion is carried away.

The *portal vein* brings to the liver blood from the stomach, spleen, pancreas, and intestine. After entering the liver, it divides into a vast number of branches, which form a plexus, the interlobular plexus in the spaces between the lobules. From this plexus the blood is carried into the lobule by fine branches that converge toward the center. The walls of these small vessels are

incomplete, so that the blood is brought in direct contact with each cell. These channels are termed sinusoids, and at the center of the lobule they empty the blood into the intralobular vein. The intralobular veins from a number of lobules empty into a much larger vein, upon whose surface a vast number of lobules rest; and therefore the name *sublobular* (under the lobule) is given to these veins. They empty into still larger veins, the *hepatic*, which converge to form three large trunks and empty into the *inferior vena cava*, which is embedded in the posterior surface of the liver.

The blood brought to the liver by the portal vein is venous blood; arterial blood is brought by the *hepatic artery*. It enters the liver with the portal vein and divides and subdivides in the same manner as the portal vein, with terminations in the sinusoids and interlobular veins. At higher pressures, more arterial blood enters the sinusoids, tending to dilute the portal supply. Blood flow through the liver has been estimated to be about 800 to 1000 ml per minute, the greater proportion coming from the portal vein.

Pressure in the liver is normally low—near zero in the hepatic vein and about 8 mm Hg in the portal vein. Increased resistance to blood flow, such as with increased amounts of connective tissue in the supporting framework, as in cirrhosis, raises the pressure in smaller vessels, and this eventually raises the portal vein pressure, with the result of increased blood flow in the accessory portal vessels.

Liver Acinus. Recent investigation of the microcirculation of the liver has resulted in the concept of the *acinus:* a mass of hepatic cells arranged around a central axis of interlobular vessels (portal, hepatic, biliary) and lying between central veins of adjacent lobules. Cells closest to the axis receive the "freshest" blood as compared to those farther away and closer to the central vein. It is known, too, that interlobular vessels do not occur at all the hepatic lobule interdigitating points; thus some cells are relatively far from their interlobular blood supply.

Functions. The liver has many meta-

---

[11] Karl William Von Kupffer, German anatomist (1829–1902).

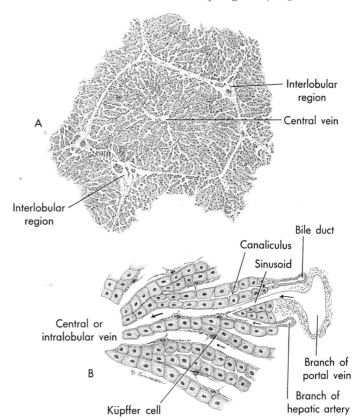

A

Interlobular region

Central vein

Interlobular region

Bile duct

Canaliculus

Sinusoid

Central or intralobular vein

B

Branch of portal vein

Branch of hepatic artery

Küpffer cell

Figure 20–24. Diagram of microscopic views of liver. *A.* Low power, showing one complete lobule in cross section and relation to other lobules. *B.* High power, cords of liver cells, bile canaliculi, and blood sinusoids. *Arrows* show direction of blood flow and bile flow.

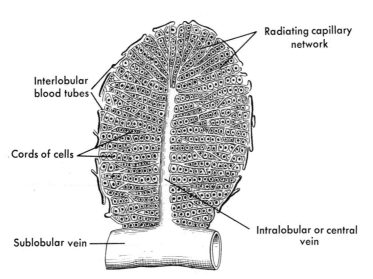

Radiating capillary network

Interlobular blood tubes

Cords of cells

Sublobular vein

Intralobular or central vein

Figure 20–25. Diagram of a hepatic lobule seen in longitudinal section.

Interlobular vessels bring blood in {Branches of hepatic artery
{Branches of portal vein

Interlobular vessels take blood out {Tributaries to hepatic vein

bolic functions of a complex nature and in various ways helps to maintain homeostasis of body fluids. Many of these functions will be discussed in detail in Chapters 21 and 22. A brief résumé is included here:

1. Bile synthesis and secretion (page 504).

2. Blood volume regulation. When hepatic venous pressure is high in the inferior vena cava, up to 400 ml of blood can be "stored" in the sinusoids. Conversely, a decrease in blood volume, as in hemorrhage, causes a marked emptying of the sinusoids to replace the blood lost. Thus, the liver is a "blood reservoir."

3. Carbohydrate metabolism (page 518). The liver forms glucose from noncarbohydrate sources (gluconeogenesis); forms glycogen from glucose; breaks down glycogen to glucose.

4. Fat metabolism (page 520). Phospholipids and other lipoproteins are synthesized in the liver; fatty acids are oxidized; fatty acids are formed from amino acids and glucose; cholesterol is formed.

5. Protein metabolism (page 524). The liver is essential for protein metabolism necessary to maintain life. Its activities include formation of urea from amino acids; deamination of amino acids; transformation of one amino acid into another; synthesis of plasma proteins (albumin, some globulin, prothrombin, and fibrinogen); synthesis of nonessential amino acids.

6. Detoxification. Harmful substances are chemically transformed to nonharmful

TABLE 20–3
ARTERIAL SUPPLY TO DIGESTIVE ORGANS OF ABDOMEN

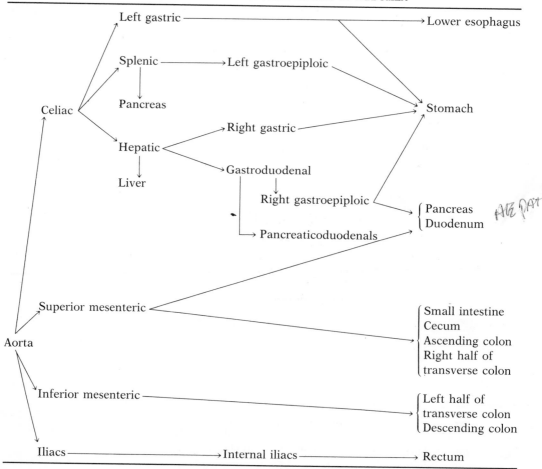

TABLE 20–4
VENOUS RETURN FROM DIGESTIVE ORGANS OF ABDOMEN

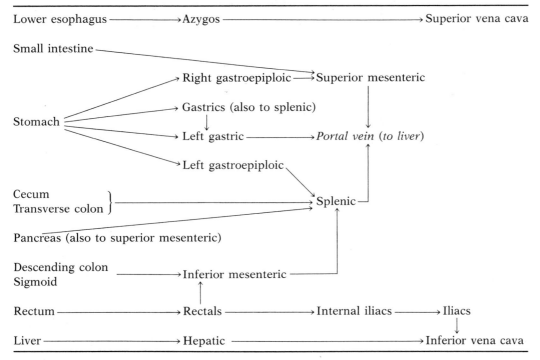

Lower esophagus ⟶ Azygos ⟶ Superior vena cava

Small intestine ⟶ Right gastroepiploic ⟶ Superior mesenteric

Stomach ⟶ Gastrics (also to splenic) ↓ Left gastric ⟶ *Portal vein (to liver)*

Stomach ⟶ Left gastroepiploic ⟶ Splenic

Cecum } Transverse colon } ⟶ Splenic

Pancreas (also to superior mesenteric)

Descending colon / Sigmoid ⟶ Inferior mesenteric

Rectum ⟶ Rectals ⟶ Internal iliacs ⟶ Iliacs ↓

Liver ⟶ Hepatic ⟶ Inferior vena cava

substances for excretion; hormones are degraded. The phagocytic activity of the Kupffer cells helps in protecting the body from harmful chemicals and from microorganisms that may have entered the portal system from the colon.

7. Storage. Iron, vitamins A, D, and B_{12}, and copper are stored in the liver.

Gallbladder (Figure 20–26). This is a pear-shaped (when full) sac lodged in the gallbladder fossa on the undersurface of the liver, where it is held in place by connective tissue. It is about 7 to 10 cm long, 2.5 cm wide, holds about 36 ml, and is composed of three coats: (1) the inner one is mucous membrane; (2) the middle one is muscular and fibrous tissue; and (3) the outer one is serous membrane derived from the peritoneum. It is only occasionally that the peritoneum covers more than the undersurface of the organ.

Most of the bile secreted continuously by the liver enters the gallbladder, where it is concentrated; thus it serves as a reservoir for bile. When required, the gallbladder contracts and expels its bile content

into the duodenum. The most potent stimuli for evacuation are the acid gastric juice and fatty foods in the small intestine. A hormone, cholecystokinin, is elaborated by cells of the small intestine in the presence of fat. This hormone causes the gallbladder to contract, thereby emptying its contents of bile into the intestine.

The sphincter of Oddi[12] is relaxed much of the time. However, when pressure in the intestine increases, the sphincter contracts to prevent ascent of intestinal contents into the biliary tract. The sphincteric mechanism must be relaxed when the gallbladder contracts; if not, the contents will not be evacuated, and the resulting distention of the bile duct will cause sharp, unbearable pain called biliary colic. This pain is also produced when the duct is obstructed by so-called "stones."

Bile Ducts (*Figure 20–24*). The surfaces of the hepatic cells are grooved, and the grooves on two adjacent cells fit together and form a passage into which the bile is

---

[12] Ruggero Oddi, Italian surgeon (late nineteenth century).

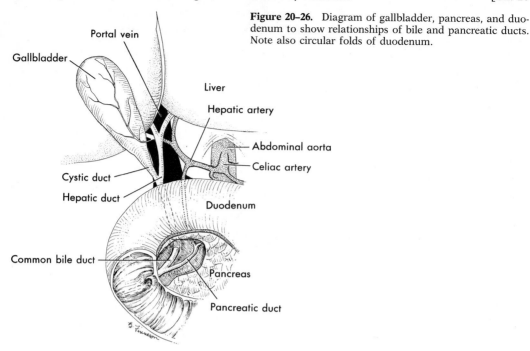

Figure 20–26. Diagram of gallbladder, pancreas, and duodenum to show relationships of bile and pancreatic ducts. Note also circular folds of duodenum.

Portal vein

Gallbladder

Liver

Hepatic artery

Abdominal aorta

Celiac artery

Cystic duct

Hepatic duct

Duodenum

Common bile duct

Pancreas

Pancreatic duct

poured as soon as it is formed by the cells. These passages form a network between and around the cells as intricate as the network of blood vessels. They are called biliary canaliculi and radiate to the circumference of the lobule, where they empty into the interlobular bile ducts. These unite and form larger and larger ducts until two main ducts, one from the right and one from the left side of the liver, unite in the portal fossa and form the *hepatic duct*.

The hepatic duct passes downward and to the right for about 5 cm and then joins (at an acute angle) the duct from the gallbladder, termed the *cystic duct.* The hepatic and cystic ducts together form the *common bile duct (ductus choledochus)*, which passes downward for about 7.5 cm and enters the duodenum about 7.5 cm below the pylorus. This orifice usually serves as a common opening for both the common bile duct and the pancreatic duct. It is very small and is guarded by a sphincter muscle, the sphincter of Oddi, which keeps it closed except during digestion.

Questions for Discussion

1. If the duodenum is removed surgically, what other procedure must be done to ensure passage of digestive fluids through the small intestine?
2. Thrombosis of the superior mesenteric artery would have an effect on what organs?
3. There is a similarity in the coats of the organs of the digestive tract. What is this?

4. Enlargement of the esophageal veins (varices) may occur in liver disease. Why?
5. Discuss the functions of the liver.
6. What hormones are secreted by the pancreas?

Summary

Digestive Processes
{ Physical and chemical breakdown of food into absorbable substances
{ Movement of food along digestive tract

Digestive System

Alimentary canal

Mouth
Pharynx
Esophagus
Stomach
Small, or thin, intestine
{ Duodenum
{ Jejunum
{ Ileum
Large, or thick, intestine
{ Cecum
{ Colon
{ { Ascending
{ { Transverse, or mesial
{ { Descending
{ { Sigmoid
{ Rectum
{ Anal canal

Accessory organs
Tongue
Teeth
Salivary glands
Pancreas
Liver
Gallbladder

Alimentary Canal

Continuous tube from mouth to anus
About 9 m (30 ft) long in cadaver
Esophagus—four coats
{ Internal, or mucous
{ Submucous, or areolar
{ Muscular
{ Serosa
From stomach to rectum—four coats
{ Mucous
{ Submucous, or areolar
{ Muscular
{ Serosa, covered with peritoneum

Mouth, or Buccal, Cavity

Roof—palate
{ 1. Hard palate { Maxillae } processes
{ { Palatine }
{ 2. Soft palate—uvula, palatine arches, and tonsils
Floor—tongue
Bounded laterally and in front by cheeks and lips
Behind it communicates with pharynx
Contains { Tonsils, orifices of ducts of salivary glands
{ Tongue, teeth

Tonsils

Masses of lymphoid tissue occupy triangular space between palatine arches on either side of throat
Function { Similar to that of other lymph nodes
{ 1. Aid in formation of lymphocytes
{ 2. Act as filters and protect body from infection

Tongue

Special organ of sense of taste
Assists in { Mastication
{ Deglutition
{ Speech

Salivary Glands

Parotid—just under and in front of ear
Submaxillary { Below jaw and under tongue
Sublingual
Function—form a secretion that, mixed with the secretion of the glandular cells of the mouth, is called saliva
Nerves—fibers from both divisions of autonomic system
Blood vessels—branches of external carotid artery

Teeth

Contained in sockets of alveolar processes of maxillae and mandible
Gums—cover processes and extend into sockets, or alveoli
Sockets—lined with periosteum { Attach teeth to sockets
{ Source of nourishment

Teeth
(cont.)

Three portions
- *Root*—one or more rootlets contained in alveolus
- *Crown*—projects beyond level of gums
- *Neck*—portion between root and crown

Composed of three substances developed from epithelium
- *Dentin*—Gives shape. Encloses pulp cavity, which contains nerves and blood vessels that enter by canal from root
- *Enamel*—caps crown
- *Cement*—covers root

Two sets
1. Deciduous— 6 months– 2 years—{ Incisors 8, Canines 4, Molars 8 } 20
   Begin to develop about the sixth week of intrauterine life
2. Permanent— 6½ years– 25 years of age—{ Incisors 8, Canines 4, Premolars 8, Molars 12 } 32
   With the exception of the second and third molars the permanent teeth begin to develop about the sixteenth week of intrauterine life

Function—to assist in the process of mastication

Pharynx

Muscular, membranous, cone-shaped tube between mouth and esophagus

Three parts
- Nasal or nasopharynx—behind posterior nares above soft palate
- Oral—extends from soft palate to hyoid bone
- Laryngeal—extends from hyoid bone to esophagus

Seven apertures
- 2 posterior nares
- 2 auditory tubes
- 1 fauces
- 1 larynx
- 1 esophagus

Nerves—receive fibers from both divisions of autonomic system

Blood vessels—branches from external carotid artery

Function
- Transmit air to larynx
- Serves as a resonating cavity
- Receives food and passes it to esophagus

Esophagus

Tube—23–25 cm (9–10 in.) long. Extends from pharynx to cardiac end of stomach

Four coats
1. Internal, or mucous
2. Submucous, or areolar
3. Muscular { Internal circular layer / External longitudinal layer }
4. External, or fibrous

Nerves
- Vagus
- Thoracolumbar system

Blood vessels
1. Inferior thyroid branch of thyrocervical trunk
2. Branches from thoracic aorta
3. Left gastric branch of celiac artery
4. Left phrenic branch of abdominal aorta

Function—receives food and passes it on to stomach

Structure of Gastro-intestinal tract

Mucous membrane
- Epithelium
- Lamina propria
- Muscularis mucosae

Submucosa

Muscularis externa
- Circular layer
- Longitudinal layer

Serosa
- Usually covered with peritoneum

Peritoneum

Layers
- Parietal—lines wall of abdominal cavity
- Visceral—reflected over abdominal organs and upper surface of pelvic organs

Fossae
- Peritoneal cavity—space between two layers
- Duodenal
- Cecal
- Pelvic
- Paracolic gutters

Folds
- Mesentery—attaches small intestine and part of large intestine to posterior abdominal wall; contains blood and lymph vessels
- Omentum—greater one hangs from stomach over intestines; may contain adipose tissue. Lesser one extends from lesser curvature of stomach to liver

Function—lubrication

Stomach, or Gaster

Dilated portion of canal, size, and shape vary
Oblique position in epigastric, umbilical, and left hypochondriac regions, under diaphragm

Openings
- Cardiac orifice—connects with esophagus
- Pyloric orifice—connects with duodenum
- Guarded by ringlike muscles known as sphincters

Curvatures
- Lesser curvature—concave border
- Greater curvature—convex border

Parts
- Fundus—blind end above entrance of esophagus
- Body—between fundic and pyloric portions
- Pyloric portion—smaller end

Four coats
1. Outer—serosa—peritoneum covering
2. Muscular
   - a. Longitudinal layer
   - b. Circular layer
   - c. Oblique layer chiefly at cardiac end
3. Submucous—vascular
4. Mucous-rugae

Glands
- Some cells of mucous membrane secrete gastrin
- Cardiac—secrete mucin
- Fundus
  - Chief cells secrete pepsinogen; parietal cells secrete H^+ and Cl^- ions, and intrinsic factor
- Pyloric—secrete pepsinogen and mucin

Nerves
- Thoracolumbar autonomic nerves from celiac plexus
- Vagus nerves

Blood vessels from three divisions of celiac
- Left gastric
- Hepatic
- Splenic

Function
1. Receives food in relatively large quantities about three times a day, holds it while it undergoes mechanical and chemical changes, then passes it on in small portions at frequent intervals into duodenum
2. Secretes mucin and gastric fluid

Small Intestine

Convuluted tube extends from stomach to colic valve. About 7 m (23 ft) long, coiled up in abdominal cavity

Three divisions
- Duodenum—about 25 cm (10 in.)
- Jejunum—about 2.2 m (7½ ft) | in the cadaver
- Ileum—about 4 m (14½ ft)

Four coats
1. Serosa covered with peritoneum
2. Muscular
   - Longitudinal layer
   - Circular layer
3. Submucous—connects muscular and mucous coats
4. Mucous
   - Circular folds
   - Villi—contain lacteals and blood capillaries

Glands and nodes
- Intestinal (glands of Lieberkühn)
  - open at base of villi
- Duodenal, or Brunner's
  - Secrete intestinal fluid
- Lymph nodules
  - Solitary
  - Aggregated lymph nodules called Peyer's patches

Nerves
- Vagi supply sensory and motor fibers
- Thoracolumbar autonomic nerves derived from plexuses around superior mesenteric artery

Blood vessels
- Branches of hepatic
- Branches of superior mesenteric
  - Distribute arched branches that lie between serous and muscular coats

Function
- Digestion
- Receives bile from liver, pancreatic fluid from pancreas
- Secretion of succus entericus
- Absorption of end products of digestion

Large Intestine

Large in width
Length, 1.5 m (5 ft); width, 6.3 cm (2½ in.) to 3.5 cm (1½ in.)
Extends from ileum to anus

Four parts
- Cecum, with vermiform appendix
- Colon
  - Ascending
  - Transverse, or mesial
  - Descending
  - Sigmoid
- Rectum—about 12 cm (5 in.)
- Anal canal—3.5 cm (1–1½ in.)
  - Internal sphincter
  - External sphincter
- Anus

Large Intestine (*cont.*)

Four coats
1. Serous, except that in some parts it is only a partial covering, and at rectum it is lacking
2. Muscular
   - Longitudinal layer — Arranged in three ribbon-like bands that begin at appendix and extend to rectum
   - Circular layer
3. Submucous
4. Mucous
   - No villi
   - No circular folds
   - Numerous
     - Intestinal glands
     - Solitary lymph nodules

Nerves—fibers from both divisons of autonomic nervous system

Blood vessels
- Superior mesenteric supplies cecum, ascending and transverse colon
- Inferior mesenteric supplies descending colon and rectum. Rectum also receives branches from hypogastric arteries

Function
- Continuance of digestion and absorption
- Elimination of waste

Pancreas

In front of first and second lumbar vertebrae, behind stomach

Hammer shape
- Head attached to duodenum
- Body in front of vertebrae
- Tail reaches to spleen

Size
- About 12.5 cm (5 in.) long
- About 5 cm (2 in.) wide

Average weight—60–90 gm (2–3 oz)

Structure
- Compound gland—each lobule consists of one of the branches of main duct, which terminates in cluster of pouches, or alveoli, is retroperitoneal
- Lobules held together by connective tissue form lobes
- Lobes form gland
- Duct from each lobule empties into pancreatic duct, also called duct of Wirsung
- Scattered throughout pancreas are islets of Langerhans
- 1. Secretes pancreatic fluid—digestive fluid
- 2. Forms hormones (insulin and glucagon)—aid in metabolism of glucose

Liver

Largest gland in body

Location
- Right hypochondriac region
- Epigastric region
- Left hypochondriac region

Convex above—fits under diaphragm

Concave below—fits over right kidney, ascending colon, and pyloric end of stomach

Four lobes
1. Right (largest lobe)
2. Left (smaller and wedge shaped)
3. Quadrate (square)
4. Caudate (taillike)

Porta hepatis
- On inferior aspect of liver; contains hepatic ducts, branches of hepatic artery, and portal vein

Five ligaments
1. Falciform
2. Coronary
3. Right lateral
4. Left lateral
   - Formed by folds of peritoneum
5. Round ligament   results from atrophy of umbilical vein

Four fossae
1. Left sagittal fossa
2. Portal, or transverse, fossa transmits
   - Portal vein
   - Hepatic artery
   - Hepatic duct
   - Lymphatics
   - Nerves
3. Gallbladder fossa
4. Fossa for inferior vena cava

Five sets of vessels
1. Branches of portal vein
2. Branches of hepatic artery
3. Hepatic veins
4. Lymphatics
5. Bile ducts

Liver (*cont.*)

Histology of liver

Cords of hepatic cells are grouped in lobules
Lobules 1.0 to 2.5 mm in diameter

Branches of portal vessel
- Interlobular veins (between lobules)
- Intralobular capillaries (within lobules)
- Sublobular veins (under lobules)
- Hepatic veins—exit at portal fossa, empty into inferior vena cava

Bile ducts
- Channels between cells form intercellular biliary passages
- Interlobular ducts
- Hepatic duct—exit at portal fossa

Branches of hepatic artery
- Interlobular arteries (between lobules)
- Intralobular capillaries (within lobules)
- Course beyond intralobular capillaries same as that pursued by blood from portal vein

Lymphatics
- Begin in lobules, form network, and run from center to periphery

Acinus
- Mass of hepatic cell surrounding interlobular vessels

Glisson's capsule invests liver
Serous membrane from peritoneum almost completely covers it

Function

Bile production
Forms and stores glycogen, which supplies glucose to the blood
Excretes bilirubin
Proteins changed into substance that can be eliminated, urea, etc.
Secretes into the bile various toxic substances
Regulation of blood volume
Manufacture of serum albumin
Manufacture of serum globulin
Manufacture of fibrinogen, prothrombin
Forms blood in embryo
Forms vitamin A from carotene
Storage of iron and copper and vitamins A, D, and B_{12}
Detoxification functions

Gallbladder

Pear-shaped sac lodged in gallbladder fossa on undersurface of liver

Size
- 7.5–10 cm (3–4 in.) long
- 2.5 cm (1 in.) wide
- Capacity about 36 ml

Three coats
1. Mucous membrane
2. Fibrous and muscular tissue
3. Serous membrane from peritoneum

Function
- Concentrates bile and serves as reservoir
- During digestion—pours bile into duodenum

Additional Readings

GUYTON, A. C.: *Basic Human Physiology: Normal Functions and Mechanisms of Disease.* W. B. Saunders Co., Philadelphia, 1971, Chapters 42, 43.

HAM, A. W.: *Histology,* 7th ed. J. B. Lippincott Co., Philadelphia, 1974, Chapters 21 and 22.

KAPPAS, A., and ALVARES, A.: How the liver metabolizes foreign substances. *Sci. Amer.,* **232**:22–31, June, 1975.

PANSKY, B., and HOUSE, E. L.: *Review of Gross Anatomy,* 3rd ed. Macmillan Publishing Co., Inc., New York, 1975, pp. 28–29, 48–49, 58–71, 300–301, 312–13, 322–23, 326–49, 360–61, 372–75.

CHAPTER 21

Food Components and Digestive Processes

Chapter Outline

Natural nutrients are in general held within cells in a nondiffusible form. The cells of plants and animals constitute a natural food supply for man. These foods are taken in periodically, digested, absorbed, used immediately by cells, or stored again within cells in a nondiffusible form for later use. Food supplies material for the manufacture of protoplasm (increasing number and size of cells) or for repair of tissues and cells.

Food Components

The food one eats and drinks may or may not be nutritious; to be nutritious food must contain the materials needed by cells. These materials, or *nutrients,* are:

Water	Proteins
Carbohydrates	Minerals
Lipids	Vitamins

Table 21–1 indicates the chemical composition of various tissues of the body, illustrating the use of nutrients ingested.

WATER   *ALL ENZYMATIC PROCESSES*

Water is an important constituent of the body tissues and is taken into the body in the form of water itself, beverages, and foods such as fruits and vegetables. In addition to this ingestion of water, water is formed in cell metabolism and is available for cell use.

CARBOHYDRATES

Carbohydrates all contain simple sugars ($C_6H_{12}O_6$) in their molecular structure,

and, depending on the number of simple sugars in the molecule, they are named monosaccharides, disaccharides, or polysaccharides (Table 21–2). Only the monosaccharides are small enough to be absorbed; the others are broken down by hydrolysis[1] in the digestive tract to their smaller subunits, prior to absorption. The primary use of carbohydrates in the body is as an energy source. Excess carbohydrates are transformed to lipids and stored in adipose tissue.

LIPIDS

Lipids are a heterogeneous group of compounds containing fatty acids, combined with an alcohol. They may be subdivided into two groups: *simple lipids* (fats, oils, waxes) and *compound lipids* (phospholipids, glycolipids, sterols). The ordinary fat of animals and vegetables are not pure compounds chemically, but are mixtures of simple lipids. *Oils* and fats are

[1] Hydrolysis is the chemical breakdown of a molecule with the addition of water.

TABLE 21–1
CHEMICAL COMPOSITION OF THE BODY

	% Total Body Wt.	H_2O	Lipids	Protein	Ash*	Ca
Skin	7.8	65	13	22	0.7	tr†
Skeleton	15	32	17	19	2.9	11
Teeth	0.06	5	3	23	70	24
Skeletal muscle	32	80	3	16	1	tr
Brain, spinal cord, nerves	2.5	73	13	12	1	tr
Liver	3	71	10	16	0.8	tr
Heart	0.7	74	9	16	0.8	tr
Lungs	4	83.7	1.5	13	1	tr
Spleen	0.2	78	1	18	1	tr
Kidneys	0.5	79	4	15	1	tr
Pancreas	0.2	73	13	13	1	tr
Alimentary tract	2	79	6	13	1	tr
Adipose	14	50	42	7	0.5	tr

*Various minerals (see Table 2–1, page 17).
†Tr = trace.

TABLE 21–2
CARBOHYDRATE, LIPID, AND PROTEIN CLASSIFICATION AND FOOD SOURCES

	Chemical Formula	Examples	Food Sources
Carbohydrates			
Monosaccharides	$C_6H_{12}O_6$	Glucose, or dextrose	Fruits, sucrose
		Fructose, or levulose	Fruits, sucrose
		Galactose	Milk (lactose)
Disaccharides	$C_{12}H_{22}O_{11}$	Sucrose	Cane sugar
		Lactose	Milk
		Maltose	Grains
Polysaccharides	$(C_6H_{10}O_5)_n$*	Starch	Grains, vegetables
		Glycogen	Meats
		Cellulose	Not digestible
Lipids			
Simple	Fatty acids, plus an alcohol	Fats, oils	Milk, butter
			Meats
			Vegetables
			Margarine (vegetable oils)
Compound	Fatty acids, alcohol, plus another compound	Glycolipids Phospholipids Sterols	Synthesized by liver
Proteins	Multiple of amino acids	Albumin	Egg white
		Casein	Milk
		Gelatin	Connective tissue of meat

*Exact formula not known; large molecules; n equals 300 or more in starch.

chemically similar, but oils are usually liquid at room temperature whereas fats are solid; oils also contain more unsaturated[2] bonds than do fats. Vegetable oils in particular tend to be highly unsaturated—a fact that may be of medical significance in regard to metabolic utilization of the fat, although the details to date have not been clarified. Experimental evidence indicates the necessity of including some of the unsaturated fatty acids such as linoleic acid and linolenic acid, which are found in most fatty foods. Waxes are indigestible and are not found in the human body. *Simple lipids* are broken down to their component parts (fatty acids and glycerol) before absorption. *Compound lipids* are composed of fatty acids, and alcohol, plus other groups (phosphate, carbohydrate, nitrogen containing). Although these complex substances may be found in the food eaten, they are broken down to subunits before absorption, and the compound lipid characteristic of human tissue is synthesized by the cells.

The primary use of all lipids is for energy. Lipids are also necessary for cell structure; excess lipids form adipose tissue.

[2] Carbon compounds having a single bond between the carbon atoms are called *saturated* compounds; those with one or more double or triple bonds between carbon atoms are termed *unsaturated* compounds.

PROTEINS

Proteins are structurally more complex than carbohydrates and lipids and contain an amino (NH_2) group in their subunits—amino acids. *Simple proteins*, such as albumin and globulin, contain only amino acids in their structure; *conjugated proteins* contain a simple protein molecule in their structure, as well as another nonprotein molecule, called a *prosthetic group*. The prosthetic group is carbohydrate in the case of glycoproteins, phosphate in the case of phosphoproteins, and lipid in the case of lipoproteins. Ingested proteins are broken down to amino acids and other constituents of their molecule before absorption, and the protein characteristic of human tissue is synthesized by the cells. The body can synthesize some amino acids, but not all. The amino acids that cannot be synthesized are called "essential" amino acids (Table 21–3); i.e., they must be taken in the daily diet. Since proteins vary in their amino acid composition, they do not all contain all the amino acids needed by the body. Those that contain all the necessary amino acids are referred to as *adequate* proteins; others are *inadequate* proteins. In general, animal proteins including milk are adequate; gelatin is an example of an inadequate protein. The primary use of proteins in the body is to form new cells and tissue. If energy sources (carbohydrate and lipid) are inadequate, tissue proteins may be broken down as a source of energy. Table 21–2 summarizes the carbohydrate, lipid, and protein classification and their common food sources.

TABLE 21–3
AMINO ACIDS FOUND IN HUMAN TISSUE

Essential	Nonessential*
Valine	Glycine
Leucine	Alanine
Isoleucine	Norleucine
Phenylalanine	Tyrosine
Threonine	Serine
Methionine	Cystine
†Arginine	Aspartic acid
Lysine	Glutamic acid
†Histidine	Hydroxyglutamic acid
Tryptophan	Proline
	Hydroxyproline

*Nonessential means that the body can synthesize in adequate amounts the amino acid listed.
†During growth.

MINERALS

The mineral elements that enter into the composition of the body are listed on page 17. Mineral constituents form about 4 per cent of the body weight and are primarily located in the skeleton. Since each element enters into the metabolism of body cells, a constant supply of each is necessary to meet the daily loss. These elements are supplied in food. On analysis many of them are classified as ash constituents, since they remain after incineration of the food. These "ash constituents" may function in the body in several ways: as constituents of bone, giving rigidity to the skeleton; as essential elements of all protoplasm associated with enzyme activity; as soluble constituents of the fluids of the body influencing the elasticity and irritability of the muscles and nerves, supplying material for the acidity or alkalinity of all body fluids, helping to maintain the acid-base equilibrium of the body fluids as well as their osmotic characteristics and solvent power; and probably in many other ways. Not only must the body be supplied with certain minerals that are important for its functions, but these must be supplied in readily absorbable form. Thus, only a small fraction, several milligrams, of the ion that is ingested is absorbed daily.

The importance of the optimum concentrations of each of these mineral salts in the tissues and fluids of the body is so great that any considerable change from the normal endangers life.

Calcium is a constituent of all protoplasm and of body fluids and is present in large proportions in bones and teeth. There is more calcium than any other cation in the body. Ninety-nine per cent of body calcium is in bones and teeth. Calcium is essential for all cellular activities. It is related to normal permeability of cellular membranes, excitability of muscle, normal heart action and nerve activity, and must be present in ionic form for normal blood clotting. Children as well as pregnant and lactating animals require large amounts. Milk is the best source of calcium, but calcium is also present in leafy vegetables. The body cannot readily adapt itself to calcium shortage; therefore, a liberal amount is needed daily. It is estimated that an intake of 1 gm of calcium would maintain the body fluids at optimum concentration. Calcium deficiency is a definite problem in the diet of Americans.

Phosphorus is essential to the normal development of bones and teeth. It is a necessary constituent of all cells, particularly nerve and muscle tissue. About 80 per cent of the total is combined with calcium in the bones and teeth. It is in organic combination such as phospholipids, phosphocreatine, and phosphorylated intermediates of carbohydrate metabolism. Blood plasma and body fluids are relatively low in phosphates. The phosphate ester is of great importance in energy transfer in the form of ATP. Phosphorylation is important in the absorption of carbohydrate from the intestine and in the reabsorption of glucose by the renal tubule. It is also concerned with maintaining enzyme systems and their functioning.

Iron in the adult body amounts to about 4.5 gm, distributed in the hemoglobin of red cells (2.5 gm), myoglobin of skeletal muscle, intracellular enzyme systems (particularly the cytochrome system), and in the tissues as ferritin. This latter is the storage form of iron—bound to protein—and is found primarily in the liver, as well as the spleen, bone marrow, and lymph nodes. In hemoglobin it is the iron-containing part of the molecule to which the oxygen is attached and which carries oxygen to the cells. Traces of copper are essential for utilization of iron in the formation of hemoglobin. Iron absorption occurs only if the iron is in a readily ionizable form. The acid of the stomach ionizes dietary iron; in gastric disease when acid production is reduced, iron absorption is usually impaired.

Storage of iron in the body is limited; therefore, foods containing it should be included in the daily diet. Recent studies indicate that iron deficiency is common among women.

At birth, a baby has a special store of iron in its body, which is used during the period of lactation. During the latter half of the first year, egg yolk and iron-bearing vegetables should be added gradually to the diet, so that as the reserve iron is depleted, fresh supplies will be available. In premature infants this special store of iron is absent; preparations of iron and copper are added to the milk.

Copper is a factor in hemoglobin formation, though it is not a part of the molecule. It must be present for the utilization of iron in hemoglobin synthesis. Since it is not a part of the hemoglobin molecule, it is believed to function as part of an enzyme-catalyzing system necessary for hemoglobin formation.

Magnesium is found in the intracellular fluid of the body and in the extracellular space of skeletal tissue, in about equal amounts in each space. Total magnesium content of the body is approximately 21 to 28 gm. Magnesium is a vital element in cell physiology because it serves as a cofactor in the metabolism of glucose, of pyruvic acid, and of adenosine triphosphate. In excessive amounts it has a depressing action on nerve impulse transmission and nerve tissue functioning in general.

Potassium content of the body is about 125 gm, found primarily inside the cells where it functions in association with various enzymes. It is intimately concerned with transmission of the nerve impulse, with skeletal muscle contraction, and with cardiac contraction. Excess or lack of the normal amount of potassium causes immediate malfunctioning of the heart.

Iodine is utilized by the thyroid gland in the synthesis of thyroxin. To maintain the body store of iodine and meet the loss in metabolism, it is estimated that a normal adult requires 0.15 to 0.30 mg of iodine daily. Regions in which the supply of iodine is insufficient in water report good results in reducing the incidence of goiter by use of iodized salt.

Cobalt is a constituent of vitamin B_{12} and is concerned with red cell formation.

Manganese is an essential element widely distributed in plant and animal tissues. Its specific function in the body is not clear, but it is known that manganese activates several important enzymes.

Zinc is essential in plant nutrition and it is evident that it is necessary in animal nutrition for it functions in enzyme systems, including carbonic anhydrase. It is universally distributed in plant and animal tissue.

Fluorine is found in bones and teeth and other tissues. In very small amounts it apparently improves tooth structure. Animal research has not provided evidence that it is an essential part of the diet.

"Trace element" is the name given to a number of other elements, i.e., molybdenum and strontium, which are found in the body in minute quantities or traces. It is believed that they have important metabolic functions.

A diet that furnishes sufficient carbohydrate, fat, and protein may be lacking in calcium, phosphorus, iron, iodine, and copper unless vegetables and fruits are added in sufficient quantities to prevent this deficiency. The amount of calcium and phosphorus needed is relatively large, and definite provision must be made for it. The amount of iron needed is minute; but since the quantities in food materials are also minute, the sources of supply must be considered in planning the diet. If the requirements for calcium and iron are met, it is probable that all other minerals will also be supplied in the same foods.

Another aspect of mineral metabolism is its relationship to water and electrolyte balance, which involves sodium, potassium, and hydrogen in particular. This will be discussed in Chapter 24.

VITAMINS

Vitamins are organic substances present in small amounts in natural foodstuffs, which are essential for growth and normal metabolism. They are considered to be essential to all cells and are needed in the diet in minute amounts. Several members of the vitamin B group are known to be components of respiratory enzymes and of other enzymes that act as catalysts for cell processes. Recent experiments emphasize the determination of optimum (as distinguished from merely adequate) amounts of the vitamins and their physiologic effects. The vitamins constitute a *nutritional factor*, which has been defined as a single substance or any group of substances performing a specific vital function in nutrition.

FAT-SOLUBLE VITAMINS

The fat-soluble vitamins require the presence of bile in the intestinal tract for absorption. Hence any defect in fat absorption may lead to deficiencies of the fat-soluble vitamins.

Vitamin A. Vitamin A and its provitamins, or precursory substances, called alpha, beta, and gamma carotenes and cryptoxanthin, constitute a nutritional factor essential to growth and to epithelial tissues in particular. The precursory substances occur in yellow pigments of plants such as paprika and carrots. These plants contain no vitamin A but do contain the precursory substances that are transformed to vitamin A, chiefly in the liver and in the wall of the small intestine. They add to the vitamin A value of the food but not to its vitamin content. Fish-liver oils contain vitamin A as well as the precursory substances; hence, they add to the vitamin content of the foods as well as to their vitamin value.

Two forms of vitamin A are detectable—an acid and an aldehyde. Vitamin A acid is responsible for the maintenance of epithelial tissues; vitamin A aldehyde makes up part of the molecule of rhodopsin, the retinal pigment. It is stored in large amounts in the liver, if the diet includes such an amount.

Deficiency results in disturbances associated with nervous and epithelial tissues: (1) retarded growth, (2) susceptibility to xerophthalmia, an eye condition conducive to subsequent infection and resulting blindness, (3) dermatosis or dry skin, (4) generally impaired epithelial tissues and resulting increased susceptibility to infections of the lungs, skin, bladder, sinuses, ears, and alimentary tract, and (5) night blindness, which results from failure of the normal regeneration of visual purple after its light-induced change, and reduced synthesis of thyroxin.

Hypervitaminosis is possible; symptoms include joint pain and loss of hair.

Vitamin D. All of the D vitamins have antirachitic properties, but there is a

TABLE 21–4
VITAMIN SOURCES AND EFFECTS

	Vitamins	Sources	Effects of Cooking	General Effects of Optimum Intake	Evidences of Deficiency
Fat-Soluble	A	Milk, butter, eggs, fish-liver oils, green vegetables, yellow vegetables (pro-vitamins)	Resists heat in absence of air; readily destroyed by oxidation	A factor in— Decreasing susceptibility to skin infections Preserving general health and vigor Effecting chemistry necessary for vision Promoting growth	Failure to gain weight, susceptibility to xerophthalmia, night blindness, dry skin, impaired epithelial tissues, increased incidence of respiratory diseases and of skin (toad skin), ear, and sinus infections, inflammations and infections of alimentary and urinary tracts, degenerative changes in nervous tissues
	D	Egg yolk, whole milk, butter, fish-liver oils	Slight; relatively stable	A factor in well-developed bone and teeth, calcium and phosphorus metabolism	Rickets (in children) Osteomalacia (in adults), bone demineralization
	E	Seeds of plants, eggs, lettuce, spinach, meat, wide distribution	Unusually heat resistant	A factor in normal gestation in rats	Sterility in rats
	K	Wide distribution, especially green leaves		A factor in normal functioning of liver and normal clotting time	Delayed clotting time

TABLE 21–4 (*continued*)

	Vitamins	Sources	Effects of Cooking	General Effects of Optimum Intake	Evidences of Deficiency
Water Soluble	B₁ Thiamine	Whole-grain cereals, legumes, eggs, pork	Destroyed by prolonged heating, by temperatures higher than boiling	A factor in— Normal carbohydrate metabolism Maintenance of normal appetite, digestion, absorption	Beriberi, polyneuritis Stunted growth of children, lowered appetite, reduced intestinal motility
	B₂ Riboflavin	Milk, eggs, green vegetables, liver, heart	Relatively heat stable	A factor in— Tissue respiration Normal growth and nutrition and vitality at all ages	Dermatitis, pellagra (in part) Well-defined eye lesions
	Niacin	Liver, milk, poultry	Destroyed by high heat	Essential in metabolic processes that release energy	Low nutritional level
	B₆ Pyridoxine Pyridoxal Pyridoxamine	Whole-grain cereals, yeast, milk, eggs, pork, liver, legumes	Unusually heat resistant	A factor in normal metabolism of fats, amino acids	Florid type of dermatitis (experimental pellagra in rats) Nervousness, irritability, and insomnia
	B₁₂ Cyanocobalamin	Liver, kidney, lean meat, milk, cheese	Relatively heat stable	A factor in red cell formation	Pernicious anemia
	Pantothenic acid	Egg yolk, kidney, liver, yeast, broccoli, lean meat, heart	Destroyed by high heat	Essential for synthesis of acetyl coenzyme A, metabolism of fats, carbohydrates, and certain amino acids	Rarely occurs in man
	Folic acid	Fresh green leafy vegetables, liver, legumes		A factor in— Functioning of enzyme systems Essential in metabolic processes— growth and development	Anemia

TABLE 21–4 (*continued*)

Vitamins	Sources	Effects of Cooking	General Effects of Optimum Intake	Evidences of Deficiency
C Ascorbic acid	Citrus fruits (raw or canned), tomatoes (raw or canned), broccoli	Readily destroyed by heat, especially slow cooking	A factor in— Red cell formation Normal integrity of capillaries Normal development of teeth and maintenance of health of gums Healing of wounds and protection against infections Normal cellular chemistry of all tissues	Low nutritional level Fragility of capillary networks Scurvy and possibly predisposition to dental caries and systemic type of pyorrhea

(Left margin bracket: **Water-Soluble** (*cont.*))

difference in potency. Irradiated ergosterol (D_2) and cholecalciferol (D_3) are powerful antirachitic vitamins. Vitamin D_3 is the most potent form.

There is a close relationship of action between vitamin D and the hormone of the parathyroid gland in calcium metabolism. Vitamin D is needed for absorption of calcium from the intestine and for reabsorption of phosphates in the renal tubules. It is needed for normal bone growth and is considered a factor in the maintenance of the normal functioning of the respiratory system, in the formation of normal teeth, and in protection against dental caries. Deficiency of vitamin D has long been known to result in rickets, which may be cured by administration of vitamin D, by direct sunlight, by ultraviolet irradiation of the body, or by administration of ergosterol or similar substances produced by irradiation. The effect of ultraviolet irradiation of the skin is through its transformation of provitamin D of skin gland secretions into vitamin D, which is absorbed by the skin. The effective rays are those that cause tanning. Ingestion of excessive amounts of vitamin

D results in elevated calcium levels in blood and tissue fluids and in abnormal calcification of soft tissues.

The Tocopherols. Vitamin E (antisterility) prevents sterility in both male and female rats, but not in humans. However, it may be important in various enzyme transformations in cell respiration.

Vitamin K. Vitamin K possesses antihemorrhagic properties. For this reason, inadequate fat absorption due to lack of bile salts is particularly significant in regard to vitamin K. It is considered a factor essential to normal clotting of blood, as it promotes the synthesis of prothrombin and proconvertin by the liver. The blood of animals having a deficiency of this vitamin shows a lowered content of prothrombin and a delayed clotting time.

Vitamin K is fat soluble and appears to be found in a great variety of foods, but knowledge regarding quantitative requirements of this vitamin is still limited. The newborn infant may have an alimentary deficiency, since the vitamin is not readily passed from mother to fetus, hence the need for giving vitamin K to the newborn and to mothers before delivery. It is also used medically to counteract bleeding effects of drugs that block vitamin K formation in the liver.

WATER-SOLUBLE VITAMINS

Vitamin B. Vitamin B is of multiple nature and is usually referred to as the "vitamin B complex." In general, this group of vitamins is necessary for formation of many enzymes, particularly those involved in (1) oxidation-reduction reactions and energy transformation; and (2) formation of red blood cells.

Thiamine (B_1). The first B vitamin, thiamine, combines with phosphate to form cocarboxylase, which plays an essential role in carbohydrate metabolism. It is essential for oxidation reactions within the cell and for transformation of the amino acid tryptophan to niacin. An adequate supply of it is necessary for normal appetite and normal motility of the digestive tract.

Beriberi occurs chiefly among Oriental people who make use of rice as food. The disease takes a variety of forms, but the symptoms are gastrointestinal disturbances, paralysis, and atrophy of the limbs. This condition is caused by limiting the diet to polished rice. If the polishings are restored to the diet, the condition disappears; or if meat or barley is used with the polished rice, the condition is avoided.

Riboflavin. Riboflavin is somewhat more heat stable than is B_1. Phosphorylation of riboflavin is essential for its absorption in the intestine. It was first isolated from milk and named *lactoflavin* and is frequently referred to as the "flavin factor." Riboflavin is the essential component of the flavoprotein coenzymes. These coenzymes catalyze hydrogen transfer in various cell reactions leading to the oxidation of hydrogen to water. Riboflavin is thus an important factor in tissue respiration and is essential to normal growth and nutrition at all ages, since it is vital for protein metabolism.

Niacin and Niacinamide (*Nicotinic Acid and Nicotinic Acid Amide*). These are constituents of two coenzymes that play vital roles in metabolism. These coenzymes function as hydrogen and electron transfer agents in oxidation-reduction reactions. They may be synthesized in man from tryptophan.

Deficiency may result in dermatitis, diarrhea, stomatitis, or glossitis. Niacin is specific for the treatment of acute pellagra.

Pyridoxine (B_6). Vitamin B_6, as pyridoxine, pyridoxal, or pyridoxamine, is important for normal growth and nutrition, and in its active form is essential for the functioning of several enzyme systems. These are the enzymes that catalyze the removal of carboxyl (–COOH) groups from amino acids and those that aid in transfer of amino (NH_2) groups from one substance to another. Pyridoxine also functions in the metabolic reactions involving fatty acids, and in the conversion of tryptophan to niacin. Deficiency produces dermatitis.

Pantothenic Acid. The vitamin pantothenic acid is a constituent of coenzyme A, which combines with acetate to form acetyl coenzyme A. It is essential for the intermediate metabolism of fats, carbohydrates, and certain amino acids (see page 526). Acetyl coenzyme A is involved in the formation of cholesterol and the steroid hormones, also of acetylcholine, which is essential to the transmission of nerve impulses. Deficiencies rarely occur in man. In animals the symptoms include growth failure, dermatitis, and nerve involvement.

Cyanocobalamin (B_{12}). Cyanocobalamin, so called because of the presence of cobalt in its complex molecule, is also named the antipernicious anemia factor. The absorption of this vitamin in the gastrointestinal tract is dependent upon the presence of a gastric factor, "intrinsic factor." The intrinsic factor is secreted by the parietal cells in the cardiac and fundic portions of the stomach. Hence patients who have a total gastrectomy will also develop vitamin B_{12} deficiency and anemia, as absence of the intrinsic factor prevents absorption of vitamin B_{12} from the intestine. It is required for normal metabolism and is essential in the formation of the red blood cell. It is possibly a growth factor for children, but this is uncertain.

Folic Acid (Pteroylglutamic Acid). This vitamin is believed to be concerned chiefly with enzyme systems involved in red cell formation. Its relationship metabolically to B_{12} is close, though as yet unclear. They both stimulate erythropoiesis. It is believed that folic acid can be synthesized by bacteria in the intestine, and that it is concerned with the use of proteins for growth and development.

Biotin, necessary in *very* minute amounts, functions in several systems as a coenzyme. It participates in carboxylation reactions as well as in deamination of certain amino acids.

Other Factors. Lipoic acid aids in decarboxylation of pyruvate before it enters the oxidative cycle. It is not a true vitamin.

Choline is not a true vitamin; however, its role in nutrition is essential. Choline is a constituent of the lecithins and is also importantly concerned in the metabolism of fats. In the form of acetylcholine, it is essential as the chemical mediator of nerve impulses. Deficiencies in experimental animals show tissue damage and abnormal accumulation of fat in the liver particularly, and also in the heart and blood vessels.

Ascorbic Acid (Vitamin C). Ascorbic acid has long been known to be essential in the prevention of scurvy. Early records of sea voyages reveal many epidemics of scurvy,

and it was reported from Austria and Russia during World War I. The cause is lack of fresh fruit and vegetables; the prevention is the use of these. Early cures were through the use of citrus fruit juices. Laboratory experiments on animals and men prove conclusively that scurvy is due to lack of vitamin C in the diet.

More recently vitamin C has been shown to be of importance in tissue respiration. Shortage of vitamin C is shown to impair general nutrition, to prevent healing of bone wounds, and to be a contributory factor in capillary fragility and the general resistance of the body to infection. Vitamin C is essential for the formation and maintenance of collagen. It is necessary for the integrity of capillary membranes and plays an important part in the formation of blood cells in bone marrow. It is also an important factor in the healing of wounds. The adrenal cortex contains a large quantity of vitamin C, which suggests its use in the metabolic synthesis of the steroid hormones. A liberal daily intake of vitamin C throughout life is recommended. The relationship of deficiency to dental caries is undetermined, but the soundness of teeth and their supporting bones and gums is believed to be dependent upon the amount of vitamin C supplied by the food.

Symptoms of scurvy are loss of weight, pallor, weakness, breathlessness, palpitation of the heart, swelling of the gums, loosening of the teeth, pains in the bones and joints, edema,

TABLE 21–5
COENZYMES AND THEIR FUNCTIONS

Coenzyme		Function	Vitamin
(NAD)	Nicotinamide adenine dinucleotide	As hydrogen acceptors in dehydrogenases	Niacinamide
(NADP)	Nicotinamide adenine dinucleotide phosphate		
(FMN)	Flavin mononucleotide	As hydrogen acceptors in aerobic dehydrogenases	Riboflavin
(FAD)	Flavin adenine dinucleotide		
	Pyridoxal phosphate	As transaminases, amino acid decarboxylases	Pyridoxine
(TPP)	Thiamine pyrophosphate	As cocarboxylase	Thiamine
(CoA)	Coenzyme A	In condensing enzymes, fatty acid utilization, acetate transfer	Pantothenic acid

nervousness, and slight hemorrhages appearing as red spots under the skin and forming hidden bleeding places in the muscles and internal organs. The heart hypertrophies and shows degenerative changes, which often cause sudden death.

As research continues, it is becoming more and more obvious that the major function of all vitamins is in relation to enzyme functioning. For many it is known that they make up part of the molecular structure of coenzymes, or cofactors, which accept atoms or groups of atoms that are removed from a substrate. These are listed in Table 21–5.

Digestive Processes

Digestion includes all the changes, physical and chemical, that food undergoes in the small and large intestines. In some instances no change is necessary; water, minerals, and certain carbohydrates in fruits are ready to be absorbed. In other instances cooking processes initiate chemical changes in food before it enters the body, for example, changing starch to dextrin, partially splitting fats into glycerol and fatty acids, and changing some proteins to the first stages of their hydrolytic products. Cooking in many instances improves the appearance, odor, and taste of food, and these changes stimulate the end organs of the optic and olfactory nerves and the taste buds, causing a reflex stimulation of the digestive mechanisms. Cooking also tends to destroy microorganisms that would be harmful to the body. At the same time cooking may destroy certain vitamins.

The digestive processes are controlled by the nervous and hormonal systems. Any strong emotion that affects the nervous system unpleasantly inhibits the secretion of the digestive fluids and interferes with digestion, often checking the appetite and even preventing the intake of food. On the other hand, pleasurable sensations aid digestion, hence the value of attractively served food, pleasant surroundings, and cheerful conversation.

Physical Digestion. This includes the various physical processes that occur in the alimentary canal. It serves the following purposes: taking food in and moving it along through the alimentary canal just rapidly enough to allow the required chemical changes to take place in each part; lubricating the food by adding the mucin and water secreted by the glands of the alimentary canal; liquefying the food by mixing it with the various digestive juices; and grinding the food into small particles, thereby increasing the amount of surface to come in contact with the digestive fluids.

Chemical Digestion. This is essentially a process of *hydrolysis* which is dependent upon the presence of enzymes. An example of hydrolysis (hydrolytic cleavage) is the splitting of maltose into glucose (also called dextrose) under the influence of maltase.

$$C_{12}H_{22}O_{11} + H_2O \rightarrow C_6H_{12}O_6 + C_6H_{12}O_6$$

Chemical digestion is necessary because foods in general cannot pass through cell membranes, and the tissues cannot use them; hence, they must be reduced to small molecules and to such substances as the tissues can use, i.e., (1) simple sugars, resulting from the hydrolysis of all carbohydrate foods; (2) glycerol and fatty acids, resulting from the hydrolysis of fats; and (3) amino acids, resulting from the hydrolysis of proteins.

Hydrolytic cleavages similar to those of digestion can be brought about in several ways. Boiling foodstuffs with acids, treating with alkali, or subjecting them to superheated steam will accomplish these changes. The *remarkable* fact is that strong acids and high temperatures, or both, are necessary to produce these changes in the laboratory, whereas in the digestive tract they take place at body temperature and are due to the enzymes present in the digestive juices.

Digestive Enzymes. The enzymes that bring about chemical digestion in the alimentary tract are exoenzymes—organic catalysts that are produced by cells, secreted into the digestive tract where they act. They may be classified as follows:

1. Sugar Splitting. The glucosidases, which hydrolyze disaccharides to monosaccharides. Examples: maltase splits maltose to glucose; sucrase splits cane sugar to glucose and fructose; and lactase splits milk sugar (lactose) to glucose and galactose.

2. Amylolytic, or Starch Splitting. Examples: salivary amylase and pancreatic amylase cause hydrolysis of starch.

3. Lipolytic, or Fat Splitting. Examples: lipase found in the pancreatic secretion causes hydrolysis of fat.

4. Proteolytic, or Protein Splitting. Examples: pepsin of gastric juice and trypsin and chymotrypsin of pancreatic juice, which cause hydrolysis of the proteins.

The digestive enzymes are found in the secretions of the various organs, in solution with a great deal of water that also contains mucus as well as electrolytes. Thus, loss of any large amount of the digestive juices may lead to dehydration and electrolyte imbalance.

Secretion of Digestive Juices. Glandular cells are well supplied with nerve fibers from the autonomic nervous system. The stimuli for varying the amount of secretory product may be chemical, nervous, or hormonal. There are secretory as well as vasomotor fibers which, when reflexly stimulated by the sight or smell of food, (1) dilate the blood vessels, increasing the volume of the vessels, and (2) cause the glands to produce a secretion that is copious in amount and watery in consistency.

DIGESTION IN THE MOUTH

Mastication. When solid food is taken into the mouth, its comminution is immedi-

ately begun. It is cut and ground by the teeth, being pushed between them by the movements of the tongue, until the whole is thoroughly crushed.

Salivation. During the process of mastication saliva is poured in large quantities into the mouth and, mixing with the food, lubricates, moistens, and reduces it to a softened mass known as a *bolus*, which can be readily swallowed.

Secretion of Saliva. The nerve supply of the salivary glands is derived in part from the craniosacral and in part from the thoracolumbar divisions of the autonomic system. Both sets of nerves carry secretory and vasomotor fibers. The craniosacral nerves cause the glands to produce a secretion that is copious in amount and watery in consistency.

The consistency of saliva depends in part on the relative number of serous and mucous cells that are secreting. Serous cells produce a thin watery secretion and mucous cells produce a thick secretion. The thoracolumbar nerves carry vasoconstrictor fibers and perhaps are relatively unimportant for controlling normal function of the salivary glands. However, if the sympathetic fibers are stimulated, vasoconstriction occurs and a scanty, viscid saliva is produced. Under normal conditions, the secretion of saliva is the result of stimulation of the secretory nerves by the smell, taste, or sight of food. Obviously, the taste buds of the tongue, and fauces are the sense organs that are stimulated by the presence of food in the mouth.

Saliva. Saliva is secreted by the salivary glands—parotid, submaxillary, and sublingual—and by the numerous minute buccal glands of the mucosa of the mouth.

It consists of a large amount of water (some 99.5 per cent) containing some protein material, mucin, inorganic salts, and *salivary amylase*. It has a specific gravity of about 1.005 and is nearly neutral in reaction (pH about 6.4 to 7.0). Although the amount of saliva secreted per day varies considerably, an average amount is from 1 to 1.5 liters. Substances in saliva include inorganic salts in solution, chlorides, carbonates, and phosphates of sodium, calcium, and potassium.

The other substances are organic, mainly mucin, salivary amylases, serum albumin and globulin, and urea. The calcium carbonate and phosphate in combination with organic material may be deposited on the teeth as tartar, especially if the saliva is alkaline and contains considerable mucin. Occasionally these salts may be also deposited in the ducts of the salivary glands.

The *functions* of saliva are to soften and moisten the food, assisting in mastication and deglutition; to coat the food with mucin, lubricating it and ensuring a smooth passage along the esophagus; to moisten or liquefy dry and solid food, providing a necessary step in the process of stimulating the taste buds, as taste sensations play a part in the secretion of gastric juice; to digest starch by means of salivary amylase.

SALIVARY AMYLASE. Salivary amylase changes starch to dextrins and maltose. The process of reducing starch to maltose is a gradual one, consisting of a series of hydrolytic changes that take place in successive stages and result in a number of intermediate compounds. The change is best effected at the temperature of the body, in a neutral solution. Boiled starch is changed more rapidly and completely than raw, but food is rarely retained in the mouth long enough for the saliva to do more than begin the digestion of starch.

Deglutition, or Swallowing. Deglutition is divided into three stages, which correspond to the three regions—mouth, pharynx, and esophagus—through which the food passes. The *first stage* consists of the passage of the bolus of food through the fauces. Contractions of the constrictor muscles of the pharynx force the bolus along. The *second stage* consists of the passage of the bolus through the laryngeal pharynx. During this stage, the respiratory opening into the larynx is closed by the approximation of the vocal folds that close the glottis, by the elevation of the larynx, and by contraction of the muscles of deglutition. The parts are crowded to-

gether by the descent of the base of the tongue, the lifting of the larynx, and the coming together of the vocal folds.

The *third stage* consists in the passage of the bolus through the esophagus. Apparently the consistency of the food affects this stage of the process. Solid or semisolid food is forced down the esophagus by a peristaltic movement and requires from four to eight seconds for passage from mouth to stomach. About half of this time is taken up in the passage through the esophagus, and the remainder is spent in transit through the cardiac orifice of the stomach. Liquid or very soft food is moved rapidly down the esophagus and arrives at the lower end in about 0.1 second. It may pass into the stomach at once or may be held in the esophagus for moments, depending on the condition of the cardiac sphincter. Repeated deglutition causes the tension of the muscles that function as a cardiac sphincter to diminish progressively, until they become completely relaxed, and food passes into the stomach. Following this, relaxation finally disappears and the sphincter becomes more contracted than usual and remains so for a considerable time.

In summary, during the process of mastication, salivation, and deglutition the food is reduced to a soft, pulpy condition, and any starch it contains may begin to be changed into sugar.

DIGESTION IN THE STOMACH

The food that enters the stomach is delayed there by the contraction of the sphincter muscles at the cardiac and pyloric openings. Within a few minutes after the entrance of food small contractions start in the middle region of the stomach and run toward the pylorus. These contractions are regular and in the pyloric region become more forcible as digestion progresses.

Weak rippling peristaltic movements, called *mixing waves,* pass over the stomach about every 15 to 25 seconds. As a result of these movements the food in the prepyloric and pyloric portions is macerated, mixed with the acid gastric fluid, and reduced to a thin liquid mass called *chyme.* At certain intervals the pyloric sphincter relaxes, and the wave of contraction forces some of the chyme into the duodenum. The fundic end of the stomach is less actively concerned with these movements but serves as a reservoir for food. The food at the fundic end may remain undisturbed for an hour or more and thus escape rapid mixture with the gastric fluid, which, therefore, penetrates slowly to the interior of the mass; hence salivary digestion may continue for a time. As the chyme is gradually forced into the duodenum, the pressure of the fundus forces the food into the pyloric end.

The time required for gastric digestion depends upon the nature of the food eaten. Liquids taken on an empty stomach pass through the pylorus promptly. Small test meals may remain from one to two hours, but average meals probably stay in the stomach from three to four and one-half hours. The shape of the stomach ("fishhook," "steerhorn") is an important factor in determining evacuation time. It has been demonstrated that emptying time was about 50 per cent faster in individuals with a "steerhorn" stomach than those with a "fishhook" (or J-shaped) stomach.

Shortly after ingestion of a meal, peristaltic waves (Figure 21–1) begin to traverse the lower stomach. These waves begin about the middle of the stomach, and sweep downward, usually passing over the pyloric sphincter and often into the duodenum. The fundic part of the stomach contains the food and functions as a storage chamber. Although it shows no peristaltic waves, it does show progressive increase in muscle tension. The increasing tension forces food toward the lower end of the stomach where peristalsis propels it foward.

Emptying of the stomach depends *almost entirely* on a pressure gradient between the stomach and duodenum. For material to leave the stomach, the *intragastric* pressure must be greater than the *intraduodenal.* The intragastric pressure is due to (1) gastric tonus changes and (2) head pressure of the peristaltic wave. The pyloric sphincter *does not* control evacuation. The sphincter may actually be relaxed during most of the digestion period *without* emptying occurring. When the intragastric pressure is adequately greater than the intraduodenal, about 2 to 5 ml of gastric content is passed into the duodenum with each peristaltic wave. Meals that are high in fat content delay the emptying time through the action of *enterogastrone,* a hormone that decreases gastric motility.

When the intraduodenal pressure exceeds the intragastric pressure, reflex constriction of the pyloric sphincter prevents the duodenal content from entering the stomach. *The major role of the pyloric sphincter is to prevent regurgitation.* With the exception of the rectal sphincter, the major role of all the sphincters of the

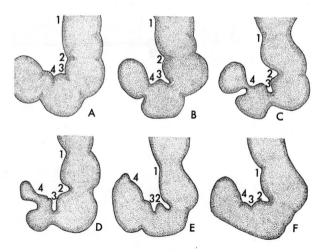

Figure 21–1. Diagrams to show peristalsis in the stomach. Locate *1, 2, 3, 4* in succeeding figures to trace a peristaltic wave over the stomach.

digestive tract is to prevent retrograde movement of digestive contents.

The secretion of gastric juice is constant. Even in the period of fasting there is a small continuous secretion, but during the act of eating and throughout the period of digestion the rate of secretion is greatly increased.

Gastric juice is produced by the mucous membrane of the stomach. The complete cycle of the activities of the gastric glands is frequently divided into three phases: the *cephalic,* the *gastric,* and the *intestinal.* The *cephalic phase* of gastric secretion refers to reflex stimulation of the gastric glands through the central nervous system by the sight, smell, or taste of food. The *gastric phase* refers to all of the activities of the gastric glands, which are brought about by conditions within the stomach itself. It includes the gastric mechanisms as well as local chemical stimulation by secretagogues (any substance that stimulates secretion by a gland), which occur as a result of the contact of the gastric mucosa with the products of digestion and the stimulation caused by distention of the stomach. During the gastric phase a hormone, *gastrin,* is secreted by cells in the antrum and transported by the blood to fundic cells; it stimulates secretion of both chief and parietal cells. Its secretion is inhibited by excessive acidity of the gastric juice (pH of 2.0).

The *intestinal phase* of gastric secretion is believed to be due to the presence of food in the small intestine, which may stimulate the gastric glands through a hormone or secretagogue. Fats inhibit gastric secretion, by stimulating the production of a hormone, *enterogastrone,* by cells in the duodenal mucosa. *Secretin,* a hormone secreted by the cells of the upper small intestine, decreases acid secretion.

Gastric Juice. Gastric juice is secreted by the gastric glands lining the mucous membrane of the stomach. It is a thin, colorless or nearly colorless liquid with an acid reaction and a specific gravity of about 1.003 to 1.008. The acid that is secreted by the parietal cell has a pH of about 0.9. However, after reaching the lumen of the stomach, the acid is partially neutralized by the gastric mucus, the more alkaline saliva, and the alkaline intestinal content, which may be regurgitated into the stomach. This reduction in acidity results in a gastric content of about pH 2.5. The quantity secreted depends upon the amount and kind of food to be digested, possibly an average of 1.5 to 2.5 liters per day. Upon analysis it is found to be a watery secretion containing some protein, some mucin (which helps protect the stomach lining from the effect of the digestive enzymes and acid), and inorganic salts; but the essential constituents for digestion are hydrochloric acid and one or possibly two enzymes—pepsin and gastric lipase.

Hydrochloric Acid. The parietal cells of the gastric glands secrete the chloride ions found in the blood and hydrogen ions formed within the cell. The chloride ion combines with the hydrogen ion in the lumen of the gland and then moves to the free surface of the stomach as hydrochloric acid. In normal gastric juice it is found in the proportion of about 0.5 per cent. It serves to activate pepsinogen and convert it to pepsin; to provide an acid medium, which is necessary for the pepsin to carry on its work; to swell the protein fibers, thus giving easier access to pepsin; and to destroy many organisms that enter the stomach.

Excessive secretion of hydrochloric acid, or gastric hyperacidity, is often found associated with peptic ulceration of the duodenum, and with some forms of gastritis, but there is little evidence that there is a cause-effect relationship between hypersecretion and ulcer disease. Hyposecretion of mucin may be a causative factor in ulcer formation. Secretion of hydrochloric acid below normal, or hyposecretion, is associated with other forms of gastritis and frequently with carcinoma of the stomach. Acid secretion is totally absent in pernicious anemia.

Enzymes. Pepsin is formed in the pyloric glands and the chief cells of the gastric glands. It is present in these cells in the form of a zymogen, an antecedent inactive substance called propepsin or pepsinogen, which is quickly changed to active pepsin by the action of hydrochloric acid.

Pepsin (gastric protease) is a proteolytic enzyme requiring an acid medium in which to function. It has the property of hydrolyzing proteins through several stages into polypeptides. This action is preparatory to the more complete hydrolysis that takes place in the intestine under the influence of trypsin (pancreatic protease) and various peptidases, for polypeptides are not absorbed but undergo a further hydrolysis to amino acids.

It is probable that the *salivary amylase* swallowed with the food continues the digestion of starchy material in the fundus for some time. Regarding the long-chain fats, it is believed that they undergo no digestion in the stomach. They are set free from their mixture with other foods by the digestive action of the gastric fluid; they are liquefied by the heat of the body and are scattered through the chyme as a coarse emulsion by the movements of the stomach, all of which prepare them for digestion. Emulsified fats such as cream may be acted upon to a limited extent by a third enzyme called *gastric lipase,* but the acid condition of the stomach contents prevents any considerable change of this sort. This enzyme is more important in the child than in the adult.

In summary, the stomach serves as a place for temporary storage and maintains a gradual delivery to the intestine; it also serves as a place for the continuation of the salivary digestion of starch, the beginning of the digestion of proteins and perhaps fats, and germicidal activity.

Inhibition of Gastric Digestion. The secretion of gastric fluid is inhibited by stimulation of the thoracolumbar system, so that various emotions and also a distaste for food may delay digestion. Secretion is also inhibited by active exercise soon after a meal, because active exercise increases the amount of blood in the skeletal muscles and decreases the supply to the stomach. When gastric digestion is much delayed, organisms are likely to cause fermentation of the sugars, producing gas, which may cause distress.

Vomiting. Vomiting is controlled by the emetic nerve center in the medulla, which can be stimulated by chemical qualities of the tissue fluid in the center and by nerve impulses that reach it. Under ordinary circumstances the contractions of the cardiac "sphincter" prevent the regurgitation of food. During vomiting the stomach, esophagus, and esophagogastric junction (so-called cardiac sphincter) are all relaxed. Spasmodic contractions of the abdominal muscles synchronously with contraction of the diaphragm cause phasic increases in intragastric pressure, which results literally in squeezing out the stomach contents in spurts. It is wrong to think that the stomach muscles contract or show reverse peristalsis; the stomach behaves passively like a water-filled rubber bulb, which spurts when it is compressed. After the stomach contents have been evacuated, the pyloric sphincter may also relax and permit the duodenal contents to be evacuated. Vomiting is usually preceded by a sensation of nausea and excessive salivation. Vomiting is a reflex act brought about by mechanical irritation of the throat or by irritating substances in the stomach and duodenum, and by pain, motion sickness, and certain emotions such as fear and repulsion.

DIGESTION IN THE SMALL INTESTINE

The chyme entering the duodenum after an ordinary meal is normally free from coarse particles of food and is acid in reaction; both the hydrochloric acid and the lactic acid produced by fermentation contribute to this condition. Much of the food is undigested. The proteins are partly digested; some progress has been made in hydrolyzing starch; fats have been liquefied and mixed with other food but probably have not been hydrolyzed themselves. If milk is part of the diet, it will have been curdled and redissolved. It is in the small intestine that this mixture undergoes the greatest digestive changes. These changes, which constitute intestinal digestion, are effected by the movements of the intestine, the pancreatic fluid, the succus entericus, or secretion of the intestinal glands, and the bile.

It is convenient to describe the secretion and digestive action of these three fluids separately, but it must be remembered that they act simultaneously. The pancreatic fluid and the bile enter the intestine about 7 to 10 cm beyond the pylorus; therefore, the foods in the small intestine throughout its length are subjected to a

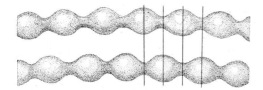

Figure 21–2. Two diagrams of a portion of the small intestine to show rhythmical movements. The small straight lines indicate the same area of the intestine at different intervals.

mixture of pancreatic fluid, bile, and small intestinal fluid.

Movements of the Small Intestine (Figure 21–2). These are described as peristaltic, rhythmical, and pendular.

Peristalsis may be defined as a wave of dilation brought about by the contraction of longitudinal muscles, followed by a wave of constriction caused by the contraction of circular muscles. The purpose is to pass the food slowly forward. *Peristaltic waves* pass very slowly along short distances of the small intestine, with an occasional rapid wave known as the *peristaltic rush*, which moves the food along greater distances. The stimulus seems to be partly mechanical, since experimental swallowing of a tube to which a small balloon is attached initiates peristalsis and propulsion of the balloon through the intestine.

The rhythmical movements consist of a series of local constrictions of the intestinal wall that occur rhythmically at points where masses of food lie. These constrictions divide the food into segments. Within a few seconds each of these segments is halved, and the corresponding halves of adjoining segments unite. Again constrictions occur, and these newly formed segments are divided, and the halves re-form. In this way every particle of food is brought into intimate contact with the intestinal mucosa and is thoroughly mixed with the digestive fluids.

Pendular movements are constrictions that move onward or backward for short distances, gradually moving the chyme forward and backward over short distances in the small intestine. These may be seen in the rabbit and other small mam-

mals, but their presence is doubted in the human.

The varied muscular movements of the small intestine increase the blood supply, bringing materials for secretion and removing absorbed materials faster. They assist the minute glands in emptying their secretion, mix the digestive fluids and food intimately, and bring fresh absorbable material constantly to the mucosa, thereby increasing absorption.

Secretion of Pancreatic Juice. Pancreatic secretion, like gastric secretion, consists of two parts: a neurally induced secretion, caused by the secretory fibers in the vagus and splanchnic nerves, and a chemically induced secretion, caused by the action of the hormones *secretin* and *cholecystokinin*. The acid gastric fluid and the products of partially digested proteins upon reaching the duodenum and jejunum stimulate the production of these hormones. They are taken by way of the blood to the pancreas where they cause secretion of large quantities of fluid rich in enzymes. It is thought that the neurally induced secretion provides pancreatic fluid in the early stages of intestinal digestion and that the chemical secretion maintains the flow until all the stomach contents reach the duodenum.

Pancreatic Juice. The *nervous secretion* of pancreatic juice is thick, and rich in enzymes and proteins. The *chemical secretion* resulting from cholecystokinin activity is thin, watery, and is also rich in enzymes. Pancreatic juice is alkaline and becomes more so with increasing rates of secretion. This is due to the effect of *secretin*, which increases secretion of bicarbonate ion (as well as decreasing gastric secretion of acid).

Pancreatic juice contains three groups of enzymes, pancreatic proteases (carboxypolypeptidase and trypsin), amylase, and lipase. The amount of pancreatic juice secreted each day varies between 600 and 800 ml.

The proteolytic enzyme trypsin under favorable conditions may hydrolyze the protein molecule to polypeptides. Trypsin is secreted in an inactive form called *tryp-*

sinogen and is activated by *enterokinase,* an enzyme that is secreted by glands of the small intestine. *Carboxypolypeptidase* breaks polypeptides into component amino acids.

Another proteolytic enzyme, *chymotrypsin,* also is present in pancreatic juice. It is secreted in an inactive form, *chymotrypsinogen,* which is activated by the enzyme trypsin.

Nuclease is a nucleic acid—splitting enzyme that results in the production of nucleotides—the subunits that form nucleic acids. There has been some question about its presence in pancreatic juice.

The amylolytic enzyme (amylase) is similar to salivary amylase in action. It causes hydrolysis of starch with the production of maltose. The starchy food that escapes digestion in the mouth and stomach becomes mixed with this enzyme and continues under its action until the colic valve is reached. Maltose is further acted upon by the maltase of the intestinal secretion and is hydrolyzed to glucose.

The lipolytic enzyme (lipase) is capable of hydrolyzing fats to monoglycerides and some to glycerol and fatty acids. The process of hydrolysis is preceded by emulsification, in which bile salts play a leading role. The lipase splits some of the fats to fatty acids and glycerol; emulsification increases the surface of fat exposed to the chemical action of the lipase and is a mechanical preparation for the further action of lipase. The glycerol and fatty acids produced by the action of the lipase are absorbed by the epithelium of the intestine. It is thought that the fatty acids form soluble and diffusible compounds with the bile salts and are absorbed in this form. After absorption the fatty acids and glycerol again combine to form fat, the triglycerides found in the bloodstream as chylomicrons. Lipases are found in blood and in many of the tissues. Although lipases are also secreted by the small intestine, that secreted by the pancreas accounts for about 80 per cent of all fat digestion. For this reason impaired fat digestion is an important result of pancreatic dysfunction.

Intestinal Secretion. This is a clear, yellowish fluid, and amounts to about 2 to 3 liters per day. Its composition varies; in the duodenum and jejunum it is slightly acid. The acidity is greatest in the duodenum, and in the ileum the secretion is practically neutral. In the duodenal bulb and the region down to the ampulla of Vater it is almost entirely mucous. Extracts of the walls of the small intestine have been found to contain four or five enzymes that influence intestinal digestion to a marked extent. The enzymes are to be found in the secretion, and their actions are as follows.

Enterokinase is an enzyme that activates the trypsin of the pancreatic fluid; aminopeptidase, carboxypeptidase, and dipeptidase are enzymes that hydrolyze peptides to amino acids, thus completing the work begun by pepsin and trypsin.

Maltase acts upon the products formed in the digestion of starches, i.e., maltose, and hydrolyzes them to glucose. *Sucrase* acts upon sucrose and hydrolyzes it to glucose and fructose. *Lactase* acts upon lactose and hydrolyzes it to glucose and galactose. This hydrolysis is necessary because disaccharides cannot be used by the tissues, but in the form of simple sugars they are readily utilized.

Nucleases act upon the nucleic acid component of nucleoproteins.

— ACTION IS PHYSICAL

Bile. Bile is formed in the liver and is an alkaline fluid, pH about 6.8 to 7.7, the specific gravity of which varies from about 1.010 to 1.050. Approximately 800 to 1000 ml are secreted daily. It is usually yellow, brownish yellow, or olive green in color. The color of bile is determined by the respective amounts of the bile pigments, (1) biliverdin and (2) bilirubin, that are present. Bile consists of water, bile pigments, bile acids, bile salts, cholesterol, lecithin, and neutral fats.

The bile acids are glycocholic and taurocholic, occurring as sodium glycocholate and sodium taurocholate. These salts are alternately poured into the duodenum, then reabsorbed, and reappear in the bile. Thus, by continued circulation, the bile salts repeat their function many times. This enterohepatic circulation helps to conserve bile salts, as each circuit is accomplished with only about 10 to 15 per cent loss. Synthesis of new bile salts from materials in the diet makes up the deficit. The mucous membranes of the bile ducts and gallbladder add a mucinlike protein called nucleoalbumin, which, together with some mucin, gives bile its mucilaginous consistencey.

Secretion of Bile. Secretion of bile is continuous, but the amount varies, increasing when the blood flow is increased and vice versa. It is thought that the presence of bile in the intestine stimulates secretion in the liver and that this is due to the bile salts, which act as a choleretic. Bile enters the duodenum only during the period of digestion. Between these periods, resistance to the entrance of bile into the duodenum is high, so that the bile is diverted into the gallbladder, where it becomes concentrated by loss of water. The ejection of chyme into the duodenum causes the contraction of the gallbladder and ejection of bile; the hormone cholecystokinin, formed by cells of the duodenal mucosa, mediates this response.

Functions of Bile. Bile salts are essential for the action of lipase. Mixtures of bile and pancreatic fluid split the fats more rapidly than pancreatic fluid alone. Bile salts lower surface tension, which aids in the emulsification of fats with concurrent production of a greater surface area, which enables lipase and other enzymes to act more effectively. Bile salts are also important for absorption of fat.

Bile is essential for the absorption of vitamin K and other fat-soluble vitamins. It also stimulates intestinal motility and neutralizes the acid chyme, creating a favorable hydrogen ion concentration for pancreatic and intestinal enzyme activity; bile salts help keep cholesterol in solution.

In addition, the bile is an excretory medium for toxins, metals, and cholesterol. The liver cells excrete the bile pigments that are brought to them by the blood, just as the kidney cells remove the urea from the blood. The cholesterol of bile is probably a waste product of cellular disintegration.

It is thought that increased putrefaction in the absence of bile is brought about by the action of bacteria on proteins and carbohydrates that have remained undigested because of the protective covering of insoluble fat that is found on them in the absence of bile.

Bilirubin Formation. When erythrocytes are broken down by the reticuloendothelial cells, the hemoglobin is also broken down, and the heme is separated from the globin. Heme is transformed first to biliverdin and then to bilirubin, which then enters the circulation. Since it is not soluble, it adsorbs to plasma albumin and eventually reaches the liver. The liver cells conjugate the bilirubin so that it is in soluble form, and the conjugated bilirubin is excreted in the bile. The first form of bilirubin is called "free" bilirubin, to distinguish it from conjugated bilirubin. Both forms are normally found in the circulation in small amounts. It is possible to identify each form by the van den Bergh test. The first-reacting (or *direct*) bilirubin is the conjugated form, since it is soluble; the second-reacting (or *indirect*) form of bilirubin is the "free" bilirubin.

Jaundice. When the flow of bile through the bile duct is interfered with, bilirubin is not removed from the blood but is carried to all parts of the body, producing a condition of jaundice, which is characterized by a yellow discoloration of the skin and of the whites of the eyes. The urine is of a greenish hue because of the extra quantity of pigment eliminated by the kidneys, and the stools are grayish in hue, owing to the lack of bile. Jaundice may also be due to the incapacity of the liver cells to conjugate bilirubin or to the presence of excessive amounts of bilirubin as in the case of too rapid destruction (hemolysis) of erythrocytes. The van den Bergh test is then a help in identifying the cause of the jaundice.

Gallstones. Abnormal composition of bile may be such that its constituents become so concentrated in the gallbladder that it tends to crystallize out, and these crystals form gallstones. Inflammatory conditions, which are often due to the typhoid and colon bacilli or to a change in the character of the bile, may cause this crystallization. Gallstones are usually formed in the gallbladder. Their passage through the cystic and common bile ducts often causes severe pain, called gallbladder colic. They may plug the duct and cause obstructive jaundice.

Microorganisms in the Small Intestine. Action of organisms in the small intestine hydrolyzes carbohydrates and proteins constantly. Fermentation of the carbohydrates gives rise to organic acids, such as lactic and acetic, but none of the products of this fermentation is considered toxic. On the other hand, the putrefaction of proteins gives rise to a number of end products that are toxic when present in large amounts. Under normal conditions and on a mixed diet, carbohydrate fermentation is the characteristic action of the organisms in the ileum, whereas protein putrefaction occurs in the large intestine. These microorganisms are in many ways beneficial to the body. They synthesize vitamins, which may be absorbed;

TABLE 21–6
SUMMARY OF DIGESTIVE SECRETIONS

Secretion	pH Volume per Day	Proenzyme/ Substance That Activates	Enzyme	Substrate	End Products
Saliva	6.8 1–1½ liters		1. Salivary amylase	Starch	Maltose
Gastric juice	2–4 1.5–2.5		1. Rennin (infants)	Casein	Paracasein
		Pepsinogen/HCl	2. Pepsin	Proteins, paracasein	Proteoses, peptones, polypeptides
			3. Gastric lipase	Emulsified fats	Fatty acids, glycerol
Pancreatic juice	8–8.4 600–800 ml	1. Trypsinogen/ enterokinase	1. Trypsin	Chymotrypsinogen	Chymotrypsin
				Proteins, paracasein, peptones, proteoses, polypeptides	Polypeptides, dipeptides
		2. Chymotrypsinogen/trypsin	2. Chymotrypsin	Proteins, paracasein, peptones, proteoses, polypeptides	Polypeptides, dipeptides
			3. Pancreatic amylase	Starch, glycogen	Disaccharides
			4. Pancreatic lipase	Emulsified fats	Fatty acids, glycerol
Bile	7.5 800–1000 ml	Bile salts (not an enzyme—action is physical)		Unemulsified fats	Emulsified fats
Intestinal juice	7–9 2–3 liters		1. Peptidase	Polypeptides, dipeptides	Amino acids
			2. Enteric lipase	Emulsified fats	Fatty acid, glycerol
			3. Sucrase	Sucrose	Glucose, fructose
			4. Maltase	Maltose	Glucose
			5. Lactase	Lactose	Glucose, galactose
			6. Enterokinase	Trypsinogen	Trypsin

their presence in the intestinal tract is not irritating and at the same time it prevents other, potentially harmful organisms that might be ingested or present in small amounts from multiplying and causing infection.

The time required for digestion in the small intestine is influenced by many factors. It depends largely on the varying proportions of the different foods included in a meal. Twenty to thirty-six hours are required for the passage of ingested food material through the gastrointestinal tract of adults who are on a mixed diet. There is considerable variation among individuals, and usually not all of the residue from a single meal is evacuated at the same time. In diarrheal conditions the time is much shortened.

According to observations made upon a patient with a fistula at the end of the small intestine, food begins to pass into the large intestine from two to five hours after eating, and it requires nine hours or more after eating before the last of a meal has passed the colic valve.

Summary of Digestive Hormones. *Gastrin* is secreted by the pyloric mucosa and excites the fundic glands of the stomach to secrete acid; it has a milder effect on the pancreas, causing it to secrete enzymes.

Enterogastrone is secreted by the duodenal mucosa and inhibits gastric secretion and motility.

Secretin is secreted by the upper intestinal mucosa and excites the pancreas to secrete bicarbonate and water, poor in enzymes. It decreases gastric secretion of acid and increases secretion of pepsinogen to a mild degree.

Cholecystokinin is secreted by cells in the upper small intestine and causes contraction of the smooth muscle of the gallbladder, therefore emptying it. In addition, it stimulates the pancreas to secrete fluid rich in enzymes.

It must be remembered that these hormones, like others, are secreted into the bloodstream and reach all cells through the circulation; only certain cells respond, however.

DIGESTION IN THE LARGE INTESTINE

Movements of the Large Intestine. The opening from the small intestine into the large is controlled by the colic valve and the colic sphincter, which is normally in a state of tone. Food begins to pass into the large intestine within two to five hours after eating. Transit of the meal through the small intestine apparently occurs at a steady rate, so that the total time for the whole meal to pass the colic valve will be determined principally by the gastric evacuation time. As food passes the colic valve, the cecum becomes filled, and gradually the accumulation reaches higher and higher levels in the ascending colon. The contents of the ascending colon are soft and semisolid, but in the distal end of the transverse colon they attain the consistency of feces.

A type of movement characteristic of the large intestine is called *haustral churning*. The pouches, or sacculations, that are present in the large intestine become distended and from time to time contract and empty themselves. Another type of movement is designated as *mass peristalsis*. It consists of the vigorous contraction of the entire ascending colon, which transfers its contents to the transverse colon. Such movements occur only three or four times a day, last only a short time, and are usually connected with eating. When food enters the stomach and duodenum, peristalsis is initiated in the colon through the autonomic nerves of the areas involved. These reflex actions are termed the *gastrocolic* and *duodenocolic reflexes*. They are most noticeable after the first meal of the day and cause increased excitability of the colon, which initiates the defecation reflexes.

Secretion of the Large Intestine. The secretion contains much mucin, shows an alkaline reaction, and secretes no enzymes. When the contents of the small intestine pass the colic valve, they still contain a certain amount of unabsorbed food material. This remains a long time in the intestine; and since it contains the digestive enzymes received in the duodenum, the process of digestion and absorption by diffusion continues.

Microorganisms in the Large Intestine. Action of microorganisms in the large intestine brings about, in an alkaline reaction, constant putrefaction of whatever proteins are present as the result of not having been digested and absorbed in the small intestine. The splitting of the protein molecules by this process is very complete; not only are they hydrolyzed to amino acids, but these amino acids are deaminized and changed to simpler groups. The list of simple substances resulting from putrefaction is long and includes various peptides, ammonia, and amino acids, and also indole, skatole, phenol, fatty acids, carbon dioxide, and hydrogen sulfide. Some of these are given off in the feces; others are absorbed and carried to the liver, where they are changed to less toxic compounds, and excreted in the urine. Some organisms of the large intestine are useful in that they are capable of synthesizing several of the vitamins needed for normal metabolism. These vitamins include several of the B group and vitamin K.

Feces. Two classes of material may be mingled in the content of the colon: (1) the residues of the diet with microorganisms and their products, and (2) the secretions of the digestive tube and its glands. The proportion existing between these two is variable. The feces consist of water; the undigested and indigestible parts of the food; pigment due to undigested food or to metallic elements contained in it and to the bile pigments; great quantities of microorganisms of different kinds; the products of bacterial decomposition, i.e., indole, skatole, etc.; the products of the secretions; mucous and epithelial cells from the walls of the alimentary tract; cholesterol or a derivative, which is probably derived from the bile; some of the purine bases; inorganic salts of sodium, potassium, calcium, magnesium, and iron.

Defecation. The anal canal is guarded by an internal sphincter and an external sphincter muscle, which are normally in a state of tonic contraction and protect the anal opening. Normally the rectum is empty until just before defecation. Various stimuli (depending on one's habits) will produce peristaltic action of the colon, so that a small quantity of feces enters the rectum. This stimulates the sensory nerve endings and causes a desire to defecate. The voluntary contraction of the abdominal muscles, the descent of the diaphragm, and powerful peristalsis of the colon all combine to empty the colon and rectum. The normal frequency of defecation varies with the individual—some persons defecating once or twice a day, and others defecating once every two or three days.

One of the commonest causes of *constipation* is the retention of feces in the rectum because of failure to act on the desire for defecation. After feces once enter the rectum there is no retroperistalsis to carry them back to the colon. Sensory adaptation occurs and often persists for 24 hours, during which time the feces continue to lose water and become harder and more difficult to expel. A certain amount of indigestible materials in the diet stimulates the lining of the intestines and promotes peristalsis, as does daily exercise.

Constipation may be due to an increased tonicity of the distal part of the colon, which causes a decrease in the lumen of the colon. Onward peristalsis is impaired, haustration is extreme, and hyperirritability and motility with delayed evacuation result. This is called *spastic* constipation. Constipation also may be due to a relaxed state of the muscle layers of the colon. The muscles fail to produce sufficient peristaltic action. The colon becomes relaxed, distended with fecal accumulation. This is called *atonic* constipation. It may follow excessive use of cathartics.

Diarrhea results from irritation of the mucosa of the intestine from bacterial infection usually, or excessive parasympathetic stimulation in stress situations. In these instances large amounts of water and electrolytes are lost in the stool, since normal absorption fails to occur and the glands of the large intestine secrete these materials along with mucus. Diarrhea may also result from pathologic conditions that cause malabsorption of certain foodstuffs, notably fats.

Questions for Discussion

1. What are the chemical digestive processes necessary for rendering the food soluble and absorbable?
2. What are the functions of each hormone in the digestive process?
3. A person has had a total gastrectomy. What physiologic process will be interfered with and what medication will the person need? Why?
4. Explain the role of two vitamins and two minerals in enzyme systems.
5. Distinguish between a proenzyme and an enzyme. Name three proenzymes and state their function.
6. What roles do the circulatory system and nervous system play in digestion?
7. Name three cranial nerves and their influence on the digestive processes.

Summary

Nutrients
Any substance taken into the body
1. To yield energy
2. To provide material for growth of tissues
3. To regulate body processes

Classification
Nutrients, or food principles
Water
Carbohydrates
Fats
Proteins
Minerals
Vitamins

Water
Enters into composition of all tissues
Sources of water content of body
Beverages
Water contained in food
Water formed in the tissues

Carbohydrates

Monosaccharides
Contain one sugar group
Glucose, or dextrose
Fructose, or levulose
Galactose

Disaccharides
Contain two sugar groups
Sucrose, or cane sugar
Lactose, or milk sugar
Maltose, or malt sugar

Polysaccharides
Contain many sugar groups
Starch
Cellulose
Glycogen

Lipids

Simple
Fats
Oils
Composed of fatty acids and glycerol

Compound
Contain groups in addition to fatty acids, an alcohol

Proteins
Contain an amino (NH_2) group in subunits—amino acids
Amino acids classed as essential, nonessential
Proteins called adequate, inadequate

Classification

Simple
Consist only of amino acids
Yield only amino acids or derivatives

Conjugated
Contain protein molecule united to some other molecule other than as a salt—yield amino acids and some other molecule
Nucleoproteins—yield amino acids and nuclein
Glycoproteins—yield amino acids and a carbohydrate
Phosphoproteins—yield amino acids and phosphates
Hemoglobins—yield amino acids and hematin
Lipoproteins—yield amino acids and a fatty substance

Minerals
- Fifteen or more elements enter into composition of body
- Five may be furnished by carbohydrates, fats, proteins, and water
- Others to be provided include:
  - Iron, calcium, sodium, potassium, magnesium, phosphorus, chlorine, iodine, fluorine, silicon
- **Function**
  - As constituents of bone
  - As essential elements of soft tissues
  - As soluble salts held in solution in fluids of body
- Calcium—bones, teeth, coagulation
- Phosphorus—teeth, bones, nerve, muscle, as phosphate ion in all cells
- Iron—hemoglobin formation, cytochome enzyme system
- Copper—erythrocyte formation
- Magnesium—bones, cofactor in glucose metabolism
- Potassium—extracellular fluid, cardiac muscle action
- Iodine—thyroxin
- Cobalt—erythrocyte formation, as part of vitamin B_{12}
- Manganese—enzyme activity
- Zinc—enzyme activity
- Fluorine—teeth, prevention of decay

Vitamins
- Essential for growth and nutrition
- Influence the metabolism of foodstuffs, as coenzymes
- Current research stresses determination of optimal amounts
- **Vitamin A**
  - Vitamin A and the related carotenes essential to growth, and to nutrition and health at all ages
  - Recent clinical observation indicates desirability of securing optimal quantity
- **Vitamin D**
  - Gives protection in childhood against rickets
  - Ergosterol transformed into vitamin D by ultraviolet light
- **Vitamin E**
  - Deficiency results in sterility in both male and female rats
  - Its value in treatment of human sterility not yet proved
- **Vitamin K**
  - Essential to normal clotting of blood
- **Vitamin B group**
  - Of multiple nature
  - Essential to health, related to composition and activity of many enzyme systems
- **Vitamin C**
  - *Ascorbic acid*
  - Essential to normal development of bones and teeth, integrity of capillary walls, and wound healing
  - Subclinical shortage frequent; attention should be given to optimal intake

Digestive Processes
- Include various physical processes that are preliminary to the more important chemical digestion
- **Physical**
  - Mastication—comminution and mixing
  - Deglutition, or swallowing
  - Peristaltic action of esophagus, stomach, intestines
  - Dissolving of nutrients in digestive juice
- **Chemical**
  - Splitting of complex substance into simpler ones
  - Process of hydrolysis that is dependent on enzymes
  - Rendered necessary by variety and complexity of foods, which must be reduced to standard and simple substances that the tissues can use, i.e.,
  - End products
    - Simple sugars
    - Glycerol and fatty acids
    - Amino acids

Enzymes
- Substances produced by living cells that act by catalysis, i.e., vary speed of reactions
- Enzyme is designated by the name of the substance on which it acts, together with the suffix, *ase*
- **Classification according to action**
  1. Sugar splitting
     - a. Hydrolytic
     - b. To yield simple sugars
  2. Amylolytic, or starch splitting
  3. Lipolytic, or fat splitting
  4. Proteolytic, or protein splitting

Secretions
- Contain electrolytes, mucin as well as enzymes
- Formed in cells from materials obtained from tissue fluid
- Nervous control is autonomic, vasomotor and secretory effects

Changes Food Undergoes in the Mouth

Mastication (chewing)—comminution and mixing

Insalivation (mixing with saliva)

Saliva

Secreted by salivary glands
- Parotid
- Submaxillary   and mucous glands of mouth
- Sublingual

Craniosacral autonomic fibers
- Carry secretory and vasodilator fibers
- Stimulated by sight or smell of food
- Causes a production of a copious amount of watery secretion

Thoraco-lumbar autonomic fibers
- Carry secretory and vasoconstrictor fibers
- Stimulated by food in mouth
- Produce a smaller amount of thicker secretion

Consists of water, some protein material, mucin, inorganic salts, and the enzyme salivary amylase

Specific gravity 1.004–1.008. Neutral in reaction—pH 6.6–7.1

Amount—1 to 1.5 liters per day

Physiology
1. Assists in mastication and deglutition
2. Serves as lubricant
3. Dissolves or liquefies the food, thus stimulating the taste buds and indirectly the secretion of gastric fluid
4. Amylase hydrolyzes starch to dextrin and maltose; maltase changes maltose to glucose

Deglutition (swallowing). Passage of food through (1) fauces, (2) pharynx, and (3) esophagus. Consistency of food affects third stage

Changes Food Undergoes in the Stomach

Time required—depends on nature of food eaten; average meal of mixed food requires 3–4½ hr

Food held in stomach by cardiac and pyloric sphincters

Cavity size of contents—never empty—always few milliliters of gastric fluid in stomach

When food enters, expands just enough to receive it; contractions start in middle region and run toward pylorus; food in prepyloric and pyloric regions macerated, mixed with gastric fluid, and reduced to *chyme*

Salivary digestion continues until gastric fluid penetrates bolus of food

Gastric juice

Periods of fasting—secreted in small amount

While eating and during period of digestion—amount increased

Secretion
- Psychic or appetite, secretion
  - Sensations of eating
  - Taste and odor of food
- Chemical
  - Food in stomach stimulates gastrin secretion, which increases glandular secretions

Secreted by glands of stomach
- Cardiac
- Fundus, or oxyntic
- Pyloric

Acid reaction due to free hydrochloric acid

Enzymes
- Pepsin
- Gastric lipase

Hydrochloric Acid

Secreted by parietal cells of gastric glands from chlorides found in blood

Chloride ions combine with hydrogen ions to form hydrochloric acid

Normal amount about 0.5%

Function
- Activates pepsinogen and converts it to pepsin
- Provides acid medium for pepsin to carry on its work
- Swells protein fibers
- Hypersecretion associated with ulcer formation
- Germicidal in action

Pepsin

Formed in pyloric glands and chief cells of gastric glands

Pepsinogen—zymogen, changed by HCl to active pepsin

Weak proteolytic enzyme—requires acid medium

Hydrolyzes proteins through several stages to peptides, which action is preparatory to more complete hydrolysis in intestine

Gastric Lipase—limited action on emulsified fats like cream

Functions of Stomach

Serves as temporary storage reservoir

Contractions promote mechanical reduction of food

Salivary digestion continues until acidity is established

Gastric digestion
- Pepsin hydrolyzes proteins
- Gastric lipase may hydrolyze emulsified fats
- HCl has germicidal action

Digestion in the Intestine	Small intestine	Movements	Peristaltic—pushes food forward slowly Rhythmical—facilitates mixing with secretions
		Secretions present	Pancreatic juice Intestinal secretions Bile
		Bacteria	Decompose carbohydrates Little or no effect on protein
		Time required	Depends on proportions of different foodstuffs Food begins to pass into large intestine 2–5¼ hr after eating, requires 9 hr or more before last of meal has passed
		Enzymes	Enterokinase—acts as coenzyme Peptidases—hydrolyze peptides to amino acids Maltase—hydrolyzes maltose to glucose Sucrose—hydrolyzes sucrose to glucose and fructose Lactase—hydrolyzes lactose to glucose and galactose Nuclease—acts upon nuclei acid portion of nucleoproteins

Pancreatic Juice

Secretion controlled by hormones secretin and cholecystokinin. Discharged into small intestine during digestion
Clear, viscid fluid, alkaline reaction

	Secretion	Neural secretion caused by secretory fibers in vagus and splanchnic. It is thick, rich in enzymes Chemical secretion due to secretin and cholecystokinin is thin, watery, rich in enzymes, alkaline
Enzymes	Proteolytic	Formerly thought one enzyme, trypsin, reduced protein to amino acids. Now thought that *trypsin* and chymotrypsin are necessary and act at different stages
	Amylolytic or amylase	Action similar to that of ptyalin Hydrolyzes starch to maltose
	Lipolytic	Hydrolyzes fats to glycerol and fatty acids Emulsification aided by bile salts occurs as soon as small amount of fat is split to fatty acids and glycerol Fatty acid combines with alkaline salts. Emulsification regarded as preparatory process

Large intestine

Movements	1. Antiperistalsis—press mass backward toward colic valve 2. Haustral churning 3. Mass peristalsis moves food from one division to another

Secretion—contains mucin, no enzymes, alkaline reaction
Digestive enzymes from duodenum continue to act

Bacteria	Mainly putrefaction of proteins with formation of relatively toxic amines or less toxic substances as indole Synthesis of certain vitamins, including some members of the B group and vitamin K

Bile

Alkaline liquid, color may be yellow, brownish yellow, or olive green
Amount secreted varies with amount of food eaten, estimated at about 800–1000 ml daily
Consists of water, bile pigments, bile acids, bile salts, cholesterol, lecithin, neutral fats, and nucleoprotein
Bile salts, i.e., sodium taurocholate and sodium glycocholate, thought to stimulate activity of liver
Secreted continuously, enters duodenum during period of digestion
Digestive secretion aids action of lipase
Excretion—eliminates toxins, metals, and cholesterol
Antiseptic—thought to limit putrefaction

Abnormal Conditions

Gallstones—concentrated cholesterol or bile salts, which crystallize out and form gallstones
Jaundice—due to absorption of bilirubin into circulation from liver; deposited in skin and whites of eyes

Feces

Consist of	Residues of diet, microorganisms, and their products Excretions of digestive tube and its glands

Contain (1) water, (2) the residues of food, (3) pigment, (4) microorganisms, (5) products of bacterial decomposition, indole, skatole, etc., (6) products of secretions, (7) mucous and epithelial cells, (8) cholesterol, (9) purine bases, and (10) inorganic salts

Defecation—term applied to the act of expelling feces from rectum

Additional Readings

DAVENPORT, H. W.: Why the stomach does not digest itself. *Sci. Amer.*, **226**:86–93, January, 1972.

GUYTON, A. C.: *Basic Human Physiology: Normal Functions and Mechanisms of Disease*, 2nd ed. W. B. Saunders Co., Philadelphia, 1977, Chapters 42, 43, 44, and 48.

ROBINSON, CORINNE, and LAWLER, M. R.: *Normal and Therapeutic Nutrition*, 15th ed. The Macmillan Publishing Co., Inc., New York, 1977.

SCRIMSHAW, N. S., and YOUNG, V. R.: The requirements of human nutrition. *Sci. Amer.*, **235**:50–64, September, 1976.

SELKURT, E. E.: *Basic Physiology for the Health Sciences*. Little, Brown & Co., Boston, 1975, pp. 219–63.

VANDER, A. J.; SHERMAN, J. H.; and LUCIANO, D. S.: *Human Physiology: The Mechanisms of Body Function*, 2nd ed. McGraw-Hill Book Co., New York, 1975, Chapter 12.

CHAPTER 22

Absorption and Metabolism

Chapter Outline

The secretory and motor activities of the gastrointestinal tract are all directed toward changing ingested food into substances appropriate for absorption into the bloodstream, which then delivers these materials to cells. Food requirements differ in amount according to the individual, although there are basic requirements common to all.

Absorption

Absorption from the gastrointestinal tract consists of transfer of materials across the cell membrane boundary and involves processes previously discussed (Chapter 2), namely, diffusion, pinocytosis, osmosis, and active transport. Particle size and concentration of the materials, as well as lipid solubility, are among the impor-

tant physical factors that influence these processes.

Conditions that determine the amount of absorption that takes place from any part of the alimentary canal are the area of surface for absorption, the length of time nutrients remain in contact with the absorbing surface, the concentration of fully

514

digested material present, and the rapidity with which absorbed nutrients are carried away by the blood.

Absorption in the Stomach. There is no active transport mechanism in the stomach, so that absorption is limited to materials already of absorbable size that can diffuse across cell membranes. Practically speaking, little absorbs except substances with high lipid solubility, such as alcohol.

Absorption in the Small Intestine. It is in the small intestine that the greatest amount of absorption takes place. The circular folds, villi, and microvilli of the small intestine increase the internal surface enormously and food remains in the small intestine for several hours; during this time the most complete digestive changes occur.

It must be remembered that blood in the capillaries is separated from the digested nutrients in the small intestine by the walls of the capillaries and the intestinal mucosa. On the intestinal side of the wall are the products of digestion and the digestive fluids. Sugars, glycerol, fatty acids, and amino acids are relatively abundant and pass into the blood. The continuous digestion of foods, the muscular activity of the intestinal wall, and the movement of the villi stir up the intestinal contents and keep relatively high the concentration of absorbable materials in contact with the absorbing membrane. These motions also increase the circulation in the villi, and therefore the absorbed materials are moved on, keeping the concentration in the blood relatively low. Absorption takes place through the membrane from a constantly higher concentration of absorbable particles to a constantly lower concentration in the circulating blood of the mucosa, until virtually all absorbable materials are absorbed.

With the exception of fatty acids and glycerol, all absorbed nutrients enter the capillaries of the mucosa and from these enter the portal circulation (see Figure 22–1). Fatty acids enter the lacteals from the epithelial cells of the mucosa and enter the lymph vessels communicating with the cisterna chyli (Figure 16–18, page 377).

When amino acid content in the lumen of the small intestine is high, some enter the lacteals as well as the capillaries.

Absorption of monosaccharides is facilitated by an active transport system, which has preference for glucose and galactose.

Absorption of amino acids also is facilitated by an active transport system in the small intestine.

Fatty acids and *glycerol* are present in the intestinal contents, dissolved in the droplets of bile salts; when the fatty acid and glycerol enter the epithelial cell of the mucosa, the bile salt remains in the lumen to combine with other fatty acids and glycerol. (*Bile salts* are absorbed in the terminal portion of the ileum.) In the epithelial cell, the fatty acids and glycerol are recombined to form triglycerides, then enter the lacteals and lymph circulation. In the lymph fat is present as tiny droplets called *chylomicrons*, groups of triglycerides with a protein coat surrounding each droplet.

Absorption of water occurs in the small intestine as a result of diffusion, but the greatest change in fluidity of intestinal contents is in the large intestine.

Absorption of electrolytes is aided by active transport for most substances. *Sodium* is actively transported out of the mucosal epithelial cells into the bloodstream, in the upper small intestine. *Chloride* ions passively diffuse along with the sodium ions; in the distal ileum and the large intestine there is active absorption, in exchange for bicarbonate ions within the epithelial cells. *Iron, potassium, magnesium,* and *phosphate* ions are all actively transported in the small intestine, as is *calcium.* Parathormone and vitamin D facilitate calcium absorption.

Absorption in the Large Intestine. When the contents of the small intestine pass into the cecum, they still contain a certain amount of unabsorbed material. Enzymes are present and digestion and absorption continue, but to a greatly reduced degree. The consistency is about that of chyme, because absorption of water in the small intestine is counterbalanced by secretion of water in the digestive juices. Water is

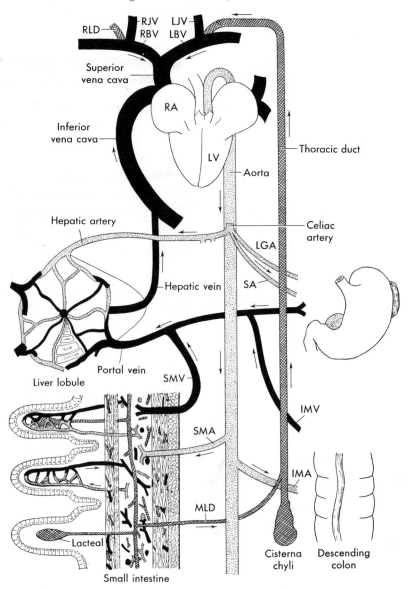

Figure 22–1. Diagram to show absorption. *IMA*, inferior mesenteric artery; *IMV*, inferior mesenteric vein; *LGA*, left gastric artery; *LBV*, left brachiocephalic vein; *LJV*, left jugular vein; *LV*, left ventricle; *MLD*, mesenteric lymph duct; *RA*, right atrium; *RBV*, right brachiocephalic vein; *RJV*, right jugular vein; *RLD*, right lymph duct; *SA*, splenic artery; *SMA*, superior mesenteric artery; *SMV*, superior mesenteric vein.

absorbed primarily in the ascending colon and the proximal portion of the transverse colon. Under usual conditions the result is the formation of semisolid or solid feces.

Cathartic Action of Salts in Solution. The "cathartic" salts are very slowly absorbed from the gastrointestinal tract. For example, the cation magnesium and the anions citrate, tartrate, sulfate, and phosphate are slowly ab-

sorbed. When taken as salts, they remain in the intestine for a comparatively long period of time. The tissues between the salts in solution and the bloodstream form a semipermeable membrane, which is readily permeable to water. Consequently fluid moves from the bloodstream to the intestinal lumen, thereby increasing the bulk of the intestinal content. The increased bulk acts as a distention stimulus to promote increased peristalsis and bowel evacuation.

It is important to remember that all materials entering the bloodstream from the intestine pass through the liver before reaching the heart and systemic circulation. (See Figure 20–22, page 473.) The liver has a significant regulatory influence on the amounts of fat, protein, and carbohydrate appearing in the blood delivered to the cells. It also clears the bloodstream of alcohol through oxidation.

Cellular Metabolism

The general term *metabolism* refers to all the changes occurring in digested foodstuffs in the body from their absorption until their elimination in the excretions. In actuality metabolism is the *sum total of the chemical changes* taking place within cells. Metabolic changes include both *anabolism,* or constructive process or synthesis, and *catabolism,* which implies the breakdown of large molecules, the products of which are of smaller molecular size.

The changes classified as anabolic include the processes by which cells take food substances from the blood and make them a part of their own protoplasm. This involves the conversion of nonliving material into living material and is a building-up, or synthetic, process. The synthesis of glycogen and of fats within the cells is also anabolism. Anabolism is accompanied by the *storage* of chemical energy.

The changes classified as catabolic consist of the processes by which cells resolve into simpler substances (1) part of their own protoplasm or (2) substances that have been stored in them. This disintegration yields simpler substances, some of which may be used by other cells, though most of them are excreted. *Release* of chemical energy accompanies this process.

In the tissues, the participation of oxygen in the chemical changes of the body forms an integral part of the processes of metabolism.

The chemical reactions occurring within the cell have two overall purposes:

1. Forming molecules that make up the cell itself, e.g., cell membrane, endoenzymes, and cytoplasm, as the cell increases in size or divides; and forming molecules to be secreted from the cell, e.g., hormones, exoenzymes and serum albumin and globulin.

2. Supplying energy for the synthetic processes listed above and for such specialized cell activities as muscle contraction, transmission of nerve impulse, ciliary movement, and sperm motility.

Fats and carbohydrates are primarily used for energy supply, although they are also necessary for synthesis of such materials as glycoproteins and cholesterol. Amino acids are primarily utilized in synthesis but may be utilized as energy sources, if fat and carbohydrate supply is inadequate. Energy supply seems to have priority over anabolism.

The chemical reactions within the cell that release energy occur in the mitochondria and are primarily oxidative reactions, that is, transfer of electrons or hydrogen from one substance to oxygen or to another substance. Oxidation may then occur aerobically (utilizing oxygen) or anaerobically. The aerobic oxidative processes yield the largest amount of energy.

Strictly speaking, with the exception of heat produced, energy is not "freed" from a substance but rather transferred. One of the most important substances in the cell in regard to energy transfer is adenosine triphosphate. To synthesize this substance, present in all cells, energy is used to form adenosine from ribose and adenine, and then to attach one, two, or three phosphate groups, adenosine monophosphate (AMP), adenosine diphosphate (ADP), and adenosine triphosphate (ATP). Seven kilocalories of energy are required to attach the second and third phosphate groups; this same amount of energy is released when the phosphate groups, through enzymatic action, are broken off during cell activity. Oxidation in the cell is thus a step-by-step reaction, producing small amounts of energy for storage in ATP molecules, which are held in reserve for utilization as needed. At the same time, utilization of ATP is a safety factor, for the release of excessive amounts of energy at one time would destroy the cell.

Muscle cells, in addition to ATP, synthesize another compound with a high-energy phosphate bond, creatine phosphate, which acts as an energy storehouse like ATP until the muscle contracts and energy is needed.

Needless to say, none of the chemical reactions within the cell, energy transferring, or of a synthetic nature could occur rapidly enough in the absence of the appropriate enzymes. Reference to the enzyme chart on page 20 and the coenzyme chart on page 496 will be helpful in the discussions to follow.

CARBOHYDRATE METABOLISM

Absorption of glucose takes place mainly into the capillaries of the small intestine. These capillaries pour their contents into the portal vein, which carries the blood, rich with glucose, to the liver. The liver cells take this glucose from the blood and convert it to glycogen, which is stored in the liver cells. By the storing up of glycogen and doling out of glucose as needed, the liver helps to maintain the normal quantity of glucose—80 to 120 mg per 100 ml of blood. From the bloodstream glucose is taken up by the skeletal muscles and stored as glycogen until needed, or it is utilized by any and all cells as a source of energy or for synthesis of appropriate cell constituents and secretions. Hence the liver functions to help maintain a constant supply of blood glucose to meet the demands of active cells. The percentage of glycogen in a muscle is small, though the total content of all the muscle cells may equal that of the liver. The maximum storage of glycogen in the body is about 400 gm, or nearly 1 lb. The need of the blood for glucose is constant, because it is constantly giving up glucose to the tissues.

Another action of the liver in relation to carbohydrate metabolism is to transform fructose and galactose to glucose.

Carbohydrate Oxidation. In all cells, the amount of glucose used is determined by its energy needs. Energy is released from glucose in the process of oxidation. The first stage of oxidation is *glycolysis*, which is outlined in Table 22-1. The first step is transferring a phosphate group to glucose (*phosporylation*), from ATP, a reaction that is catalyzed by the enzyme *hexokinase*. All cells contain this enzyme. Glucose phosphate then may be transformed to glycogen (*glycogenesis*) in the case of liver or skeletal muscle cells; or glycolysis may occur—the breakdown of glucose for energy release.

Glycolysis. This first stage of glucose oxidation is aerobic. The six-carbon glucose molecule is slightly modified in each step, splitting into three-carbon molecules at the midpoint: 3-phosphoglyceraldehyde and dihydroxyacetone phosphate. Only the first progresses to pyruvic acid, with the production of two molecules of ATP. Dihydroxyacetone phosphate is transformed to 3-phosphoglyceraldehyde, and again the sequence continues to pyruvic acid, with the production of two more molecules of ATP. In this way, there is a net gain of two molecules of ATP, with replacement of the two ATP molecules used in the early stage.

During glycolysis, the enzyme *nicotinamide adenine dinucleotide* (NAD) is reduced, by the addition of two hydrogen atoms, to $NADH_2$. The hydrogen will be transferred to an enzyme system in the mitochondria, called the cytochrome system, and combined with oxygen to form water. In the absence of oxygen, the hydrogen can be used in forming lactic acid. With release of hydrogen NAD is reformed, to participate in further glycolysis.

The pathway of glycolysis from glucose-6-phosphate to pyruvic acid is reversible, and any of the intermediates can be used to form glucose-6-phosphate, if this should be needed by the cell. However, only the liver has the enzyme *glucose*

TABLE 22–1
OXIDATION OF GLUCOSE

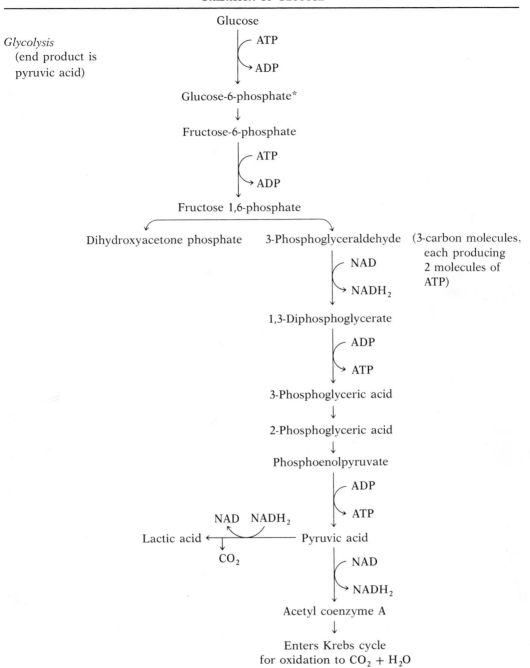

Glucose

Glycolysis
(end product is
pyruvic acid)

ATP → ADP

Glucose-6-phosphate*

↓

Fructose-6-phosphate

ATP → ADP

Fructose 1,6-phosphate

Dihydroxyacetone phosphate ← → 3-Phosphoglyceraldehyde (3-carbon molecules, each producing 2 molecules of ATP)

NAD → NADH$_2$

1,3-Diphosphoglycerate

ADP → ATP

3-Phosphoglyceric acid

↓

2-Phosphoglyceric acid

↓

Phosphoenolpyruvate

ADP → ATP

Lactic acid ← Pyruvic acid
NAD   NADH$_2$
CO$_2$

NAD → NADH$_2$

Acetyl coenzyme A

↓

Enters Krebs cycle
for oxidation to CO$_2$ + H$_2$O

*Numbers refer to the carbon atoms that carry the phosphate group. Glucose has six carbons, numbered as follows:

$$H-\overset{O}{\underset{H}{C}}-\overset{OH}{\underset{OH}{C}}-\overset{H}{\underset{H}{C}}-\overset{OH}{\underset{H}{C}}-\overset{OH}{\underset{H}{C}}-\overset{OH}{\underset{H}{C}}$$

1   2   3   4   5   6

phosphatase, which will convert glucose-6-phosphate to glucose.

The next stage in glucose metabolism is the oxidation stage, in which pyruvic acid is changed to acetylcoenzyme A and enters the citric acid (Krebs) cycle (Figure 22–2, page 526), a pathway for oxidation of certain amino acids, fatty acids, and glycerol.

Hormones Concerned with Carbohydrate Metabolism. The complex processes by which glucose is utilized—glycolysis, glycogenesis, glycogenolysis, and gluconeogenesis—require regulation. Certain hormones are known to be important.

Insulin is concerned with the diffusion of glucose across tissue cell membranes from the extracellular fluids. In the liver, insulin inhibits glycogen formation from noncarbohydrate sources. *Glucagon* promotes the conversion of glycogen to glucose and gluconeogenesis in the liver.

Somatotropin and *adrenocorticotropin* are insulin antagonists and raise the blood sugar level; i.e., they inhibit glucose phosphorylations by inhibiting hexokinase and glucose utilization in the tissues.

The *glucocorticoids* decrease the use of tissue glucose, increase blood sugar, increase gluconeogenesis from amino acids, and increase the production of glucose from glycogen.

Epinephrine causes a rapid conversion of liver glycogen to glucose and increases the rate of use of glycogen in muscle tissue.

Thyroxin influences glucose metabolism by increasing the rate of oxidation in tissue cells. In the liver it increases the rate of glycogen formation from noncarbohydrate sources. At the same time it increases glucose absorption from the intestine, so that its immediate effect is to raise blood sugar level.

Functions of Carbohydrates. The oxidation of glucose serves the following purposes: It furnishes the main source of energy for all cells. The nerve cells use glucose only for energy; thus, it is critical for brain functioning. Glucose furnishes an important part of the heat needed to maintain the body temperature. The oxidation of each gram of glucose yields 4 Cal[1] of heat; and since the carbohydrates form the largest part of our diet and are easily oxidized, they must be regarded as specially available material for keeping up body heat. Glucose prevents oxidation of the body tissues, because it constitutes a reserve fund that is the first to be drawn upon in time of need. It is "protein sparing," for protein will be used as the energy source if glucose is not available. Carbohydrates, in excess of the amount that can be stored as glycogen in the liver and muscles, are converted into depot or stored fat.

End Products of Carbohydrate Metabolism. Eventually the glucose derived from the glucose of the blood or from the glycogen of the cell is oxidized by the cell, via the citric acid cycle, to *carbon dioxide* and *water.*

Derangements of Carbohydrate Metabolism. Ingestion of a larger amount of sugar than the liver and muscles can store results in an increased amount in the blood (*hyperglycemia*). A higher percentage of glucose than normal in the blood is excreted in the urine. This is designated as temporary glycosuria.

DIABETES MELLITUS. The most common derangement of carbohydrate metabolism is diabetes mellitus, in which the pancreas fails to produce insulin, or produces inadequate amounts. As a result, glucose level in the bloodstream is high (hyperglycemia); it is excreted in the urine (*glycosuria*); it causes loss of large amounts of water to be excreted in the urine. The person is therefore thirsty and drinks large amounts of water (*polydipsia*). Since carbohydrates are not metabolized, fat and protein stores are broken down, the fat metabolism leading to production of ketone substances that are acidic in nature and lead to acidosis—*diabetic acidosis.* The treatment, of course, is to give exogenous insulin.

LIPID METABOLISM

When the lymph enters the bloodstream, the triglycerides enter also and

[1] The *large* calorie, abbreviated Cal, is the amount of heat required to raise 1 kg of water 1° C. It is the same as the kilocalorie (kcal).

TABLE 22–2
FACTORS THAT INFLUENCE BLOOD GLUCOSE LEVEL*

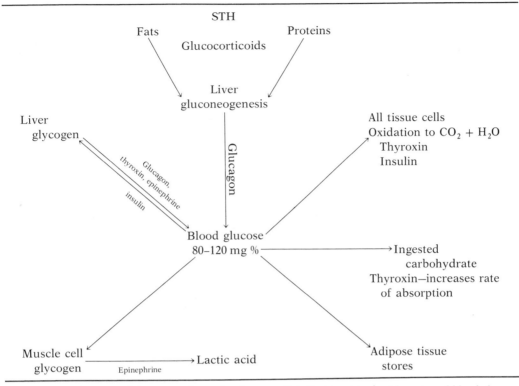

*Note that the direction of the arrows (inward and outward) indicates raising and lowering of blood glucose.

reach all cells. (In the blood they are referred to as *neutral fats.*) Complete oxidation may occur, or adipose tissue is formed as a means of storing energy-supplying material.

As with glucose, the liver has many functions related to lipid metabolism. It synthesizes triglycerides from carbohydrate sources, and to a lesser extent from amino acids; it synthesizes other lipids from fatty acids, notably cholesterol and phospholipids; it is a major source of triglyceride degradation for energy.

Triglyceride Oxidation. All cells (with the exception of brain cells, which employ only glucose) use triglycerides for energy. The first step in oxidation of triglycerides is the splitting of glycerol from the fatty acid portion of the molecule, with the assistance of the enzyme *lipase.* Glycerol is a three-carbon molecule that, under the influence of intracellular enzymes, enters the glycolytic pathway.

The long chains of fatty acids are broken down sequentially in the mitochondria by the breaking-off of two-carbon fractions, each of which is used to form acetylcoenzyme A—which enters the Krebs cycle of oxidation. The enzymes NAD and FAD (flavine adenine dinucleotide) are necessary for this oxidation, as is one molecule of ATP. Both NAD and FAD are reduced (to $NADH_2$ and $FADH_2$), which subsequently release the energy to form ATP. The breakdown of a gram of neutral fat releases about three times the amount of energy that a gram of glucose does.

In the liver triglyceride oxidation progresses at such a rapid rate that excess fatty acids are produced—too many for the metabolic processes of the liver cells. As a result some fatty acids are metabolized to acetoacetic acid, beta-hydroxybutyric acid, and acetone (the "ketone bodies"), which enter the circulation and are metabolized in other tissue cells, notably

TABLE 22–3
SUMMARY—IMPORTANT STEPS IN CARBOHYDRATE METABOLISM

Sources of Carbohydrates	Role of the Liver	Tissue Cells
Digestion of carbohydrates to monosaccharides in small intestine: glucose fructose galactose	→ Taken to liver via portal vein and converted to glycogen, which is stored and subsequently converted to glucose as needed by cells	→ To all cells of the body for oxidation for energy by way of the citric acid cycle
Products of muscle activity lactic acid pyruvic acid	→ Conversion to glycogen	→ Stored as glycogen in muscle cells
Liver activities: Conversion of monosaccharides into glycogen Conversion of fatty acids into glucose and glycerol Conversion of amino acids into glucose	Synthesis of glycogen from noncarbohydrate sources (fats, amino acids) Stored or converted to glucose as needed by cells	→ Taken to cells by bloodstream
	Synthesis of fatty acids from glucose	→ Stored as adipose tissue
	Glucose	→ Synthesis of lactose in mammary glands
		→ { Glycolipids / Glycoproteins / Nucleic acids } All cells

skeletal muscle. When carbohydrate metabolism is inadequate, as in starvation and in lack of insulin (*diabetes mellitus*), more triglycerides are metabolized, and the "ketone bodies" are produced too rapidly for utilization; they then appear in large amounts in the bloodstream and in urine.

Certain lipotropic factors, such as choline, function in the liver to prevent excessive accumulation of fat in this organ, mainly through the formation of lipoproteins. Choline is obtained in the diet, as well as synthesized from the amino acid methionine.

Hormonal Influences on Lipid Metabolism. *Glucocorticoids, somatotropin, ACTH, epinephrine,* and *norepinephrine* cause fat mobilization and increase the rate of lipid oxidation. *Thyroxin* mobilizes fat by increasing the overall energy metabolism of cells.

Functions of Lipids. Much of the fat absorbed from the intestine is deposited in fat-storage cells, which are widely distributed throughout the body. From these cells fat is constantly being withdrawn to meet the demands of active cells. Some is oxidized to provide energy, some is used for synthesis of lipids for cell use. Each gram of fat yields 9 Cal of heat; thus fat is a valuable source of body heat.

When the food eaten is in excess of the energy requirements of the individual, fat accumulates in the storage cells, thus increasing the amount of adipose tissue. The fat stored in adipose tissue is in a dynamic state; i.e., it is continually formed (*lipogenesis*) and continually being broken down (*lipolysis*). Adipose tissue is a valuable reserve of nutrient and also provides protection and support for organs and insulation for the animal. It is used to form essential tissue and cell components, particularly in the nervous system (myelin and the phospholipids).

Obesity results from ingestion of excessive amounts of nutrients, particularly carbohydrate and fat. It imposes demands on the heart, which may be harmful, and in general shortens the life-span. As with many diseases, prevention is easier than cure. Habits formed early in life that match food ingestion to energy needs and the decrease of caloric intake when adulthood is reached (and there is less physical activity) help to prevent this condition.

Experimental work has recently been devoted to the relationship of heart attacks and deposits of lipids, including cholesterol, in the walls of arteries (*atherosclerosis*). It is known that high levels of serum lipids lead to increased deposits of lipids in the arterial wall.

How to decrease these high levels has not been reliably established, although fatty acids containing many "unsaturated" or double bonds between carbon atoms seem more effective than fatty acids that are "saturated" (no double bonds). The essential fatty acids (those that must be present in the diet) are all unsaturated—linoleic, linolenic, and arachidonic acids. Although heart attacks do appear to be related to the deposits of lipids in arterial walls, there still is no concrete evidence that changing from a diet high in "saturated" fatty acids to one high in "unsaturated" fatty acids is beneficial.

TABLE 22–4
SUMMARY—IMPORTANT STEPS IN LIPID METABOLISM

Sources of Fat	Bloodstream	Role of the Liver	Tissue Cells
Foods Digestion of fats to fatty acids and glycerol in the intestine	Found as neutral fats in form of chylomicrons (fat droplets visible under the microscope)	Stores phospholipids and glycerides	Good source of energy by way of citric acid cycle to CO_2 and H_2O—9 Cal per gm of fat
		Oxidation of fatty acids Formation of unsaturated fatty acids	Necessary constituent of all cells
	Phospholipids	Phospholipids formed at a higher rate than in any other organ	Phospholipids and cholesterol essential constituent of all cells Lecithin—essential for formation of myelin sheaths
Cells Mobilization of fat from adipose tissue cells	Cholesterol	Formation of cholesterol	Cholesterol found in all living cells. It is essential for the formation of steroid hormones in the adrenal cortex and gonads
	Cholesterol esters	Cholesterol esters formed and distributed to the cells or excreted in the bile	Cholesterol enters into basic structure of the cell and is essential for the normal permeability of cell membranes
Liver Synthesis of lipids from: Glucose Acetic acid Amino acid Pyruvic acid	Ketone bodies	Ketone bodies formed in the liver in the course of oxidation of fatty acids	Ketone bodies oxidized in muscle and other cells. Excess excreted by kidney

PROTEIN METABOLISM

Amino acids, which are absorbed by the blood capillaries of the villi, pass into the portal vein, are carried through the liver into the blood of the general circulation, and are distributed to the tissues. The tissues select and store certain of these substances; and in each organ they are either synthesized into new tissue or used to maintain and repair tissue. Such synthesis of tissue protein, or of materials to be secreted from the cell such as lipoproteins, is under the direction of RNA.

Amino acids not used in synthesis of protoplasm are broken down or deaminized in the liver. In *deaminization,* the amino groups are removed from amino acid molecules; *transamination* involves the transfer of the amino group from an amino acid to another organic acid. These NH_2 groups may be used for synthesis of other amino acids or for formation of urea by the liver. In the kidney, ammonia is formed from certain amino acids and is used for the production of ammonium salts, thus sparing the sodium ions of the blood in maintaining acid-base balance (page 566).

In skeletal muscle amino acids are utilized to form creatine and creatine phosphate, substances that have an active role in muscle contraction.

The nonnitrogenous portion of the amino acid molecule that is left after deamination is oxidized to furnish energy (see Krebs cycle, page 526) or is synthesized into glycogen or into fat. Therefore, this portion of the amino acid molecule may be regarded as a source of energy. Oxidation of amino acids is not complete, for nitrogenous substances appear in the urine: urea, uric acid, and creatinine. Thus, some energy is lost.

Certain amino acids have physiologic roles that are distinctive. *Glycine* is used by the liver in detoxification of benzoic acid, a food preservative, and in formation of hemoglobin, one of the bile acids, purines, and fatty acids. *Methionine* reacts with niacin and participates in formation of phospholipids, ergosterol, and histamine and in the detoxification of certain poisons such as chloroform and carbon tetrachloride. *Tryptophan* is a precursor of niacin, one of the B-complex vitamins.

Hormones Influencing Protein Metabolism. *Somatotropin* stimulates cell and tissue formation. *Androgens,* particularly testosterone, have a similar action; this is noticeable at puberty when the muscles increase in size, and when the overall size increases. *Insulin* is necessary for somatotropin effect, and it increases protein synthesis by promoting glucose utilization for energy in the cell, thus sparing amino acids. *Glucocorticoids* decrease the amount of protein in extrahepatic tissues, but increase liver and plasma proteins. *Thyroxin,* by increasing cell metabolic rate, increases the rate of protein synthesis, *provided* that adequate fat and carbohydrates are available for energy. Otherwise, protein is broken down to provide energy for the increased metabolic rate.

Nitrogen Equilibrium. Nitrogen continues to be excreted in the urine even though the diet is devoid of nitrogen. This represents a condition in which the body is oxidizing its own tissues to supply its needs. The nitrogenous portion of the protein molecule of ingested proteins is not stored in the body but is eliminated chiefly in the urine in the form of urea and to a limited extent in the feces. It is therefore important that the body receive daily an amount of protein nitrogen equal to the amount eliminated in the excreta. When this condition exists, the body is said to be in nitrogen equilibrium. If there is a *positive balance,* it means that protein is being made into body protoplasm; and this is an ideal condition during the period of growth or convalescence from wasting illness. If the balance is *negative,* it means that the body is oxidizing its own protein. Minimum nitrogen equilibrium can be maintained on about 40 gm of protein or less per day, but it is thought that higher protein intake results in greater resistance to disease and a higher state of physiologic

TABLE 22–5
SUMMARY—IMPORTANT STEPS IN PROTEIN METABOLISM

Sources of Protein	Role of the Liver	Tissue Cells
Digestive processes Digestion of protein foods to amino acids —→	→ Taken to liver via portal vein Amino acids —————————→	⌐ To all cells of body for maintenance, growth, and repair of tissues
		Nucleoproteins in cytoplasm and in nucleus of all cells
	Synthesis of: Serum albumin Serum globulin Fibrinogen Prothrombin	
		⌐ Synthesis of certain hormones and enzymes
	Synthesis of essential nitrogen containing nonproteins: Choline Purines Creatine Pyrimidines	
	Deaminization of amino acids, conversion of amino acids into glucose and fats	
	Synthesis of amino acids from other amino acids	Excess amino acids oxidized by way of citric acid cycle to carbon dioxide and water
	Formation of keto acids ——→	Oxidized by way of citric acid cycle to carbon dioxide and water

efficiency. It is customary to add 50 per cent to the average indicated as the actual requirement based on laboratory experiments, thus bringing the amount up to 60 to 100 gm of protein per day, which is somewhat more than 1 gm per kilogram of body weight.

Functions of Proteins. The major function of protein absorbed from the small intestine is the formation of new protein: new cells, tissue repair, and replacement of protein components of the cell structure. Secretions from the cell (hormones, digestive enzymes, mucin), intracellular enzymes, many blood coagulations factors—these all require amino acids for their synthesis.

Protein oxidation yields 4 Cal of heat; this oxidation occurs only if carbohydrate and lipid sources are inadequate. Thus, adequate protein intake is a necessity, and it must be accompanied by carbohydrate and fat.

KREBS[2] (CITRIC ACID) CYCLE

The final stage in oxidation of glucose, fatty acids, glycerol, and amino acids occurs in the mitochondria. It is a complex series of chemical reactions in which two-carbon fragments combine with coenzyme A to form acetyl coenzyme A, which in turn helps to form citric acid (Figure 22–2), and the coenzyme A is released to function again. By a series of steps the three carbon atoms in the pyruvic acid molecule are converted to three molecules of carbon dioxide, and hydrogen is transferred to carriers. When the hydrogen combines with oxygen in the

[2] Hans Krebs, English biochemist (1900–00).

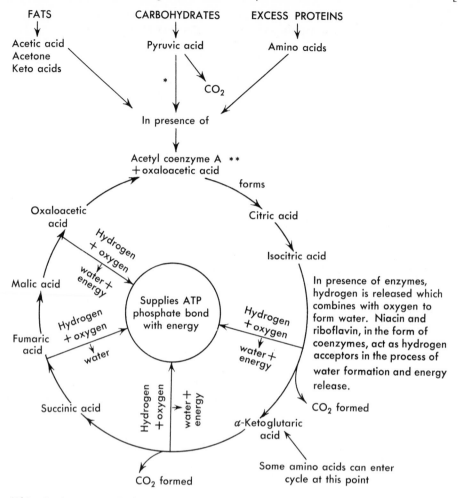

FATS　　　　CARBOHYDRATES　　EXCESS PROTEINS

Acetic acid　　　Pyruvic acid　　　Amino acids
Acetone
Keto acids

CO_2

In presence of

Acetyl coenzyme A **
+oxaloacetic acid

forms

Oxaloacetic
acid

Citric acid

Isocitric acid

Hydrogen
+ oxygen

water +
energy

Malic acid

Supplies ATP
phosphate bond
with energy

Hydrogen
+ oxygen

In presence of enzymes,
hydrogen is released which
combines with oxygen to
form water. Niacin and
riboflavin, in the form of
coenzymes, act as hydrogen
acceptors in the process of
water formation and energy
release.

Hydrogen
+ oxygen

water +
energy

Fumaric
acid

water

CO_2 formed

Hydrogen
+oxygen

water +
energy

Succinic acid

α-Ketoglutaric
acid

Some amino acids can enter
cycle at this point

CO_2 formed

*Thiamine is necessary in the coenzyme which participates in pyruvic acid transformation.
**Coenzyme A contains pantothenic acid in its molecule.

Figure 22–2. Citric acid cycle. Using fats, carbohydrates, and excess amino acids as fuel, the citric acid cycle supplies energy-rich bonds to adenosine triphosphate. A series of acids are converted one to another by enzymes with release of carbon dioxide and transfer of hydrogen to oxygen, forming water. The energy released is used to form *ATP* and to provide body heat. *ATP* is essential for all muscle contraction and is also found in all cells of the body. The B vitamins in the form of coenzymes are an integral part of the cycle.

cytochrome system of the mitochondria, ATP is formed: each two hydrogens lead to a synthesis of three molecules of ATP. Each glucose molecule oxidized yields a total of 38 ATP molecules; each glycerol molecule oxidized yields a total of 22 molecules of ATP. The ATP yield from fatty acid oxidation depends on the length of the chain, but it is much greater than that for glucose as the chains are very long. For example, stearic acid, the fatty acid found in beef fat, contains *18* carbons.

Not all the energy released in cellular metabolism is transferred to ATP. In the complete oxidation of glucose, for example, 39 per cent of the potential energy is used to form ATP; 61 per cent of the energy is released as heat.

Interconversion of Nutrients. Amino acids, triglycerides, and glucose can enter Krebs cycle through some metabolic intermediate. Glucose can be converted to triglycerides or amino acids by means of such common intermediates as pyruvic acid, acetyl coenzyme A, and alpha-keto-glutaric acid. Some amino acids can be converted into glucose. Glycerol can be converted to glucose, but fatty acids cannot because of the irreversibility of pyruvic acid → coenzyme A reaction.

Basal Metabolism

Energy released from breakdown of amino acids, lipids, and glucose appears as heat, or is incorporated in ATP and subsequently used for work: muscle contraction, cell secretion, and so forth. (During growth periods there is storage of energy.) Some of the energy in ATP will appear as heat also; e.g., as muscles contract heat is produced, and friction of blood flowing against resistance produces heat. Heat, of course, is necessary in maintaining the body temperature, but much of the heat is lost from the surface of the body.

In measuring the metabolic rate, it is simplest if the individual is at complete rest—not contracting muscles, not digesting and metabolizing food. The energy expenditure then is equivalent to heat production. *Basal metabolic rate* is determined when the subject is resting quietly, in a room of comfortable temperature, 12 to 18 hours after a meal. The *direct method* involves placing the subject in a special chamber, called a calorimeter, for a specified period of time; the temperature change in the calorimeter is equivalent to the heat produced by the subject. The *indirect method* measures the amount of oxygen used during a period of time. Heat calculation is based on the assumption that catabolism of food in the body produces the same amount of energy (heat) as in the laboratory. The average is 4.8 Cal/liter of oxygen, averaging the specific values for fat, carbohydrate, and protein, and assuming an average diet. This is a method that is accurate enough for usual purposes. The basal metabolic rate is determined by multiplying the oxygen, in liters, consumed times 4.8.

What the normal basal metabolic rate should be can be determined from tables. Calculations based on surface area of the body are more accurate than those based on body weight, since heat loss increases in proportion to surface. Using Figure 22–3, a woman 5 ft 4 in. tall, weighing 56 kg, has a surface area of 1.6 sq m. The normal basal metabolic rate for this woman should be 36.9 Cal/sq m/hr.

Variations in Basal Metabolism. A number of factors, such as age, sex, sleep, and thyroid hormone, influence the basal metabolic rate. In women the rate is a little lower than in men; it gradually decreases with age; it may be increased by systematic exercise over a long period; prolonged undernourishment reduces it. All the preceding relate in great degree to the amount of muscle tissue in the individual, which has a higher metabolic rate than adipose tissue; certain races appear to have a slightly higher rate than others. Increased thyroxine secretion raises the basal metabolic rate; decreased thyroxine level lowers the rate.

Emotional tension increases it, but the individual then is not basal—i.e., not completely relaxed. Temperature has a marked effect on the metabolic rate. Fever causes an increase—often beyond the energy supplied in the diet, and body tissues are utilized for energy. Similarly, a decrease in body temperature decreases metabolic rate and therefore decreases the amount of oxygen required.

Daily Caloric Requirement. In order to maintain weight, with neither gain nor loss, the daily output of calories (as determined by calorimeter tests) must be balanced by calories taken in. On such data the daily food requirement in terms of calories is based. Since basal metabolism represents the heat given off when physiologic work is at a minimum (in the morning after a comfortable night's rest, re-

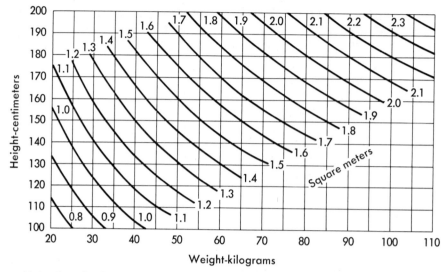

Figure 22–3. Chart for determining surface area of adults from weight and height. Height in inches divided by 0.393 gives height in centimeters. Weight in pounds divided by 2.2 gives weight in kilograms. (Courtesy of Dr. E. Du Bois, and the *Archives of Internal Medicine.*)

laxed in bed, before breakfast), it is obvious that any increase in physiologic activity, even the slightest (sitting up under the same conditions), increases the metabolic rate over the minimum. Comparison of the basal metabolic rate with the metabolic rates during various types of work shows that work results in an increment in the amount of heat eliminated. Such increases have been estimated and graded for the average individual according to the calories required, as shown in Table 22–6.

Total Caloric Requirement. The metabolic rate is reduced about 10 per cent beyond the basal level during sleep. Allowance is made for this in estimating the total daily calorie requirement of an individual. On the other hand, the process of digestion of food itself brings about a need for an increase of 6 to 10 per cent of the calorie intake. This is called the *specific dynamic action of food* (S.D.A.), and allowance must be made for it. Protein S.D.A. is higher than that for fat or carbohydrate; thus weight-losing diets contain a high proportion of protein.

TABLE 22–7
DAILY CALORIC REQUIREMENT
FOR ADULT MALE

	Calories
8 hours of sleep at 65 Cal	520
2 hours of light exercise at 120 Cal	240
8 hours of moderate work at 175 Cal	1400
6 hours sitting at rest at 100 Cal	600
	2760
6–10 per cent for S.D.A.	250
Total requirement for 24 hours	3010

TABLE 22–6
EFFECT OF WORK ON CALORIC REQUIREMENTS

	Calories Per Hour
Very light work, or sitting at rest	100
Light work	120
Moderate work	175
Severe work	350

TABLE 22–8
APPORTIONMENT OF 3000-CALORIE DIET

	Calories	Approximate % of Total Calories	Grams
Carbohydrate	1440	48	380
Fat	1200	40	133
Protein	360	12	90

After consideration of all these factors, the total calorie requirement of the average man will be about 3000 Cal, as shown in Table 22–7. The main constituents should be apportioned as shown in Table 22–8.

Body Temperature Regulation

Source of Body Heat. As explained earlier in this chapter, heat is produced in the catabolic processes within the cells, resulting in warming of body tissues and the blood. A small amount of heat is introduced into the body when hot liquids are drunk. Blood is warmest in the inner parts of the body, particularly in the liver, where a great deal of metabolism occurs, and is coolest at the surface, the skin; however, this variation is minor—1° or 2° C. The constant temperature of the body is maintained by means of a balance between heat production, *thermogenesis*, and heat loss, *thermolysis*. Overall control of the balance lies in the hypothalamus and its influence on the sympathetic nervous system.

Thermogenesis. Since heat is produced in catabolic chemical reactions, more heat is generated when these processes are occurring at a greater rate. Thus, during exercise, digestion, and metabolism following a meal there is greater heat production. The most important influence on the rate of cellular metabolism is the hormone *thyroxin*. It speeds up all metabolic processes, and when there is excess of this hormone (as in thyroid disease) the excess heat production is noticeable to the individual, who has a decreased tolerance for external heat.

Thermolysis. Normal means of eliminating heat from the body are shown in Table 22–9: skin, expired air, and urine and feces. As indicated, by far the most important means is the skin. *Radiation* of heat occurs from the skin, as it does from any warm object (including the sun). *Conduction* of heat is the movement of heat directly from one molecule to another, or one object to another, as from the body to the clothes worn or to the air. It also oc-

curs within the body, from one organ to another and to the blood. *Convection* of heat is the movement of warm air from the area next to the body, cooler air replacing it. Warm air is less dense and therefore rises, so it is a continual process. Radiation and convection of heat are particularly dependent on the difference in temperature between the body and the external environment. When air is cool, more radiation and convection of heat are possible.

Another means of heat loss from the body is *evaporation* of sweat from the skin. Evaporation removes heat as the liquid is transformed to a gaseous state; therefore, it removes heat from the skin and cools it in the process. Evaporation of sweat occurs even when we are not aware (*insensible* sweating), at a rate of about 600 ml per day; on hot days and during and after exercise, the amount is greatly increased—to as much as 2 to 3 liters per hour. Evaporation from the skin is less when the moisture content of the air is high (high humidity).

Heat loss in respiration is minor, although continuous; when respirations are rapid, as in exercise, heat loss is greater. Heat loss in the urine and feces is minimal.

Temperature Regulation. The hypothalamus has two areas that control the balance of heat loss and heat production. One is sensitive to heat—the heat loss center; one is sensitive to cold—the heat conservation center (Figure 22–4). Neurons from one center to the other control the balance of heat loss and heat production.

During muscular activity, or when the external temperature is high and *heat loss* must be increased, skin receptors are stimulated; impulses are transmitted to the heat loss center of the hypothalamus. At the same time the warm blood stimulates neurons of this center directly. The

TABLE 22–9
HEAT LOSS FROM THE BODY IN 24 HOURS'
FROM AN INDIVIDUAL DOING LIGHT WORK

Lost from Skin	2156 Cal, or 87.5%
1792 Cal, or 73.0%, by radiation and conduction	
364 Cal, or 14.5%, by evaporation of perspiration	
Lost in Expired Air	266 Cal, or 10.7%
182 Cal, or 7.2%, vaporization of water	
84 Cal, or 3.5%, warming air	
Lost in Urine or Feces	48 Cal, or 1.8%
Total heat loss per 24-hour day	2470 Cal, or 100.0%

result is sympathetic stimulation of sweat glands to secrete more sweat; vaporization of sweat then removes heat from the skin. Blood in the skin capillaries is cooled, and since blood is continually circulating, warmer blood brought to the surface loses heat.

Central depression of the vasoconstrictor center, as a result of hypothalamic influence, results in dilation of skin arterioles; more blood dilates the skin capillaries, and blood cooling progresses at a faster rate.

When it is necessary to *conserve heat*, as in a cold environment, arterioles to the skin and the capillaries they supply are constricted; there is less loss of heat from the blood. Sweat glands secrete less sweat (down to the minimal insensible level), and heat production increases. These responses are due to stimulation of the cold receptors in the skin, which initiate impulses for transmission to the central nervous system and the hypothalamic center for heat conservation. Efferent impulses to skeletal muscle increase muscle tone, and soon shivering occurs—rhythmical oscillations of muscle contraction and relaxation. Both increased muscle tone and shivering increase the metabolic production of heat.

A more important means of increasing heat production is the hypothalamic secretion of TSH, causing secretion of *thyroxin*, and a generalized increase in metabolic rate of all cells.

It is in heat conservation that adipose

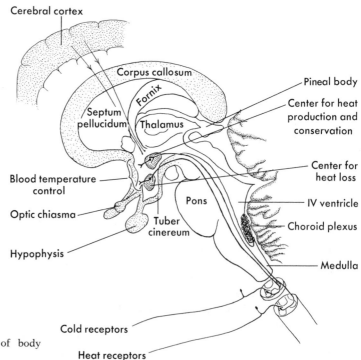

Figure 22–4.   Nervous control of body temperature.

tissue underlying the skin becomes important—it serves to insulate the inner parts of the body. This insulating effect may be deleterious when it is necessary to lose heat: a fat person must sweat more than a thin person.

In summary, heat conservation involves constriction of skin blood vessels, decrease in sweating to the minimal level, increased muscle tone, shivering, and increased metabolic rate as a result of secretion of thyroxin. Heat loss involves dilation of skin blood vessels and sweating; radiation, conduction, and convection are continuous. Central control of temperature regulation lies in the hypothalamus.

Normal Variations in Temperature. Normal body temperature varies with age and sex. As age increases, the normal body temperature decreases slightly; children have a slightly higher body temperature, partly related to their increased activity and to the increased metabolic rate in synthesizing new tissue, as they grow. Women may have a slightly lower body temperature than men as a result of their slightly lower metabolic rate.

Daily variations in temperature are normal: the temperature is usually lowest between 3 and 5 A.M. It rises slowly during the day, reaches its maximum at about 4 P.M., and falls again during the night. This corresponds to the usual temperature ranges in fever, when the minimum is in the early morning and the maximum is in the evening. Muscular activity and food cause a slight increase in temperature. This probably accounts for the increase in temperature during the day. In the case of nightworkers who sleep during the day, the increase in temperature occurs during the night, which is the period when food is eaten and work performed.

The *normal body temperature* is 37° C (98.6 F) when taken by mouth; it is 1° higher when the thermometer is inserted in the rectum.

Temperature of Infants. Infants, and in particular premature infants, have difficulty regulating their body temperature, because the hypothalamic centers are not yet fully developed. For this reason, they must be carefully protected from environmental changes in temperature. Premature infants are cared for in incubators, with careful regulation of the temperature inside the incubator.

Effect of Clothing. Clothing serves to insulate the body from the external environment by trapping air between the clothing and the skin. This interferes with radiation, conduction, and convection of heat.

Subnormal Temperature. In order to carry on the activities essential to life, the body must maintain a normal temperature. If the temperature falls much below normal, to about 35° C (95° F), life may be threatened. Subnormal temperature may be due to excessive loss of heat, profuse sweating, hemorrhage, and lessened heat production, as in starvation. In cases of starvation the fall of temperature is very marked, especially during the last days of life. The diminished activity of the tissues first affects the central nervous system; the patient becomes languid and drowsy, and finally unconscious; the heart beats more and more feebly, the breath comes more and more slowly, and the sleep of unconsciousness passes insensibly into the sleep of death.

HYPOTHERMIA. It is possible to decrease body temperature by giving sedatives to depress the hypothalamic thermostat and then using ice packs or a cooling mattress. Temperature can be maintained considerably below 32.2° C (90° F). Artificial cooling of the body is used during heart surgery so that the heart can be stopped for several minutes at a time without apparent untoward physiologic results, since the metabolic rate decreases and oxygen need is less.

Fever. An elevation of body temperature above the normal range is a fever. This may be caused by an abnormality of the brain, such as a tumor, or manipulation of the brain during surgery; or it may be due to toxins produced by bacteria and by certain breakdown products of protein. These substances, known as *pyrogens*, act by "resetting" the hypothalamic center for temperature, so that temperature regulation is at a higher level. Destruction of leukocytes, as in infection, releases a substance that is known to be pyrogenic. Dehydration is also a cause of fever, presumably by its effect on the hypothalamus. A temperature above 41° C is extremely hazardous to life. During the fever, particularly the early stages, "chills" often occur. Since the "thermostat" is set at a higher level than the temperature of the circulating blood, the normal mechanisms for thermogenesis come into play—excessive shivering, peripheral vasoconstriction—and the person feels cold. This continues until the blood and tissues of the body are at the elevated temperature of the thermostat.

Fever has a beneficial effect in many infectious diseases because the high temperature kills the causative organism. Inability to produce a fever during an infection has a negative effect as far as recovery is concerned.

Heat Stroke. Heat stroke occurs when the individual is exposed to extremely high atmospheric temperature for a long period, and when the humidity is high so that heat loss is difficult. As the body temperature rises, the hypothalamic center becomes damaged, and heat regulation is impossible.

Questions for Discussion

1. What are the normal end products of digestion, and how do they reach the bloodstream?

2. If the quantity of food from a meal is in excess of immediate energy needs, what will become of the excess? Through what metabolic process might this be accomplished?

3. Since the body does not use the food directly for energy, what is the function of the citric acid cycle?

4. Which hormones influence the level of blood glucose? What other factors are involved?

5. What is the role of the liver in relation to metabolism of carbohydrates, fats, and proteins?

6. Differentiate between metabolism and basal metabolism and explain three factors that will affect each one.

7. How can you determine your metabolic needs? Why is "calorie counting" important (a) for growing children, (b) for energy production, (c) for individuals who have a tendency to put on weight, and (d) for "senior" individuals with a "heart condition"?

8. How is heat lost from the body? How is it conserved?

9. What does the hypothalamus do in relation to temperature control?

10. How does fanning help to cool the body?

Summary

Absorption

Determining conditions
{ Passage of digested food material from the cavity of the alimentary canal to the blood
1. Area of surface for absorption
2. Length of time food is in contact with absorbing surface
3. Concentration of digested material present
Above conditions realized in small intestine:
1. Circular folds villi and their microvilli increase internal surface
2. Food remains for several hours
3. Products of digestion higher in intestine, lower in blood

Paths of absorption
1. Capillaries of villi absorb sugars, amino acids, and some of the glycerol and fatty acids, carry them to portal vein, then to liver
2. Central lymph channel of villus absorbs glycerol and fat, empties into larger lymph vessels, then into thoracic duct, superior vena cava, and right atrium of heart

Stomach
Diffusion only of substances with high lipid solubility

Small intestine
Monosaccharides: active transport with preference for glucose, galactose
Amino acids: active transport
Fatty acids, glycerol: recombined in epithelial cell to form triglyceride; enter lacteals as chylomicrons; bile salts necessary for solution of fat
Bile salts: in terminal ileum
Water: by diffusion
Electrolytes: mostly active transport

Absorption (*cont*)	Large intestine	Limited absorption by diffusion; marked water absorption
Metabolism	Consists of	**Anabolism**—processes by which living cells take food substances from the blood and make them into protoplasm and stored products **Catabolism**—processes by which living cells change into simpler substances (1) part of their own protoplasm, (2) stored products, or (3) absorbed nutrients
		Catabolic processes: 1. Simple splitting of complex molecules into simpler ones 2. Hydrolysis, or the splitting of complex molecules into simpler ones through reaction with water 3. Oxidation, with the production of carbon dioxide and water
	Purpose	Growth and repair of tissue Release of chemical energy in the form of heat, nervous activity, muscular activity, etc.
Metabolism of Carbohydrates	Processes	Glycogenesis, or the production of glycogen in the liver Glycogenolysis, or the conversion of glycogen to glucose according to body needs Glycolysis, or the anaerobic oxidation in the tissues Gluconeogenesis, formation of glycogen or glucose from noncarbohydraté sources ATP formation for storage of energy
	Hormones	Pancreas—produces *insulin* and *glucagon* Insulin a. Concerned with phosphorylation of glucose b. Inhibits glycogen formation in liver from noncarbohydrate sources c. Decreases plasma glucose level Glucagon—favors glycogenolysis and gluconeogenesis in the liver Somatotropin—opposes action of insulin in glucose phosphorylation, increases glucose formation from noncarbohydrate sources Adrenal cortex—glucocorticoids decrease tissue utilization of glucose, increase blood glucose and gluconeogenesis Adrenal medulla—epinephrine stimulates glycogenolysis in liver and muscle Thyroid—thyroxin increases oxidative rate, therefore, utilization of glucose, increases glyconeogenesis
	Functions	Furnish main source of energy for muscular work and all the nutritive processes Help to maintain body temperature Protect body tissues by forming reserve fund for time of need (glycogen) Excess carbohydrates are converted into depot fat May be used in synthesis of cell structure
	End products	When completely oxidized, the waste products are carbon dioxide and water
	Derangements	1. High glucose intake, giving rise to alimentary glycosuria 2. Plasma glucose level rises and renal threshold is exceeded (diabetes mellitus)
Metabolism of Lipids		**Reconstruction**—in act of passing through epithelial cells of villi, glycerol and fatty acids combine to form triglycerides; move into lacteals
	Functions	Yield heat and other forms of energy Stored as adipose tissue Synthesized to form compound fats and fatlike substances; lecithin, cholesterol Glycerol may be converted to glucose
	Terms	Lipogenesis—formation of lipids Lipolysis—breakdown of lipids
	Oxidation	Fatty acids split; 2-carbon units enter citric acid cycle; eventually oxidized to carbon dioxide and water
	Factors affecting	Choline and methionine essential for synthesis of phospholipids in the liver Hormones—epinephrine and norepinephrine, ACTH, and glucagon stimulate release of fatty acids from depot fats
	Body fat	Formed from fats, carbohydrates, and proteins of food in order named
	Obesity	May be caused by eating more food than body needs, by lack of exercise, or both

Metabolism of Proteins

- All cells — Amino acids used as cell constituents, to form new cells, for secretions
- Liver — Transamination; deamination; nonnitrogenous portion enters Krebs cycle for oxidation; gluconeogenesis
- Skeletal muscle — Forms creatine, creatine phosphate
- End products — CO_2, water, urea, uric acid, creatinine
- Nitrogen equilibrium maintained on 40 gm protein per day
- Nutritive value
  - **Adequate** proteins contain all the amino acids for maintenance and growth of tissue
  - Proteins may yield energy
- Hormones
  - Somatotropin, testosterone, insulin are anabolic
  - Glucocorticoids, excessive thyroxin are catabolic

Citric Acid Cycle — A complex series of chemical reactions whereby glucose, fatty acids, and amino acids enter the final stage of oxidative reactions and energy is released, to form ATP. Formation of carbon dioxide and water is incidental

Basal Metabolism

- Calorie
  - Unit of measurement of heat production
  - Large calorie = quantity of heat necessary to raise temperature of 1 kg (2.2 lb) of water 1° C; also called kilocalorie
- Carbohydrates 1 gm—4 Cal
- Fat 1 gm—9 Cal
- Protein 1 gm—4 Cal
- **Basal metabolism**—energy production at minimal level; heat production
- **Rate**—measured by direct, indirect methods
  - Surface area necessary
- **Variations** in basal metabolism influenced by age, sex, level of thyroid hormone

Daily Calorie Requirement

- **Factors to consider**—body surface, work, sleep
- Distribution
  - Proteins—10–15%
  - Fats—about 25–35%
  - Carbohydrates—depends upon form in which taken, also on amount of fat

Body Heat

- Derived from
  1. In the process of **oxidation** every body cell produces heat, but the most important heat-producing organs are the **muscles** and **liver**
  2. Minor sources
     - Friction of muscles, blood
     - Hot substances ingested
- Distributed—by the blood circulating through the blood vessels
- Lost by
  - Skin—2156 Cal, or 87.5%
    - Offers large surface for radiation, conduction, convection, and evaporation of sweat
    - Contains large amount of blood
  - Lungs—266 Cal, or 10.7%, is lost by warming the inspired air and in the evaporation of the water of respiration
  - Urine and feces—48 Cal, or 1.8%, is lost in the urine and feces
- Regulation
  - Hypothalamus
    - Heat loss center
      - Sweating
      - Dilation skin capillaries, arterioles
    - Heat conservation center
      - Constriction skin capillaries, arterioles
      - Increased muscle tone
      - Shivering
    - Secretion of TSH; result is increase in metabolic rate

Variations in Temperature

- The normal temperature by mouth is about 37° C (98.6° F)
- Normal
  1. Depends on where temperature is taken
     - Mouth
     - Axilla
     - Rectum
  2. Depends on time of day
     - Lowest in early morning, between 3 and 5 A.M.
     - Highest in late afternoon, about 4 P.M.
  3. Slightly increased by muscular activity and the digestive processes
  4. Age. Higher and more variable in infants; lower in aged
- Abnormal
  - Hypothermia
  - Fever

Additional Readings

GUYTON, A. C.: *Basic Human Physiology: Normal Functions and Mechanisms of Disease,* 2nd ed. W. B. Saunders Co., Philadelphia, 1977, Chapters 45, 46, and 47.

KOSHLAND, D.: Protein shape and biological control. *Sci. Amer.,* **229:**52–64, October, 1973.

SCHACTER, S.: Obesity and eating. *Science,* **161:**751–56, August 23, 1968.

SELKURT, E. E.: *Basic Physiology for the Health Sciences.* Little, Brown & Co., Boston, 1975, Chapters 11 and 13.

VANDER, A. J.; SHERMAN, J. H.; and LUCIANO, D. S.: *Human Physiology: The Mechanisms of Body Function,* 2nd ed. McGraw-Hill Book Co., New York, 1975, pp. 77–89, 410–29.

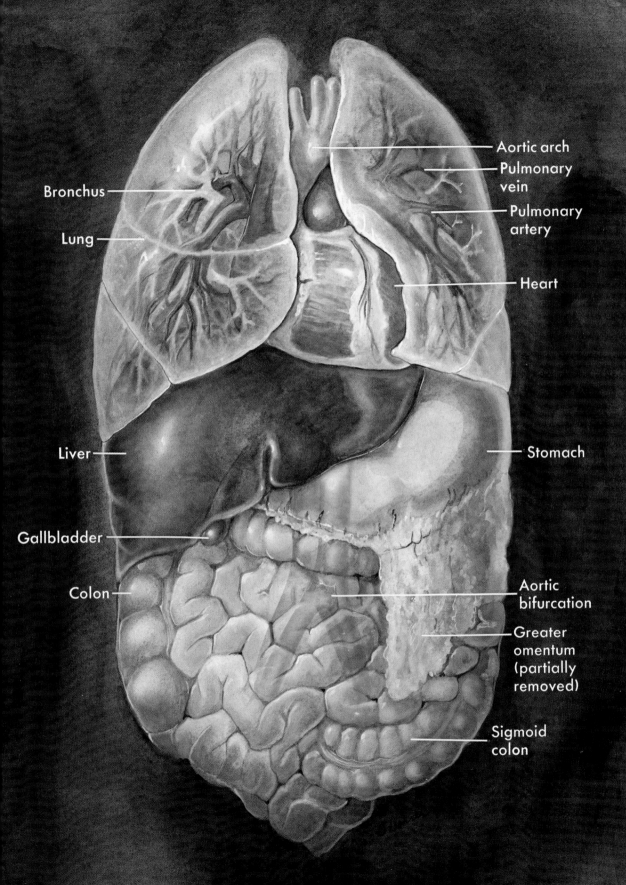

Bronchus

Lung

Aortic arch

Pulmonary vein

Pulmonary artery

Heart

Liver

Stomach

Gallbladder

Colon

Aortic bifurcation

Greater omentum (partially removed)

Sigmoid colon

Plate I. Anatomic relationships in thorax and abdomen.

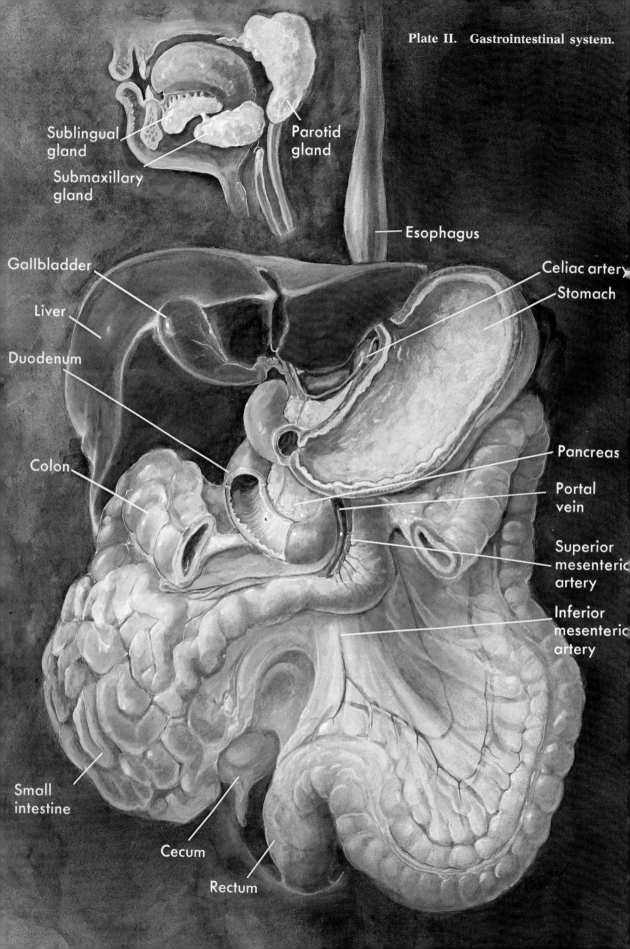

Plate II. Gastrointestinal system.

Sublingual gland

Submaxillary gland

Parotid gland

Esophagus

Gallbladder

Liver

Duodenum

Colon

Small intestine

Cecum

Rectum

Celiac artery

Stomach

Pancreas

Portal vein

Superior mesenteric artery

Inferior mesenteric artery

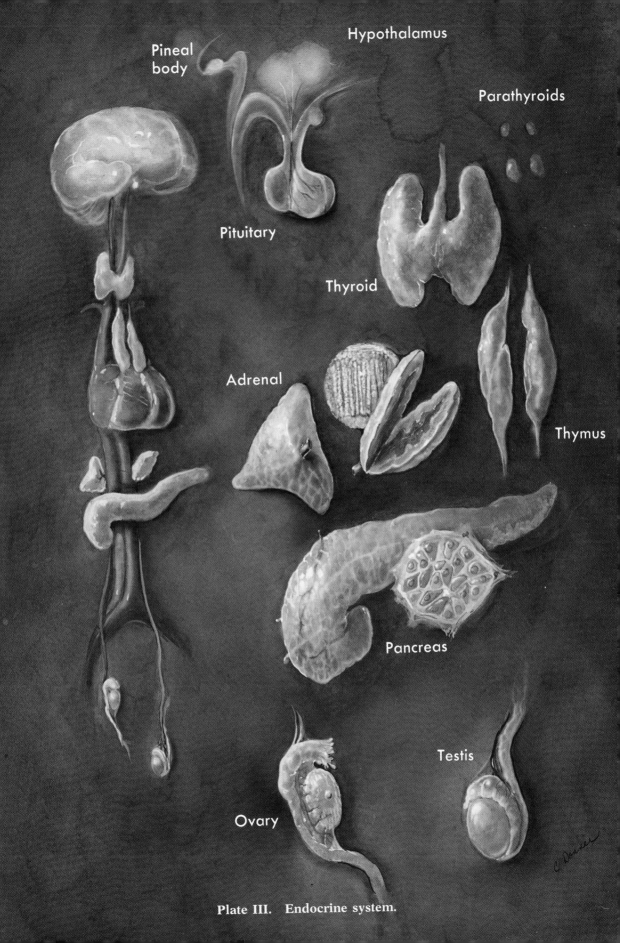

Pineal body

Hypothalamus

Parathyroids

Pituitary

Thyroid

Adrenal

Thymus

Pancreas

Ovary

Testis

Plate III. Endocrine system.

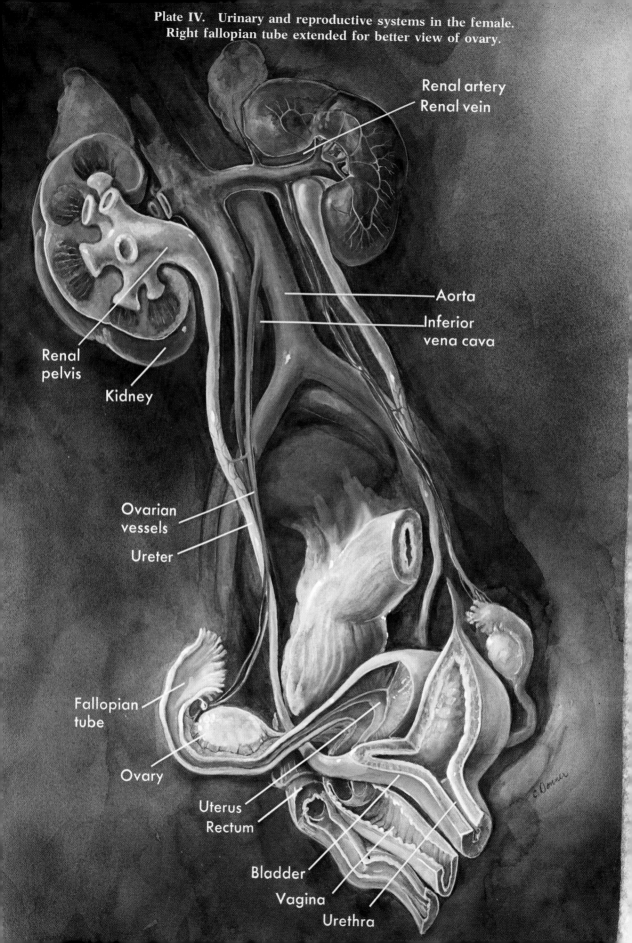

Plate IV. Urinary and reproductive systems in the female.
Right fallopian tube extended for better view of ovary.

Renal artery
Renal vein

Aorta

Inferior
vena cava

Renal
pelvis

Kidney

Ovarian
vessels

Ureter

Fallopian
tube

Ovary

Uterus

Rectum

Bladder

Vagina

Urethra

Body Maintenance: Homeostasis of Body Fluids

Claude Bernard, a French physiologist, first stated in the late nine-teenth century that the constancy of the internal environment was the prime essential for cell life. When materials enter the body in excessive amounts, the excess must be removed as must the waste products of cell metabolism. Only thus can the chemical character-istics of the internal environment remain within the narrow limits of efficient cell functioning. This vital role belongs to the kidney.

The Urinary System

Chapter Outline

Nutrients added to the bloodstream by the digestive organs and oxygen from the lungs are utilized by cells of the body for growth and repair, for synthesis of hormones or other secretions, and as a source of energy for these and other cell activities. As a result of the complex chemical reactions taking place within the cell, certain products are formed that tend to alter the normal internal and external environment of the cell. Unless these are kept within the normal range, cell functioning will deteriorate, causing eventual death of the cell and possibly of the individual. The kidney is the organ most responsible for maintaining homeostasis of body fluids.

Materials for Elimination

The materials to be eliminated from the body include nonabsorbed substances, which appear in the feces; metabolic production of carbon dioxide, eliminated during expiration; water, eliminated by various means, but primarily by the kidney; metabolic end products of protein—urea, uric acid, creatinine, eliminated by the kidney; and heat, eliminated by the skin. Materials to be eliminated are called *excreta*, and the process of elimination is called *excretion*. The excretory organs are listed in Table 23–1. With the exception of the kidneys, all of these excretory organs have been discussed in previous chapters.

TABLE 23–1
EXCRETORY ORGANS AND PRODUCTS EXCRETED

Excretory Organ	Essential	Incidental
Lungs	Carbon dioxide	Water, heat
Kidneys	Water and soluble salts, resulting from metabolism of proteins, neutralization of acids, etc.	Carbon dioxide, heat
Alimentary canal	Solids, secretions, etc.	Water, carbon dioxide, salts, heat
Skin	Heat	Water, carbon dioxide, salts

Components of Urinary System

The urinary system includes two kidneys, which form urine, each with a ureter to transport urine to the urinary bladder, and the urethra, which conducts urine from the bladder to the external environment (Figure 23–1).

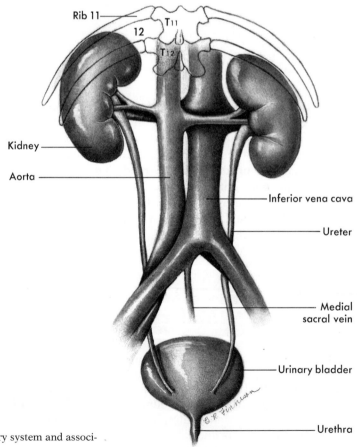

Figure 23–1. Diagram of the urinary system and associated blood vessels, posterior view.

Kidneys

The kidneys are bean-shaped organs, placed at the back of the abdominal cavity, one on each side of the spinal column and behind the peritoneal cavity. They correspond in position to the space included between the upper border of the twelfth thoracic and the third lumbar vertebrae. The right kidney is a little lower than the left, because of the large space occupied by the liver.

Each kidney with its vessels is embedded in a mass of fatty tissue termed an *adipose capsule*. The kidney and the adipose capsule are surrounded by a sheath of fibrous tissue called the *renal fascia*. The renal fascia is connected to the fibrous tunic of the kidney by many trabeculae, which are strongest at the lower end. The kidney is held in place partly by the renal fascia, which blends with the fasciae on the quadratus lumborum and psoas major muscles and also with the fascia of the diaphragm, and partly by the pressure of neighboring organs.

The kidneys are bean shaped, with the medial or concave border directed toward the medial line of the body. Near the center of the concave border is a fissure called the *hilum* (hilus), which serves as a passageway for the ureter, and for the blood vessels, lymph vessels, and nerves going to and from the kidney. Each kidney is covered by a thin but rather tough envelope of fibrous tissue. At the hilum of the kidney the capsule becomes continuous with the outer coat of the ureter. If a kidney is cut in two lengthwise (Figure 23–2), it is seen that the upper end of the ureter expands into a basinlike cavity, called the *pelvis* of the kidney. The substance of the kidney consists of an outer portion called the cortical substance (cortex) and an inner portion called the medullary substance (medulla). Between cortex and medulla are the arterial and venous arches.

Medulla. The medulla is red and consists of from 8 to 18 radially striated cones, the

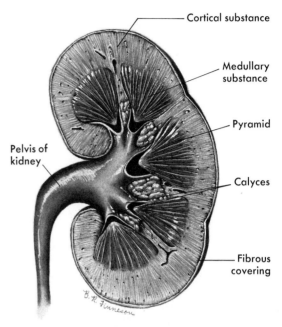

Figure 23–2. Diagram of kidney. Longitudinal section.

renal pyramids, which have their bases toward the circumference of the kidney, whereas their apices converge into projections called papillae, which are received by the cuplike cavities, or calyces, of the pelvis of the kidney.

Cortex. The cortex is reddish brown and contains the glomeruli and convoluted tubules of the nephron and blood vessels. It penetrates for a variable distance between the pyramids, separating and supporting them. These interpyramidal extensions are called the *renal columns* and support the blood vessels. A glance across the shiny surface of a freshly cut kidney discloses both a granular appearance and also areas showing radial striations. In general, these alternate with each other. The granular areas contain the renal corpuscle, glomeruli, and convoluted areas of the tubule. The radially striated areas contain other parts of the tubule.

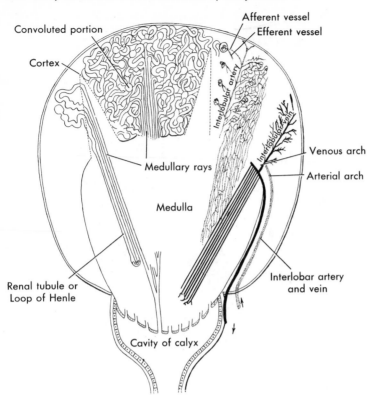

Figure 23–3. Diagram of a longitudinal section of a lobe of the kidney showing the arrangement of tubules and blood vessels in the lobe. The calyx embraces the apex of the pyramid. It is lined with epithelium, which continues from it over the apex, the latter being perforated with many apertures of collecting tubules. Note the arrangement that gives the granular and radial striations in the cortex. Nephron, arteriole, and venous units are diagrammed separately; in actuality they all occur in each pyramid and cortex section. (Modified from Gerrish.)

Nephron (Figures 23–3 and 23–4). The bulk of kidney substance, in both cortex and medulla, is composed of minute tubules closely packed together and having only enough connective tissue to carry a large number of blood vessels, lymphatics, and nerve fibers. The appearance of the cortex and medulla is due to the arrangement of these tubules, the *nephrons*, the functional unit of the kidney.

The nephron consists of a glomerulus, its tubule, and its blood supply. Tubules vary in length, and the glomeruli vary in size. The largest ones are found nearest the medulla.

The *renal tubule* begins as a closed, invaginated layer of epithelium, the *glomerular capsule,* or capsule of Bowman.[1] The inner layer of this globelike expansion closely invests a capillary tuft called a *glomerulus* (Figure 23–5). The glomerulus consists of capillary loops branching from a single *afferent arteriole;* the capillaries

unite to form a single efferent vessel, the *efferent arteriole.* It was formerly thought

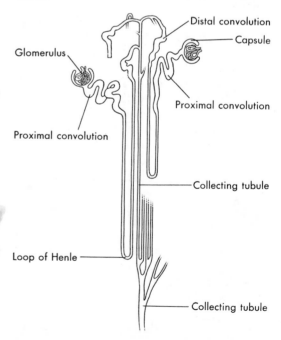

Figure 23–4. Diagram of the course of two renal tubules.

[1] Sir William Bowman, English anatomist and ophthalmologist (1816–1892).

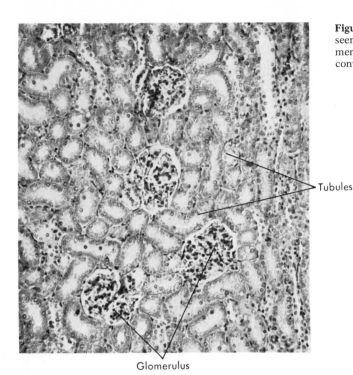

Figure 23–5. Section of renal cortex, as seen with the light microscope. Four glomeruli are visible and proximal and distal convoluted tubules, cut in cross section.

Tubules

Glomerulus

that the glomerular capillaries did not anastomose; however, recent electron micrographs demonstrate anastomoses. The glomerulus is completely encapsulated, except at the point where the afferent and efferent arterioles enter and leave the capillary tuft. The glomerulus, its basement membrane, and the enveloping capsule make up a renal corpuscle, or malpighian[2] body. A million or more glomeruli are found in the cortex of each kidney.

The cells in the visceral layer of Bowman's (glomerular) capsule have numerous large arms of cytoplasm, called major processes, extending from its cell body, which terminate as "feet" on the membrane of the glomerular capillary wall. These cells are called *podocytes* (Figure 23–6). Smaller processes and "feet" interdigitate with the major processes around the glomerular capillaries, leaving tiny spaces between them. It has been suggested that these podocytes may have contractile or elastic power.

As seen in Figure 23–7, the renal capsule

joins the rest of the tubule by a constricted *neck;* the *tubules,* each composed of a proximal convolution, loop of Henle, and distal convolution, open into *collecting ducts,* which pour their contents through their openings on the pointed ends, or papillae, of the pyramids into the calyces of the kidney. About 20 of these collecting ducts empty into the calyces from each papilla.

The epithelial lining of the renal tubule varies in different parts of the tubule. A distinctive characteristic of cells of the proximal convoluted tubule is the presence of microvilli forming a "brush border," which projects into the lumen. In the convoluted tubules and ascending limb of Henle's[3] loop the cells are columnar, and in the glomerular capsule and descending limb of Henle's loop they are thin, squamous cells. The collecting tubules have well-defined columnar cells, definitely resembling those of excretory ducts. The ascending limb of the loop of Henle re-

---

[2] Marcello Malpighi, Italian anatomist (1628–1694).

[3] Friedrich Gustav Jakob Henle, German anatomist (1809–1885).

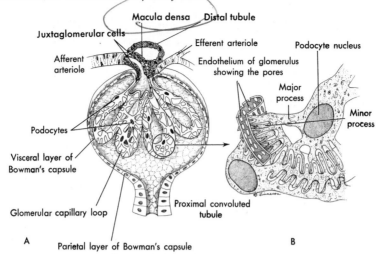

Figure 23–6. Schematic diagram from an electron micrograph of the glomerulus. *A.* Glomerulus, juxtaglomerular cells, and macula densa. Parietal and visceral layers of endothelium of Bowman's capsule are continuous. Note that the afferent arteriole is wider than the efferent arteriole. (Modified from Bailey.) *B.* Enlarged microscopic section of *A.* Note the many processes of podocytes and their relation to the glomerular capillary pores. (Modified from Harris.)

turns to the glomerulus at the point of the afferent and efferent vessels before continuing as the distal convoluted portion. At this point it is in contact with both blood vessels and particularly with the afferent because there is no basement membrane between cells of the tubule and those of the afferent arteriole. The tubular cells

in contact with the blood vessels are heavily nucleated and appear thicker than other tubular cells; they are collectively called the *macula densa.* At the same time the smooth muscle cells of the afferent arteriole at this point are larger than in other parts of the afferent arteriole and contain granules composed of inactive *renin* (see

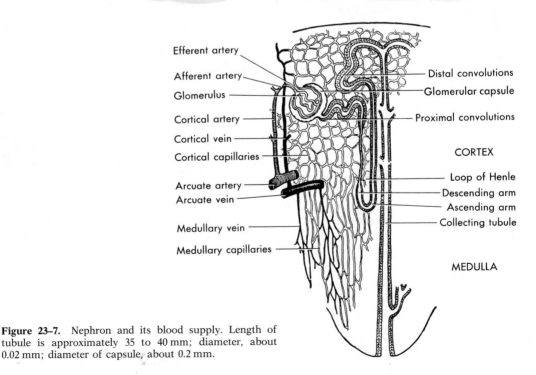

Figure 23–7. Nephron and its blood supply. Length of tubule is approximately 35 to 40 mm; diameter, about 0.02 mm; diameter of capsule, about 0.2 mm.

p. 548). These granular cells are termed *juxtaglomerular* cells (Figure 23–6). The macula densa cells are peculiarly different in that the Golgi apparatus is between the nucleus and the juxtaglomerular cell, rather than toward the lumen as is usual in the tubular cell.

Blood Supply of the Kidney (Figure 23–7). The kidney is abundantly supplied with blood by the *renal artery*, which is a branch of the abdominal aorta. Before or immediately after entering the kidney at the hilum, each artery divides into several branches, which enter the renal parenchyma separately. These branches travel up the renal columns as interlobar arteries.

When these arteries reach the boundary zone between the cortex and medulla, they divide laterally and form the *arch*, or *arcuate arteries*. From the convexity of these arches, the *interlobular arteries* (cortical) enter the cortex, giving off at intervals minute *afferent arterioles*, each of which branches out as the *capillaries of a glomerulus*. These capillaries reunite to form an *efferent arteriole* much smaller than the afferent arteriole. The efferent arteriole breaks up in a close meshwork, or *plexus*, of capillaries, which are in close approximation with both the convoluted tubule in the cortex and the loop of Henle in the medulla. These capillaries unite to form the *interlobular veins* (cortical) and *medullary veins*, which pour their contents into the *arcuate veins* lying between the cortex and the medulla. The arcuate veins converge to form the *interlobar veins*. These merge into the *renal vein*, which emerges from the kidney at the hilum and opens into the inferior vena cava.

Nerves of the Kidney. These are derived from the *renal plexus*, which is formed by branches from the celiac plexus, the aortic plexus, and the lesser and lowest splanchnic nerves. They accompany the renal arteries and their branches and are distributed to the blood vessels. They contain sympathetic fibers as well as parasympathetic fibers from the vagus nerve. The sympathetic nerve fibers are vasomotor

and, by regulating the diameters of the small blood vessels, control the pressure and amount of blood flowing through the kidney and glomeruli. The function of the parasympathetic fibers is not known.

Urine Formation. Three processes are involved in urine formation: filtration by the glomerulus, reabsorption, and secretion by the tubular cells.

The *glomerulus* functions as an ultrafilter, permitting particles smaller than the size of the endothelial pores to escape, thus filtering small colloidal and noncolloidal substances of plasma into the plasma. There is no selectivity to this filtration, other than that of particle size; so composition and concentration of materials in the filtrate are the same as those of plasma. The filtrate is generally spoken of as "protein free," although small amounts of low-molecular-weight plasma proteins do filter through the glomerular membrane and are reabsorbed in the tubules by pinocytosis.

The filtering process in the glomerulus is directly related to blood pressure. Hydrostatic pressure in the glomerulus is about 75 to 80 mm Hg and the protein osmotic (oncotic) pressure is about 25 mm Hg. This means that the effective hydrostatic pressure, or filtration pressure, is about 50 or more mm Hg. This pressure is essential for the filtration of 170 to 180 liters of the protein-free filtrate. The pressure must also be adequate to overcome resistance to movement of fluid within the tubule. This back pressure has been estimated to be about 7 mm Hg. The total filtration pressure is therefore equal to the glomerular hydrostatic pressure, minus the oncotic pressure and the back pressure within the tubules. Back pressure in the tubules is increased by intraureteral pressure, which is transmitted upward through the renal pelvis and inhibits the onward movement of filtrate. Pressure in the ureter rises if urine flow is impeded by obstruction of the ureteral openings into the bladder, or of the urethra.

Although systemic blood pressure may vary, the pressure in the glomerulus is maintained by alterations in the lumen of the efferent arteriole as compared with the afferent arteriole. Thus filtration continues until blood pressure falls to very low levels; and when hypertension is present, filtration continues at the normal rate. The average rate of filtration is about 120 ml per minute. This means that about 170 to 180 liters are filtered in 24 hours into the tubules. Urine output is about 1 ml per minute, or about $1\frac{1}{2}$ liters in 24 hours. As the protein-free filtrate passes through the tubule, it is changed in both volume and composition by the reabsorption of about 168 or 169 liters of water and certain of its constituents.

The *tubular cells* are responsible for reabsorbing materials to approximate normal blood levels of each. In addition, secretion of electrolytes occurs, particularly in relation to maintaining acid-base balance (see next chapter), although secretion of other materials, such as creatinine, also occurs. Much of the reabsorption is an active process involving carriers within the tubular cell to aid in transporting the substance across the membrane. There is a limit to the rate at which the active transport system can operate; when this limit is reached, the material, e.g., glucose, which is present in excessive

amounts in filtrate, then fails to be reabsorbed and is excreted in the urine. The term *threshold* is used in relation to reabsorption of transport substances. The threshold of a particular substance is that point above which the substance will appear in urine, and below which it will not. In addition to active transport, there is passive transport across the membrane by diffusion and osmosis. All three are involved in reabsorption.

Tubular reabsorption varies in different parts of the tubule, as shown in Table 23–2.

Countercurrent Mechanism. The ability of the kidney to conserve water for the body by forming concentrated urine, and to permit water loss by forming dilute urine, is a function of the loop of Henle and its associated capillaries. The mechanism for effecting dilution and concentration has been termed the *countercurrent mechanism*. In the ascending limb of the loop sodium is actively transported out of the filtrate into the surrounding interstitial fluid, with chloride ions moving out also. Since the ascending limb is highly impermeable to water without the influence of the antidiuretic hormone, the filtrate becomes more and more dilute as sodium is

---

TABLE 23–2
TUBULAR REABSORPTION

1. Proximal tubule and descending limb	Glucose, amino acids, vitamins, and sodium (largely under influence of aldosterone) are actively reabsorbed; chloride, sulfate, phosphate ions, and urea are passively reabsorbed; bicarbonate is actively reabsorbed in relation to systemic pH; water is reabsorbed with these substances, leaving the filtrate osmotic pressure unchanged. Eighty per cent of the water is reabsorbed in this way and is known as obligatory water reabsorption (a must)
2. Loop of Henle	Sodium is actively transported from the filtrate in the ascending limb into the medullary interstitial fluid, thus raising its osmotic pressure. This causes more water to be reabsorbed from the descending limb and the collecting duct and results in the concentration of the urine
3. Distal tubule and collecting ducts	Active reabsorption of sodium (in exchange for secreted potassium or hydrogen) and reabsorption of water. Filtrate is progressively more concentrated and volume greatly reduced as water continues to be reabsorbed, under the influence of ADH, depending on body needs. This is known as facultative reabsorption (optional). About 10 to 15 per cent of the water may be absorbed in this way

pumped out and the water remains. As sodium increases in the interstitial fluid, it diffuses back into the descending limb, increasing the osmolality of the filtrate moving down toward the tip of Henle's loop, hence countercurrent. A similar phenomenon occurs in the capillary loops nearby. Sodium diffuses into the section of capillary near the ascending limb and out of the capillary loop near the descending limb. Blood flow through the capillaries in the medulla, where the countercurrent mechanism is operating, is sluggish, and only about 5 per cent of the total renal blood flow is in this area. Ninety-five per cent of the blood flow is in the cortex where the glomeruli lie and where the distal convoluted tubules are functioning to reabsorb (or fail to reabsorb) water. Thus the effect of the countercurrent mechanism is to alter filtrate concentration before it reaches the blood supply of the cortex. In addition, body electrolyte concentrations are not appreciably altered.

Hormonal Influences on Tubular Cells. The reabsorption of water in the distal tubules of the kidney is promoted primarily by the antidiuretic hormone *ADH*, also called vasopressin. ADH increases the size of the pores in the epithelial cells of the distal tubule and collecting ducts, permitting reabsorption of water from the filtrate. The regulation and production of the ADH are under the influence of the hypothalamus. There are receptor cells in the hypothalamus called *osmoreceptors* that are sensitive to the osmotic pressure of the

plasma. A rise causes an increased secretion of ADH and inhibition of normal water diuresis. This is represented in Figure 23–8, which at the same time illustrates how water may be held in the tissues. The capacity of the kidney to concentrate substances in the filtrate (Table 23–3) is very marked and is a critical capacity. In renal disease, this capacity is lost.

The *mineralocorticoids* of the adrenal cortex (aldosterone) promote the excretion of potassium ions and the reabsorption of sodium ions, which then electrostatically attract chloride ions and cause their reabsorption throughout most of the tubule. In this way, water and sodium reabsorption and elimination by the kidneys are rigorously controlled. Eighty per cent of the tubule water is reabsorbed by osmosis as a result of decreased osmolality created by the reabsorption of these electrolytes and, to a lesser extent, other solutes.

KNOW 6 FUNCTIONS

Kidney Functions. By means of filtration, reabsorption, and secretion, the kidney accomplishes many functions:

1. *Regulation of Osmotic Pressure of Extracellular Fluids.* This is accomplished by the relative amounts of water and sodium chloride excreted. If large quantities of fluids are ingested, more water will be eliminated, and the ratio between water and electrolytes will change. This causes a decrease in specific gravity of urine. Conversely, ingestion of excess sodium chloride without an increase of fluid intake will raise the specific gravity of urine.

2. *Regulation of the Electrolytic Pattern of Extracellular Fluids.* The kidneys regulate not only the total concentration of water and electrolytes but also the concentration of each electrolyte. The regulation is complex and is accomplished by tubular reabsorption and tubular secretion, under the influence of ADH and aldosterone.

3. *Excretion of Metabolic Wastes.* The kidney excretes metabolic wastes, particularly those arising in protein metabolism (urea, uric acid, creatinine, and ammonia).

4. *Regulation of pH.* By regulating the rate of excretion of hydrogen ions and electrolytes, the kidney helps to keep the

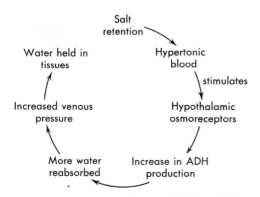

Figure 23–8. Diagram of effect of salt retention.

TABLE 23–3
CONCENTRATING CAPACITY OF THE KIDNEY

Substance	Blood Plasma %	Urine %	Number of Times Concentrated by Kidney
Water	90–93	95	—
Proteins, fats, and other colloids	7–9	—	—
Glucose	0.1	—	—
Urea	0.03	2	60
Creatinine	0.001	0.075	75
Uric acid	0.004	0.05	12
Sodium ions	0.32	0.35	1
Potassium ions	0.02	0.15	7
Ammonium ions	0.001	0.04	40
Calcium ions	0.008	0.015	2
Magnesium ions	0.0025	0.006	2
Chloride ions	0.37	0.6	2
Phosphate ions	0.009	0.15	16
Sulfate ions	0.002	0.18	90

pH of the plasma within normal limits. This is discussed more fully in Chapter 24.

5. *Regulation of the Volume of Extracellular Fluid.* The kidneys excrete concentrated or diluted urine as required to maintain normal blood volume.

6. *Secretion of Renin.* Certain cells located in the juxtaglomerular apparatus (JGA) of the afferent arterioles respond to pressure changes and secrete an enzyme substance called *renin.* Renin enters the blood and functions as a proteolytic enzyme that activates a plasma globulin known as angiotensinogen to form angiotensin I. Under the influence of another plasma enzyme, two amino acid units are split off from angiotensin I to form angiotensin II, an extremely potent vasopressor substance. In addition, angiotensin II stimulates the release of aldosterone from the adrenal cortex.

Recent research indicates excess renin production may be a factor in causing hypertension in some individuals.

When the kidney is unable to secrete nitrogenous wastes and to regulate pH and electrolyte concentrations of the plasma (as in kidney disease), artificial control may be instituted. *Hemodialysis* by means of an artificial kidney involves removal of blood from an artery, transporting it through a lengthy coiled cellophane tube immersed in a "water bath," and returning the blood to the body via a vein. The radial artery and the cephalic or basilic veins are often used. The cellophane tube functions as a semipermeable membrane, and the mixture of ions in the water bath is calculated so that materials *diffuse out* of the plasma. If, for example, the patient has a high serum potassium or urea level, there will be little or no potassium and *no* urea in the water bath. Since the substances are in higher concentration in blood plasma than they are in the water bath, they dialyze *out* of the plasma into the bath, thereby lowering plasma concentrations. Conversely, low calcium blood levels may be raised by adding calcium to the water bath.

Peritoneal dialysis may be used instead of the artificial kidney. In this technique the *peritoneal cavity* is used, the dense capillary network of the entire peritoneal cavity functioning as a dialyzing membrane. A sterile solution of the desired mixture of substances is introduced into the abdominal cavity by means of a tube inserted through a small incision in the abdominal wall. The fluid usually remains for 30 to 45 minutes to allow for sufficient time for dialysis and is then permitted to flow out through the tube. The process may be repeated any number of times until normal plasma concentrations of the various substances have been attained.

An understanding of the physiology of urine formation, with filtration of deproteinized blood plasma through the glomerular capsule and concentration in the renal tubule, helps to explain symptoms of *glomerular nephritis.*

When the glomerulus is impaired, proper filtration does not occur; nonprotein nitrogenous substances are retained in the blood, edema may be present, and a small amount of highly concentrated urine containing albumin may be passed.

Diuretics are substances that increase the volume of urine excreted, causing a condition known as *diuresis*. All diuretics act by preventing reabsorption of sodium ions by the tubular cells, which in turn causes water to remain in the filtrate. Careful regulation of these drugs is necessary to prevent fluid and electrolyte imbalance.

Physical Characteristics of Urine. Normal urine is usually a yellow, transparent liquid with an aromatic odor. Cloudy urine is not necessarily pathologic, for turbidity may be caused by mucin secreted by the lining membrane of the urinary tract; this, however, if present in excess does denote abnormal conditions. If urine is alkaline, especially after a meal, turbidity may be due to phosphates and carbonates. The color of urine varies with the changing ratios of water and substances in solution and may, of course, be affected by the presence of abnormal materials such as those produced by disease or certain drugs.

In health the specific gravity of urine may vary from 1.008 to 1.030, depending upon the relative proportions of solids and water. When the solids are dissolved in a large amount of water, the specific gravity will naturally be lower than when urine is more concentrated. The ability to concentrate urine when water is lost from the body by other means, as in sweating, is an important characteristic of the kidney and depends on adequate numbers of tubular cells. When disease destroys them, the individual excretes dilute urine.

The average quantity of urine excreted by a normal adult in 24 hours varies from 1200 to 1500 ml (40 to 50 oz). Much wider variations may occur for short periods of time without pathologic significance, as, for example, when a rise in environmental temperature or unusual muscular activity increases perspiration and so lessens the urinary output. The quantity of urine may be affected by the amount of fluid taken in by the body; the amount of fluid lost in perspiration, respiration, or in vomiting, diarrhea, or hemorrhage; the health of the organs concerned, the kidneys, heart, blood vessels; and the action of certain drugs such as diuretics.

The amount of urine excreted by children in 24 hours is great in proportion to their body weight (see Table 23–4). Thus, fluid intake in children must be closely observed for adequacy.

Chemical Characteristics of Urine. Urine is usually slightly acidic, although its *pH* may vary between 5.5 and 7.5. Diet affects this reaction; a high-protein diet increases acidity, and a vegetable diet increases alkalinity. This variation is due to the difference in the end products of metabolism in each case. If human urine is allowed to stand, it will eventually become alkaline owing to the decomposition of urea with production of ammonia, and many solutes will precipitate.

Water forms about 95 per cent of urine. The solutes (on chemical examination, precipitated as solids—some 60 gm in 1500 ml of urine) are organic and inorganic waste products.

Sodium chloride is the chief inorganic salt, about 15 gm being excreted daily by the kidneys; however, sodium chloride excretion will vary with intake.

End products of protein metabolism appearing in urine are called *nonprotein nitrogen* (NPN) substances. These are creatinine, creatine, urea, and ammonia. The amounts depend on the dietary intake of protein (Table 23–5). *Creatinine* is always present in the urine, and in amounts independent of the proteins of the diet. It is thought, therefore, to be an endogenous substance resulting from cellular metabolism of certain protoplasmic constituents.

TABLE 23–4
EFFECT OF AGE ON QUANTITY OF URINE

Age	Weight	Amount
6 months–2 years	17–26 lb	540–600 ml
2–5 years	26–38 lb	500–780 ml
5–8 years	38–55 lb	600–1200 ml
8–14 years	55–103 lb	1000–1500 ml

TABLE 23–5
NITROGEN OUTPUT AS INFLUENCED
BY LEVEL OF PROTEIN INTAKE

	High-Protein Diet (Free from Meat)	Low-Protein Diet* (Starch and Cream)
Total nitrogen	16.8 gm	3.6 gm
Urea nitrogen	14.7 gm, or 87.5%	2.2 gm, or 61.7%
Ammonia nitrogen	0.49 gm, or 2.9%	0.42 gm, or 11.3%
Uric acid nitrogen	0.18 gm, or 1.1%	0.09 gm, or 2.5%
Creatinine nitrogen	0.58 gm, or 3.6%	0.60 gm, or 17.2%
Undetermined nitrogen	0.85 gm, or 4.9%	0.27 gm, or 7.3%
Water output	Normal	Diminished

*Note that urea nitrogen decreased 12.5 gm when the low-protein diet was used. The creatinine nitrogen varied only 0.02 gm. These diets were similar in all respects, except for protein intake.

About 1 to 2 gm of creatinine are excreted in the urine daily, the amount being stable for the individual. Thus, it serves as a check for the adequate recording of urine output. The source of creatinine in urine is creatine and creatine phosphate of muscle. It is not yet known whether this change from creatine to creatinine, which involves a loss of water, is accomplished in the blood or in the kidney.

Creatine is not excreted as much in the urine of the adult male, but it is constantly present in the urine of children and often in that of women after menstruation and during and shortly after pregnancy. Creatine is also present in the urine during starvation or fever, probably because the body tissues are being utilized at a rate too high for all the creatine to be changed to creatinine. It is markedly increased in the urine of individuals who have muscular diseases.

Urea constitutes about one half (30 gm daily) of all the solids excreted in the urine. It is made by the liver cells from NH_2 radicals released on the deaminization of amino acids. Normally 27 to 28 mg of urea are contained in each 100 ml of blood. The kidneys constantly remove the urea as it is formed, keeping the amount in the bloodstream at its normal level.

Ammonia in the urine is formed by the kidney from amino acids, especially glutamine. The amount of ammonia produced by the kidney may depend upon the general need of the body for conserving sodium ions to offset acid substances in the blood and tissues.

Hippuric acid is thought to be the means by which benzoic acid, a toxic substance occurring in food and from body processes, is eliminated from the body. A vegetable diet increases the quantity of hippuric acid excreted, probably because fruit and vegetables contain benzoic acid.

Purines are derived from foods containing nucleic acid (exogenous) and from the catabolism of the body cells (endogenous). The exogenous purines excreted depend upon the quantity eaten of purine-containing foods such as meat; the endogenous purine waste depends upon the health and activities of the body and is normally fairly constant.

Abnormal Constituents of Urine. Serum *albumin* is a normal constituent of the blood plasma, but it is not usually filtered into the renal capsule. Its presence in the urine is spoken of as *albuminuria* and is usually due to increased permeability of the glomerular membrane.

Normal urine contains so little *sugar* that for clinical purposes it may be considered absent. In health the amount of glucose present in the blood varies from 80 to 120 mg per cent. When the blood level of glucose rises above 180 mg per cent, it "spills" into the urine; i.e., the tubules fail to reabsorb it.

Indican (indoxyl potassium sulfate) is a potassium salt that is formed from indole. Indole results from the putrefaction of protein food in the large intestine. It is absorbed by the blood and carried to the liver, where it is probably changed to indican. Traces of indican are found in normal urine, but its presence in larger amounts is abnormal.

The *ketone bodies*, acetoacetic acid, beta-hydroxybutyric acid, and acetone, normally appear in the urine in very small amounts. However, when excessive quantities of fatty acids are oxidized in the liver, as in acute starvation, or in insulin deficiency of diabetes, these ketones are excreted in the urine in appreciable quantities. In normal individuals they may appear in the urine during periods of fasting.

In some abnormal conditions the kidney tubules become lined with substances that harden and form a mold or *cast* inside the tube. Later these casts are washed out by the urine, and their presence can be detected with the aid of a microscope. They are named either from the substances composing them or from their appearance. Thus, there are pus casts, red cell casts, epithelial casts from the walls of the tubule, granular casts from cells that have

decomposed and form masses of granules, fatty casts from cells that have become fatty, and hyaline casts that are formed from coagulable elements of the blood.

Mineral salts in the urine may precipitate and form *calculi*, or stones. Calculi may be formed in any part of the urinary tract from the tubules to the external orifice of the urethra. The causes that lead to their formation are an excessive amount of salts, a decrease in the amount of water, and abnormally acidic or abnormally alkaline urine.

In suppurative conditions of any of the urinary organs, *pus cells* are present in the urine.

In cases of acute inflammation of any of the urinary organs *blood* may be present in the urine, a condition known as *hematuria*. Blood imparts a smoky or reddish color to urine.

Bile pigments in the urine may be caused by obstructive jaundice, when bile has been reabsorbed from the biliary tract into the bloodstream, or by diseases in which an abnormal number of erythrocytes are destroyed. Bile pigments give the urine a greenish-yellow or golden-brown color.

Ureters

The ureter is a tube that conveys the urine from the renal pelvis to the bladder. Each commences as a number of cuplike tubes, or *calyces*, which surround the renal papillae. The calyces (varying in number from 7 to 13) join and form two or three short tubes, which unite and form a funnel-shaped dilatation called the *renal pel-*

vis. The ureter is about 25 to 30 cm long, about 4 to 5 mm in diameter, and consists of three coats: an outer fibrous coat, a muscular coat, and an inner mucous lining. The contractions of the muscular coat produce peristaltic waves, which commence at the kidney end of the ureter and progress downward.

Bladder

The bladder is a hollow muscular organ situated in the pelvic cavity behind the pubes, in front of the rectum in the male, and in front of the anterior wall of the vagina, and the neck of the uterus, in the female. It is a freely movable organ but is held in position by folds of peritoneum and fascia. During infancy it is conical and projects above the upper border of the pubes into the hypogastric region. In the

adult, when quite empty, it is placed deeply in the pelvis; when slightly distended, it has a round form; but when greatly distended, it is ovoid and rises to a considerable height in the abdominal cavity. It has four coats: (1) The *serous* coat is a reflection of the peritoneum and covers only the superior surface and the upper part of the lateral surfaces. (2) The *smooth muscle* coat has three layers, an inner lon-

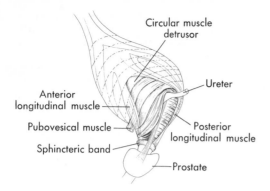

Figure 23–9. Lateral view of male bladder. (Modified from McCrea.)

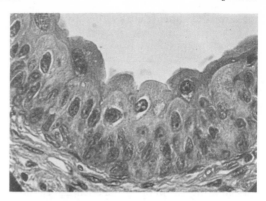

Figure 23–10. Transitional epithelium of the bladder, as seen with light microscope. Note the change of columnar cells at the base to squamous cells at the top.

gitudinal, middle circular, and outer longitudinal. The fibers of the outer layer are arranged in a more or less longitudinal manner, up the inferior surface of the bladder, over its vertex, and descending along the fundus. They are attached in the male to the prostate, and in the female in front of the vagina. At the sides of the bladder the fibers are arranged obliquely and intersect one another. This layer is called the *detrusor* muscle (Figure 23–9). The circular fibers are collected into a layer of some thickness around the opening of the bladder into the urethra. These circular fibers form a sphincter muscle, which is normally in a state of contraction, relaxing only when the accumulation of urine within the bladder renders its expulsion necessary. (3) The *submucous* coat consists of areolar connective tissue and connects the mucous and muscular coats. (4) The *mucous* membrane of transitional epithelium lining the bladder is like that lining the ureters and the urethra (Figure 23–10). This coat is thrown into folds, or rugae, when the bladder is empty, with the exception of a small triangular area formed by the two orifices and the internal orifice of the urethra where the mucous membrane is firmly attached to the muscular coat. This area is called the *trigone* (Figure 23–11).

There are three openings into the bladder. The two ureters open into the corners of the trigone, 1.25 cm ($\frac{1}{2}$ in.) from the midline. The ureters take an oblique course through the wall of the bladder, downward and medialward. The urethra

leads from the bladder, its vesical opening lying in the midline below and in front of the openings of the ureters.

Nerve Supply (Figure 23–12). The bladder is supplied by nerves from both the craniosacral and thoracolumbar divisions of the autonomic nervous system. The sacral fibers bring about contraction of the muscles of the bladder and relaxation of the internal sphincter. It is believed that the reflex centers for these fibers are located

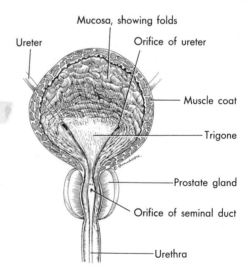

Figure 23–11. Diagram of the bladder opened ventrally to show bladder wall, orifices of ureters, and urethra. The orifices of the ureters and urethra form a triangle—the trigone. Note that the ureters enter the bladder wall from the posterior and run obliquely downward for about 2 cm. (Modified from Pansky and House.)

in the midbrain, anterior pons, and posterior hypothalamus. The function of the thoracolumbar fibers seems to be primarily related to blood supply to the bladder wall. Spinal nerves are important as well, because they innervate the muscles of the pelvic floor and perineum and aid in conscious control of the external sphincter.

Blood Supply. The superior, middle, and inferior vesical arteries (branches of the hypogastric artery) supply the bladder. In the female, branches of the uterine and vaginal arteries also supply the bladder.

Function. The bladder serves as a reservoir for urine. Its capacity varies. When moderately distended, it holds almost one liter in the adult, but this volume is never attained physiologically. The usual maximum volume is about 600 ml.

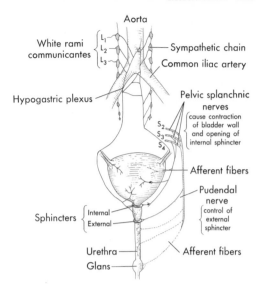

Figure 23–12. Innervation of bladder and urethra. (Modified from Grant.)

Urethra

In the *female* the urethra is a narrow membranous canal that extends from the bladder to the external orifice, the meatus. It is placed behind the symphysis pubis and is embedded in the anterior wall of the vagina. Its diameter, when undilated, is about 6 mm ($\frac{1}{4}$ in.), and its length is about 3.8 cm ($1\frac{1}{2}$ in.). Its direction is obliquely downward and forward, its course being slightly curved, with the concavity directed forward and upward. The meatus is the narrowest part and is located between the clitoris and the opening of the vagina.

The walls of the urethra consist of three coats: (1) an outer muscular coat, which is continuous with that of the bladder; (2) a thin layer of spongy tissue, containing a plexus of veins; and (3) a mucous coat, which is continuous internally with that lining the bladder and externally with that of the vulva.

The *male* urethra is about 20 cm long. It is divided into three portions: (1) the prostatic, which runs vertically through the prostate; (2) the membranous, which extends between the apex of the prostate and the external sphincter; and (3) the cavernous portion, which extends from the membranous to the urethral meatus.

The male urethra is composed of (1) mucous membrane, which is continuous with the mucous membrane of the bladder and is prolonged into the ducts of the glands that open into the urethra; and (2) a submucous tissue, which connects the urethra with the structures through which it passes.

Micturition

The act by which the urine is expelled from the bladder is called micturition. The desire to urinate is due to stimulation of pressure receptors in the bladder itself caused by the volume of urine. As the bladder fills, pressure is exerted against the detrusor muscle. This reflexly causes the muscle to contract, and the internal sphincter to relax. The reflex is sustained until the bladder is completely empty. The act of micturition is started by voluntary relaxation of the external sphincter and surrounding muscles of the perineum.

Although the emptying of the bladder is reflexly controlled, it may be initiated voluntarily and may be started or stopped at will.

Involuntary Micturition, or Incontinence. In young infants incontinence of urine is normal. The infant voids whenever the bladder is sufficiently distended to arouse a reflex stimulus. Children vary markedly in the ease with which they learn to control micturition and defecation. During the first year, some children can be taught to associate the act with the proper time and place. By the second year, regular training in habit formation and proper feeding should enable the child to inhibit the normal stimulus and to control micturition, at least during the day. Control of micturition at night is a habit requiring longer practice and is usually formed by the end of the second year. When involuntary voiding occurs at night with any degree of regularity after the third year, it is called enuresis. If this is not caused by emotional stress or irritation of the bladder, it will usually yield to proper training.

Involuntary micturition may occur as the result of a lack of consciousness or as the result of injury to the spinal centers that control the bladder. If the spinal cord is transected, all bladder controls are abolished and loss of voluntary control is permanent. If there is injury to the sympathetic fibers, no significant effects result, but frequency may be present for a short time. There may be difficulty in initiating the act of voiding, where there is injury to the brain cortex.

Involuntary micturition may also result from some irritation due to abnormal substances in the urine, or to disease of the bladder (*cystitis*). Emotional stress may provoke the desire to urinate when there is only a small amount of urine in the bladder, owing to failure of the detrusor muscles to relax. This desire may also be aroused by visual and auditory impressions, such as the sight and sound of running water.

Retention of Urine. Retention, or failure to void urine, may be due to: (1) some obstruction in the urethra or in the neck of the bladder, (2) nervous contraction of the urethra, or (3) lack of sensation to void. In the last two conditions retention is often overcome by measures that induce reflexes, i.e., pouring warm water over the vulva, or the sound of running water. If micturition does not occur and the bladder is not catheterized, distention of the organ may become extreme, and there is likely to be constant leakage, or involuntary voiding of small amounts of urine without emptying the bladder. This condition is described as retention with overflow.

Anuria. A far more serious condition than retention is the failure of the kidneys to form urine. This is spoken of as suppression, or *anuria*. Unless anuria is relieved, a toxic condition known as uremia will develop. When the formation of urine is decreased below the normal amount, the condition is spoken of as *oliguria*.

Questions for Discussion

1. Explain the functions of the glomerulus and renal tubule in relation to volume and composition of urine.
2. Which hormones affect kidney function? What is the specific action of each?
3. Explain the relationship of arterial blood pressure to kidney function.
4. What is the trigone? How does it differ from the rest of the bladder? Is this important?
5. Explain how the parasympathetic nervous system affects micturition.
6. What is the normal composition of urine?
7. Would it be possible to remove excessive fluid from an edematous patient by use of the artificial kidney? Explain.

Summary

Excretion
- Lungs—carbon dioxide, water, heat
- Kidneys—water, electrolytes, protein metabolites
- Alimentary canal—unabsorbed materials
- Skin—heat, water

Urinary System
- **Kidneys** (2)—form urine
- **Ureters** (2)—ducts that convey urine from kidneys to bladder
- **Bladder** (1)—reservoir for urine
- **Urethra** (1)—tube through which urine is voided

Kidneys

Location
- Posterior part of lumbar region, behind peritoneum
- Placed on either side of spinal column and extend from upper border of twelfth thoracic to third lumbar vertebra

Covering and support
- Embedded in a mass of fatty tissue, adipose capsule
- Surrounded by fibrous tissue called renal fascia
- Held in place by renal fascia, which blends with fasciae on { quadratus lumborum, psoas major, and diaphragm
- Also by pressure and counterpressure of neighboring organs

Size and shape
- About 11.25 cm (4½ in.) long, 5–7.5 cm (2–3 in.) broad, and 2.5 cm (1 in.) thick
- Weight about 135 gm (4½ oz)
- Bean-shaped organs
- Concave border directed toward median line of body
- Hilum—fissure near center of concave side serves for vessels to enter and leave

Gross structure
- **Pelvis**—upper expanded end of ureter
- **Calyces**—cuplike cavities of the pelvis that receive papillae of pyramids
- **Medulla**—inner striated portion, made up of cone-shaped masses
  - **Pyramids**—8–18
  - **Bases** directed toward circumference of kidney
  - **Papillae**—apices of pyramids, directed into pelvis
- **Cortex**—outer portion of kidney
  - **Renal columns**—extensions of cortical substance between pyramids

Unit
- **Nephron**—consists of a renal tubule and glomerulus
- **Renal tubule**—begins as capsule enclosing capillary tuft—glomerulus—in the cortex, includes proximal convoluted tubule, loop of Henle, and distal convoluted tubule, opens into straight collecting tube, which pours contents into calyx of kidney pelvis

Blood supply
- **Renal artery**—direct from aorta
- Before entering kidney divides into several branches
- **Arterial arches**
  - Lateral branches at the boundary zone between cortex and medulla
  - Send branches to cortex; afferent arteriole supplies glomerulus; efferent arteriole transmits blood from glomerulus to medullary capillaries
- **Venous arches**
  - Lateral branches at level of base of pyramids
  - Receive blood from cortex
  - Receive blood from medulla
- Veins empty into renal vein, leave kidney at hilus, and empty into inferior vena cava

Autonomic nerves
- Nerves derived from renal plexus, and from the lesser and lowest splanchnic nerves
- Vasomotor, by regulating size of blood vessels, regulates blood flow and pressure in glomerulus

Urine Formation
- Efferent vessel is smaller than afferent, making blood pressure in glomerulus high
- Deproteinized plasma filters through glomerulus into tubule
- Efferent vessel, with others, forms plexuses about tubule
- Urine is concentrated in tubule by diffusion of water and some salts back into the blood
- Secretion of H^+, NH_4^+, or K^+, reabsorption of Na^+
- Countercurrent mechanism functions in loop of Henle to alter concentration of filtrate
- **Hormonal control**
  - ADH—reabsorption of water in the kidney tubule
  - Mineralocorticoids (aldosterone)

Functions of kidneys
- Regulation of osmotic pressure of extracellular fluids
- Regulation of electrolytic pattern of extracellular fluids
- Excretion of metabolic wastes
- Regulation of acid-base ratio
- Regulation of the volume of extracellular fluid

Urine

Physical characteristics
- An aqueous solution of organic and inorganic substances
- **Color** and **transparency** depend upon concentration, diet, etc.
- **Reaction,** usually slightly acid, pH 5.5–7.5

Urine (cont.)

Physical characteristics (cont.)
Specific gravity 1.010–1.030, depending upon proportions of solids and water
Quantity—1200–1500 ml daily. Affected by:
Amount of fluid ingested
Amount of fluid lost in other excretions than urine, e.g., perspiration
Ability of organs to function—heart, blood vessels, kidneys, etc.; age is a factor
Action of specific substances, as diuretics

Normal Constituents
95% water; remainder organic, inorganic solutes
Exogenous wastes—vary with diet (urea)
Endogenous wastes—constant rate of excretion (creatinine)

Inorganic
Chlorides
Sulfates of
Phosphates
Sodium
Potassium
Magnesium
Calcium

Organic
Creatinine—metabolic derivation from creatine in skeletal muscle
Urea—from catabolism of protein; varies in amount with dietary protein intake; usually one half of urinary solutes
Ammonia—end product of protein metabolism; varies with need to conserve sodium
Purines—end product of nucleoprotein catabolism; appears in urine as uric acid

Abnormal Constituents
Albumin
Sugar
Casts
Erythrocytes (hematuria)
Leukocytes (pus cells)
Bile pigments

Ureters
Excretory ducts. Connect kidneys with bladder and serve as passageway for urine
Commence as calyces that surround renal papillae. These join to form two or three short tubes, and these unite to form renal pelvis
Duct is 25–30 cm long
Three coats
1. Mucous—lining
2. Muscular
  Inner, longitudinal layer
  Outer, circular layer
3. Fibrous—carries blood vessels and nerves
Function—conduct urine to the bladder

Bladder
Hollow muscular organ
Situated in pelvic cavity behind the pubes
  In front of rectum in male
  In front of anterior wall of vagina and neck of uterus in female
Freely movable. Held in position by folds of peritoneum and fascia
Size, shape, and position depend upon age, sex, and whether bladder is full or empty
Four coats
1. Mucous—lining
2. Submucous—connects mucous and muscular
3. Muscular
  Inner layer—longitudinal
  Middle layer—circular
  Outer layer—longitudinal, forms detrusor muscle
4. Serous—partial covering derived from peritoneum
Three openings
Ureters run obliquely downward through the bladder wall opening into lower part, about 1.25 cm ($\frac{1}{2}$ in.) from median plane
Urethral opening is below and in front of the opening of the ureters
Nerve Supply
Cranio-sacral
  Contraction of muscle of bladder
  Relaxation of internal sphincter
Thoracolumbar—primarily related to blood supply of bladder wall
Spinal nerves—conscious control of external sphincter
Function
Serves as a reservoir for the reception of urine
When moderately distended, holds about 500 ml

Urethra
Membranous canal, extends from the bladder to the urinary meatus
3.8 cm long in female, about 20 cm long in male
In female behind symphysis pubis, and embedded in the anterior wall of vagina
Three coats
1. Mucous—lining
2. Submucous—supports network of veins
3. Muscular
  Inner—longitudinal
  External—circular
Meatus urinarius—external orifice located between clitoris and vagina

Micturition
Act of expelling urine from bladder
Reflex act—controlled by voluntary effort

Retention ⎰ Failure to void urine
⎱ Due to ⎰ 1. Obstruction in urethra or neck of bladder
⎱ 2. Nervous contraction of urethra
3. Lack of sensation
May be accompanied by constant leakage, or involuntary voiding of small amounts

Anuria—Failure of the kidneys to secrete urine
Oliguria—Deficient secretion of urine

Additional Readings

BARAJAS, L.: Renin secretion: an anatomical basis for tubular control. *Science,* **172:**485–87, April 30, 1971.

GUYTON, A. C.: *Basic Human Physiology: Normal Functions and Mechanisms of Disease,* 2nd ed. W. B. Saunders Co., Philadelphia, 1977, Chapter 24.

SELKURT, E. E.: *Basic Physiology for the Health Sciences.* Little, Brown & Co., Boston, 1975, Chapter 21.

VANDER, A. J.; SHERMAN, J. H.; and LUCIANO, D. S.: *Human Physiology: The Mechanisms of Body Function,* 2nd ed. McGraw-Hill Book Co., 1975, pp. 321–31.

CHAPTER

Water, Electrolyte, and Acid-Base Regulation

Chapter Outline

WATER
    TOTAL BODY WATER
    DISTRIBUTION OF WATER IN THE BODY
    FLUID MOVEMENT BETWEEN COMPARTMENTS
    REGULATION OF BODY WATER
    WATER BALANCE
    FLUID EXCHANGE
    RESERVE BODY WATER

ELECTROLYTES
    HORMONAL REGULATION OF ELECTROLYTES
ACID-BASE REGULATION
    MECHANISMS FOR REGULATION
        Buffer Systems of the Body
        Respiratory Regulation of pH
        Renal Regulation of pH
    ABNORMALITIES OF pH

Water is the universal medium in which all of the complex metabolic processes of life take place. Water, electrolyte concentration, and pH are interrelated, and it is only by the careful regulation of extracellular fluid volume and concentration that normal cell activity can occur.

Water

Total Body Water. Water enters into the composition of all the tissues: within the cell as well as in the tissue fluid. Most cells contain between 75 and 90 per cent water, by weight, as shown in Table 21–1 (page 487). However, there are exceptions. Bone makes up much of the body weight but contains less than 20 per cent water. Adipose tissue water content is also low; as this tissue increases in the body, the overall body water decreases, and this accounts for the sex differences in adults. Women characteristically are less muscular than men and have a thicker layer of subcutaneous fat; body water content is about 50 per cent for women and 60 per cent for men.

Age is another variable. Seventy to sev-

enty-five per cent of infant body weight is water (extracellular water content is higher); after the first few months of life water content gradually decreases. As tissues continue to age, they become relatively dehydrated, so that in those over 65 the total body water may be 40 to 50 per cent of body weight.

Distribution of Water in the Body. Water is distributed in the body in three main compartments; figures are in terms of adult male (Figure 24–1):

1. Blood plasma   5% of body weight
2. Interstitial fluid 11% of body weight
3. Intracellular fluid

                44% of body weight

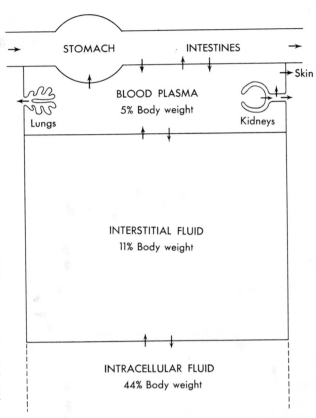

Figure 24–1. Diagram illustrating distribution of water in the body and exchange of water between compartments, normal sources of body fluid, and fluid loss. Figures are average for lean muscular adult. *Arrows* indicate direction of fluid movement. (Reprinted by permission of the publishers from James Lawder Gamble, *Chemical Anatomy, Physiology and Pathology of Extracellular Fluid: A Lecture Syllabus.* Harvard University Press, Cambridge, Mass., Copyright, 1942, 1947, 1954, by The President and Fellows of Harvard College.)

This means that if an individual weighs about 70 kg there are about 3.5 liters of water in the plasma compartment, 7.7 liters of water in the interstitial compartment, and 30.8 liters of water within the cells. This makes a total of 42 liters of water. The amount of fluid in each compartment may be measured with relative accuracy, and the measurement of plasma volume is done frequently prior to surgery.

Plasma volume may be measured by giving a known quantity of the blue dye T 1824 (Evans blue). It is nontoxic, soluble in water, but not in fat; it does not readily pass through the capillaries, for it unites with plasma proteins; hence it may be used as a measure of plasma volume.

The dye is given intravenously, and after it has had an opportunity to be mixed well in the bloodstream, samples of blood are taken. Plasma is separated from the cells and matched against samples of similar known dye dilutions. These values are considered along with hematocrit readings, i.e., the ratio of cells to plasma.

Radioactive chromium also is used to meas-

ure blood volume; in this case the red cells are the significant component, not the protein. A small amount of blood is removed, mixed with radioactive chromium for a half-hour; then the cells are washed, the degree of radioactivity is determined (from the chromium that has entered the cell), and the cells are reintroduced into the bloodstream. The degree to which these cells have been diluted, as determined by the degree of radioactivity of a second sample, gives an estimate of the total blood volume.

The *extracellular fluid volume* can be measured with radioactive sodium; the *total body water,* with "heavy water" (containing deuterium, the isotope of hydrogen). In these instances the amount remaining in a plasma sample is the index of measurement.

Fluid Movement Between Compartments. Water moves between compartments relatively freely, since cell membranes are completely permeable to water. The net gain or loss of water is determined by the osmotic pressures on either side of the membrane. Water movement is enormous, but the amount in each compartment is stable and changes only when there are extreme changes in the osmotic pressure, as in a decrease in plasma proteins (failure of the liver to produce them, or loss in the urine in glomerular disease), or in a decrease in plasma volume (hemorrhage and excessive sweating).

Regulation of Body Water. Controls for total fluid volume are incompletely understood, but involve both hormonal and neural regulation.

A *decrease in blood volume,* such as in hemorrhage, (1) stimulates secretion of renin by the juxtaglomerular cells of the kidney; the resultant increase in formation of angiotensin II causes release of aldosterone from the adrenal cortex, which, acting on the tubular cells of the kidney, increases sodium and water reabsorption. (2) The decreased blood volume stimulates cells in the hypothalamus to secrete ADH, which also acts on the tubular cells, increasing reabsorption of water from the filtrate. (3) Decreased blood volume causes decreased cardiac output and blood pressure, both of which decrease the glomerular filtration rate, thus decreasing water loss. (Similarly an increase in blood volume and blood pressure can increase filtration rate, if they are excessive.) Since large amounts of fluid are normally filtered, only slight increases or decreases of the rate effect changes in the urinary output. (4) A sensation of thirst accompanies a decrease in blood volume; voluntary increase in amount of fluid intake usually follows. The end result of these four mechanisms is a return of the blood volume to normal.

In severe stress situations, such as burns and hemorrhage, ACTH secretion and its effect on the adrenal cortex also play an important part in conservation of body fluids.

Changes in osmolality of plasma accompany changes in the plasma protein level, and the overall concentration of electrolytes in plasma. Increased sweating and dehydration from lack of fluid intake *increase* the osmolality; compensatory water conservation due to ADH secretion by the hypothalamic cells results. *Decreased* osmolality results primarily from a decrease in plasma proteins and has the opposite affect.

Water Balance. Water balance is that of intake and output. *Fluid intake* is in the form of water, other liquids or beverages, and foods. Normally water furnishes about a third or so of the required need, and the rest is furnished by other liquids and foods. Some foods have a much higher water content than do others. Another source of water to the body fluids is the water of metabolism. An ordinary mixed diet yields about 250 ml in 24 hours. On the whole, 100 gm of starch yields 55 gm of water, 100 gm of fat yields 107 gm of water, and 100 gm of protein yields 41 gm of water.

The amount of fluid needed is related to size, weight, and activities of an individual. This means that fluid intake may vary from 1800 to 3000 ml in 24 hours.

When excess amounts of fluids are taken by mouth, water is rapidly absorbed into the plasma compartment. If large quantities are taken with no food, absorption is complete in about 30 to 40 minutes. Blood volume may be temporarily increased, as well as cardiac output. However, there is rapid adjustment to the increased water load by the opening of capillary networks to increase the vascular bed; the sinusoids of the liver and spleen open to hold more blood, and fluid is rapidly transferred to the interstitial compartment. The kidneys eliminate excess water, and balance is restored within two or three hours.

Normally *fluid output* is directly related to fluid intake. Water loss via the kidneys is rigorously controlled and varies from 1000 to 1500 ml in 24 hours. As a rule water loss through feces is about 100 to 150 ml. Through insensible perspiration about 450 to 800 or so milliliters of water is lost in 24 hours and via the lungs about 250 to 350 ml. There is a relationship between water loss through kidneys and through the skin. Usually when large quantities of water are lost via the skin, urine volume is decreased.

It is readily understandable that excessive loss of fluid by diarrhea, vomiting, or

TABLE 24–1
FLUID MOVEMENT DURING A 24-HOUR PERIOD (EXCHANGE AND LOSS)

Average Amounts

Blood plasma	→ Saliva	1500 ml	
	→ Gastric juice	2000 ml	
	→ Intestinal juice	3000 ml	→ Villi and large intestine
	→ Bile	1000 ml	
	→ Pancreatic juice	800 ml	→ Return to blood
	→ Kidney: filtrate, 170 L	{ 168 L reabsorbed	
		1500 ml *lost* as urine	
	→ Lungs: moisture	300 ml *lost*	
	→ Skin: insensible moisture	600 ml *lost*	

From the above figures it can be seen that the kidney has an important function in maintaining water and electrolyte balance.

perspiration, excess loss by the kidneys, loss due to burns, or loss of fluid through hemorrhage causes fluid loss from the plasma and interstitial compartments. Failure to ingest sufficient quantities of fluid may also deplete the fluid in the interstitial compartment, and eventually the plasma compartment may be disturbed.

Fluid Exchange. There is a tremendous exchange of fluids between plasma and secreting cells and back to plasma again (Table 24–1). Normally fluid enters the body by mouth. In the stomach, fluid moves from the blood plasma to the cells of the gastric glands for the manufacture of about 1500 ml of gastric juice in 24 hours. This fluid moves, along with ingested foods and fluids, to the intestine. Here again large quantities of water move from the blood plasma to gland cells for the manufacture of large quantities of digestive fluids, and finally the fluids and end products of digestion are returned to the bloodstream.

Reserve Body Water. It is useful in considering water balance to think in terms of the individual's water "reserve"—his ability to withstand water loss. That is, should he be faced with a stress such as lack of water for drinking, or unusual loss as in vomiting and diarrhea, or excessive perspiration, how good is his reserve? Obviously the person with the greatest percentage of body water, other things being equal, has the best reserve. Obesity is the commonest cause of decreased body water. The very obese person's body water content may be as little as 25 to 30 per cent of his body weight.

Infants also have poor reserve. Although they have more body water in terms of percentage than most adults, they also lose relatively more in urinary output each 24 hours; thus their reserve is poor. Under normal conditions this large water loss is matched by a large water intake. Stressful situations such as those mentioned above easily cause severe dehydration in infants.

Electrolytes

Acids, bases, and salts are known as electrolytes because, when in water solution, they conduct an electric current. Such solutions owe their chemical activity to the presence of dissolved ions. In general, the higher the concentration of the ions, the more active the solution. Some electrolytes ionize freely and provide a high concentration of ions; these are known as strong electrolytes. Some electrolytes maintain a reserve of neutral molecules in the solution, consequently, a low ion concentration; these are known as weak electrolytes. Both weak and strong electrolytes are important physiologically.

Within the body, electrolytes (especially sodium, potassium, chloride, and bicarbonate ions) function in various ways to hold fluid within compartments, and to maintain the acid-base relationship essen-tial for enzyme activities. Figure 24–2 illustrates the distribution of various ions within the blood plasma, interstitial fluid, and cells. The concentrations of the various ions are expressed in milliequivalents per liter, the measuring unit that provides a basis for comparing relative concentrations of ions.

To transpose values recorded in milligrams per 100 ml of blood to milliequivalents per liter, it is necessary to know the equivalent

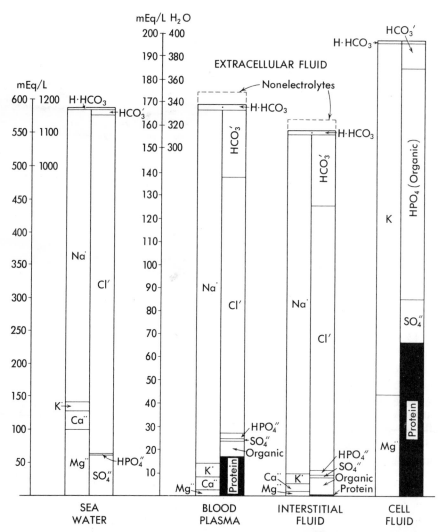

Figure 24–2. Chart illustrating the distribution of electrolytes and nonelectrolytes in the body. It will be noted that blood plasma and interstitial fluid are almost identical. Comparison with sea water shows differences. (Reprinted by permission of the publishers from James Lawder Gamble, *Chemical Anatomy, Physiology and Pathology of Extracellular Fluid: A Lecture Syllabus.* Harvard University Press, Cambridge, Mass., Copyright, 1942, 1947, 1954, by the President and Fellows of Harvard College.)

weight of the ion. This is obtained by dividing the atomic weight of the element or formula weight by its valence. Thus, for sodium with an atomic weight of 23 and a valence of one, the equivalent weight of the sodium ion is $^{23}/_1 = 23$. To transpose from milligrams per 100 ml to milliequivalents per liter:

Na^+ 330 mg per   100 ml blood plasma
        3300 mg per 1000 ml blood plasma
                                    (1 liter)

Therefore, $^{3300}/_{23} = 143$ milliequivalents of Na^+ per liter, expressed 143 mEq/L.

Sodium and chloride ions are in higher concentration in plasma and interstitial fluid than in intracellular fluid. Sodium ion concentration, to a large degree, controls the movement of water between cells and the interstitial fluids. Potassium ions are in much higher concentration within cells than in the extracellular fluids. The distribution of negatively charged ions is also of significance. Phosphates, sulfates, and proteinates predominate within cells where their presence is attributed to metabolic activities in the cells. The electrolytes function in the control of osmotic pressures in the various compartments and thereby help to regulate water balance. They help to maintain the acid-base balance essential for normal cell activities. In addition, many of these electrolytes have specific functions, which have been discussed in Chapter 21.

Hormonal Regulation of Electrolytes. Certain electrolytes are regulated by hormones. A high potassium level of extracellular fluid and a low sodium level prompt the secretion of *aldosterone* (via the renin-angiotensin system); this causes tubular reabsorption of sodium and loss of potassium. A high serum calcium level prompts secretion of *calcitonin;* low calcium stimulates release of *parathyroid hormone.* These together help keep calcium levels within normal limits: calcitonin, by inhibiting bone resorption; parathyroid hormone, by increasing bone resorption, increasing gastrointestinal absorption of calcium, and increasing tubular reabsorption of calcium. Parathyroid hormone also influences phosphate level by decreasing its tubular reabsorption. This is an important effect, since the ionic level of calcium is related to the calcium: phosphate ratio—as phosphate rises, calcium decreases, and vice versa. It is the *ionic* calcium level that stimulates hormone secretion.

Acid-Base Regulation

Within the body as a whole, the balance between acid and alkaline components is maintained within a remarkably constant range. This is strikingly illustrated by the range within blood plasma, which normally varies only between pH 7.35 and 7.45, with an average value of pH 7.4. (See Figures 24–3 and 24–4.) Extremes beyond 7.0 and 7.8 are life threatening.

The *sources* of acids and bases in the body are (1) the cells and (2) the ingestion of food. The cells continuously form carbon dioxide through the metabolism of carbohydrates, fats, and proteins. This forms carbonic acid by combination with water. Lactic acid is a product of carbohydrate metabolism; acetoacetic acid is formed in the liver from fatty acids; the sulfate ion is formed from certain amino acids and is a potential source of sulfuric acid. These substances escape from the cells and, when ionized, tend to increase the hydrogen ion concentration of blood and tissue fluids.

Various organic acids enter with the food, for instance, acetic, citric, and tartaric acids. Fatty acids are also absorbed from the intestine as end products of fat digestion. These tend to increase the hydrogen ion concentration of blood. Plant cells contain salts, chiefly potassium salts of weak acids. Sodium salts, such as so-

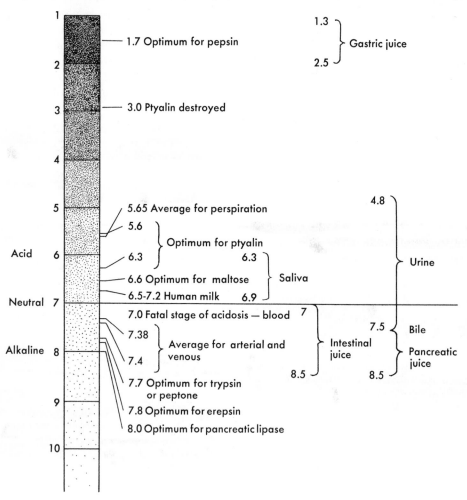

Figure 24–3. The pH scale runs from 1 to 14. In physiology the pH range is comparatively narrow; figures represent average values.

dium chloride and sodium bicarbonate, are frequently added to foods during preparation.

Mechanisms for Regulation. The constancy with which the hydrogen ion concentration of blood is regulated is due to three well-integrated mechanisms. These are the presence of buffer systems in extracellular fluids and within cells, the removal of carbon dioxide in the lungs, and renal regulation of the bicarbonate buffer system.

Buffer Systems of the Body. These consist of weak acids accompanied by their sodium or potassium salts. The three important buffer systems are:

In Plasma:

$$\frac{\text{Carbonic acid}}{\text{Sodium bicarbonate}} \quad \frac{H_2CO_3}{NaHCO_3}$$

$$\frac{\text{Sodium dihydrogen phosphate}}{\text{Sodium monohydrogen phosphate}} \quad \frac{NaH_2PO_4}{Na_2HPO_4}$$

$$\frac{\text{Hydrogen proteinate}}{\text{Sodium proteinate}}$$

The efficiency of the buffers is determined by the ratio of acid to its salt.

In *cells* the same buffers occur but the salts are potassium salts.

A buffer system acts through an exchange of ions that results in reduction of H^+ ion concentration through the forma-

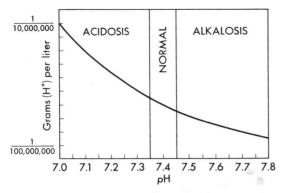

Figure 24-4. Diagram illustrating the range of hydrogen ion concentration compatible with life. It will be noted that the physiologic range lies on the alkaline side of neutrality. In health the hydrogen ion range is between pH 7.35 and 7.45. (Reprinted by permission of the publishers from James Lawder Gamble, *Chemical Anatomy, Physiology and Pathology of Extracellular Fluid: A Lecture Syllabus.* Harvard University Press, Cambridge, Mass., Copyright, 1942, 1947, 1954, by The President and Fellows of Harvard College.)

tion of molecules of a weaker electrolyte. This may be illustrated by reference to the functioning of the carbonic acid–sodium bicarbonate buffer system.

When any acid stronger than carbonic acid enters the blood, it will be buffered by the reaction with the sodium bicarbonate salt. Hydrogen ions will be removed to form molecules of carbonic acid and a sodium salt of the stronger acid. For example:

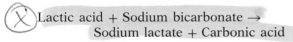

Lactic acid + Sodium bicarbonate →
Sodium lactate + Carbonic acid

The carbonic acid thus formed can be buffered by both the phosphate salt buffers or the protein. In fact, the most important buffer for carbonic acid is the potassium salt of hemoglobin within the erythrocytes. Most of the carbon dioxide that enters the plasma goes into the red blood cells. Part of it combines with water within the cell and forms carbonic acid, which in turn reacts with the hemoglobin salts to make the hemoglobin acid, a much weaker acid than is carbonic acid. In this way, respiration plays an important role in buffer systems.

Strong bases such as sodium hydroxide and potassium hydroxide are not usual components of food, nor are they products of cell metabolism. Nevertheless, intake of substances that tend to increase the salt of the buffer pair will influence the pH because the relative amount of each salt determines the degree of ionization of its appropriate acid. For example, as the sodium bicarbonate increases, there is less ionization of the carbonic acid, and vice versa. The various buffering mechanisms of cells, tissue fluids, and plasma thus serve to maintain the pH within rather narrow limits whether metabolic end products tend to lower or raise the pH.

The bicarbonate buffer system is the most significant aspect of acid-base regulation because the body has efficient means of altering the relative concentrations of carbonic acid and sodium bicarbonate.

Respiratory Regulation of pH. The normal ratio of carbonic acid to sodium bicarbonate is 1:20. At this ratio the ionization of carbonic acid is such that the free hydrogen ions are equivalent to a pH of 7.4. When the ratio changes to increase carbonic acid, there is more free hydrogen; when the relative amount of carbonic acid is less, there is less free hydrogen. *Increasing* the hydrogen ion concentration lowers the pH (and increases the acidity); *decreasing* the hydrogen ion raises the pH (increasing the alkalinity, or decreasing the acidity).

From this it is obvious that as respiratory rate and depth increase, more carbon dioxide is lost, thus decreasing carbonic acid concentration in the blood. As respiratory rate and depth decrease, as with shallow breathing, less carbon dioxide is removed, more carbonic acid is retained in the blood, and there is a change in the ratio of carbonic acid to sodium bicarbonate. Since low pH is a stimulus to the respiratory center, it is difficult to alter the pH voluntarily. However, drugs that depress the respiratory center, such as barbiturates, prevent normal removal of carbon dioxide since respirations are shallow and slow and cause a decrease in blood pH. Voluntary hyperventilation has the opposite effect but is self-limiting and is followed by a period of apnea during which the carbon dioxide content (and carbonic acid level) of the blood rise to a normal level. Normal elasticity of lung tissue is a requisite for efficient removal of carbon dioxide from the lungs. Any disease that decreases this elasticity results in retention of carbon dioxide and of carbonic acid, thus causing acidosis—a lower-than-normal pH.

Renal Regulation of pH. In the tubular cells of the kidney, as carbon dioxide is formed during cell activity in the citric acid cycle, it combines with water under the influence of carbonic anhydrase as in other cells, and carbonic acid is formed. Hydrogen ions from this acid are secreted into the filtrate in exchange for a sodium ion (Table 24–2). The secreted hydrogen ion replaces the sodium in the phosphate molecule and is lost in the urine. If the sodium were combined in sodium bicarbonate, the resulting H_2CO_3 in the filtrate is not all lost in the urine, for carbonic acid breaks down to CO_2 and water, and the carbon dioxide diffuses back into the tubular cell and is returned to the capillaries as bicarbonate ion or sodium bicarbonate.

Secretion of ammonia is a second means the tubular cell uses to regulate pH (Table 24–3). When the acidity of body fluids is low, glutamine is metabolized, from which NH_3 results. When ammonia is secreted into the filtrate, again a sodium ion is returned to the tubular cell and bloodstream. Ammonia secretion does not occur when body fluids are not acidic.

TABLE 24–2
CONSERVATION OF SODIUM BY THE KIDNEY*

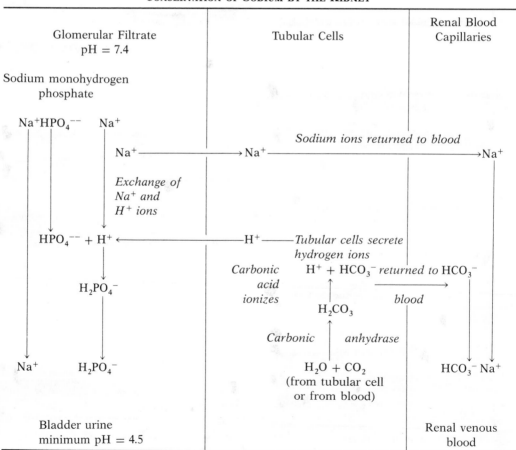

Glomerular Filtrate pH = 7.4	Tubular Cells	Renal Blood Capillaries
Sodium monohydrogen phosphate		

*Modified from Davenport.

TABLE 24–3
FORMATION OF AMMONIA BY THE KIDNEY*

Glomerular Filtrate	Tubular Cells	Renal Blood Capillaries

Cl^- Na^+

Na^+ ————— *enters tubular cell* —→ Na^+ ——— *returns to blood* ——→ Na^+

Exchange of ions

leaves

H^+ ←——————— H^+ and HCO_3^- ——— *returns to blood* ——→ HCO_3^-

tubular cell

Tubular cells remove
NH_2 ions
from glutamine and
some other amino acids

$H^+ + NH_3$ ←————————————— NH_3

Cl^- NH_4^- HCO_3^- Na^+

Urine   *Renal venous blood*

*Modified from Davenport.

Secretion of hydrogen ions and of ammonia in exchange for sodium in the filtrate results in return to sodium bicarbonate to the blood, thus tending to increase the bicarbonate fraction of the carbonic acid: sodium bicarbonate buffer pair.

Abnormalities of pH. In spite of the remarkable adjustments thus made among the electrolytes, if abnormal demands are made on the system, disturbance of the acid-base balance does occur and acidosis or an alkalosis may result, although blood pH in these conditions is still alkaline, i.e., above 7.0, except in extreme circumstances.

In *acidosis* the lower-than-normal blood pH may be of metabolic or respiratory origin. Respiratory acidosis results from inadequate removal of carbon dioxide from the body, thus preventing the usual regulation of carbonic acid level and altering the 1:20 ratio. Metabolic acidosis (see Figure 24–5) results from production of substances that are acidic and "tie up" temporarily sodium ions so that the net effect is to lower the sodium bicarbonate portion of the 1:20 ratio. Such substances are produced in excessive fat oxidation, as in starvation or diabetes mellitus. Another means of decreasing sodium bicarbonate is the loss from the body of positively charged ions, potassium and sodium both, as occurs in diarrhea.

Alkalosis is present when the pH of body fluids is higher than normal, and less free hydrogen ion is present in the bloodstream. Voluntary hyperventilation has already been mentioned as a possible cause. Others include excessive vomiting, with loss of hydrogen ion in the hydrochloric acid of the gastric juice, gastric drainage, which has a similar effect, and excessive ingestion of sodium bicarbonate as a medicament.

Acidosis and alkalosis occur only when the compensatory mechanisms of the body have been exhausted, or when they

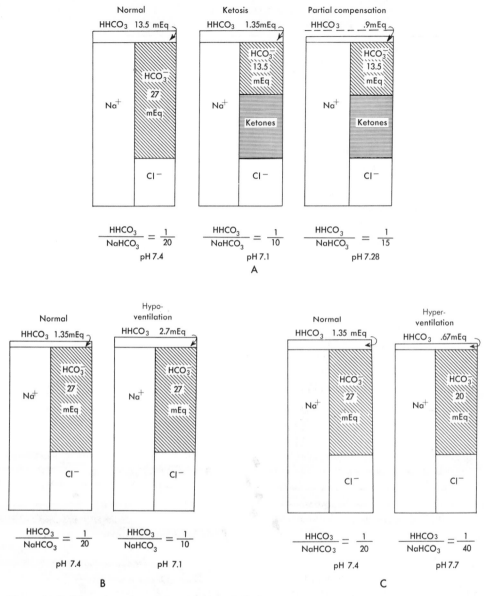

Figure 24–5. Electrolyte changes in acidosis and alkalosis. *A.* Diabetic acidosis with partial compensation. Note the decrease in bicarbonates and the increase of ketones. *B.* Comparison of the ratios of carbonic acid and bicarbonate in respiratory acidosis. *C.* Ratios of carbonic acid and bicarbonate in the development of respiratory alkalosis caused by hyperventilation.

are incompetent from disease processes involving the lung and kidney. *Compensatory mechanisms* in acidosis include hyperventilation, and renal excretion of hydrogen and ammonia with reabsorption of sodium ions. The urine may be more acid than normal (pH 5.5 to 7.5) depending on the amount of ammonium salts present. In alkalosis the kidney does not secrete hydrogen ions, permits sodium to be lost, and ventilation is shallow. By these means the blood pH is kept within its narrow range, in spite of varying dietary pattern and cellular activity.

Questions for Discussion

1. An adult female was admitted to the hospital with a history of vomiting for two days. Urine output was scant. The doctor's orders included intravenous injection of physiologic saline with 5 per cent glucose.
   a. What electrolytes were being lost by vomiting?
   b. What is the usual amount of gastric juice formed each 24 hours?
   c. Would there be a tendency for this patient to develop acidosis or alkalosis? Explain.
   d. Why was the urine output decreased in amount?
   e. Discuss why sodium chloride and glucose solution was ordered.

2. When an individual perspires profusely (e.g., during a baseball game on a hot day), how is total body fluid kept within normal limits?

3. Which hormones are essential for maintaining the constancy of body fluids in relation to electrolytes?

4. When faced with acute lack of water, which would suffer most: the mother, the father, or a six-month-old infant? Which least? Explain.

5. When an individual takes one large teaspoonful of bicarbonate of soda in water for indigestion, what happens to the blood pH?

Summary

Water Balance

Distribution of water in the body
1. Blood plasma 5% of body weight
2. Interstitial fluid 11% of body weight
3. Intracellular fluid 44% of body weight

Movement of water
Water and solutes move freely between compartments

Water intake, 1800 to 3000 ml in 24 hours
Water
Other fluids
Foods
Water metabolism 300–350 ml in 24 hours

Water output
Feces   100 to 150 ml
Skin   450 to 800 ml
Lungs   250 to 350 ml
Urine   1000 to 1500 ml

Adjustment to excess fluid intake
Fluid rapidly absorbed into bloodstream
Blood volume may be temporarily increased
Cardiac output may be increased
Capillary networks open, vascular bed increased
Sinusoids of liver and spleen hold more blood
Fluid rapidly transferred to interstitial compartment
Kidney eliminates excess water and balance is restored in several hours

Regulation by hormones
ADH regulates water reabsorption in the kidney tubule

Stress situations
ACTH and the adrenal cortex hormones respond to stress situations and play a role in maintaining homeostasis of the body fluids

Reserve body water
Ability of individual to withstand water loss from any cause
Person with greatest percentage of body water has best reserve
Obesity reduces water reserve
Infants have poor reserve

Electrolyte Balance

An electrolyte may be an acid, base, or salt that, in solution, has the power to conduct an electric current. Some ionize more than others

Distribution
In extracellular fluids — Chiefly Na^+, Cl^-, HCO_3^-; small quantities of K^+, Ca^{++}, HPO_4^{--}, SO_4^{--}, Mg^+
Intracellular fluids — Chiefly K^+, Mg^{++}, HPO_4^{--}; small quantities of SO_4^{--}, HCO_3^-, Na^+

Milliequivalents
Concentrations expressed in terms of milliequivalents — Shows relative magnitudes and interrelationships

Electrolyte Balance *(cont.)*		Important cations—	Important anions—

Important cations—
Na$^+$ 142 mEq/L
Ca^{++} 5 mEq/L
Mg^{++} 3 mEq/L
K$^+$ 5 mEq/L

Important anions—
Cl$^-$ 103 mEq/L
HCO$_3^-$ 27 mEq/L
HPO$_4^{--}$ 2 mEq/L
SO$_4^{--}$ 1 mEq/L
Organic acids   6 mEq/L
Protein   16 mEq/L

Electrolyte regulation
Kidneys most important for regulation
Aldosterone controls sodium, potassium levels
Parathyroid hormone {
Calcitonin { Control calcium level

Acid-Base Balance

Normal pH range: 7.35–7.45
Depends on degree of hydrogen ion concentration, which is determined by carbonic acid: sodium bicarbonate ratio of 1:20

Buffering
Potassium salts buffer within the cells
Sodium salts buffer in the extracellular fluids
Cell proteins, especially hemoglobin, are effective buffers
Buffering is accomplished through exchange of ions
Acids react with buffer salts to form weaker acids
Bases react with buffer acids to form water and salts

pH regulated by
1. Secretion of acid urine, removing hydrogen ions
2. Production of ammonia in kidney cells with conservation of sodium ions in blood
3. Elimination of carbonic acid in lungs

Sources of acid and base
1. The cell. Products of metabolism of {
  Carbohydrates
  Fats
  Proteins
} {
  Carbon dioxide and water, lactic acid, acetoacetic acid, and from protein, sulfate and phosphate ions
}

2. From foods. The average diet provides various acids and salts of organic acids {
  Citrate
  Lactate
  Acetate } Ions
  Fatty acid
  Potassium salts
  Sodium bicarbonate
}

Acidosis, alkalosis
Pathologic conditions when compensatory mechanisms are inadequate

Additional Readings

DAVENPORT, H. W.: *The ABC of Acid-Base Chemistry,* 6th ed. University of Chicago Press, Chicago, 1973.

GUYTON, A. C.: *Basic Human Physiology: Normal Functions and Mechanisms of Disease,* 2nd ed. W. B. Saunders Co., Philadelphia, 1977, Chapters 25 and 26.

SELKURT, E. E.: *Basic Physiology for the Health Sci-* ences. Little, Brown & Co., Boston, 1975, Chapters 22, 23, and 24.

VANDER, A. J.; SHERMAN, J. H.; and LUCIANO, D. S.: *Human Physiology: The Mechanisms of Body Function,* 2nd ed. McGraw-Hill Book Co., 1975, pp. 331–54.

Perpetuation of the Species

Human sexual functions have as a goal reproduction of the individuals involved, and additionally a larger goal—perpetuation of the human species. The male is responsible for sex determination; the female is responsible for nourishment of the fertilized ovum following conception. Male and female together contribute inherited traits and characteristics, which are determined by the chromosomes in the spermatozoon and the ovum. During intrauterine life, cell division and differentiation result in formation of distinctly different tissues and organs typical of the human. In this process characteristics of one individual are transmitted from generation to generation.

CHAPTER 25

The Reproductive System

Chapter Outline

Perpetuation of the human species requires development of separate male and female *gametes, or sex cells*; the union of these cells; and a hospitable environment for development of the product of the union prior to birth. The specialized organs for accomplishing this constitute the male and female reproductive systems.

Sexual Maturation

Although sex is determined at the time of union of the male and female gametes (*conception*), the reproductive organs do not acquire their morphologic characteristics until the end of the second month of development. Functional maturity is attained at the time of puberty. *Puberty* may be defined as the period when the *gonads or sex glands attain normal adult function*. In temperate climates, the age at which boys usually attain puberty is between 14 and 16 years; in girls, puberty is signaled by the beginning of the menses and occurs between the ages of 11 and 14. In warmer

573

climates, puberty often occurs earlier, and in the Arctic regions, one or two years later. However, the time of puberty varies from individual to individual. Psychologic factors producing stress at the time of puberty may cause physiologic changes that induce early puberty or impede its progress. The onset of puberty is initiated by hypothalamic releasing factors, which stimulate secretion of follicle-stimulating hormone (FSH) and luteinizing hormone (LH). (See Chapter 14, page 317.)

Physical Changes During Puberty. The onset of puberty is signaled in the male by production of functional spermatozoa and in the female by the beginning of ovulation and menstruation. At birth, the testes contain thousands of immature spermatocytes and the ovaries contain thousands of partially developed germ cells, but these cells do not mature until the onset of puberty. Puberty is also marked by the gradual appearance of the secondary sex characteristics induced by the increasing amounts of circulating sex hormones in the body.

In the *male*, increased secretion of testosterone by the testes causes the increased development of skeletal muscle, enlargement of the external genitalia, and hair growth in the axillae and on the face, pubes, and to a varying degree on the extremities. The male larynx increases in size and accentuates the thyroid prominence called the "Adam's apple." The vocal folds thicken, and the male voice becomes lower.

The *girl* undergoes a gradual change of figure; the pelvis widens, and fat deposits increase around the hips, thighs, and buttocks. Hair grows on the pubes and in the axillae. Subcutaneous fat deposition increases in the breasts. The menstrual cycle is initiated, and cyclical changes begin to occur in the ovaries, uterus, and vagina. Increased production of estrogens and progesterone by the ovaries and their feedback effects upon pituitary gonadotropic secretion bring about these changes in the female figure and reproductive organs.

These secondary sex alterations begin at puberty and continue to develop over a number of years. This period is known as *adolescence*, and it extends from puberty until the age of 17 to 20 years in the female and until the age of 18 to 21 years in the male. At the age of 20 and 21, in the average man and woman the rapid increase in height characteristic of adolescence comes to an end owing to the closure of the epiphyseal plates of bones. The complete development of all secondary characteristics has also been achieved.

Male Organs of Reproduction

The male reproductive organs include two *testes* (producing the male sex gametes, spermatozoa, and the male hormone, testosterone); the bilateral accessory organs—*seminal vesicles, ductus deferens* (*vas deferens*), *ejaculatory ducts, epididymides* (singular *epididymis*), and *bulbourethral glands*; as well as the following single structures—*prostate gland, penis, urethra,* and *scrotum*. (Figures 25–1 and 25–2 illustrate these organs.)

Testes and Epididymides. The two *testes* are glandular organs lying within the scrotum. Each testis is about the size and shape of a small egg and is attached to an overlying structure called the epididymis. The testes are covered by fibrous tissue, the tunica albuginea, which sends incomplete partitions into the central portion of each gland, dividing it into communicating cavities. In these cavities are winding seminiferous tubules surrounded by blood vessels and supported by interstitial tissue. The seminiferous tubules, which produce spermatozoa (discussed on page 579), intertwine and join together in a meshwork of exiting small ducts called the *rete testis* and finally all unite in the epididymis.

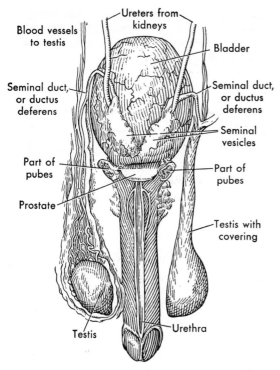

Blood vessels to testis

Ureters from kidneys

Bladder

Seminal duct, or ductus deferens

Seminal duct, or ductus deferens

Seminal vesicles

Part of pubes

Part of pubes

Prostate

Testis with covering

Testis

Urethra

Figure 25–1. Male reproductive organs, dorsal view.

Cells in the interstitial tissue produce the male hormone, testosterone, which is secreted in relatively steady amounts during adult life as a result of the negative feedback mechanism with LH (ICSH). When plasma testosterone level falls, more LH is secreted; when it rises, less LH is secreted. The interstitial cells are often called the *cells of Leydig.*[1]

The *epididymides* are long bilateral narrow bodies that lie on the superior portions of the testes and are composed of 15 to 20 tortuous tubules that eventually open into a single convoluted tubule. These tubules contain smooth muscle cells in their wall and are lined with mucous membrane. They connect the testes with the *ductus deferens* and serve as areas for final maturation of the spermatozoa.

Descent of the Testes. The testes are formed and develop in the abdomen slightly inferior to the kidneys, behind the peritoneal cavity. During growth and de-

[1] Franz von Leydig, German histologist (1821–1908).

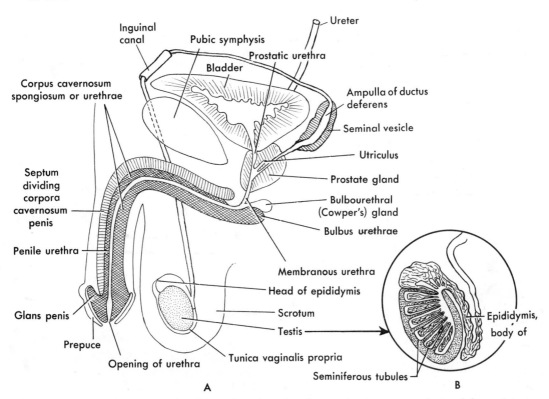

Figure 25–2. *A.* Diagram of midsagittal section of male reproductive organs; ductus deferens is curving laterally from midline. *B.* Detail of testis structure.

velopment of the fetus, they migrate downward and through the inguinal canal into the scrotum. Shortly before birth or soon afterward, the testes are in the scrotum. Sometimes, particularly in premature infants, a testis has not descended and is found in the inguinal canal or even in the abdominal cavity. As a rule, it soon descends; however, if it does not, the condition is referred to as *cryptorchidism* and may be either unilateral or bilateral. If the testis remains inside the abdomen after puberty, spermatogenesis is depressed owing to the slightly higher intraabdominal temperature, and eventually it ceases altogether. It can be treated successfully prior to puberty by a simple surgical procedure.

Scrotum. The scrotum, covering testes and epididymides, is a thin pouch of skin, smooth muscle, and fascia. The skin is in folds so that the testes are mobile within it. The smooth muscle contracts reflexly to raise the testes closer to body warmth in a cold environment, and when the skin of the upper thigh is briskly stroked, e.g., with a tongue blade (the *cremasteric reflex*). Originally two separate pouches, developmentally the medial walls fuse to form right and left halves. The wall in each half represents outpouching of the abdominal wall, with layers of fascia continuous with the fascia from three muscles forming the abdominal wall (transversalis, internal oblique, and external oblique). These layers cover the spermatic cord as well. Skin and subcutaneous tissue of the scrotum are continuous with that of the perineum and groin.

Ductus Deferens, Seminal Vesicles, and Ejaculatory Ducts (Figure 25–2). The right and left *ductus deferens,* continuations of the epididymides, are important storage sites for spermatozoa and are the excretory ducts of the testes. Each ductus runs from the epididymis up through the inguinal canal, then downward and backward lateral to the bladder. It then curves medially above the ureter, and down to join the duct of the seminal vesicle on the poste-

rior aspect of the bladder. The distal portion of the duct, the *ampulla,* is enlarged.

The *seminal vesicles* are two membranous tubes, giving the appearance of two pouches, but in reality they are long, convoluted tubes with blind superior endings. Inferiorly the duct joints the ductus deferens to form the ejaculatory duct on either side. The seminal vesicles produce secretions containing fructose, amino acids, mucus, prostaglandins, and small amounts of ascorbic acid. During ejaculation, these substances are added to the semen at the time spermatozoa are transported to the ejaculatory ducts from the vas deferens. The fructose and other substances contained in the seminal fluid provide nutrients and protection for the spermatozoa.

The *ejaculatory ducts* are short and very narrow; they begin at the point of union of the ductus deferens and the duct of the seminal vesicle, bilaterally, descending into the prostate gland to join the ureter, into which they discharge their contents.

Spermatic Cords. The fibrous tissue running from the abdominal fascia to the scrotum forms the two spermatic cords. They contain the ductus deferens, arteries and veins to and from the scrotum, lymph vessels, and autonomic nerve fibers. Beginning at the point of the deep inguinal rings, they pass through the inguinal canals to the scrotum.

Inguinal Canals. These bilateral short canals are formed by the descent of the testes to the scrotum, which creates openings in the inguinal fascia of the abdominal muscles. Each *deep (internal) inguinal ring* is a small opening in the lower border of the transversalis fascia, just above the midpoint of the inguinal ligament (Figure 1–6, page 15). Each *superficial (external) inguinal ring* is a small opening in the external oblique aponeurosis, just lateral to the pubic tubercle. The inguinal canals are formed by fibrous connective tissue between these rings and continuous with that of the scrotum and that of the abdominal muscle fascia. They run obliquely above the inguinal ligament in a posterolateral direction. Since the inguinal canals are weak spots in the lower abdominal wall, excessive abdominal pressure, as in lifting heavy objects, may push a loop of small intestine (and peritoneum) into the tract. This condition is called a *hernia*.

Penis (Figure 25–1, and 25–2). The penis, or organ of copulation of the male, is a short, cylindrical, pendulous body that is suspended from the front and sides of the pubic arch. It is composed of three cylindrical masses of cavernous erectile tissue bound together by fibrous strands and covered with skin. The lateral two masses are known as the *corpora cavernosa penis.* The third, known as the *corpus cavernosum urethrae* or *spongiosum*, makes up the ventral surface and contains the urethra. The term *cavernous* is used because of the relatively large venous spaces present within its structure. It is also described as erectile tissue because the venous spaces may become distended with blood during sexual excitement, and the penis then becomes firm and erect.

At the end of the penis there is a slight enlargement known as the *glans penis*, which is part of the corpus spongiosum. It forms a cap over the ends of the corpora cavernosa. The glans contains the external urethral orifice (meatus) and the sensory end organs that are stimulated during sexual intercourse. The loose skin and fascia of the penis form a fold over the glans—the *prepuce*, or *foreskin*. Sometimes the foreskin may cover the glans too tightly causing restricted circulation to the area and collection of the sebaceous secretion, called *smegma*, which provides a good environment for microbial multiplication and inflammation. Constriction of the foreskin is known as *phimosis* and is often prevented and treated by the operation known as *circumcision*, surgical removal of the foreskin.

The *male urethra* is an S-shaped tube lined with mucous membrane, about 17.5 cm long, extending from the internal to the external urethral orifice. It serves at separate times as conveyor of both urine and semen to the exterior. There are three parts: the *prostatic*, the *membranous*, and the *cavernous*, or *penile* (Figure 25–2).

Prostate Gland. The prostate gland is situated immediately inferior to the bladder and internal urethral orifice. It surrounds the first portion of the urethra, referred to as the *prostatic urethra*, and is comparable to a chestnut in shape, size, and consistency. The prostate is covered by a dense fibrous capsule and consists of glandular units surrounded by fibromuscular tissue that contracts only during ejaculation. The glandular tissue consists of tubules that communicate with the urethra by minute orifices. The function of the prostate gland is to secrete a thin, milky alkaline fluid that enhances spermatic motility. Fluid in the ductus deferens is quite acidic owing to the presence of the metabolic end products of the stored sperm. The secretions of the female vagina are also quite acidic. Therefore, it is probable that alkaline prostatic fluid neutralizes the acidity of the semen and vaginal secretions and thereby greatly increases the motility and fertility of the spermatozoa.

Age Changes in the Prostate. The prostate enlarges during adolescence along with the other reproductive organs owing to the effect of androgens secreted by the interstitial cells of the testes. It attains full size during the twenties. In older age, for reasons not yet understood, frequently the prostate increases in size so that two out of every three men reaching the age of 70 suffer from some degree of obstruction to urination. This obstruction is due to the decreased size of the lumen of the prostatic urethra, which can result in a degree of obstruction to urine flow.

Bulbourethral Glands. These are two small bodies about the size of peas situated on either side of the membranous portion of the urethra a little inferior to the prostate gland. Each small gland is provided with a short duct that empties its mucous secretion, a viscid alkaline fluid, into the urethra.

Blood Supply to the Male Pelvis (Figure 16–10, page 366). When the descending aorta reaches the body of the fourth lumbar vertebra, it divides into the two *common iliac* arteries. These arteries pass downward and outward for about 5 cm (2 in.), and then each divides into the hypogastric, or internal, and external iliac arteries.

The *internal iliacs* are short, thick vessels that descend to the greater sciatic foramen, where they divide into anterior and posterior divisions. The *posterior division* passes backward and supplies parietal branches, the *iliolumbar* and *lateral sacral* arteries, to the pelvic wall; and the *superior gluteal artery* to the gluteal muscles and hip joint. The *anterior division* is a direct continuation of the internal iliac. It divides into three branches to the thigh and perineum: the *obturator, internal pudendal,* and *inferior gluteal arteries,* as well as branches to the pelvic viscera: the *umbilical, inferior vesical,* and *middle rectal arteries.* A summary of the regions supplied by the branches of the anterior division is as follows:

Obturator	Urinary bladder, pubis, hip joint
Internal pudendal	Perineum, scrotum, penis
Inferior gluteal	Lower portions of gluteal muscles, skin and tissues over coccyx, back of thigh
Umbilical	Ductus deferens, superior surface of urinary bladder
Inferior vesical	Prostate, inferior portion of urinary bladder
Middle rectal	Middle portion of rectum

The external iliacs extend from the bifurcation of the common iliacs to a point halfway between the anterior superior spines of the ilia and the symphysis pubis. They enter the thigh and become the femoral arteries.

The external iliacs send small branches to the psoas major muscles and to the neighboring lymph nodes, and each gives off the *inferior epigastric* and the *deep iliac circumflex.* The *cremasteric artery,* which supplies the pubes and the cremasteric muscle of the scrotum, is a branch from the inferior epigastric arteries.

The testicular arteries arise from the front of the aorta, a little below the renal arteries. They give off branches to the ureters and supply the testes.

The veins of the pelvis correspond to the arteries with one exception. The right spermatic vein empties into the inferior vena cava; the left, into the left renal vein.

Ejaculation. The sensory nerve endings of the glans penis initiate the male sexual act; impulses are transmitted to the sacral plexus of nerves, to the spinal cord, and up the cord to the cerebrum. Integration of impulses occurs at lower levels in the brain, including autonomic control centers, as well as in the spinal cord.

Erection is the first effect of male sexual stimulation, and it is brought about by parasympathetic impulses that pass from the sacral portion of the spinal cord via the *nervi erigentes* (pelvic splanchnic nerves) to the penis. These impulses cause dilatation of the penile arteries, which results in compression of the exiting veins. Increased blood supply that is under high pressure and unable to leave the area results in filling of the venous spaces and erection of the organ.

Ejaculation, or the discharge of semen to the exterior, is initiated by peristaltic waves moving along the tubes leading from the testes and by rhythmical contractions of the smooth muscle layers of the testes, epididymides, seminal vesicles, and prostate gland. Increased pressure upon all these structures causes expulsion of the semen. The bulbourethral glands discharge additional quantities of mucus into the urethra at this time. The process to this point is called *emission* and is brought about by rhythmical sympathetic impulses, which leave the spinal cord at L_1 and L_2 and then pass through the hypogastric plexus to the genital organs. Ejaculation proper is brought about by contraction of the skeletal muscles that encase the base of the erectile tissue of the penis and that are innervated by fibers traveling in the pudendal nerves. In some stages of sexual life, especially during the teens, the male may have nocturnal emissions during dreams.

Semen is the fluid that is ejaculated during the male sexual act. It is composed of the combined fluids from the testes, epididymides, seminal vesicles, prostate gland, and bulbourethral glands—a grayish-white viscid liquid that contains carbohydrates, mucin, proteins, salts, and about 100,000,000 spermatozoa per milliliter. At each ejaculation, 2 to 5 ml of semen are usually expressed through the urethra.

Spermatogenesis. At the time of puberty, the onset of secretion of follicle-stimulating hormone (FSH) from the adenohypophysis, under the influence of hypothalamic releasing factor, initiates the development of the spermatozoa in the seminiferous tubules (Figure 25–3) of the testis. From this time, FSH is produced at a relatively steady rate (in contrast to the female, in which it is cyclical), and the production and maturation of germ cells, or spermatogenesis, continues throughout adult life. Germ cells called spermatogonia, formed during fetal life, now begin to proliferate and differentiate through def-inite stages to form spermatozoa. As these cells divide, increase in number, and move toward the center of the tubule, they become *primary spermatocytes*. Primary spermatocytes then go through a meiotic, or reduction, division, during which the number of chromosomes of the developing cells is halved from 46, the *diploid* number, to 23, the *haploid* number. (See Figure 25–5.)

Each primary spermatocyte divides into two *secondary spermatocytes*, each of which contains 23 double chromosomes. The secondary spermatocytes each divide again to form two *spermatids* with 23 single chromosomes contained within each one. In summary, then, from every primary spermatocyte, four spermatids are derived. The process of meiosis apportions the sex chromosomes so that each spermatozoon receives one. One half of the spermatozoa carry the X chromosome, and one half carry the Y chromosome. The sex of the fertilized ovum is determined by the sex chromosome contributed by the spermatozoon because all ova normally contain one X chromosome. When a spermatozoon enters an ovum at fertilization, the original complement of 46 chromosomes is reestablished.

The *spermatid* develops into a *spermatozoon* when a tail, or flagellum, is formed, and the cytoplasm has contracted around the cell nucleus to form the head. The mature spermatozoon (Figure 25–4), then,

Figure 25–3. Seminiferous tubules, as seen with light microscope. In the center two tubules are seen cut in cross section; *L* indicates lumen of tubules; rounded, darkly stained bodies are the nuclei of spermatozoa at different stages of development.

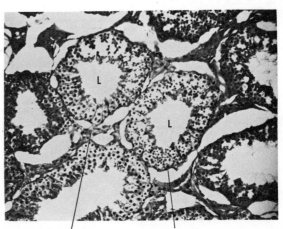

Interstitial tissue

Seminiferous tubule

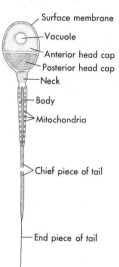

Surface membrane
Vacuole
Anterior head cap
Posterior head cap
Neck
Body
Mitochondria
Chief piece of tail
End piece of tail

Figure 25–4. Diagram of spermatozoon.

has four sections: a head, neck, body, and tail. The body contains many mitochondria, producing ATP, the source of energy for movement of the tail. The structure of the tail is similar to that of the cilia, with paired microtubules in the center surrounded by nine pairs of microtubules. The action of the tail, waving back and forth, propels the spermatozoon forward at a speed up to 20 cm per hour. Hyaluronidase and various proteases contained in the head, in the modified Golgi body, are important in entry of the sperm into the ovum at the time of fertilization.

Normal spermatogenesis requires FSH and LH, the gonadotropic hormones, testosterone, and the action of Sertoli[2] cells. Sertoli cells are found in the seminiferous tubules; they extend from the base of the epithelium to the lumen of the tubule. The spermatids attach themselves to the Sertoli cells; these cells provide nourishment or, possibly, hormones or enzymes that change the spermatids into mature spermatozoa. FSH is responsible for the changing of primary spermatocytes into

[2]Enrico Sertoli, Italian histologist (1842–1910).

secondary spermatocytes; without FSH this change does not occur. LH promotes the formation and secretion of testosterone by the interstitial cells. Testosterone action in spermatogenesis is poorly understood, but it seems to be necessary. During puberty it causes the testes to enlarge, as well as the scrotum and penis; and, as described earlier in the chapter, it is responsible for the secondary sex changes.

Spermatozoa are formed at the edge of the lumen of the seminiferous tubules and reach the epididymis via a series of excretory ducts called the *tubili recti,* the *rete testes,* and the *efferent ductules.* They are conveyed by the motion of ciliated epithelium, for at this stage of their development they are nonmotile. Sperm most probably mature within the epididymis, since after they have remained there for one-half day or more, they have developed the power of motility and the capability of fertilizing the ovum. They are stored in the ductus deferens, where they remain dormant within the acid medium resulting from their own metabolism. Viable sperm may be stored in the epididymis and vas deferens for as long as six weeks, but once ejaculated, they survive at normal body temperature for only 24 to 72 hours.

Male Fertility. Approximately 100 million spermatozoa are contained in each milliliter of ejaculated semen. These are highly uniform in size and shape, with occasional sperm having two heads or two tails or being otherwise abnormal. When the percentage of abnormal spermatozoa is greater than 25 percent, or the sperm count is less than 20 million per milliliter, fertility is greatly decreased. Even though only one spermatozoon fertilizes each ovum, a large number of normal ones is necessary to ensure fertilization.

Male Climacteric. Hormonal function usually does not significantly decline at a particular period in the life of the male as it does in the female. If and when testosterone production by testicular Leydig cells does decrease, symptoms may appear that are similar to those observed in menopausal women.

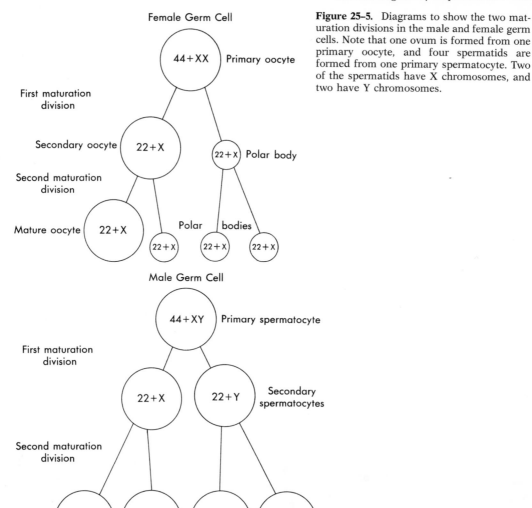

Female Germ Cell

44+XX | Primary oocyte

First maturation division

Secondary oocyte | 22+X

22+X | Polar body

Second maturation division

Mature oocyte | 22+X

Polar | bodies

22+X 22+X 22+X

Male Germ Cell

44+XY | Primary spermatocyte

First maturation division

22+X   22+Y | Secondary spermatocytes

Second maturation division

22+X 22+X 22+Y 22+Y | Spermatids

Figure 25–5. Diagrams to show the two maturation divisions in the male and female germ cells. Note that one ovum is formed from one primary oocyte, and four spermatids are formed from one primary spermatocyte. Two of the spermatids have X chromosomes, and two have Y chromosomes.

Female Organs of Reproduction

The female organs of reproduction include the internal organs—bilateral *ovaries* and *fallopian* (*uterine*) *tubes* or *oviducts*—and the *uterus* and *vagina*. The external genitalia are *bilateral labia majora, labia minora, glands of Bartholin,*[3] the *clitoris,* and *mons pubis.* (See Figure 25–6; see also Plate IV, inserted between pages 536–37.) Certain female reproductive organs are

homologous to male organs, since they develop from similar embryologic structures. The clitoris is comparable to the penis, the labia majora to the scrotum, and the ovaries to the testes.

Ovaries. Each ovary is a slightly flattened, almond-shaped body measuring from 1.5 to 3 cm in width and about 8 mm in depth. One is located on each side of the pelvis, lateral to the uterus, below the fallopian

[3]Caspar Bartholin, Danish anatomist (1655–1738).

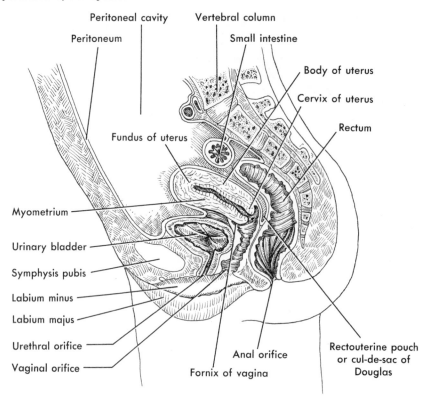

Peritoneal cavity   Vertebral column

Peritoneum   Small intestine

Body of uterus

Cervix of uterus

Rectum

Fundus of uterus

Myometrium

Urinary bladder

Symphysis pubis

Labium minus

Labium majus

Urethral orifice

Vaginal orifice

Anal orifice

Fornix of vagina

Rectouterine pouch
or cul-de-sac of
Douglas

Figure 25–6. Median sagittal section of female pelvis.

tubes and attached to the posterior surface of the broad ligament. The *broad ligament* is a reflection of peritoneum that
supports the uterus and extends from each
side of the uterus to the lateral pelvic wall.
In addition, each ovary is attached to the
lateral angle of the uterus by a short *ovarian ligament*, a fibrous cord within the
broad ligament, and to the end of the fallopian tube by the largest of its fringelike
process. The ovaries produce the ova and
the female sex hormones, estrogen and
progesterone. (Ovulation is discussed on
page 586.) The ovaries, like the testes, develop during fetal life in a position below
the kidneys, but they descend a shorter
distance, remaining within the pelvic
cavity.

Fallopian (Uterine) Tubes (Figure 25–7).
These are bilateral muscular ducts, lined
with mucous membrane, that pass from
the upper angles of the uterus in a curve
between the folds and along the superior
margin of the broad ligament toward the

sides of the pelvis. They are about 10 cm
long, and the margin of the dilated end, or
ampulla, is surrounded by a number of
fringelike processes called *fimbriae*. The
lumen of the tubes varies in diameter,
being narrow in the portion closest to the
uterus where the muscle wall is thickest,
and wider in the terminal portion (ampulla
and *infundibulum*). The mucous membrane of the wall, arranged in longitudinal
folds, is covered with ciliated epithelium.

Since the ovum is not motile, unlike the
spermatozoon, the fallopian tubes convey
the ova from the ovaries toward the uterus
by peristaltic action of the smooth muscle
of the walls, and by the action of the ciliated epithelium.

Uterus (Figure 25–7). The uterus is a hollow, thick-walled, pear-shaped muscular
organ about 7 cm long, situated in the pelvic cavity between the rectum and the
bladder. Three parts of the uterus can be
distinguished: (1) the fundus, the portion
above the level of entrance of the uterine

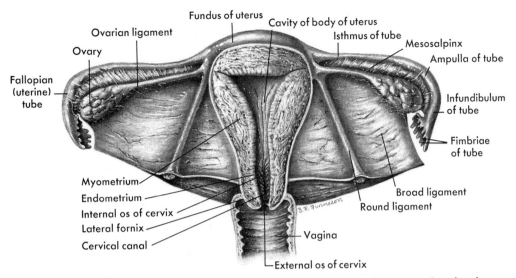

Figure 25–7. Uterus in section showing relation to ovary, uterine tube, and ligaments. Anterior view.

tubes; (2) the body, or middle portion; and (3) the cervix, or cylindrical lower part, which surrounds the *cervical canal* and projects into the vagina. The short (2.5-cm) cervical canal extends from the *internal orifice,* or *os,* of the uterus to the *external os* at the termination of the cervix.

The cavity of the uterus is small because of the thickness of its walls. The part of the cavity within the body is triangular and has three openings, one very small one at each upper angle communicating with the fallopian tubes, and the third, the *internal os,* opening into the cervical canal. The uterus is the organ of the reproductive tract in which the embryo grows and develops.

Structure of Uterus. The thick wall of the uterus has three layers. The external layer is the *serosa.* The peritoneum covers the fundus and much of the posterior surface of the uterine body and is reflected from the body to the bladder as the *vesicouterine* pouch. Posteriorly it may descend as far as the upper part of the vagina before reflection to the rectum as the *rectouterine pouch* (*cul-de-sac of Douglas*). (See Figure 25–6.) Laterally it becomes the broad ligament.

The middle, *muscular layer* is about 2 cm thick and is called the *myometrium.* It consists of three ill-defined layers of smooth muscle disposed longitudinally, circularly, and spirally. Peristaltic-like movements of the muscles are thought to be increased around the time of ovulation. It is believed that these contractions assist the ascension of spermatozoa by the suction they produce within the organ. The larger blood vessels are found deep within the myometrium.

The inner, *mucous layer* is called the *endometrium.* It is continuous with the mucous membrane that lines the vagina and uterine tubes. The endometrium is highly vascular and has numerous glands. The surface is ciliated and secretory columnar epithelial cells, except for the lower third of the cervical canal, where it gradually changes to stratified squamous epithelium similar to that lining the vagina. In the sexually mature, nonpregnant female, the uterine mucosa is subject to cyclical menstrual changes that are influenced by ovarian hormonal secretions.

Position of Uterus. The uterus is not firmly attached or adherent to any part of the skeleton. Its main support is the muscles of the perineum (page 586). Fibrous cords and folds of peritoneum, called *ligaments,* suspend the uterus in the pelvic cavity. Normally the fundus is inclined forward, and the general line of the uterus is almost at a right angle to that of the vagina. A full bladder tilts it backward, and a full rectum pushes it forward. During pregnancy the uterus becomes enormously enlarged, extending high up into the abdominal cavity.

If the uterus becomes fixed or rests habitually in a position beyond the limits of normal variation, it is said to be displaced. *Retroversion* signifies a backward turning of the whole uterus without a change in the relationship of the body to the cervix. *Retroflexion* signifies a bending backward of the body on the cervix at the level of the internal os. *Anteversion* means a forward turning of the whole uterus, and *anteflexion*, a forward bend of the body at the isthmus, which brings the fundus under the symphysis pubis.

Ligaments (Figure 25-7). Several fibrous cords covered with mesothelium and peritoneal folds, all called *ligaments*, assist in holding the internal reproductive organs in normal position and in anchoring them to the wall and floor of the pelvis. The ligaments of the uterus are two *broad ligaments*, two *round ligaments*, two *cardinal ligaments*, two *uterosacral*, and one *anterior* and one *posterior ligament*. The *broad ligaments* are the largest and the most important, and they provide the reproductive organs with the greatest support. They are wide peritoneal folds that are slung over the anterior and posterior surfaces of the uterus, ovaries, and uterine tubes, and extend laterally to the walls of the pelvis. Other important structures found between the layers of the broad ligaments are the ligaments of the ovary, nerves, blood vessels, and lymphatics and the cardinal and round ligaments. The fold of broad ligament extending between the ovary and upper uterus is called the *suspensory ligament;* it contains ovarian blood vessels.

The *round ligaments* are flattened fibrous cords that extend from the upper, lateral borders of the uterus, within the broad ligaments, to the connective tissue and skin of the labia majora via the inguinal canals of their respective sides. They help to hold the fundus forward in a slightly anteflexed position. The *uterosacral ligaments* are peritoneal folds that pass from the cervix to the sacrum, one extending on each side of the rectum.

The *cardinal ligaments* are less important supporting structures of the uterus. They are enveloping bands of fascia that surround the uterine blood vessels as they pass to the vagina and cervix from the lateral pelvis. Folds of peritoneum that contribute little or no support to the pelvic viscera, but which are frequently included in the list of ligaments, are the *anterior* and *posterior ligaments*. The sheath of peritoneum that extends from the urinary bladder to the anterior surface of the uterus is called the *anterior ligament*. The *posterior ligament* is the reflection of the peritoneum from the anterior surface of the rectum upon the posterior surface of the uterus and vagina. The *cul-de-cac*, or *pouch of Douglas*, is the deep recess formed by this peritoneal reflection between the uterus and the rectum. The recess formed by the peritoneal reflection between the uterus and bladder is the *vesicouterine pouch.*

Vagina. The vagina is a fibromuscular tube, 7.5 to 10 cm in length, situated anterior to the rectum and anal canal and posterior to the bladder and urethra. It is parallel to the direction of the urethra; that is to say, it is directed upward and backward. It is the organ of copulation, for the deposition of semen in the female, and during the birth process (*parturition*) it serves as the exit from the uterus. The cervix projects into the vault of the vagina, and the vaginal recesses are formed around it. These recesses are known as the *anterior, posterior,* and *lateral fornices* (singular, *fornix*).

The vaginal wall consists of fibrous, muscular, and mucous coats. The mucous coat is composed of stratified squamous epithelium with glycogen stored within its cells. Estrogen secretion during the menstrual cycle and pregnancy seems to cause an increase in glycogen stores and keratinization of the surface epithelium. The inner surface of mucous membrane is thrown into two longitudinal folds and numerous transverse folds, or rugae. The circular and longitudinal smooth muscle layers hypertrophy during pregnancy, and these layers, together with the rugae of the mucous coat and the interstitial elastic connective tissue, allow for extreme distensibility of the canal during parturition. Striated muscle fibers form a ring-shaped sphincter around the introitus, or external orifice of the vagina. This opening may be partially occluded in the virgin by a fold of mucous membrane containing squamous epithelium with a thin connective tissue core called the *hymen.*

The vagina normally has a pH of between 4 and 6. This acidic environment impedes the growth of microorganisms and thus functions to prevent infection of the pelvic organs. The mucus that lubricates the vagina originates from the glands of the cervix. This mucus is acidified by the fermenting action of the vaginal bacteria, mainly lactobacilli, upon the glycogen from the vaginal epithelium.

Vaginal Cells. The cells found in the vaginal secretion may be studied microscopically to determine hormonal balance of the individual, since there is a cyclical change during the menstrual cycle, and to identify cancerous cells. These vaginal cells are derived from the endometrium, the inner and outer surfaces of the cervix, and the lining of the vagina.

External Female Genitalia. The external reproductive organs of the female, often called the *vulva*, include the *mons pubis, labia majora, labia minora, clitoris, vestibule* of the vagina, and the *greater vestibular (Bartholin's) glands.* These are shown in Figure 25–8.

The mons pubis is a pad of fat making up an eminence situated anterior to the symphysis pubis. After puberty it is covered with hair.

The labia majora are two prominent longitudinal folds that begin at the mons pubis anteriorly and extend posteriorly to within 1 in. of the anus. The labia majora are homologues of the scrotum in the male and, like the scrotum, contain large sebaceous glands and become pigmented after puberty. They protect the vagina and help to maintain its secretions. Within their substance lie the terminations of the round ligaments.

The labia minora are two thin longitudinal folds of skin bordering the vestibule of

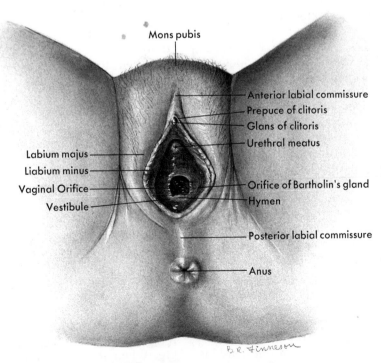

Figure 25–8. External female genitalia.

the vagina. They are situated between the labia majora, are united anteriorly in the hood or prepuce of the clitoris, and form the boundaries of the vestibule, which is the area between them.

The clitoris is a small protuberance more or less hidden by the folds of skin called the *prepuce*, situated at the apex of the triangle formed by the junction of the labia minora. It is the homologue of the penis in the male and, like it, contains erectile tissue, venous cavernous spaces, and specialized sensory corpuscles that are stimulated during coitus.

The vestibule of the vagina is the area situated posterior to the clitoris and between the labia minora. The urethra opens into this space anteriorly, and the vagina opens into it posteriorly. Several glands open into the floor of the vestibule.

Bartholin's glands, or the *greater vestibular glands,* open on either side of the vaginal orifice. These glands together with the smaller lesser vestibular glands and paraurethral glands have a moistening function and are of clinical diagnostic importance because they may become infected with microorganisms, particularly the gonococcus, which causes the venereal disease known as gonorrhea.

Perineum. The perineum is the external surface of the floor of the pelvis, including the underlying muscles and fascia. Figure 25–8 shows the female perineum. It is a diamond-shaped area extending from the pubis to the coccyx, and from the ischial tuberosity on one side to that on the other. If the diamond is divided by a line transversely, the anterior half is the *urogenital triangle;* the posterior half is the *anal triangle.* The urogenital triangle in the male contains the root of the penis and the membranous urethra; in the female it contains the clitoris and the urethra. The anal triangle in both sexes contains the anal canal and its orifice. A wedge-shaped upward extension of the perineum anterior to the rectum is called the perineal body, which in the female forms the septum between vagina and rectum.

The perineum is distensible and is stretched to a remarkable degree during parturition. Nevertheless, it may be lacerated during delivery and, if not surgically repaired, may cause weakening of the muscular and fascial supports of the pelvic floor. Without adequate support the bladder may prolapse with the anterior vaginal wall to form a hernia known as a *cystocele.* The herniation of the bowel through prolapse of the posterior vaginal wall is known as a *rectocele.*

Blood Supply to Female Pelvis. Arteries and veins of the female pelvis are similar to those in the male (page 578).

The ovarian arteries in the female arise from the same portion of the aorta as the spermatic arteries in the male. They supply the ovaries and send small branches to the ureters and uterine tubes. One branch unites with the uterine artery (a branch of the hypogastric) and assists in supplying the uterus. During pregnancy the ovarian arteries become considerably enlarged. The right ovarian vein empties into the inferior vena cava, and the left into the left renal vein.

The *blood supply of the uterus* is abundant and reaches it by means of the *uterine arteries,* branches from the internal iliac (hypogastric) arteries, and the *ovarian arteries,* which are branches from the abdominal aorta. The arteries are remarkable for their tortuous course (hence called *spiral* arteries) and frequency of anastomoses. The veins are large and correspond to the arteries in size and frequency of anastomoses. The uterine veins empty into the internal iliac (hypogastric veins.)

Development of Ova. During fetal life primordial ova develop (probably in the yolk sac, see page 604) and migrate to the ovarian cortex; here they rapidly divide and form the primitive female germ cell, the oogonia. About 2 million oogonia are present at this time in both ovaries, but the number is decreased to about 400,000 at the time of puberty. Each oogonium differentiates into a larger cell, known as a *primary oocyte,* during the period from the fourth through the seventh month of fetal life. When the primary oocyte becomes surrounded with a single layer of epithelioid cells, called *granulosa cells*, the body thus formed is termed a *primary follicle*. The primary oocytes enter the prophase of their first meiotic division; at birth they are still resting in this stage.

During the sexual life of the female only about 400 of the primary follicles become mature; the rest regress. It is believed that during the years before puberty a few follicles do mature, with formation of small amounts of estrogen. The hypothalamus is thought to be very sensitive to estrogen at this time, so that FSH releasing factor is not produced and, therefore, few follicles mature. However, at the time of puberty this sensitivity changes and FSH begins to be secreted in adult amounts.

Menstrual Cycle. During the sexual life of the female beginning with puberty, there is maturation of one ova on an average of each 28 days (Figure 25–9). At the same time there are cyclical changes in the endometrium of the uterus, which occur in three phases: the *proliferative phase*, lasting from the end of the first phase until ovulation occurs; the *secretory phase*, from ovulation to the beginning of menstruation; and the *menstrual phase*, lasting about four days.

Proliferative (Estrogenic) Phase. The adenohypophysis begins secreting more and more hormones beginning at the age of about seven years until the beginning of adult sexual life at puberty. At this time

follicle-stimulating hormone (FSH) from the anterior pituitary begins to be secreted in large quantity. This causes growth and development of the *primordial follicle* and of its surrounding layer of cells, the *theca interna*, which then begins to produce increasing amounts of *estrogens*. FSH causes the ovum to enlarge with concurrent development of follicular fluid encompassing it (Figure 25–10). Several follicles begin to develop under this stimulation, but normally only one fully matures under the influence of the estrogen secreted by the theca interna.

Estrogenic hormones cause increased development of the inner lining of the uterus, the endometrium, and increase its vascularity and glandular secretions. This stage continues over approximately one half of the menstrual cycle and is often referred to as the *proliferative*, or *estrogenic, phase*. As the blood level of estrogens rises, this has a feedback effect on the hypothalamus, causing FSH secretion to decrease. At the same time estrogens are thought to stimulate secretion of the luteinizing hormone (LH). It is possible that this hormone helps in the final extrusion of the ovum from the follicle.

No matter how long the menstrual cycle, ovulation is believed to occur between the thirteenth and fifteenth day before the beginning of menstrual bleeding. *Ovulation* can be defined as the extrusion of the ovum surrounded by a mass of granulosa

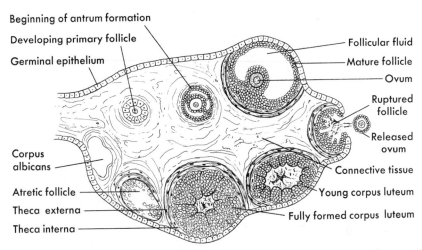

Figure 25–9. Diagrammatic view of the ovary showing ovum in various stages of maturation.

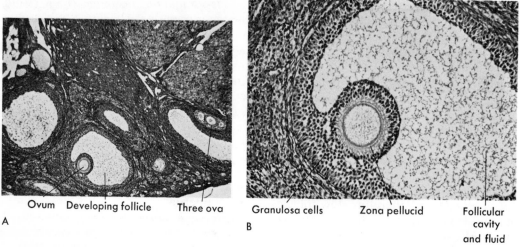

Ovum Developing follicle Three ova Granulosa cells Zona pellucid Follicular cavity and fluid

A   B

Figure 25–10. Developing follicle, as seen with light microscope. *A.* Low power. *B.* Higher power and at a later stage of development.

cells, into the peritoneal cavity. As the fluid content of the follicle increases, the blood supply is decreased; and it is perhaps, although by no means certain, for this reason that the follicle ruptures and releases the ovum. In its mature state the follicle is known as the *graafian follicle* and is about 10 to 12 mm in diameter.

The mature ovum is globular, almost 0.2 mm in diameter. It is much larger than the spermatozoon, whose elliptical head is only about 0.003 mm long and whose tail measures about 0.06 mm in length, or one half the diameter of the human ovum. If it is not fertilized within approximately 24 hours, it is no longer viable. *Meiosis* continues in the ovum, under the influence of FSH. As seen in Figure 25–5, the *primary oocyte* finishes its first maturation division, which was started a long time previously during prenatal life; and a thick membrane, the *zona pellucida*, develops around it. From the first meiotic division two daughter cells of unequal size, each with 23 double chromosomes, are formed. The larger of the two cells is known as the *secondary oocyte*, and the second, much smaller cell, containing very little cytoplasm, is called the *first polar body.* The secondary oocyte then begins a second maturation division. A *mature ovum* containing 23 single chromosomes and another polar body result from this division.

The secondary oocyte is extruded from the ovary before the second maturation division is completed. This division may never be terminated unless fertilization takes place. The first polar body may or may not undergo a second division. Three or four polar bodies may be formed during the two maturation divisions, but they are reabsorbed and do not serve any reproductive function.

Secretory (Progestational) Phase. After ovulation the remaining granulosa and theca cells of the follicle and the cells of the theca interna undergo a process of luteinization, or cellular accumulation of yellow, lipid inclusions. Under the continued stimulation of LH the mass of lutein cells becomes the *corpus luteum,* a secretory organ producing a large quantity of progesterone and, to a lesser degree, other progestational substances. This phase of the menstrual cycle, usually comprising the 13 to 15 days before the commencement of menstrual bleeding, is most commonly referred to as the *progestational,* or *secretory, phase.*

The function of *progesterone* is to increase the secretory function of the endometrium, and to bring about the formation of glycogen and lipid stores within its structure. It also inhibits contractility of the uterine smooth muscle layers, the myometrium, thereby preventing expulsion of the embryo. In these ways progesterone functions to prepare the uterus for the implantation, early growth, and development of the fertilized ovum or zygote. High blood levels of progesterone also prepare the breasts for lactation.

The increasing production of progesterone toward the end of the cycle has a feedback inhibitory effect on the anterior pituitary production of LH. This causes the corpus luteum to involute and become a *corpus albicans*, or scar, and results in decreasing production of progesterone and estrogens.

Menstrual Phase. About two days before the end of the cycle the ovarian hormone secretion decreases sharply to low levels and *menstruation* ensues. Inadequate amounts of ovarian hormones probably are the cause of vasospasm of blood vessels to the mucosal layers of the endometrium, and necrosis, or death, of the inner layers of endometrium. The dead tissues and released blood initiate uterine contractions which expel the sloughed-off uterine contents. Approximately 40 ml of blood and about the same amount of serous fluid are lost during menstruation. Small amounts of *fibrinolysin* released from the desquamated tissues prevent clotting of this blood and fluid. Menstruation usually ceases after four to six days and the endometrium becomes completely reepithelialized. FSH secretion, no longer inhibited by estrogens at this time, resumes in increasingly larger amounts, the endometrium proliferates, another follicle begins to mature, and a new menstrual cycle has begun.

If fertilization of the ovum occurs, menstruation does not take place. The fertilized ovum completes its second maturation division, and male and female pronuclei unite, forming a zygote that implants within the uterus after about a week following fertilization. The developing placenta secretes a hormone called *chorionic gonadotropin*, which has much the same function as the pituitary hormone, LH. It prevents the involution of the corpus luteum at the end of the menstrual cycle, causing it to enlarge considerably and to continue to secrete large amounts of progesterone and estrogens. Desquamation of the endometrium, or menstruation, is thus prevented if pregnancy occurs.

Summary of Hormonal Changes and Influences During the Menstrual Cycle (*Figures 25–11 and 25–12*).

1. The anterior pituitary begins the secretion of *FSH* during the time of menstrual flow. This causes follicles to grow and develop in the ovary. At the end of about two weeks *one follicle* reaches maturity. Under the influence of FSH the *theca interna* begins to secrete gradually increasing amounts of *estrogens*. This is often referred to as the *estrogenic* or *proliferative phase of the cycle* and it lasts, usually, about two weeks.

2. The estrogens aid in the repair and proliferation of the endometrium of the uterus. They cause an increased thickening, keratinization, and glycogen storage of the vaginal epithelium and also probably aid *libido*. The rising blood levels of estrogens stimulate the secretion of *LH* toward the end of the two-week period and decrease production of FSH.

3. LH in some ways helps to cause ovulation or rupture of the mature follicle and release of the ovum about 13 to 15 days before the beginning of the next menstrual flow. It causes the remains of the ruptured follicle to be transformed into an endocrine gland known as a *corpus luteum* that secretes increasingly large amounts of *progesterone* during the second, *progestational* or *secretory*, phase of the cycle.

4. *Progesterone* has a secretory effect on the uterine endometrium. That is to say, it causes increased secretions and a nutrient storage within the uterine mucosa in preparation for the reception and nourishment of a fertilized ovum. It also appears to have a quieting effect upon uterine musculature, which favors implantation of the ovum. As the level of *progesterone increases* toward the end of the cycle, it appears to have a feedback effect upon the anterior pituitary by causing it to secrete *less* and *less LH*. When the production of LH is thus inhibited, the *corpus luteum involutes* and ceases to secrete progesterone.

5. *Menstruation* is believed to be caused

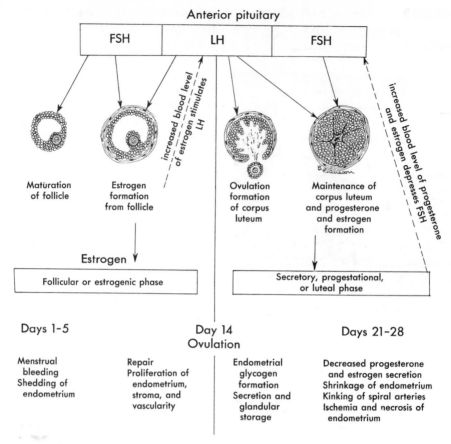

Figure 25–11. Summary of hormonal effects upon the ovary and uterus.

by the resulting low levels of progesterone and estrogens. Lack of sufficient amounts of estrogens is thought to cause vasospasm of the spiral arteries to the endometrium. This results in inadequate nutrition of the superficial layers of endometrium, which then become necrotic and slough off.

6. The *decreased* blood level of estrogen is no longer sufficient to *inhibit* the anterior pituitary production of *FSH;* so it is *again secreted* and the cycle resumes.

7. If *fertilization* and *implantation* occur, the *corpus luteum persists* and continues to secrete large amounts of *estro-*

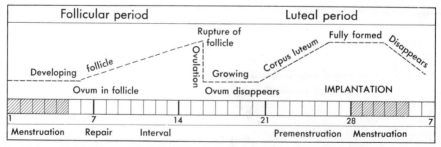

Figure 25–12. Diagram to show sequence of events in ovary and in uterus in human menstrual cycle. (Modified from Corner.)

gens and *progesterone* under the stimulating influence of *chorionic gonadotropin* produced by the developing placenta.

Fertilization, or Conception. Fertilization occurs when the nucleus of the spermatozoon, or male gamete, fuses with the nucleus of the ovum, or female gamete, to form the *zygote.* Soon after extrusion into the peritoneal cavity, the ovum, surrounded by the follicular cells that make up the corona radiata, passes into the fallopian, or uterine, tube. It is apparently carried into the infundibulum, or funnel, of the tube by currents in the peritoneal fluid created by the fimbriae and the cilia of the infundibulum and tubular mucosa. It is possible that an ovum from one ovary can pass to the opposite uterine tube. Rhythmical muscular contractions of the uterine tube also aid the passage of the ovum toward the uterus.

Fertilization normally occurs in the ampulla of the uterine tube or at least in the distal third. It is believed that tubular and uterine musculature contractions cause aspiration of spermatozoa into the uterus and tubes. The flagella, or tails, of the sperm probably contribute only slightly to their ascent into the female reproductive tract, since they enable them to move only from 1 to 4 mm per minute. The main function of these flagella seems to be to aid in the penetration of the corona radiata and the surrounding membranes of the ovum, the zona pellucida, and the vitelline membrane.

Only one spermatozoon of the 200 to 300 million spermatozoa deposited in the female reproductive tract is necessary for fertilization of the ovum. It is believed that other surrounding spermatozoa release enzymes such as hyaluronidase that detach the layer of corona radiata and thereby aid the penetration of the fertilizing cell. In the human being both the head and tail enter the ovum. If two ova mature at the same time and both are fertilized, fraternal twins result.

As soon as the spermatozoon enters the ovum, the female germ cell finishes its second maturation division, and its 22 plus one X chromosomes make up the *female pronucleus.* The spermatozoon moves toward the female pronucleus, and its swollen head becomes the *male pronucleus.* The two pronuclei meet and join, thus restoring the complete 46, or diploid number of, chromosomes in the human somatic cell. The tail that had been detached from the head immediately following fertilization contributes to the formation of the centrosome, which soon divides into two halves, each half moving to the opposite pole of the spindle, thus forming two new cells. Each cell of the rapidly forming individual contains 46 chromosomes, that is, the original diploid number.

Fertilization determines the sex of the zygote, restores the diploid number of chromosomes, and causes the initiation of mitotic, cleavage division, all of which result in the formation of the embryo.

Female Fertility. The ovum is capable of being fertilized for only a short period, probably about 24 hours. Sperm can remain viable up to about 72 hours; therefore, the period of possible conception is only a few days, if the time of ovulation is known. Ovulation is thought to occur 14 days prior to menstruation, whether the menstrual cycle is 28 days or longer or shorter, and is accompanied by a sharp rise of about a half degree in temperature (F). The most common cause of sterility in women is the failure to ovulate. This may be diagnosed by tests for presence of progesterone breakdown products in the urine, and by microscopic examination of the uterine lining, cervix, and vaginal epithelium for cellular changes due to lack of progesterone.

Menopause. The menopause, or female climacteric, is the period during which there is a physiologic cessation of the menstrual flow, the termination of development of the follicles in the ovaries, a decrease in estrogen production, and consequently the end of the childbearing period. It is usually marked by atrophy of the breasts, uterus, uterine tubes, and ovaries. The onset of menopause usually occurs somewhere between the ages of 45 and 50 and may or may not be indicated by a number of troublesome symptoms that include insomnia, irregular menstruation with its eventual cessation, nervous irritability, palpitations, increased sweating, periods of depression, and intolerance to heat. The exact cause of these symptoms has been ill defined, although they presumably follow the reduction of blood and tissue levels of estrogens. It is probable that the estrogen deficiency itself is not the sole cause of emotional disturbances that may become evident at this time, but rather that the events associated with the climacteric amplify or bring about latent potentialities.

Other menopausal problems resulting from estrogen deficiency may include obesity, believed to result from decreasing caloric expenditure, and *osteoporosis*, due to decreasing protein anabolism, causing a loss of protein matrix particularly of the vertebral column with resultant softening and decalcification.

Mammary Glands (Figure 25–13). The two mammary glands, or breasts, are structurally and developmentally closely related to the integument but function as accessory organs of the reproductive system since they secrete milk for nourishment of the infant.

The mammary gland is contained entirely within the superficial fascia and is composed of 15 to 20 glandular tissue lobes, divided by connective tissue bands and arranged radially about the centrally located *nipple*. The glandular tissue occupies only a small portion of the breast in the nonpregnant or nonnursing breast. A variable but usually considerable amount of adipose tissue is contained between and around the lobules and makes up most of the peripheral part of the structure.

Each breast covers a nearly circular space anterior to the pectoralis muscles extending from the second to the sixth ribs and from the sternum into the axilla. The increase in the size and the shape of the mammary glands at the time of puberty and adolescence is due to increased amount of glandular tissue and adipose tissue brought about under the influence of estrogens and progesterone. The glandular tissue remains underdeveloped unless conception takes place.

The nipple is perforated at the tip by the 15 to 20 minute openings of the lactiferous ducts, each one of which is an excretory duct from one of the lobules to the surface. The skin of the nipple extends outward on the surface of the breast for 1 to 2 cm to form a pink or brown *areola*. The areola contains numerous large sebaceous glands that secrete a lipoid material that protects and lubricates the nipple during nursing of the infant.

During pregnancy *estrogens* probably

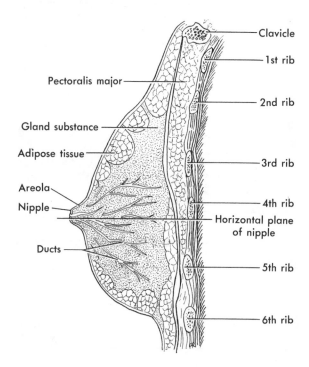

Pectoralis major

Gland substance

Adipose tissue

Areola

Nipple

Ducts

Clavicle

1st rib

2nd rib

3rd rib

4th rib

Horizontal plane of nipple

5th rib

6th rib

Figure 25–13. Right breast in sagittal section.

stimulate the growth of the glandular duct system, whereas *progesterone* is considered to be the stimulus for glandular cell or acini formation. Complete development of the mammary system for lactation requires the concerted action of estrogens, progesterone, lactogenic hormone, and somatotropin. Before parturition lactation is presumably held in abeyance by the high titers of placental sex steroids, which suppress secretion of *prolactin*. After delivery and the expulsion of the placenta, the anterior pituitary is no longer inhibited by progesterone and estrogens, and lactogenic hormones begin to be secreted in increasingly large amounts. These hormones are especially important for their major role in the initiation and maintenance of lactation.

Oxytocin is secreted by the posterior pituitary by a reflex action induced by the infant's sucking on the breast. The milk letdown principle of this hormone causes contraction of the myoepithelial cells of the mammary alveoli and ductules, which forces the milk into collecting ducts to be expelled.

The mammary glands are well supplied with blood brought to them by the thoracic branches of the axillary, internal mammary, and intercostal arteries. The nerve fibers are derived from the ventral and lateral cutaneous branches of the fourth, fifth, and sixth thoracic nerves.

The *lymphatic drainage* of the breasts is of the greatest importance in relation to malignant conditions of the breast. Malignant cells may readily spread from the affected breast to other areas of the body via the lymphatic vessels and nodes that drain the adjacent regions. Direct paths of lymph drainage follow the blood vessels and are primarily to the nodes within the pectoral muscles, the axillary nodes, and the internal mammary nodes. Some lymph also drains into the subcutaneous plexus and into the deep cervical nodes posterior to the clavicle.

In the surgical removal of malignant disease of the breast, the entire breast, all of the underlying subcutaneous tissue, the pectoralis muscles of that side, and all of the aforementioned lymph nodes are removed with the hope of preventing the spread of the tumor to other areas of the body, either by direct extension into adjacent tissue, or by spread of some of the cells via lymph or blood systems.

Questions for Discussion

1. Define the terms *puberty* and *adolescence*. What are the changes occurring at this time in the male? In the female?
2. Describe the descent of the testes. If the testis remains in the abdominal cavity, how will its function be affected?
3. What stimulates the production of LH and FSH?
4. Where is progesterone produced? What are its functions?
5. Describe the endometrial changes occurring in the three phases of the menstrual cycle.
6. Describe the pathway of the spermatozoa from the seminiferous tubule to the urinary meatus.
7. How does meiosis differ from mitosis?
8. What is fertilization? Where does it usually occur?

Summary

Puberty
{
Initiated by hypothalamic production of releasing factors for LH and FSH
Age at which sex organs attain normal adult function
Male begins to produce functional spermatozoa in temperate climates between ages 14–16 years
Female begins to ovulate and menstruation commences in temperate climate between ages 11–14 years
Period marked by gradual appearance of secondary sex characteristics in both sexes

Adolescence	Period from puberty to early twenties during which complete development of all secondary sex characteristics have been achieved—epiphyseal plates close at age of 20–21 years Physical growth accompanies psychologic growth in maturity
Male Organs of Reproduction and Functions	2 testes—contained in the scrotum—produce spermatozoa and testosterone 2 epididymides (singular, epididymis)—lie superior and posterior to testes—spermatozoa mature here 2 ductus deferens—storage sites for spermatozoa—convey them from epididymides to seminal vesicles 2 seminal vesicles—membranous tubes—lie posterior to bladder—produce nutritious secretions added to semen 2 ejaculatory ducts—short passageways within prostate gland—convey semen from seminal ducts to urethra Prostate gland—size of chestnut situated immediately inferior to internal urethral sphincter—surrounds proximal urethra—secretes alkaline fluid—increases motility of spermatozoa—prostate may increase in size in old age causing urinary retention 2 bulbourethral glands—small bodies on either side of membranous urethra—secrete lubricating and protective mucus prior to ejaculation Penis—organ of copulation suspended from front and sides of pubic arch—consists of 3 bodies of cavernous tissue—2 lateral corpora cavernosa penis—and 1 corpus cavernosum urethrae—these erectile bodies become distended with blood to produce erection of organ—glans penis, expansion at lower portion of penis, contains sensory end organs and urethral orifice—covered by foreskin or prepuce Blood supply—testicular arteries, internal and external iliacs. Corresponding veins Spermatic cords—bilateral; from abdominal fascia to scrotum; contain ductus deferens, arteries, veins, lymph vessels, autonomic fibers; pass through inguinal canal
Ejaculation	Erection—caused by filling of venous spaces in penis Ejaculation—discharge of semen

Ejaculation	**Semen**	Fluid derived from the various sex glands in the male—contains approximately 100,000,000 spermatozoa per milliliter. Fertility depends on number and quality of sperm
	Nervous control	Parasympathetic nerves (sacral 2, 3, 4) control erection of penis Sympathetic nerves (lumbar 1, 2, 3) control emission and ejaculation of semen

Male Urethra	S-shaped tube—extends from internal urethral orifice to external urethral orifice of glans penis—17.5 cm in length—divided into 3 parts: prostatic, membranous, and penile urethra
Maturation of Germ Cells	Process whereby male and female reproductive cells grow, develop, reduce their number of chromosomes to 23, and are prepared for fertilization **Oogenesis**—process of forming female sex gamete—ovum **Spermatogenesis**—process of forming male gamete—spermatozoon
Spermatogenesis	Begins at puberty; continues throughout adult life Spermatogonia formed during fetal life; mature under influence of FSH Primary spermatocytes—at puberty primary spermatocytes develop from spermatogonia Primary spermatocytes undergo first maturation division of meiosis—46 or diploid number of chromosomes reduced to 23 double chromosomes in each of two secondary spermatocytes formed Secondary spermatocytes—each secondary spermatocyte undergoes second maturation division of meiosis to form a spermatid containing 23 single chromosomes Spermatids—a spermatid develops into a spermatozoon when it develops a head and tail, or flagellum Four spermatozoa formed from each primary spermatocyte Two of these cells carry an X chromosome Two of these cells carry a Y chromosome Two maturation divisions of meiosis take place in the formation of spermatozoa from primary spermatocytes Male fertility depends on adequate numbers of healthy spermatozoa
Female Organs of Reproduction and Functions	Homologous with male organs: testis and ovary; clitoris and penis; labia majora and scrotum

Female Organs of Reproduction and Functions	**2 ovaries**	Almond-shaped bodies measuring from 2.5–5 cm (1–2 in.) in length located one on each side of pelvic cavity—are attached to posterior surface of broad ligament, to uterus by ligament of the ovary, to ovarian tubes by fimbriae—functions are to produce ova and the sex hormones, progesterone and estrogens

Female Organs of Reproduction and Functions (*cont.*)

2 uterine or fallopian tubes (oviducts)

Muscular ducts lined with mucosa—containing ciliated epithelium—enclosed in layers of broad ligament—extend from upper angles of uterus to sides of pelvic cavity

Divisions
- Isthmus—inner constricted portion near uterus
- Ampulla—dilated portion that curves over ovary
- Infundibulum—trumpet-shaped extremity—fimbriae

Coats
- External, or serous
- Middle, or muscular
- Internal, or mucous, arranged in longitudinal folds—lined with ciliated epithelium

Function
- To convey ova to uterus, provide environment for fertilization and early development of fertilized ovum

Uterus

Hollow, pear-shaped, muscular organ placed in pelvic cavity between bladder and rectum

Divisions
- Body—superior part, upper rounded portion above entrance of tubes called fundus
- Isthmus—middle or slightly constricted portion
- Cervix—lower and smaller portion extends into vagina—contains internal and external os

Three coats
- External, or serosa, derived from peritoneum, covers intestinal surface and anterior surfaces to beginning of cervix
- Muscular or myometrium
  - Circular layer
  - Longitudinal layer
  - Spiral layer
  } Interlaced
- Mucous membrane, or endometrium { Lines internal aspect

Ligaments
- Broad—two large layers of serous membrane—from uterus to walls of pelvic cavity
- Round—two fibromuscular cords from sides of uterus to labia majora
- Anterior—peritoneal fold from bladder to uterus
- Posterior—peritoneal fold from uterus to rectum
- Uterosacral—two partly serous, partly muscular ligaments from cervix to sacrum
- Cardial—fascia surrounding uterine vessels from pelvic wall to cervix and vagina

Function
- To receive fertilized ovum, provide for implantation, nourishment, and environment for growth and development of fetus until parturition

Vagina

Extends from uterus to vulva—about 7.5–10 cm (3–4 in.)

Coats
- Internal mucous lining arranged in rugae
- Layer of submucous connective tissue
- Muscular coat

Location—placed anterior to rectum, posterior to bladder and urethra

Function—for deposition of semen, serves as exit of birth canal

External genitalia

Mons pubis—cushion of areolar, fibrous, and adipose tissue in front of pubic symphysis, covered with skin and after puberty covered also with hair

Labia majora—two folds that extend from mons pubis to within an inch of anus—protect vagina—maintain secretions

Labia minora—two folds situated between labia majora

Clitoris—small protuberance at apex of triangle formed by junction of labia minora—well supplied with nerves and blood vessels

Vestibule—cleft between labia minora

Hymen—fold of mucous membrane partly covering and surrounding vaginal orifice

Glands—greater vestibular or Bartholin's—oval bodies situated on either side of vagina—secretions have moistening function

Blood Supply

Similar to male; uterine arteries from hypogastrics; ovarian arteries from aorta; uterine veins empty into internal iliacs; ovarian vein (right) empties into inferior vena cava, left into left renal vein

Perineum

External surface of floor of pelvis; includes underlying fascia and muscles; extends from pubis to coccyx and between ischial tuberosities; includes urogenital and anal triangles

Oogenesis

Primitive female germ cells, oogonia, formed in embryo

Oogonia located in clusters in ovarian cortex, surrounded by layer of flat epithelial cells

Fourth to seventh month of fetal life—oogonium differentiates into primary oocyte, which enters prophase of meiosis

Primary oocyte, present at birth, finishes first maturation division of meiosis under influence of FSH, and zona pellucida develops

From this division, secondary oocyte and first polar body develop, each with 23 double chromosomes

Secondary oocyte undergoes second maturation division forming mature ovum with 23 single chromosomes and another polar body

Ovum is extruded from ovary before second maturation division is completed

If ovum is not fertilized, division may not terminate; ovum is viable for 24 hours

One follicle matures during each menstrual cycle, from puberty until time of menopause

Menstrual Cycle

Menstruation or desquamation of the endometrium is thought to be due to ischemia of this tissue following drop in blood levels of progesterone and estrogens prior to beginning of cycle. In menstruation about 40 ml of blood and about the same amount of serous fluid are lost—cycle begins on first day of menstruation

FSH secreted at beginning of cycle causes development of a graafian follicle and production of estrogens by follicle and theca interna

Estrogenic follicular or proliferative stage

Estrogens cause increase in thickness, vascularity, and secretions of endometrium—rising blood levels of estrogens stimulate production of LH; decrease production of FSH

LH assists production of ovulation or extrusion of ovum into peritoneal cavity usually between 13th–15th day before beginning of menstrual bleeding. One mature ovum derived from each primary oocyte. Each ovum contains 22 + one X chromosomes. Polar bodies disintegrate

Progestational, luteal, or secretory stage

LH stimulate luteinization of follicle after ovulation with formation of a corpus luteum, a secretory body that produces large quantities of progesterone and smaller amount of estrogens

Progesterone increases secretory function of endometrium, causes formation of glycogen, and lipid stores within endometrium, inhibits contractility of uterine smooth muscle. High levels decrease production of LH. Corpus luteum involutes as a result of diminished LH, and menstruation ensues.

Hormonal changes in pregnancy

If fertilization occurs, menstruation does not take place—placenta secretes hormone called chorionic gonadotropin, which causes continuation and enlargement of corpus luteum—large amounts of estrogens and progesterone are secreted, and thick, vascular secretory endometrium is maintained for implantation and development of embryo

Fertilization

Ovum conveyed to uterine tube by currents created in peritoneal fluid by fimbriae and cilia of tubular mucosa

Flagella of spermatozoa propel them only 1–4 mm per minute, their ascent in female tract believed to be assisted by aspiration following tubular and uterine contractions

Only one out of 200–300 million deposited sperm necessary for fertilization—penetration of ovum assisted by enzyme, hyaluronidase, released by one or more sperm

Head and tail of sperm enter ovum

Fertilization—normally occurs when nucleus of spermatozoon fuses with nucleus of ovum to form *zygote;* occurs in distal third of uterine tube

Pronuclei of sperm and ovum join to restore 46 or diploid number of chromosomes

Menopause

Female climacteric—the period of physiologic cessation of menstrual flow, termination of follicular development, decrease in sex hormones, and the end of the childbearing period—occurs between the ages of 45–50 and may or may not cause troublesome symptoms such as increased irritability, hot flushes, insomnia, depression, and osteoporosis

Mammary Glands

Composition — Made up of 15–20 lobes of glandular tissue divided by connective tissue bands and embedded in superficial fascia

Location — Anterior to pectoralis muscles, extending from 2nd–6th ribs and from sternum into the axilla

Size and shape — Increased size at puberty under influence of increased hormones produced, especially estrogens and progesterones

Mammary Glands
(*cont.*)

Pregnancy and postpartum	Increase in size and numbers of glands mainly under influence of pituitary hormones and estrogens and progesterone. Lactation depends on concerted action of progesterone, estrogen, prolactin, and somatotropin
	After delivery and expulsion of placenta, decrease in female sex steroids allows prolactin to be released from anterior pituitary. Milk letdown principle of oxytocin causes expulsion of milk

Blood supply—from thoracic branches of axillary, internal mammary, and intercostal arteries

Nerve supply—from lateral and cutaneous branches of fourth, fifth, and sixth thoracic nerves

Lymphatic drainage—follows pathway of blood vessels—drains into nodes within pectoral muscles, into axillary nodes and internal mammary nodes. Some lymph drains into deep cervical nodes

Additional Readings

BASMAJIAN, J. V.: *Grant's Method of Anatomy*, 9th ed. Williams & Wilkins Co., Baltimore, 1975, Chapters 19, 20, and 21.

GUYTON, A. C.: *Basic Human Physiology: Normal Functions and Mechanisms of Disease*, 2nd ed. W. B. Saunders Co., Philadelphia, 1977, Chapters 54 and 55.

HAM, A. W.: *Histology*, 7th ed. J. B. Lippincott Co., Philadelphia, 1974, pp. 852–60, 900–910.

PANSKY, B., and HOUSE, E. L.: *Review of Gross Anatomy*, 3rd ed. Macmillan Publishing Co., Inc., New York, 1975, pp. 264–65, 320–26, 366–99.

VANDER, A. J.; SHERMAN, J. H.; and LUCIANO, D. S.: *Human Physiology: The Mechanisms of Body Functions*, 2nd ed. McGraw-Hill Book Co., New York, 1975, Chapter 14.

26

Development of the Human Organism

Chapter Outline

The human organism begins life as a single cell derived from the fusion of two parental cells: the ovum and the spermatozoon. The fertilized cell, or *zygote*, undergoes mitotic cell division and cells become differentiated into tissues, organs, and systems—the multicellular, highly organized replica of the species. This chapter presents a brief résumé of the development from zygote to human infant.

Inheritance

Development of the zygote is directed by information stored in the chromosomes received from the parents. Because of meiosis, however, the zygote chromosomal complement is not identical with that of either the mother or the father.

Chromosomal Replication in Meiosis. Human cells contain 46 chromosomes, of which 44 are *autosomes* (nonsex chromosomes) and two are sex chromosomes.

Each chromosome has a partner, with similar morphologic characteristics, so that they may be paired (Figure 26–1). The two sex chromosomes (X and Y) do not look alike, except in females when the sex chromosomes are both X, but are considered a pair in spite of this. One chromosome of each pair is derived from the mother, and one from the father.

Prior to puberty, the germ cells have entered the prophase of the first stage of

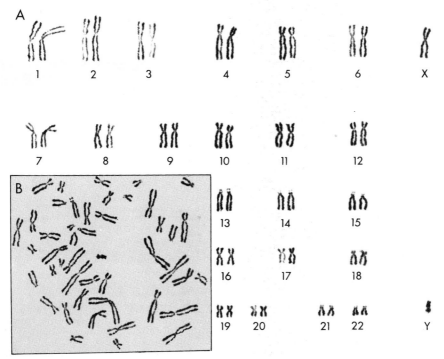

Figure 26–1. Normal human male chromosomes as they appear during metaphase. Cell division has been arrested by treatment with colchicine; hypotonic salt solution then is added to swell and disperse the chromosomes and make them more visible. *B*. Inset: chromosomes as they appear under the microscope following this treatment. *A*. Karyotype: chromosomes are paired and arranged according to a standard classification based on the size, position of the centromere, and other characteristics. The normal human has 22 somatic pairs plus two sex chromosomes (an X and Y in males, two X's in females). (Courtesy of Dr. J. L. German, III, Cornell University Medical College, New York, N.Y.)

meiosis. The DNA, which replicates in interphase, has already done so and the chromosomes have doubled. Such double chromosomes are held together at a point known as the *centromere*, or *kinetechore*; the individual member of the doubled chromosome is called a *chromatid*. These doubled chromosomes then align as pairs, twining around one another in a process called *synapsis*. In late prophase these paired, doubled chromosomes are visible in the cell as four chromatids. Under the influence of FSH at the time of puberty meiosis continues.

The first maturation division (Figure 25–5, page 581) results in the close approximation of the chromatids of homologous chromosomes. At this time there is exchange of segments of the homologous chromosomes, and at the end of the first maturation division of meiosis, when the doubled chromosomes of each pair sepa-

rate from one another, there has been a *crossing over* of genetic material. After the first meiotic division each daughter cell has 23 doubled chromosomes—one of each chromosome pair.

When the second maturation division begins, there has been no further synthesis of DNA; the doubled chromosomes divide at the centromere and each daughter cell has the haploid number of chromosomes: 23 single chromosomes. Each has half as much DNA as the normal human cell. In the ovum, the sex chromosome is always an X chromosome; in spermatozoa, it may be either X or Y.

When fertilization occurs, the full complement of 46 chromosomes is achieved. Males then have an X and a Y chromosome; females have two X chromosomes. As the zygote divides and eventually tissues and organs are formed, *each* cell has the same chromosomal makeup.

[However, cells vary in which portions of particular chromosomes (and DNA) are active, e.g., skeletal muscle cells produce enzymes for formation of glycogen but smooth muscle cells do not.] Thus, it is possible to determine genetic sex when, in exceptional cases, the sex of an infant is in doubt, by examining the squamous cells of the oral mucosa. Two X chromosomes are identifiable in a cell by staining techniques: the extra X chromosome appears as a rounded body attached to the inner nuclear membrane.

Genetic Advantage of Meiosis. The reducing divisions of meiosis separate each pair of chromosomes; thus it halves the number of chromosomes, but chance alone decides the actual distribution of the maternal or paternal member of any pair to any particular daughter cell. Meiosis also provides an opportunity for re-shuffling of the genes from one chromosome to another as the result of the *crossing over* or interchange of chromosome parts. In man, reduction to the haploid number of chromosomes during meiosis makes possible in each gamete one of 8 million final combinations of chromosomes. A further increase in new heredity combinations is made possible by the phenomenon of crossing over. Each child carries the genes inherited from his parents and their ancestors, but each child has a different inheritance and different appearance from every other person in the world because of the infinite variety provided by reduction and the recombination of chromosomes during meiosis.

At this point it is well to consider the meaning of the term *gene*. As discussed in Chapter 2, the codon determines the amino acid sequence in protein synthesis and, therefore, the kind of enzymes formed. Since the metabolic activities of the cell are biochemical, the enzymes have a significant influence on the cellular activity. The term *gene* is presently defined as the segment of DNA (one or more codons) responsible for formation of a particular protein, or segment of protein. In a more general sense, the term *gene* is still used to refer to particular traits, such as color of eyes, stature, or intelligence.

Heredity. Heredity is a term applied to the transmission of potential traits, physical or mental, from parents to their offspring. Each autosome carries the same general set of hereditary genes as its mate. For example, each of the paired chromosomes may carry a gene for eye color, for hair form, or for length of fingers. These genes may be the same or different, depending on the genetic contribution of each parent. If the offspring has received a gene for brown eyes from each parent, we say he is *homozygous* for brown eyes; i.e., the gene for eye color is the same on each of the chromosomes of the pair. On the other hand, if he has received a gene for brown eyes from one parent and a gene for blue eyes from the other, we say he is heterozygous (different genes) for eye color.

Two genes are *alleles* when they carry the same trait (such as eye color) and occupy the same position or locus on the chromosome. Alleles may be alike alleles, producing the same effect, or nonalike alleles, producing differing effects.

The genes for some traits, such as brown eye color, are *dominant* under usual conditions and are bound to cause their effect. Others are *recessive*, such as the genes for blue eye color, and will not cause their effect unless two recessive genes are received by the offspring.

The term *genotype* means the type of genes present in the individual. *Phenotype* refers to the expression of the genes. The person with one gene for blue eyes and one for brown eyes might be said to have the heterozygous genotype for brown color, but the phenotype of brown, since this is the color of this individual's eyes due to the dominance of the brown gene.

The system of using letters as symbols for genes as well as the whole development of the gene concept was devised by the Austrian monk Gregor Mendel,[1] the "father of modern genetics." A small letter

---

[1] Gregor Johann Mendel, Austrian monk and geneticist (1822–1884).

stands for a recessive gene, the capitalized form of the same letter for the gene that is dominant over this recessive. For instance, B might represent the dominant brown eye color and b the recessive color blue.

Dominance may be complete, partial, or absent. A gene that is dominant to one gene may be recessive to another one. Certain abnormalities are inherited as dominants over the normal condition, e.g., extra digits or excessively short digits. Others are recessive to the normal; e.g., albinism is a recessive gene, and albino individuals must have homologous genes. Some abnormalities appear as *sex-linked recessive* (gene on the X chromosome), becoming evident in males with a single recessive factor but in females only when there are two, e.g., red-green color blindness and hemophilia.

The genetic makeup is powerful in its influence on the development of the individual. However, environmental factors play an important part also—genes for height cannot be fully expressed without adequate nutrition.

Developmental Periods

Although this chapter focuses on the period before birth, it is important to remember that development continues after this time. Prenatal (before birth) development is divided into two periods: the *embryonic period* and the *fetal period*. The embryonic period begins about the end of the second week after fertilization and lasts until the end of the seventh week, when all the major structures have begun to form. The fetal period then begins and lasts until birth. Developmental changes are not so dramatic in the fetal period as in the embryonic, but there is a fast rate of growth in the third and fourth months, and great increase in weight in the last months of pregnancy. The period of pregnancy usually lasts for 280 days.

PREEMBRYONIC AND EMBRYONIC PERIOD

Formation of Morula and Blastocyst. Cleavage consists of a number of rapid mitotic divisions that result in the production of a number of increasingly smaller cells known as *blastomeres* (Figure 26–2). Cleavage is not really a growth process since there is no increase in protoplasmic volume despite the progressive increase in cell number. As the zygote passes down the fallopian tube, cleavage continues. When the 16- to 20-cell stage is attained, it is known as a *morula*.

The morula is a solid ball of cells, which contains the inner cell mass, a group of centrally located cells, and cells in the periphery, known as the outer cell mass. The inner cell mass gives rise to the tissues that make up the embryo itself, but it also contributes to the formation of embryonic membranes called the amnion and yolk sac. The outer cell mass forms the *trophoblast*, from which the outer embryonic membranes known as the *chorion* and the *placenta* are developed. The morula reaches the uterine cavity about three to four days after ovulation, the zona pellucida disappears, and the zygote produces a fluid-filled cavity between the inner and outer cell masses, resulting in the formation of the *blastocyst* (Figure 26–3).

Occasionally, as the inner cell mass develops in this early period prior to tissue differentiation, it separates into two groups of cells, each of which then matures in the normal fashion as described below. Should this occur, identical twins result—"identical" because each has developed from one fertilized ovum, and therefore they have the same genetic inheritance.

Implantation of Blastocyst (Figure 26–4). *Implantation*, or uterine attachment, of the

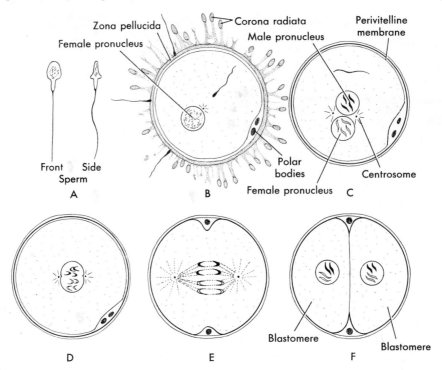

Figure 26–2. Diagrams showing the fertilization of the ovum, the joining of the male and female pronuclei, the chromosomes organized on the spindle, and the two-cell stage of blastomeres.

blastocyst probably occurs between the seventh and ninth days after ovulation. It is thought that the penetration and erosion of the endometrium necessary for implantation result from the combined effects of proteolytic enzymes produced by the trophoblast and by the vascular changes in the endometrium. In any case, the blastocyst normally implants in the endometrium of the body of the uterus. The ero-

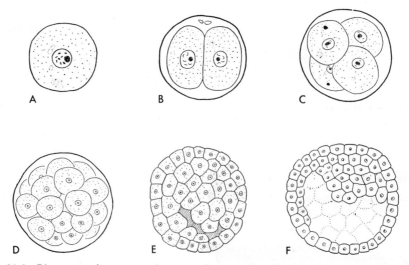

Figure 26–3. Diagram to show very early stages of mammalian development. *A*, One-celled embryo; *B*, two-celled embryo; *C*, four-celled embryo; *D*, berrylike ball of cells or *morula*; *E*, beginning formation of the blastocyst; *F*, well-developed blastocyst, consisting of a hollow ball of *trophoblast* cells and an inner mass of cells known as formative cells.

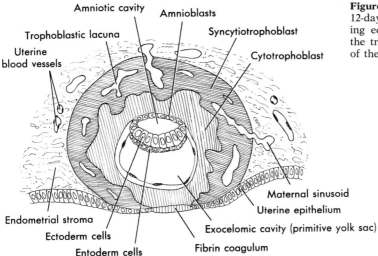

Amniotic cavity
Amnioblasts
Trophoblastic lacuna
Syncytiotrophoblast
Uterine blood vessels
Cytotrophoblast

Endometrial stroma
Maternal sinusoid
Uterine epithelium
Ectoderm cells
Exocelomic cavity (primitive yolk sac)
Entoderm cells
Fibrin coagulum

Figure 26–4. Diagram of a 9- to 12-day blastocyst to show developing ectoderm, entoderm, layers of the trophoblast, and the beginning of the uteroplacental circulation.

sion in the uterine mucosa brought about by the penetration of the blastocyst is gradually obliterated by the growth of adjacent epithelium and by the formation of a *fibrin coagulum*. If the blastocyst implants abnormally in close proximity to the cervical internal os, a condition known as *placenta previa* occurs and causes severe bleeding in the latter part of pregnancy and during delivery. Implantation in the uterine tube results in a tubal pregnancy, which is dangerous because it causes rupture of the tube, severe internal hemorrhage, and death of the embryo during the second or third month of pregnancy. Implantations anywhere outside the uterus are known as *extrauterine*, or *ectopic*, *pregnancies*.

During pregnancy, the uterine mucosa is highly modified and is called the *decidua*. The glands become extremely convoluted and hyperactive, and stromal cells become differentiated into decidual cells that possess variable amounts of glycogen, lipids, and increasing numbers of mitochondria. After implantation of the blastocyst and until the fourth month of gestation, three parts of the decidua can be recognized: the *decidua basalis* underlying the implanted ovum, the *decidua capsularis* that surrounds the surface of the implanted chorionic sac, and that portion that lines the rest of the uterus, the *decidua parietalis*, or *vera*.

Major Events in the Second and Third Weeks (Figure 26–5). The blastocyst becomes firmly embedded in the uterine mucosa during the second and third weeks of development and its two parts, the *embryoblast* and the *trophoblast*, begin to grow and differentiate. The embryoblast gives rise to the three basic layers of the embryo proper, the *ectoderm*, the *entoderm*, and the *mesoderm*. The cells of the trophoblast grow deeply into the endometrium and form the *placenta*.

By the eighth day of development, the embryoblastic cells differentiate into two distinct cell layers, the inner *entodermal* germ layer, which is composed of flattened polyhedral cells facing the lumen of the blastocyst, and the outer *ectodermal* germ layer, which is composed of a layer of tall columnar cells.

The trophoblast forms an inner pale layer, the *cytotrophoblast*, and an outer darker zone referred to as the *syncytiotrophoblast*, or *syncytium*, which has many small open spaces called *lacunae*. The endometrial stroma at the implantation site is highly vascular and edematous, and its enlarged glands secrete mucus and glycogen. Between the ectoderm cells and the trophoblast an opening begins to form, called the *amniotic cavity*, and its outer portion is lined with flattened cells, the *amnioblasts*. The amniotic cavity becomes filled with a thin, clear fluid, which serves

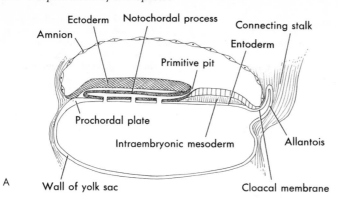

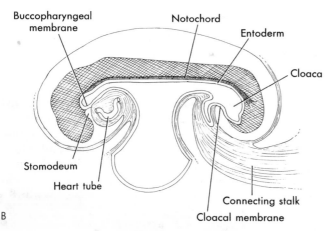

Figure 26–5. Diagrams of midsagittal sections through embryos. *A.* Extension of notochordal process and fusion with the entoderm in 18-day embryo. The newly formed intermediate mesodermal layer is shown. It is thought to be derived from modified ectodermal cells. *B.* Fourteen-somite embryo with developing heart and buccopharyngeal and cloacal membranes.

as a protective cushion to absorb shock, to maintain fetal environmental temperature, to prevent adherence of the embryo to the surrounding surface, and to allow for fetal movements. The gradually enlarging cavity extending toward the lumen of the uterus forms the *celomic* cavity, or *primitive yolk sac.*

Maternal capillaries become dilated and congested and begin to extend into the syncytiotrophoblast. Maternal blood begins to enter the *lacunae* of the syncytium as the syncytial cells begin to erode the endothelial lining of the maternal sinusoids. Thus the future uteroplacental circulation begins to be established. Maternal blood begins to flow through the trophoblastic lacunar system. Bleeding may occasionally occur at the implantation site around the thirteenth day of development owing to increased blood flow into the trophoblastic lacunar spaces at this time, although the epithelial surface has usually healed by then. This bleeding

may be confused with normal menstrual bleeding since it occurs at about the twenty-eighth day of the menstrual cycle. Therefore, the anticipated delivery date may be estimated inaccurately.

Cells of the inner surface of the cytotrophoblast delaminate and differentiate to form the loose network of tissue known as the *extraembryonic mesoderm.* This mesoderm fills the expanding space between the amnion and primitive yolk sac internally and the trophoblast externally. When large cavities develop in this tissue and become confluent, a new space, the *extraembryonic celom,* is formed. This cavity surrounds the blastocyst except between the germ disk and trophoblast, where the attachment remains.

The extraembryonic celomic cavity enlarges, and by the twentieth day the embryo is attached to the surrounding trophoblast by a narrow *connecting body stalk,* which later develops into the *umbilical cord,* attaching the embryo to the placenta.

During the second and third weeks, ectodermal cells in the caudal region of the germ disk begin to multiply and to migrate toward the midline, forming a narrow groove, the *primitive streak*. It is believed that modified ectodermal cells proliferate and migrate between the ectodermal and entodermal germ layers and spread laterally at this time, forming the intermediate cell layer known as the *mesoderm*. The *notochord* now is formed along a longitudinal axis; it is along this axis that the vertebral column will develop. The entoderm layer establishes firm contact with the ectoderm making up the *posterior cloacal membrane* from which *urogenital* and *anal membranes* are derived. The *buccopharyngeal* membrane is later developed from the *prochordal* plate, which now begins to appear at the anterior ectodermal attachment. An outpocketing called the *allantois*, or *allantoenteric diverticulum*, appears about the sixteenth day and extends from the posterior wall of the yolk sac into the connecting stalk. This structure in some lower vertebrates becomes a large reservoir for urine storage, but in man it is normally rudimentary and gradually disappears during further embryonic development.

Primary Germ Layers. During the embryonic period the shape and the appearance of the embryo are greatly altered, and by the end of the second month of development, all the important features of the external body may be recognized. Each germ layer starts a course of differentiation into specific tissues, organs, and systems, and by the end of the embryonic period, all the major body systems have begun to form.

Ectodermal Layer. The ectoderm gives rise to the formation of the *central nervous system*. During the third and fourth week the *neural* plate, posterior to the *notochord*, is formed from ectoderm cells, and it soon invaginates to form the *neural groove* lined by the *neural fold*. When the folds approach each other and fuse, the groove becomes the *neural tube*. This fusion begins in the future neck region and proceeds simultaneously in cephalic and caudal directions. The tube does not close off entirely at this time but temporarily remains open for some time at the anterior

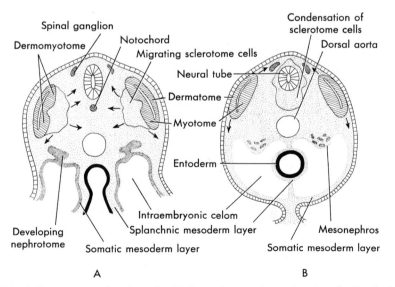

Figure 26–6. *A*. Transverse section through a 26-day embryo to show migration of cells of *sclerotome*, or ventromedial part of the somite. The remaining cells form the *dermomyotome*. The intermediate mesoderm has proliferated to form *nephrotomes*, the excretory units of the urinary system. The *arrows* indicate the direction of the migrating cells of the *dermatome* and *sclerotome*. *B*. Transverse section through a 28-day embryo to show condensation of sclerotome cells around the neural tube to form the axial skeleton. (From Langman, J.: *Medical Embryology*. Courtesy of Williams & Wilkins Co.)

and posterior neuropores. The brain begins to enlarge and develop at the cephalic end of the neural tube, and the spinal cord and peripheral nerves develop from the remainder of the tube. At the end of the first month the *otic vesicle* and *optic vesicle*, outpocketings of the brain, are formed. From the former, parts of the ear are derived, and from the latter, the retina and optic nerve develop.

From the ectodermal germ layer the following parts of the body are derived: (1) the central nervous system and hypophysis, (2) the peripheral nervous system including the autonomic nervous system, (3) the sensory epithelium of the sense organs, (4) the enamel of the teeth, and (5) the epidermis including hair, nails, and subcutaneous glands.

Mesodermal Layer. The mesoderm is formed between the ectoderm and the entoderm. The mesodermal cells form a thin layer on each side of the midline, until the end of the third week when the mass begins to thicken immediately lateral to the notochord, to become the *paraxial mesoderm* (the future somites), the *intermediate mesoderm* (the future excretory units), and the *lateral plate*, which splits into somatic and visceral layers.

By the end of the third week, the paraxial mesoderm becomes segmented into approximately 40 pairs of somites. These *somites* mold the contours of the embryo, and from them are formed the *mesenchyme*, which gives rise to connective tissue, cartilage, and bone, and *myoblasts*, which give rise to striated muscle cells. Some of the cells from the somites become mesenchymatous and spread under the ectoderm to form the subcutaneous tissue of the skin (integumentary system).

The *cardiovascular system* is derived from the mesodermal germ layer during the third week. Blood islands, lined by endothelial cells, become arranged in isolated clusters, which then fuse and give rise to small blood vessels. During the fourth week, a single *primitive heart tube* is formed and is suspended in the pericardial cavity. *Extraembryonic blood vessels* are also formed in a similar manner during

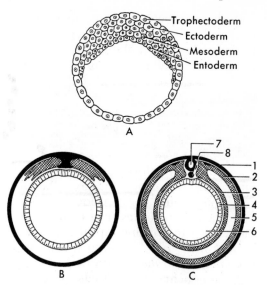

Figure 26–7. Diagram of a section of an embryo, showing the beginning of tissue formation. A. The embryonic disk shows the cells arranged in layers. B,C. Two diagrams of sections of embryos showing later stages of tissue formation. *Ectoderm* is shown in black, *entoderm* as cells; the two layers of *mesoderm* are cross hatched. *1*, Ectoderm; *2*, parietal mesoderm; *3*, visceral mesoderm; *4*, entoderm; *5*, future body cavity; *6*, enteron; *7*, neural tube; *8*, notochord.

this time and become the *umbilical* and *vitelline vessels*, which, as they develop, begin to penetrate the embryo proper where they reach the independently developing intraembryonic vascular system. The pharyngeal arches give rise to the maxillary, mandibular, hyoid bones, and ossicles of the ear, as well as other ligaments, muscles, and bones.

The lateral plate of the mesoderm separates into two layers, the *somatic, or parietal, layer* lying next to the ectoderm, and the *splanchnic, or visceral, layer* lying next to the entoderm. The ectoderm and the somatic mesoderm form the *somatopleure*, and from this the body wall is developed. The entoderm and the splanchnic mesoderm form the *splanchnopleure*, and from this the viscera are developed. The *celom* is a cavity between the two layers of mesoderm that develops into the body cavity. The peritoneal, pleural, and pericardial cavities develop from the celom and are lined with mesothelium derived from mesoderm encompassing them.

The important structures derived from

the mesodermal layer are (1) cartilage, joints, and bones; (2) connective tissue; (3) blood and lymph cells, walls of blood and lymph vessels, and the heart; (4) the spleen; (5) serous membranes; and (6) kidneys, gonads, and their ducts.

Entodermal Layer. As the embryo folds and its head comes closer to the tail, a portion of the yolk sac, lined with entoderm, becomes incorporated into the embryo proper and forms the *primitive foregut* and *hindgut*, which are lined with epithelium of entodermal origin. The *buccopharyngeal* membrane ruptures at the end of the third week, and thus an open connection between the primitive gut and the amniotic cavity is established. The primitive gut located between the fore- and hindgut remains temporarily in open connection with the yolk sac by way of a wide duct, the *omphalomesenteric* or *vitelline duct*.

Important structures that are subsequently derived from the entodermal germ layer are (1) the epithelial lining of the digestive and respiratory tracts and part of the bladder and urethra, (2) the epithelial lining of the tympanic cavity and eustachian tube, and (3) the main cellular portions of the tonsils, parathyroids, thymus and liver, pancreas, and gallbladder.

Development of the Extremities (Figure 26–8). The fore- and hindlimbs appear as buds at the beginning of the second month. The *forelimb buds* arise quite high, at the level of the fourth cervical to the first thoracic somites, and this explains their subsequent innervation by the *brachial plexus*. The *hindlimb* buds appear a day or two later than the arm buds at the level of the lumbar and sacral somites just below the attachment of the umbilical stalk and are later innervated by nerves from the *lumbosacral plexus*. As the limb buds grow, they undergo a 90-degree rotation but in opposite directions, so that the elbow points dorsally and the knee points ventrally. Hands and feet begin as paddle-shaped plates at the ends of the buds; then ridges appear in the plates that become the separated fingers and toes.

Development of the Sex Organs. The genetic sex is determined at the time of fertilization, but there is no indication of sex until the seventh week. The early genital system is similar in both sexes, and the embryo is potentially bisexual. In males, the Y chromosome stimulates development of testes, and the potentially female structure regresses. As the testes develop, they produce androgens that stimulate

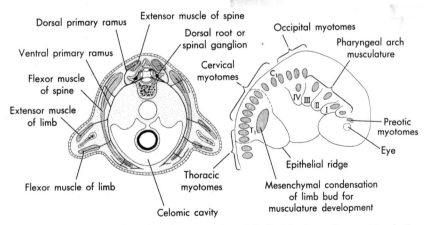

Figure 26–8. *A.* Cross section through embryo in region of the limb bud attachment. Muscle tissue has penetrated limb bud and has divided into ventral (flexor) portion and dorsal (extensor) components. The spinal nerves follow a similar orientation but eventually unite to form large dorsal (radial) and ventral (median and ulnar) nerves in upper extremity. *B.* Longitudinal view to show *myotomes* in head, neck, and thoracic regions of embryo at seven weeks. The tissue of the somite that remains after the migration of the cells of the *sclerotome* and dermatome makes up the *myotome*. *Myoblasts* of three of the four *occipital myotomes* migrate forward to form muscles of the tongue. The upper extremity is budding opposite the lower cervical and upper thoracic segments. (Modified from Langman.)

growth and development of male genitalia. The XX complement of chromosomes in the female stimulates development of the ovaries, and there is regression of the male embryonic structure. Maternal and placental estrogens circulating in the fetus stimulate development of the female reproductive organs.

Critical Period. The embryonic period is a time of great sensitivity because it is a period of differentiation. Factors that may cause malformation are called *teratogens*, and include certain drugs, the German measles virus, and x-rays. Exposure to such a factor, particularly just before the visible appearance of a structure, will result in derangement of the development of that structure. Exposure in the first two weeks results in complete failure to develop and the embryo is naturally expelled from the uterus.

FETAL PERIOD

During the fetal period some differentiation of tissues does continue, but the major changes are brought about by the rapid growth of the body. At the beginning of the third month the head constitutes approximately one half of the crown-rump (CR) length, or sitting height, but at birth the proportion has diminished to one fifth. The eyes are initially directed laterally, but during the third month the eyes are located on the ventral aspect of the face; the ears, originally formed in the neck region, reach their final position; and the limbs

their relative length in comparison to the rest of the body, although the lower limbs remain a little shorter and slightly less developed than the upper extremities. The *external genitalia* have usually developed by the end of the third month, so that the sex of the fetus can be determined by external appearance.

During the fourth and fifth months, at the end of the first half of intrauterine life, the fetus lengthens rapidly and its CH, or crown-heel length, is approximately 23 cm. This is about one half the total length of the full-term newborn.

There are many presumptive signs of pregnancy, but the three positive signs occur during the fetal period and include hearing the fetal heart between the eighteenth and twentieth weeks, visibility of the fetal skeleton by x-ray during the fourteenth to sixteenth weeks, and the physician's observance of fetal movement during the fifth month. The mother is usually able to discern fetal movements during the fifth month (quickening).

The weight of the fetus increases considerably during the second half of intrauterine life, from 500 gm at the end of the fifth month to 3200 gm by the end of the ninth month. Subcutaneous fat is formed during the last months before birth so that the fetus loses much of its former wrinkled appearance. At birth the fetus is approximately 50 cm long, and the skull still has the largest circumference of the body. At birth the testes have usually descended through the inguinal canal and into the scrotum.

Table 26–1 outlines the development during embryonic and fetal periods.

TABLE 26–1
DEVELOPMENTAL TIMETABLE

First Week
Fertilization, usually in fallopian tube:
  Second maturation division occurs, polar body expelled from ovum, tail separates from spermatozoon
  Male pronucleus combines with female pronucleus to form zygote
As zygote passes down tube cleavage occurs, forming several small blastomeres
After three days, morula (ball of 16 or more blastomeres) enters uterus

TABLE 26-1 (*Continued*)

Cavity forms in morula, converting it to blastocyst:
    ⎧ inner cell mass, or embryoblast
    ⎨ blastocyst cavity
    ⎩ outer layer of cells, or trophoblast
By sixth day blastocyst adheres to endometrium, invading epithelium
Embryonic entoderm begins to form on ventral surface of inner cell mass
Second Week
Rapid proliferation and differentiation of trophoblast
    Lacunar network develops, fills with maternal blood
    Trophoblast erodes endometrium
    Primary villi form on outer surface of chorionic sac, implantation complete, conceptus embedded within endometrium
Extraembryonic mesoderm forms from inner surface of trophoblast, reducing size of blastocyst cavity; blastocyst cavity becomes primitive yolk sac
Extraembryonic celom forms from extraembryonic mesoderm
Primitive yolk sac decreases in size; becomes secondary yolk sac
Inner cell mass differentiates into bilaminar embryonic disk:
    ⎧ embryonic ectoderm adjacent to amniotic cavity
    ⎨ embryonic entoderm adjacent to blastocyst cavity
    ⎩ prochordal plate develops from embryonic entoderm

Embryonic Period

Third Week
Formation of head, tail, and lateral folds
Dorsal part of yolk sac enclosed during folding to become primitive gut
Transverse folding forms lateral and ventral body walls
Gut pinches off from yolk sac, forming yolk stalk
Amnion expands
Primitive streak appears in ectoderm
Third primary germ layer (embryonic mesoderm) forms, except at oropharyngeal and cloacal membranes and in midline (notochord)
Notochord formed
Neural tube formed from neural groove in ectoderm
Somite formation begins cranially by end of third week
Celom forms (to become body cavity)
Blood islands form
Two heart tubes formed
Cytotrophoblastic shell formed, increasing surface area for embryonic-maternal exchange and firmly anchoring chorionic sac to endometrium
Longitudinal and transverse folding have converted trilaminar embryonic disk into **C**-shaped cylindrical embryo
Fourth Week
Arm and leg buds develop
Otic pits visible
Heart is ventral prominence
Head and tail folds give curved shape to embryo
Fifth Week
Head grows markedly due to growth of brain
Elbow and wrist regions identifiable; digital ridges appear in hand plates
Eye developing retinal pigment
Auricle of ear begins to form with groove in center (to become auditory meatus)

TABLE 26-1 *(Continued)*

Embryonic Period (*Cont.*)

Sixth Week
Head larger than trunk, is bent over heart prominence
Head and trunk begin to straighten at end of week
Yolk sac is small, now called vitelline duct
Intestines enter extraembryonic celom
Arms increase in length
Toe ridges appear in foot plates
Germinal tissues develop
Seventh Week
Head more rounded and erect
Ears not fully developed
Neck forms
Eyelids more obvious
Extremities lengthen, fingers and toes well differentiated
Abdomen less protuberant
Umbilical cord decreases in size
Intestine still within umbilical cord
At end of this week embryo has developed from undifferentiated cell mass to recognizable human shape

Fetal Period

Third Month
Head is half of fetal length; body grows to double crown-rump length
Arms grow, almost reaching final length; have separated fingers
Legs grow; have separated toes
Heart begins to beat
Intestinal coils visible in umbilical cord until tenth week, when they reenter abdomen
Eyelids closed; eyes widely separated and ears low set, begin to move to correct position
External genitalia similar at ninth week; different by twelfth week and have begun their descent
Placenta develops to final form (two layers)
Fourth Month
Growth rapid, legs lengthen
Ossification of skeleton begins
Ears and eyes in mature position
Fifth Month
Growth rate slows; legs achieve final length
Fatty tissue forms (special type: brown fat)
Hair grows on head, eyebrows visible, fine hair (lanugo) covers body
Skin develops covering of fatty, cheeselike material (vernix caseosa) that remains until birth
Fetus moves extremities
Sixth Month
Gain in weight
Skin wrinkled and translucent
Organs fairly well developed, but respiratory system still immature
Body is lean, better proportioned
Subcutaneous adipose tissue forming
Seventh Month
Nervous system matures—body temperature control and rhythmical respiration are possible
Eyelids reopen
Eighth and Ninth Months
Skin smooth, color same in white and dark-skinned races (melanin produced on exposure to light)
Testes have descended to scrotum
Finishing off of development and growth
Fetus usually plump at birth

Placental Development and Functions

The placenta is composed of an embryonic portion, the *chorion frondosum*, and a maternal portion, the *decidua basalis*. Each portion has its own blood supply, and there is no direct connection between them. Exchange of substances between the two systems takes place by diffusion.

The *trophoblast* develops a great number of *secondary villi* or *cytotrophic projections* that extend into and are attached to the *maternal decidua* (Figure 26-9). They are now referred to as the *chorionic villi*. The portion of the chorion that contains the expanding villi and is adherent to the decidua basalis is called the *chorion frondosum* (bushy chorion), and the portion projecting into the lumen of the uterus is smooth and almost nonvascular and is known as the *chorion laeve*.

The *decidua basalis* (Figure 26–10), the endometrial layer adjacent to the embryonic *chorion frondosum*, consists of a compact layer that is tightly connected to the chorion and a spongy layer that contains dilated glands and the spiral arteries. The compact layer is often referred to as the *decidual plate*. The *decidua capsularis* is the decidual layer adjacent to the chorion laeve, which projects into the uterine lumen. During the third month, as the fetus increases in size, the decidua capsularis degenerates, and the *chorion laeve*

fuses with the *decidua parietalis* on the opposite side of the uterus. Most of the uterine cavity is now obliterated and the chorion frondosum remains the only functional part of the chorion. The fully mature *placenta* is composed of the chorion frondosum, its fetal portion, and the decidua basalis, its maternal portion. The placenta enlarges greatly as the fetus and uterus increase in size, and it amounts to about 25 per cent of the internal surface of the uterus. The main functions of the placenta are the exchange of gaseous and metabolic products as well as nutrients between the maternal and fetal bloodstreams, and the production of hormones to maintain pregnancy.

There are intervillous spaces between the chorionic and decidual plates (Figure 26–11). These spaces are filled with maternal blood and are lined with *syncytium* of fetal origin. Fetal villous capillaries from umbilical veins project into these intervillous spaces, where they are bathed by approximately 150 ml of oxygenated blood from the maternal spiral arteries. Blood from the intervillous lakes is returned to the maternal circulation via venous openings from these spaces. It is estimated that from 500 to 600 ml of maternal blood circulates through the intervillous spaces every minute, thus permitting a replace-

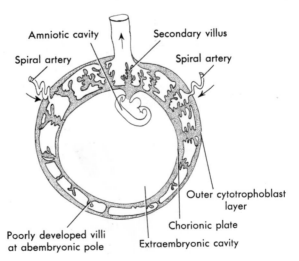

Figure 26–9. Diagram of embryo at beginning of second month of development to show numerous well-formed secondary villi. (Modified from Langman.)

Amniotic cavity

Secondary villus

Spiral artery

Spiral artery

Outer cytotrophoblast layer

Chorionic plate

Poorly developed villi at abembryonic pole

Extraembryonic cavity

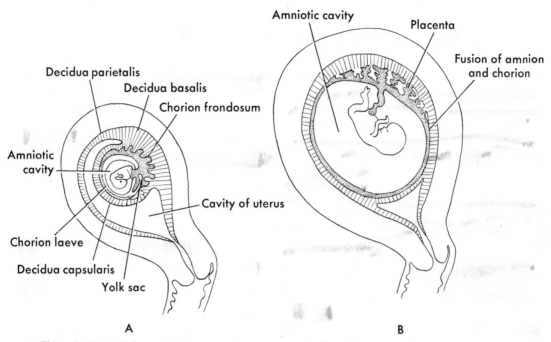

Figure 26–10. *A.* Schematic drawing to show the decidua and fetal membranes in embryo at the end of the second month. Villi have disappeared at the abembryonic pole. *B.* Schematic drawing of embryo at the end of the third month to show obliteration of the decidua capsularis and chorion laeve. Fusion of the amnion and chorion has also occurred.

ment of 150 ml in the intervillous spaces every two to three minutes.

The placental barrier, or dividing membrane, is made up entirely of fetal tissue. Until the fourth month the barrier is composed of four layers, but then it becomes much thinner and retains only two, the endothelial lining of the capillaries and the syncytial covering that lies in intimate contact with them. The thinner layer allows for more rapid exchange of substances such as nutrients, gases, and other metabolic products between the two blood systems. Hormones and antibodies also pass across the placental barrier, although it is not known just how the high-molecu-

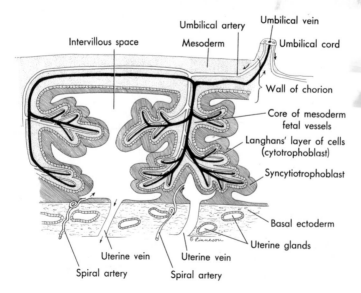

Figure 26–11. Schematic drawing of the structure of the villi at an early stage of development. The capillaries of the fetal circulation are separated from maternal blood in the intervillous spaces by surrounding layers of mesoderm, cytotrophoblast, and syncytiotrophoblast. After the fifth month only a single layer, the syncytiotrophoblast, lies between the fetal capillary wall and the maternal blood. Note the umbilical cord containing one umbilical vein and two arteries. Two uterine arteries and two veins are shown.

lar-weight substances such as proteins and maternal gamma globulins can pass through the barrier. The fetus acquires some of the antibodies that the mother has produced against such infectious diseases as scarlet fever, measles, smallpox, and diphtheria. The precise mechanism by which these antibodies reach the fetus is unknown, but it is probably pinocytosis.

The placenta manufactures sufficient amounts of gonadotropins, progesterone, and estrogens by the end of the fourth month of pregnancy so that the ovarian corpus luteum is no longer needed and it therefore begins to degenerate. Estrogenic hormones are produced in increasing amounts until a maximum level is reached just before the end of pregnancy.

The presence of chorionic gonadotropin in the urine early in pregnancy is used as an indicator in some varieties of pregnancy tests. When this urine is injected into the immature mouse, it causes ovarian hyperemia, and in the young female rabbit it causes ovulation. If young male frogs are injected with urine or blood serum containing these gonadotropins, it will cause their ejection of spermatozoa. More recently immunologic tests have been used—gonadotropin in the urine showing agglutination when antiserum is added.

Circulation

Fetal Circulation Before Birth (Figure 26–12). The umbilical cord unites the placenta with the navel of the fetus. The cord is made up of two arteries and one large vein. These vessels are surrounded and protected by soft mucous connective tissue known as *Wharton's jelly*. Nutrients and oxygenated blood are conveyed to the fetus via the umbilical vein, which travels to the liver within the anterior peritoneal attachment, the *falciform ligament*. The main portion of the blood to the fetus bypasses the liver. It flows from the umbilical vein into the short vessel called the *ductus venosus* and from there into the inferior vena cava. Since the liver, at this time, is only partially functional, there is no need for the major portion of blood to perfuse it. Only a small amount of blood enters the sinusoids of the liver and mixes with the blood from the portal circulation. There is a sphincter mechanism in the ductus venosus near the entrance of the umbilical veins. When venous return is too great because of the additional pressure caused by a uterine contraction, this sphincter, it is believed, closes so that the heart will not be overloaded with blood.

Blood from the inferior vena cava enters the right atrium and is directed toward the *foramen ovale* by the valve of the inferior vena cava. As a result the largest portion of the returning blood bypasses the non-functioning lungs and passes directly into the left atrium. This blood will supply the coronary vessels of the heart and the carotid arteries to the brain with well-oxygenated blood.

A small portion of blood from the inferior vena cava joins the desaturated blood from the superior vena cava, which flows into the right ventricle and out into the pulmonary artery. During fetal life resistance in the pulmonary vascular system is very high, causing the blood to pass through the *ductus arteriosus* (Figure 26–12) into the descending thoracic aorta, where it mixes with blood from the left heart. The *umbilical arteries*, branches of the hypogastric arteries, return the fetal blood to the placenta, where it is reoxygenated, receives nutrients diffused from the mother's blood, and discharges the excess carbon dioxide and nitrogenous metabolic wastes into her circulation. The capillary networks of fetal and maternal circulation are not directly connected, as has previously been stated, but are in close association within the placenta so that diffusion of products between the two separate systems is feasible. At the end of pregnancy a small portion of blood may be carried by the pulmonary artery through the fetal lungs.

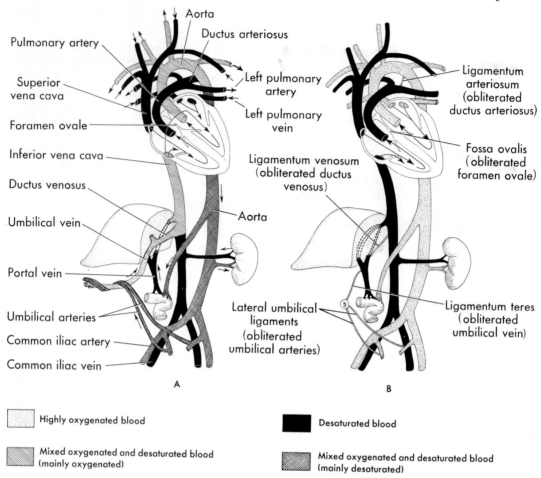

Figure 26–12. *A.* Fetal circulation. *B.* Circulation after birth.

Circulatory Changes at Birth. At birth a number of changes occur in the newborn due to the cessation of the placental flow and the beginning of lung respiration. When the amniotic fluid in the alveolar sacs and bronchial tree is replaced by air at birth and when the pressure in the right atrium decreases as a result of interruption of the placental blood flow, many of the fetal vascular structures are no longer needed and they cease functioning. The *ductus arteriosus* closes because of muscular contraction of its wall. The amount of blood flowing through the lung increases as the fluid is expelled and as breathing fills the lungs with air. Since the blood pressure within the right atrium has decreased as a result of the interrupted placental flow and the pulmonary pressure increases as a result of the increased pulmonary blood flow, the pressure becomes equalized on both sides of the *foramen ovale*. The *septum primum* is then apposed to the *septum secundum*, and the interatrial opening closes functionally. In 20 per cent of all adults perfect anatomic closure of the foramen ovale may never be obtained, although there may be no flow of blood from one atrium directly into the other. Anatomic fusion of the two septa is usually completed at the end of the first year of life.

The umbilical arteries, umbilical vein, and ductus venosus close shortly after birth owing to contraction of smooth muscle of the vessel walls as a result of the ligation of the umbilical cord and the thermal and mechanical stimuli, as well as a change in oxygen tension. Following their obliteration, the umbilical arteries become the *lateral vesicoumbilical ligaments*, and the umbilical vein becomes the connective tissue band, the *ligamentum teres hepatis*, in the margin of the *falciform ligament*. The ductus arteriosus closes almost immediately after birth following the contraction of its muscular wall and becomes the fibrous band known as the *ligamentum arteriosum*. If either the foramen ovale or the ductus arteriosus does not become entirely obliterated, the affected child may suffer from inadequate oxygenation of his blood, resulting in a bluish appearance of the skin, known as cyanosis, especially after exertion.

Complete anatomic closure of all these structures by proliferation of vessel wall intima and fibrous tissues takes anywhere from several months to a year.

CONGENITAL MALFORMATIONS

Congenital abnormalities, which are those present at birth, may be due to teratogenic influences, as mentioned earlier, or they may be due to specific genes within a chromosome, as in hemophilia. Another cause is abnormal chromosome number. Individuals with mongolism have three number 21 chromosomes. Mental retardation, congenital heart defects, deafness, cleft lip and palate, or other defects may be found in the child with extra chromosomes numbers 13 to 15 and 17 to 18.

Abnormalities in the sex chromosomes may cause congenital defects. If, during meiosis of the female sex cell, the two homologous X chromosomes fail to separate and move instead into one daughter cell, the resultant ovum has either two X chromosomes or none. If an ovum with two X chromosomes combines with a sperm containing a Y chromosome, the result is a male with an XXY complement (Klinefelter's syndrome). Some of the important features of this syndrome are testicular atrophy, sterility, and mental retardation. If the ovum without sex chromosomes is fertilized by a sperm containing an X chromosome, the result is an individual with an XO complement (Turner's syndrome). This condition found in women is characterized by the absence of ovaries. Last, if an ovum containing two X chromosomes combines with an X sperm, the result is an XXX individual. Women with the triple-X syndrome have some degree of mental retardation, are infantile, and have scant menses.

Amniocentesis. It is possible to identify the presence of an abnormal chromosome number by a procedure known as *amniocentesis*. In this procedure a sterile needle is inserted through the abdominal wall and through the uterine muscle into the amniotic sac. Several milliliters of amniotic fluid are withdrawn and examined. Cells from the fetus are found floating in this fluid—epithelial cells from the skin and respiratory tract. These cells are then cultured; sex, abnormality of chromosomes, and enzyme defects can then be determined.

Questions for Discussion

1. Explain why no two children in the same family (unless they are identical twins) have exactly the same physical appearance despite the fact that every child has received 23 chromosomes from each of his parents.
2. Explain how the diploid number of chromosomes is reduced to the haploid number of single chromosomes during meiosis.
3. How many spermatozoa are formed from one primary spermatocyte? What percentage of the spermatozoa will carry X chromosomes?
4. What is crossing over and when does it occur?
5. How many ova are produced from each primary oocyte? When do the first and second maturation divisions occur?

6. If one parent is homozygous for brown eyes (B brown, B brown) and the other parent is heterozygous for brown eyes (B brown, b blue), can any of their children have blue eyes? Explain.
7. What important structures are developed from the embryoblast and from the trophoblast of the blastocyst?
8. Explain the development and functions of the placenta. What is the placental barrier?
9. Describe fetal circulation before birth. Explain the causes of circulatory changes in the newborn infant.
10. What important parts of the body are developed from ectoderm? The mesoderm? The entoderm?

Summary

Meiosis	**First Maturation Division**	Each chromosome duplicates itself and becomes doubled Kinetocore or centromere—point of union of doubled chromosome Chromosomes begin to pair; synapsis—entwining of homologous chromosomes in center of cell Four chromatids in each pair of chromosomes Crossing over—exchange of segments between homologous chromosomes Cell completes metaphase, anaphase, and telophase Cell division results in 23 doubled chromosomes, held together at chromomere
	Second Maturation Division	No synthesis of new DNA Doubled chromosomes separate Cell division results in each cell having 23 single chromosomes; ovum has X chromosome; spermatozoon has either X or Y
	Significance	Halves number of chromosomes Provides opportunity for crossing over and reshuffling of genes Provides offspring with infinite variations resulting in greater individuality

Heredity

Heredity—term applied to transmission of potential traits, physical or mental, from parents to their offspring; offspring receives one chromosome of a homologous pair from each of his parents
Autosome—chromosome that carries same general set of genes as its mate
Heterozygous—having alternate genes for the same trait, e.g., one gene for blue and one for brown eyes
Homozygous—having two similar genes for the same trait, e.g., two genes for brown eyes
Dominant gene—gene whose trait is bound to appear under normal circumstances
Recessive gene—gene whose trait will not appear usually unless two of them are present or dominance is not complete
Capital letter used to denote dominant trait, small letter for recessive trait
Dominance may be complete, partial, or absent
Alleles—genes that carry the same trait and occupy the same position or locus on the chromosomes—come together during synapsis
Gregor Mendel—"father of modern genetics," responsible for whole development of the gene concept
Genotype—refers to type of genes present in individual
Phenotype—refers to expression of genes present
Total final expression of individual's characteristics due to blending of hereditary and environmental factors

Genetic abnormalities	Abnormalities may be inherited as dominants over normal condition, e.g., extra digits or excessively short digits, or may be recessive to the normal, e.g., albinism Some abnormalities appear as sex-linked recessives evident in males with only one factor, but in females only when two factors are present, e.g., red-green color blindness or hemophilia Abnormal number of chromosomes, especially of sex chromosomes, results in abnormality

Developmental Periods

Prenatal—before birth
— First two weeks
— Embryonic period—first two months
— Fetal period—third through ninth month
Postnatal—after birth; growth continues especially during first year and during puberty

Preembryonic and Embryonic Period

Cleavage—consists of a number of rapid mitotic divisions, which result in an increasing number of smaller cells known as *blastomeres*
Morula—16–20-cell stage of cleavage resulting in solid ball of cells—inner cell mass forms embryo and embryonic membranes, the amnion and yolk sac—outer cell mass forms the trophoblast from which the chorion is developed
Morula reaches uterine cavity 3–4 days after ovulation
Blastocyst—formed when cavity appears between inner and outer cell masses

Implantation of Blastocyst — Implantation—uterine attachment of the blastocyst—occurs 7–9 days after ovulation and normally in the endometrium

Preembryonic and Embryonic Period (*cont.*)	**Implantation of Blastocyst** (*cont.*)	Decidua—modified uterine mucosa during pregnancy—can be divided into three parts: *decidua basalis* at base of placenta, *decidua capsularis* portion surrounding surface of chorionic sac, and *decidua parietalis* lining the rest of the uterus
	Major Events in Second and Third Weeks	Blastocyst—made up of *trophoblast* and *embryoblast* Embryoblast gives rise to *ectoderm*, *entoderm*, and *mesoderm* Trophoblast has inner layer, *cytotrophoblast*, and outer layer, *syncytiotrophoblast* Amniotic cavity—forms between trophoblast and ectoderm—fills with fluid to protect and nourish the embryo Connecting stalk—later becomes umbilical cord, attaches embryo to placenta Primitive streak and notochord (embryonic backbone) formed in midline from modified ectodermal cells; notochord later replaced by segmented vertebral column Posterior cloacal membrane—formed from joining of ectoderm and entoderm, urogenital and anal membranes formed from it Prochordal plate—appears at anterior ectodermal attachment—buccopharyngeal membrane formed from this plate Allantois and yolk sac—pocket-like extensions of ventral side of embryo—they can be seen in umbilical cord of young embryo Neural tube—developed from ectoderm—gives rise to brain, spinal cord, autonomic and spinal nerves Primary germ layers formed—ectoderm, mesoderm, entoderm
	Portions of the Body Derived from Ectoderm	1. Central nervous system and hypophysis 2. Peripheral nervous system 3. Sensory epithelium of sense organs 4. Enamel of teeth 5. Epidermis including hair, nails, and subcutaneous glands
	Portions of the Body Derived from Mesoderm	1. Cartilage, joints, and bones 2. Connective tissue 3. Heart, blood, and lymph cells and vessels 4. Spleen 5. Serous membranes 6. Kidneys, gonads, and their ducts
	Portions of the Body Derived from Entoderm	1. Epithelial lining of digestive and respiratory tracts and parts of bladder and urethra 2. Epithelial lining of tympanic cavity and eustachian tube 3. Main cellular portions of tonsils, parathyroids, thymus, liver, pancreas, and gallbladder
	Development of Extremities	Fore- and hindlimbs appear as buds at beginning of second month—as buds grow, they undergo 90° rotation in opposite directions—elbow points dorsally, knee ventrally. Hands and feet begin as plates at end of buds; ridges become separated fingers and toes
	Development of Sex Organs	Sex determined by sex chromosome; early stage similar; hormones influence development of external genitalia
		Critical Period—embryonic period is one of differentiation; at end of embryonic period all major organ systems have begun to develop
Fetal Period		Interval from beginning of third month to end of intrauterine life—major changes are brought about by rapid body growth—at end of 4th–5th month the CH length is approximately 23 cm and the weight 500 gm At birth the fetus is approximately 50 cm long and the weight is about 3200 gm
Placental Development and Functions		Chorion frondosum—embryonic portion of placenta Decidua basalis—maternal portion of placenta Chorion laeve—portion of chorion that is smooth, almost nonvascular, and projects into uterus Main functions of placenta—exchange of gaseous and metabolic products between fetus and mother, and production of hormones
Development of Blood Circulation Within Placenta		Intervillous spaces between chorionic and decidual plates are filled with maternal blood and are lined with syncytium of fetal origin Fetal villous capillaries derived from umbilical arteries and veins project into these spaces where they are bathed with oxygenated blood from maternal circulation Placental barrier—is made up of fetal tissue starting with four layers, but retaining only two after fourth month—endothelial lining of capillaries and syncytial covering—thinner layer allows for more rapid exchange of nutrients, gases, metabolic products, hormones, and antibodies

Fetal Circulation	Umbilical cord—unites the placenta with the navel of the fetus—contains *two umbilical arteries*, extensions from fetal hypogastric arteries, and one umbilical vein that travels in falciform ligament of fetus to the ductus venosus—these vessels are surrounded by soft mucous connective tissue called *Wharton's jelly* Direct communication between right and left atrium by means of *foramen ovale* Direct communication between pulmonary artery and aorta known as *ductus arteriosus* Direct communication between umbilical vein and inferior vena cava through *ductus venosus* Oxygen and nutritive substances diffuse from maternal blood in placenta across placental barrier to fetal blood. Carbon dioxide and metabolic wastes diffuse from fetal capillaries in placenta across the same barrier into maternal blood
Changes in Circulation of Infant Following Birth	Infant respiration stimulates pulmonary circulation; this causes a rise in blood pressure in left atrium—pressure is equalized across atrial septum and connective tissue begins to close over foramen ovale—will become *fossa ovalis* Ductus arteriosus becomes fibrous cord—*ligamentum arteriosum* Umbilical vein becomes fibrous cord—*ligamentum teres*, or round ligament, in edge of falciform ligament Ductus venosus becomes fibrous cord—*ligamentum venosum* Umbilical arteries are obliterated to become *lateral vesicoumbilical ligaments;* their most proximal portions remain as *hypogastric* arteries Complete anatomic closure of all these structures takes from several months to a year

Additional Readings

BEARN, A., and GERMAN, J.: Chromosomes and disease. *Sci. Amer.*, **205**:66–76, November, 1961.

DAVIS, B.: Prospects for genetic intervention in man. *Science*, **170**:1279–83, December 18, 1970.

EDWARDS, R., and FOWLER, R.: Human embryos in the laboratory. *Sci. Amer.*, **223**:44–54, December, 1970.

FRIEDMANN, T.: Prenatal diagnosis of genetic disease. *Sci. Amer.*, **225**:34–42, November, 1971.

LANGMAN, J.: *Medical Embryology*, 3rd ed. Williams & Wilkins Co., Baltimore, 1975, pp. 19–106.

TUCHMANN-DUPLESSIS, H.; DAVID, G.; and GAEGEL, P.: *Illustrated Human Embryology*, Vol. 1. Springer-Verlag, New York, 1972, pp. 14–54.

Glossary

Terms adequately defined in the body of the text are not always included in the Glossary. They may be located through the Index.

ab'duct. To draw away from the median line or from a neighboring part or limb

acetab'ulum. "Shallow vinegar cup"; the depression in the innominate bone that receives the head of the femur

ac'etone. $(CH_3)_2CO$. A simple ketone with a sweetish odor, present in blood and in body excretions whenever fats are used in metabolism without the presence of sufficient carbohydrate

acetylcho'line. A reversible acid ester of choline, $CH_3 \cdot CO \cdot O \cdot CH_2 \cdot CH_2 —N—(CH_3)_3 \cdot OH$, normally present in many parts of the body and having important physiologic functions, such as the transmission of a nerve impulse across a synapse

acido'sis. Condition of pH of body fluids lower than normal; decrease in alkaline reserve

acro'mion. Point of the shoulder. The spine of the scapula terminates in the acromion process

ad'duct. To draw toward a center or toward a median line

adrener'gic. Acting like epinephrine; sympathetic nerve fibers secreting norepinephrine

adsorption. The attachment of one substance to the surface of another

adventi'tia. The outermost layer of the organs that are not bounded by a serous coat, the outer areolar connective tissue being continuous with that of the other organs

af'ferent. Toward; applied to nerve impulses and fibers (toward the brain or spinal cord) and to arterioles of nephron (toward the glomerulus)

agglu'tinins. Substances that induce adhesion or clumping together of cells

agglutin'ogen. Substances (ABO) in the red blood cells that when in contact with agglutinins (Aa, Bb) will cause agglutination

al'binism. Congenital absence of the pigment melanin in the skin, iris, and hair. It may be partial or complete

albu'mins. Proteins; soluble in water, dilute acids, dilute salines, and concentrated solutions of magnesium sulfate and sodium chloride. They are coagulated by heat. Examples: egg albumin and serum albumin of blood

alimen'tary. Petaining to diet or food

alkalo'sis. Condition of pH of body fluids higher than normal; increase in alkaline reserve

ampul'la. A flasklike dilatation of a canal

amyg'daloid. Shaped like an almond

anaero'bic. Lacking in oxygen; applied to microorganisms that grow in an environment with decreased or absent oxygen

anastomo'sis. Joining of a blood vessel with another (arterial or arteriovenous); joining of one nerve to another; surgical joining of a part (e.g., stomach with jejunum)

an'drogen. Hormone causing masculinization, having effects similar to testosterone

an'ion. An electrically charged ion bearing a negative charge

619

an'kylose. Immobilization of a joint by pathologic or surgical process

anom'aly. Anything unusual, irregular, or contrary to the general rule

an'tibody. Globulin that confers immunity by inactivating or destroying foreign proteins introduced into the body

an'tigens. Name given to foreign proteins and certain other substances that upon entering the bloodstream cause the formation of antibodies in the serum

antimetab'olite. A substance that resembles chemically a normal metabolite but is foreign to the body and competes with, replaces, or antagonizes the latter. These substances may be used to prevent growth of cancer cells

a'pex. The top or pointed extremity of a body

aponeuro'sis. A white, flattened, or ribbonlike tendinous expansion, serving mainly as an investment for muscle or connecting a muscle with the parts that it moves

arach'noid. Resembling a cobweb

arboriza'tion. A branching distribution of venules or of nerve filaments, especially the branched, terminal ramifications of neurofibrils

artic'ular. Of or pertaining to a joint

au'ricle. Ear. The *pinna* of the ear that with the *external meatus* constitutes the external ear

autonom'ic. Performed without the will; automatic

az'ygos. Without a fellow; single; unpaired

bas'al metab'olism. Rate of energy metabolism of a person at rest 12 to 18 hours after eating, as measured by the calorimeter or a BMR machine

bas'ilar. Pertaining to the base of an object; e.g., basilar artery located at the base of the brain

bicip'ital. Referring to the biceps

blas'tula. A hollow sphere of embryonic cells; the last stage in development before the embryo divides into two layers

brachiocephal'ic. Pertaining to both the upper arm and head, as the brachiocephalic (innominate) artery and veins

buc'cal. Pertaining to the cheek

cal'cify. Harden by deposit of salts of calcium; petrify

cal'culus; pl., cal'culi. A stone

cal'orie. The Calorie is the amount of heat required to raise the temperature of 1 kg of water 1° C. The small calorie is the amount of heat required to raise the temperature of 1 gm of water 1° C

canalic'ulus; pl., canalic'uli. A minute channel or vessel

carbohy'drate. Organic compound composed of carbon, hydrogen, and oxygen in a specific proportion $(CH_2O)n$. Among important carbohydrates are starches and sugars

car'dia. The heart. The esophageal orifice of the stomach

carot'id. The principal artery of the neck

castra'tion. Removal of the testes in the male or the ovaries in the female

catacrot'ic. Referring to an irregularity of the pulse in which the beat is marked by two or more expansions of the artery. A tracing of this pulse shows one or more abnormal elevations on the downward stroke

catal'ysis. A changing of the speed of a reaction, produced by the presence of a substance that does not itself enter the final products

cat'ion. An electrically charged ion bearing a positive charge

celluli'tis. Inflammation of connective tissue, most commonly of superficial fascia

chi'asm. An X-shaped crossing or decussation, especially that of the fibers of the optic nerve

choa'na. Any funnel-shaped cavity, such as the posterior nares

cholere'tic. Any substance that stimulates secretion and flow of bile

choles'terol. A sterol, $C_{27}H_{43}OH$, found in small quantities in the protoplasm of all cells, especially in nerve tissue, blood cells, and bile. Same as cholesterin

choliner'gic. Having the same action as acetylcholine; nerve fibers that secrete acetylcholine

chor'da tym'pani. The tympanic cord, a branch of the facial (seventh cranial) nerve, which traverses the tympanic cavity and joins the gustatory (lingual) nerve

chro'maffin. Certain cells, occurring in the adrenal medulla, along the sympathetic nerves, and in various organs, that stain deeply with chrome salts (brownish-yellow)—hence the name chromaffin. The whole system of such tissue throughout the body is named the chromaffin or chromaphil system

chro'matin. Portions of the nucleus that stain deeply with basic dyes; e.g., methylene blue

chro'mosomes. Rodlike strands of DNA and protein, visible during mitosis in the nucleus; contain the genes

cica'trix. The mark or scar left after the healing of a wound

cister'na mag'na or **cister'na cerebel'lo med'ullaris.** That portion of the subarachnoid cavity between the cerebellum and the medulla

clea'vage. The process of division of the fertilized ovum before differentiation into layers occurs

co'don. A triplet of three bases in a DNA or RNA molecule that codes for a specific amino acid

coen'zymes. Nonprotein substances produced by living cells, which are essential for action of certain enzymes

col'lagen. An albuminoid, the main supportive protein of skin, tendon, bone, cartilage, and connective tissue

col'loid. A state of matter in which particles having a diameter of 0.1 to 0.001 μm are dispersed in a medium such as water

comminu'tion. Grinding into small particles

com'missure. A joining. A bundle of nerve fibers passing from one side of the brain or spinal cord to the other side. The corner or angle of the eyes or lips

concep'tus. Embryo (or fetus) and its membranes

congen'ital. Existing from or before birth

con'jugate. Join together; unite chemically

cor'acoid. Shaped like a crow's beak; a process of the scapula

cor'pus. A body or mass

cor'tex. The bark or outer layer; the outer portion of an organ

cre'nated. Notched on the edge (indented)

crepita'tion. A grating or crackling sound or sensation, like that produced by fragments of a fractured bone rubbing together, or when tissues contain small air pockets

crib'riform. Perforated like a sieve

cu'neiform. Wedge shaped

cyano'sis. Blueness of the skin, resulting from insufficient oxygenation of the blood

cysti'tis. Inflammation of the bladder

deaminiza'tion. Removal of the amino group, NH_2, from an amino compound

dec'ibel. Unit of loudness, tenth of a bel. For measuring loudness, differences of sounds

decid'uous. That which falls off; not permanent

dehydra'tion. Removal of water, as from a tissue

delamina'tion. Separation of cells into layers

del'toid. Triangular; resembling in shape the Greek letter Δ, delta

detri'tus. Waste matter from disintegration

diapede'sis. Passing of any of the formed elements of the blood through vessel walls without rupture

diath'esis. A predisposition to certain kinds of disease

dichot'omous. Divided into two; consisting of a pair or pairs

diffu'sion. Continual movement of molecules among each other in liquid or in gases. The passage of a substance through a membrane. The diffusing substance always diffuses from an area of high concentration to an area of lesser concentration (of that particular substance)

dip'loid. Having the full complement of chromosomes, 46 in the human

dor'sal. Pertaining to the back surface, same as posterior

dura mater. "Hard mother"; the tough outer membrane enveloping the brain and spinal cord

dyne. Unit of force that, when acting on a mass of 1 gm for 1 second, will cause an acceleration of 1 cm per second

dyspha'gia. Difficulty in swallowing

ec'toderm. Outermost of three primary germ layers from which skin and neural tube develop

ectop'ic. Out of place. *Ectopic gestation* refers to pregnancy when the fecundated ovum, instead of entering the uterus, remains in either a fallopian tube or the abdominal cavity

ede'ma. Swelling due to abnormal effusion of serous fluid into the tissues

ef'ferent. Away from; opposite of afferent

elec'trolyte. Atom or radical that in solution conducts an electric current

electrophore'sis. Movement of charged particles through a liquid medium under the influence of electrical potential; technique of recording this movement

empir'ical. Founded on experience; relating to the treatment of disease according to symptoms alone, without regard to scientific knowledge

emul'sion. A mixture of two fluids insoluble in each other, where one is dispersed through the other in the form of finely divided globules

endergon'ic. A reaction characterized by absorption of energy; this is characteristic of many anabolic processes, such as synthesis

endochon'dral ossifica'tion. Ossification in which cartilage is formed first and then is gradually replaced with bone

endog'enous. Originating within the organism. Opposite of exogenous

endother'mic. In a reacting system, if heat is absorbed the reaction is endothermic

en'ergy. Capacity or ability to do work, activity, exertion of power

en'toderm. Innermost of three primary germ layers that forms lining of gastrointestinal, respiratory, and urinary tracts; accessory digestive organs

epicrit'ic. Relating to or serving the purpose of accurate determination. Applied to cutaneous nerve fibers that serve the purpose of perceiving fine variations of touch or temperature

equilib'rium. The balanced condition resulting when opposing forces are exactly equal. The term may refer to the maintenance of correct concentration of constituents of the body fluids or to the harmonious action of the organs of the body as in standing

erg. A unit of work. The work done in moving a body 1 cm against a force of 1 dyne

es'trogen. Group of hormones, natural or synthetic, causing glandular development in endometrium and secondary sex changes of puberty

eth'moid. Resembling a sieve

evagina'tion. Protrusion of some part or organ from its normal position

evapora'tion. The changing of a liquid into a vapor. Heat is necessary for evaporation, and if not otherwise applied it is taken from near objects. Thus, the heat necessary for the evaporation of perspiration is taken from the body

ever'sion. A turning outward or inside out

exergon'ic. A reaction characterized by release of free energy; this is characteristic of catabolic reactions

exog'enous. Originating outside the organism; opposite of endogenous

exophthal'mic. Pertaining to abnormal protrusion of the eyeball

exother'mic. In a reacting system if heat is formed, the reaction is said to be exothermic

exten'sile. Capable of being stretched

exten'sion. A movement that brings the members of a limb toward a straight condition

extrin'sic. Originating outside of a part or cell

ex'udate. A fluid or semifluid that has oozed through the tissues into a cavity or upon the body surface

fal'ciform. Sickle shaped

fal'ciform ligament. Fold of peritoneum between the liver and the anterior abdominal wall in which the umbilical vein (ligamentum teres) is contained in fetus

fas'cia. A sheet or band of fibrous tissue that covers the body under the skin and invests the muscles and organs

fascic'ulus; pl., fascic'uli. A bundle of close-set fibers surrounded by connective tissue, usually muscle or nerve fibers

fecunda'tion. Fertilization; impregnation

fenes'trated. Having windowlike openings; perforated

fibril'la; pl., fibril'lae. A small fiber or filament

fim'bria; pl., fim'briae. A fringe

flat'ulence. Distention due to generation of gases in the stomach and intestine

flex'ion. The act of bending; condition of being bent

fontanelle'. "Little fountain"; membranous space between bones of cranium in infants

fos'sa. A pit, depression, trench, fovea, or hollow

gastrocne'mius. "Belly of the leg"; one of the calf muscles

gas'troepiplo'ic. Pertaining to the stomach and greater omentum

gen'erative. Having the power or function of reproduction

genes. The factors on chromosomes that determine certain hereditary characteristics believed to be the sequence of bases in the DNA molecule

genu. Any structure with the shape of the flexed knee

ger'minal. Cells or tissue from which new ones arise

gesta'tion. Pregnancy

glob'ulins. Protein substances (myosin, fibrinogen, etc.) similar to albumins but insoluble in water and soluble in dilute solutions of neutral salts

glossi'tis. Inflammation of the tongue

glucocor'ticoid. Steroid hormone from adrenal cortex affecting glucose metabolism

gluconeogen'esis. Formation of glucose or glycogen from noncarbohydrate substances

glycoca'lyx. Carbohydrate-rich coat located on external surface of cell membrane

glycogen'esis. The production of glycogen

glycogenol'ysis. Splitting of glycogen into glucose by the liver, or into pyruvic or lactic acid in other tissues

glycol'ysis. Splitting of glucose within the cell; a series of enzyme-aided reactions that yield pyruvic or lactic acid

gon'ad. Gamete-producing gland; e.g., testis, ovary

gy'rus; pl., gy'ri. One of the tortuous elevations (convolutions) of the surface of the brain caused by enfolding of the cortex and separated by the fissures and sulci

hap'loid. Having half the full number of chromosomes

hem'atin. An iron-containing compound derived from heme, the colored nonprotein constituent of hemoglobin

hemorrhoi'dal. Pertaining to hemorrhoids, varicosities caused by dilation of the veins of the anal region

hep'arin. A mucopolysaccharide acid occurring in various tissues. Used as an anticoagulant

hi'lus or **hi'lum.** The depression, usually on the concave surface of a gland, where vessels and ducts enter or leave

his'tamine. An amine, beta-imidazolylethylamine, occurring in animal tissues. It is a powerful dilator of the capillaries and a stimulator of gastric secretion

homeosta'sis. Constancy of the internal environment

homoge'neous. Of the same kind or quality throughout; uniform in nature; the reverse of heterogeneous

hu'moral. Pertaining to substances (hormones, ions) within the body fluids

hy'aline. Glassy; translucent

hyaluron'idase. Enzyme causing breakdown of polysaccharide materials; "spreading factor," produced by sperm and by certain bacteria

hydrostatic pressure. A pressure exerted by a liquid resulting from a force applied to it, e.g., blood pressure due to cardiac muscle contraction

hy'oid. Y or U shaped

hyperglyce'mia. An abnormally high amount of glucose in the blood

immu'nity. Condition of a living organism whereby resistance to infection is conferred. *Active* immunity is due to individual producing own antibodies; *passive* immunity is due to receiving antibodies produced by another person or animal

in'guinal. Pertaining to the groin

interme'diary meta'bolism. Refers to metabolism taking place after absorption and before excretion from the excretory organs occurs in cells

internun'cial. Connecting, applied to neuron between two others in a nerve pathway

intersti'tial. In the interspaces of a tissue; refers to the connective tissue framework

intrin'sic. Originating within a part; belonging to the essential nature of a part

intro'itus. An entrance to a cavity or space

inver'sion. A turning inward, inside out, upside down, or other reversal of the normal relation of a part

ionize. To form ions that are atoms, or groups of atoms, bearing electric charges

ische'mia. Local anemia due to mechanical obstruction (mainly arterial narrowing) to the blood supply

is'chium; pl., **is'chia.** The lower portion of the os innominatum; that upon which the body is supported in a sitting posture

isoagglu'tinin. A substance present in the blood serum that can agglutinate or clump together the erythrocytes of other individuals of the same species

isoagglutin'ogen. A substance in blood cells that stimulates the action of agglutinins

i'somer. One of two or more chemically alike substances that have different relative positions of the atoms within

i'sotope. Isotopes are atoms that have different numbers of neutrons in their nuclei, but have the same number of protons. They weigh differently but behave alike chemically. Some of them are radioactive and can be used as tracers in the body; e.g., iron, iodine

ker'atin. A scleroprotein that is the principal constituent of epidermis, hair, nails, horny tissues, and organic matrix of teeth enamel

ketogen'ic. Tending to produce "ketone bodies," acetone, acetoacetic acid, and beta-hydroxybutyric acid

kinesthe'sia. The sense by which muscular motion, weight, and position are preceived

lacer'tus fibro'sus. Aponeurotic band from biceps tendon to fascia of forearm

lacta'tion. The secretion of milk

lacu'na; pl., **lacu'nae.** A minute, hollow space

lambdoi'dal. Resembling the Greek letter Λ, lambda

lamel'la; pl., **lamel'lae.** A thin plate, or layer

lam'ina. A thin plate; a germinal layer

laryn'goscope. The instrument by which the larynx may be examined in the living subject

lemnis'cus. A band of longitudinally arranged sensory fibers in the medulla and pons extending upward from the decussation, passing along the outer surface of the superior cerebellar peduncle, and terminating in the thalamus

libi'do. Conscious or unconscious sexual desire

lim'bic. Pertaining to a limbus, or border, margin, edge

lin'ea as'pera. A rough, longitudinal line on the back of the femur

lip'id. Organic substances that are insoluble in water; e.g., fats, waxes, phospholipids, and sterols

lipotro'pic. Applied to substances that prevent or reverse the condition of fatty liver

lymphangi'tis. Inflammation of a lymphatic vessel

ly'sin. Lysis, "to dissolve." An antibody that can dissolve cells

macera'tion. The softening of the parts of a tissue by soaking

mac'rophage. Phagocytic cell of reticulo-endothelial system

macroscop'ic. That which can be viewed with the naked eye

manom'eter. An instrument for measuring the pressure or tension of liquids or gases

maras'mus. Progressive wasting and emaciation, especially in young infants

mas'toid. Nipple shaped

matura'tion. Cell division in which the number of chromosomes in the germ cells is reduced to one half the number usual for the species

me'dial. Pertaining to the middle; nearer the median plane

medul'la. The central portions of an organ. Marrow. The medulla oblongata above the spinal cord

menis'cus. The curved surface of a liquid column; a crescentic interarticular fibrocartilage

mesenter'ic. Pertaining to the mesentery

mesoco'lon. A process of the peritoneum by which the colon is attached to the posterior abdominal wall

mes'oderm. One of three primary germ layers, lying between ectoderm and entoderm, from which develop connective tissue, muscle, bone, vascular and urogenital system

metabol'ite. Product of chemical interactions within the cell; intermediary substance in metabolism

methemoglo'bin. A transformation product of oxyhemoglobin found in the circulating blood after poisoning with acetanilid, potassium chlorate, etc.; the iron is oxidized from ferrous to ferric form; this compound does not carry oxygen

microceph'alus. Congenitally defective fetus with a very small head

micro'villi. Fine protoplasmic extensions at the apical surface of a columnar epithelial cell

mineralocor'ticoid. Steroid hormone from adrenal cortex affecting sodium and potassium excretion

mononu'clear. Having but one nucleus

mo'tor. Producing or subserving motion. A muscle, nerve, or center that affects or produces movement

mu'cin. A glycoprotein, a constituent of mucus

muta'tion. A distinctive character appearing for the first time in a pure line that is transmitted through succeeding generations. It is due to some change in the chromosomes

my'elocyte. A bone marrow cell giving rise to granulocytes

myogen'ic. Originating in muscular tissue

myoneu'ral. Pertaining to both muscle and nerve

my'osin. A globulin, chief protein substance of muscle

na'ris; pl., na'res. A nostril

navic'ular. Boatlike

ner'vus er'igens; pl., ner'vi erigen'tes. A nerve fiber supplying the bladder, genitalia, and rectum; derived from the second and third sacral nerves

neurogen'ic. Originating in nerve tissue

norepineph'rine. A primary amine that differs from epinephrine in the absence of an N-methyl group. Also called noradrenaline

no'tochord. The primitive backbone in the embryo

nu'cleotide. One of the repeating units that make up nucleic acids (DNA, RNA); composed of sugar, phosphate group, and nitrogenous bases

nystag'mus. Rhythmical oscillation of the eyeballs, either horizontal, rotary, or vertical. A symptom seen sometimes in disease of the inner ear or cerebellum

odon'toid. Toothlike

olfac'tory. Pertaining to olfaction, or the sense of smell

olige'mia. Condition of decreased quantity of blood (total or in a specific anatomic area)

onco'tic pressure. Osmotic pressure due to colloids in solution, as blood proteins in plasma or tissue fluid

o'ocyte. The primitive ovum in the cortex of the ovary, before maturation takes place

os; pl., o'ra. A mouth

os; pl., os'sa. A bone

osmo'sis. The passage of fluids and solutions, separated by a membrane or other porous septum, through the partition, so as to become mixed or diffused through each other

os'sa innomina'ta, pl. of os innomina'tum. "Unnamed bones." The irregular bones of the pelvis, unnamed on account of their nonresemblance to any known object

os'sicle. Any bonelet or small bone

os'teoblasts. The cells forming or developing into bone

os'teoclast. A large cell found in the bone marrow, believed to be capable of absorbing bone

o'tic. Pertaining to the ear

o'toliths. Particles of calcium carbonate and phosphate found in the internal ear on the hair cells

oxidize. To increase in positive valence or decrease in negative valence; to combine with oxygen

papil'la; pl., papil'lae. A small eminence; a nipplelike process

paranas'al si'nuses. Sinuses that communicate with the cavity of the nose. They are often called air sinuses of the head

paren'chyma. The essential or specialized part of an organ (e.g., hepatic cell), as distinguished from the supporting tissue

parie'tal. Pertaining to the walls of a cavity

parturi'tion. Act of giving birth

patel'la. A small pan; the kneecap

ped'icle. A stalk

pedun'cle. A narrow part acting as a support

pep'tide. Two or more amino acids, intermediate in synthesis of protein

pet'rous. Stonelike

phagocyto'sis. The engulfing of particulate matter by cells and microorganisms and foreign particles by phagocytic cells; e.g., leukocytes

phlebot'omy. The surgical opening of a vein; venesection

phren'ic. Pertaining to the diaphragm

pi'a ma'ter. "Tender mother"; the innermost membrane closely enveloping the brain and spinal cord

pinocyto'sis. Phenomenon in which fluid passes into cell by being incorporated in invagination of cell membrane that is pinched off to form cytoplasmic vacuole

pir'iform. Pear shaped

pis'iform. Pea shaped

plex'us; pl., **plexus** or **plex'uses.** A network or tangle, especially of nerves, veins, or lymphatics

polar'ity. Tendency of a body to exhibit opposite properties in opposite directions; referring to the possession of positive and negative poles

poles. Points having opposite properties, occurring at the opposite extremities of an axis. Either end of a spindle in mitosis

precip'itins. Antibodies in the blood serum that are capable of precipitating antigens

proges'terone. Hormone produced by corpus luteum that causes secretion of endometrial glands

pro'tein. Complex organic nitrogenous compounds, widely distributed in plants and animals, that form principal constituents of cell protoplasm. Proteins are combinations of α-amino acids and their derivatives

psy'chic. Pertaining to the mind

ptery'goid. Wing shaped

pyogen'ic. Producing pus

pyrex'ia. Elevation of temperature; fever

quadrigem'inal. Consisting of four parts

race'mose. Resembling a bunch of grapes, as in racemose gland

ra'mus; pl., **ra'mi.** A branch, as of an artery, bone, nerve, or vein

rec'tus; pl., **rec'ti.** Straight. Name given to certain straight muscles of the eye and abdomen

re'flex. Involuntary, unchanging response to a stimulus

regurgita'tion. The casting up of undigested food from the stomach. A backward flowing of blood through a cardiac valve because of imperfect closure of a valve leaflet

resis'tance. Opposition to force, as in force of blood flow; general term meaning capacity to counteract infection

retic'ular. Resembling a fine network (reticulum)

rhe'obase. Minimal electric current required to excite a tissue; e.g., nerve or muscle

rhom'boid. A quadrilateral figure whose opposite sides and angles are equal but which is neither equilateral nor equiangular

saliva'tion. Secretion of saliva

saphe'nous. Superficial, as the saphenous veins of the body

saponifica'tion. Alkaline hydrolysis; when fats are thus hydrolized, soaps are produced

sarco'mere. The structural unit of muscle confined between two Z-lines

semilu'nar. Resembling a half-moon in shape; e.g., valves of aorta

sig'moid. Shaped like the letter S

si'nus. A recess, cavity, or hollow space; a dilated channel for venous blood, found chiefly within the cranium; an air cavity in one of the cranial bones, especially those communicating with the nose

ska'tole. A strong-smelling crystalline substance from human feces, produced by decomposition of proteins in the intestine

slough. Mass of necrotic tissue in process of separating from a wound

sol'ute. A dissolved substance

sol'vent. A substance, usually liquid, that is capable of dissolving another substance

somat'ic. Pertaining to the body, especially the body wall

so'mite. Series of paired segments of mesoderm on either side of neural tube

specif'ic grav'ity. A comparison between the weight of a substance and the weight of an equal volume of some other substance taken as a standard. The standards usually referred to are air for gases and water for liquids and solids. For instance, the specific gravity (s.g.) of carbon dioxide (air standard) is 1.5, meaning that it is 1.5 times as heavy as an equal volume of air. Again, the specific gravity of mercury (water standard) is 13.6, meaning that mercury is 13.6 times as heavy as an equal volume of water. The specific gravity of solutions, as a salt solution, will necessarily vary with the concentration

sphe'noid. Wedge shaped

sphinc'ter. A circular muscle that contracts the aperture to which it is attached

splanch'nic. Pertaining to the viscera

stereogno'sis. The faculty of perceiving and understanding the form and nature of objects by the sense of touch

ster'oid. Class of organic chemicals having rings of six carbon atoms similar to cholesterol; e.g., bile acids, sex hormones, and adrenal cortex hormones

stomati'tis. Inflammation of the soft tissues of the mouth

sub'strate. A substance upon which an enzyme acts

sul'cus. A groove, trench, or furrow, especially as seen on the surface of the brain

summa'tion. Addition; finding of total or sum

suppura'tion. Formation of pus

sur'face ten'sion. The force that exists in the surface film of liquids that tends to bring the contained volume into a form having the least superficial area. It is due to the fact that the particles in the film are not equally acted on from all sides but instead are attracted inward by the pull of molecules below them

su'ture. That which is sewn together, a seam; the synarthrosis between two cranial bones

syn'apse. The anatomic relation of one nerve cell to another; the region between the processes of two adjacent neurons, forming the place where a nervous impulse is transmitted from one neuron to another

syner'gic or **synerget'ic.** Acting in harmonious cooperation, said especially of certain muscles

syno'via. A viscid fluid containing mucin and a small proportion of mineral salts. Secreted by synovial membranes, it is found in joint cavities, called *synovial fluid*

tac'tile. Pertaining to the sense of touch

ten'do achil'lis. "Tendon of Achilles." The tendon attached to the heel, so named because Achilles is supposed to have been held by the heel when his mother dipped him in the river Styx to render him invulnerable

tet'any. A disease characterized by painful tonic and symmetric spasm of the muscles of the extremities

the'nar. Mound at base of thumb

thermogen'esis. The physiologic process of heat production in the body

thermol'ysis. Loss of body heat by evaporation, radiation

thermotax'is. The normal adjustment of the bodily temperature. The movement of organisms in response to heat

thresh'old. Lower limit of stimulus applied to nerve that evokes a response; limit above which substances appear in the urine

tocoph'erols. Group of organic chemical substances, among which vitamin E is one

to'nus. The slight, continuous, and maintained contraction of muscle, which in skeletal muscles aids in the maintenance of posture and in the return of blood to the heart

trabec'ula; pl., **trabec'ulae.** A supporting fiber; a prolongation of fibrous membrane that forms septa or partitions

troch'lear. Pertaining to a pulley. The trochlear nerve supplies the superior oblique muscle of the eye

u'vula. "Little grape"; the soft mass that projects downward from the posterior middle of the soft palate

vac'uole. A space or cavity within the protoplasm of a cell, containing nutritive or waste substances

vas'cular. Latin, *vasculum*, a small tube. Refers to tubes conveying liquids, as blood vascular and lymph vascular systems

ve'na co'mes; pl., **ve'nae com'itantes.** A deep vein following the same course as the corresponding artery

ven'tral. Pertaining to the front or belly surface; anterior

ven'tricle. Any small cavity, especially either of the lower cavities of the heart or one of the several cavities of the brain

ver'miform. Worm shaped

ves'icle. A small sac containing liquid or gas

vis'ceral. Pertaining to a visceral or internal organ

vo'lar. Pertaining to the palm of the hand or the sole of the foot

xiph'oid. Shaped like a sword

zygoma'tic. Pertaining to the zygoma, the arch formed by the zygomatic process of the temporal bone and by the zygomatic or cheek bone

Index

Numbers in **boldface** refer to illustrations; *n.* indicates a footnote; *t.*, a table.